Builder's Comprehensive Dictionary

Robert Putnam

Craftsman Book Company
6058 Corte del Cedro
Carlsbad, CA 92009

Library of Congress Cataloging-in-Publication Data

Putnam, Robert E.
Builder's comprehensive dictionary / Robert Putnam.
p. cm.
Originally published: Reston, Va. : Reston Pub. Co., c1984.
ISBN 0-934041-50-4
1. Building—Dictionaries. I. Title.
TH9.P83 1989
690'.03—dc20 89-27846
CIP

Second Printing 1991

preface

The *Builder's Comprehensive Dictionary* is designed as an introduction to the current state of the art in all areas of building and construction. Basic trade terms are defined for both residential and commercial construction. Tools, equipment, and materials used in all phases of construction are covered. Definitions are clearly written so the non-professional can easily grasp their meaning, while the book's technical accuracy meets the needs of the building tradesman, both apprentice and journeyman. Related fields—such as surveying, site preparation, environmental planning, solar use, legal usage, and drafting—are covered.

How to Use the Dictionary

All dictionaries can be easily used if you know the term you want defined. The *Builder's Comprehensive Dictionary* is unique in that it is designed to lead to a term you need even though you may not remember the term or the spelling. Terms are heavily cross referenced so you can follow from a general or related term to a more specific term. For example, *solar heating* is defined and then six cross references are given. If you followed just one of these, for example *passive solar heating*, you would find not only a definition but a further cross reference to the very crucial concept of *orientation*. By following multiple cross references you can branch out to cover a whole subject area or type of construction. In the case of multiple terms for one building part or concept, cross references lead to the more common or more widely used term. You can also quickly discover construction areas of interest by thumbing through until you find illustrations that show the general area you're interested in. You can then go to the term that references the illustration and start following cross references if appropriate.

Technical Terms and Legal Terms

The dictionary is designed to cover not only technical terms but also the related legal, financial, and management terms. "Legal, Real Estate and Management Terms" are covered at the end of the book. The more recent concepts of construction management are introduced.

Standards and Codes

Definitions have been written to conform to national standards that are widely used in the construction industry, and many of the illustrations have been taken from these standards. Local codes, in as far as they are consistent with national codes, should be compatible with the definitions. Of course, always check the latest standards and codes in your area.

technical terms

abrasion The wearing away of material by friction, as by sanding, grinding, smoothing, or polishing.

abrasive A hard substance used for abrading or wearing down another material. Several common materials are used: flint (crushed quartz), garnet, emery, aluminum oxide, silicon carbide, and tungsten carbide. The crushed abrasives are glued to a backing material such as paper or cloth. Backings are graded *A* (lightweight paper), *C* and *D* (intermediate weight paper), *E* (heavy paper—used for mechanical sanding), and *F* (extra heavy paper—heavy-duty mechanical finishing). Cloth backing has a light *J* weight and a heavy *X* weight. Abrasive papers come in 9″ × 11″ sheets or in rolls. Cloth abrasive belts or disks are used for power sanders. Abrasive papers or abrasives are graded by grit numbers from coarse to fine. Abrasive papers are also graded as to *open coat* or *closed coat.* The open coat has fewer abrasive grains on the backing and is used with softer gummy woods (which tend to clog the paper). Abrasive grading is based on grit size. The standard method used today is the *mesh system.* Mesh size is determined by a mesh cloth screen with sized openings calculated by the number of openings per inch. An older system of describing abrasives was the *aught system,* which grades abrasives from 0 as a base down to 3 (coarse) and up to 7/0 (very fine). Figure 1 compares the mesh system to the aught system. Nominal grit sizes describe the roughness or fineness of the abrasive.

abrasive cut-off saw Heavy-duty power saw that uses an abrasive blade. Used for cutting metal and for tuckpointing masonry. (Figure 2.)

abrasive disks Disks that are used with power *disk sanders* or with a special attachment with power drills. Various grades or *meshes* are available.

DESCRIPTIVE RATING	MESH SYSTEM	AUGHT SYSTEM
SUPER FINE	600 500 400	12/0 11/0 10/0
EXTRA FINE	360 320 280	– 9/0 8/0
MED. FINE	240 220	7/0 6/0
FINE	180 150 120	5/0 4/0 3/0
MEDIUM	100 80	2/0 1/0
COARSE	60 50	1/2 1
MEDIUM COARSE	40 36	1-1/2 2
EXTRA COARSE	30 24 20	2-1/2 3 3-1/2
SUPER COARSE	16 12	4 4-1/2

Figure 1 Abrasive ratings: new mesh system and old aught system

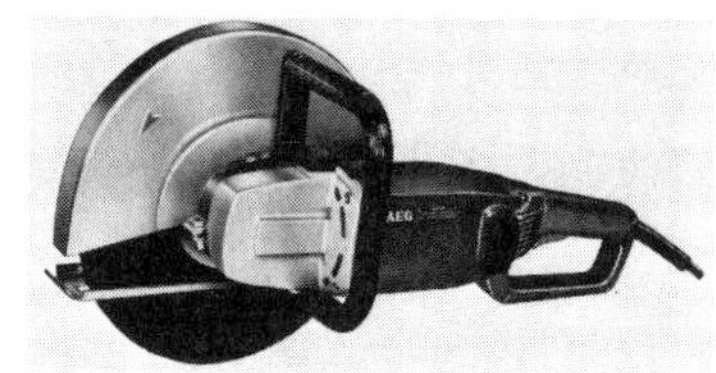

Figure 2 Abrasive cut-off saw

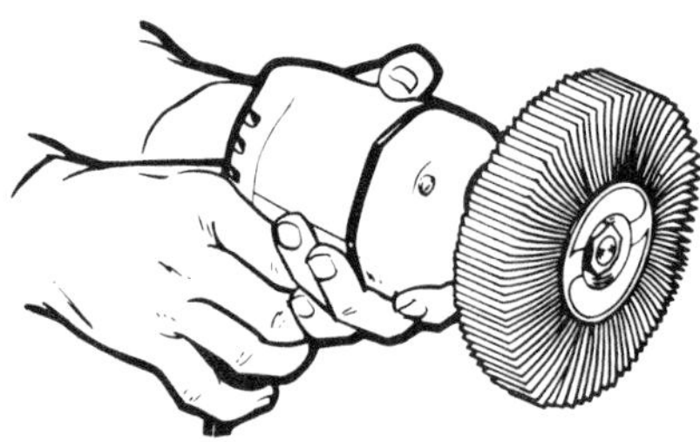

Figure 3 Abrasive flap wheel

abrasive flap wheel An abrasive wheel made up of many flaps of abrasive paper or cloth. (Figure 3.) Used in an electric drill for cutting and sanding.

abrasive paper Paper with abrasives glued to it. Common abrasive papers are sandpaper, garnet paper, and emery paper. Cloth is also used as a backing. See *abrasive.*

abrasive stones Shaped stones used for grinding, smoothing, or polishing. Common abrasive stones are: emery wheels, used to quickly grind down metal; grindstones, used to sharpen tools; and whetstones, used to put a fine edge on cutting tools. See *oilstone, waterstone.*

ABS A black plastic (acrylonitrile-butadiene-styrene) pipe widely used for drainage and venting. See *plastic piping.*

absolute zero Point at which all molecular activity stops; −458.8°F or −273.1°C.

absorber plate In solar heating, the black surface inside a solar collector that absorbs solar radiation and converts it to heat.

absorption The taking of liquid, heat, light, or the like into a material; the process of a liquid's filling in the pores of another material, as water being drawn into the ground. In acoustics, the ability of a material, such as acoustical tile, to take in and trap sound rather than reflect it.

absorption field A private waste system of perforated pipe or tile, or plastic tubing, laid out to receive the overflow of a septic tank. The overflow or effluent runs through the pipe and seeps (percolates) into the soil, where it is purified. Used for house waste when no public sewer is available. (Figure 4.) Also called a *distribution field.* Figure 5 shows how the absorption field is laid out.

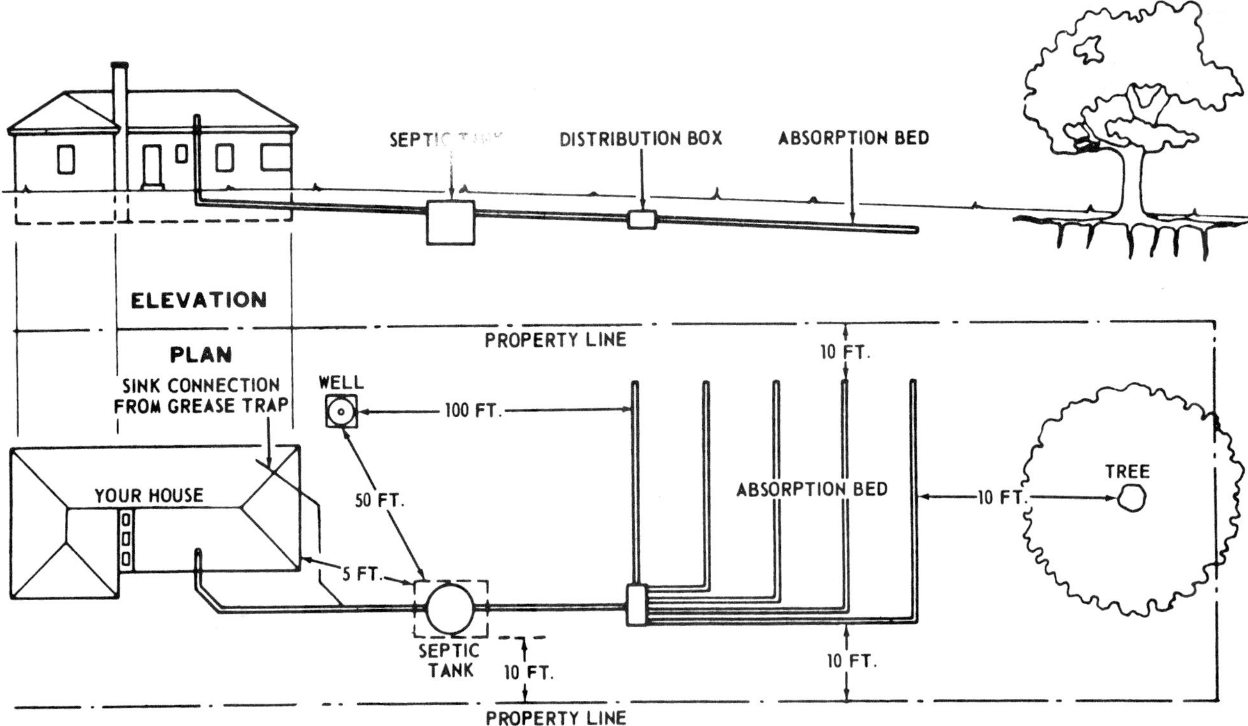

Figure 4 Absorption or distribution field

abut To end against, as when an interior partition hits an outside wall at a right angle.

abutment A support that receives thrust at the end of an arch or bridge.

AC See *alternating current.*

accelerator Admixture added to concrete, plaster, or mortar to speed up setting, such as calcium chloride.

access A passageway into a building; a door or panel that allows mechanical equipment to be reached for repair or inspection.

access court Open area that allows entrance to a group of residences or apartments. Often found in a U-shaped apartment complex.

access door Small door opening into attic and wall areas, often a panel that is screwed to the opening.

access panel Panel that allows entrance to mechanical equipment, such as plumbing pipes.

accordion doors Folding doors with small folds, used for closets and as room separators. (Figure 6.) See *door.*

acetylene A highly combustible gas that is mixed with oxygen to produce a hot flame used in *oxyacetylene welding (oxyfuel welding).*

acid-core solder Solder that has an acid-core flux; it is suitable for joining metal pieces together but cannot be used for soldering electrical wires. See also *rosin-core solder.*

acoustical board Sound-absorbing insulating board or tile made of pressed fiber. Holes are made into the surface to further absorb sound. Often used as tile on ceilings.

acoustical materials Sound-absorbent materials used to cover floors, walls, and ceilings; cuts down transmission of sound into or out of living area.

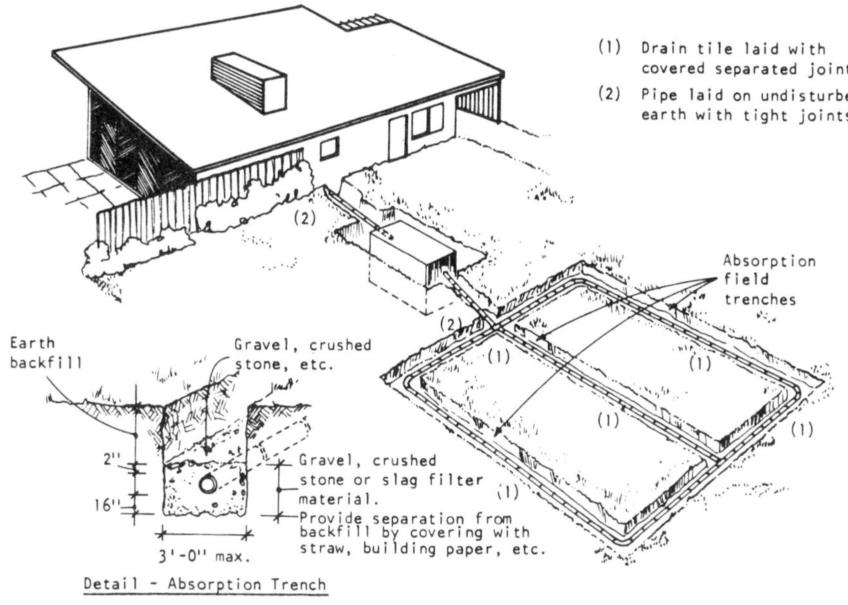

Figure 5 Absorption-field installation and layout

acoustical plaster Finishing plaster with a sound-absorbent aggregate added to the mix.

acoustical tile Special sound-absorbing tile, often made of pressed fiber, used in walls and ceilings. The tile has a roughened or perforated surface that absorbs sound. Tiles are installed individually.

acoustics The science of sound. The study of building materials for sound-transmission qualities.

acre Land area containing 43,560 square feet (160 square rods); if square, each side is 208.71 feet long.

active Use of mechanical means to move heat, as opposed to *passive.*

active solar heating Heating of a structure by indirect use of sunlight. *Solar collectors* or *panels* are located, usually on a roof, looking southward to collect the sunlight; fluid or air in the solar collectors is heated by the sunlight and pumped (by "active" mechanical means) into the structure to heat the building. See *solar heating.*

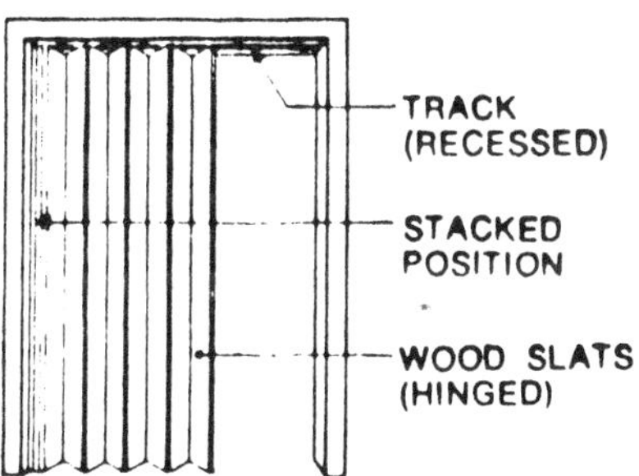

Figure 6 Accordian door

Figure 7 Air-operated adhesive gun. The adhesive is pumped from a bulk container

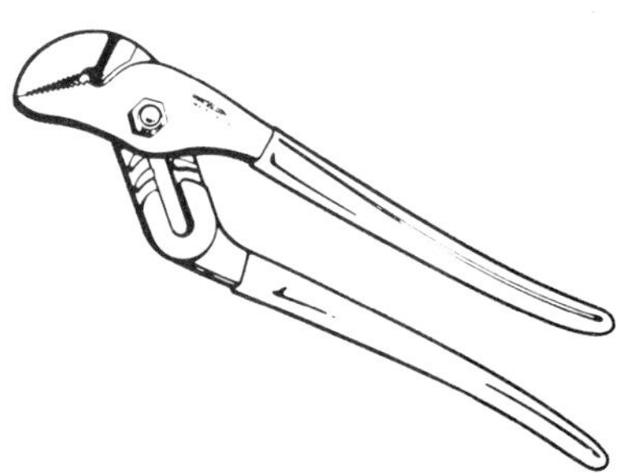

Figure 8 Adjustable-joint pliers

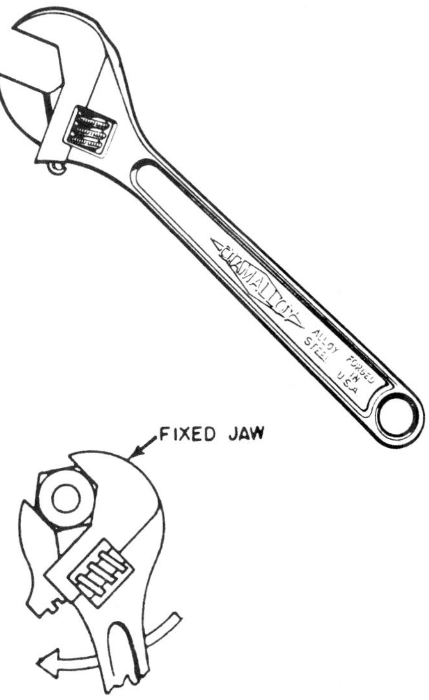

Figure 9 Adjustable wrench

AC transformer welder See *transformer, welding machine.*

actual size lumber Finished lumber size after dressing and shrinkage. A nominal 2″ × 4″ would have an actual size of 1½″ × 3½″. (*Figure 578*, p. 246.) See *board, dimension lumber, nominal lumber.*

acute angle Angle less than 90°. See *angle.*

addition Expansion of a building, normally by the adding of a room or rooms. Also, material added to cement at time of manufacture to improve a specific physical property, such as curing under low temperatures.

adhesion Property of holding fast; adhering.

adhesive Liquid material used to bond parts together by application to the surface of one or both the parts to be bonded. See *glue.*

adhesive, cold-setting Adhesive that sets at temperatures below 87°F.

adhesive, contact Adhesive that forms an instant bond upon contact between the two treated surfaces.

adhesive, gap-filling Adhesive that fills cracks and gaps of the joining surfaces.

adhesive gun A hand-powered or air-powered gun used to deposit a bead of adhesive. (Figure 7.) See *caulking gun.*

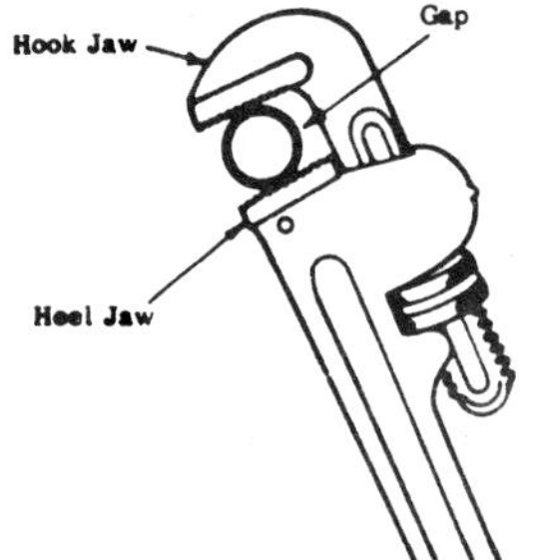

adhesive, heat-activated A dry adhesive film that is made fluid and active by the application of heat.

adhesive, hot-melt Adhesive that is applied hot and hardens into a bond when cool.

adhesive, hot-setting Adhesive that sets at temperatures over 212°F.

adhesive, intermediate-setting Adhesive that sets at a temperature from 87° to 211°F.

adhesive, separate-application Adhesive that is applied to each surface to be joined. The bond is made when the two surfaces are brought together.

adjacent Near by or touching; next to, as the adjacent lot.

adjustable clamp Holding device that can be loosened or tightened. Also, a clamping device used to hold concrete forms, such as column forms.

adjustable-joint pliers Pliers that have jaws with several adjusting grooves for changing the jaw bite or capacity. (Figure 8.) Used in electrical work for turning fittings, in plumbing for grasping pipe or couplings. Various sizes are available. See also *pliers.*

adjustable wrench Turning tool that has adjustable jaws. (Figure 9.) See *wrench.*

admixture Something added to concrete (other than cement, water, or aggregate) to change its properties or change setting time. See *accelerator, retarder.* Admixtures are added to change strengths, to allow for drying under high winds, to allow for low temperatures, high temperatures, and so on.

adobe A clay soil used for making sun-dried bricks.

adobe brick A large brick made using an adobe clay; brick is formed in a mold and sun-dried.

adobe construction Houses and other structures made using sun-dried adobe brick. Walls are very thick. The structure may be washed with a whitewash or cement paste. Used for building in the southwestern part of the United States and in Central and South America. (Figure 10.) See also *architectural styles.*

adze A cutting tool with the blade at a right angle to the handle. Used for dressing timber. (Figure 11.) Also called a *carpenter's adze.*

aerated concrete A lightweight concrete that has air bubbles in it. Used as a sound-deadening material and to cut down heat transmission.

aerial survey Survey of land areas by photographing the area from an airplane. Aerial surveys are used in making topographical maps. This process is called *photogrammetry.*

A-frame Wood-frame house constructed with an "A" shape when seen in end views. (Figure 12.) Often used for summer or vacation homes.

AGC Associated General Contractors of America.

aggregate In concrete, a part of the mix: sand, crushed stone, or gravel. Sand is a fine aggregate; crushed stone and gravel are coarse aggregates. In plastering, sand or other lightweight aggregates such as vermiculite are used. The term refers also to exposed crushed rock on the plaster or concrete surface.

aggregate gun Machine for spraying aggregate on a surface.

air acetylene welding (AAW) Gas welding that uses a mixture of air and acetylene; a type of *oxyfuel welding.*

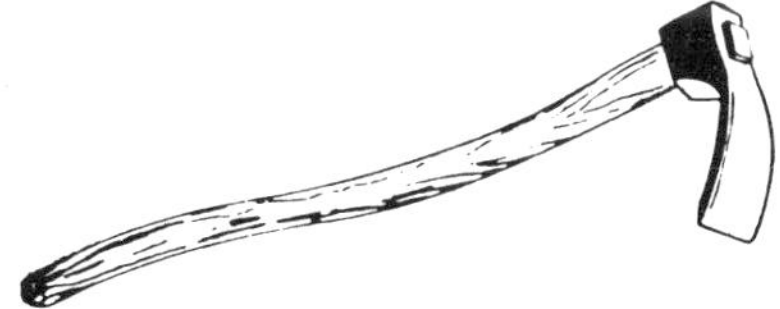

Figure 11 Adze

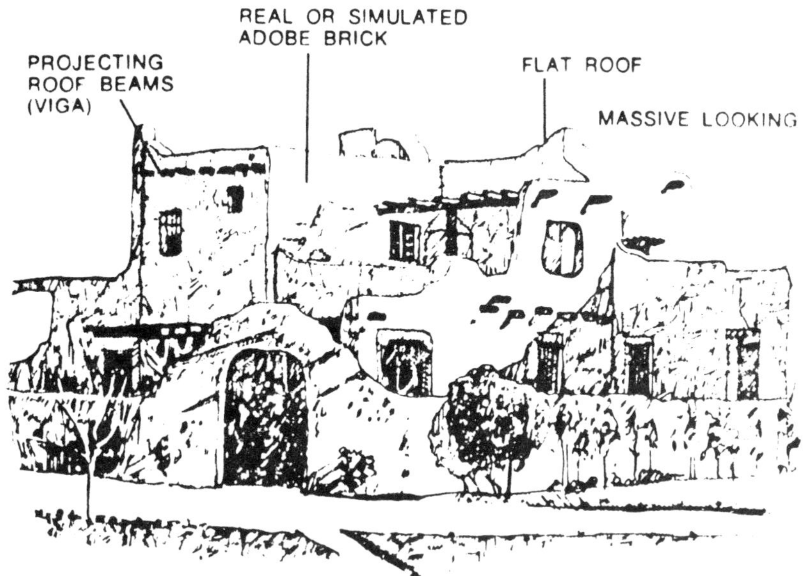

Figure 10 Adobe construction

Figure 12 A-frame

ELECTRONIC AIR CLEANER
CONTROL PANEL
COOLING COIL
FLUE VENT
BLOWER
HUMIDIFIER
HEAT EXCHANGER
BURNER CONTROLS

Figure 13 Air conditioning system: furnace, humidifier, cooling coil and dehumidifier, air cleaner and blower. Cools and dehumidifies in summer and heats and humidifies in winter

Figure 14 Air-driven nailer

Figure 15 Alidade being used on a plane table

air brick A metal or perforated brick used to allow air to enter into a foundation or wall.

air change In ventilation, the number of times per hour air is changed in a room.

air conditioner Device for cooling air inside a structure. Conditioning of air includes both heat and moisture control.

air conditioning The circulating of cooled, cleansed, and moisture-controlled air in a building. The concept of conditioned air includes heating as well as cooling. Figure 13 shows a typical system. See *furnace, heat pump.*

air-dried Lumber that has been dried by exposure to natural air currents. The lumber is stacked so air can flow between the individual pieces. Air-seasoned. Opposed to *kiln-dried.*

air-driven nailer Air-driven (pneumatic) nailing gun used in all forms of rough construction. Nails for the gun come in coils or sticks; sizes vary from 2″ to 3½″. (Figure 14.) See also *nailer.*

air duct Formed sheet metal or a manufactured pipe that carries warm air from the furnace or cooled air (cool air return) back to the furnace. See *ductwork.*

air entrainment Increase in the amount of air in a concrete mix by addition of a special admixture that creates small bubbles throughout the concrete. Improves the concrete's workability and ability to withstand the freezing and thawing cycle.

air gun A device used for spraying adhesive insulating material on a wall or ceiling.

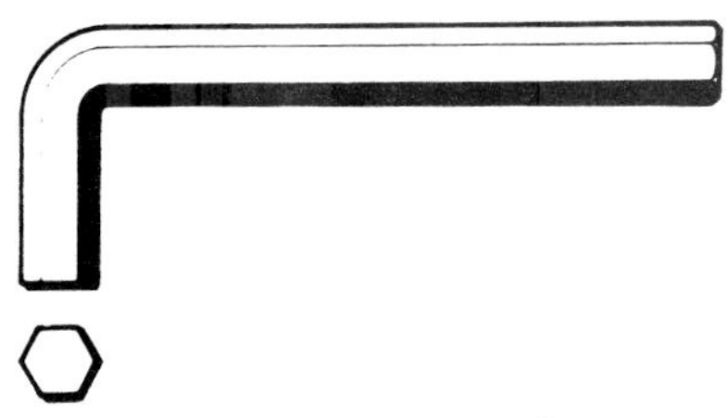

Figure 16 Allen wrench

air lock The stopping of water flow by air in the lines.

air pocket A space with air trapped in it, especially an air void in concrete.

air space An open area or void, as in a wall or floor or between building members.

air trap A water seal in piping that prevents sewer gas from seeping out of plumbing fixtures or drains. See *P trap, S trap.*

air-type solar system An active solar heating system that uses air to collect and carry solar heat. Heated air may run directly into the house. See *fluid-type solar system.*

aisle A passageway between or leading to chairs or seating, as between the pews in a church.

alcove A recess in a wall.

alidade Sighting instrument used on a *plane table* to take bearings on distant objects. Used for mapping small areas and for completing details of a survey. (Figure 15.)

align To line up something in relation to something else.

aligning punch Punch used to line up holes in two different pieces of metal. See *punch.*

alignment Locating of building members so they line up to a set line.

Allen wrench L-shaped wrench that has a hexagonal end, used to fit into the recessed hexagonal slots or sockets of Allen screws. (Figure 16.) See also *wrench.*

alley Road running behind rows of buildings.

alligatoring Cracking of a surface in a close pattern of random cracks resembling rough scales, as on an alligator hide. Occurs on a paint surface or a bitumen built-up roof surface. Caused by the top coat separating to expose a lower surface, as a new paint coat separating to expose the old paint coat.

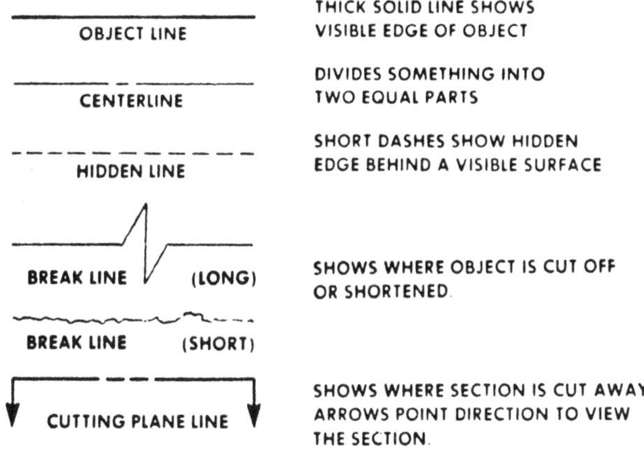

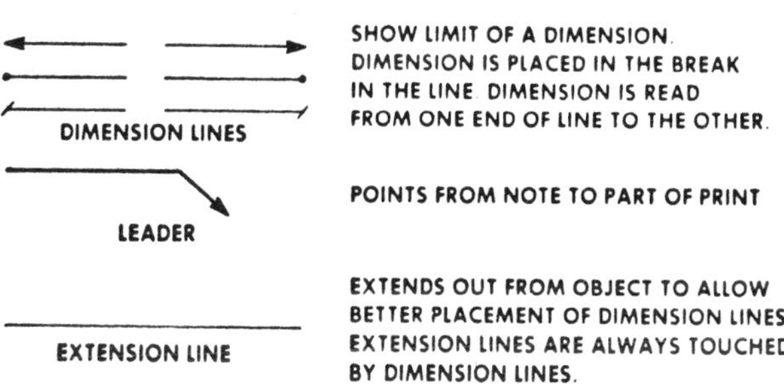

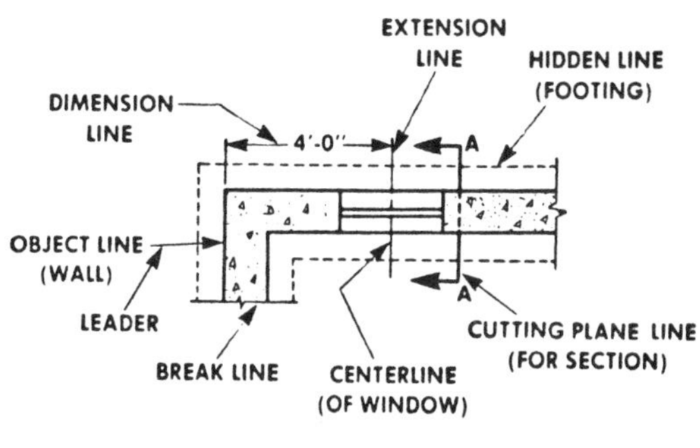

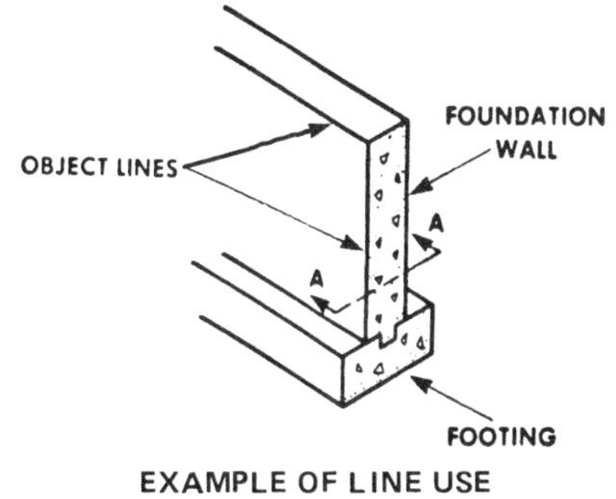

EXAMPLE OF LINE USE

Figure 17 Alphabet of lines

alloy A metallic substance made by combining two or more metals, as brass is a combination of copper and zinc. A combination of a metal and a nonmetal, as steel is a combination of iron and carbon.

alluvial Material deposited by flowing water.

alphabet of lines Lines used on blueprints to show different parts of a structure. (Figure 17.) Also called *line conventions.*

alteration A change in an existing building.

alternating current (AC) Electrical current that constantly reverses direction of the flow of electricity (electrons) through the circuit. House current is 60-cycle (60-Hertz) AC. The 60-cycle current goes in one direction 60 times per second and in the opposite direction 60 times per second. See also *direct current.*

altitude Angle of the sun above the horizon. See also *solar azimuth.* Also, the elevation of something above sea level.

aluminum nails Rust-resistant nails with a high tensile strength; sizes from 6*d* to 20*d*.

aluminum oxide A hard, synthetic, abrasive mineral widely used in all grits, wet and dry abrasive papers, belts, and discs. Aluminum oxide in a pure, natural state is a gemstone, sapphire or ruby. See *abrasive.*

ambient temperature The temperature of the surrounding air.

American architecture See *Colonial architecture, nineteenth-century American, post-World War II American.* See also *architectural styles.*

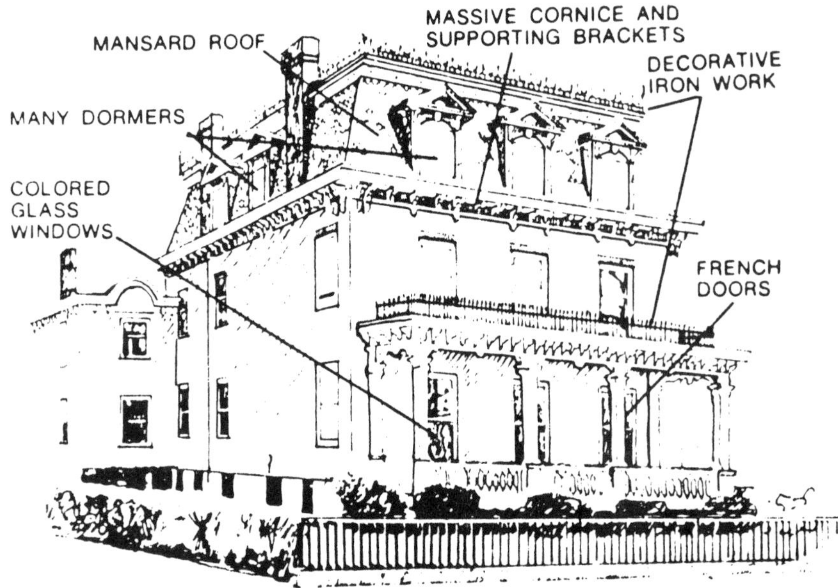

Figure 18 American mansard or Second Empire style

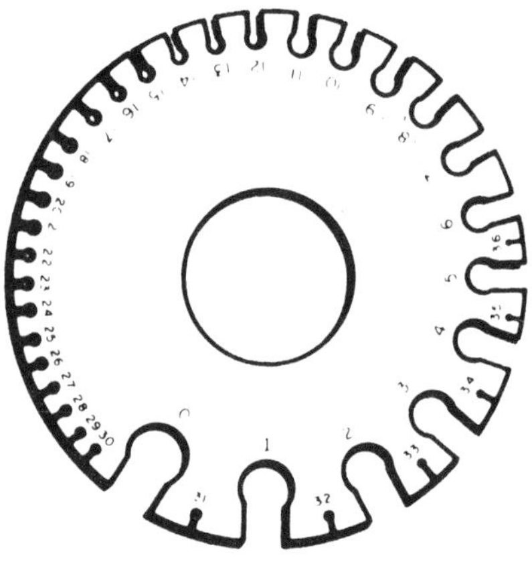

Figure 19 American standard wire gauge

American bond Masonry bond pattern, a *common bond.*

American Institute of Architects (AIA) Professional society of architects.

American mansard Multistory house with ornate mansard roof. (Figure 18.) Developed from the French Second Empire (1852 to 1870). See also *architectural styles.*

American National Standards Institute (ANSI) An association that publishes standards in various areas of building construction, including safety and drafting.

American Society for Testing and Materials (ASTM) Establishes voluntary standards for materials and products, often referenced in specifications.

American Softwood Lumber Standards A recognized standard (PS20-70) for the grading of softwood lumber. Determines classifications, nomenclature, grades, sizes, and description. See *lumber grades.*

American standard wire gauge A recognized grading system for wire sizes (diameter). Used for copper and aluminum wire. Gauge is specified by number. Figure 19 shows the circular gauge used to measure and check wire diameters. The wire end is inserted in the slot on the gauge. It should have a snug fit. Note the hole at the base of the slot. This is used for easy removal of the wire. See also *wire sizes.*

ammeter Electrical device for measuring current or amperes.

ampacity Current-carrying capacity of a circuit, calculated in amperes.

ampere Rate of electrical flow; the current in a circuit.

$$\text{amperes} = \frac{\text{watts}}{\text{voltage}}$$

$$\text{amperes} = \frac{\text{volts}}{\text{ohms}}$$

Amperes are generally referred to as *amps.*

ampere-hour Quantity of electricity produced by a one-ampere current flowing for one hour.

ampere-hour meter Device for recording ampere-hours.

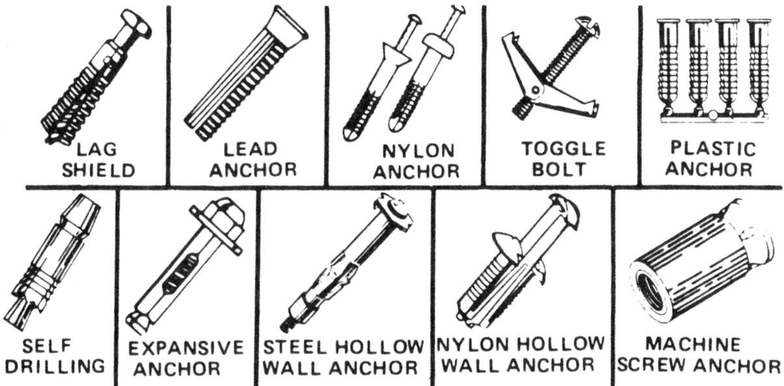

Figure 20 Anchors

amps Amperes (A).

anchor Shaped metal fastener used to hold building parts together. A *framing anchor* holds timber or lumber pieces together. A *masonry anchor* or *tie* fastens masonry together or holds wood structural members to the masonry. A device used to hold something to a concrete or masonry floor or a concrete, masonry, plaster, or gypsum panel wall. (Figure 20.) See *fiberplug, hollow-wall anchor, lag shield, lead anchor, machine-screw anchor, nylon anchor, nylon expansion anchor, plastic anchor, plastic toggle, rawplug, sleeve anchor, snap-off anchor, stud anchor, toggle bolt, wedge anchor.*

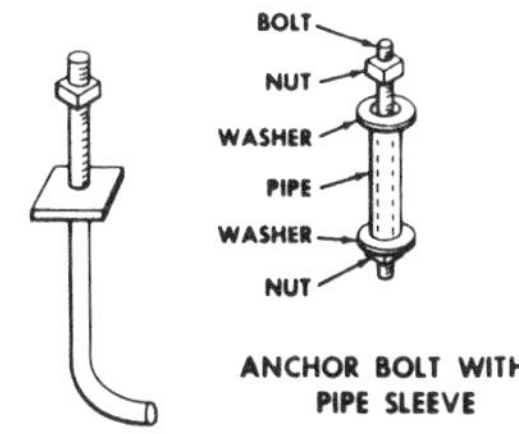

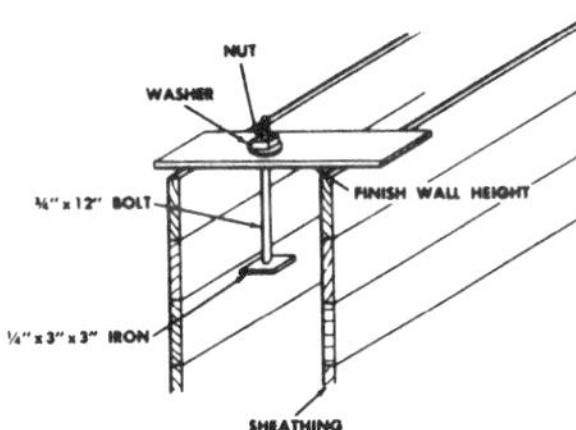

Figure 21 Anchor bolts

anchorage Securing or fastening a structural part.

anchor block Wood block built into masonry walls, used as a base for fastening walls and partitions.

anchor bolt Bolts used to fasten sills, columns, or beams to the foundation. The bolts are set in concrete or in concrete blocks (with concrete). Figure 21 shows anchor bolts. Also called a *foundation bolt.*

anchor bolt plan Plan drawing showing location of all anchor bolts that will be placed in the foundation or in column footings.

angle Relation of two connected straight lines, measured in degrees. Figure 22 shows the common types of angles. An *angle iron.*

angle bead See *corner bead.*

angle brace Support piece that runs across the inside angle of the framework.

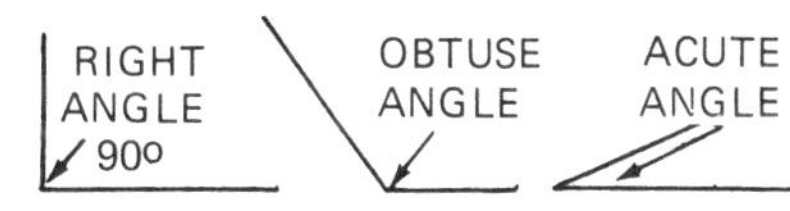

Figure 22 Angles

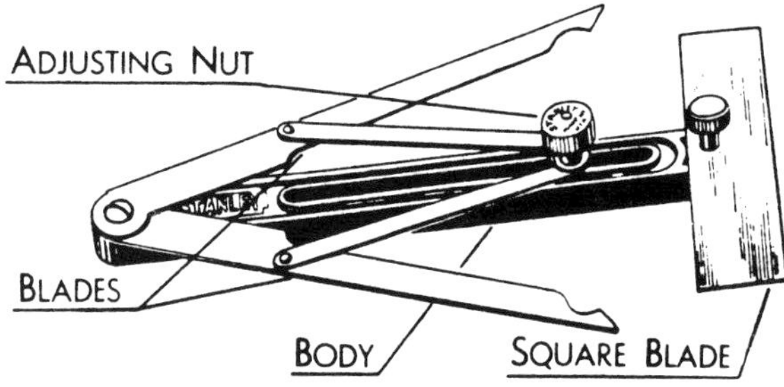

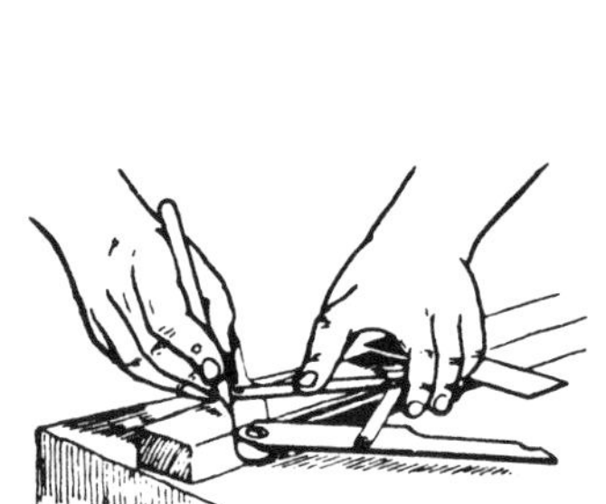

Laying off a miter with an angle divider. The square blade may be used for a try square when working on a surface.

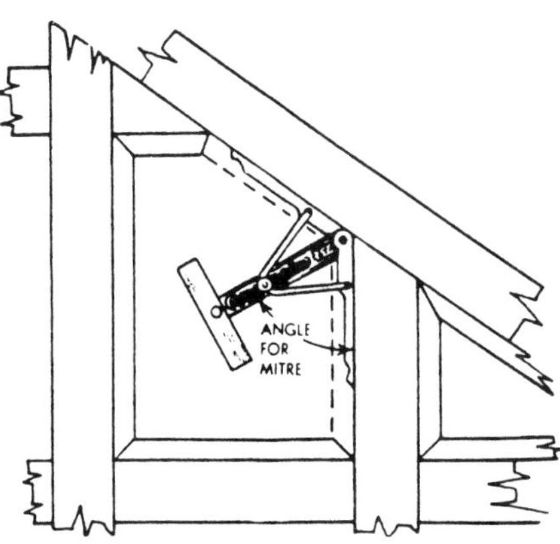

The angle divider is a double bevel. It is used to take off and divide angles for the miter cut in one operation. The handle is graduated on the back for laying off 5, 6, 8 and 10 sided work.

Figure 23 Angle dividers

EQUAL LEG ANGLES	L	3½	x 3½	x ¼	x 5'-6
	Group Symbol	Leg Width in inches	Leg Width in inches	Thickness in inches	Length in ft. and in.

Length — Leg — Heel — Toe — Fillet — Leg — Toe — Thickness

UNEQUAL LEG ANGLES	L	6	x 4	x ⅜	x 10'-3
	Group Symbol	Long Leg in inches	Short Leg in inches	Thickness in inches	Length in ft. and in.

Length — Short Leg — Long Leg — Thickness

Figure 24 Angle-iron terminology and specifications

angle dividers Layout tool for measuring or bisecting angles. (Figure 23.) Used for laying out miters and laying out identical angles.

angle float In plastering, an angled trowel which is used to finish inside walls. See *angle trowel, corner trowel, floats, trowels.*

angle iron Right-angle bar, commonly used over masonry openings for reinforcement. Figure 24 shows the terms associated with angles and gives specification examples.

angle of inclination Angle a plane or surface makes in relation to the horizon.

angle of repose Slope at which a pile of loose material, such as sand or rocks, or an earthen bank, will hold without sliding. Also called *angle of rest.* (Figure 25.)

angle paddle In plastering, a paddlelike tool used to clean out or finish an inside corner where two walls meet.

angle plane In plastering, a tool used to even the brown coat by removing high spots, cleaning angles, and scraping down walls. Used before the application of the finish coat. (Figure 26.)

angle plow See *corner trowel* (plastering).

angle post A newel located at a stairway angle.

angle rafter A *hip rafter.*

angle trowel In plastering, a shaped metal trowel used for cutting sharp angles, as in corners (Figure 27.) See *trowels.*

angle valve A globe valve that has its inlet and outlet openings at a 90° angle to one another. See *globe valve.*

angular perspective A three-dimensional drawing that has two vanishing points. Also called *two-point perspective.* See *perspective drawing.*

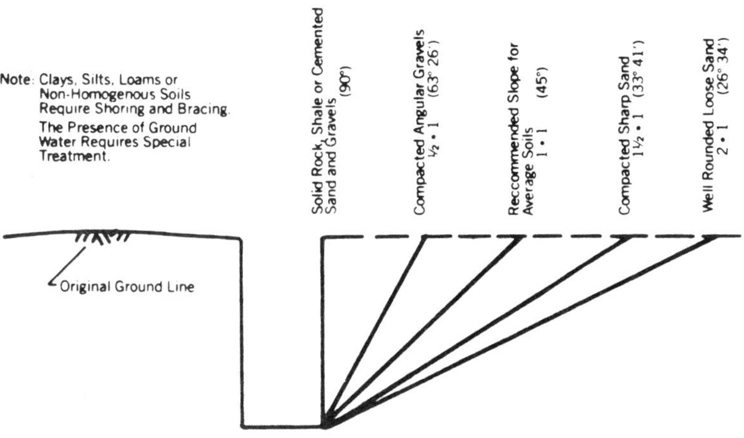

Figure 25 Angle of repose

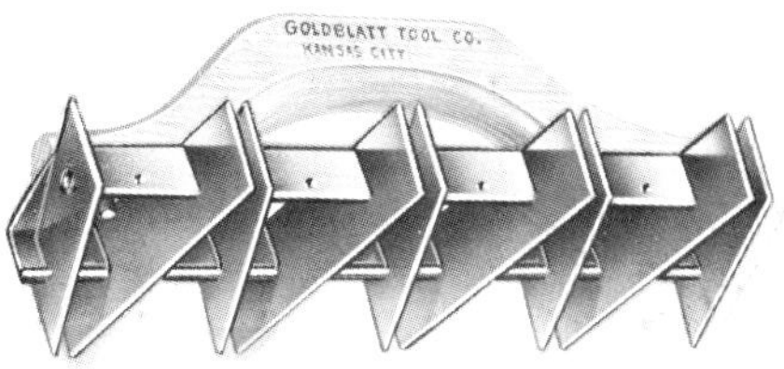

Figure 26 Angle plane

anhydrous lime Unslaked lime (no water added); same as quicklime.

annealing Softening of metal.

annual ring Yearly growth ring on a tree. The ring is formed of light-colored *springwood* and dark-colored *summerwood.* When the trunk is cut, the age of the tree can be determined by counting the annual rings. See *tree.*

annular Ring-shaped or with ring shapes on it, as an annular nail.

annular bit A hollow bit which cuts out a plug. See *plug cutter.*

annular nail A nail that has raised circular ridges on the shank. (Figure 28.) Also called a ring-shank nail. See *nail.*

antishort bushing Vinyl device installed on the end of armored cable for safety. (Figure 29.)

anvil Shaped iron or steel block used as a base for working metal. (Figure 30.) The front rounded part is called a horn or beak; the back part is called the tail. The square hardy hole in the tail is used to hold anvil tools such as a *hardy* or *swage.*

apex Top or tip of something, as the highest point of a tower or the top of a church spire.

appliance Electrical household device such as an air conditioner, stove, furnace, vacuum cleaner, coffee pot, toaster.

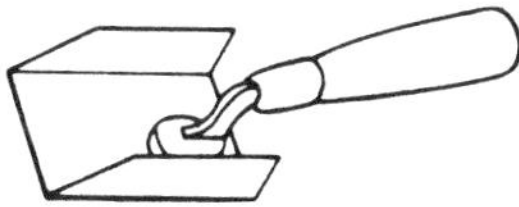

Figure 27 Angle trowel

Figure 28 Annular nail

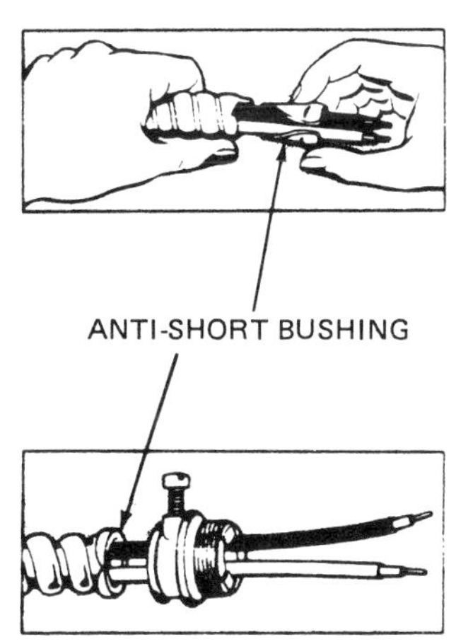

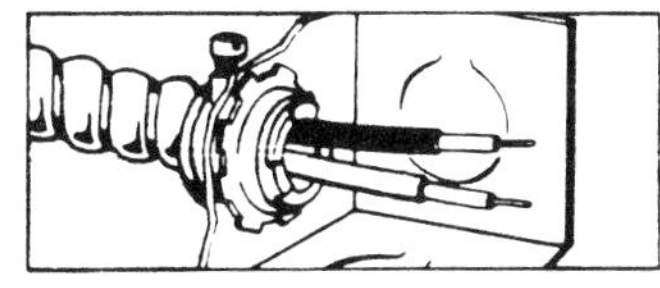

Figure 29 Antishort bushing

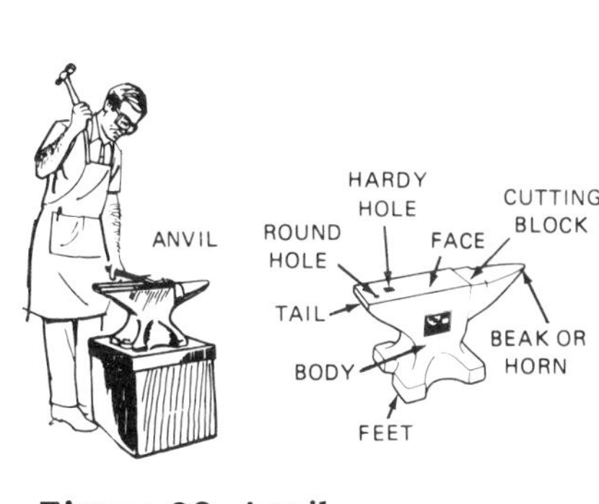

Figure 30 Anvil

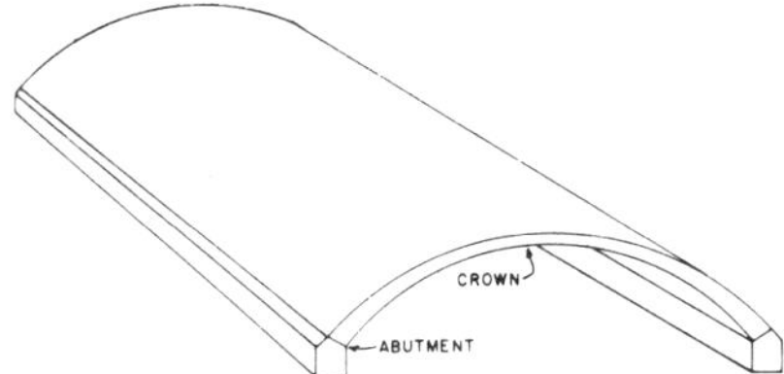

Figure 31 Arch (fixed foundation: concrete)

applied molding A cast plaster molding that is affixed (applied) to a flat surface to form a raised design.

apprentice Someone serving an apprenticeship in a trade, usually for three or four years. Usually the apprentice signs an agreement with a *Joint Apprenticeship and Training Committee* (JATC). After the training is satisfactorily completed, the apprentice becomes a qualified *journeyman.*

apprenticeship A period of training, usually three or four years, in a skilled craft or trade. Both theory and hands-on practice are covered.

apron Horizontal wood board that runs around the edge of the foundation just below the drip cap. An interior piece placed below a window to cover any rough edges. A wide concrete or asphalt slab just outside of a garage or other building opening.

arbor saw See *table saw.*

arc A segment of the circumference of a circle.

arcade Open passageway, often with arches and an overhead covering.

arc cutting Cutting metal using an electric arc. Both metallic-arc and carbon-arc processes are used to melt the metal.

arch A curved structural support over an opening or open area. Arches may be fixed or hinged. (Figures 31 and 32.) A large arch over an open area and supported at each end by flexible anchorage is designated as two-hinge. An arch supported at both ends by flexible anchorage and joined at the top center by a pivot is called three-hinge. Figure 33 shows various arch shapes.

arched beam Beam with an arched shape.

architect A person who designs buildings and other structures; a person skilled in methods of construction and in planning buildings and who is licensed to practice.

architectural concrete Concrete that is exposed and may need special treatment so as to present an attractive view.

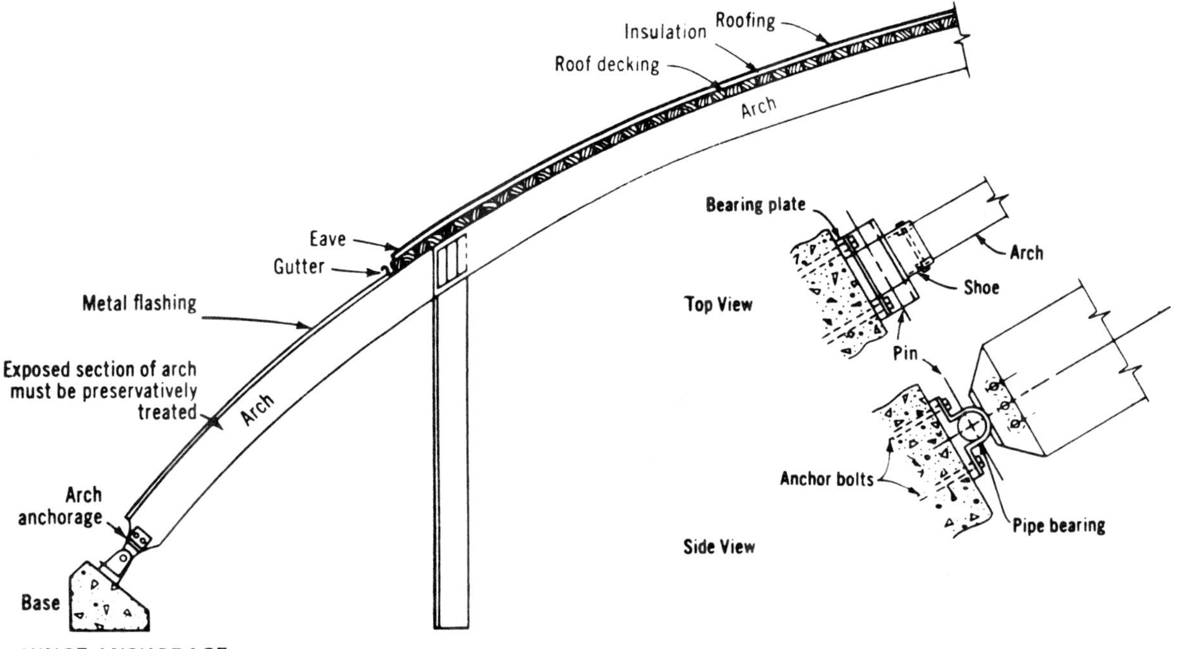

Figure 32 Arch (hinged: timber)

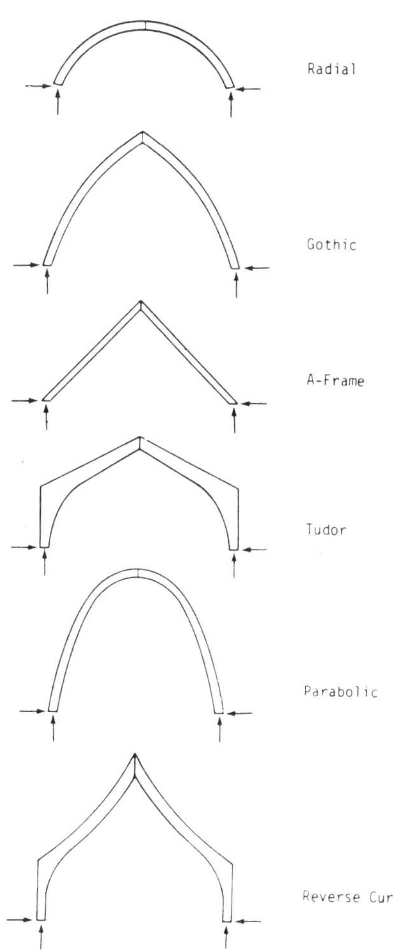

Figure 33 Arch shapes

Scale	Relation of Scale to Object
16	Full Scale
3	3" = 1'-0"
1 1/2	1 1/2" = 1'-0"
1	1" = 1'-0"
3/4	3/4" = 1'-0"
1/2	1/2" = 1'-0"
3/8	3/8" = 1'-0"
1/4	1/4" = 1'-0"
3/16	3/16" = 1'-0"
1/8	1/8" = 1'-0"
3/32	3/32" = 1'-0"

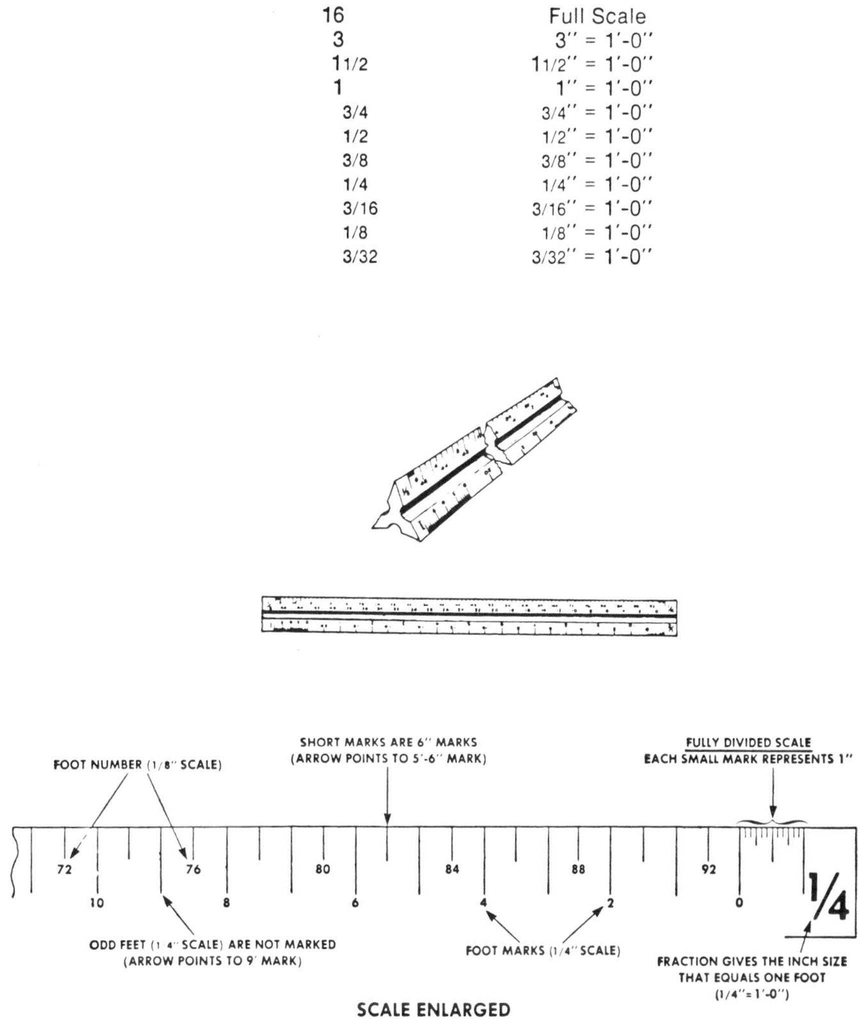

Figure 34 Architectural scale

architectural drawings Blueprints that show the basic nonstructural part of a building, as basic framing, windows, doors, schedules, trim, and so on. Identified by an A in the title box. Opposed to *structural drawing, mechanical drawing,* or *electrical drawing.* In a small set of working drawings, as for a house, all drawings are assumed to be architectural drawings and no separate structural, mechanical, or electrical drawings are made.

Architectural Graphic Standards A widely used guide or standards encyclopedia prepared by the American Institute of Architects (AIA). It provides drafting and practice standards.

architectural orders See *orders.*

architectural scale Scale with drawing reductions built into the separate sides. The triangular scale (Figure 34) has 15 separate scales built in. For example, the ¼" = 1'–0" scale has divisions representing measurements figured at that reduction.

architectural style Type of building design that has accepted and recognizable features. See *Colonial, English, French, nineteenth-century American, twentieth-century American, post-World War II.*

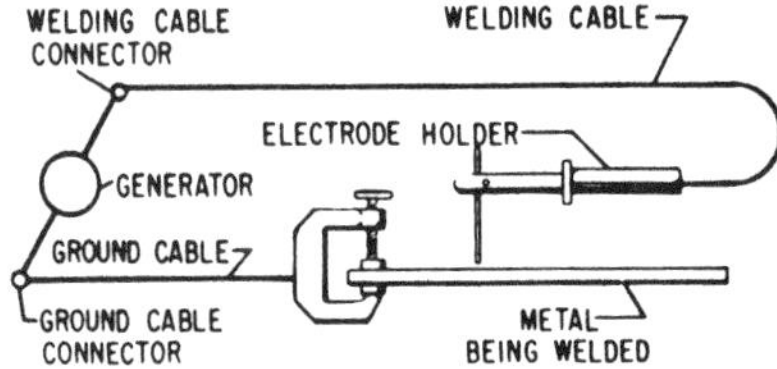

Figure 35 Arc-welding circuit

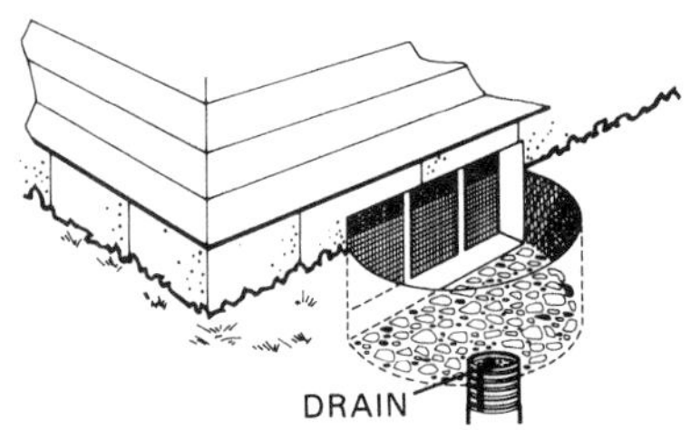

Figure 36 Areaway or window well

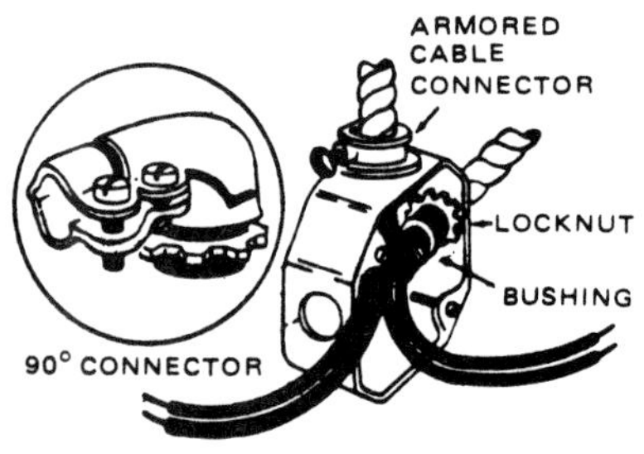

Figure 37 Armored cable

architecture The art and science of designing and building structures, including houses, apartment buildings, schools, commercial structures, bridges, churches, and so on.

architrave Molding above and on the sides of an opening; casing or trim.

arch stone A wedge-shaped stone used in building a masonry arch. A *keystone*.

arc length In arc welding, the distance the arc jumps from the end of the electrode to the work surface.

arc voltage In arc welding, the voltage used for a specific job; the voltage of the actual arc.

arc welding Electric welding process that uses the heat of an electric arc between an electrode and the work piece. Figure 35 shows a typical welding circuit. See *gas metal-arc welding (GMAW), gas tungsten-arc welding (GTAW), shielded metal-arc welding (SMAW)*. Electrical power is supplied by a *welding machine*. See also polarity.

area An open space, as a dining area, generally assigned a specific use or living purpose. The square footage or square-inch surface of a space. See also *cubic measure*.

area plan Preliminary site development plan. See *plot plan*.

areaway A sunken opening outside a basement window designed to admit light and provide an emergency exit; a window well. (Figures 36; *72*, p. 28; *784*, p. 344)

armor Metal wrapping used on electrical cables.

armored cable Flexible metal-clad cable with two or three rubber-insulated conductors. (Figures 37; *232*, p. 88.) Also called BX. Four types are used: AC, ACT, ACL, and MC. Type MC is used in commercial construction. Types AC and ACT are used in exposed or concealed work in dry locations. Type ACL can be run underground or imbedded in concrete or masonry units. See *conduit*.

arris A sharp edge formed when two surfaces meet at a sharp angle. (*Figure 380*, p. 152)

arris fillet Small strip of wood used at the eaves to raise roof tiles or slate so as to throw off water.

arriswise Sawing of square timber or lumber pieces at an angle.

artificial stone Odd-shaped cement-based colored building units made in molds. Designed to resemble field stone. Also a stucco-like colored building finish that is crafted to resemble set rough stone.

artisan A craftsman or artist.

asbestos A mineral fiber sometimes used as insulation. It is resistant to high temperatures. Less used today because of the harmful effect of the fibers.

asbestos cement A Portland cement with asbestos fibers mixed in.

asbestos shingles A fire-resistant roof shingle with asbestos added.

as-built drawings Drawings reflecting the actual construction with all final changes; *record drawings*.

ash dump A covered opening in the floor of a fireplace. It opens so the ashes can fall into an opening or *ash pit* under the fireplace. See *fireplace*.

ashlar Squared stone used in constructing walls.

ashlar brick A rough brick scored to resemble quarry stone.

ashlar masonry Masonry work of evenly shaped stone. (Figure 38.) See also *rubble masonry.*

ash pit Opening under a fireplace for holding ashes. Ashes are pushed into the ash pit through the *ash dump* in the floor of the fireplace. Ashes are removed at the base of the pit. See *fireplace.*

asphalt Bitumen; a heavy petroleum-based substance used in various forms for hot-mopped roofing, dampproofing or waterproofing, and for driveway and highway slabs.

asphalt felt A felt saturated or covered with asphalt. Used in built-up roofing. See *felt.*

asphalt lute A type of blunt-toothed *asphalt rake* used for spreading and smoothing asphalt. The toothed side is used for spreading and the opposite, flat side is used for smoothing.

asphalt mastic Asphalt mixed with aggregate used in bonding roofing. May be poured when hot.

asphalt paint A dampproof paint with an asphalt base.

asphalt rake Long-toothed rake used for spreading and smoothing asphalt. The toothed side is used for spreading. The opposite, flat side is used for smoothing. (Figure 39.) See also *asphalt lute.*

asphalt roofing Roofing consisting of layers of felt hot-mopped with asphalt. (Figure 40.)

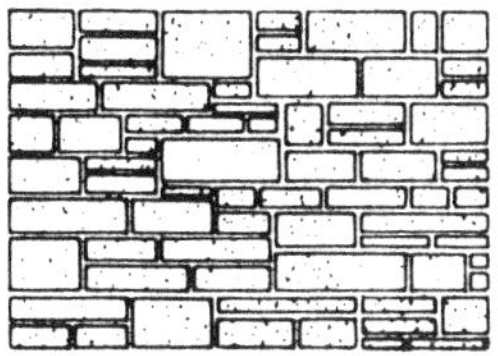

Random Stone

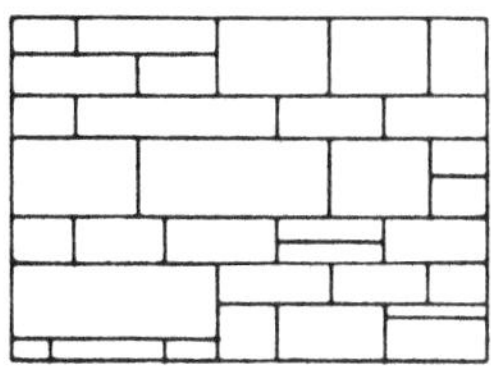

Coursed Stone

Figure 38 Ashlar masonry

Figure 39 Asphalt rake

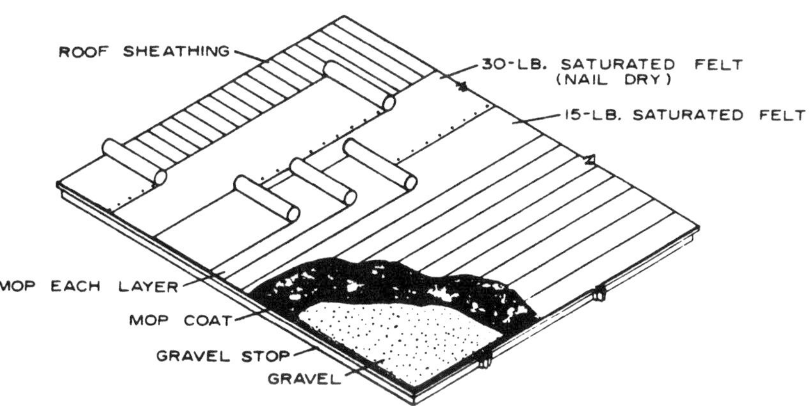

Figure 40 Asphalt roofing

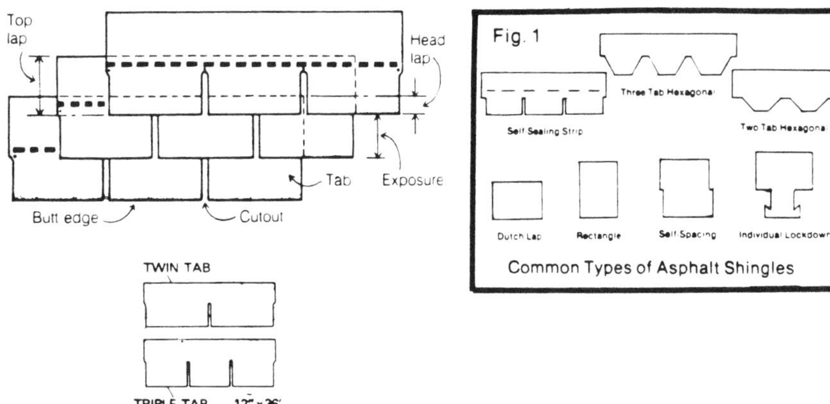

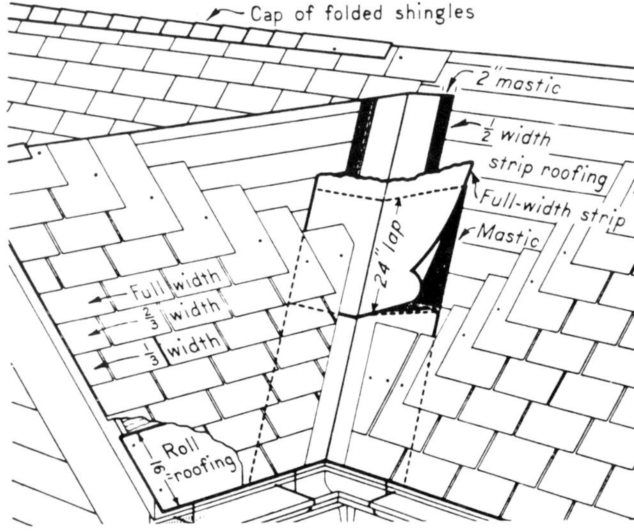

Figure 41 Asphalt shingle layout

asphalt shingle Asphalt-based roofing shingle topped with crushed rock. Figure 41 shows installation techniques.

assembly drawing Drawing that shows how the parts of an assemblage, structure, or machine go together.

Associated General Contractors of America An association of contractors engaged in construction.

ASTM American Society for Testing and Materials.

astragal On double-door construction, a vertical member attached to the lock stile of one door so as to overlay the second door when closed. A bead molding. See *molding.*

as welded The weldment after welding but before any treatment or finish, such as filing or grinding.

atmospheric pressure Pressure of the surrounding air. At sea level 14.7 pounds per square inch.

attachment plug Two- or three-pronged device at the end of a flexible cord running from electrical equipment, appliances, or lamps. Used to plug into a receptacle to obtain electrical current.

attenuation Thinning down or tapering to a point.

attic Unheated space above the ceiling of the top floor and below the roof rafters.

attic exhaust fan Large fan installed in the attic floor so it draws air directly from the top floors of the house. Air pulled in from the house is vented out the attic vents.

attic ventilation Circulation of air into and through the attic area. Louvers are used at the gable ends of the roof, often assisted by a ventilator fan. (Figure 42.) Fresh air is drawn into the attic through eave vents, screens in the cornice plancier or soffit. *Roof vents* or a *ridge ventilator* may also expel attic air. Attic ventilation is critical in both summer and winter. In the summer, heat build-up will cause very high temperatures in the attic (much higher than outside), which in turn leads to overheating of the house living area or to unneeded use of the cooling system. Without adequate ventilation in the winter, moisture will collect, causing rotting. Freezing of the moisture can cause structural parts to separate from each other.

auger See *drain auger.*

auger bit Spiral-shanked, single or double twist bit. The tip has a feed screw that draws the bit into the wood. The outside cutting lip or spur starts the cut. (Figure 43.) The auger bit is used in a hand brace or power drill. Sizes and lengths vary. See also *bit, ship auger.*

auger bit file Very fine, flat file used for sharpening the spur and cutting edge of an auger bit or ship auger bit. See *file.*

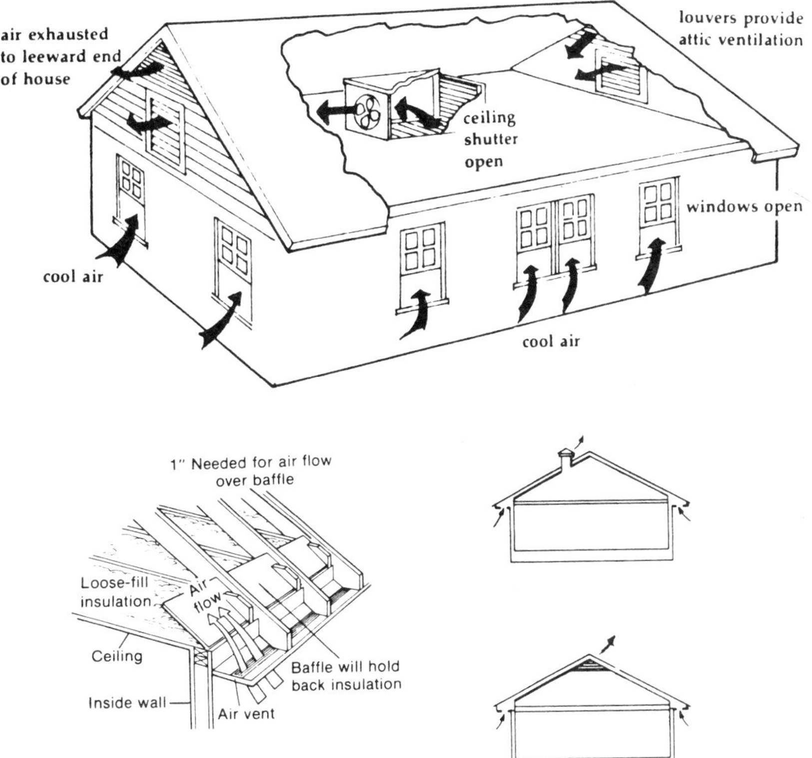

Figure 42 Attic ventilation

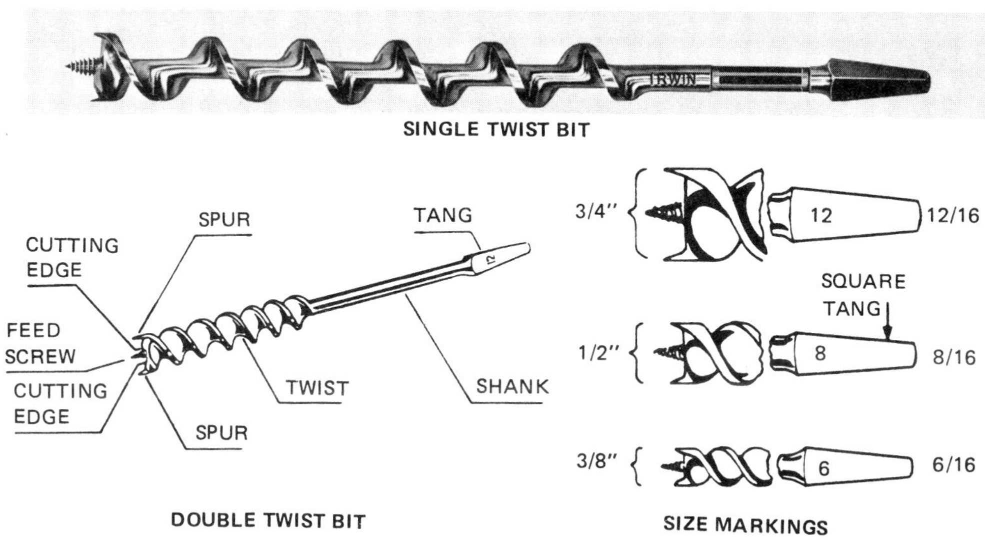

Figure 43 Auger bit

Figure 44 Automatic level

Figure 45 Automatic taper

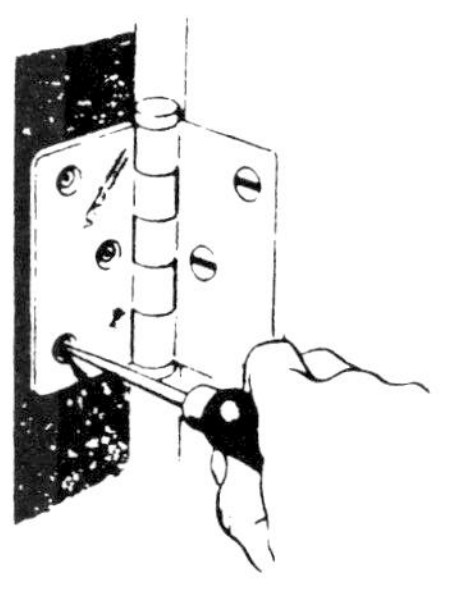

Figure 46 Awl

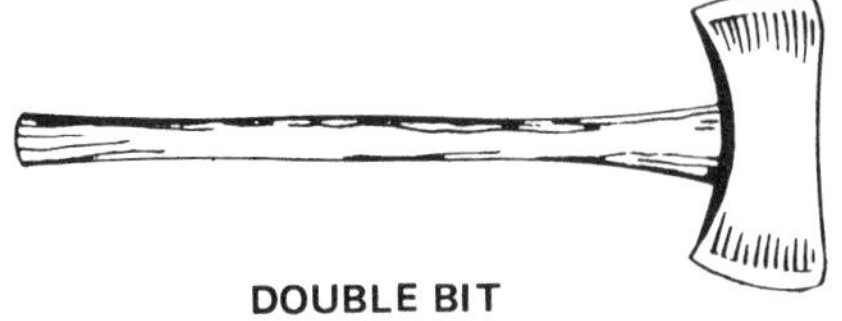

Figure 47 Axes

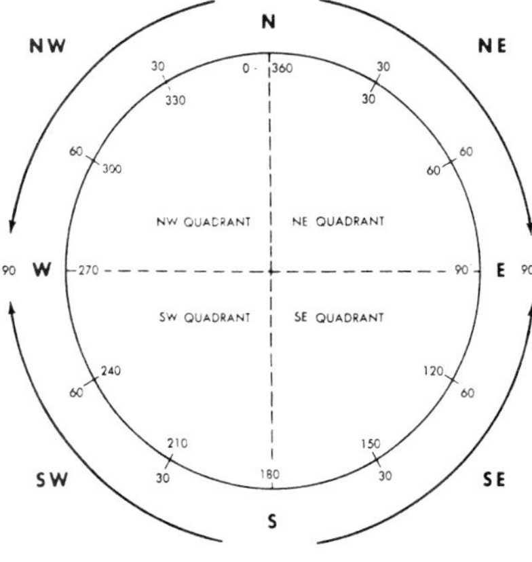

Figure 48 Azimuth (inside circle) and bearings (outside circle)

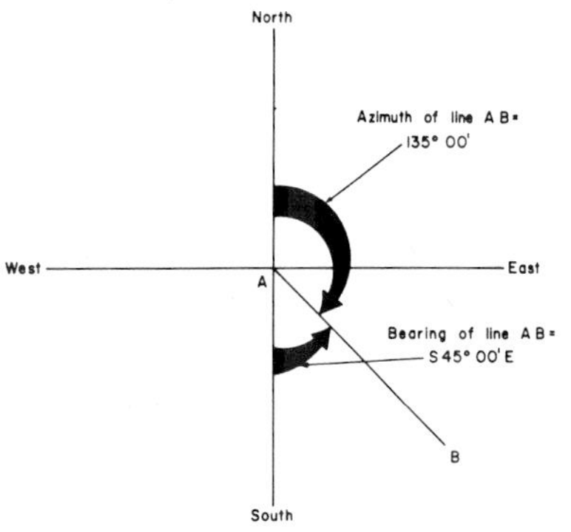

Figure 49 Azimuth and bearings

automatic level Builder's level that automatically levels itself and needs no manual adjustments. (Figure 44.)

automatic taper Long-arm drywall taper that simultaneously applies joint compound and tape to the joint. (Figure 45.)

automatic vent damper A damper in the furnace vent that automatically closes when the furnace burner shuts off. Designed to prevent the loss of heated air up the flue pipe. See *barometric damper.*

automatic welding Welding that is completely automatic; no operator control or adjustment is involved. See *machine, manual, or semiautomatic welding.* See *welding process* for a comparison of the processes.

awl Sharp-pointed tool used for making holes in wood to start a screw. (Figure 46.)

awning window Outswinging window with hinges at the top. See *window.*

AWS American Welding Society.

axe Long-handled, wide-bladed tool used for chopping and cutting wood. (Figure 47.) See also *hatchet.*

axial force Compression or tension along the length of a member.

azimuth Identification of a direction using a 360° circle; measurement is clockwise in degrees. Normally, in-plane surveying azimuths are read from 0° north clockwise around the circle. (Figure 48.) For example, east has an azimuth of 90°, south is 180°, west is 270°. Figure 49 shows a north azimuth for line *AB* of 135°. South is used as a 0° reference point in geodetic survey and astronomy; measurement is clockwise. See also *bearing, magnetic meridian, solar azimuth, true meridian.*

b

back The rear side, as the rear of a panel.

back cut Cut made into the back side of a wood piece so the material will lie flat, as on a wood flooring piece.

backfill Sand, gravel, pea stone, crushed stone, slag, or cinders used for filling around foundations or piping. In general, to back fill is to replace earth in a trench or around a foundation.

backfire In oxyacetylene welding, a condition where the torch suddenly goes out with a loud snap or pop.

backflow In plumbing, an undesirable reversal in the direction of water flowing in a pipe; the flow of fluids other than pure water into the water supply system.

backhand welding A welding technique where the welding torch slants in the opposite direction to the process of the weld. (Figure 50.) Opposed to *forehand welding.*

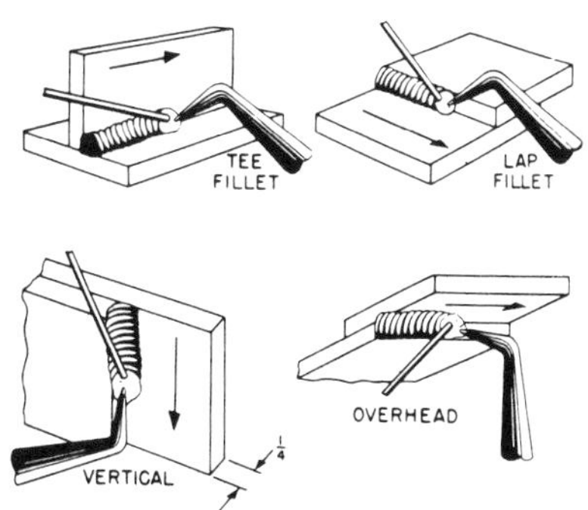

Figure 50 Backhand welding

Figure 51 Backhoe

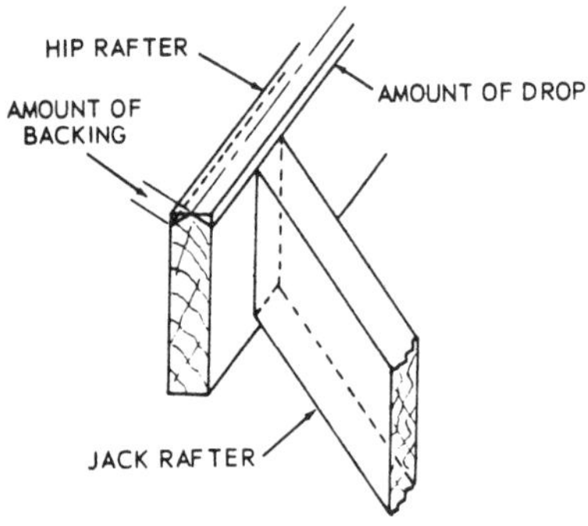

Figure 52 Backing

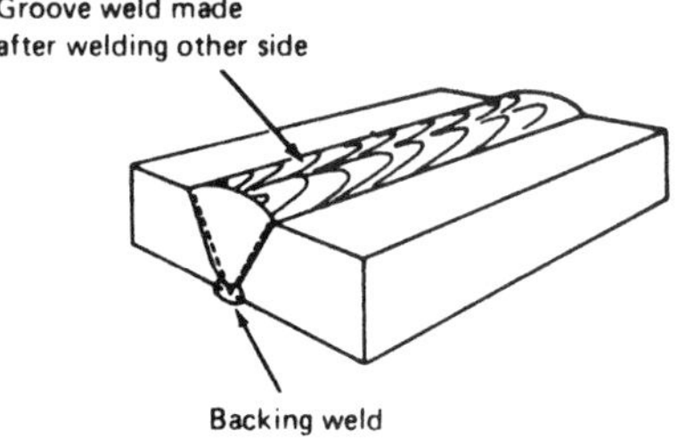

Figure 53 Backing weld

back hearth The back area of a hearth where the fire is built. See *fireplace.*

backhoe An excavating machine having a bucket in front that digs by being drawn downward and back toward the machine. (Figure 51.) See also *excavator.*

backing In masonry: base for the first plaster coat; random rubble stone; bricks behind the facing bricks. In carpentry: small wood strips used to provide nailing base for finish at partition corners; shims used on joists to provide a level surface; a bevel on a hip rafter made to fit the adjacent roof structure. (Figure 52.) In welding: a material placed behind the work to support the molten weld metal.

backing up Laying of backing—bricks or concrete block—on the inner side of a wall.

backing weld Weld bead made on the back side of a single-groove weld before welding on the face side. (Figure 53.)

backnailing Nailing of roof plies in built-up roofing; also, *blindnailing.*

back pressure In a plumbing system, pressure that causes water in a system to reverse its direction.

backsaw Fine-toothed saw used in cabinet work and for fine cutting. The saw gets its name from the metal strip running along the back edge for reinforcement. The backsaw is available in sizes from 12″ to 28″ with 11 to 14 teeth to the inch. The backsaw is used in a *miter box.* See also *dovetail saw.*

backset Horizontal distance from door lock to keyhole or doorknob.

backsight In surveying, a rod reading taken to a previous known point from a new position, as opposed to *foresight,* which is a rod reading taken at an unknown location. Also called a *plus sight.* Abbreviated as B.S. or +S. (Figure 54.)

back-to-back Placement of plumbing fixtures together and connected at the same level, as two bathrooms are commonly placed back-to-back and share the same vent stack.

backup Withe of brick or concrete block behind the exterior withe or facing masonry; *backing.*

backup brick Brick used on inside withe.

backup strips Wood strips used to provide nailing base.

backup system A system that parallels a functioning system but is activated only when the first system fails—for example, a conventional heating system that backs up a *solar heating* system or a *heat pump.*

back weld A weld made on a back of a single-groove weld. A *backing weld.*

badger Instrument for cleaning mortar out of drain joints.

baffle Plate or surface for deflecting or deadening sound to provide better acoustics; a plate for regulating fluid or gas flow.

bagtie Wire ties used to fasten reinforcing bars together at intersections. See *bar tie.*

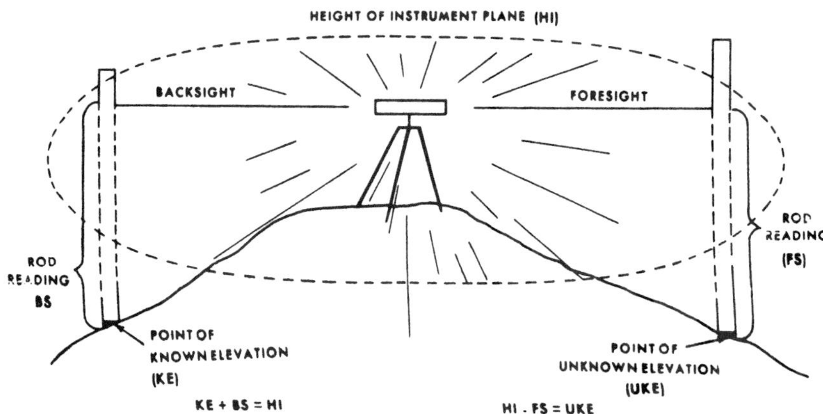

Figure 54 Backsight

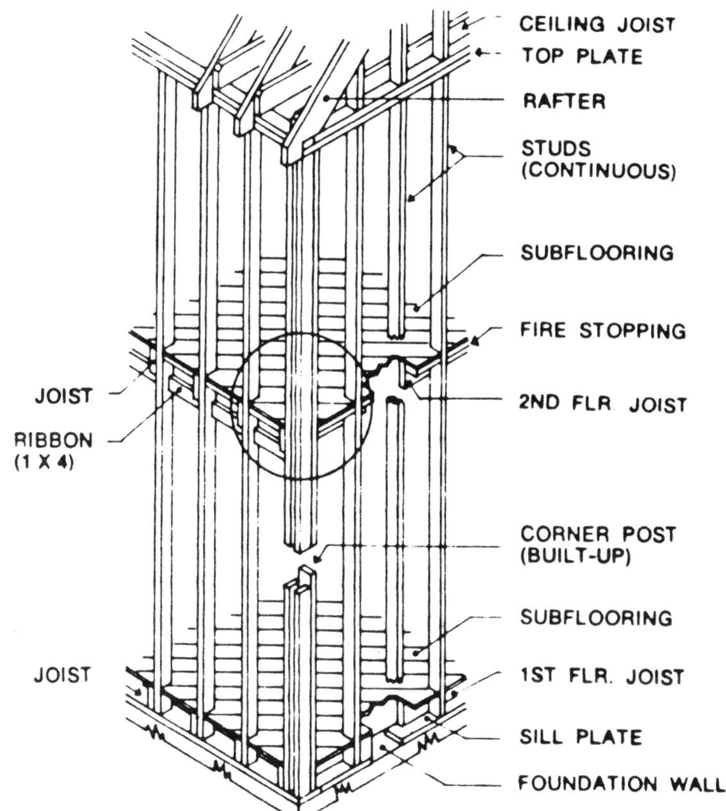

Figure 55 Balloon framing

balcony An outside platform, with railings around the three sides, that projects out from the face of a wall. Entrance onto the balcony is made from a door leading from the inside of the structure. May be supported by columns or by brackets.

ballast A current-limiting device used to stabilize and maintain a constant current in vapor lamps, such as a fluorescent lamp.

balloon framing A type of frame (wood) construction where the studs are continuous from foundation sill to roof line; joists at each floor are nailed to the side of studs (Figure 55) and rest on a ribbon or ledger board (*Figure 397*, p. 160).

ball peen hammer Hammer used in metalworking: the striking surfaces are especially tempered for striking metal. The hammer has a ball-shaped striking head and a peen. Various weights are available. (Figure 56.) See also *cross peen*, *straight peen*.

baluster Small upright pillars in a balustrade; supports for a stair railing. (Figure 57.) See *stairs*.

balustrade A railing; balusters held at the top by a rail and seated at the bottom in a stair or floor.

bandage Metal band or strip secured around a structure, such as a tower, to give strength.

band clamp See *weld clamp*.

band joists Joists that run on the outside of the floor platform. (*Figure 81*, p. 32) Also called *rim joists*, *sill joists*. See specifically *end joists*, *header joists*. See also *platform framing*.

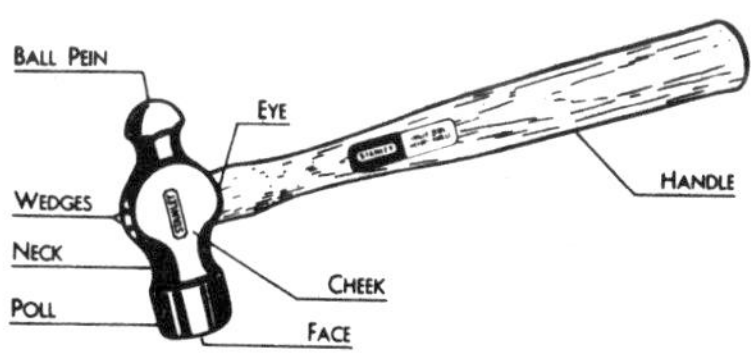

Figure 56 Ball peen hammer

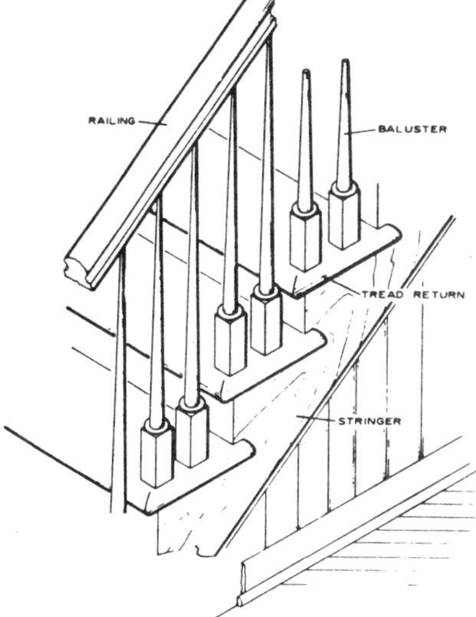

Figure 57 Balusters supporting a stair railing

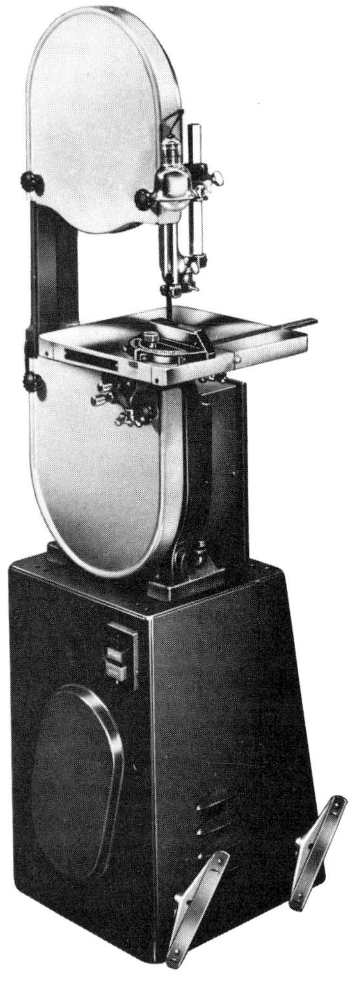

Figure 58 Band saw

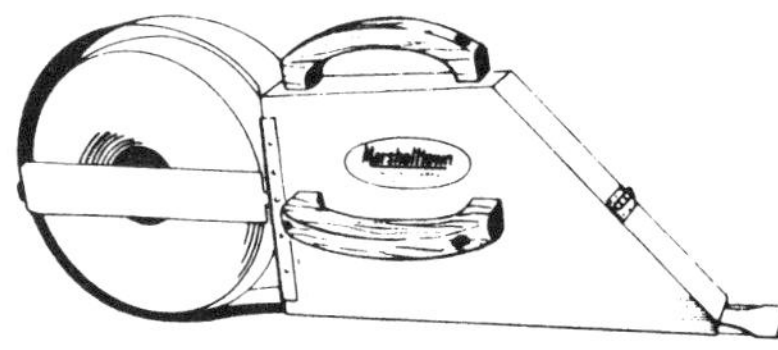

Figure 59 Banjo drywall taper

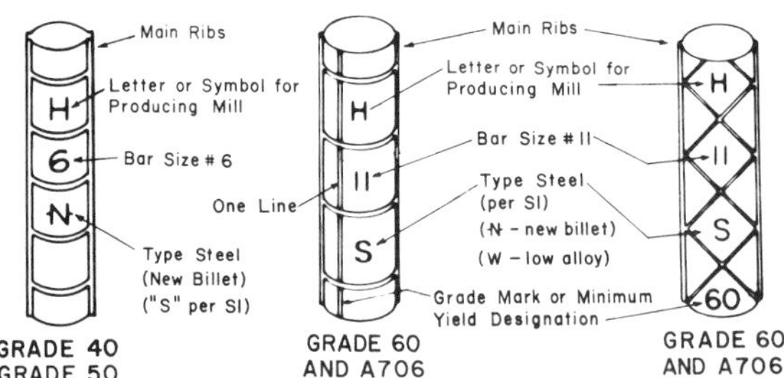

1st — Producing Mill (usually a letter)
2nd — Bar Size Number (#3 through #18)
3rd — Type Steel: N for New Billet
S for Supplemental Requirements A615
A for Axle
I for Rail
W for Low Alloy

Figure 60 Bars (deformed)

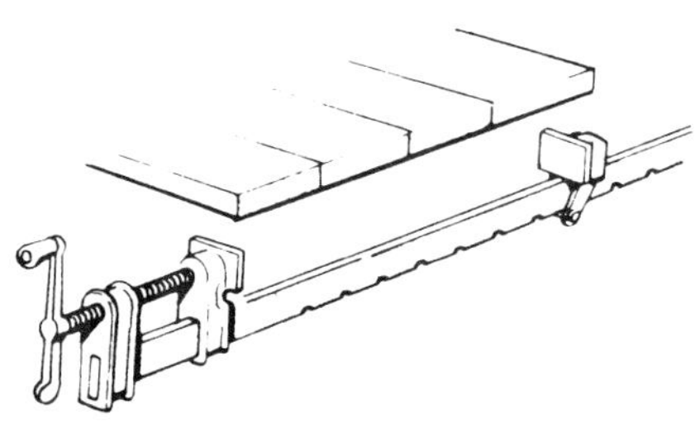

Figure 61 Bar clamp

band saw Woodworking saw with a continuous revolving blade running around two wheels. (Figure 58.) Widely used for cutting curved lines or for contour cutting.

banister Balustrade.

banjo drywall taper In drywall construction, a tool for holding and dispensing joint tape. (Figure 59.) See also *automatic taper, tape corner tool.*

banker Workbench used by bricklayers and stone masons when shaping arches.

bar Round rod used for reinforcing concrete, also called a *rebar.* (Figure 60.) Deformed bars (with raised ribs) are normally used. See also *hook, truss bar, reinforcing bars.* Also, the wooden *sash bars* used for separating window panes. See *double-hung window, window.*

bar chair Shaped wire device for holding up reinforcing bars or welded-wire fabric so concrete will flow evenly around the steel reinforcement. See *bar support, bolster, chair.*

bar chart Graph used by builders to schedule and chart progress of construction work. Horizontal bar represents time from beginning to completion.

bar clamp Woodworking clamp with jaws that slide on a long bar. (Figure 61.) Similar in usage to a *pipe clamp.*

bare conductor Electrical conductor with no covering.

bare electrode Welding electrode that has no coating, used as filler metal. See *covered electrode.*

barefaced tenon A tenon shouldered on one side only.

bare-metal arc welding (BMAW) Arc-welding process that uses a consumable bare, uncoated electrode.

bar folder Sheet metal machine used to bend edges. (Figure 62.) See also *brake.*

barge board Board that covers gable end rafters. A *verge board.* Fascia board on the cornice.

barge course Course of bricks laid on edge to form wall coping.

barge rafter Rafter at a gable end overhang. The barge rafters butt against each other at the ridge past the end of the ridgeboard. (Figure 63.) See *fly rafter,* which frames into the ridgeboard at the same location.

bar hanger Hanger used to mount ceiling electrical outlet boxes. (Figure 64.)

bar joist See open-web joist.

bark Rough outer covering of a tree trunk. See *tree.*

bark pocket Bark partially or wholly enclosed in the tree wood.

bar list Schedule giving reinforcement bar sizes, types, lengths, and bending information. See also *bar schedule, bending schedule.*

bar number Size number rolled onto the surface of reinforcement bars. Roughly represents $\frac{1}{8}''$. For example, a 4 denotes a bar with a diameter of $\frac{4}{8}''$ or $\frac{1}{2}''$. (Figure 65.)

barometric damper Damper on furnace flue that is designed to automatically open and close for better draft control. See *automatic vent damper.*

bar-placing subcontractor Subcontractor responsible for correct placement of reinforcing bars before concrete is placed; also called a bar placer or placer.

barrel arch An arch shaped like the curve of a barrel side.

barrel bolt Door fastener with a manually operated sliding bolt. (Figure 66.)

barrel roof Roof shaped like the inside of a barrel. (Figure 67.)

Figure 62 Bar folder

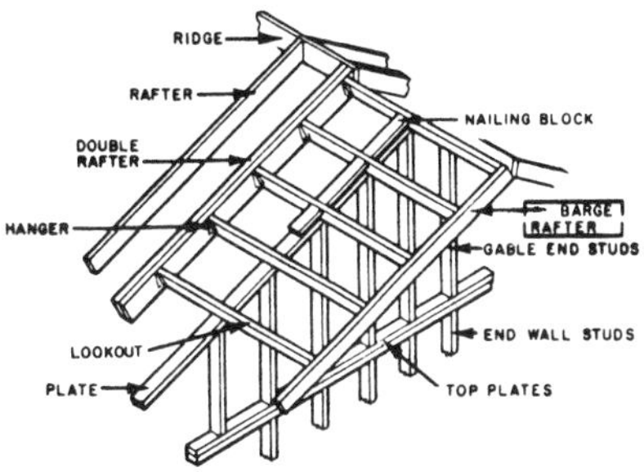

Figure 63 Barge rafter

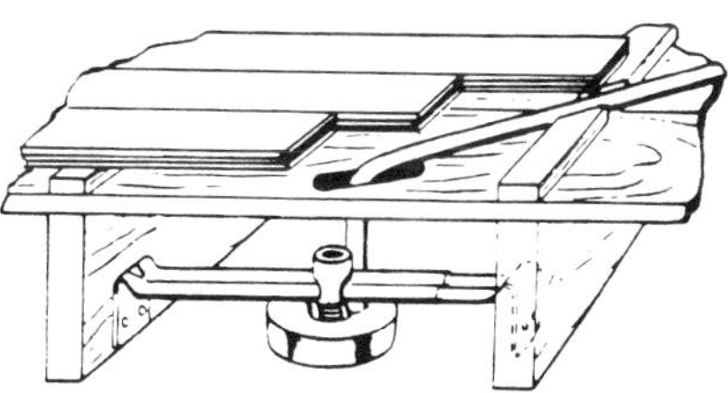

Figure 64 Bar hanger and ceiling outlet box

ASTM STANDARD REINFORCING BARS				
BAR SIZE	WEIGHT	NOMINAL DIMENSIONS—ROUND SECTIONS		
DESIGNATION	POUNDS PER FOOT	DIAMETER INCHES	CROSS SECTIONAL AREA SQ. INCHES	PERIMETER INCHES
#3	.376	.375	.11	1.178
#4	.668	.500	.20	1.571
#5	1.043	.625	.31	1.963
#6	1.502	.750	.44	2.356
#7	2.044	.875	.60	2.749
#8	2.670	1.000	.79	3.142
#9	3.400	1.128	1.00	3.544
#10	4.303	1.270	1.27	3.990
#11	5.313	1.410	1.56	4.430
#14	7.65	1.693	2.25	5.32
#18	13.60	2.257	4.00	7.09

Figure 65 Bar numbers: #3 to #18

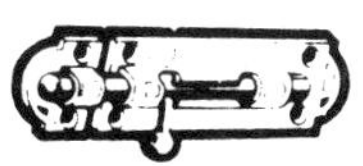

Figure 66 Barrel bolt

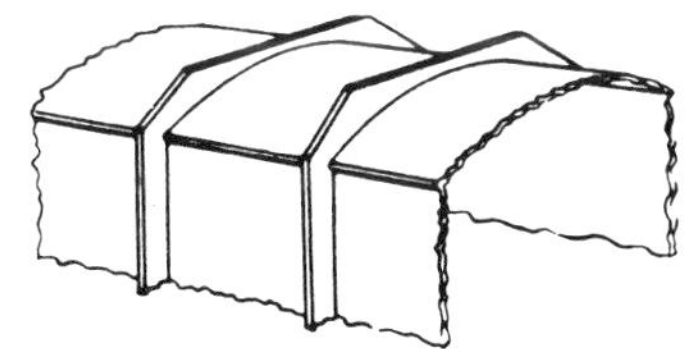

Figure 67 Barrel roof

SYMBOL	BAR SUPPORT ILLUSTRATION	BAR SUPPORT ILLUSTRATION PLASTIC CAPPED OR DIPPED	TYPE OF SUPPORT	SIZES
SB	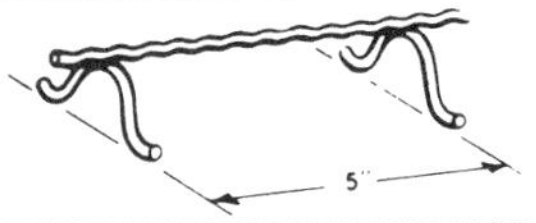 5"	 CAPPED 5"	Slab Bolster	¾, 1, 1½, and 2 inch heights in 5 ft. and 10 ft. lengths
SBU*	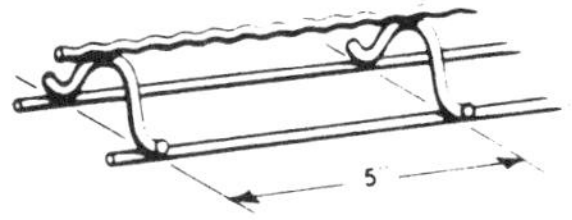 5"		Slab Bolster Upper	Same as SB
BB	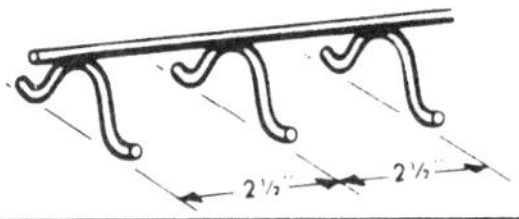 2½" 2½"	 CAPPED 2½" 2½"	Beam Bolster	1, 1½, 2, over 2" to 5" heights in increments of ¼" in lengths of 5 ft.
BBU*	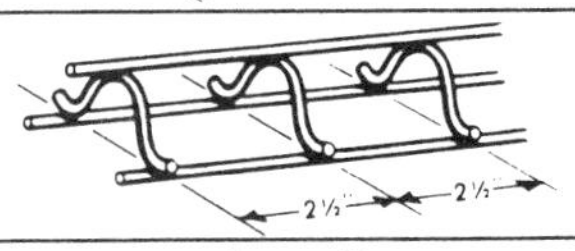 2½" 2½"		Beam Bolster Upper	Same as BB
BC	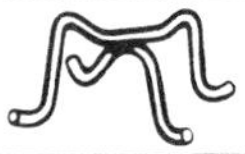	DIPPED 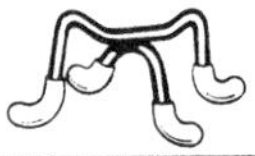	Individual Bar Chair	¾, 1, 1½, and 1¾" heights
JC	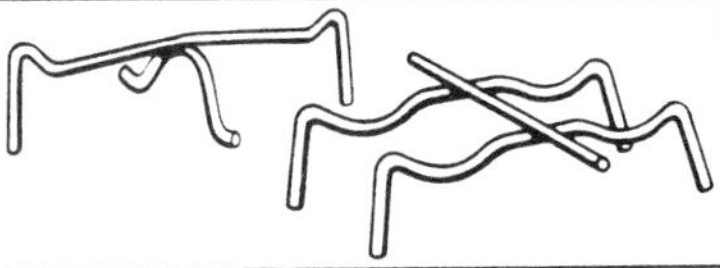	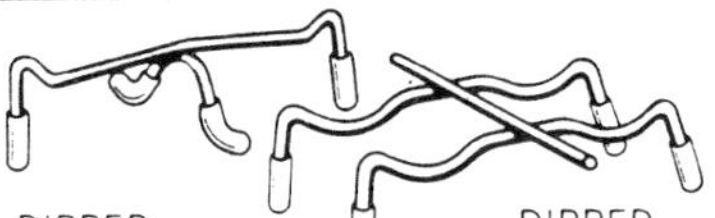 DIPPED DIPPED	Joist Chair	4, 5, and 6 inch widths and ¾, 1 and 1½ inch heights
HC	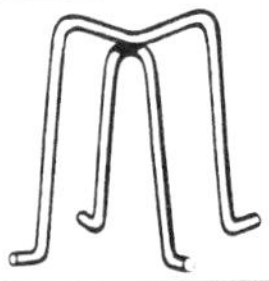	CAPPED 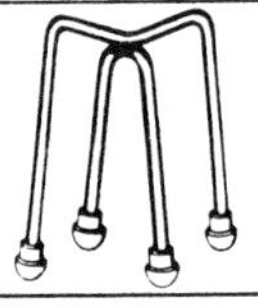	Individual High Chair	2 to 15 inch heights in increments of ¼ inch
HCM*	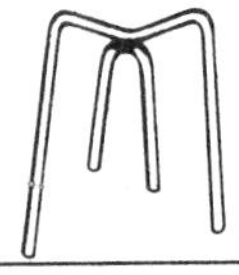		High Chair for Metal Deck	2 to 15 inch heights in increments of ¼ in.
CHC	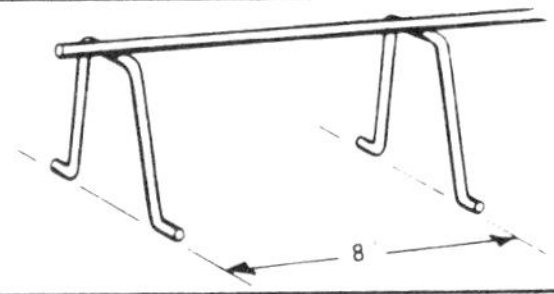 8"	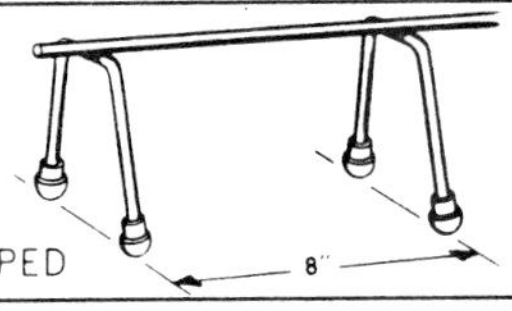 CAPPED 8"	Continuous High Chair	Same as HC in 5 foot and 10 foot lengths
CHCU*	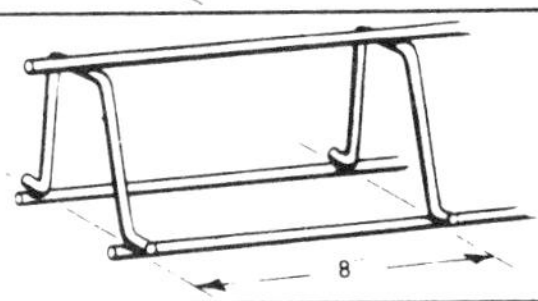 8"		Continuous High Chair Upper	Same as CHC
CHCM*	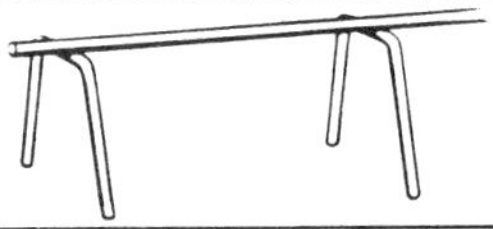		Continuous High Chair for Metal Deck	Up to 5 inch heights in increments of ¼ in.
JCU**	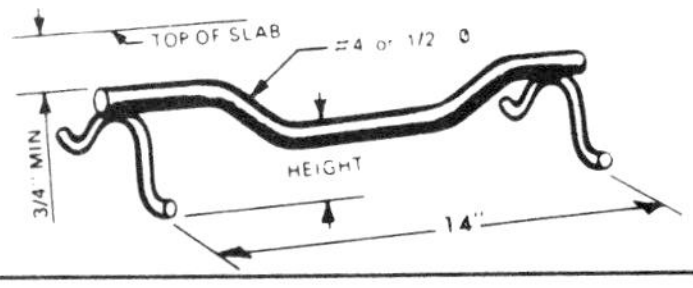 TOP OF SLAB, #4 or 1/2 Ø, 3/4 MIN, HEIGHT, 14"	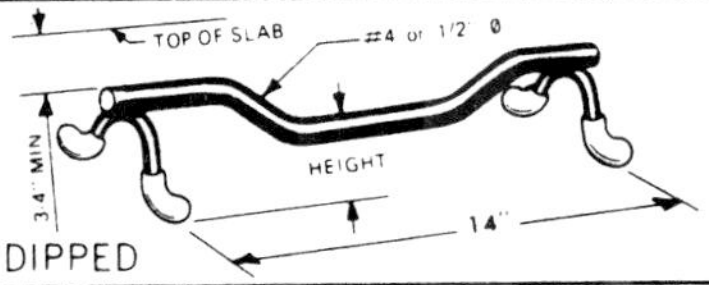 TOP OF SLAB, #4 or 1/2 Ø, 3/4 MIN, HEIGHT, 14", DIPPED	Joist Chair Upper	14" Span. Heights −1" thru +3½" vary in ¼" increments

*Usually available in Class 3 only, except on special order.
**Usually available in Class 3 only, with upturned or end bearing legs.

Figure 68 Bar support

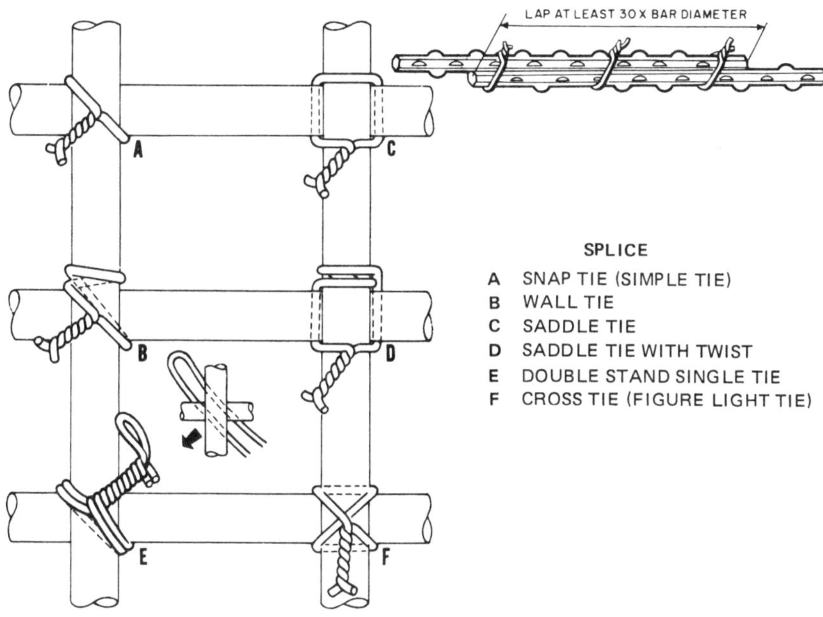

Figure 69 Bar ties

bar schedule List giving reinforcement bar size, type, length, and bending information. See *bar list, bending schedule.*

bar spacing Placement of reinforcing bars measured center-to-center of bars. Bars run parallel to each other and at right angles to each other; they are wired together to form a *mat.*

bar supports Formed wire supports used to hold up reinforcing bars so concrete can flow around the bars. (Figure 68.) See *bar chairs, bolsters, chairs.*

bar tie Wire ties used to fasten reinforcing bars together. (Figure 69.)

basal angle Angle measured from the base of something.

base Lowest part, as the base of a wall.

baseboard Board at the base; finishing board at the bottom of a wall next to the floor. (Figures 70; *178,* p. 65.)

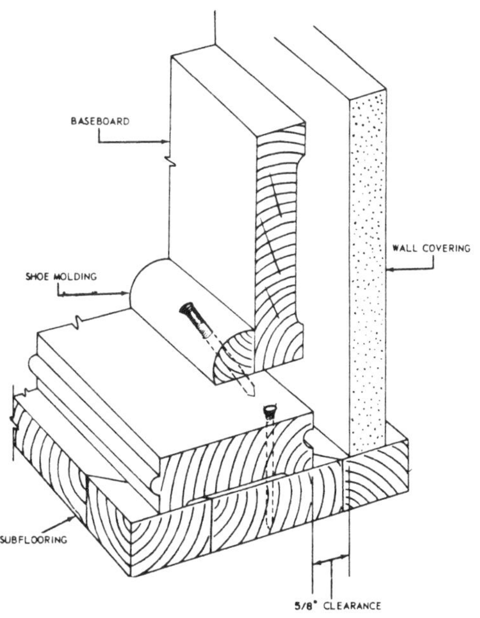

Figure 70 Baseboard

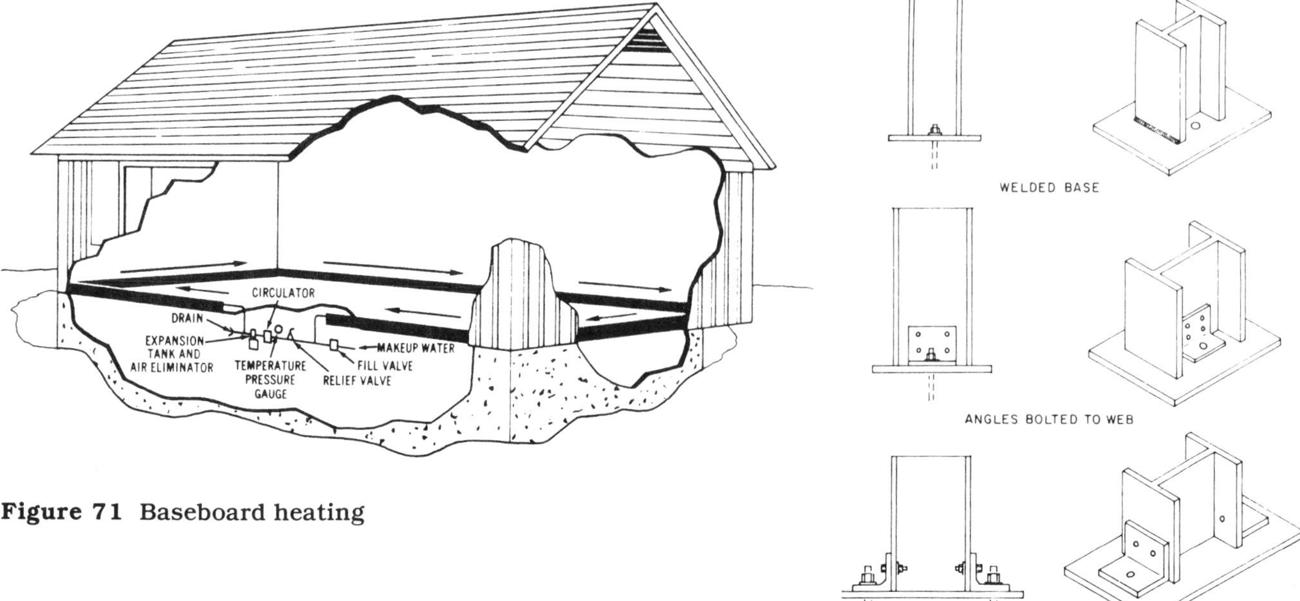

Figure 71 Baseboard heating

Figure 73 Base plates

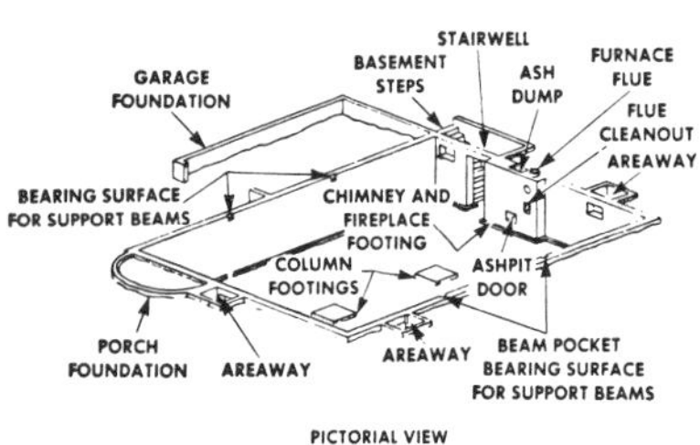

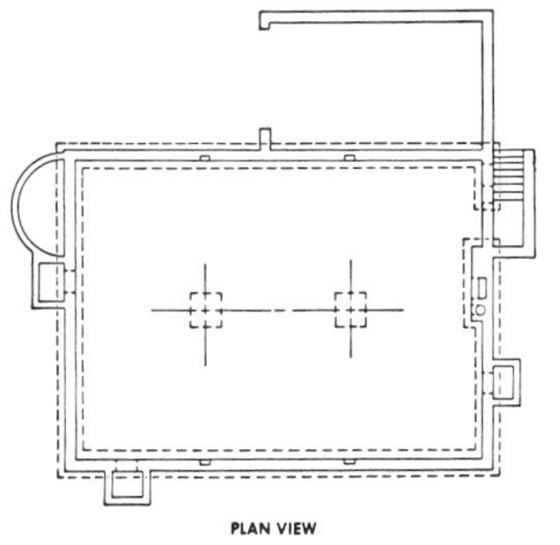

Figure 72 Basement and plan view

baseboard heating Perimeter heating with radiators or convectors located at the base of the walls. (Figure 71.) The heating unit is located where the baseboard would normally be. Hot water is circulated by pump through the system. See also *convector.*

base coat Stain or paint applied to raw wood as a base for final finish. See also *body coat.*

base course First or lowest course in a masonry wall.

base line Established building line or reference line from which other measurements are taken. Often the top of foundation is assigned an elevation of 100.0′ and other elevations and measurements are taken from there. In *perspective* the intersection of the *ground plane* and the *picture plane.* In surveying, an east-west latitude line that is used to establish a local line for establishing townships. See *principal meridian, township numbers.* Also, in surveying, a line of known length used as a starting point in a survey. See also *triangulation.*

basement The story of a building below the main floor; a story partly or wholly below the ground level; the finished portion of a building below the main floor or section. (Figures 72 (top); *529,* p. 227.)

basement plan Plan view that shows the layout of the basement, including location of stairs, doors, windows, and partitions. Symbols are commonly used showing location of the furnace, plumbing fixtures, and electrical outlets and switches. Figure 72 (bottom) shows a simplified plan view. See also *foundation plan.*

basement wall foundation A continuous outer wall that supports the structure forming a room-high inside space.

base metal In welding, the material to be welded, brazed, or cut.

base molding Molding at the bottom of a wall at the intersection with the floor. See *chair rail, molding.*

base plate Steel plate on which a column rests. (Figure 73.)

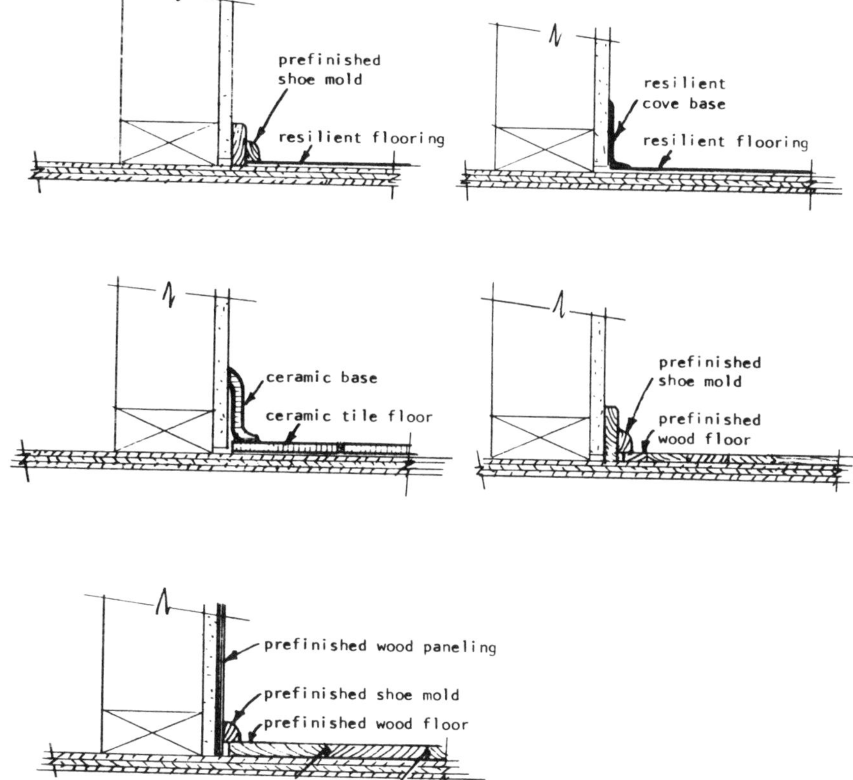

Figure 74 Base trim: details shown with different finish floors

base ply In roofing, the first ply of a built-up roof membrane; a *base sheet.*

base sheet In roofing, the first sheet or ply of asphalt felt placed in a built-up roof; a *base ply.*

base shoe Molding strip nailed to the bottom of the baseboard. See *molding.*

base trim Finish made at the base of the wall over the finish floor. (Figure 74.)

basil A *bezel;* beveled cutting edge.

bas relief Sculpture that projects out slightly from the background; a *low relief.*

bastard-sawed Lumber sawed at 30° to 60° angle to the annual rings. Between *quarter-sawed* and *plainsawed.*

bat Brick piece that has been broken off the whole brick. See *cut brick.* One end is still complete; the other end is broken.

batch An amount of concrete mixed at any one time.

batching Proportioning out the ingredients for a batch of concrete or mortar and mixing together.

batching plant Manufacturing plant where concrete mix is prepared. See *central-mixed concrete.*

bath-shower module An enclosed one-piece fiberglass plumbing unit that includes both a shower and bath.

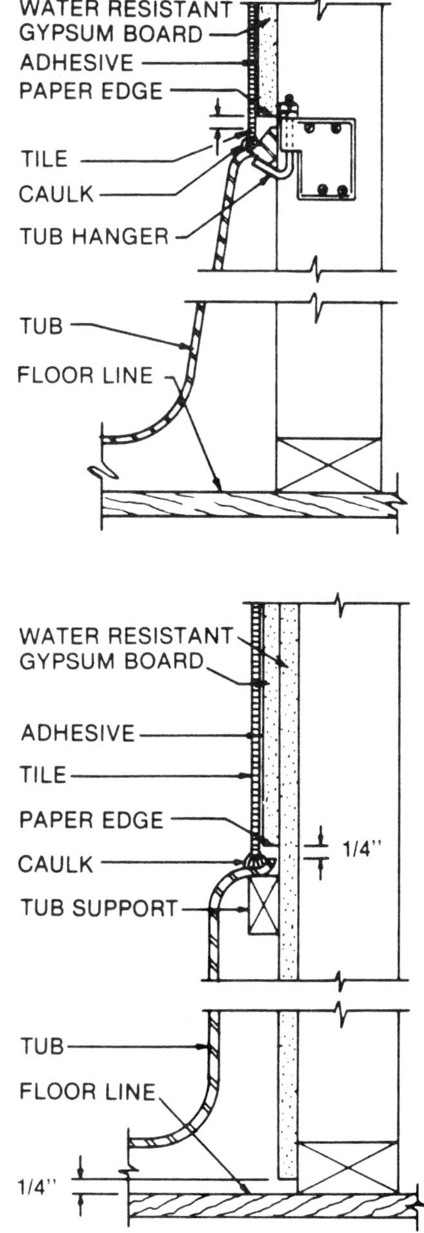

Figure 75 Bathtub support

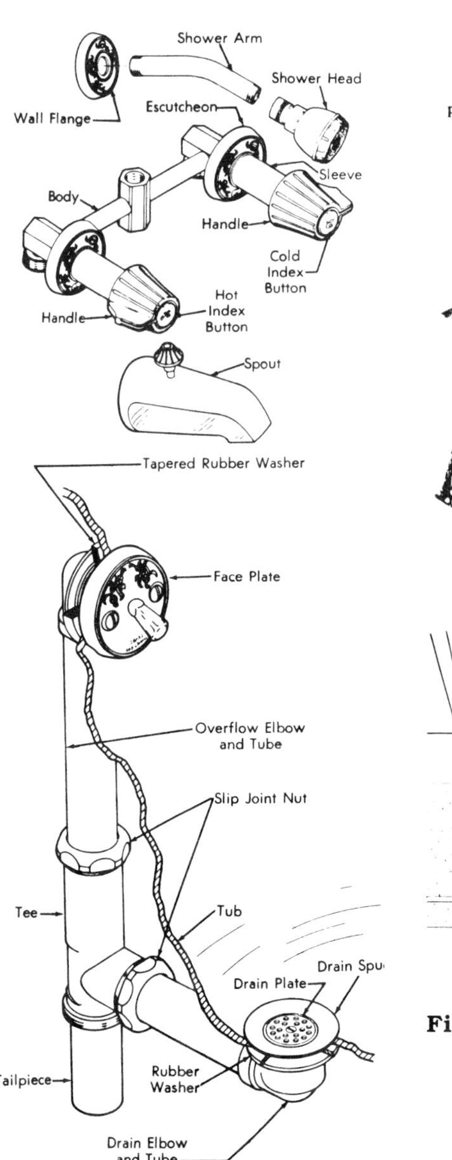

Figure 76 Bathtub: hot- and cold-water fixtures and drain

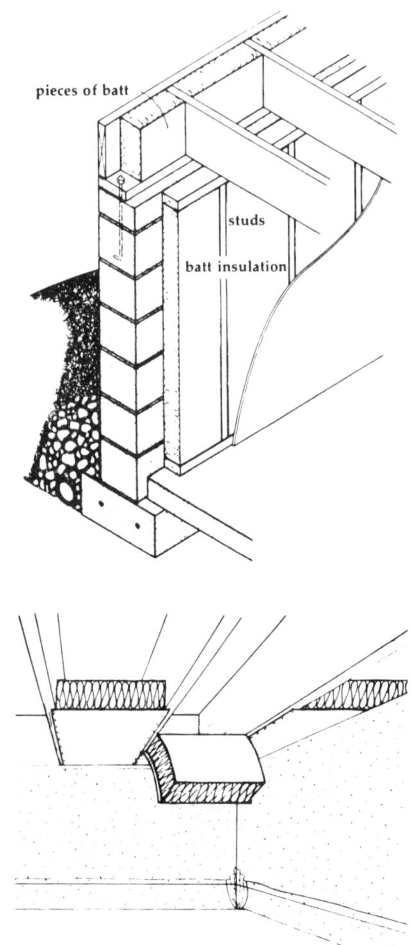

Figure 77 Batt insulation

bathtub Large receptable for washing the whole body in. There are two varieties: righthand and lefthand. A righthand bathtub has the drain on the right (as you face it); a lefthand bathtub has the drain on the left (as you face it). Most bathtubs are recessed or built into the bath area. (Figure 75.) Figure 76 shows a hot- and cold-water inlet and drain for a tub.

batt Batt insulation; insulation, usually fiberglass, compacted together in continuous sections or blankets. Sizes are from $3\frac{1}{2}''$ to $5\frac{1}{2}''$ thick and from $15\frac{1}{4}''$ to $23\frac{1}{4}''$ wide. Batt insulation is placed between studs, joists, or ceiling rafters. (Figures 77; *512*, p. 221.) See *insulation*.

batten A narrow board or wood strip used to cover openings between other boards or panels. (Figure 78.) A type of siding called *boards and battens.* A long narrow board used across several boards or planks as a support and reinforcement; a long cleat. See also *molding.*

batten plate In steel construction, a steel plate used to strengthen the member or a joint.

batter A wall or slope that inclines or recedes away from the viewer; a wall out of plumb (slanting away from true vertical); a wall or *retaining wall* that is perpendicular on the intersurface but has a sloping outer face: narrower at the top, wider at the base. A slant surface or *draft.*

batter boards Position boards erected to locate corners of new buildings—the horizontal boards are nailed at right angles to each other and supported on posts. (Figure 79.) They locate the corners and, with cords stretched between different corners, the location of the excavation and the location of footings and foundations, in plan and elevation.

battering wall Wall with a sloping outer face used for holding or retaining earth or water. A *retaining wall.*

battery Device for storing electricity.

bay Area projecting out from a wall with windows. A space between open columns or piers; space between two rigid support systems in a structure. (Figure 80.) Also, open storage areas or compartments in a warehouse or barn. See *rigid frame construction.* (*Figure 792,* p. 249) In rigid frame construction, two *bents* make a bay.

bayonet saw See *saber saw.*

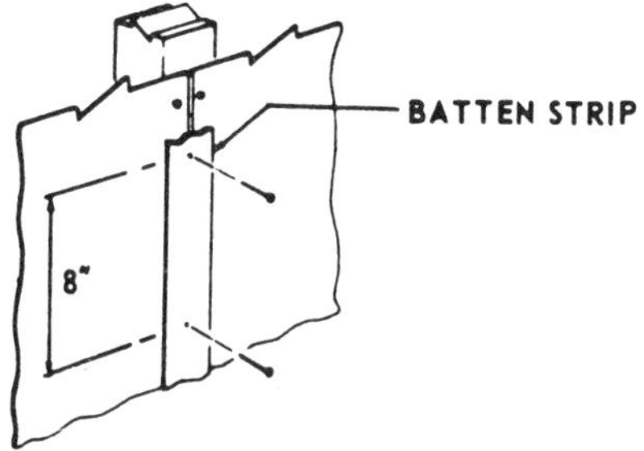

Figure 78 Batten strip

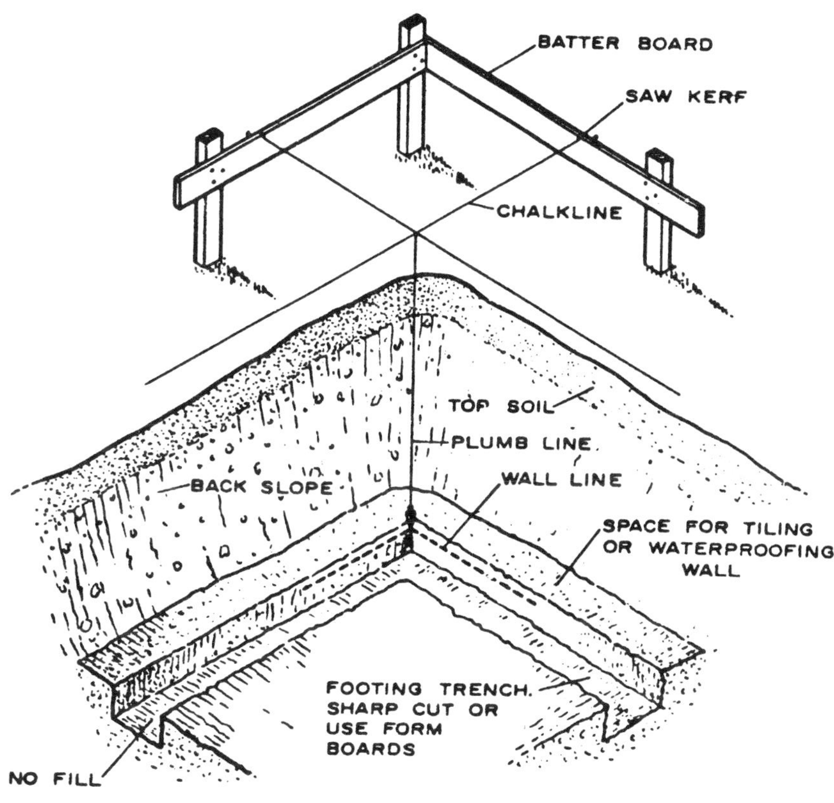

Figure 79 Batter boards

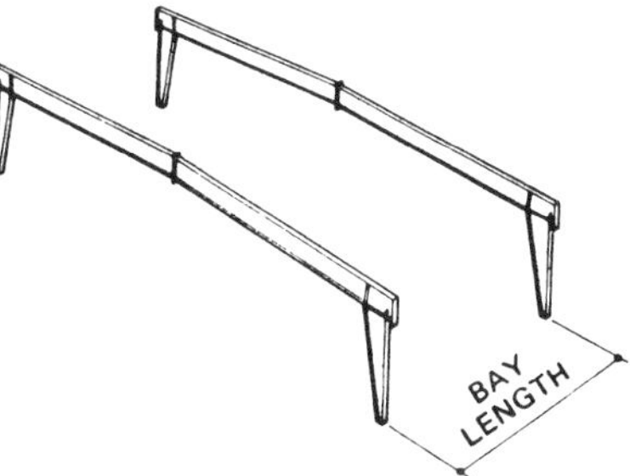

Figure 80 Bay (rigid frame construction)

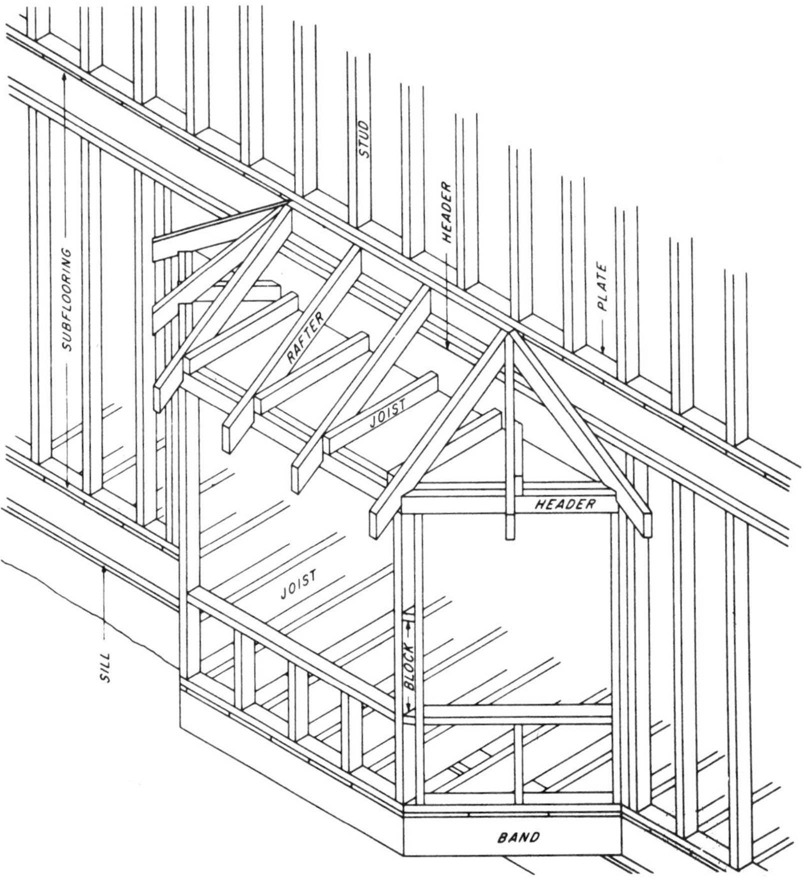

Figure 81 Bay window framing

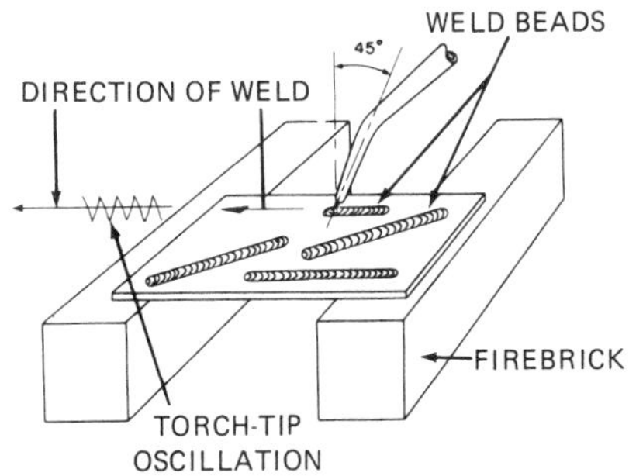

Figure 82 Beads (weld)

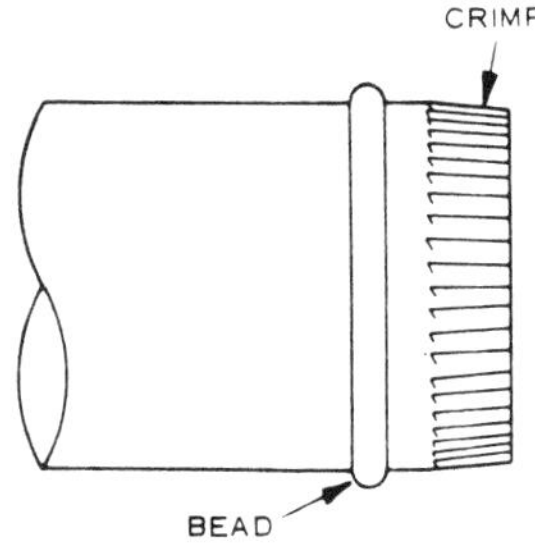

Figure 83 Bead (sheet metal)

bay window Window and framing that projects outward from the wall of a building. (Figure 81.)

bead A molding; a rounded projecting band, curved out to form a circular ridge. See *bead molding.* In welding, a narrow deposit of metal made by the welding rod or electrode. (Figure 82.) A long, pencil-shaped line of glue or adhesive used to hold panels to joists, studs, or other surfaces. In sheet metal work, a raised, rounded projection in the metal, sometimes called beading. (Figure 83.) In stucco work and plastering, a metal edging with a short section of wire mesh used at corners (corner bead), around windows, and at the sheathing edge by foundation (stop bead). (*Figure 251,* p. 95.) In drywall work, a narrow, right-angle corner bead. See *corner bead, cornerite.*

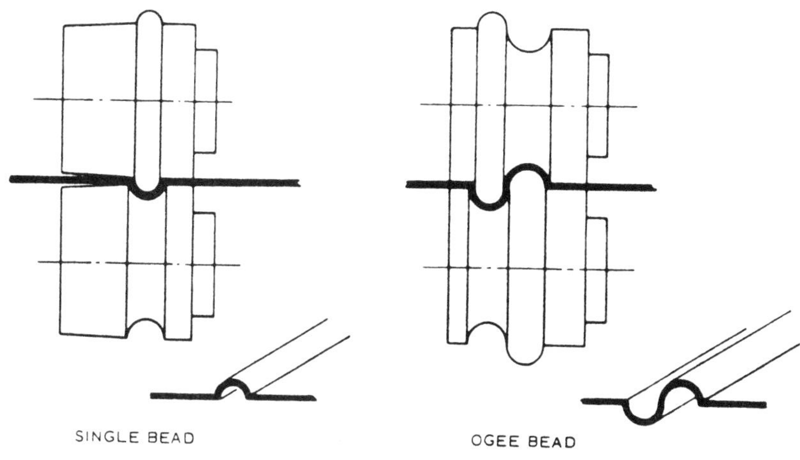

Figure 84 Beading: using a rotary machine

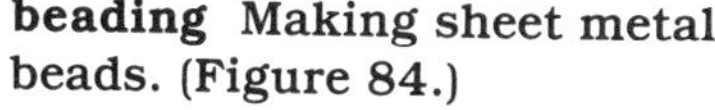

beading Making sheet metal beads. (Figure 84.)

bead molding A molding curved to form a ridge or raised groove. See *molding.*

bead plane A special plane used for cutting beads.

bead weld A single weld made on the surface of the metal. (Figure 85.) See *surfacing weld.*

beam Any horizontal structural member, supported at each end, that supports a load. Examples are girders, steel beams, rafters, joists, and purlins. Beams may be solid timber, laminated wood, steel, or reinforced concrete. Figure 86 shows various wood beams. *Simple beam:* supported at both ends. *Fixed beam:* rigidly supported at each end, as beams embedded at each end in concrete or a masonry wall. *Cantilever beam:* supported at one end only. *Continuous beam:* supported at three or more points. A beam is sometimes called a *girder,* although in heavy construction a girder is a heavy support member that may support a beam. See also *framed beam construction, seated beam construction, timber construction.*

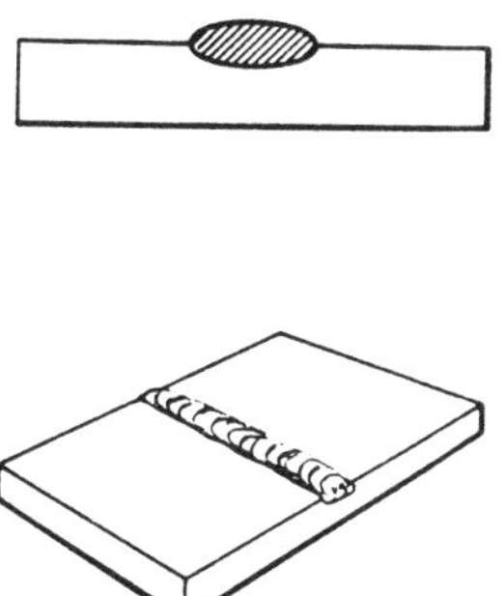

Figure 85 Bead weld

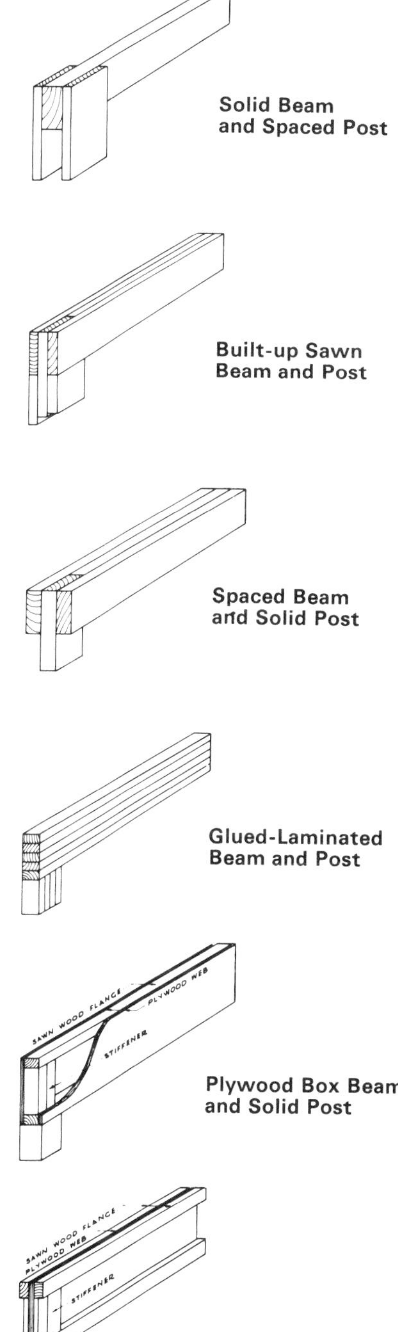

Figure 86 Beams (wood)

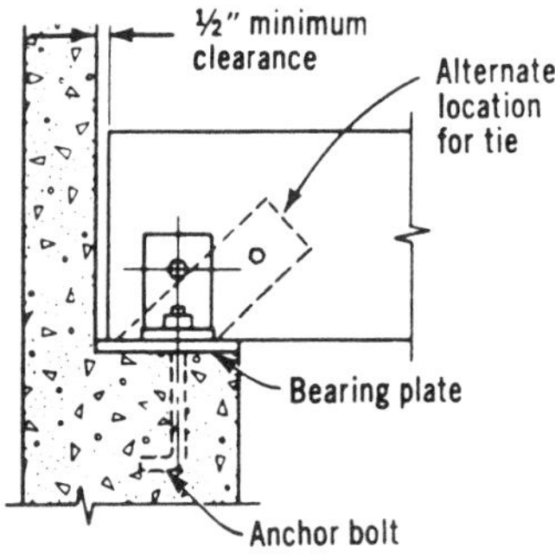

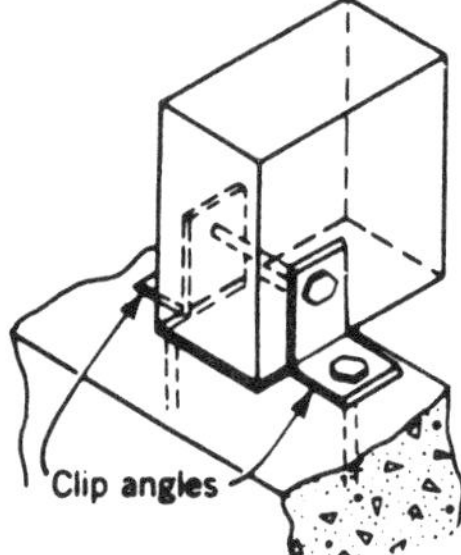

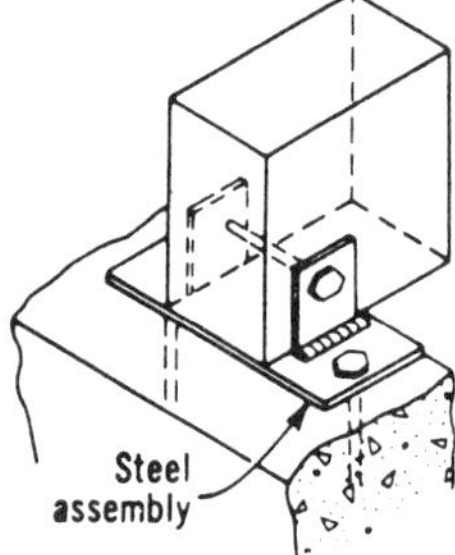

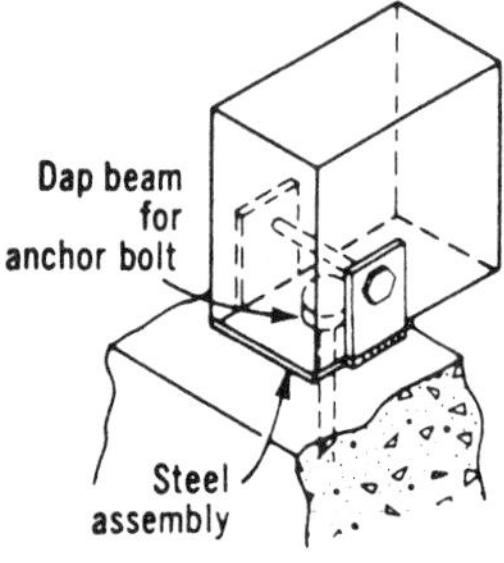

Figure 87 Beam anchors

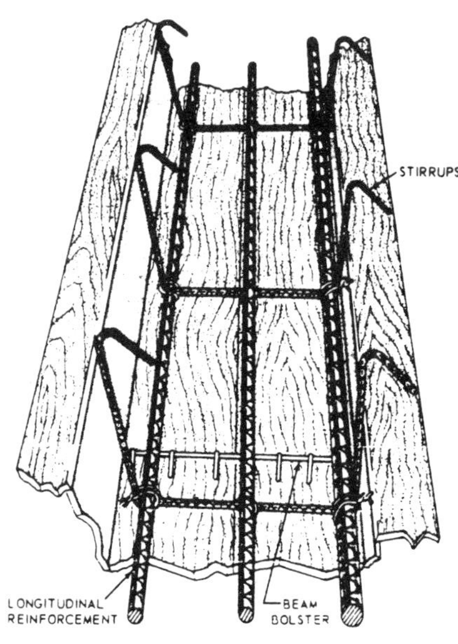

Figure 88 Beam bolster

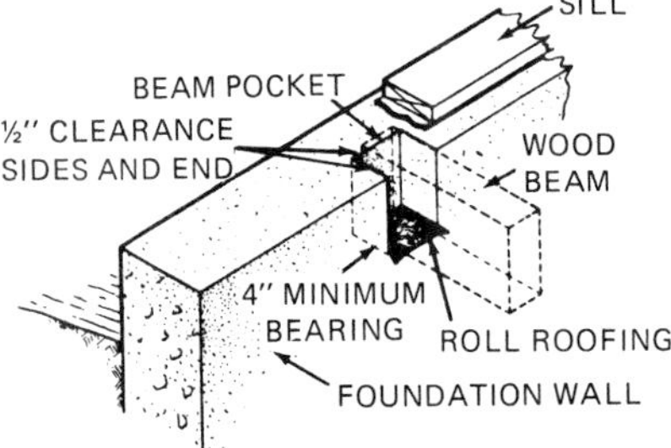

Figure 89 Beam pocket

beam anchor Metal support at the end of a beam. (Figure 87.) S[illegible]n *anchor.*

beam-and-slab floor construction Reinforced concrete slab floor supported by integral reinforced parallel beams (running one way only).

beam bolster Continuous reinforced bar support consisting of *bolsters* welded together with a wire or bar, used to support beam reinforcing bars. (Figure 88.) See *bolster.*

beam ceiling Open ceiling with exposed beams; the beams may be solid or false.

beam clip See *post cap.*

beam compass Drafting instrument used for drawing large circles or arcs. Two legs slide on a bar or beam or create the needed width.

beam fireproofing Enclosing of steel beams with noncombustible material to increase the fire rating. Application of gypsum paneling around the exposed beam. See *caged beam, column fireproofing.*

beam girder In steel construction, two or more beams fastened together to form a stronger unit; a built-up girder.

beam hanger See *joist hanger.*

beam number Number assigned to a beam on a framing or erection plan.

beam pocket Opening left in foundation top or wall for the end of a beam. (Figures 89; *683,* p. 295.) Opening in a column or girder to receive the end of a beam.

beam schedule Table giving number, size, type, and location of steel beams.

bear To rest on another member and transfer load.

bearer Structural member that supports other structural members or load.

bearing That surface area of a girder, beam, joist, rafter, or truss that rests on a support; the support surface for any structural member. In surveying, the angle in degrees from a north or south reference point. (Figures, 48 and 49, *48,* p. 20; *49,* p. 20.) There are two 90° quadrants to the left and right of the North 0° reference point, and two 90° quadrants to the left and right of the South 0° reference point. A bearing, for example, 45° into the northwest quadrant, would be noted as N 45° W. A bearing 45° into the southeast quadrant would be noted S 45° E. See also *azimuth, quadrant.*

bearing area Square-inch surface area of contact between a load-bearing member and a load-transferring member, as between the end of a beam and its support.

bearing capacity The load the soil will hold without significant movement.

bearing plate Plate placed under a load-bearing structural member to distribute the load on the support surface.

bearing-type connection In steel construction, a bolted connection where load is carried by bearing rather than friction; shear load resistance is by body of the high-strength bolt(s) against sides of the holes in connected member. See *friction-type connection.*

bearing wall Wall or partition that holds part of the load of the building. See also *nonbearing wall.*

bed Cement or mortar in which masonry units are laid. The lower surface of the masonry unit. See *bed joint.*

bed joint Horizontal masonry joint, opposed to a vertical masonry joint (*head joint*). Also called *beds.*

bed molding Molding in an angle, as the intersection of cornice or eaves with the end wall of a building. See *molding.*

bedrock Solid, natural rock below the soil or cover.

beds *Bed joint.*

bell In plumbing, the enlarged end of a pipe into which another pipe end fits to form a joint, also called a *hub* or a *socket.* In heavy concrete construction, the bell-shaped part of a pier, column, or caisson.

bell-and-spigot cast-iron soil pipe A heavy cast-iron pipe used for drainage and vent pipe. Sizes vary from 2″ to 15″ in diameter. (Figure 90.) The plain end is called a spigot; the bell-like opening is called a hub. The spigot or plain end of one pipe fits into the bell or hub opening of another. See also *no-hub soil pipe.*

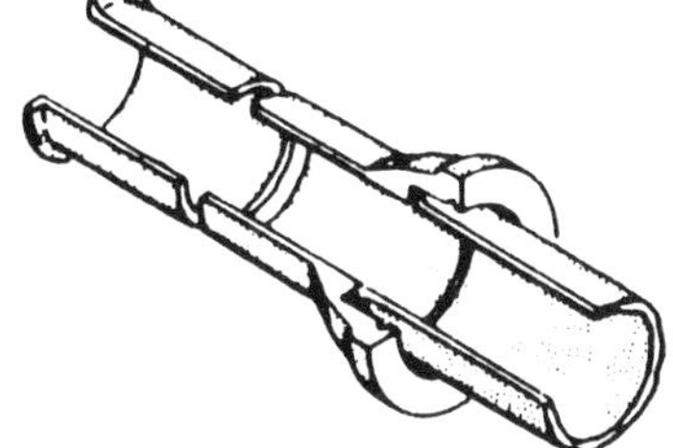

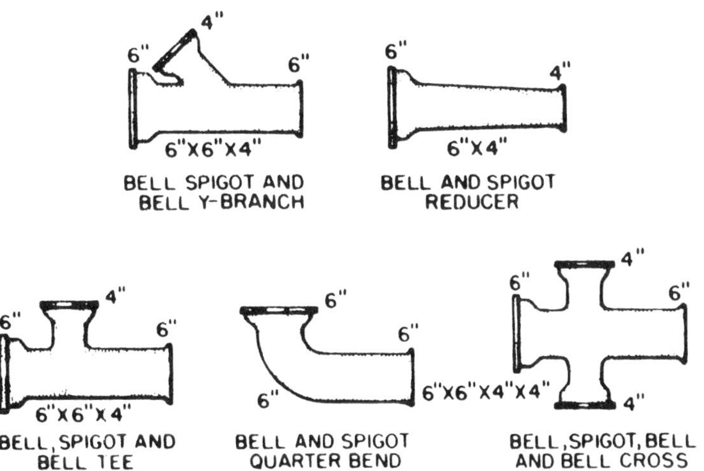

Figure 90 Bell-and-spigot piping

belled excavation Shaft or footing excavation that is enlarged or bell-shaped at the bottom.

bell wire General term for 18-gauge wire used for making doorbell or chime and thermostat connections.

belt course A band or layer (course) of masonry units or molded plaster at the same level, running around a building.

Figure 91 Belt sander: portable

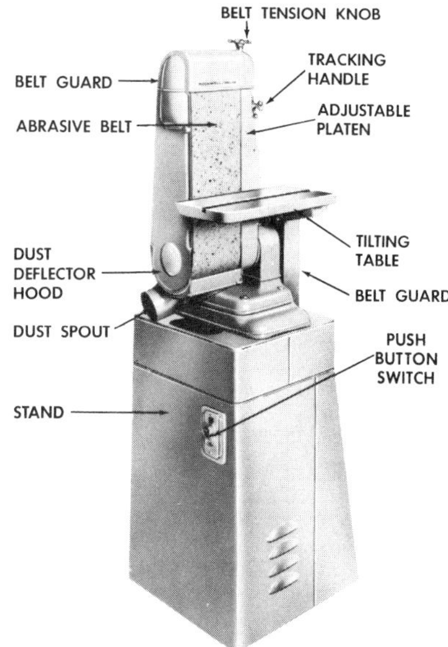

Figure 92 Belt sander: stationary

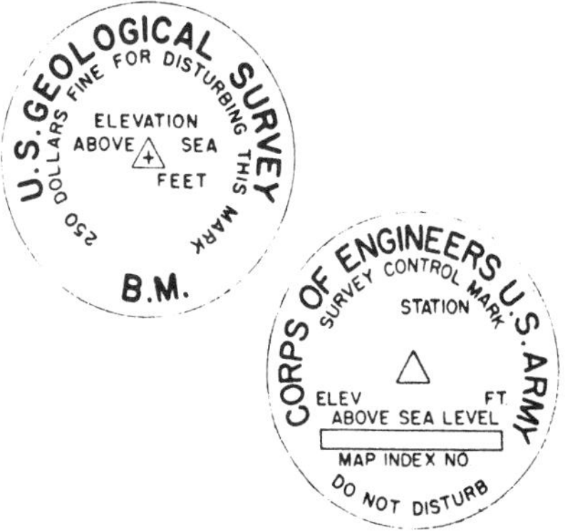

Figure 93 Bench marks

belt sander Power sander used for sanding large flat surfaces. (Figures 91, 92.) The sanding belt revolves around two drums. The belt is replaced when worn.

bench dog Peg placed in a workbench hole to secure work.

bench lumber High-grade, clear lumber used in trim and cabinet work.

bench mark (B.M.) A permanent survey marker giving accurate elevation above sea level as determined by U.S. Geological Survey. (Figure 93.) Often used to determine local datum point.

bench saw See *table saw.*

benchstones In woodworking, sharpening stones used in the shop. See *oilstone, waterstone.*

bench stop Metal device used on a workbench to hold work while planing or otherwise working on it.

bench table Course of stone or brick projecting out at the base of a wall or building; course is wide enough to form a seat.

bench vise Ordinary vise which is attached to a workbench.

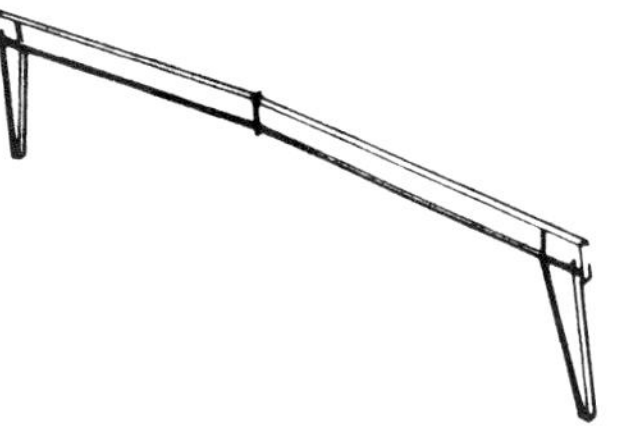

Figure 94 Bent (rigid frame construction)

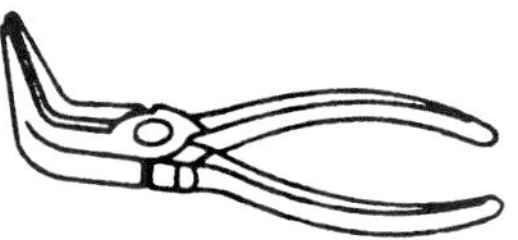

Figure 95 Bent-nose pliers

bench work Work that is done at a workbench.

bend A short curved pipe; an *elbow.*

bender Device for bending thin-wall tubing and rigid conduit. See *conduit bender, hickey.*

bending moment Force acting perpendicular to the length of a member. Equal to force times length. See *stress.*

bending schedule Table giving information on type and length of bends on reinforcement bars. See *bar schedule.*

bent In steel construction, a single rigid support system running from one side of the building to the other, including columns and roof support. (Figure 94.) Two bents make a *bay.* A bent normally has two or more columns braced together to form a support. In highway or bridge construction, however, a single support member, often in the form of a reinforced concrete "T", is common. Also, a row of piles connected together. See *bay, pile* (*Figure 684,* p. 295), *rigid frame construction.*

bent bar Reinforcing bar bent to specified shape, such as with a hook or stirrup. See *reinforcement symbols.*

bent cap In reinforced concrete or heavy timber construction, a support beam or surface running across the top or head of a concrete or timber *bent,* as for a highway bridge support. (*Figure 1026,* p. 452.)

bent-nose pliers A *long-nosed plier* that has curved jaws. Used for wire bending and for close, small work. (Figure 95.) Also called *curved-nose.* See *pliers.*

bentwood Wood or plywood permanently bent to a desired shape or curve. A form may be used to give the needed shape; steam and heat are used to give flexibility to the wood.

berm A built-up earthen slope or bank around a structure, used to provide insulation; an earthen dike or long mound. (Figure 96.)

bevel An edge that is not at a right angle, as on the edge of a hip rafter; the cutting edge of a chisel.

bevel board Board with bevel angles laid out for use as a template for cutting needed bevels.

bevel siding Beveled (flat wedge-shaped) boards used as horizontal siding. Also called *lap siding.* See *siding.*

bevel weld A weld joining pieces with one- or two-angled edges. A weld of two beveled edges is called a *V weld.*

bezel Sloping edge of a cutting tool.

bib Water faucet onto which a hose may be screwed. Also called a *bibcock, hose bibb,* or *sill cock.*

bibcock Water faucet onto which a hose may be screwed. Also called a *bib, hose bibb,* or *sill cock.*

bifold door Hinged door that runs on an overhead track.

bight A loop in a rope or wire cable.

bill of materials List of materials or parts needed for a specific project or to construct something.

binder Something that holds materials or parts together, such as the mortar to hold masonry units together.

binder bar Long metal piece used for holding down carpeting edges when the carpet stops at a doorway or meets a tile or wood floor.

binding Condition where a moving or sliding part catches or freezes and does not move freely.

bird screen Screen installed behind attic louvers or on chimneytops to prevent entry of birds.

Figure 96 Berming around house

bird's-eye figure In wood, a figure resembling a bird's eye, often found in hard maple.

bird's mouth Cut-out portion or notch of a rafter where it rests against the top plate. (*Figure 763,* p. 335.) The top horizontal cut is called the seat cut and rests on the top plate. The vertical cut is called the plumb cut.

bisect To divide, as a line or plane, into two equal parts.

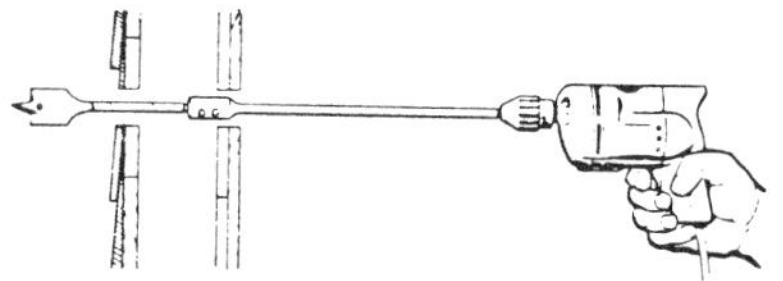

Figure 97 Bit extension used with a spade bit

Figure 98 Bit gauge

bit A boring tool used for drilling holes. Various bits are used in hand and power drills. A *spade bit* is commonly used in power drills. *Masonry bits* are used for drilling into concrete or other masonry units. Bits usually come in sets. See also *auger bit, bit extension, counterbore, countersink, door lock bit, drill, drill and counterbore bit, expansion bit, Forstner bit, hole saw, lockset bit, masonry bit, plug cutter, power bore bit, power drill, push drill, screw-bit, self-feed bit, ship auger, spade bit.*

bit brace A device for holding bits. See *brace.*

bit extension Shaft used to extend the length of a bit. (Figure 97.) Lengths and sizes vary.

bit gauge Device that clamps onto a bit or drill to control the hole depth. (Figure 98.)

bitumen A tarlike substance used in roofing; asphalt.

black knot A black or dark knot in the wood, caused by a dead branch or limb on the tree.

blacktop Asphalt paving or topping used for driveways or highways.

blade A cutting implement. The longest arm on a *framing square.*

blanket insulation A roll insulation with or without backing, usually $15\frac{1}{4}''$ or $23\frac{1}{4}''$ wide, from $3\frac{1}{2}''$ to $5\frac{1}{2}''$ thick; usually fiberglass. Today the term often means the same as *batt.*

bleach In woodworking, an oxidizing agent used to remove color or dark blemishes on fine hardwood.

bleaching Using an oxidizing agent on wood to make it light in color.

bleed To ooze out; the breakout of a lower paint coat through an overcoat or finish coat.

bleed hole In concrete placement, a hole in a material, such as an angle or plate, to allow entrapped air or water to escape during the concrete placement.

bleeding In painting and wood finishing, the oozing out of an undercoat through the finish coat. In lumber, the discharge of a preservative or a resin out of the wood. In cement work, moisture or free water that rises to the surface of fresh concrete, often carrying a white lime or cement concentrate. In plumbing and hot-water heating, the process of purging air from water pipes.

blemish Visual irregularity in lumber that is unsightly but does not affect the strength of the wood. See *defect* (which does affect the wood's strength). See also *blue stain.*

blending Mixture of two or more different things, as (in painting) the mixture of different-colored pigments. The application of paint to a painted surface so as to match the existing paint.

blind doweling Joining two pieces of wood with dowels so the dowels are hidden, as at a miter joint. See *wood joints.*

blind edge Sheet metal edge that is lapped back to cover fasteners, used when applying sheet metal to a wood surface.

blind hinge Concealed hinge.

blind joint In welding, soldering, or woodworking, a joint that is hidden and cannot be seen.

blind nailing Driving nails so that the nail head or hole (from sunken nails) is not visible.

blind rafter Rafter that runs from the ridge of a dormer to the ridge of the main roof.

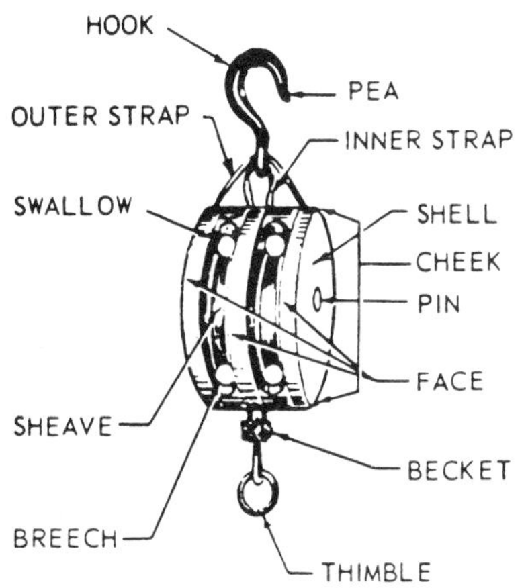

Figure 99 Block

blind valley A valley that is almost flat, not readily apparent from the ground.

blister Painting defect; the paint coat raises or puffs out, leaving an air pocket, apparently when direct heat, such as sunlight, causes moisture expansion under the paint coat. Blisters also sometimes occur in plaster.

block Short piece of wood used to space out or separate building members and to give support; block used in an interior angle of a joint to give support. A *concrete block.* A device for raising loads. (Figure 99.) See *block and tackle.*

block and tackle Hoist for lifting loads. The line (rope) threads through the pulley or sheave in the block. There can be several sheaves in the block; double and triple blocks are common. (Figure 100.) A mechanical advantage is realized using the blocks; the power increase is equal to the number of lines running through the blocks. Two blocks with single sheaves have two lines: it takes only half the power to lift a load. This gives a mechanical advantage of two. A block with two sheaves working with a single sheave block takes one-third the power to lift a load, or a mechanical advantage of three. See also *reeving blocks, tackle.*

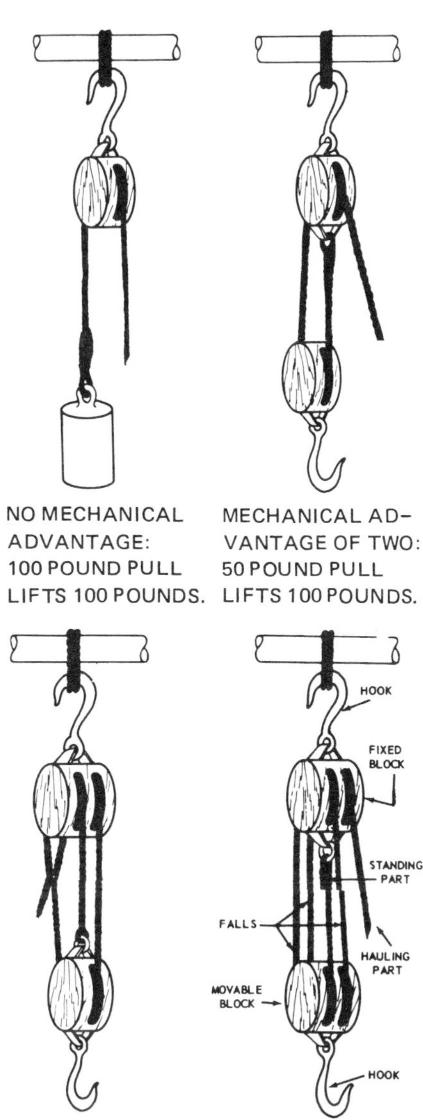

Figure 100 Block and tackle

block bond See *stack bond.*

block bridging Short pieces of lumber used between joists or studs to give support and brace the members. Short wood pieces used between framing members, such as studs in balloon framing, to retard the spread of fire, called *fire blocks.*

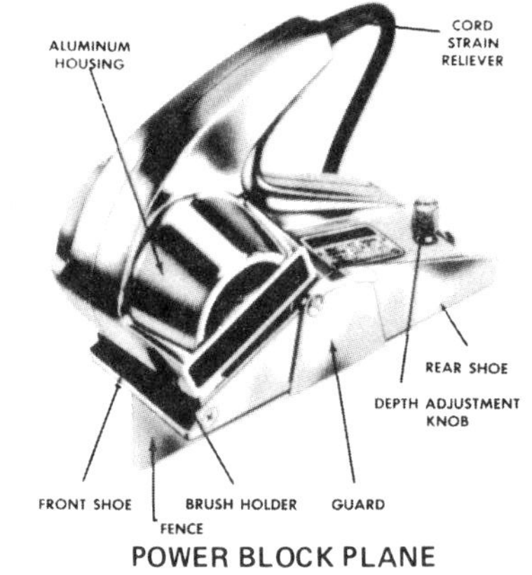

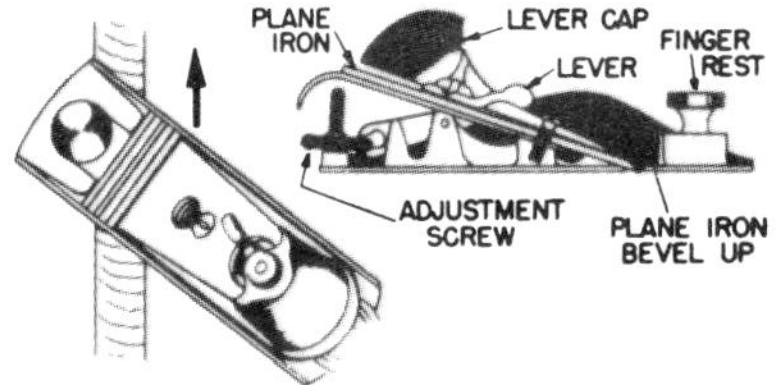

Figure 101 Block planes

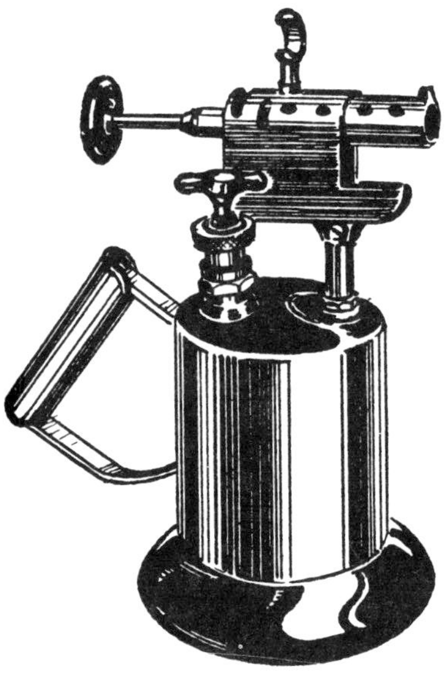

Figure 102 Blow torch

blocking Using wood blocks as filler pieces between other wood members; strengthening the interior angle of a joint by gluing in a wood block.

block plane Small hand plane, normally about 6″ long with a $1\frac{5}{8}$″ cutter. Used for smoothing small pieces and end grain. It is operated with one hand. (Figure 101.) See also *plane.*

block tin Pure tin.

bloom An efflorescence of lime on a masonry wall; flower-shaped excretion of cement onto a concrete wall. Normally occurs in the first year after concrete is placed, or in damp areas, such as basements.

blowdown On a boiler, the difference between original opening pressure and final closing pressure on a safety relief valve.

blower and vacuum fish line Method of forcing a line through an electrical conduit. A small plastic foam plug is blown or vacuumed through the system. A string attached to the plug is used to pull a wire and then the actual conductors through the conduit.

blowhole In welding, an older term for *porosity.* In concrete, a hole on the face of the hardened concrete caused by trapped air. In metal casting, a hole left by trapped gas.

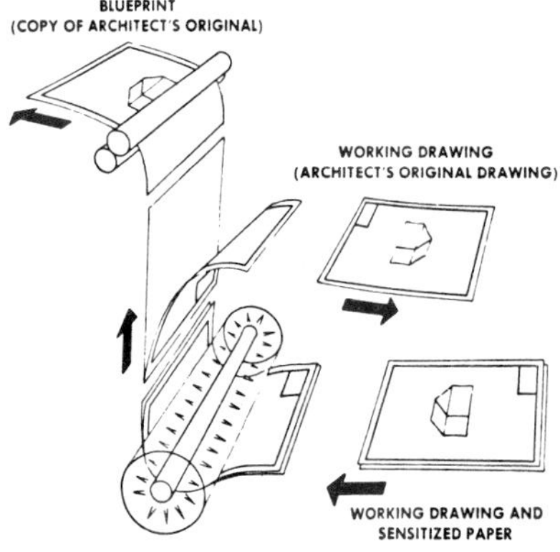

Figure 103 Blueprint copied from original drawing

blowpipe Welding torch.

blowtorch A gasoline-fueled torch used for heating metal, removing paint, and sweating joints on copper tubing. A hot flame is produced by vaporized gasoline. (Figure 102.)

blued nail Nail with a blue surface caused by heating to produce a hard, oxidized, rust-resistant surface.

blue stain Blue coloring found in lumber; it is caused by a fungus growth and does not impair the strength of the lumber.

blueprint Copy of the original architect's pencil *working drawings.* Blueprints (or *prints*) are drawings that show how to build the structure. Figure 103 shows how a blueprint is copied. Originally, blueprints had white lines on a blue ground. Today, the lines on a blueprint are commonly blue or black. Blueprints, along with *specifications,* give the information needed to build the structure. See also *drafting sheets.*

blushing In painting, a gray cloudy film that appears on fresh paint; it may be caused by high humidity.

board Lumber sized from 1 × 4 up to 1 × 12 inches:

Nominal size	*Actual size*
1 × 4	$\frac{3}{4} \times 3\frac{1}{2}$
1 × 6	$\frac{3}{4} \times 5\frac{1}{2}$
1 × 8	$\frac{3}{4} \times 7\frac{1}{4}$
1 × 10	$\frac{3}{4} \times 9\frac{1}{4}$
1 × 12	$\frac{3}{4} \times 11\frac{1}{4}$

boards and battens Vertical siding that has battens nailed between spaced boards. See *siding.*

board foot equivalent of a board one foot square and one inch thick or 144 cubic inches. Abbreviated b.f. A 2″ × 4″ that is 18″ long is a nominal board foot (144 cu in.), as is a 1″ × 6″ that is 24″ long.

board measure System for measuring lumber; the basic unit is a *board foot* (b.f.). Lumber prices are often quoted in thousand board feet. A board foot is equivalent to 1″ × 12″ × 12″.

board-measure scale Scale on the back of a steel framing square; also called *essex board measure.*

board rule Special scale used for quickly determining the number of board feet in lumber. (Figure 104.)

boaster Chisel used to smooth stone.

boasting Dressing a stone with chisel and mallet.

body coat Paint coat between the priming coat and the finish coat.

boil board Support board nailed to a concrete form to stiffen the form.

boiler Closed vessel for heating water to create hot water or steam. See *high-pressure boiler, low-pressure boiler.*

bolster Shaped wire device for holding up reinforcing bars or welded wire fabric so the concrete can flow around the support. (Figure 105.) See also *bar chairs, bar support, chairs.* Also, another term for a *brick set.* Also, a short horizontal timber or steel beam on top a column, designed to give a larger-area support base for a beam or girder.

bolt Fastening device provided with a head on one end and threaded at the other end for a nut. (Figure 106.) Machine bolts have either a square or a hex head and nut. Carriage bolts have a round curving head. Stove bolts have a round head with a slot for a screwdriver. An eye bolt has a closed hook on one end instead of a head. A U bolt is shaped like a U with two threaded ends. See also *expansion bolt, handrail bolt, high-*

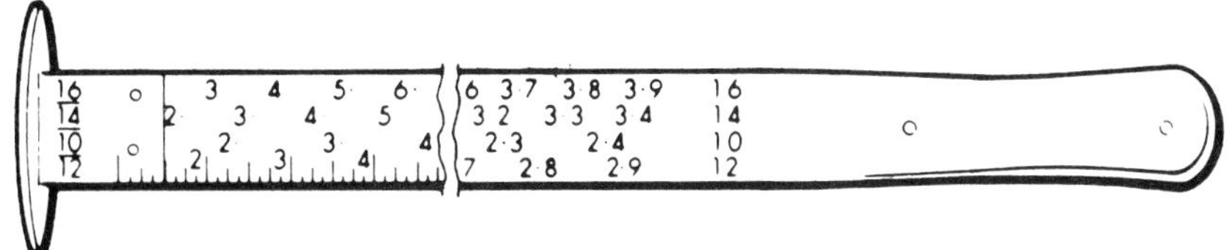

Figure 104 Board rule

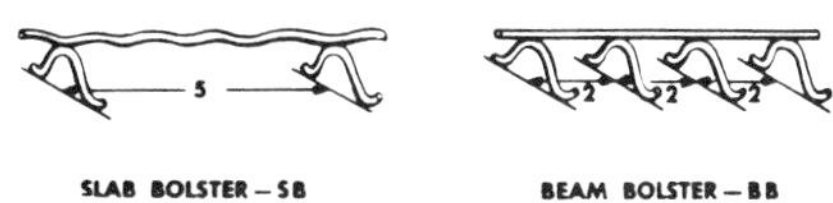

Figure 105 Bolsters

RIBBED NECK

CARRIAGE BOLTS

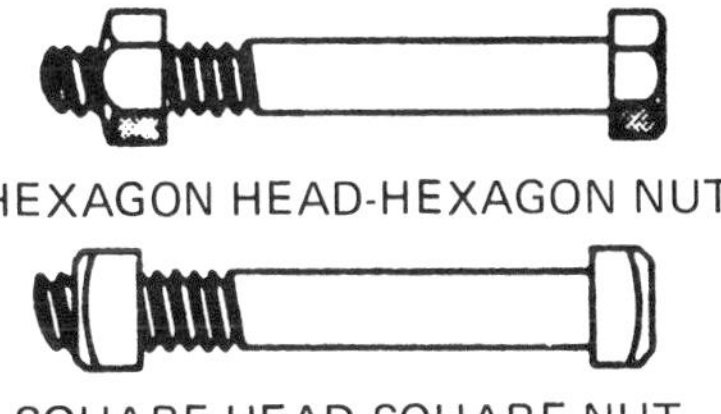

SQUARE HEAD-SQUARE NUT

MACHINE BOLTS

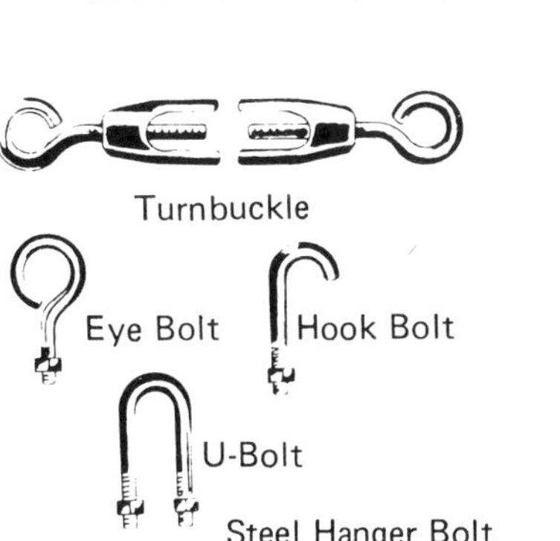

SPECIAL BOLTS

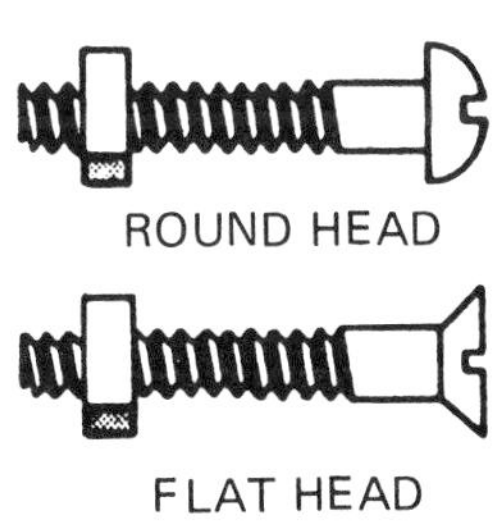

STOVE BOLTS

Figure 106 Bolts

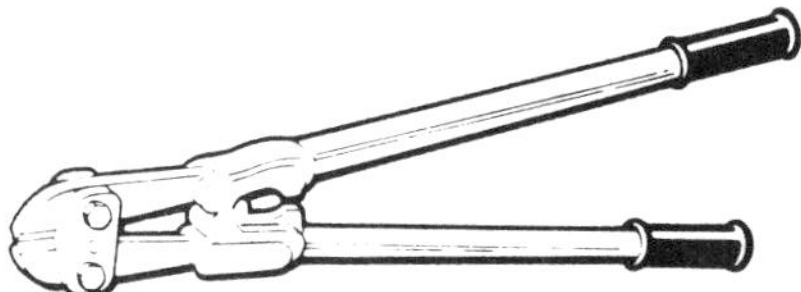

Figure 107 Bolt cutters

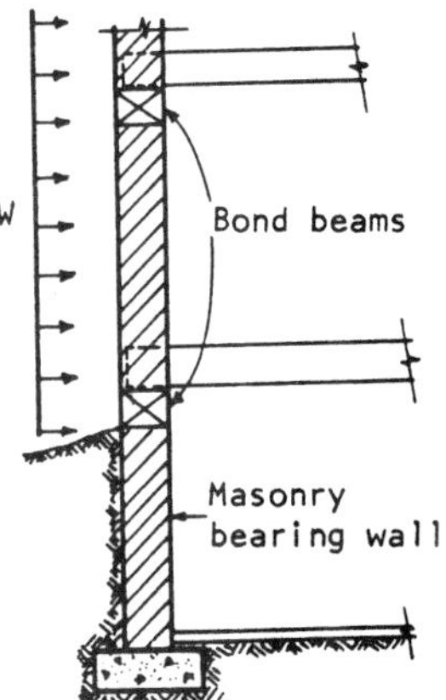

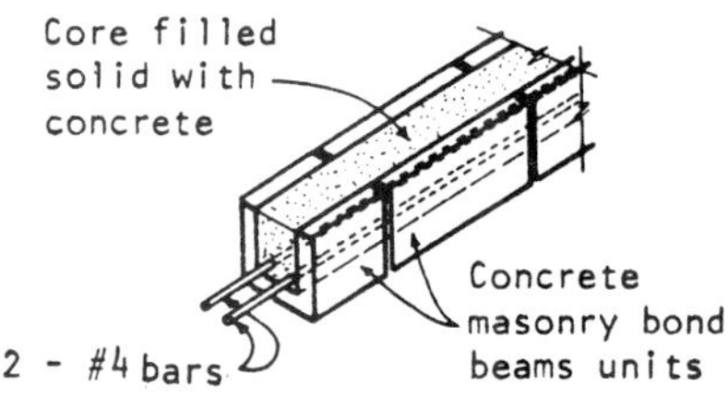

Figure 108 Bond beams

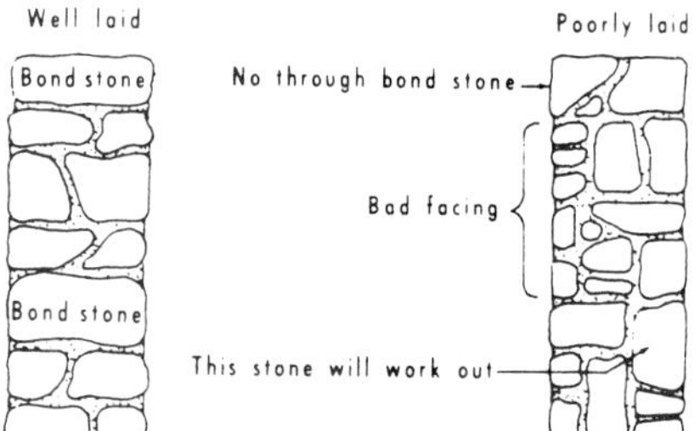

Figure 109 Bond stone used in well laid rubble wall (left). Poorly laid wall (right) has no through stones to hold it together

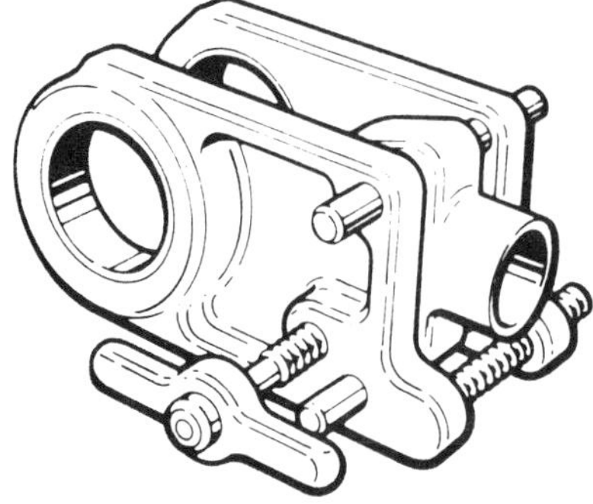

Figure 110 Boring jig

strength bolt, hollow-wall anchor, lag screws, machine screws, nut, stud, toggle bolt, turnbuckle. Also, in veneer cutting, a short log or flitch that is used to cut a continuous strip of veneer. See *rotary-cut veneer.*

bolt cutters Heavy, high-strength steel tool used for cutting bolts, steel dowels, chains, and the like. (Figure 107.)

bolt extractor Threaded device used for removing broken bolts or screws. A hole is drilled in the broken bolt end, then the bolt extractor is screwed in counterclockwise to remove the bolt. Also called a *screw extractor.*

bolt symbols Symbols that show how shop bolts and field bolts are to be finished. See *rivet symbols.*

bond A holding or gluing together of separate parts; glue joint; mortar joint between masonry units. The pattern made by masonry units in place. See *brick bond.* In electrical wiring, a wiring connection. A wiring connection between separate conductors; between conductors and equipment; a connection to ground. Also called a *bond wire* or *bonding wire.*

bond beam Horizontal masonry members bonded with reinforced concrete to form a continuous unit or beam. (Figure 108.) Designed to withstand lateral pressures from wind and earthquake.

bond breaker Substance used to prevent one surface from adhering to another, such as an oil used on forms to prevent the surface from sticking to the hardened concrete.

bond course In masonry, a course of headers.

bonded joint In concrete work a *construction joint.* Also a glued joint.

bonding wire Wire used for making a ground; a *bond wire.*

bond stone A stone or masonry unit that projects back and ties the face units to a backing. Similar to a header except it may not project entirely through the backing (as a header does). A stone that goes through a rubble wall. Figure 109 shows bond stones used in a rubble wall.

bond wire Wire used to make an electrical ground from conduit or cable terminal end to within service equipment, as from the conduit bushing grounding lug to a terminal within a panelboard. (*Figure 477,* p. 195.) Also called *bonding wire.* See *conduit bushing, grounding wire.*

bonnet Cover for the top of a ventilating pipe or chimney.

boom Spar projecting from the tower of a *tower crane,* used for lifting or guiding weights; lifting arm of a *derrick;* lifting arm of a *crane.*

border Outside edge; finishing strip.

bore A drilled hole; the internal diameter (I.D.) of a hole or pipe.

boring Process of making a hole.

boring jig Device used to align and guide bit when boring cylinder lock holes. (Figure 110.)

borrow In site development, the removal (borrowing) of earth from an off-site source for use as fill at a construction site. See also *cut and fill, filling, top dressing.*

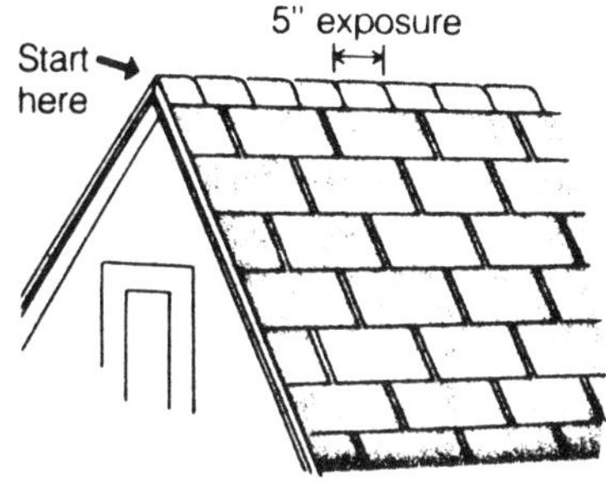

Figure 111 Boston ridge

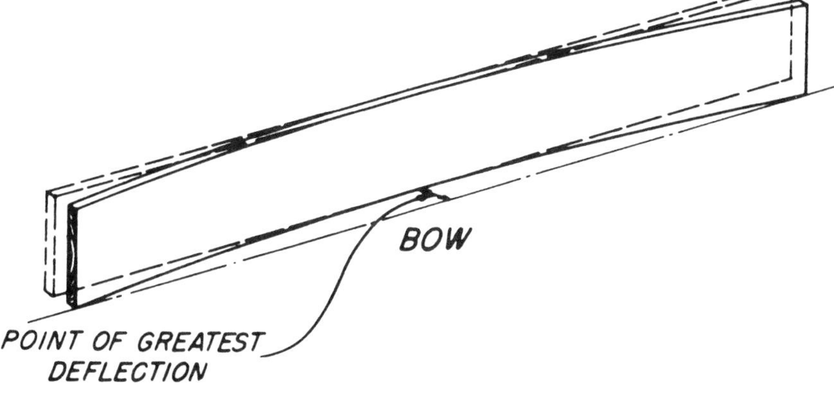

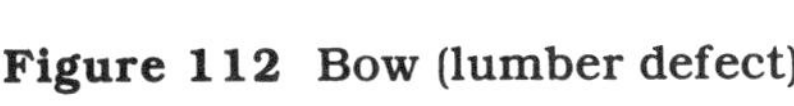

Figure 112 Bow (lumber defect)

Boston ridge Special method of shingling the ridge of a roof; the shingles are saddled over the ridge. (Figure 111.)

bottle See *gas cylinder.*

bottom chord Horizontal bottom support member of a truss. See *truss.*

bottom plate *Sole plate.*

bottom rail Bottom horizontal member of a window sash or panel door. See *door, window sash.*

boulder A large stone, generally over a foot in diameter.

boulder wall Rough wall made by setting field stones or boulders together.

bow A flat arc or curve. In lumber, a defect where a board curves slightly along its long axis. (Figure 112.)

bow compass In drafting, a small instrument used for drawing circles or arcs. One leg has a needle point, the other leg holds a pencil lead or pen point. The legs are held under tension at the top; an adjusting screw is used to set the width. (Figure 113.) See *compass.*

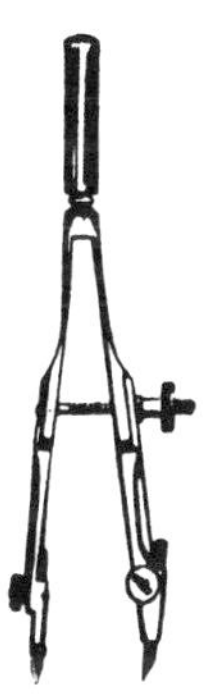

Figure 113 Bow compass with pencil lead

bow dividers In drafting, a small instrument with needle points on each leg. Used for transferring measurements. (Figure 114.) See *dividers.*

bow window Bay window that forms a curve or bow.

box and pan brake Sheet metal brake used for bending the four sides of a box. (Figure 115.) See also *brake, cornice brake.*

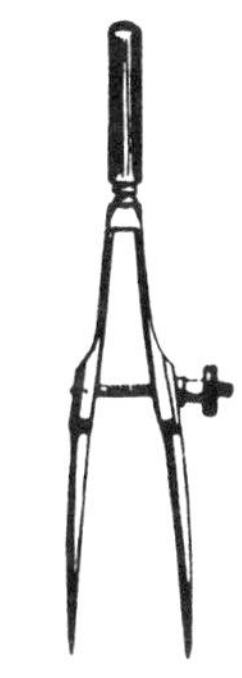

Figure 114 Bow dividers

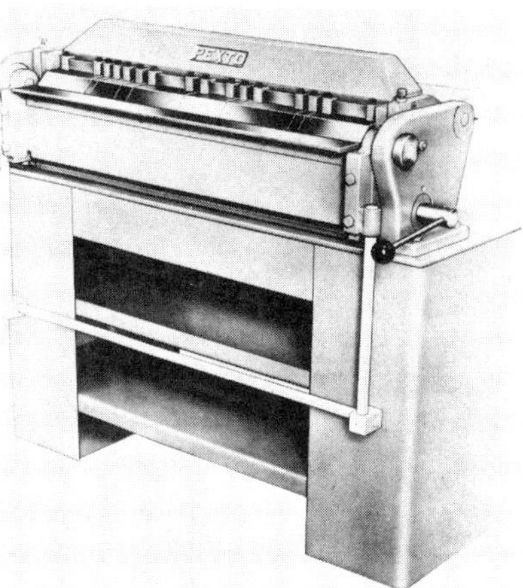

Figure 115 Box and pan brake

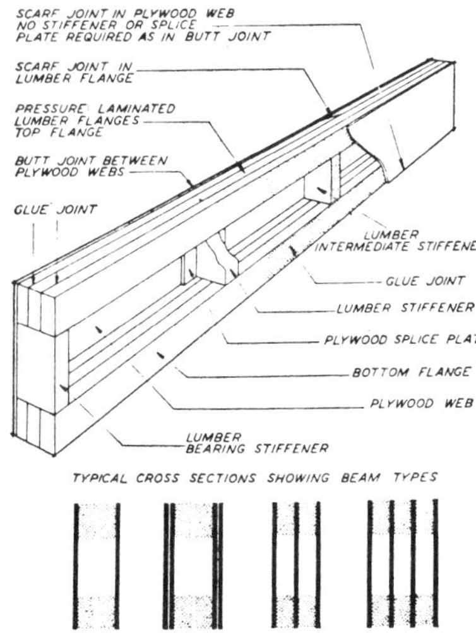

Figure 116 Box beam

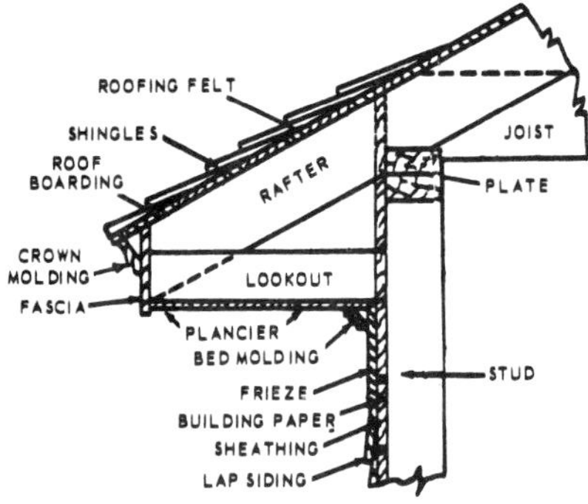

Figure 117 Box cornice

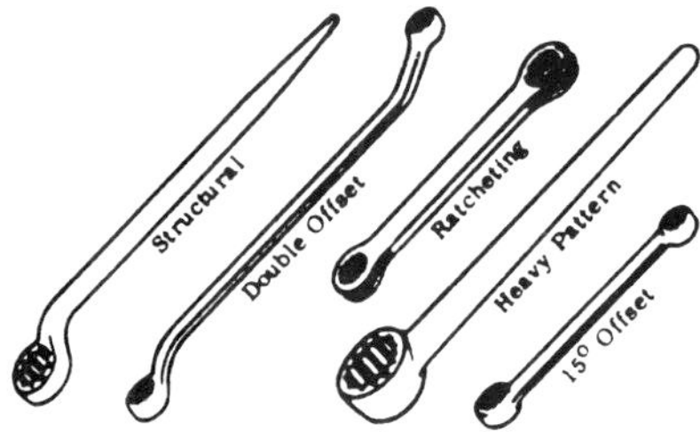

Figure 118 Box wrenches

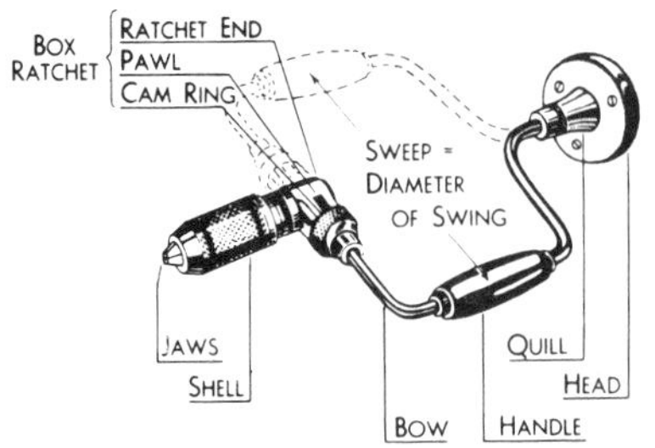

Figure 119 Brace (with ratchet)

box beam Built-up, hollow boxlike beam or girder; plywood is fastened over a wood framework. (Figure 116.) See *wood beam.*

box casing Inside casing of a *window frame.*

box column Built-up, hollow boxlike column.

box cornice Completely closed cornice. (Figure 117.) Also called a *closed cornice.*

boxed cornice A *box cornice, closed cornice.*

boxed pith Lumber piece that contains tree *pith.*

box form A box-shaped opening built into formwork, designed to leave openings for windows, doors, ductwork. See *buck.*

box frame construction In reinforced concrete construction, a framework or skeleton of joined vertical and horizontal slabs. The structural base looks like open-sided boxes.

box girder Built-up, hollow boxlike girder. A compound steel girder, as a girder made up of two I beams bolted together with cover plates. See *box beam.*

boxing up Closing or framing in a building; the rough covering is on, with doors and windows in place.

box nail Thin-shank nail, similar to the common nail. See *nails.*

box out To form an opening, as for a window, in a concrete wall by using a boxlike form.

box sill A foundation where a header or joist rests on the sill plate, and the sole plate rests on the subflooring over the header. (*Figure 282,* p. 105; *428,* p. 173.) Used in *platform framing.*

box stairs Stairs built between two walls, usually supported by the outside stringers.

box wrench Wrench with an enclosed head. The head may have six or twelve points. The six-point box is satisfactory for most work; the twelve-point establishes a much firmer seat and can often be used to turn a stripped nut or bolt head that would slip with a six-point. (Figure 118.) See *wrench.*

brace Framing member used to strengthen or stiffen some part of a structure. See *cut-in brace, diagonal brace, L brace, let-in brace, T brace, tension strip, wall brace, wind brace.* A hand boring tool that holds bits for drilling holes in wood. (Figure 119.)

brace and bit A brace that holds bits for boring. See *brace.*

braced frame Wood or steel frame that has diagonal bracing to resist lateral forces.

brace measure A table on the back of the tongue of a steel framing square. Also called a *brace scale* or *brace table.* See *rafter square.*

bracing Metal or wooden support braces used anywhere in a structure. See *cut-in brace, let-in brace, tension straps.*

bracket A support, usually metal, projecting out from a wall, used to hold some architectural feature such as guttering or an outside light. A right-angle steel support used for connecting steel beams, or beams and columns.

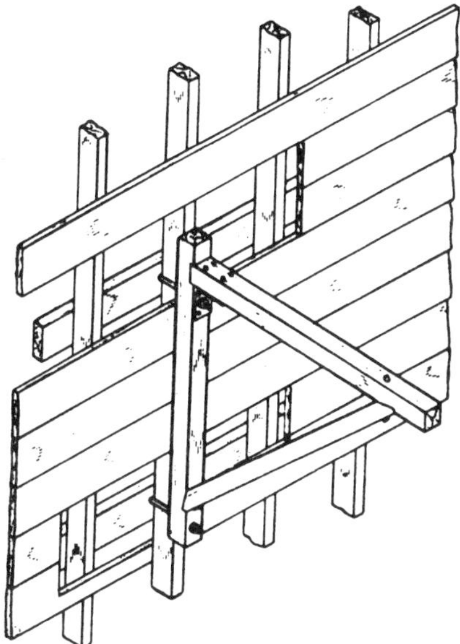

Figure 120 Bracket scaffold

bracket scaffold Scaffolding bracket supports that fasten directly to the wall. (Figure 120.) See *ladder jack.*

brad pusher Small screwdriver-shaped tool with a hollow magnetic tip. Used to start and drive small nails or brads into wood. The hollow tip retracts back as the brad is driven in. (Figure 121.)

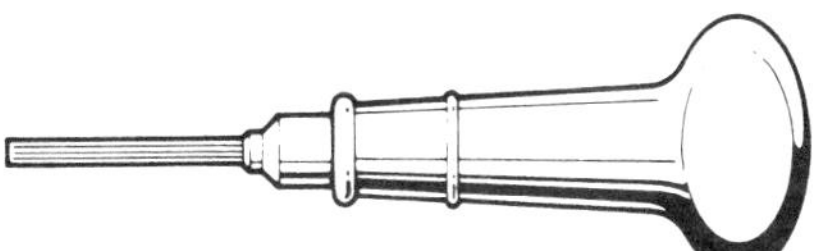

Figure 121 Brad pusher

brake Machine used for bending sheet metal. (Figure 122.) See *bar folder, box and pan brake, cornice brake, rotary machine, slip roll forming machine.*

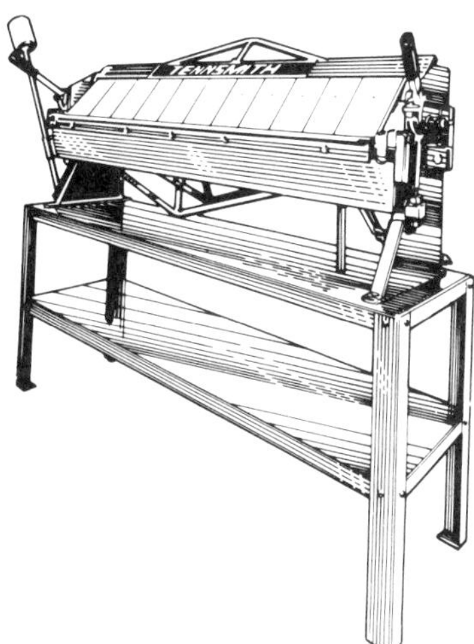

Figure 122 Brake

branch Something moving off from a larger whole; a horizontal duct pipe, or wiring run. In plumbing, a smaller outlet or pipe leading off from a larger pipe. The whole series of plumbing fixtures connected together in one run to receive fresh water or to discharge waste water into the main drain. In heating, a smaller air duct leading away from a larger duct; a series of heating outlets connected to a run of ductwork or one hot water or steam piping run. In electrical work, one continuous circuit with fixtures and outlets connected to one circuit breaker or fuse; a *branch circuit.*

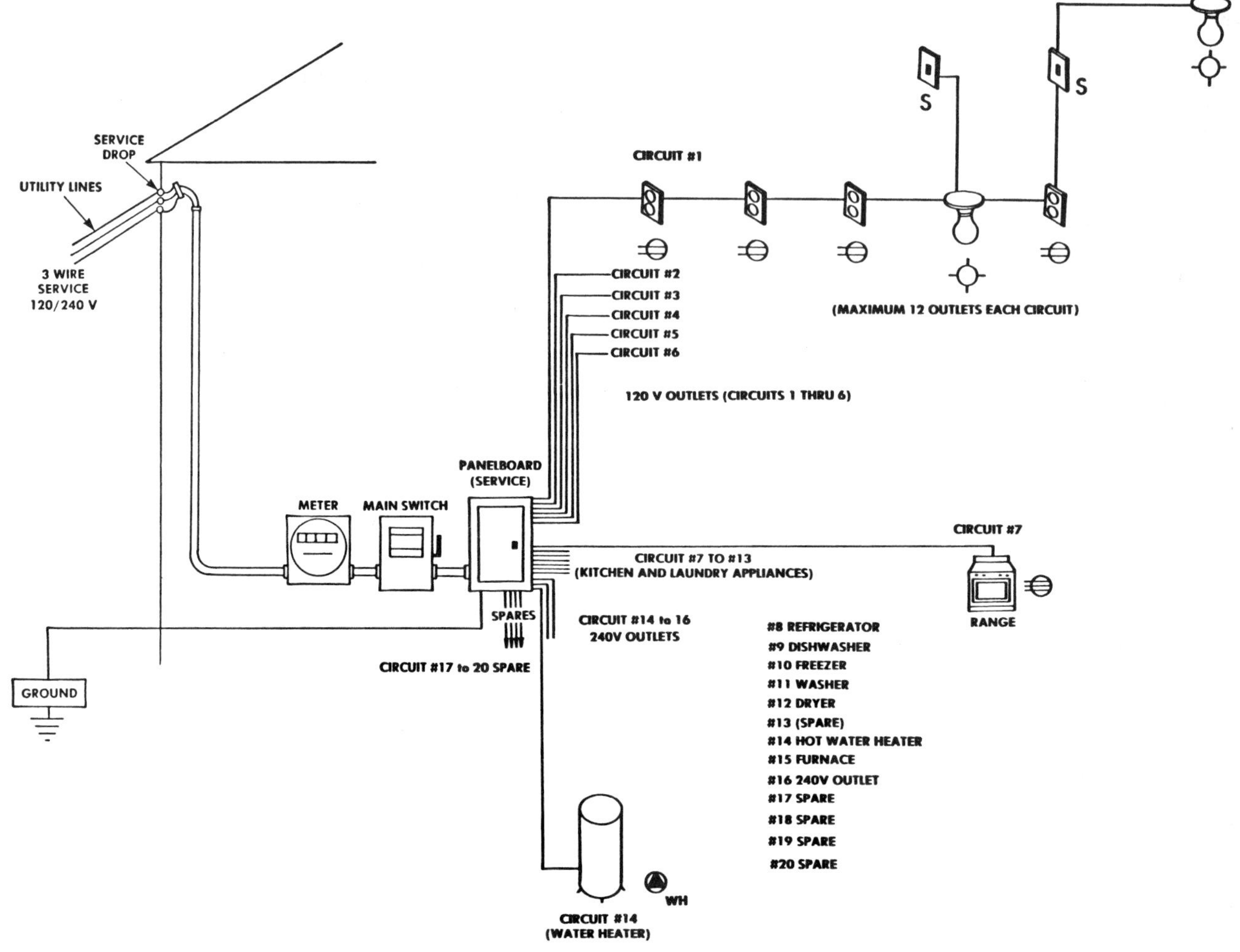

Figure 123 Branch circuits run from panelboard

branch circuit Circuit from the final circuit breaker or fuse to the outlets and back again. One continuous wiring run protected by one circuit breaker or fuse. There would be several convenience outlets on one circuit. (Figure 123.) With a heavy-duty circuit there would be only one special outlet.

branch drain Sanitary drainage pipe from fixture to soil stack.

branch vent Horizontal vent pipe that connects two or more individual vents to a main vent.

brandering In plastering, the nailing down of furring strips to provide a base for the metal lath and to provide a space between the lath and a solid surface, such as a masonry wall.

brashness Wood defect causing major loss of strength. Fiber integrity is lost and wood cracks or fragments easily without splintering.

braze To perform a kind of high-temperature soldering, similar to welding, that uses a bronze filler rod to make a strong, durable joint. Metals are joined with the rod filler metal. See *brazing*.

brazing A welding made by heating the base metal and melting a filler metal into the hot joint. (Figure 124.) The filler metal has a melting point lower than the metal to be brazed. The molten filler metal flows into the joint by *capillary action*. Brazing differs from soldering by using a filler metal with a higher melting point (over 840°F).

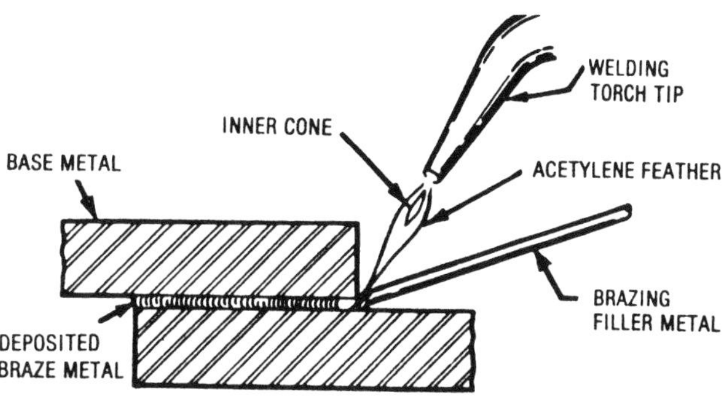

Figure 124 Brazing

break A coming apart. An abrupt change in the direction of a wall; a projection out from a wall, as a room add-on; a discontinuity in a wall, such as an opening for a gate, doorway, or driveway.

breaking radius Limit of curvature a piece of wood or plywood can be bent without breaking.

break joint A breaking, staggering, or offsetting of masonry joints, shingles, board ends, or panels to avoid having joints line up in a vertical line, which would create a structural weakness.

breezeway Passageway open at both ends, under or through a structure; used in high-density areas where buildings abut directly against each other.

brick Masonry building unit. Bricks are made in a mold, then hardened in a kiln. The basic brick size is $2\frac{1}{4}'' \times 3\frac{3}{4}'' \times 8''$. Figure 125 shows the standard terms used in referring to a brick (or to other similarly shaped masonry units such as concrete block). Brick commonly is made with two rows of holes; this is called a *cored brick*. The position or pattern of a brick (or other masonry unit) in place also has a specific terminology. Figure 125 shows these specific terms: *course, header, soldier, stretcher* and *wythe*. A rowlock header is called a *bull header*, a rowlock stretcher is called a *bull stretcher*. A brick for a 6″ wall, called an *SCR brick*, is also available. Bricks are specified and ordered as SW (severe

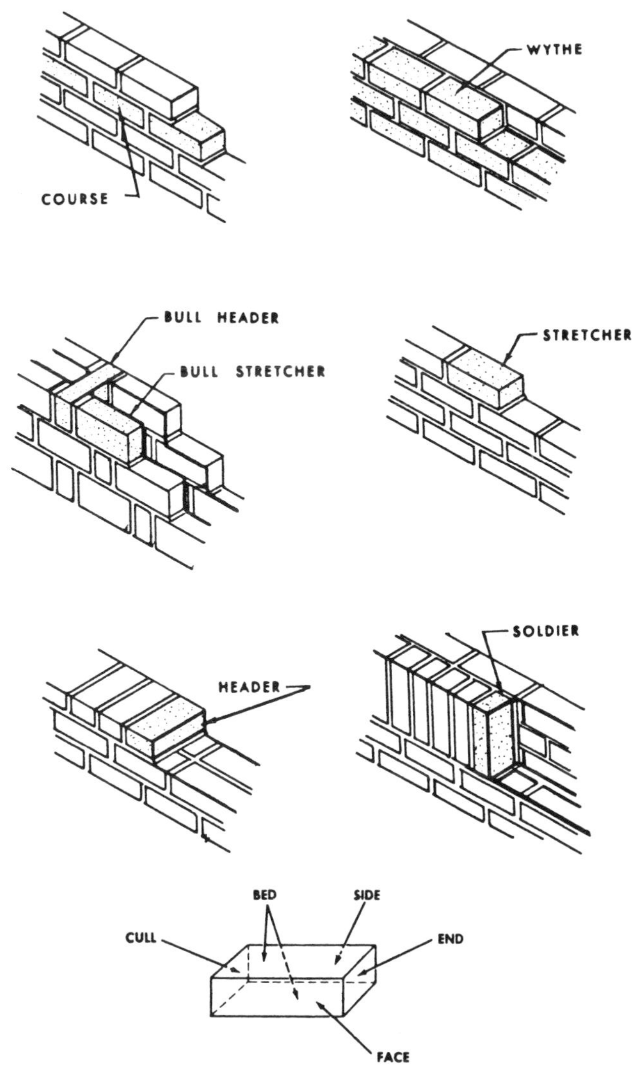

Figure 125 Brick terminology

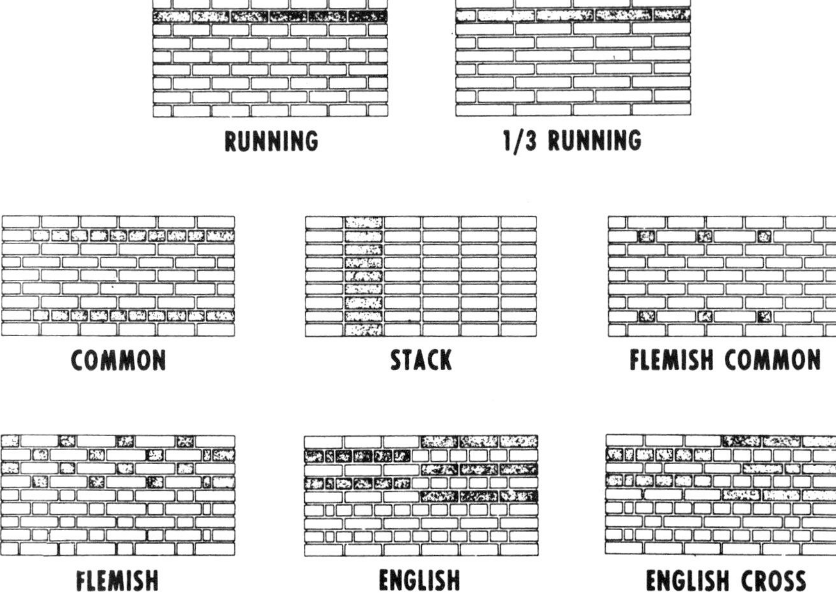

Figure 126 Brick bonds

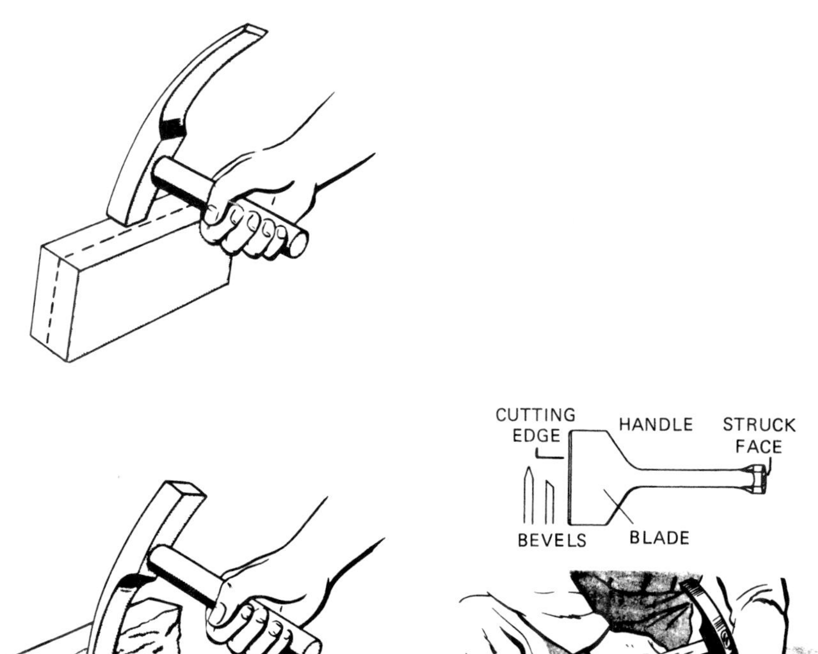

Figure 127 Brick hammer

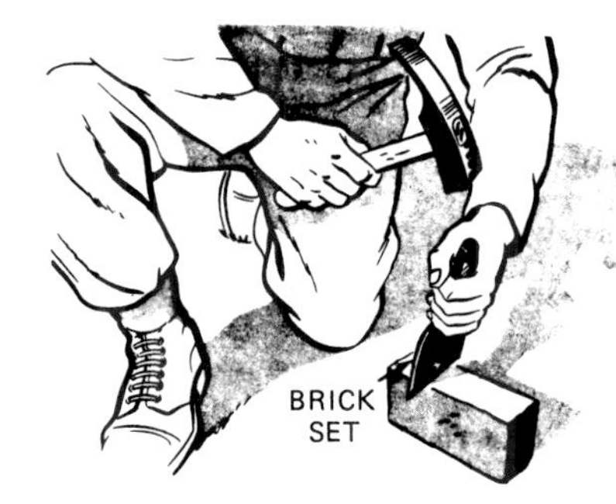

Figure 128 Brick set used with brick hammer

weathering) for outside, exposed use; MW (moderate weathering) for inside or sheltered use; and NW (no weathering) for inside use. See also *brick bonds, building brick, clinker brick, cut brick, English brick, face brick, fire brick, glazed brick, Norman brick, Roman brick.*

brickbat Broken bricks.

brick bond The mortar joint between bricks. See *masonry joints.* The style or pattern of course layout. Figure 126 shows common brick patterns: *running or stretcher, common, stack, Flemish common, Flemish, English,* and *English cross.* See also *chimney bond, header bond, herringbone bond, quarry stone bond, raking bond, stack bond, structural bond.*

brick construction Construction where outside bearing walls are solid brick.

brick facing See *brick veneer.*

brick hammer Tool for dressing or breaking brick. The head is flat on one end and on the other has a sharpened blade. (Figure 127.)

bricklayer One who builds walls, partitions, fireplaces and chimneys, and other structures of brick or concrete block. Also builds with other masonry units, such as structural tile, terra cotta, glass block, or cinder block.

bricklayer's hammer See *brick hammer.*

bricklayer's scaffold Temporary framing on exterior of wall that supports the mason, bricks, and mortar. See *mason's scaffold.*

brick masonry A general term used to describe laying of bricks in mortar.

brick seat Shelf or ledge sufficient to support a course of masonry.

brick set Wide-bladed chisel used for cutting brick. (Figure 128.) Also called a *bolster.*

brick trowel Flat, triangular trowel used to pick and spread brick mortar. See *trowel.*

brick veneer Single-wythe brick wall attached to a wood frame structure, to a concrete block wall, or to a concrete wall. The brick veneer is mainly decorative and carries none of the building load. (Figure 129.)

brick veneer construction Type of structure that has an exterior brick veneer.

brickwork Anything constructed of brick.

bridge Structure made for passage over a depression or waterway; something that goes over an opening.

bridgeboard A *stair stringer* or *carriage.*

bridging Short pieces of wood or metal pieces used between framing members, such as joists, studs, and flat roof trusses, to stiffen and give lateral support. Figures 130 and 529, p. 227, show bridging between joists. See also *subfloor.*

British thermal unit (Btu) Heat needed to raise one pound of water one degree Fahrenheit. Btu's are used to express energy used in heating.

broken joints Joints staggered so they will not fall in a straight vertical line. See *brick bond.*

bronze Alloy of copper and tin.

broom closet A small compartment or cabinet used for storing brooms and other cleaning materials.

broom finish In concrete flat work or plaster, a rough finish made by sweeping a broom over the surface before final curing.

brooming In built-up roofing, use of a broom to embed and smooth out a roof ply of asphalt felt. Use of a broom on fresh concrete flat work or fresh plaster to give a texture.

brown coat In plastering, a rough plaster coat to which the finish coat is applied. The brown coat may be the base coat that goes directly on the metal lath (two-coat plastering). It may be the second or intermediate coat that goes on the base coat, then followed by the finish coat (three-coat plastering).

browning brush Brush used to throw water on applied mortar to cause the mortar to soften up so it can be worked and straightened.

brown rot A brownish fungus decay in wood produced when cellulose decays, leaving the lignin.

brushes See *paint brushes.*

buck The framing around a masonry wall opening, especially the metal door frame. Also, the wooden box form in a concrete wall to make an opening when concrete is placed.

bucket trowel Stub-nose trowel used for mixing small batches of mortar in a bucket. (Figure 131.) See *trowel.*

buckle To fail structurally by bowing outward, as does an overloaded column, or by twisting or crumpling. A term usually associated with steel that fails by severe overloading or heat from a fire.

buffalo box An iron boxlike shaft that holds the *curb cock* that controls fresh water flow into a building. A small iron cover at ground level allows entry into the buffalo box. Also called *stop box.* See *water service.*

builder's hardware Metal parts used in finish construction, as hinges, catches, pulls, locks, and so on.

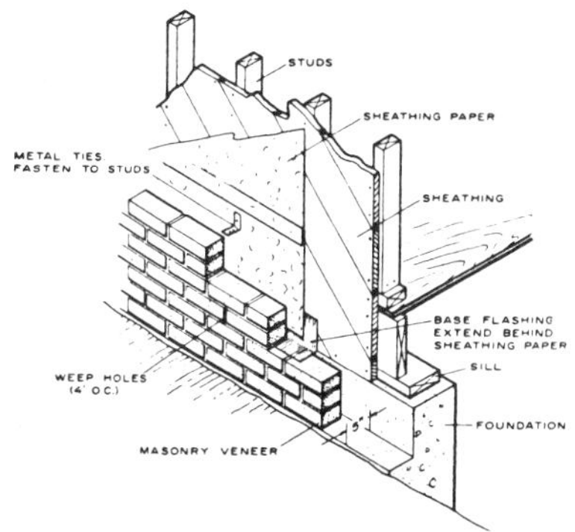

Figure 129 Brick veneer

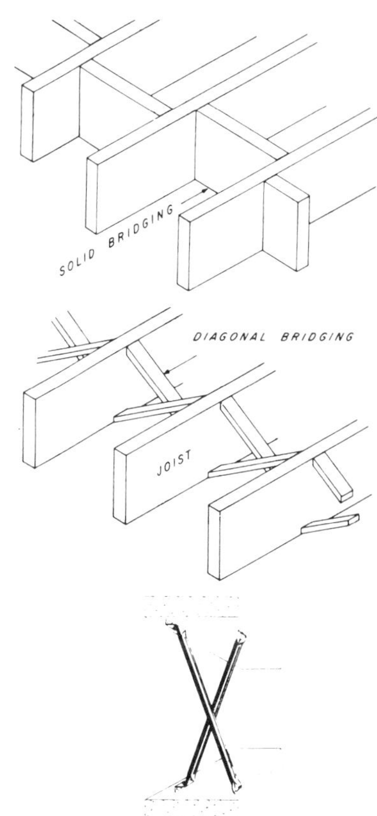

Figure 130 Bridging

Figure 131 Bucket trowel

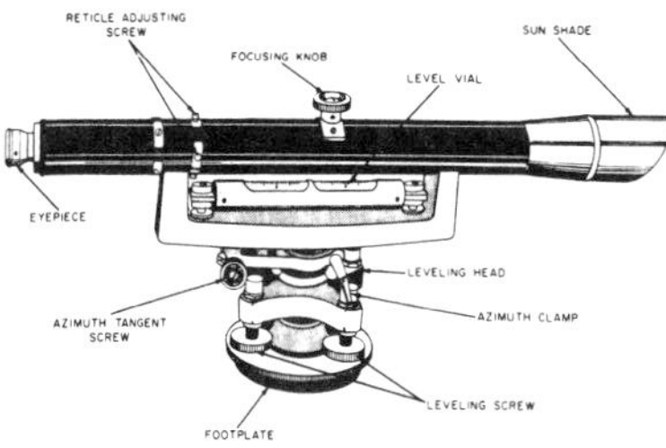

Figure 132 Builder's level

builder's level A surveying instrument with a 20× to 32× telescope mounted on a tripod. (Figure 132.) The level can turn only in the horizontal plane. A 360° horizontal circle and vernier allows angles to be read. Leveling screws are used to level the instrument. The level is located over an exact point with a plumb bob and line for measuring exact angles. See *alidade, automatic level, differential leveling, leveling, leveling rod, theodolite, transit, transit laser, vernier.*

builder's tape Steel measuring tape that comes in lengths of 50 and 100 feet; used on the job for determining longer distances. See *steel tape.*

building block *Concrete block.*

building board A manufactured, pressed insulating board.

building brick The standard brick used in construction. Also called hard or kiln-run brick. In the past called *common brick.*

building code A collection of regulations adopted by a city, county, or other governmental unit for the construction of buildings and to protect the health, safety, and general welfare of those within or near the buildings.

building drain Drainage piping that carries building waste to the building sewer. A building drain may connect to the sewer system or to a septic tank and private *absorption field.* Also called a *house drain.* See *plumbing system.*

building inspector One who checks structures to see that they conform in every way to codes and minimum building and safety standards.

building line The line established by the local municipality controlling the location of structures. The building line usually parallels the property line. The building may not extend beyond the building line.

building paper A heavy water-resistant construction paper often used over house sheathing or over rough flooring. Designed to prevent air passage and inhibit moisture passage.

building permits Authorization given by local governing body to construct a building.

building restrictions Prohibitions by a local governing body against, or controlling construction of, certain types of structures.

building sewer Drainage system that receives discharge from building drain and carries waste to the public sewer or private disposal field. Also called a *house sewer.*

builds *Head joint.*

built-in Referring to parts of a house, such as kitchen cabinets, bookcases, and shelves, which are constructed on-site by a journeyman, rather than being factory-built components.

built-up Refers to a structural member made up of several parts (lumber pieces, steel beams) fastened together (with spikes or bolts, or welded) to work as a unit, such as a built-up beam or girder. (Not the same as a glued or laminated member.) See *wood beam.*

built-up beam Beam made by nailing or bolting planks together.

built-up column In steel construction, a column or stud made up of two or more steel structural elements welded, riveted, or bolted together to form a structural unit. A wood column made of two or more structural wood pieces fastened together.

built-up roof Roof cover made by sealing together several waterproof layers, usually with asphalt-covered felt; the roofing material is put down one layer at a time and is mopped with hot asphalt, which serves as a sealer. Used on flat or low-sloped roofs. (Figure 133.)

built-up roof membrane A continuous roof cover, as asphalt sheets or plies bonded together with bitumen.

built-up timber Lumber pieces nailed or bolted together to make a larger structural unit.

bulkhead A boxlike raised structure on a roof used to cover an elevator shaft or stairwell. Also, in concrete placement, a vertical divider placed in the forms to stop the flow of concrete. (Figure 134.) See *construction joint*. A retaining wall made of sheet pilings used to retain earth by river banks or lake shores.

bulldozer Steel-treaded tractor with a large concave blade on the front. (Figure 135.) Widely used for clearing building sites and loosening dense soil. Also called a dozer, crawler tractor, tractor. Figure 136 shows a bulldozer being used to spread material on the site.

bullfloat A long-handled float used for finishing concrete flatwork. It embeds the large aggregate into the concrete mass. (Figure 137.)

bull floating In concrete work, a type of finishing of flatwork after the *screening* levels out the concrete. A long-handled *bullfloat* is used. See *darbying*.

bull header Masonry unit, such as a brick or concrete block, laid on face, that runs between two wythes in a cavity wall and ties the wythes together. See *brick*.

bull nose Blunt, rounded edge. In masonry, a brick, concrete block, or ceramic tile with a rounded corner used at the edge of openings. See *concrete block*.

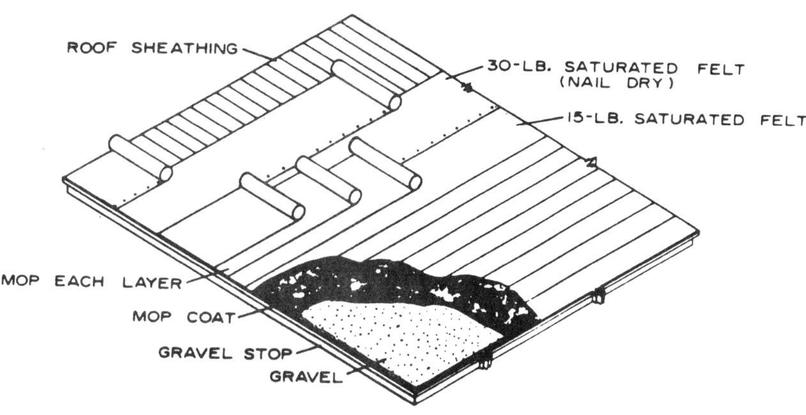

Figure 133 Built-up roofing

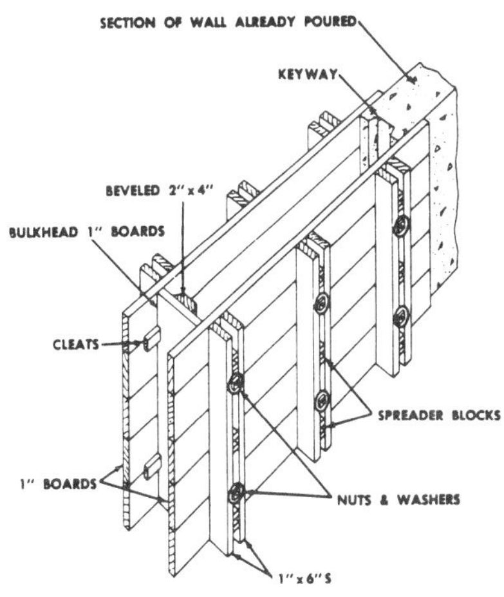

Figure 134 Bulkhead

Figure 135 Bulldozer

Figure 136 Bulldozer used to spread material

Figure 137 Bullfloat being used

Figure 138 Bull-nose plane

bull-nose plane A small plane with the blade located at the forward edge. Commonly 4″ long with a $1\frac{3}{32}''$ cutter. Can be used to plane into corners. Named for its rounded, bull-like nose. (Figure 138.) See also *plane.*

bull pin Steel, tapering pins used for aligning holes between structural steel members so they can be bolted together. See also *drift pin, erection wrench.*

bull stretcher Brick laid on face in a wythe. See *brick.*

bumping In fine metal work, the raising out of flat metal by striking with a special hammer to make various curved metal shapes.

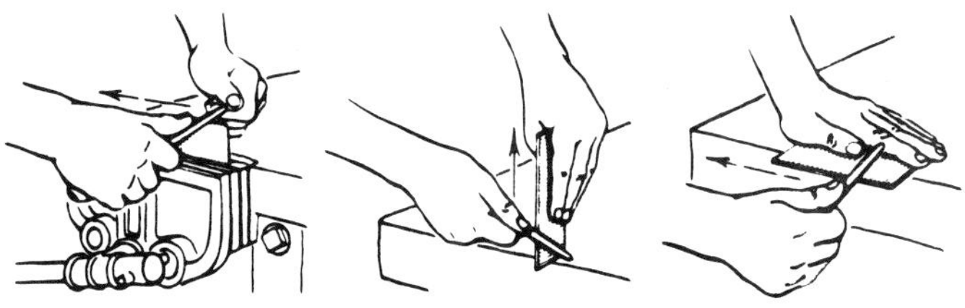

TURN THE EDGE WITH A FEW STROKES OF THE BURNISHER. THE SCRAPER CAN BE HELD IN ANY OF THE THREE WAYS SHOWN ABOVE. DRAW THE BURNISHER TOWARD YOU THE FULL LENGTH OF THE BLADE, WITH A SLIDING STROKE

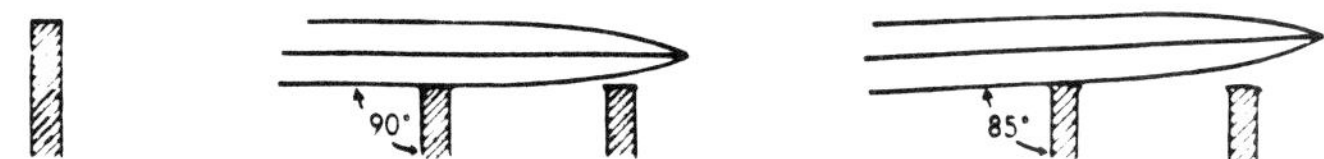

TO TURN THE EDGES OUT, THE BURNISHER IS HELD AT 90° TO THE FACE OF THE BLADE FOR THE FIRST STROKE. FOR EACH OF THE FOLLOWING STROKES TILT THE BURNISHER SLIGHTLY UNTIL AT THE LAST STROKE IT IS HELD AT ABOUT 85° TO THE FACE OF THE BLADE. A DROP OF OIL ON THE BURNISHER HELPS

Figure 139 Burnisher

burl In hardwood, a dense wood mass with many twists and swirls of wood grain, often found near a crotch in the tree. Used in decorative veneer and for carving and turning.

burning Cutting with a welding torch. See *oxygen cutting.*

burnisher A hardened steel rod used to turn over an edge on a hand scraper. (Figure 139.)

burnishing Glazing of a wood surface from tool friction or use of a clogged abrasive sheet or belt.

bush hammer Hammer for finishing rough architectural stone. (Figure 140.) Different fineness of finish is obtained by using a finer cut of hammer. A four-cut and six-cut bush hammer is commonly used. Air-operated or electric bush hammers are also used. See also *boasting.*

bushhammering Finishing of rough stone with a bush hammer. Also, chipping of concrete to give an attractive, rough finish similar to natural stone; normally a pneumatic tool is used.

bushing Finishing of rough architectural stone with a *bush hammer.* A four-cut or six-cut finish is common, depending on the fineness of the teeth cut on the hammer.

bus way Grounded metal enclosure in which factory-assembled electrical conductors run.

butt In roofing, the bottom, thick end of a wood shingle; the part that is exposed to weather when in place on the roof.

butt chisel Short, wide-blade wood chisel used for cutting gains for butt door hinges. See *wood chisel.*

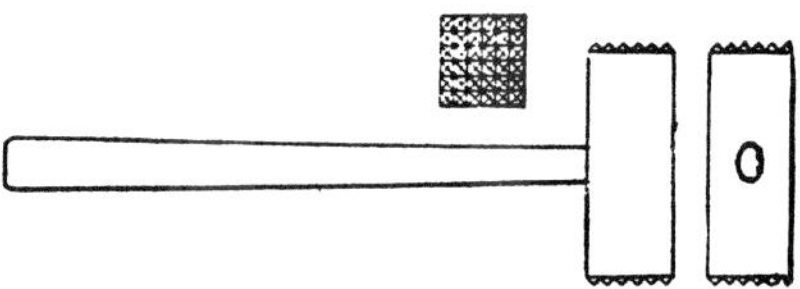

Figure 140 Bush hammer

butterfly roof A roof with a valley in the middle and the two roof sides forming an open V.

buttering In brick or concrete block masonry, the spreading of mortar on the building unit before laying.

buttering trowel Small trowel used for spreading mortar on a brick.

butt gauge Marking gauge used to cut or scribe lines to indicate location of a mortise. (Figure 141.) See also *mortise gauge.*

butt hinge Hinge commonly used for attaching doors to the frame. The hinge is recessed into the wood. (Figures 142; *46*, p. 20.) See *hinges, piano hinge.*

butt joint Lumber pieces joined end-to-end; an *end joint.* Also, the square end of one piece framed against the face of another. See *wood joints.*

buttress An exterior support for a wall that is structurally part of the wall, designed to receive lateral force.

butt weld Weld where two pieces are joined together end-to-end and fused.

BX cable *Armored cable.*

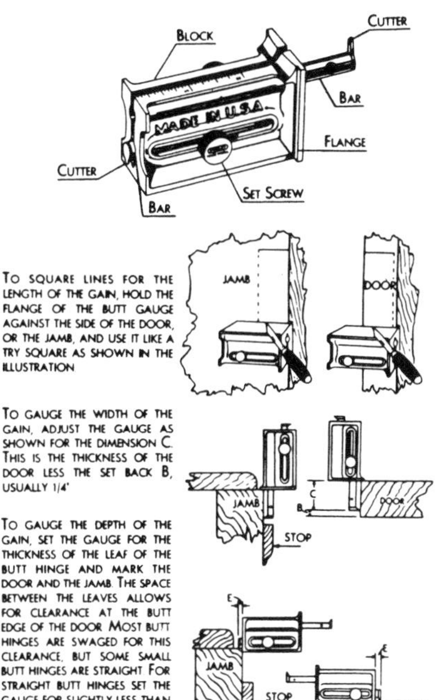

Figure 141 Butt gauge

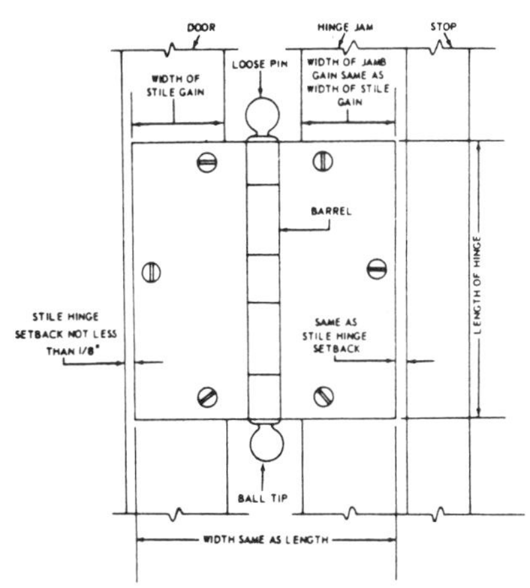

Figure 142 Butt hinge

C

cabinet clamp Lightweight bar clamp used for holding finished wood surfaces while gluing. (Figure 143.)

cabinet drawing Method of showing a solid object with the front view parallel to the viewer (orthographic view) in full scale and the receding sides drawn up at an angle, usually of 30° or 45°, and at one-half scale. (Figure 144.) This method allows you to see the front view in correct proportion. Widely used to show cabinet work. See *pictorial drawings.*

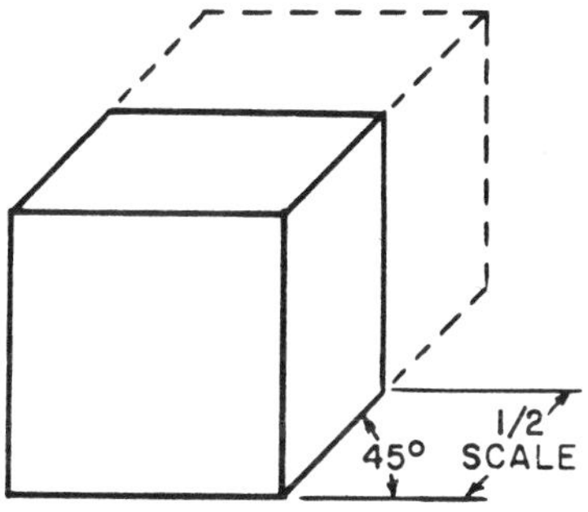

Figure 144 Cabinet drawing of a cube

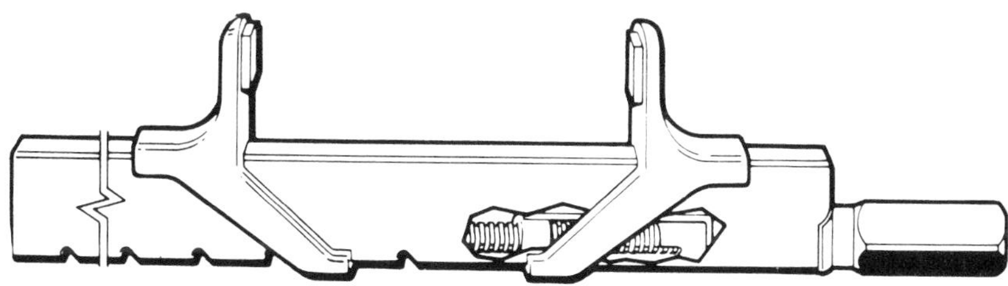

Figure 143 Cabinet clamp

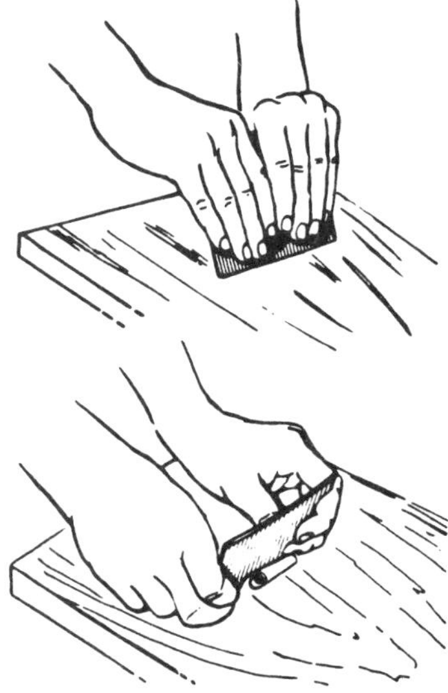

Figure 145 Cabinet scraper

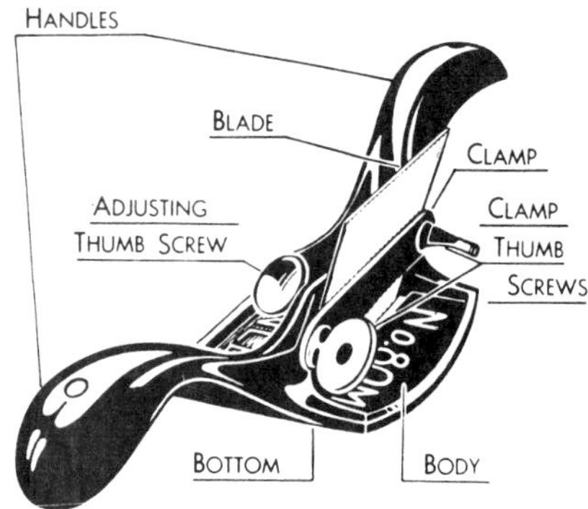

Figure 146 Cabinet scraper

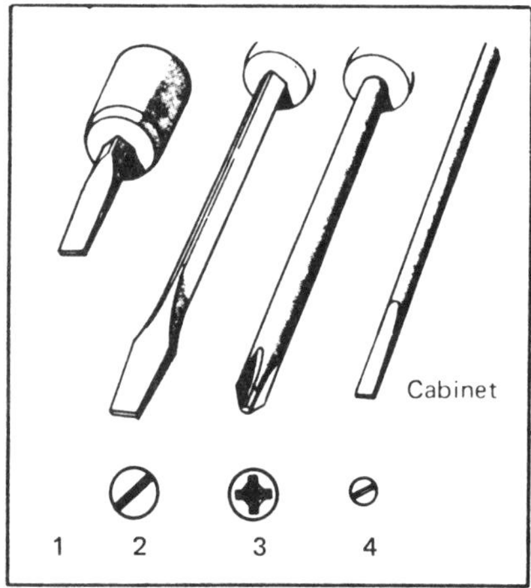

Typical screwdrivers
1. Stubby screwdriver for working in close quarters.
2. Screwdriver with a square shank to which a wrench can be applied to remove stubborn screws.
3. Screwdriver for Phillips screws.
4. Cabinet screwdriver has a thin shank to reach and drive screws in deep, counter-bored holes.

Figure 147 Cabinet screwdriver (right) compared to other screwdrivers

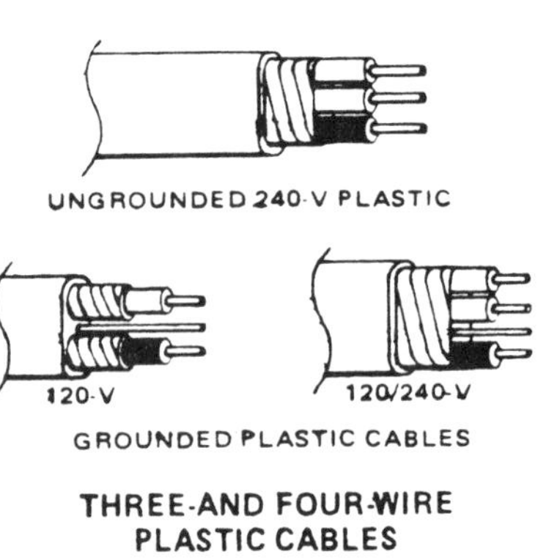

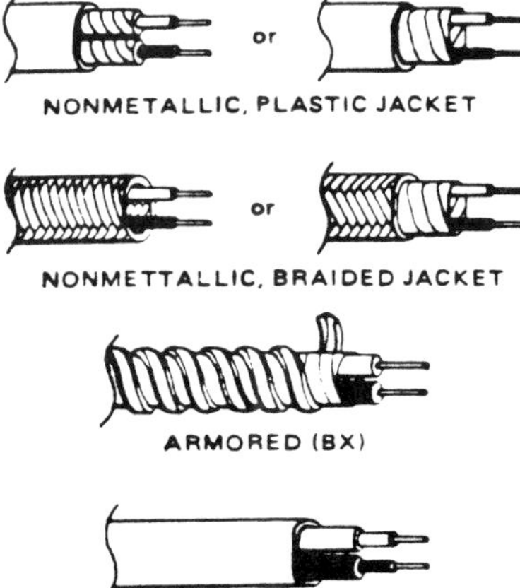

Figure 148 Cable

cabinetmaker One who builds and installs cabinets and shelving. May also install factory-built cabinets.

cabinet rasp A half-round file or rasp that is used for rounding out. See *file, rasp.*

cabinet scraper Flat piece of steel with a thin edge that is turned over with a *burnisher* to form a fine edge. The scraper can be pulled toward you or pushed away to smooth the wood. (Figure 145.) Various shapes are available. Also, a two-handed planing tool that is used for finishing; it is pulled toward you. (Figure 146.)

cabinet screwdriver A long, straight-shanked screwdriver used in turning recessed screws; widely used in cabinetwork. The blade and tip run parallel to the shank to allow entrance into holes to reach recessed screws. Figure 147 compares a cabinet screwdriver shank to other types.

cabinetwork Construction or installation of kitchen or bathroom cabinets; the cabinets themselves.

cable In electrical wiring, the protected conductors that carry the electrical power. Conductors are bound together in a sheathing or casing. The main types of cable commonly used are: *armored cable,* also called BX; *nonmetallic sheathed cable; plastic sheathed cable;* and *service entrance cable.* Cable is specified by the number of wires or conductors: two-wire, three-wire. (Figure 148.) See *conduit, wire sizes.*

cable attachments Method of attaching a steel cable to the load. (Figure 149.) See *rigging.*

cable rack Frame used to hold electrical or telephone cables.

cable ripper Special electrician's tool for stripping nonmetallic sheathed cable. (Figure 150.) See also *wire stripper.*

cable slitter See *wire stripper.*

cable splice Joining wire cables by splicing or weaving together the wire strands.

cable winch A ratchet-activated hoist that uses a steel cable through the block. See *ratchet-lever hoist.*

CAD Computer-assisted drafting.

cadastral survey Survey of land boundaries and subdivisions. Industrial land tracks are identified.

CADD Computer-aided design and drafting.

cage An assembly of reinforcing bars wired together, as the reinforcement for a concrete column. (*Figure 210,* p. 78.)

caged beam Steel beam that is enclosed on three exposed sides in double or triple layers of gypsum paneling. (Figure 151.) Designed to increase fire rating of the member. See *beam (fireproofing), column fireproofing.*

caisson A deep, tubelike foundation support. A containerlike structure used to seal out water for construction work under water; box, open, and pneumatic caissons are used. A box caisson is sunk in place and filled to create a support. An open caisson is sunk in place, sealed at the bottom with concrete, pumped dry, and filled with concrete. A pneumatic caisson has a sealed chamber with air locks; compressed air holds back water and mud so underwater work and construction can be done. Also, a sunken ceiling panel.

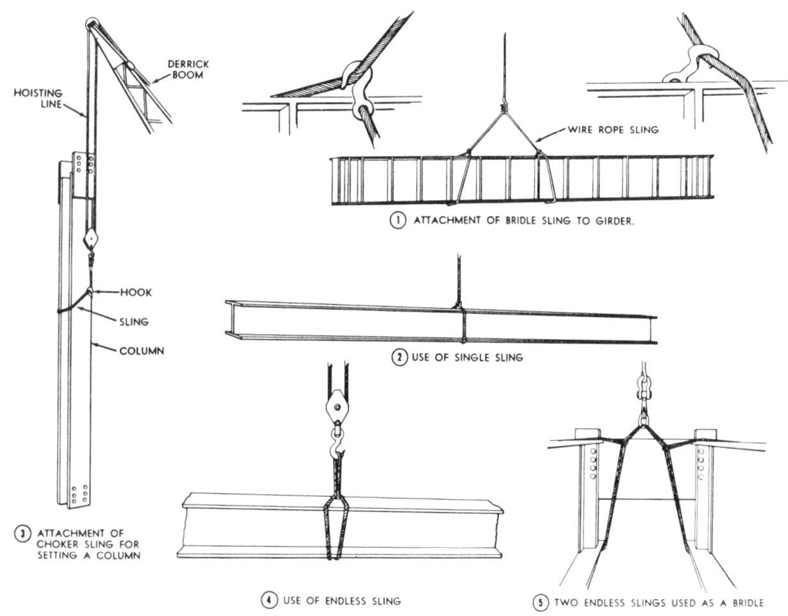

Figure 149 Cable attachments

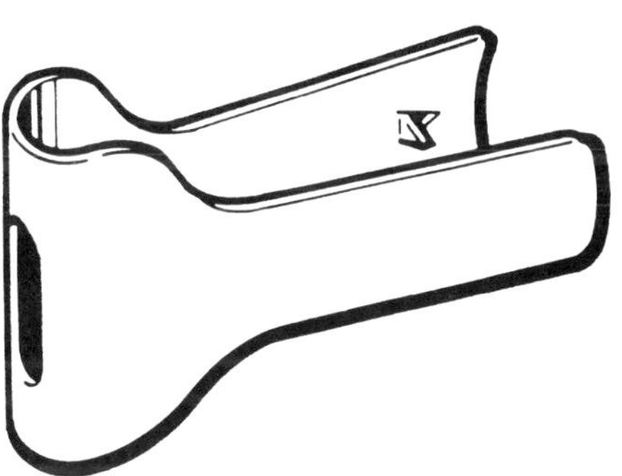

Figure 150 Cable ripper

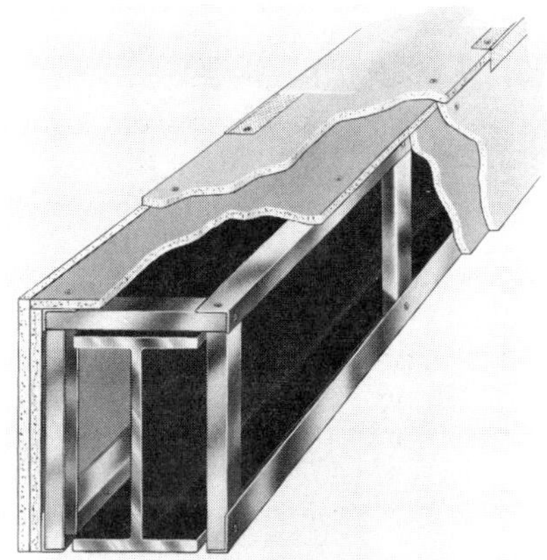

Figure 151 Caged beam

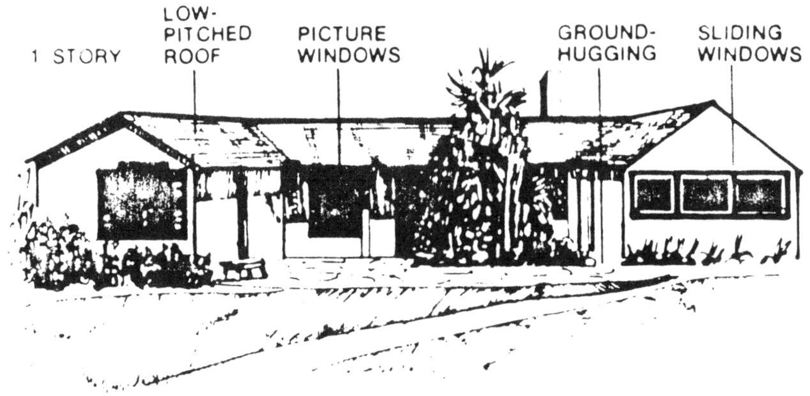

Figure 152 California ranch style

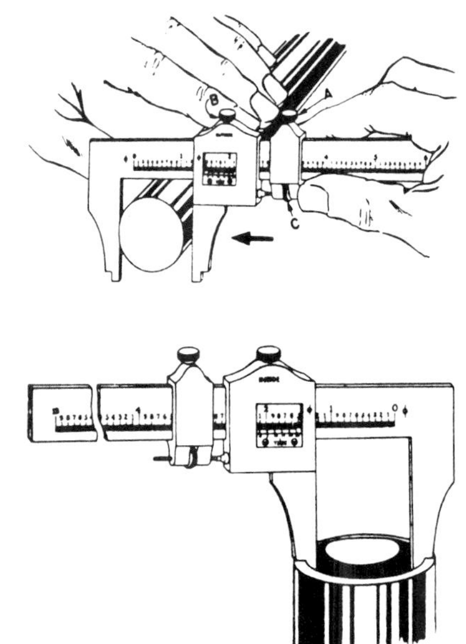

Figure 153 Caliper rule

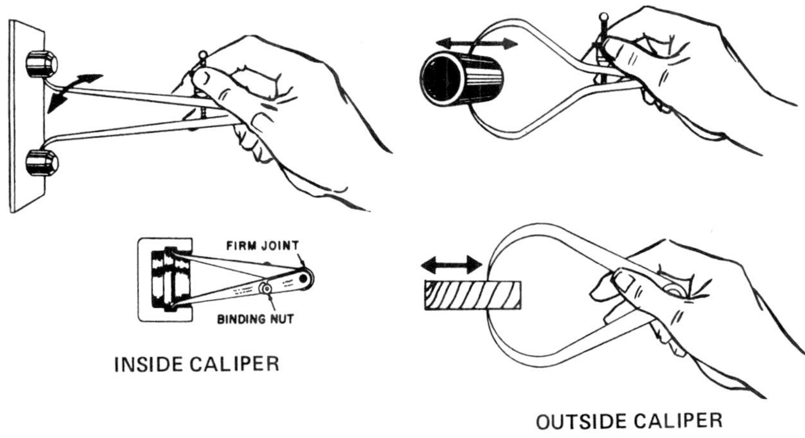

Figure 154 Calipers

California ranch Low, one-story house with a low-pitched roof. (Figure 152.) See also *architectural styles.*

caliper rule Rule for taking both inside and outside measurements. (Figure 153.)

calipers Two-legged measure used where it is awkward to use a rule, as on pipes. (Figure 154.) The *outside caliper* takes outside measurements; the *inside caliper* takes inside measurements, as inside a pipe.

camber A slight upward (convex) or downward (concave) curve in a horizontal support member. Positive camber is upward, negative camber is downward. Positive camber is designed to counteract loading (downward pressure). Positive camber is common in reinforced prestressed concrete beams and steel bridge girders. Positive camber is also used on steel boiler support girders and on cantilevered girders. Large steel trusses, over 80′ wide, may also employ positive camber.

cambium Soft layer of living tissue just below the bark of a tree where new wood is formed. See *tree.*

cant To incline, tilt, set at an angle.

canted beam In steel construction, the framing of one steel beam into the web of another so the faces of the webs are perpendicular but the flanges of the one are tilted in relation to the flanges of the support beam. (Figure 155.) See *skewed beam, sloped beam.*

cant hook Wooden lever with a movable steel hook at the end. (Figure 156.) Used for rolling or turning logs or heavy timber. See also *peavy.*

cantilever To extent beyond a support by using short members that are only supported on one end. A projecting floor, joist, or beam supported only at one end. (*Figure 576,* p. 244.)

cantilevered beam Beam extended beyond support.

cantilevered joists Short joists that run at right angle to the floor joists. They project out away from the floor to provide support for a bay window or balcony. See also *lookouts.*

cant strip Cut lumber piece (triangular in cross section) used on a flat roof where the roof meets a wall or parapet. (Figures 157; *247,* p. 94.) Designed to eliminate the sharp angle and prevent cracking. Flashing and roofing are applied over the strip. Also called a *canting strip.* Also, a molding near the bottom of an outside wall designed to allow water to flow away from the house foundation; a water table. See *molding.*

cap Top part or piece. Threaded fitting over the end of a pipe. A steel plate used on the top of a structural member. See also *bent cap, capping, cap plate, post cap.*

Cape Ann Colonial Symmetrical one-and-a-half story American Colonial house with a gambrel roof; variety of *Cape Cod Colonial.* (Figure 158.) See also *architectural styles.*

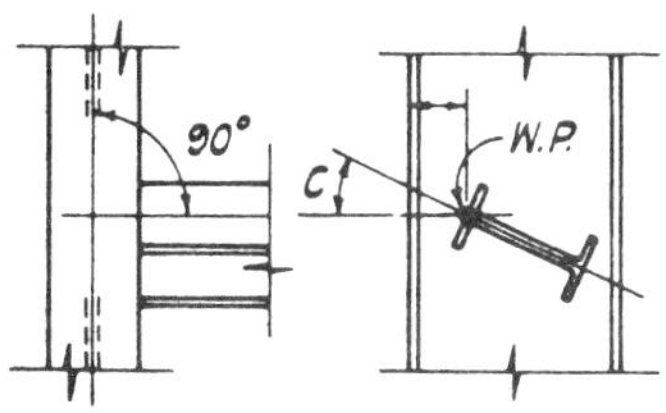

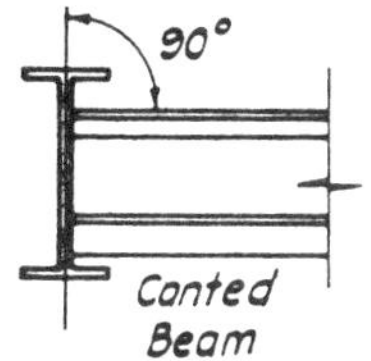

Figure 155 Canted beam

Figure 156 Cant hook

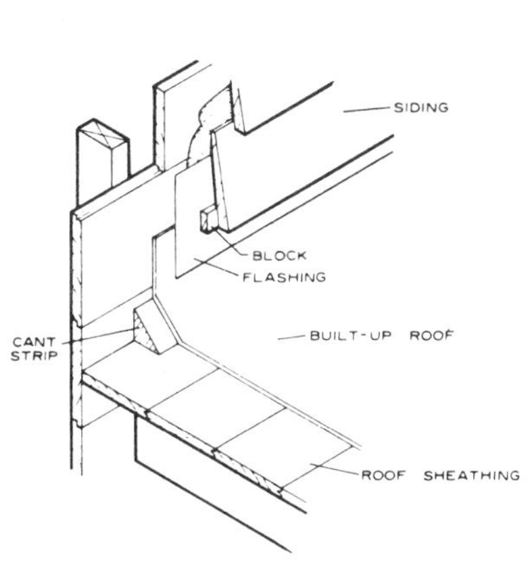

Figure 157 Cant strip

Figure 158 Cape Ann Colonial

Figure 159 Cape Cod Colonial

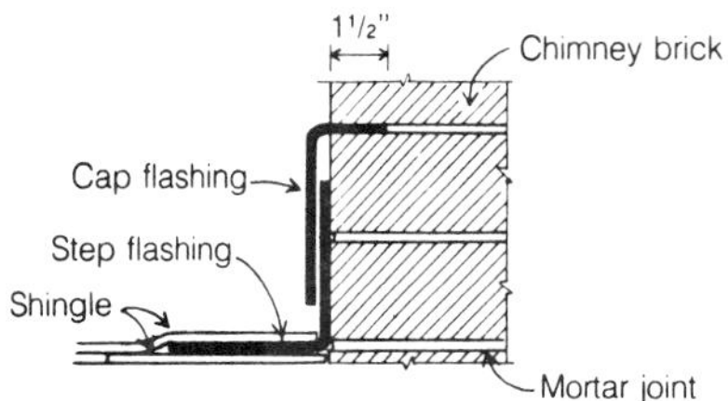

Figure 160 Cap flashing

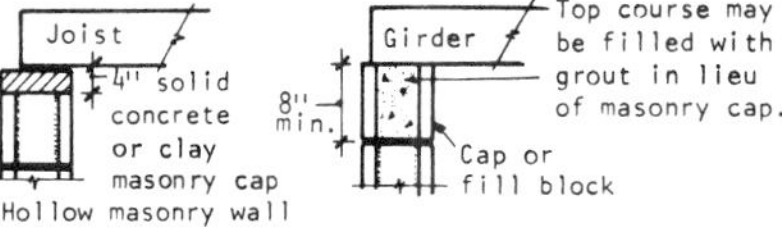

Figure 161 Capping

Cape Cod Colonial An East Coast symmetrical house with a full first floor and a partial second floor under a steep gable roof. (Figure 159.) Also referred to as *one-and-a-half story construction.* See also *architectural styles, Cape Ann Colonial, story.*

cap flashing Flashing placed over base flashings; a counterflashing. Figure 160 shows a cap flashing set into the brickwork of a chimney. See *counterflashing.*

capillary action Process of flowing of a liquid (water, molten solder, or molten filler metal) into fine cracks or openings by molecular attraction.

capital Upper part of a column or pier. (*Figure 784,* p. 344.) In the classical orders, the top part of a column. See *orders, drop panel.*

capping Top of a part; the crown. Solid masonry cap at top of masonry wall to support joist or girder. (Figure 161.)

capping brick Special brick used at the top of a wall, coping brick.

capping in In roofing, the placement of roofing felt on the roof. Also called *drying in.* (*Figure 40,* p. 17)

cap plate Angle supports at the top of a column. (Figure 162.) They attach column to girder or beam. See *post cap.*

cap sheet Aggregate-coated finish sheet used in built-up roofing.

cap strip seam See *drive slip.*

carbon-arc welding Electric-arc welding process where a nonconsumable carbon electrode is used to produce the electric arc between the electrode end and the work piece. A filler rod is used if additional metal is needed for the weld. (Figure 163.)

carbon electrode Graphite or baked carbon electrode used in carbon-arc welding.

carborundum Very tough abrasive made from carbon and silicon. Used in abrasive sheets, belts, and stones.

carborundum cloth Cloth with finely crushed carborundum glued on, used for sanding. Carborundum paper is also used.

carborundum stone Abrasive stone made of carborundum, used for sharpening tools.

carcase Cabinet or chest frame; a *carcass.*

carcass In woodworking, the boxlike frame of a cabinet, chest, or furniture piece, without doors, drawers, or hardware installed. (Figure 164.) Also, a house frame.

carpenter A journeyman who constructs houses, buildings, and other structures; a craftsman who works with wood to build structures. The rough carpenter erects wood framework of buildings, including joists and studs for floors, walls, and partitions, and rafters or trusses for roofs. Also builds forms and scaffolding. The finish carpenter installs molding, paneling, cabinets, window sashes, door frames, hardware, and finish floor; also builds stairs.

carpenter-built Something that is built on-site in the building by a carpenter, as a kitchen cabinet or stairs.

carpenter's level A bubble level used to determine level horizontal and vertical surfaces. See *level.* Also called a *spirit level.*

carpenter's pencil Flat wooden pencil with a heavy, wide lead, used for marking on the job.

carpenter's square Steel square used for laying out, for testing for straightness, for squaring. (Figure 165.) See also *rafter square,* which includes a rafter table. See *steel square.*

carport Open structure attached to the side of a house for sheltering cars; has a roof but is open on the back or side.

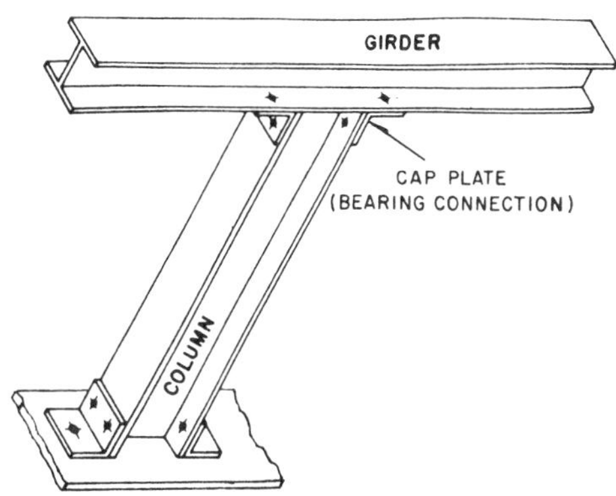

Figure 162 Cap plate

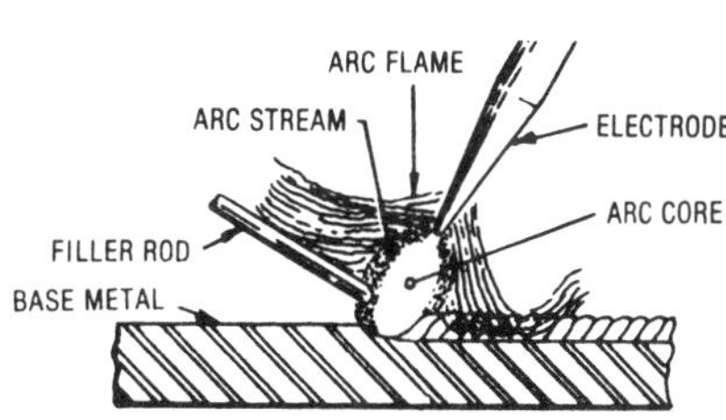

Figure 163 Carbon arc welding

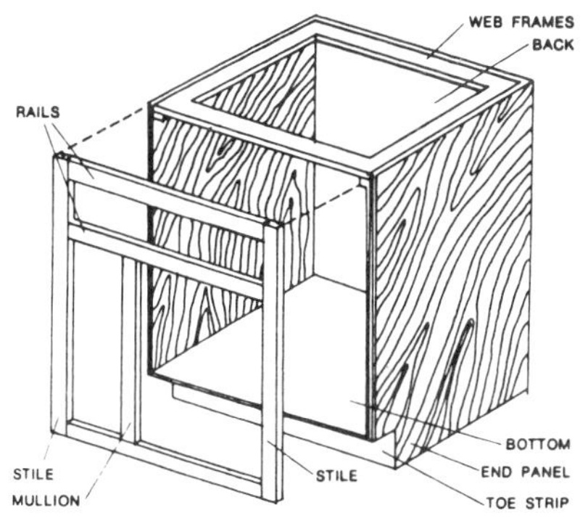

Figure 164 Carcass

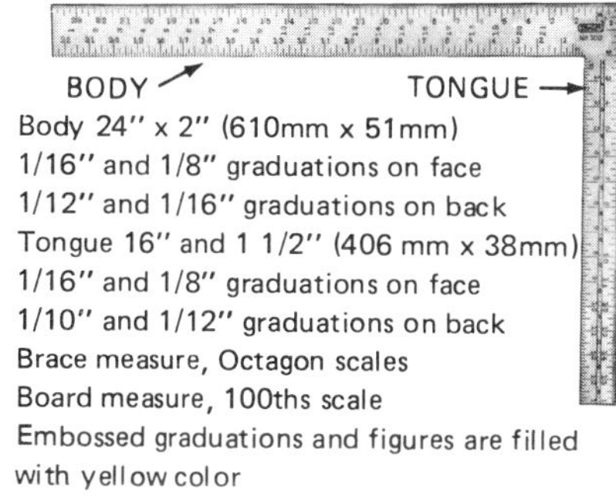

Figure 165 Carpenter's square

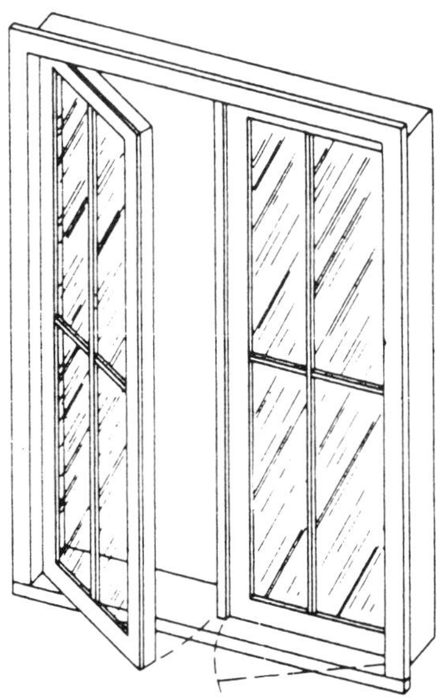

Figure 166 Casement window

carriage A support for stair steps; two or three carriages are used under the steps. Also called a *stringer*. See *stairs*.

cartridge fuse Circuit protection used for circuits up to 600 volts. The fuse is inserted into a special fuse clip. Fuseholders are designed to make it difficult to put in cartridges designed for a lower or higher voltage. Two types are available: blade contact and ferrule contact. Cartridge fuses are called pull-out fuses. A special *fuse puller* is used to remove the fuse. See *fuse*.

carving An ornament cut on woodwork or furniture.

case Structural framework of a building; external masonry facing; the *carcass* or framework of a cabinet.

cased opening Finished window or door opening with casing—ready to install.

casement window Outswinging window with hinges at the side, usually operated with a small hand crank. (Figure 166.) See *window*.

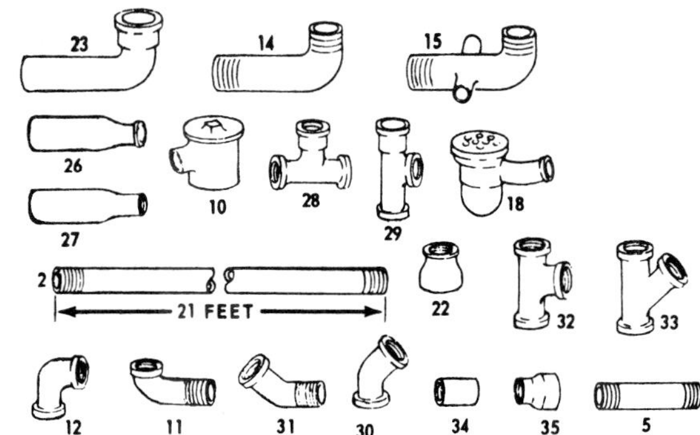

23 SANITARY LONG ¼ BEND
14 TOILET BEND
15 TAPPED TOILET BEND
26 SPIGOT END INCREASER
27 THREADED END INCREASER
10 DRUM TRAP
28 VENT TEE
29 WASTE TEE
18 FLOOR DRAIN WITH TRAP
2 THREADED STEEL PIPE
22 THREADED SPIGOT
32 DRAINAGE TEE
33 DRAINAGE Y
12 90° DRAIN ELL
11 90° DRAIN STREET ELL
31 45° DRAIN STREET ELL
30 45° DRAIN ELL
34 COUPLING
35 REDUCER
5 THREADED NIPPLE

Figure 167 Cast iron fittings

casing Trim around a window or door. (*Figure 336*, p. 129.) See *door, molding, window*.

casing nails Special nails used for installing casing or other wood members where the head should be flush to the wood face. See *nails*.

casting plaster Fine plaster used for casting architectural ornaments and pieces of art.

cast in place Concrete that is placed (poured) at the construction site in concrete forms, as opposed to off-site *precast concrete*. *Reinforced concrete* is cast in place with reinforcing bars located in the concrete. See *reinforced concrete floors*. See also *prestressed concrete*.

cast-iron fitting Plumbing fittings used for vent piping and for sewer and storm drainage piping. Figure 167 shows standard cast-iron threaded fittings commonly used. One type of cast-iron fitting has a wide or recessed shoulder and is used with drainage piping. The recessed shoulder provides a strong joint and a smoother interior surface. See also *galvanized malleable iron fittings*.

cast-iron soil pipe Heavy iron pipe used for sewer and drainage pipe, and for vent piping. Two types are used: *bell-and-spigot* and *no-hub*. Bell-and-spigot piping varies from 2″ to 15″ in diameter. No-hub varies from $1\frac{1}{4}$″ to 10″ in diameter.

catch basin In plumbing, a trap or reservoir used in a sewer system to catch and retain sediment and greasy or oily materials and prevent them from reaching the sewer. An *interceptor*. In site drainage, an opening to allow the inflow of storm sewer into a culvert.

catches Devices designed to hold a cabinet door closed. (Figure 168.) They may be magnetic or mechanical. A magnetic hinge simply holds the door closed by attraction between a steel plate on the inner surface of the door

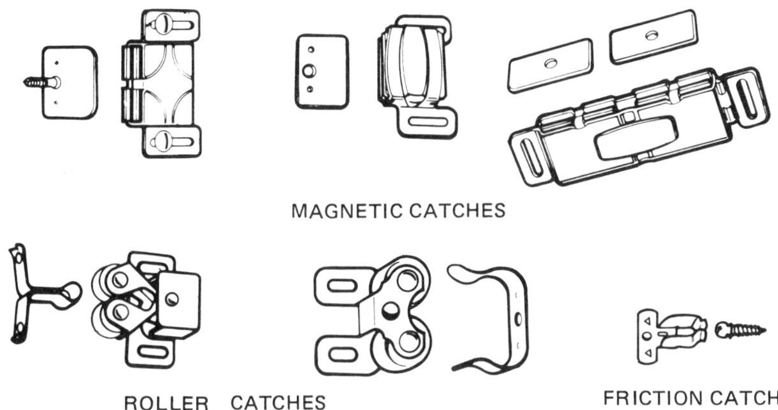

Figure 168 Catches

and a magnet attached to the carcass of the cabinet. Mechanical catches hold by friction, by a head or roller being caught. See *friction catch, roller catch.*

cat head A frame and sheave, through which a lifting cable, as on a crane, is operated.

catslide A *salt box Colonial*

caulk In plumbing, to fill joints with oakum. In painting, to fill cracks or holes with a caulking material. Several caulks are available: oil base, vinyl latex, butyl rubber, acrylic latex, silicone, and polysulfide. Caulks often come in cartridges for use in a *caulking gun.*

caulking In plumbing, driving lead and oakum into pipe joints. See *leaded joint.* In construction, using a caulking gun to force caulking compound into cracks and other openings in the house, as around window and door frames. (Figure 169.) The soft waterproof material used to seal cracks, usually available in a tube for use in a caulking gun. See *caulk.*

caulking anchor *Machine-screw anchor.*

caulking gun Hand-operated device used to force caulking out of a caulking tube. (Figure 169.) See also *adhesive gun.*

caulking trowel See *joint filler.*

cavalier drawing Oblique projection with the receding or oblique planes in the same scale as the front plane. Compare to *cabinet drawing.* See also *pictorial drawings.*

cavil Heavy mason's hammer used for rough-dressing stone at a quarry; has a pointed end for dressing and a flat hammer head. (Figure 170.)

cavity wall A two-part masonry wall with an air space, usually 2″, between the parts or wythes. (*Figure 1072,* p. 473.) Also called a *hollow wall.*

C clamp Steel woodworking clamp shaped like a "C", used for holding small wood pieces together while they are being glued or worked on. (Figure 171.) See *clamp, edging clamp.*

cedar shingles Manufactured cedar shingles used for roofing and siding.

ceiling Top surface of a room.

ceiling fan Exhaust fan mounted in the ceiling to vent air directly to the outside, used in kitchen areas and bathrooms. See *range hood fan, wall fan.*

ceiling fixture box A circular outlet box used in the ceiling for lamp wiring. (*Figure 64,* p. 25.)

Figure 169 Caulking gun

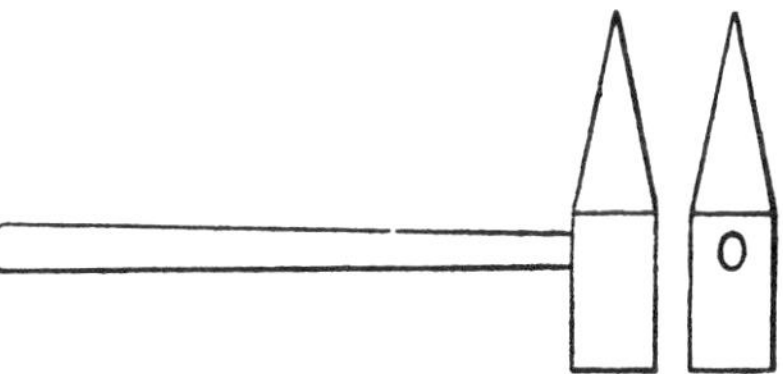

Figure 170 Cavil

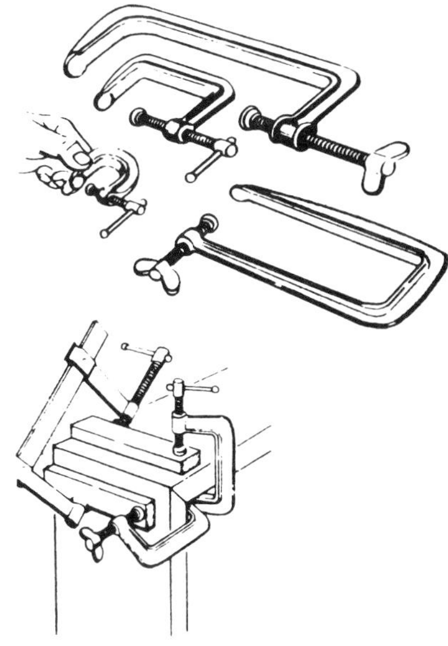

Figure 171 C clamps

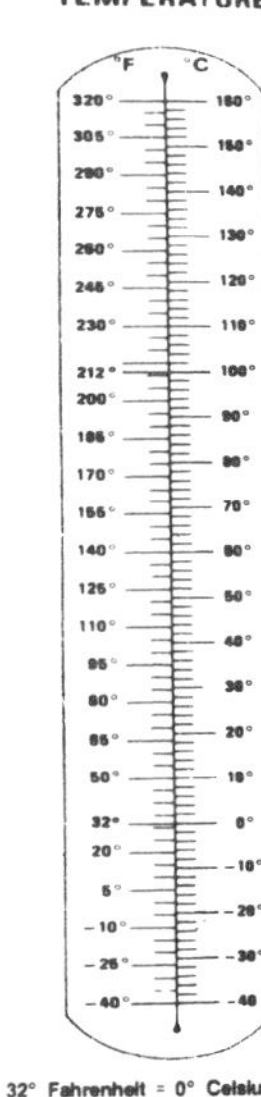

Figure 172 Celsius and Fahrenheit scales

ceiling joists Support joists over a ceiling. If there is a floor above, then they also serve as floor joists (to the floor above).

cellular concrete Low-density concrete with a high volume of air and a low weight (less than 55 pounds per cubic foot).

cellulose Wood ingredient that forms the major part of wood; the wood cell framework. See *lignin.*

Celsius Metric temperature measure. On this scale the freezing point of water is 0°C and the boiling point is 100°C. Abbreviated C. Figure 172 compares the Celsius to the Fahrenheit scale.

cement Hydrated lime (Portland cement) which is mixed with water, sand, and aggregate to harden into a strong, solid body (concrete). Five types of Portland cement are commonly used: *Type I:* Standard or normal (a type 1A is air-entrained). *Type II:* Used in large structures, has a lower hydration heat. *Type III:* High early strength, has an accelerated hydration and is used in cold weather. *Type IV:* Low heat, used in large structures. *Type V:* Sulfate-resistant, used in areas or for structures exposed to high sulfate concentrations. Various special cements are available, including *hydraulic* and *epoxy.* Vinyl and acrylic cements are also available. See *accelerator, admixture, expansive cement, retarder.*

cement-coated nails Nails coated with cement to give greater holding power; the greater strength is temporary.

cement coloring Special pigments mixed with cement to give permanent color to concrete, used for decorative effect.

cement gun Special nozzle which uses pneumatic pressure to spray concrete or mortar on a surface.

cement mason One who works on concrete construction: places and finishes concrete for footings, walls, floors, steps, and sidewalks. Installs preformed decorative concrete panels.

cement mix Different proportions of cement ingredients are used to obtain different strengths or types of concretes.

cement mortar Paste made of Portland cement, sand, and water; used for binding masonry units together or for making stucco walls.

cement trowel Rectangular, flat trowel for smoothing concrete. Also called a *finishing trowel.* See *trowel.*

center bearing truss Truss with center support in addition to heel support (at each side).

centerline The center of something that can be divided into equal parts. A dot-and-dash line on a blueprint that shows the center of something; the parts of each side of the line are the same. See *alphabet of lines.*

center punch A steel tool used to make a dent in metal for starting a drill hole. See *punches.*

center to center Measurement from the center of one thing to the center of another, as studs are spaced 16″ center to center; same as *on center.*

centimeter A metric measure equal to 0.394″. There are 2.54 centimeters per inch. See *metric.*

central-mixed concrete Concrete that is mixed at the factory and then transported to the job site in a truck.

ceramic tile Fired clay tile used on floors and walls. The tile is cut with a *tile cutter.* It may also be cut by scoring with a glass cutter and then breaking by snapping along the scored line.

cesspool Covered pit into which raw sewage runs; the raw untreated liquid part seeps out into the surrounding soil. Prohibited in most areas and replaced by a *septic tank.*

chain Metal hoisting or pulling device made of flexible metal links. In surveying, a metal measure with flexible links; a chain has 100 wire links each 7.92 inches long for a total of 66 feet or 4 rods, called a Gunter chain. Ten square chains equal one acre; 80 chains equal one mile. An engineer's chain is 100 feet long with one-foot links. (Survey chains are rarely used today and have been replaced with steel tapes.)

chain guard Safety chain latch used inside a door. (Figure 173.)

chain hoist Hoist that uses a chain to lift the load.

chaining In surveying, measuring of linear distance with a surveyor's chain. Today, measuring is done with a steel tape.

chaining pin In surveying, a metal pin used to mark a 100′ length of a distance being measured. A pin is used at every 100′ mark. Each 100′ mark is called a station. (Figure 174.) See also *taping pin.*

chain-nose pliers Long needle-nose pliers with side cutters. Used in wiring.

chain saw Gasoline-powered saw used for rough sawing. A chain with cutting teeth revolves around a guide bar. (Figure 175.)

chain tongs A kind of pipe wrench that uses a chain around the open end. Used for turning large plumbing pipe. See also *chain wrench, strap wrench.*

chain wrench In plumbing, a wrench with a chain on the end that fits over a pipe; used for turning a pipe or electrical conduit. (Figure 176.) See also *chain tongs, strap wrench.*

chair Bent rod supports used for holding up reinforcing bars and welded-wire mesh so concrete can flow evenly around the steel. (Figure 177.) See *bar support, bolster, continuous high chair, high chair, joist chair.*

chair rail Molding around a room at chair-back height; designed to protect the wall finish from chairs being pushed against it. (Figure 178.) See *molding, panel molding.*

chalk line A long, heavily chalked string used for marking lines. The string is secured at both ends exactly over the line to be marked, then pulled up and released, leaving a chalk line on the material where the string is snapped against it. (Figure 179.)

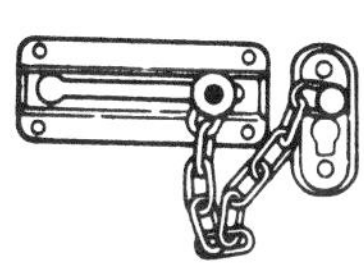

Figure 173 Chain guard

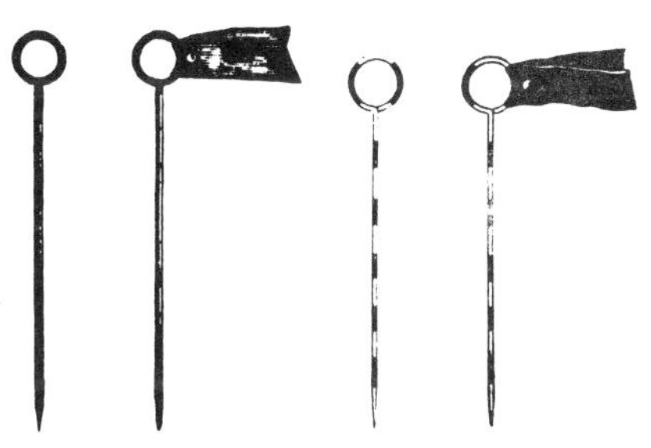

Figure 174 Chaining or taping pins

Figure 175 Chain saw

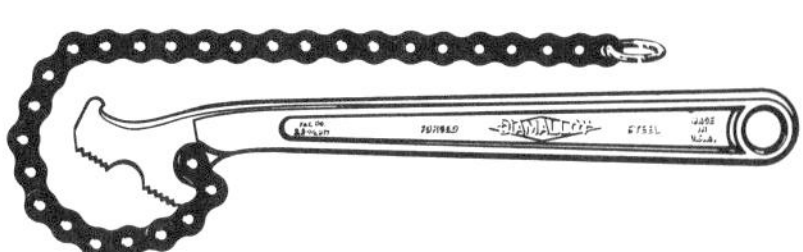

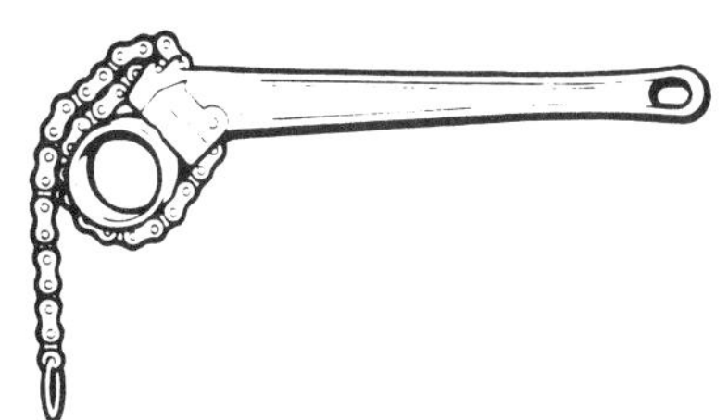

Figure 176 Chain wrench

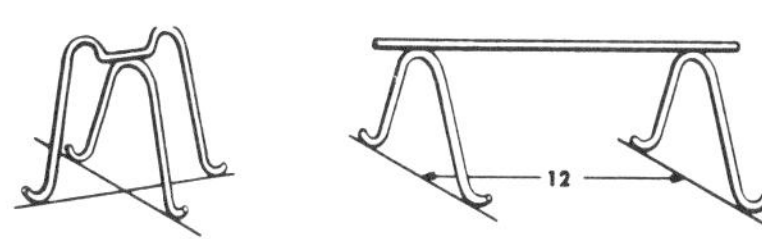

Figure 177 Chairs

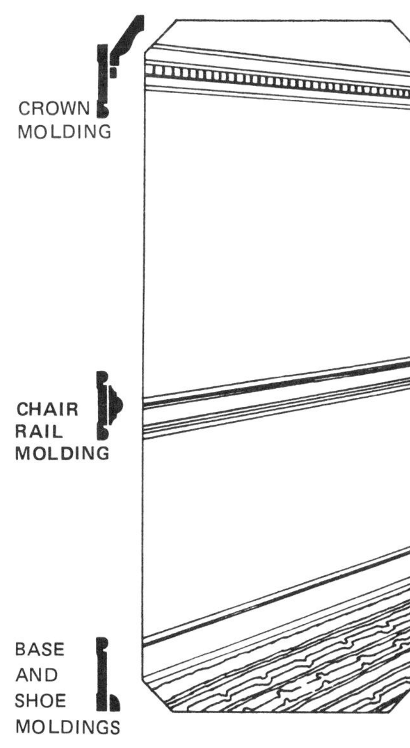

Figure 178 Chair rail molding

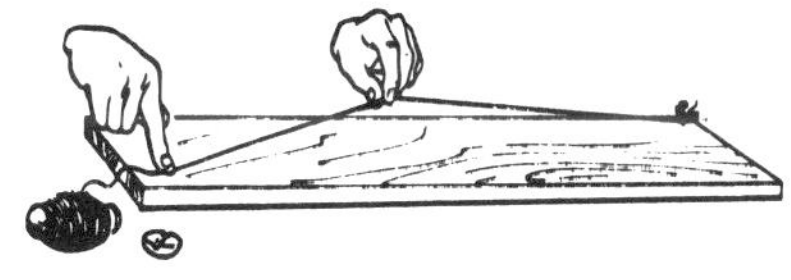

Figure 179 Chalk line

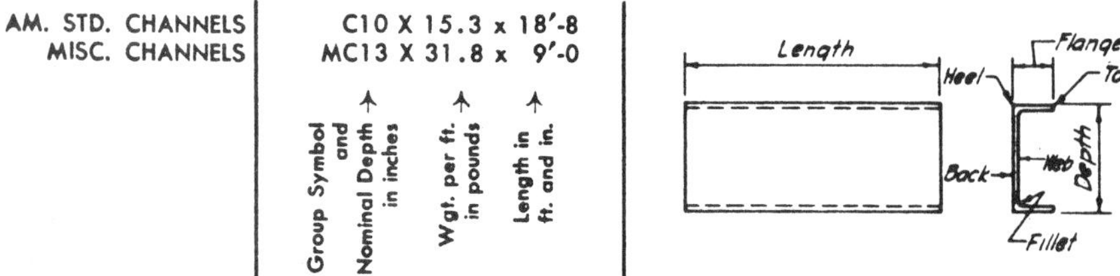

Figure 180 Channel-iron terminology and specifications

Figure 181 Channel shear

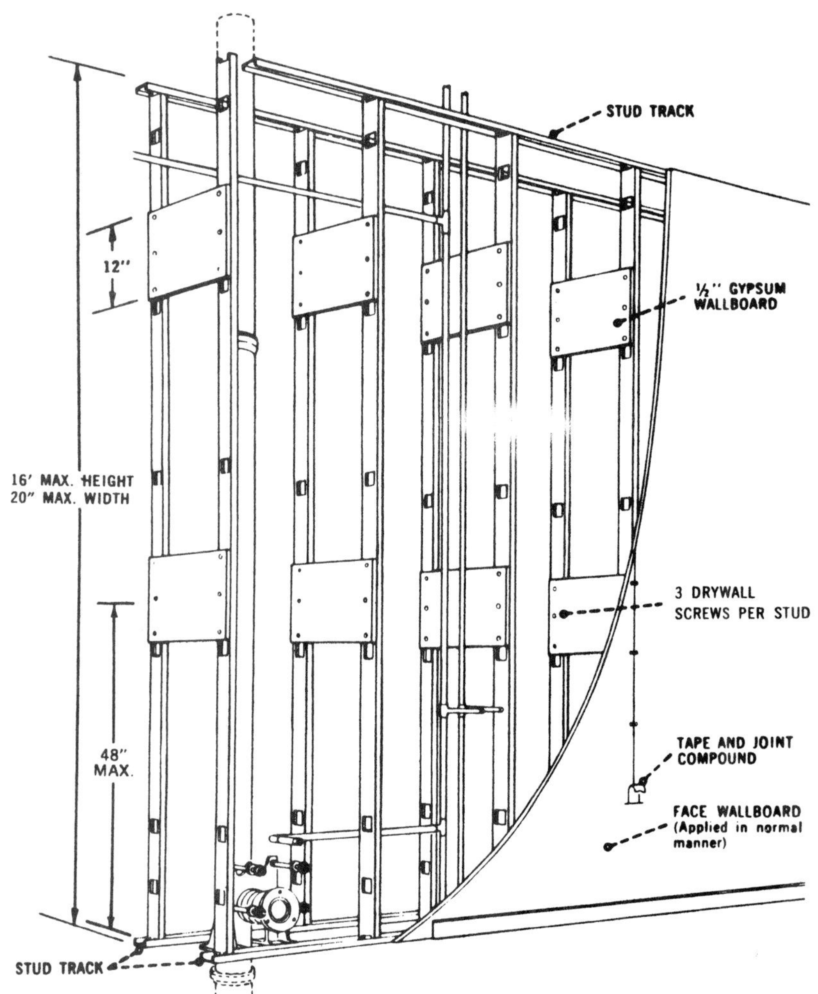

Figure 182 Chase wall

chamfer Beveled edge; an edge with material cut away at a 45° angle. See *molding.*

channel A groove cut into something. A *channel iron.*

channel iron Steel member used in construction. Figure 180 shows the parts of a channel iron and gives specification examples. Widths vary from 3" to 18".

Channellock® Proprietary term sometimes used to refer to an *adjustable-joint pliers.*

channel shear Lever hand shear for cutting through metal stud and runner channels. (Figure 181.) Also called *stud shear.*

charging Putting concrete, mortar, or plaster materials in a mixer.

chase A channel or metal sleeve in a masonry wall or floor to receive pipes or ductwork. See *sleeve. A channel* or recess on or inside a masonry wall to receive piping, conduit, ductwork.

chase wall A double or cavity wall that contains piping or ductwork. (Figure 182.)

chasing Indenting metal to form decorative patterns.

chat Crushed, gravel-size limestone, used for road topping and under-slab fill.

chatter A rough, uneven sound, as in the sound of power tools when a problem occurs.

check Wood defect where the wood fibers separate along the wood rays. See also *heart shake.*

check dam Small retaining dam built in a gully or other small waterway to check waterflow or runoff and control erosion and sediment loss.

checking Cracks opening in wood or in a paint coating.

check valve A valve that automatically closes to prevent water backflow; liquid flows only in one direction.

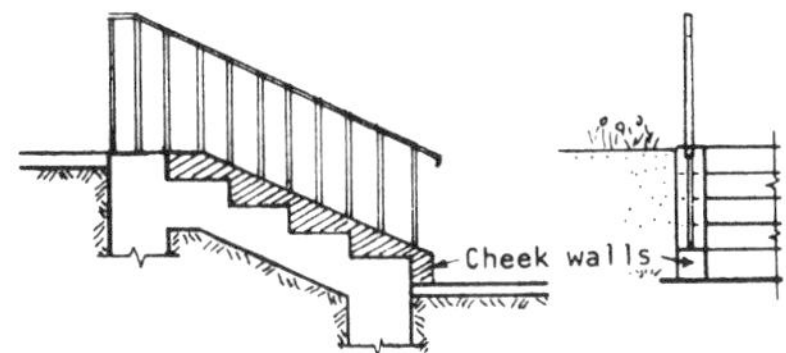

Figure 183 Cheek wall

cheek Side of something, side of a mortise cut, sides of a tendon, the flat side of a hammer. Bevel cut on a rafter edge or end.

cheek cut Side angle cut on the end of a rafter, as for a jack that connects to a hip or valley. A *side cut.* See *rafter cuts.*

cheek wall Low walls on the sides of exterior concrete steps. (Figure 183.)

chimney Brick or metal structure that contains one or more flues for drawing off fumes from the furnace or fireplace. See *fireplace.*

chimney bond A *running bond* used in constructing brick chimneys.

chimney cap Metal covering for a chimney, designed to improve the draft, cut down entry of rain and snow. Also, a *concrete cap* or cover at the chimney top, around the flue.

chimney cricket A small V-shaped arch built behind a chimney on the roof, designed to divert water runoff. See *cricket.*

chimney effect Term used to describe effect of heated air or gas rising, as in a chimney. Heated air or gas rises because of its lower density. In a building, heated air rises to be replaced by cooler outside air.

chimney flashing Flashing and counterflashing laid against the side of a chimney.

chimney throat Area above the fireplace where the sides converge together. See *fireplace.*

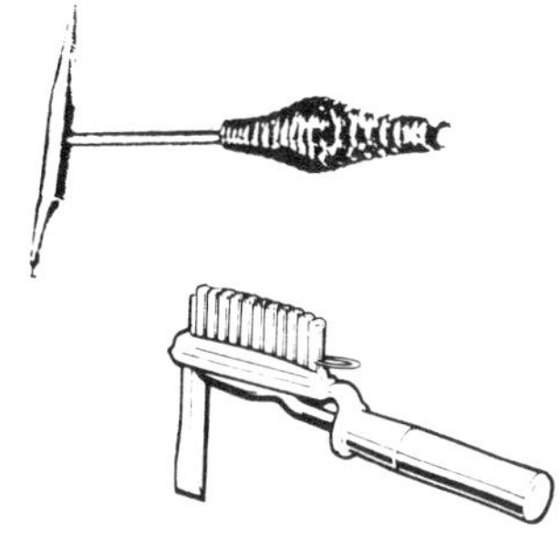

Figure 184 Chipping hammers

chipping hammer In welding, a hammer used to remove slag or spatter from the weld area. (Figure 184.)

chisel A tool used to cut metal or wood. Cold chisels are used to cut metal. They have a double bevel and are classified by the shape of their points. They are struck with a ball peen hammer. See *cold chisels.* Woodworking chisels have a single-bevel blade that fits in a wooden or plastic handle. They are struck with a wooden mallet or are manipulated by hand. See *wood chisels.* See also *brick set, glazier's chisel, stone cutter's chisel.*

choke Clogging of a hard plane mouth with shavings.

chop saw Heavy-duty saw used on the job for cutting metal studs, angle iron, conduit, rebar, and the like. (Figure 185.)

chord Major support members in a truss. See *truss.*

Figure 185 Chop saw cutting rebars

TO OPEN

SPINDLE OF DRILL

REMOVABLE CHUCK KEY

JAWS

CHUCK

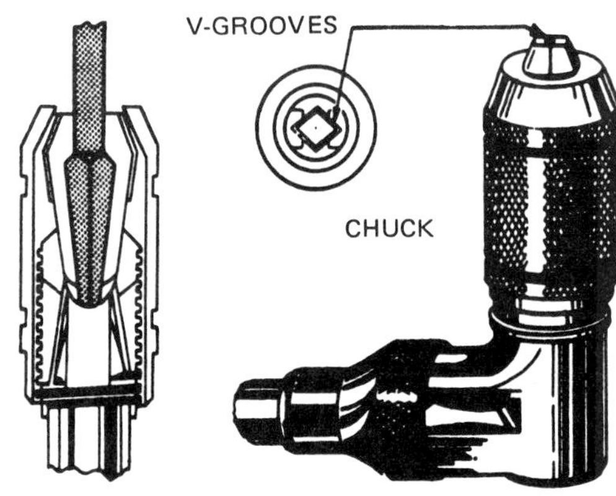

Turn key to open chuck. Drill fits in chuck. Remove key before turning drill on.

POWER DRILL

HAND BRACE

Figure 186 Chuck

chuck A device with closing jaws that hold a drill bit or saw blade in place. Figure 186 shows a chuck in a hand brace and chuck and key for a power drill.

chute Long metal trough used for placing concrete. See *placing.*

cinder block Building block made of cement and cinders.

circle cutter Device used for cutting circular holes in gypsum wallboard; arm has sliding knife that can be adjusted to the needed diameter. (Figure 187.) Also, a drafting instrument for cutting circles.

circuit A wiring loop that runs through a house to the various outlets. A circuit runs from the power source, in a loop to the outlets, and back to the power source. A complete circle or circuit is made. Also called *branch circuit.*

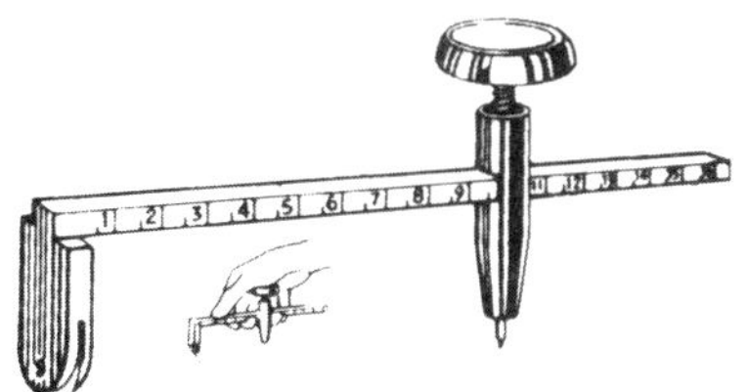

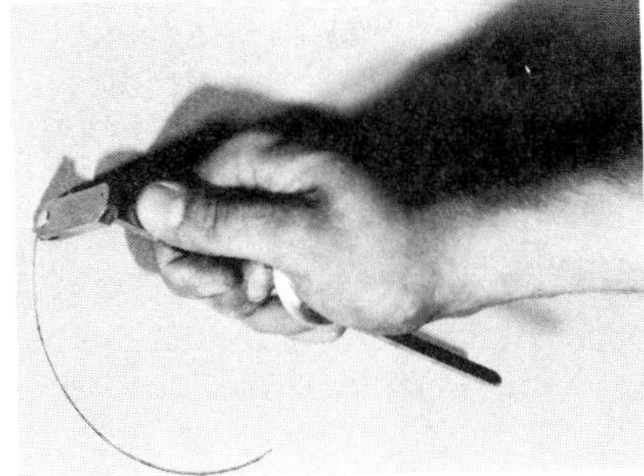

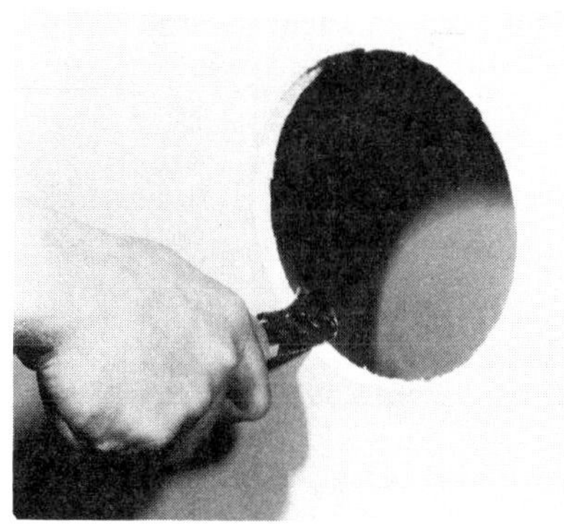

Figure 187 Circle cutter

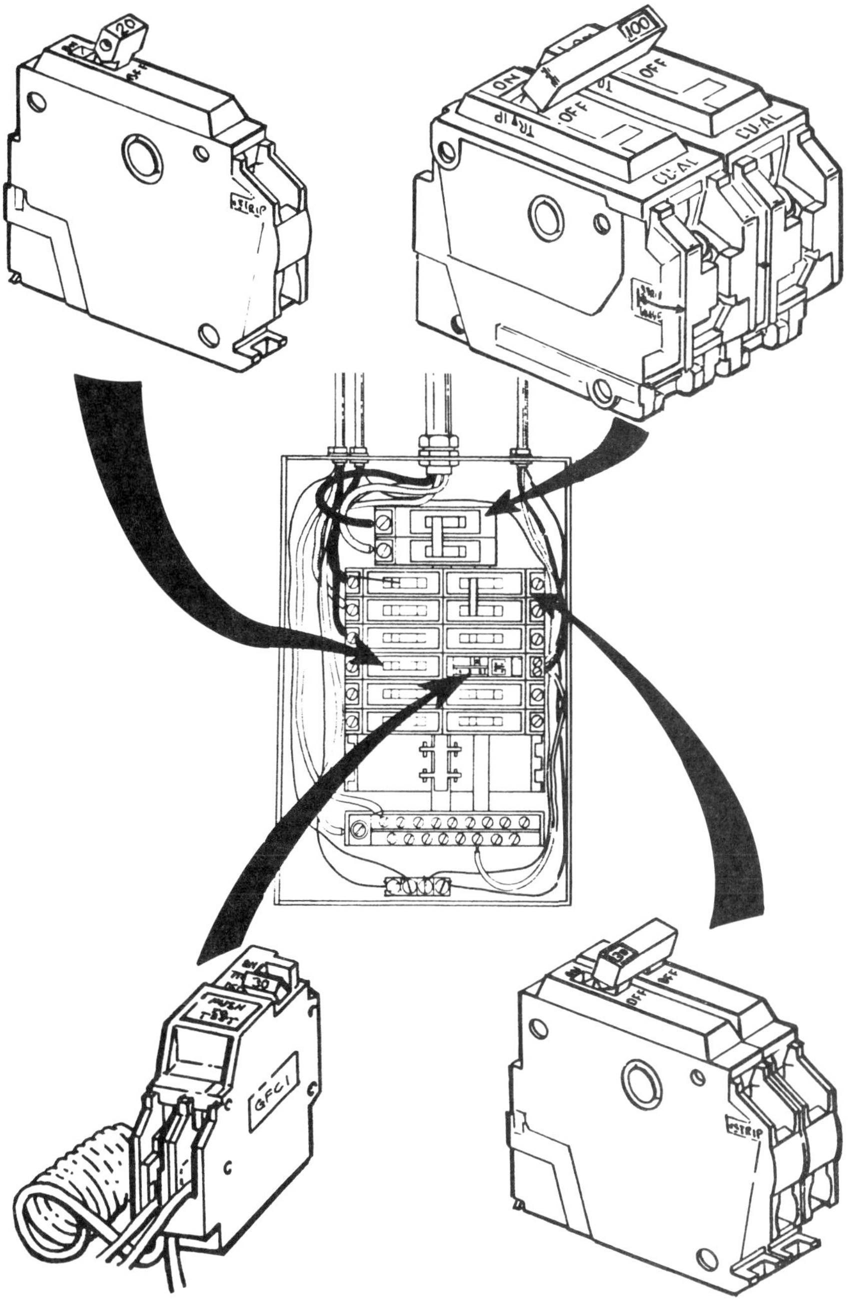

Figure 188 Circuit breakers

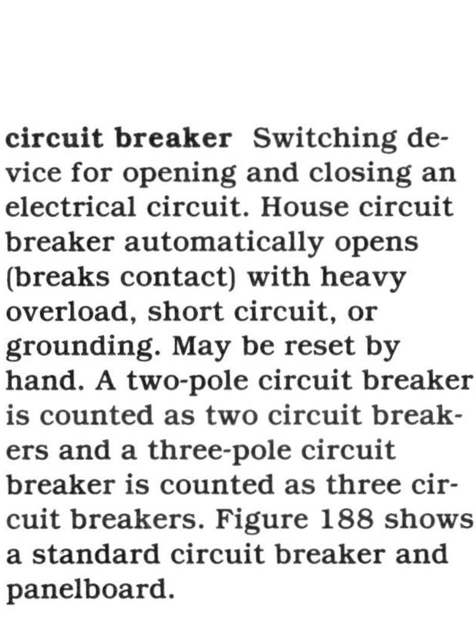

circuit breaker Switching device for opening and closing an electrical circuit. House circuit breaker automatically opens (breaks contact) with heavy overload, short circuit, or grounding. May be reset by hand. A two-pole circuit breaker is counted as two circuit breakers and a three-pole circuit breaker is counted as three circuit breakers. Figure 188 shows a standard circuit breaker and panelboard.

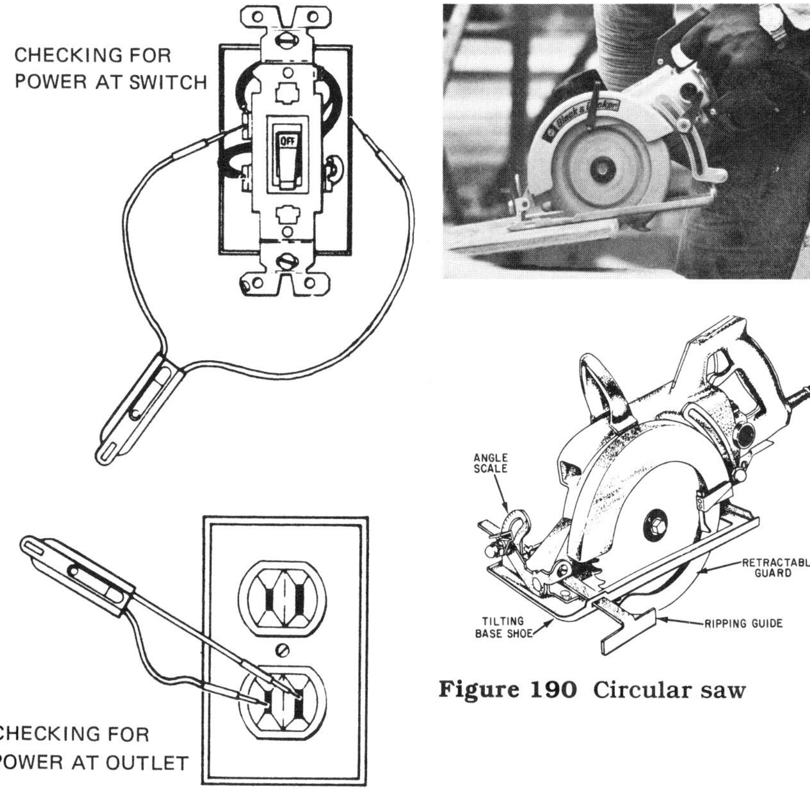

Figure 190 Circular saw

Figure 189 Circuit tester

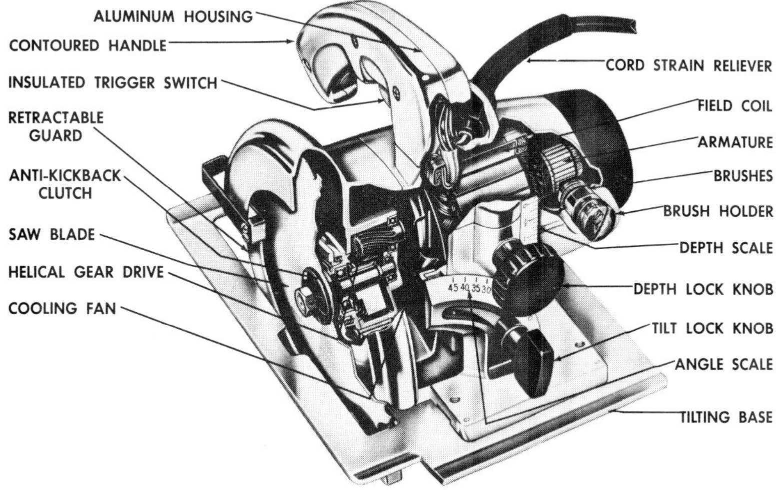

Figure 191 Circular-saw terminology

circuit tester Device used for testing for live circuits from 90V to 300V. (Figure 189.) When the ends of the tester are touched to hot and neutral wires, the bulb will glow if the circuit is live. Ends can also be inserted into the openings of a receptacle, or into the small (hot) receptacle opening and a ground; again, if the circuit is live the bulb will glow. Also called an *electric tester*, or *voltage tester*. See also *continuity tester*.

circuit voltage The voltage in a circuit: 120V or 240V are used in the home. See *convenience circuits, special circuits*.

circular and angular measure A measure in degrees, minutes, and seconds:

60 seconds (″)	=	1 minute (′)
60 minutes (′)	=	1 degree (°)
90 degrees (°)	=	1 quadrant
4 quadrants	=	1 circle or circumference
360°	=	1 circle

circular mil Measurement in electrical wiring: an area based on a circle one mil ($\frac{1}{1000}$ inch) in diameter. Cross-sectional area is determined in this system by multiplying the wire diameter (in mils) by itself. For example, a wire with a diameter of 32 mils (20 gauge) has a circular mil area of 32 × 32 = 1024 circular mils.

circular saw Portable power saw widely used in construction. It takes circular blades of various sizes: $7\frac{1}{4}''$, $7\frac{1}{2}''$, $8''$, and $8\frac{1}{4}''$ are widely used in construction. (Figure 190.) A retractable guard protects the blade. Figure 191 shows the parts of a circular saw. See *saw blades* for the types of circular blades used.

circular stairs Stairs that turn around in a circle as they go upward, often made as one steel unit. See *winding stairs*.

circulation Passage of air from one area to another; flow of hot water or steam in piping.

circumference The outside perimeter of a circle.

circumscribe To encircle; to enclose; to outline; to delineate something by drawing its outside boundary.

cistern Underground reservoir for storage of rainwater.

cladding In heavy construction, an exterior covering of a steel or concrete building. See *composite cladding, panel.*

clamp Woodworking holding device screwed together to hold objects against each other for gluing or for working on. Several types of clamps are used: *C clamp, corner clamp, cross clamp, edging clamp, parallel* or *hand screw clamp, pipe* or *bar clamp, spring clamp, web clamp.* See also *cabinet clamp, framing clamp, miter clamp, picture framing vise.*

clamping time In gluing, the amount of time the wood pieces must be held together before the glue sets.

clamshell Crane attachment used for scooping large amounts of earth. (Figure 192.) Figure 193 shows the clamshell bucket.

clap board A board siding that is overlapped. See *siding.*

clasp nails Two-shanked nail shaped like a square letter U.

classical Type of architecture from Greek and Roman times (classical period); any architecture based on classical themes or design. See *architectural styles, orders.*

classical orders. See *orders.*

claw Forked end of a hammer head or crowbar.

claw bar A *crowbar.*

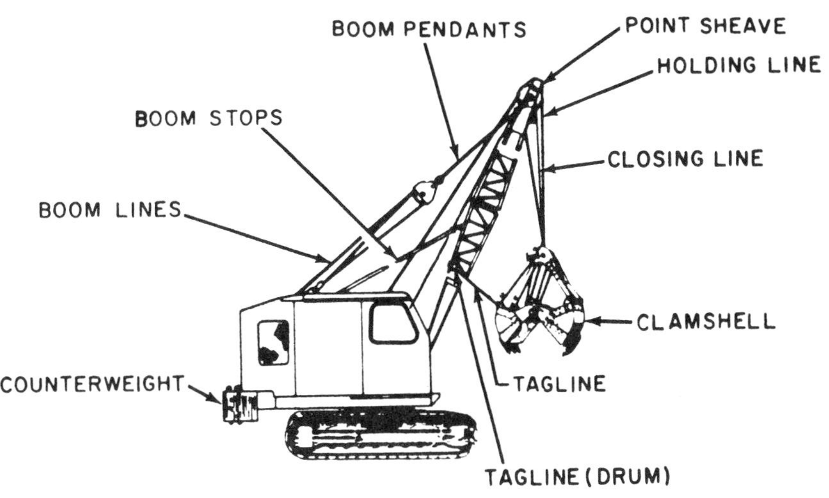

Figure 192 Clamshell attachment on crane

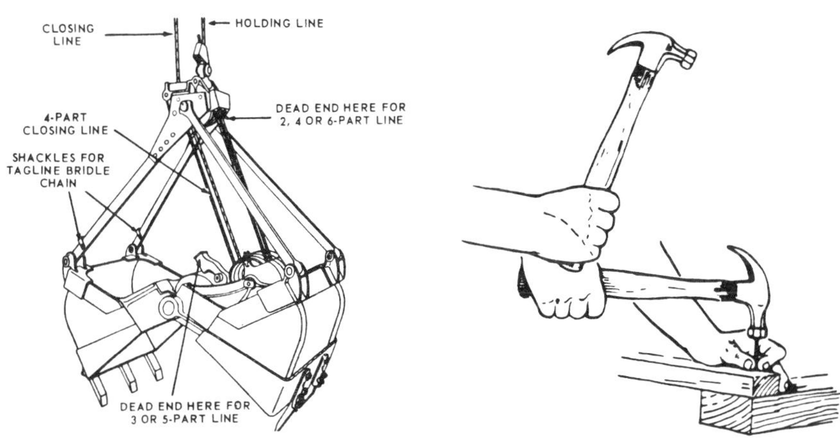

Figure 193 Clamshell bucket

claw hammer Carpenter's hammer with a split claw for pulling nails. The curved claw is most commonly used; a straight claw is used for laying wood flooring (the claw is used as a sort of blade or hatchet for cutting and breaking). (Figure 194.) See *hammer.*

claw hatchet Hatchet with a nail notch on the side of the blade, for pulling nails.

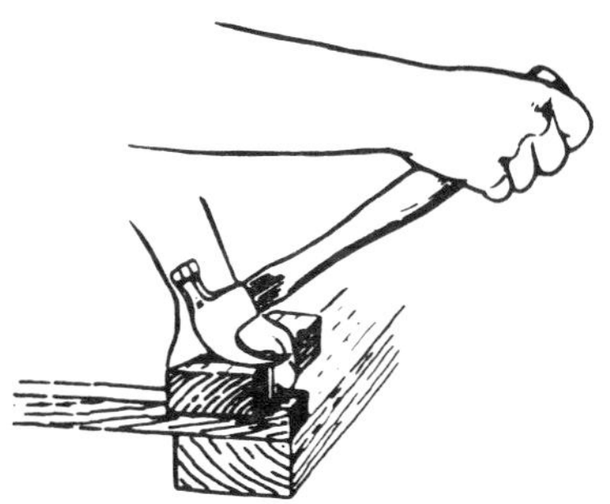

Figure 194 Claw hammer

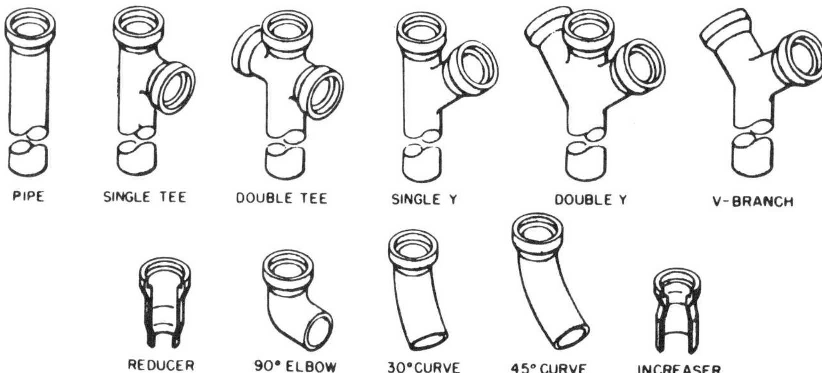

Figure 195 Clay tile

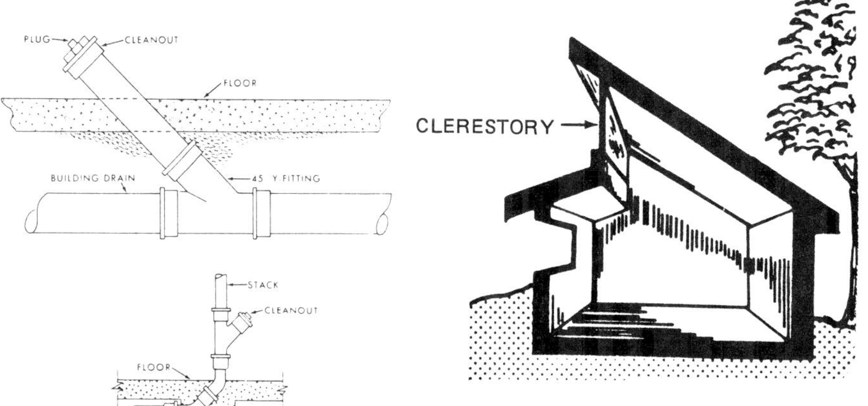

Figure 196 Cleanout

Figure 197 Clearstory

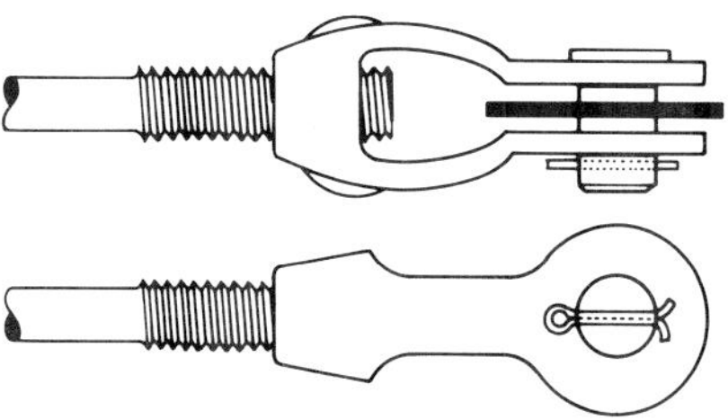

Figure 198 Clevis

clay tile Vitrified clay tile used for drainage and sewer systems. (Figure 195.)

Clean Air Act Federal legislation passed in 1970 requiring individual states to develop pollution control standards.

cleanout In plumbing, a short section of pipe that juts above the floor with a removable plug on the end; allows access to the building drain for cleaning or rodding. (Figure 196.) In cement masonry, an opening in the forms so refuse can be removed before the concrete is poured. Also, an opening into the ash pit of a fireplace used to remove ashes.

clearance Distance between two surfaces. Space left between framing and reinforced concrete members to allow for inaccuracies or shrinkage and expansion. Open distance between two things, as between a hot water heater and the wall.

clearing and grubbing Removal of debris, stumps, and rocks from a building site.

clear lumber Lumber free of knots or other defects.

clear span Horizontal distance from the face of one support to the face of the opposite support. See *effective span, span.*

clearstory Break in a roof system where two roofs meet at different levels, creating a vertical space that often has windows. (Figure 197.) Window area under an open roof area. Also spelled *clerestory.*

cleat Small wood piece used to strengthen a wood joint. Small board used as a brace in framework.

clevis End connection for a threaded rod. (Figure 198.) Threaded rod is tightened into the clevis connector to create the tension needed. See also *turnbuckle.*

climbing crane In heavy construction, a tower crane on the top of the building that moves up as the building goes up. (Figure 199.)

clinch To bend over the end of a nail.

clinker brick A rough, hard brick that is often irregular in shape. Clinker bricks are overburned in the kiln.

clip Metal fastener used to help secure panels or other framing members, as the clips used to hold plywood roof panels. (Figure 200.) Also, a brick cut to length; part of a brick.

clockwise Moving or rotating in the same direction as the hands of a clock.

closed circuit In electrical wiring, a circuit that is continuous and has a flow of electricity; a live circuit. Opposed to an *open circuit.*

closed construction Building that cannot be inspected at site without damage or disassembly.

closed cornice A boxed cornice; plywood or other material is nailed to the exposed rafters; lookouts may be used. (Figure 201.)

closed-loop solar system Solar heating active system that circulates a heated fluid through a closed pipe loop. Heat from the circulating fluid is then transferred (using a *heat exchanger*) to heat the household water. This system is called an indirect system. See *open-loop solar system, thermosiphoning.*

closed traverse In surveying, a series of lines with direction and distance that ends at the starting point. See also *open traverse.*

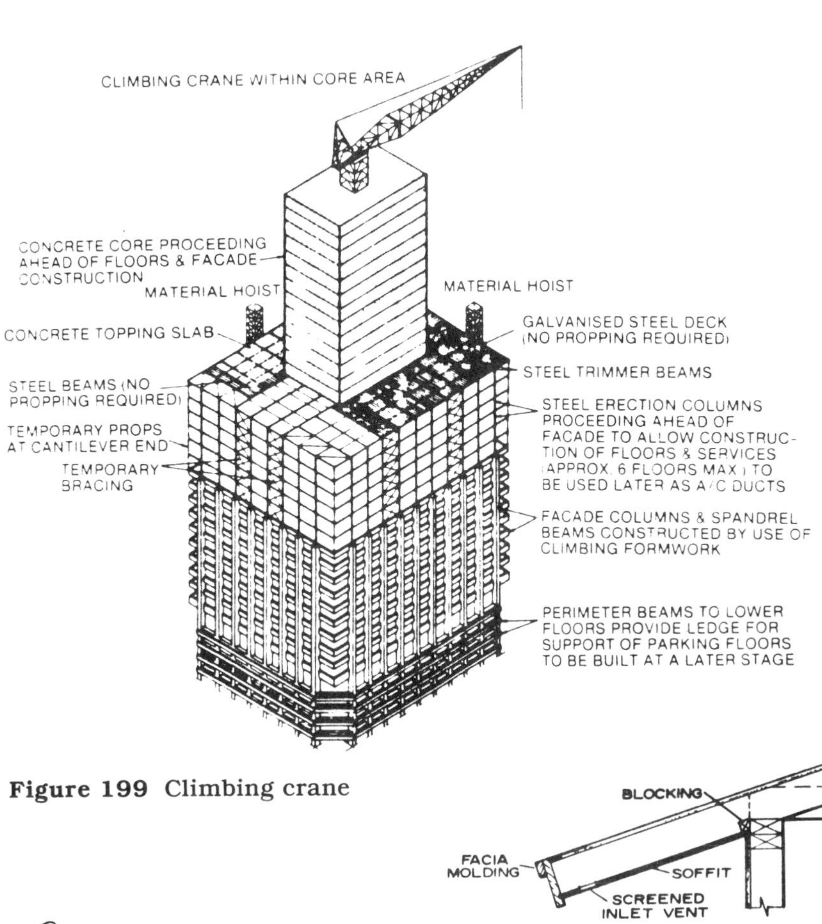

Figure 199 Climbing crane

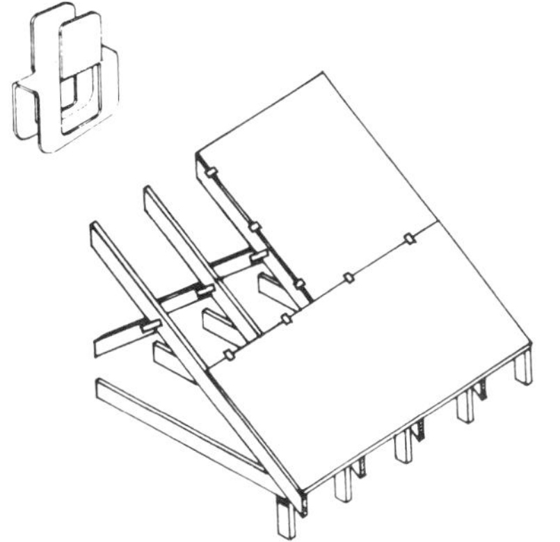

Figure 200 Clips used to secure roof panels

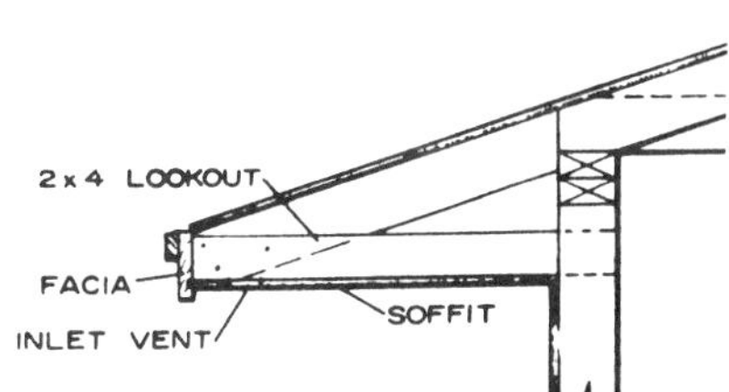

Figure 201 Closed cornice

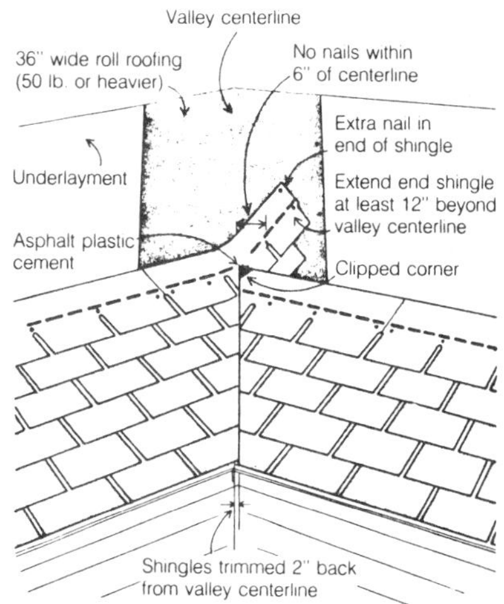

Figure 202 Closed valley

closed valley Roof valley at which shingles are run into each other. (Figure 202.) See *open valley.*

close grain Pattern of small and tight annual rings in wood.

closer Shortened brick that is used to close up or finish a brick course or to allow vertical joints to be staggered. See *king closer* (approximately three-fourths normal brick length) and *queen closer* (half brick).

close string A stair that is closed in; the outside stringers (carriages) are covered or encased; also called a *closed string stair.*

closet Small room used for hanging clothing and for storing small articles.

closure block The last concrete block laid in a course.

closure brick The last brick laid in a course.

cluster A group of similar things arranged together, as several lights in one fixture.

clustering Arranging a group of houses closely together, often around an open area. Common in planned developments.

coarse grain Pattern of large and conspicuous annual rings in wood.

coarse threads Screw threads commonly used on bolts and nuts, specified as UNC (Unified Coarse).

coat A layer or covering, as a layer of paint or plaster.

cobble A small stone.

code Set of rules regulating construction or installation, as the building code, plumbing code, or electrical code. Rules regarding safety, health, or fire prevention. Major model codes in the United States are:

Basic Building Code Building Officials and Code Administrators International, 17926 S. Halsted St., Homewood, IL 60430.

Energy Conservation in New Building Design (ASHRAE 90A-1980) American Society of Heating, Refrigerating and Air-Conditioning Engineers, 345 East 47th St., New York NY 10017.

National Building Code American Insurance Association, 85 John St., New York NY 10038.

National Electrical Code National Fire Protection Association, Battery March Park, Quincy, MA 02269.

National Fire Prevention Code National Fire Protection Association, Battery March Park, Quincy, MA 02269.

National Plumbing Code International Association of Plumbing and Mechanical Officials, 5032 Alhambra Ave., Los Angeles, CA 90032.

One- and Two-Family Dwelling Code Council of American Building Officials, 5201 Leesburg Pike, Suite 507, Falls Church, VA 22041.

Standard Building Code Southern Building Code Congress International, 900 Montclair Rd., Birmingham, AL 35213.

Uniform Building Code International Conference of Building Officials, 5360 South Workman Mill Rd., Whittier, CA 90601.

coffer Recessed panel in a ceiling or soffit.

cofferdam Open, watertight enclosure made of piles and earthwork to hold back water while constructing foundations, as for a bridge in a body of water.

coign Wedge; a projecting corner.

coil tie Concrete form tie used in heavy construction. (Figure 203.) A special bolt screws into the tie and is removed during stripping. Forms are removed with a crane. See *snap tie.*

cold air duct Ductwork returning cooled air back to the furnace for reheating.

cold chisel A hardened solid steel chisel used for cutting metal. It is struck with a ball peen hammer. A cold chisel has a double-bevel edge and is classified by the cutting edge or point. (Figure 204.) The standard flat cold chisel is used to cut bolts, rivets, metal sheets, and so on. The cape chisel is used for cutting narrow grooves and square corners. The half-round and round-nose chisels are used to cut circular grooves and to chip inside corners. The diamond-point chisel is used for cutting V grooves and sharp corners. Various sizes of cold chisels are available. See also *chisel.*

cold glue Any glue that is used cold, such as a prepared glue that is used directly from the container.

collapse Area in the wood that caves in or flattens during drying or pressure treatment.

collar beam Horizontal brace in a roof system; braces opposing rafters. (Figure 205.)

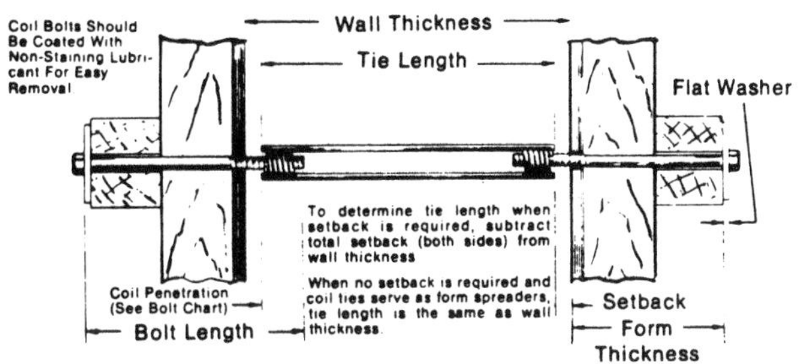

Figure 203 Coil tie

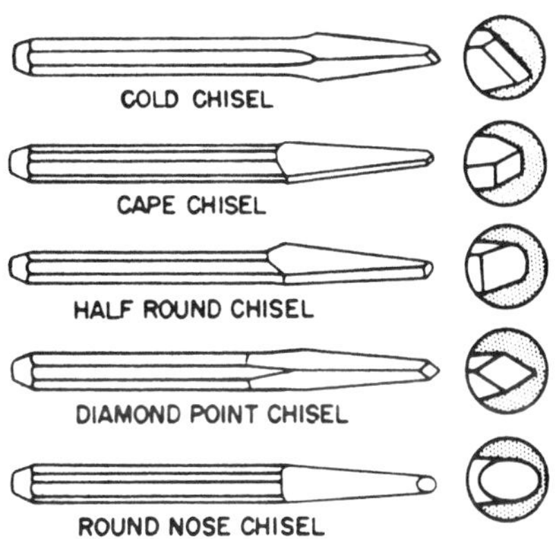

Figure 204 Cold chisels

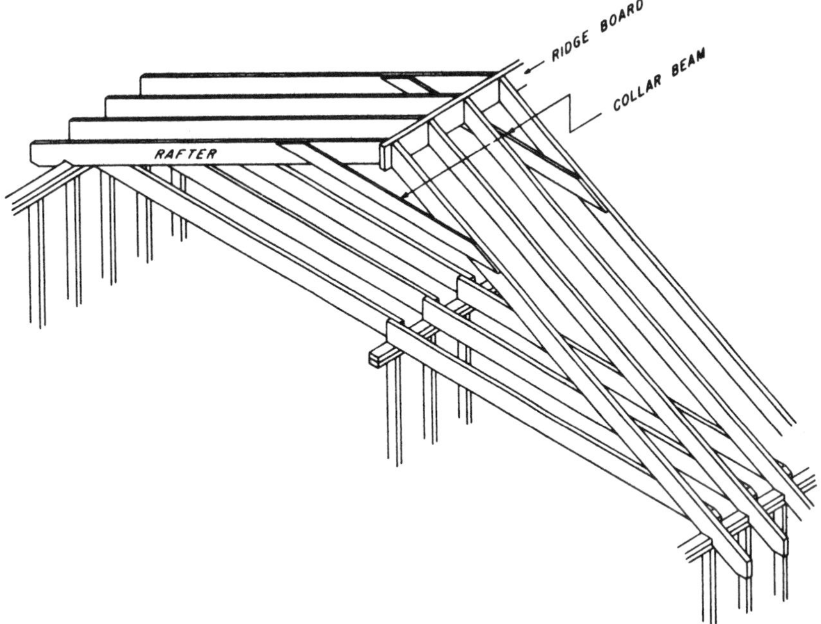

Figure 205 Collar beam

Figure 206 Collector bank

collar joint Masonry joint running between wythes.

collector bank Series of solar collectors that function together and provide solar heat for several buildings. (Figure 206.)

collector coolant Fluid used in a solar fluid-heating system, such as water, antifreeze, or oil.

collector panel Flat-plate solar collector.

collect Device used to hold a bit, as on a *router.* See also *chuck.*

Colonial architecture Architectural style used during American Colonial times before the American Revolution. Usually a two-story house with a balanced facade. Attic area may have dormers. See *Cape Ann, Cape Cod, New England Colonial, salt box, Southern Colonial, Federal.* See also *architectural styles.*

color Color is based on the reflection of light off something. White is a reflection of the continuous spectrum of color. Black is an absence of light or light reflection. White reflects sunlight and is used on the exterior of buildings to create a cool interior. Black absorbs sunlight and is used with solar collectors to absorb and collect heat. A limitless number of colors can be made by mixing various colors and tints together.

color codes for electrical wiring See *electrical wiring color codes.*

color codes for safety See *safety color codes.*

colorimetric test In concrete work, a test made to determine the presence of organic impurities in sand.

column Vertical support member. (Figures *784,* p. 344, *162,* p. 61, *683,* p. 295, *792,* p. 349, *1066,* p. 471.) See also

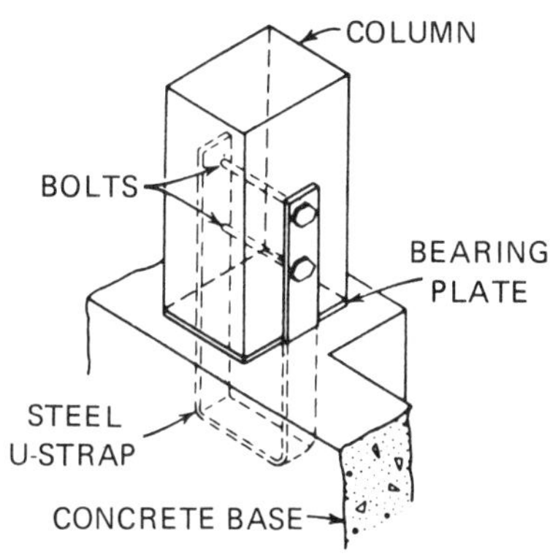

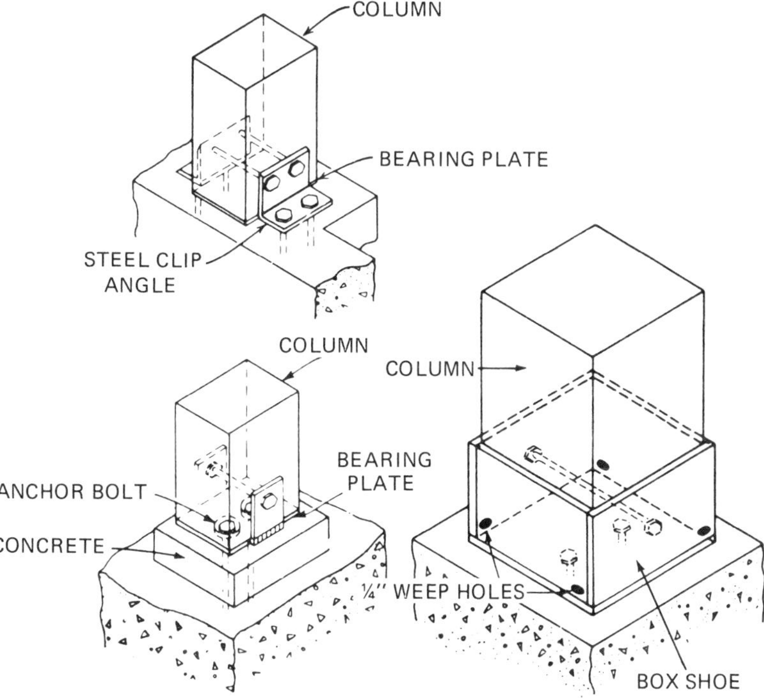

Figure 207 Column anchors

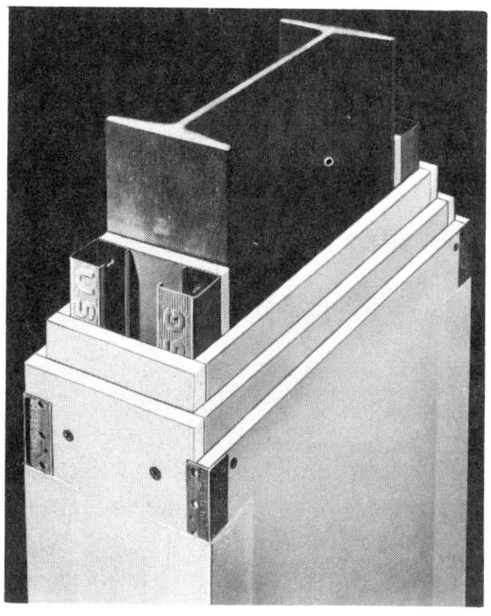

Figure 208 Column fireproofing

column anchor Metal support used at the base of a column. (Figure 207.) See *base plate, beam anchor.*

column fireproofing Enclosing of steel column with noncombustible material to increase the fire rating. Application of gypsum panel layers, concrete, masonry, or the like around the outside of the column. (Figure 208.)

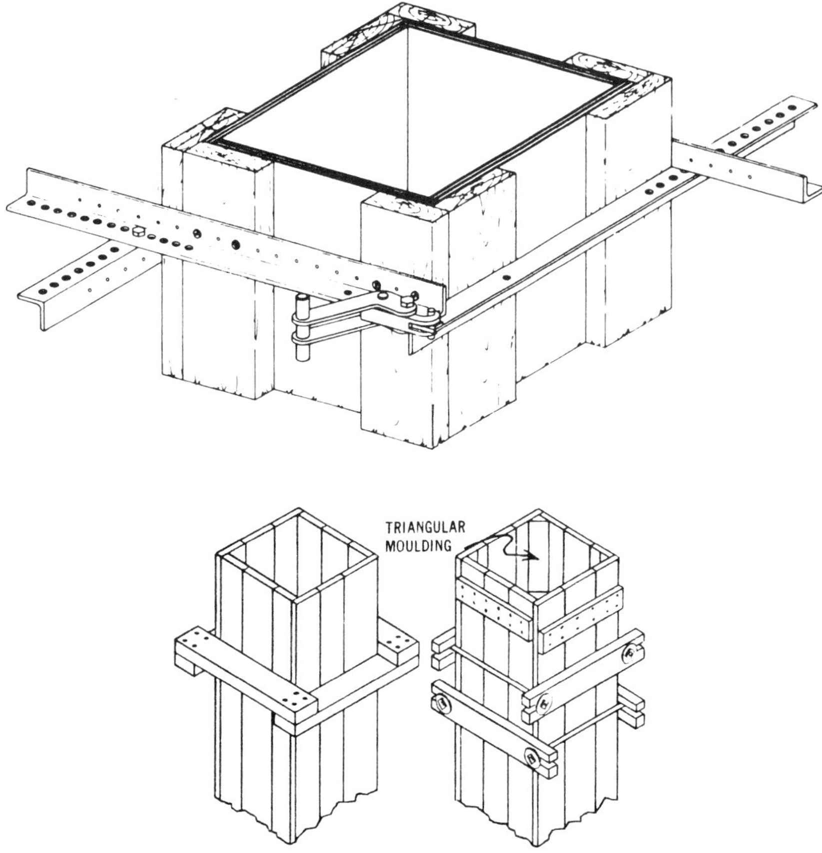

Figure 209 Column form clamps or yokes

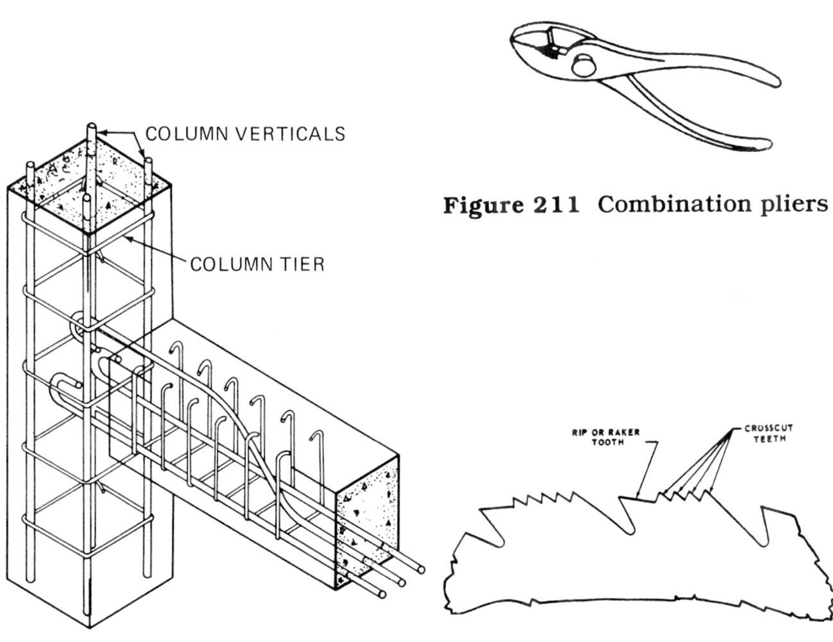

Figure 211 Combination pliers

Figure 210 Column and beam reinforcement

Figure 212 Combination saw blade

column form clamp In reinforced concrete, horizontal bracing or adjusting clamps used to hold the formwork for a column. (Figure 209.) Also called a *yoke*.

column forms Forms used to hold column concrete. Reusable forms are used in multistory construction. (Figure 209.) See *forms*.

column number Number assigned to a column on a framing or erection plan.

column schedule Table giving the number, size, and steel placement for columns in a building.

column tier In reinforced concrete, bars bent around the vertical reinforcing bars to hold them in place. (Figure 210.)

column verticals Vertical reinforcing bars in a concrete column. (Figure 210.)

comb Ridge of a roof. A masonry tool used for dressing stone.

combed plywood A finished plywood with small, irregular parallel grooves or striations. Used for exterior siding.

combination blade Circular power saw blade that has teeth cut for both ripping and crosscutting. See *combination saw blade*.

combination door Door used both in summer and winter; part of the door is removable; screens can be inserted for warm weather, storm windows for cold weather.

combination pliers Adjustable pliers used for bending and cutting. Also called slip-joint (Figure 211.) See *pliers, slip-joint pliers*.

combination rotary machine See *rotary machine*.

combination saw blade Circular saw blade used both for ripping and cross cutting. (Figure 212.)

combination screw bit See *screw bit*.

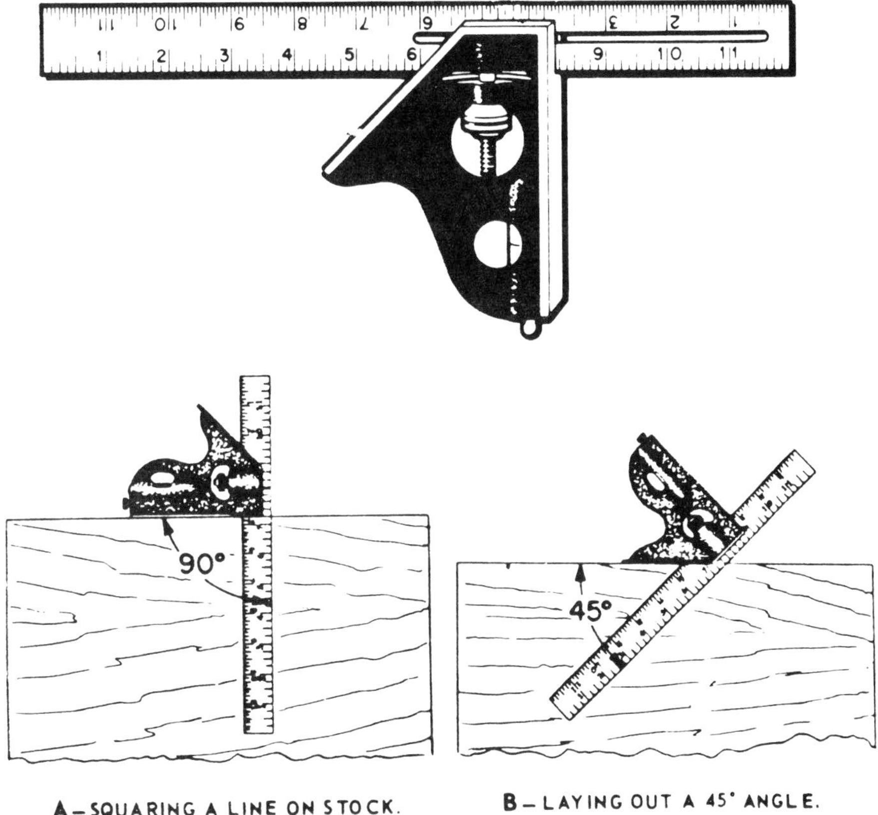

A–SQUARING A LINE ON STOCK.

B–LAYING OUT A 45° ANGLE.

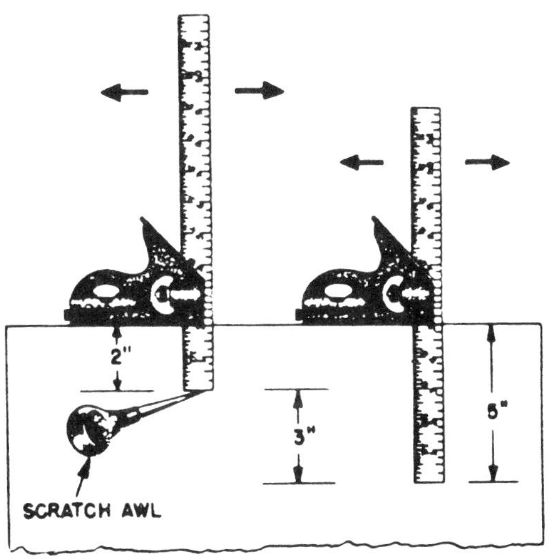

C–DRAWING PARALLEL LINES.

Figure 213 Combination square

combination sewer Sewer that carries both sanitary (waste) and storm flow. See *combined sewer, sanitary sewer.*

combination square Square combining the uses of several tools. (Figure 213.) Used for 45° and 90° measure, as a depth gauge, and as a level.

combination switch/receptacle Electrical switch and receptacle mounted in the same unit. (Figure 214.)

combination window Window used both in summer and winter; part of the window is removable: screens can be inserted for warm weather, storm windows for cold weather.

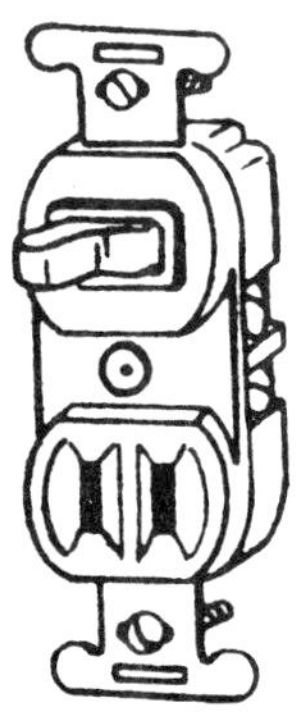

Figure 214 Combination switch/receptacle

Figure 215 Combination wrench

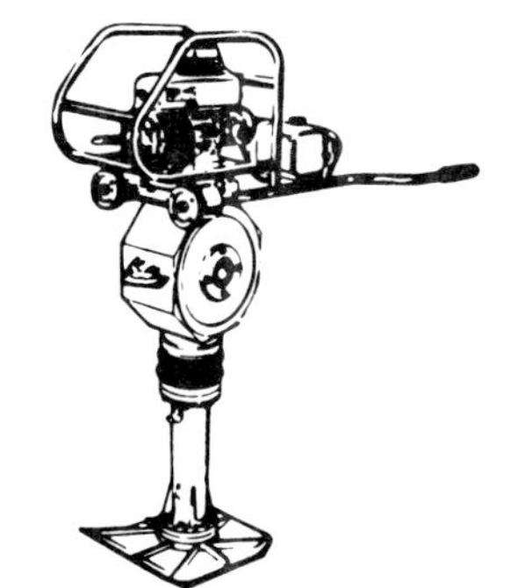

Figure 216 Compactor

combination wrench A wrench with one open end and one closed (box) end. (Figure 215.) See *wrench.*

combined sewer Sewer system that receives storm water or other runoff in addition to building waste. A *combination sewer.*

combing Raking of damp plaster with a special toothed tool (scarifier) to produce an uneven base for the finish coat.

commercial construction Buildings designed for commerce and industry.

commercial property Property used for business; property used to produce income.

common bond Masonry bond pattern with stretcher courses alternated with a header course every fifth, sixth, or seventh course. See *brick bond.*

common brick Ordinary *building brick.*

common nail Most widely used building nail; comes in lengths from 2*d* to 60*d*. Various materials are used, such as steel, aluminum, and copper. See *nail.*

common rafter Rafter extending from top plate to ridge. See *rafters.*

common wall A *party wall* between two living units; the wall is shared between the two units and is common to both.

compact To pack down, as to pound the earth to develop a firmer base of dense soil.

compacting In site development, the compressing of the soil to make it firm; rollers are used for large sites, small hand-held compactors are used over a small area and for top soil compaction. In concrete work, the removal of voids and the working down of freshly placed concrete.

compactor Gasoline- or air-operated tool for compacting down earth. (Figure 216.)

compass An instrument for determining magnetic north. Also, the full 360° circumference of a circle. (Figure 217.) Each degree (°) has 60 minutes ('). Each minute has 60 seconds (''). See also *quadrant*. In drafting, a two-legged instrument used for drawing circles and curves. One leg has a needle point, the other leg holds a pencil lead or pen point. (Figure 218.)

compass saw Saw with narrow, tapering, interchangeable blades, used for cutting small openings such as a keyhole; a *keyhole saw*.

complete specification Very detailed technical specifications. The *Construction Specification Institute* (CSI) format is often followed. See also *specifications*.

component A manufactured or prefabricated part of a house, delivered to the job site as a complete unit.

composite beam Beam made up of different materials, as a steel beam encased in concrete; a steel beam structurally integrated to a concrete slab.

composite cladding In heavy construction, an exterior covering of a steel or concrete building that is integrated to the frame so as to develop a structural value.

composite column Column made by combining different materials (such as a steel beam or pipe and surrounded by concrete) to form a support unit that works together as a whole.

composite construction Construction that combines different building materials (such as steel and precast concrete) to create a structure that works together as a whole.

composite order A Roman column and entablature based on a synthesis of earlier Greek styles. See *orders*.

composition roofing A roll roofing laid down with asphalt.

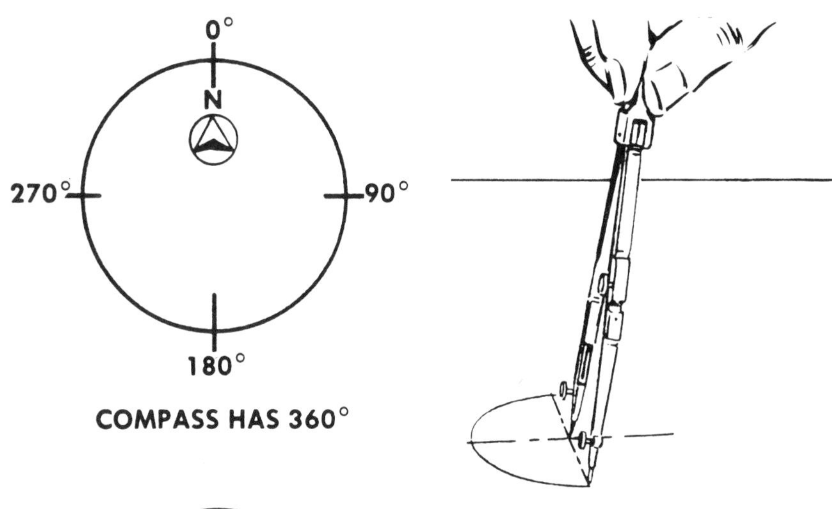

Figure 218 Compass: drafting

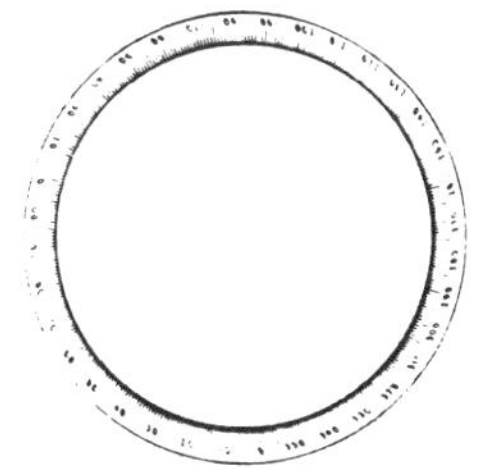

Horizontal Circle, numbered 0-360

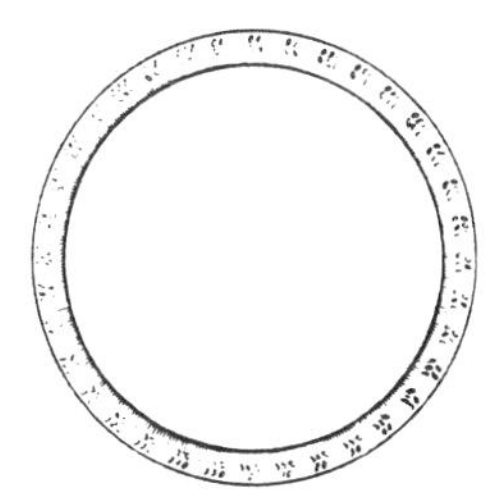

Horizontal Circle, numbered 0-360 and 360-0

Figure 217 Compass

compound beam Built-up beam.

compreg Bonded wood plies saturated with liquid resin; compregnated wood.

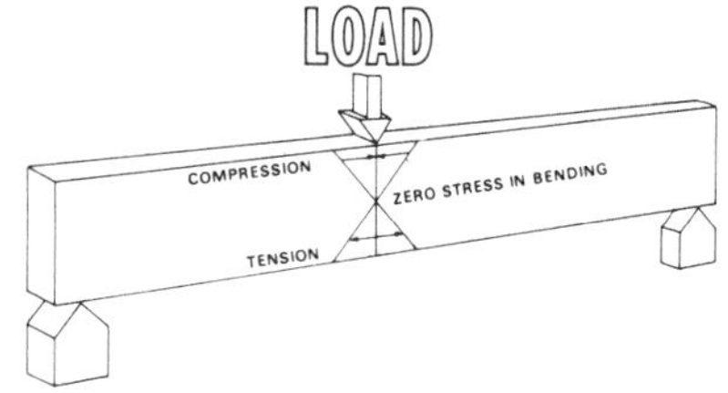

Figure 219 Compression and tension

Compression perpendicular to grain "Fc"

Where a joist, beam or similar piece of lumber bears on supports, the loads tend to compress the fibers. It is therefore necessary that the bearing area is sufficient to prevent side grain crushing.

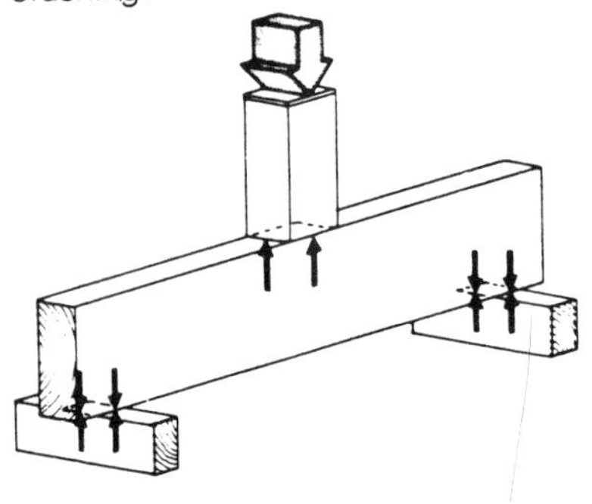

Figure 221 Compression perpendicular to grain

Compression parallel to grain "Fc"

In many parts of a structure, stress-grades are used with the loads supported on the ends of the pieces. Such uses are as studs, posts, columns and struts. The internal stress induced by this kind of loading is the same across the whole cross-section and the fibers are uniformly stressed parallel to and along the full length of the piece.

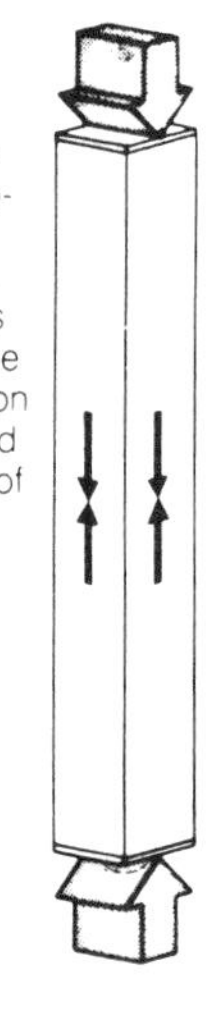

Figure 220 Compression parallel to grain

compression A stress or squeezing together. Stress on a column is an example of compression parallel to the wood grain. Stress on a beam end from the building load is an example of compression perpendicular to the wood grain. (Figure 219.) See also *compression parallel to grain, compression perpendicular to grain, shear, tension, torsion.*

compression joint Plumbing joint made in hub-and-spigot piping using a neoprene gasket inside the hub. The spigot is forced into the hub and gasket to form a tight waterproof fit.

compression parallel to grain **(F_c)** Internal stress running along the length of a vertical support member, such as a stud or post. (Figure 220.) Design values are available for different woods. See also *extreme fiber stress in bending.*

compression perpendicular to grain ($F_{c\perp}$) A crushing of wood fiber at support areas; the load compresses the fibers at these points. (Figure 221.) Design values are available for different woods. See also *extreme fiber stress in bending.*

compression wood Abnormal wood found on the underside of bent softwood. The wood has an unusual amount of summerwood and shrinks lengthwise excessively.

compressive strength Maximum compressive stress a material can bear before failure.

compressor In an air conditioning system, a device that compresses the heated refrigerant vapor to cause it to turn back into a liquid under high pressure. The compressed, heated liquid refrigerant is then forced into the *condensor,* where heat is lost to the outside air. See *heat pump.*

computer-assisted drafting (CAD) Use of a computer to produce final drawings or to design a structure to specifications. Also called *computer-aided drafting* or *computer-aided design and drafting.*

computer drafting Use of a computer to produce final drawing or to design the structure: called *computer-assisted drafting* (CAD). See also *CRT, interactive drafting.*

concave Having the shape of a curving depression, bowl-like hollow. Reverse of *convex.*

concealed gutter Gutter constructed so it is not easily seen.

concentrated load Load concentrated at one point or small area.

concentrating solar collector Solar heat collector that uses lenses or reflectors to concentrate the sun's rays. A concentrated high temperature is obtained. See *flat-plate solar collector.*

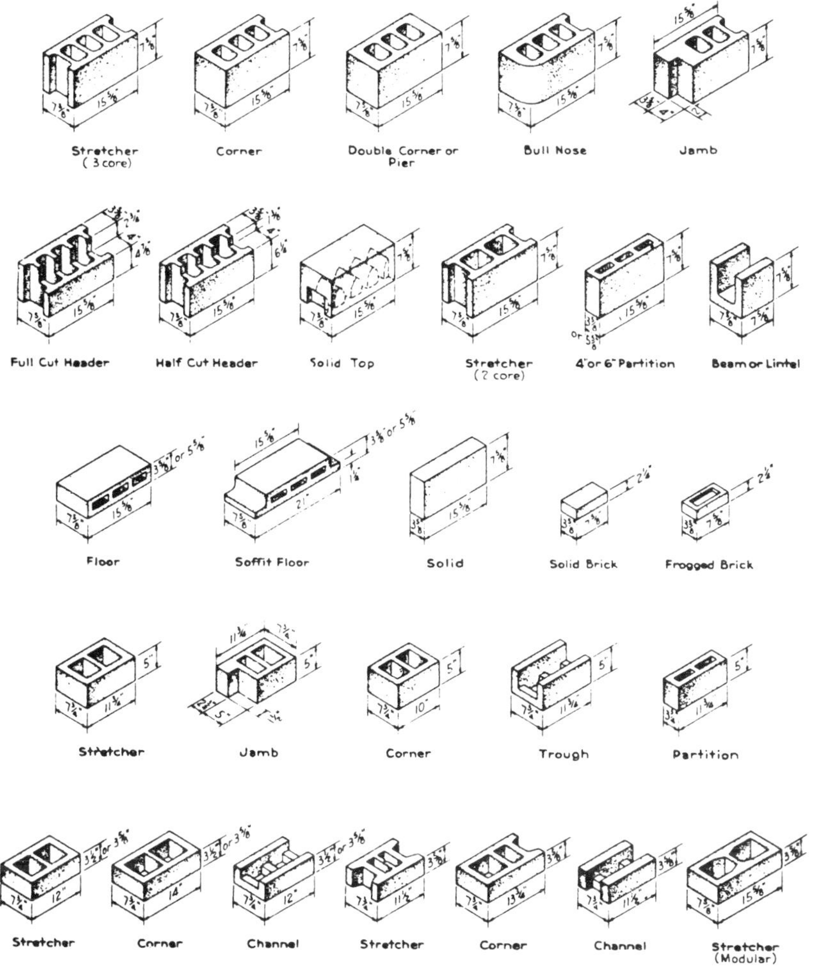

Figure 222 Concrete blocks

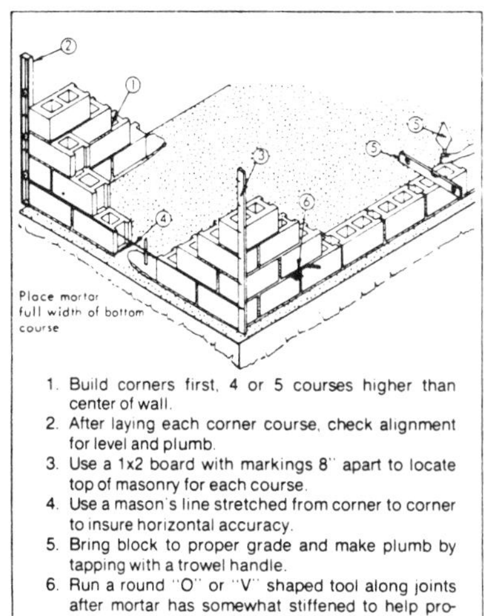

Figure 223 Concrete block layment

concentric Having a common center, as annual rings on a tree.

concrete Mixture of cement, sand, aggregate (gravel or crushed rock), and water; sets into a hard, strong mass. Fresh concrete is placed in forms at the site, in other words it is *cast in place.* Reinforcement (steel bars or welded-wire mesh) is used in the concrete to give it tensile strength. See *cellular concrete, plain concrete, prestressed concrete, reinforced concrete, reinforced concrete floors.*

concrete, aerated Concrete with air bubbles, used for sound deadening.

concrete block Masonry unit made of concrete, nominal size 8″ × 8″ × 16″. Figure 222 shows some of the many concrete blocks used. Both cored and solid units are available. The parts of a concrete block have a standard terminology; also the position or pattern of concrete block in place is identified by special terms. (*Figures 125*, p. 47 , and *126*, p. 48 , show the terminology for brick; the same terms are used for concrete block.) See *masonry joints* (*Figure 588*, p. 253) for the finished joints made between concrete block. Concrete blocks are specified and ordered as N (outside exposed) and S (outside not directly exposed). Figure 223 shows the sequence of laying concrete block.

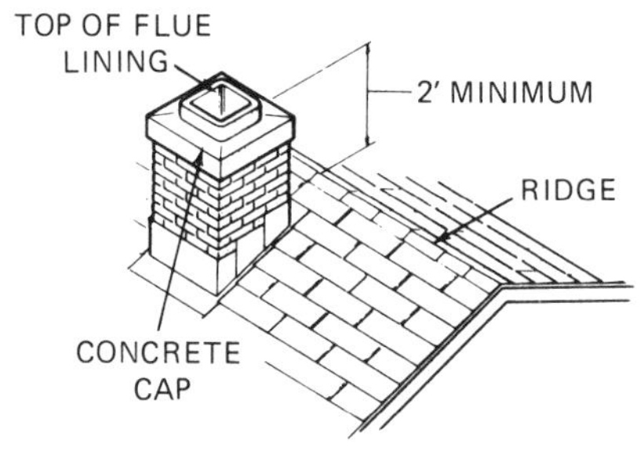

Figure 224 Concrete cap that finishes chimney top

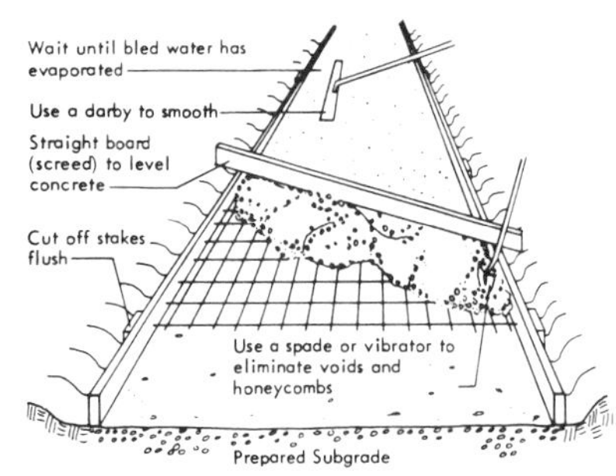

Figure 225 Concrete finishing

Figure 226 Concrete mixer

Figure 227 Concrete placer

concrete block bar support In reinforced concrete, precast concrete blocks used to support reinforcing bars.

concrete brick Solid brick-size concrete block.

concrete cap A concrete cover and finish at the chimney top, around the flue. (Figure 224.)

concrete cover In reinforced concrete, the distance from the surface of the concrete to the reinforcing steel.

concrete expansion Size expansion in cured concrete caused by high temperature.

concrete finishing In concrete flatwork the freshly placed concrete is first struck, leveled out with a screed, floated with a darby or bull float, and then finally troweled with a steel trowel and a wood or aluminum float. (Figure 225.) Edges are finished with an edger. Cross joints or control joints are finished with a groover. See also *control joint.* Vertical surface finish is provided by the *forms.*

concrete form A mold constructed to hold the concrete until it sets. Forms are removed after the concrete is set. See *forms.*

concrete masonry A general term used to describe laying of concrete blocks in mortar.

concrete mixer On-site gas or electric power concrete mixing machine. Concrete ingredients are placed in the revolving drum for mixing as needed at the site. (Figure 226.)

concrete nail Special, hardened nail used for driving into set concrete. See *nails.*

concrete paint Special cement-based paint used to seal and finish exposed concrete.

concrete paving surfacing a road with aggregate and concrete. See *rigid paving.*

concrete placer Long-handled, wide-bladed tool for spreading and smoothing concrete after placement. (Figure 227.) See also *concrete rake, kumalong, mobile concrete placer.*

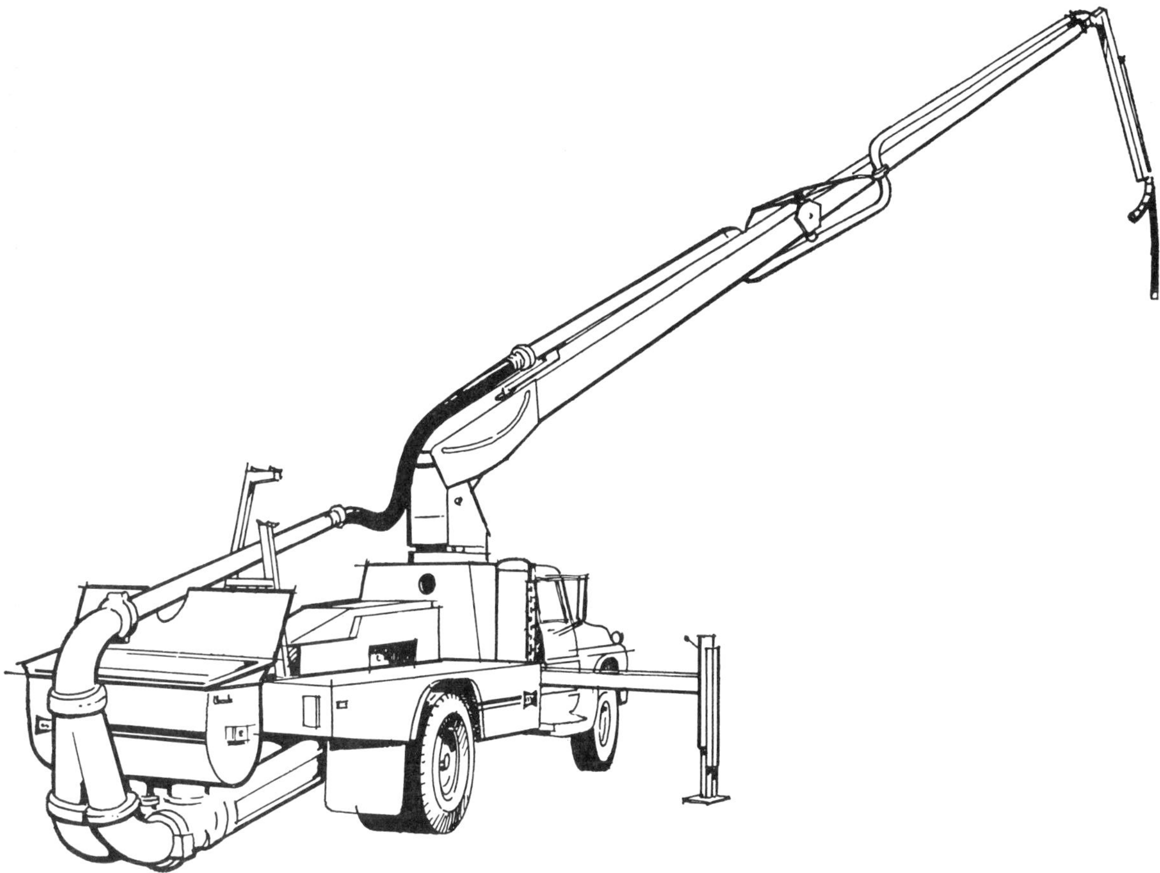

Figure 228 Concrete pump

concrete pumping Method of placing concrete by pumping concrete through pipes or flexible hoses to the exact place needed. (Figure 228.) See also *mobile concrete placer.*

concrete rake Long-handled, wide-bladed tool for spreading and smoothing concrete. One side has wide blunt teeth. Similar to a *concrete placer* or *kumalong,* which are used for the same purpose but have no teeth.

concrete, reinforced See *reinforced concrete.*

concrete saw Heavy-duty, gasoline-powered or electric saw used for cutting grooves in concrete flatwork or for cutting concrete blocks. (Figure 229.) See *abrasive cut-off saw.*

Figure 229 Concrete saw

concrete shell structures Various cast-in-place reinforced thin-shell roofing systems are used: the folded plate (W shape), barrel (cylindrical shape), and dome are common. See *barrel roof, dome.*

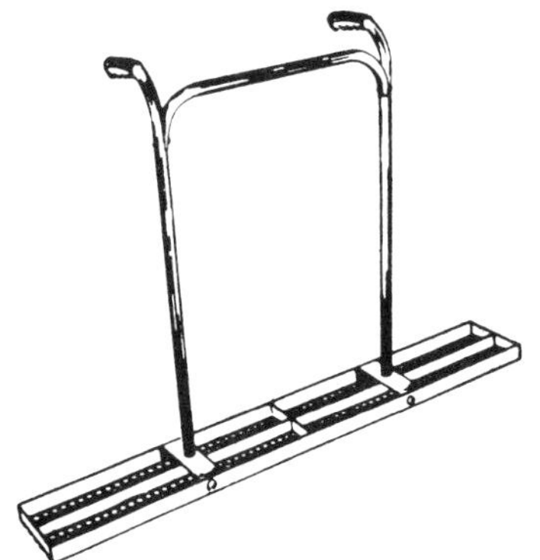

Figure 230 Concrete tamper

concrete shrinkage Size reduction in concrete during the setting period, caused by water loss; size reduction caused by lower temperature.

concrete slab Horizontal layer of reinforced concrete laid on the ground, as a floor, driveway, or sidewalk.

concrete tamper Tool for tamping concrete slab to consolidate and bring fine material to the surface for easy finishing. (Figure 230.)

concrete truck See *ready-mix truck.*

concrete wall Vertical reinforced concrete wall, as a basement wall.

condensate Water condensed out of steam in a steam heating system; liquid formed out of a vapor.

condensation Changing of a vapor to a liquid because of cooling; also, deposit of moisture from moisture-saturated air, as in improperly ventilated attics or crawl spaces.

condensor In air conditioning, a unit that expells heat out into the atmosphere. A condensor may use a cooling tower that allows warm water to fall and transfer heat to the air. Most small systems circulate the refrigerant to an air-cooled condensor where a fan circulates the air to carry away heat. See *compressor, heat pump.*

conditioned air Year-round heating and cooling of air. See *air conditioning.*

condominium Individual ownership of a living unit in a larger multiunit structure; common areas and facilities are shared, and mechanical and maintenance expenses are shared.

conduction Direct flow of heat from a warm area to a cooler area; heat flows through some material, such as an uninsulated wall.

conductor In electricity, wires through which the current flows. They are sized by gauge number; the higher the number, the smaller the conductor. (Figure 231.) Both copper (CU) and aluminum (AL) wire are used as conductors. Three types of conductors are recognized by the National Electrical Code: *bare* (no covering), *covered* (covered by material or a thickness not recognized by the Code), and *insulated* (covered by material and a thickness recognized by the Code). See *wire sizes.* Also, a *downspout* or *leader.*

TW

Type TW conductors may be used in conduit or other approved raceways in commerical wiring, as specified by the National Electrical Code. Suitable for use in wet or dry locations at conductor temperature not to exceed 60° C.

THW*

Type THW conductors are primarily used for feeders and branch circuit wiring in conduit or other recognized raceways, as specified in the National Electrical Code. THW conductors may be used in wet or dry locations where conductor temperatures do not exceed 75°C.

THHN

Type THHN conductors are primarily used in conduit as feeder or branch circuit wiring in commercial or industrial applications, as specified in the National Electrical Code. All Southwire THHN, except solid conductors, carries a triple rating (THHN or THWN or MTW) and may be used at conductor temperatures not to exceed 90°C in dry locations, 75°C in wet locations, and 60°C when used as machine tool wiring and exposed to oil or coolant. Solid conductors carry a dual rating only and may be used at conductor temperature not to exceed 90°C in dry locations and 75°C in wet locations. Stranded sizes AWG 14 through 8 carry a quad rating and may be used as appliance wiring material at conductor temperatures not to exceed 105°C in dry locations, in addition to all applications for triple-rated THHN.

XHHW*

Type XHHW conductors are especially designed for power distribution and branch circuit wiring in commercial and residential applictions, as specified in the National Electrical Code. XHHW conductors may be used in wet locations at temperatures not to exceed 75°C, 90°C in dry locations.

USE**

Type USE is primarily used as underground service entrance cable, direct burial, at conductor temperatures not to exceed 75°C. This product is also suitable for use in conduit or other recognized raceways per the National Electrical Code as type RHH for use in dry locations at conductor temperatures not to exceed 90°C and as type RHW for use in wet or dry locations at conductor temperatures not to exceed 75°C.

NM*

Type NM is primarily used in residential wiring as branch circuits to outlets, switch legs, and other loads. NM may be used for both exposed and concealed work in normally dry locations at temperatures not to exceed 60°C as specified in the National Electrical Code.

UF*

Type UF is generally used as underground feeder, including direct burial, to outside post lamps, pumps, and other load apparatus fed from a distribution point in an existing building as specified by the National Electrical Code. Multiple conductor UF cable may be used for interior branch circuit wiring in wet, dry or corrosive locations at conductor temperatures not to exceed 60°C.

SEU*

Type SE, Style U service entrance cable is used to convey power from the meter base to the distribution panelboard, to connect a masthead to a meter base (where overhead service drops are employed), and as branch circuits as permitted by the National Electrical Code. SEU phase conductors are suitable for use at temperatures not to exceed 90°C in dry locations, 75°C in wet locations.

SER*

Type SE, Style SER service entrance cable is primarily used as panel feeder in multiple unit dwellings; however, it may be used in all applications where Type SE cable is permitted. SER may be used in wet locations at temperatures not to exceed 75°C, 90°C in dry locations.

*Available in copper and Triple E® aluminum alloy.
**Available in copper and aluminum alloy #1350.

Voltage rating for all conductors is 600 volts.

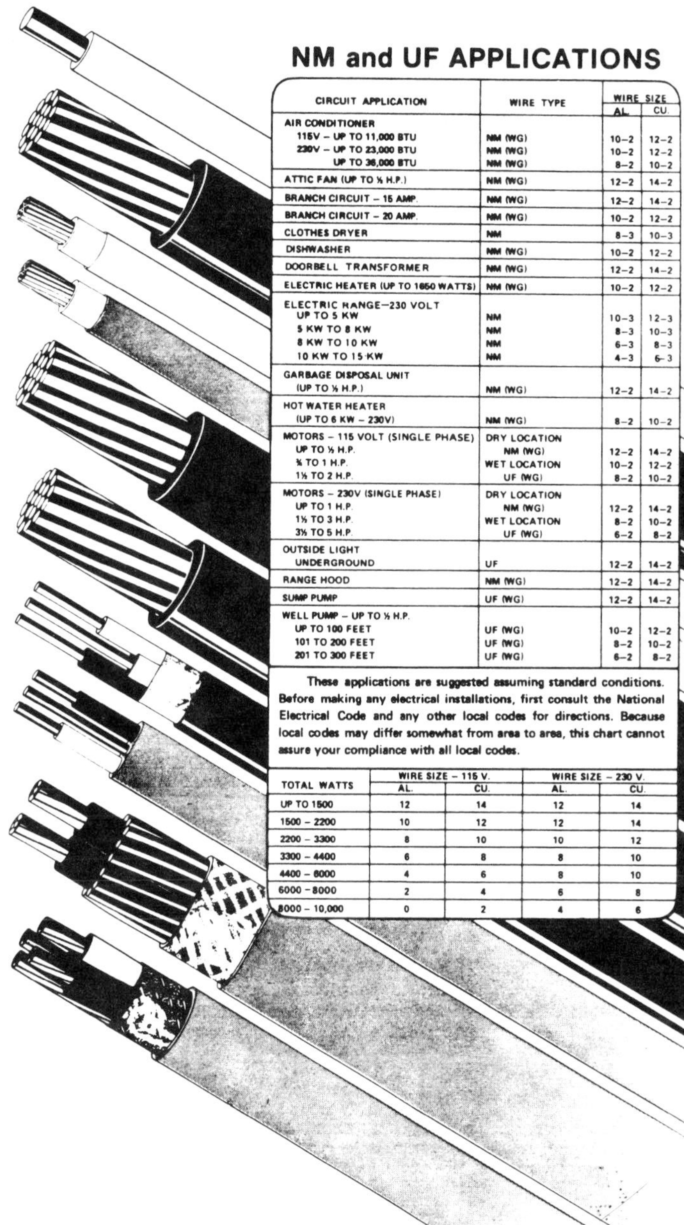

NM and UF APPLICATIONS

CIRCUIT APPLICATION	WIRE TYPE	WIRE SIZE AL.	WIRE SIZE CU.
AIR CONDITIONER			
115V – UP TO 11,000 BTU	NM (WG)	10–2	12–2
230V – UP TO 23,000 BTU	NM (WG)	10–2	12–2
UP TO 36,000 BTU	NM (WG)	8–2	10–2
ATTIC FAN (UP TO ½ H.P.)	NM (WG)	12–2	14–2
BRANCH CIRCUIT – 15 AMP.	NM (WG)	12–2	14–2
BRANCH CIRCUIT – 20 AMP.	NM (WG)	10–2	12–2
CLOTHES DRYER	NM	8–3	10–3
DISHWASHER	NM (WG)	10–2	12–2
DOORBELL TRANSFORMER	NM (WG)	12–2	14–2
ELECTRIC HEATER (UP TO 1650 WATTS)	NM (WG)	10–2	12–2
ELECTRIC RANGE—230 VOLT			
UP TO 5 KW	NM	10–3	12–3
5 KW TO 8 KW	NM	8–3	10–3
8 KW TO 10 KW	NM	6–3	8–3
10 KW TO 15 KW	NM	4–3	6–3
GARBAGE DISPOSAL UNIT (UP TO ½ H.P.)	NM (WG)	12–2	14–2
HOT WATER HEATER (UP TO 6 KW – 230V)	NM (WG)	8–2	10–2
MOTORS – 115 VOLT (SINGLE PHASE)	DRY LOCATION		
UP TO ½ H.P.	NM (WG)	12–2	14–2
¾ TO 1 H.P.	WET LOCATION	10–2	12–2
1½ TO 2 H.P.	UF (WG)	8–2	10–2
MOTORS – 230V (SINGLE PHASE)	DRY LOCATION		
UP TO 1 H.P.	NM (WG)	12–2	14–2
1½ TO 3 H.P.	WET LOCATION	8–2	10–2
3½ TO 5 H.P.	UF (WG)	6–2	8–2
OUTSIDE LIGHT UNDERGROUND	UF	12–2	14–2
RANGE HOOD	NM (WG)	12–2	14–2
SUMP PUMP	UF (WG)	12–2	14–2
WELL PUMP – UP TO ½ H.P.			
UP TO 100 FEET	UF (WG)	10–2	12–2
101 TO 200 FEET	UF (WG)	8–2	10–2
201 TO 300 FEET	UF (WG)	6–2	8–2

These applications are suggested assuming standard conditions. Before making any electrical installations, first consult the National Electrical Code and any other local codes for directions. Because local codes may differ somewhat from area to area, this chart cannot assure your compliance with all local codes.

TOTAL WATTS	WIRE SIZE – 115 V. AL.	WIRE SIZE – 115 V. CU.	WIRE SIZE – 230 V. AL.	WIRE SIZE – 230 V. CU.
UP TO 1500	12	14	12	14
1500 – 2200	10	12	12	14
2200 – 3300	8	10	10	12
3300 – 4400	6	8	8	10
4400 – 6000	4	6	8	10
6000 – 8000	2	4	6	8
8000 – 10,000	0	2	4	6

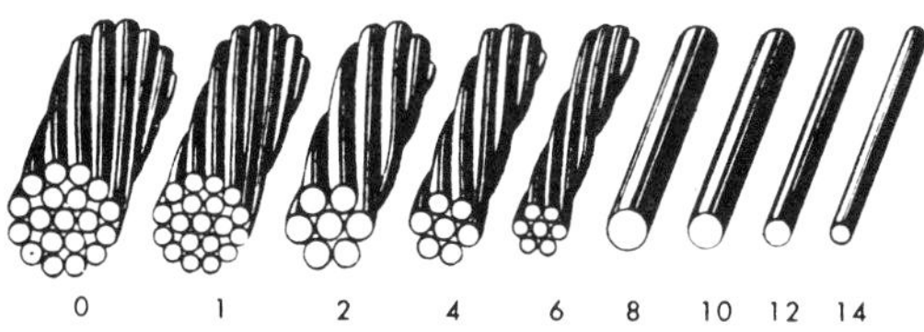

Actual size of copper conductors. Note the larger the gauge number the smaller the diameter of the wire.

Figure 231 Conductors

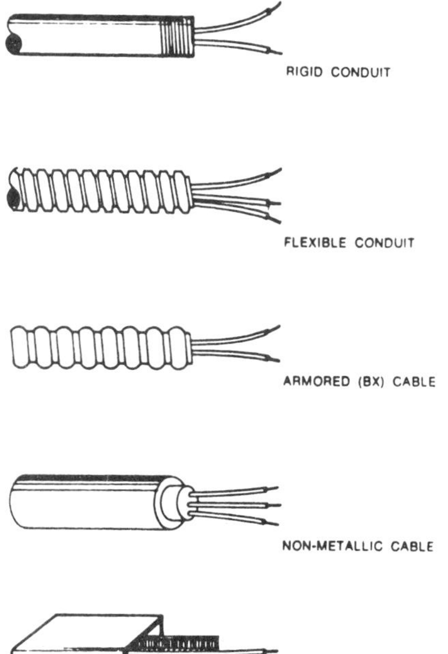

Figure 232 Conduit

Figure 233 Conduit bender

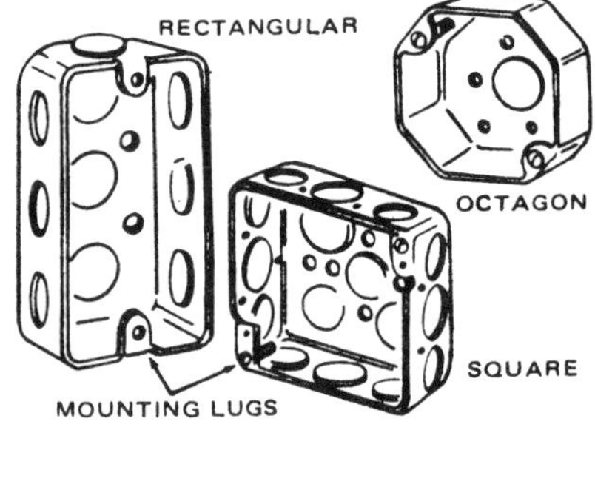

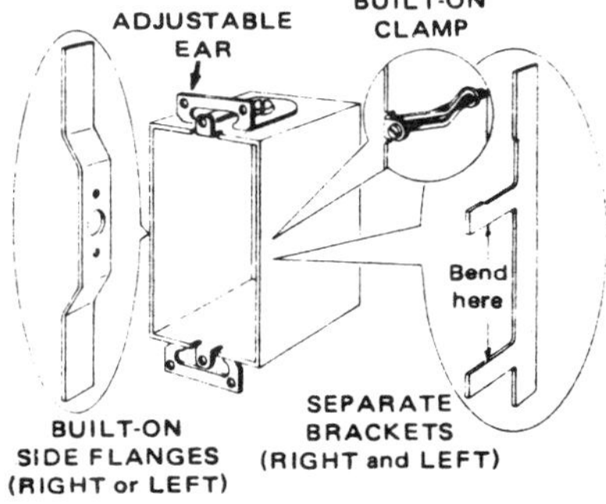

Figure 234 Conduit boxes

conduit In electrical wiring, a pipe through which conductors or cables are drawn. (Figure 232.) Also, any pipe or channel which can carry fluids. See *flexible metal conduit, rigid metal conduit, thinwall conduit.*

conduit bender Hand bender used to bend rigid conduit or EMT (electrical metallic tubing). (Figure 233.) The bender is sometimes referred to as a *hickey.* See *bender.*

conduit box In electricity, a metal box for receiving the ends of the conduit; conductor wires run into the box and are attached to the electrical fixture. (Figure 234.) See *junction box, outlet box.*

conduit bushing Metal fitting or sleeve used to fasten the end of conduit to a panelboard or outlet box. (*Figures 37*, p. 16, *998*, p. 441 .) Note the grounding lug for grounding the equipment.

conduit coupling Short conduit piece with internal threading, used to connect two threaded conduit ends.

conduit fittings In electrical wiring, the parts needed for joining conduit.

conduit straps Straps used for supporting conduit. (Figure 235.)

conduit wiring Conductors (wires) which are run through a *conduit.*

cone In gas welding, the small conical part of the flame just outside the tip opening.

conifer *Softwood* or *evergreen* tree with needle-like leaves and seed cones. The leaves do not fall off with the cold season.

consistency In concrete masonry, the firmness of the concrete; the *slump.*

construction Erection of structures. Often noted by the type of material, as frame, timber, masonry, veneer, concrete, steel, prestressed concrete.

construction adhesive Adhesive, commonly a mastic type, used on the job to bond construction members or building parts together, with or without nailing. See *adhesive.*

construction classifications Classifications of building into types based on fire properties of walls, floors, roofs, ceilings, and other elements. The U.S. Department of Housing and Urban Development in the *Minimum Property Standards* has developed these classifications:

Type 1, fire-resistive construction That type of construction in which the walls, columns, floors, roof, ceiling and other structural members are noncombustible with sufficient fire resistance to withstand the effects of a fire and prevent its spread from one story to another. **Type 2, noncombustible construction** That type of construction in which the walls, columns, floors, roof, ceiling and other structural members are noncombustible but which does not qualify as Type 1, fire-resistive construction. Type 2 construction is further classified as Type 2a (1 hr protected) and Type 2b, which does not require protection for certain members. **Type 3, exterior protected construction** That type of construction in which the exterior walls are of noncombustible construction having a fire resistance rating as specified and which are structurally stable under fire conditions and in which the interior structural members and roof are of protected combustible construction, or of unprotected heavy timber construction in public areas only. Type 3 construction is divided into two subtypes as follows:

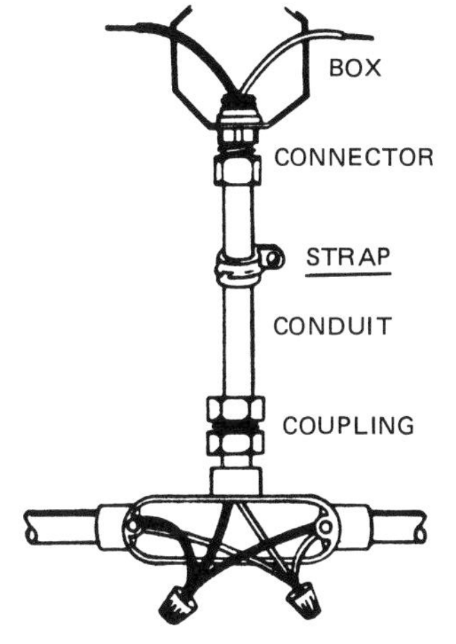

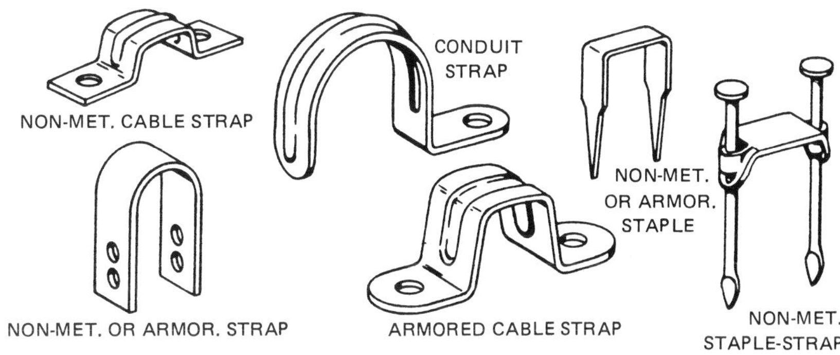

Figure 235 Conduit straps

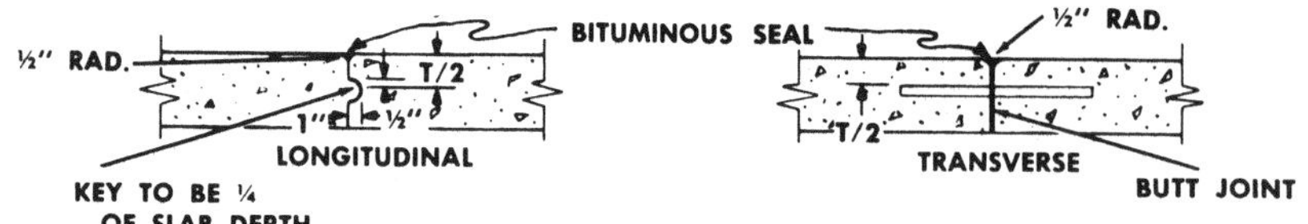

Figure 236 Construction joint

Type 3a Exterior protected construction in which the interior exitways, columns, beams and bearing walls are noncombustible in combination with the protected floor system roof construction and non-load-bearing partitions of combustible construction.

Type 3b Exterior protected construction in which the interior structural members are of protected combustible materials, or of heavy timber unprotected construction.

Type 4, protected wood frame construction That type of construction in which the exterior walls, partitions, floors, roof and other structural members are wholly or partly of wood or other combustible materials.

construction joint Concrete joint made between separate sections or batches of concrete placements; a joint made between the placing of one day's concrete and the next. A vertical divider called a *bulkhead* is used. A key is formed between the two sections. (Figure 236.) The interface between the foundation footing with key and the wall is also called a construction joint. See also *control joint, isolation joint, lift.*

Construction Specifications Institute (C.S.I.) An organization that has developed standardized construction specifications. There are four major groupings: *Bidding Requirements, Contract Forms, General Conditions,* and *Specifications (Technical).* There are 17 permanent divisions under the specifications format. These are used to write standard technical specs:

0 Bidding and contract requirements
1 General requirements
2 Site work
3 Concrete
4 Masonry
5 Metal
6 Wood and plastics

7 Thermal and moisture protection
8 Doors and windows
9 Finishes
10 Specialties
11 Equipment
12 Furnishings
13 Special construction
14 Conveying systems
15 Mechanical
16 Electrical

construction survey Survey of a site; lines, grades, and distances are laid out as a guide to construction. See also *plane survey, topographical survey.*

contact cement Adhesive that forms an instant bond upon contact between the two treated surfaces.

contemporary architecture Functional architecture that uses modern materials to create a functional and unified whole. Glass is widely used and living areas are integrated to include outside space. Influenced by Frank Lloyd Wright and the Bauhaus school. (Figure 237.) See also *architectural styles, international architecture.*

continuity tester Device used for testing electrical circuits and devices to see if they are operating properly. The tester is self-energizing and is used only on dead or disconnected circuits or devices. (Figure 238.) See also *circuit tester.*

continuous Structural member running over three or more supports, as a continuous beam or girder.

continuous beam Beam supported at more than two points.

continuous high chairs Continuous reinforced bar support consisting of *high chairs* welded together with a wire or bar, used to support reinforcing bars. See *chair.*

continuous weld Weld bead that is unbroken across or around the entire joint.

contour Outline of an object; also, a *contour line.*

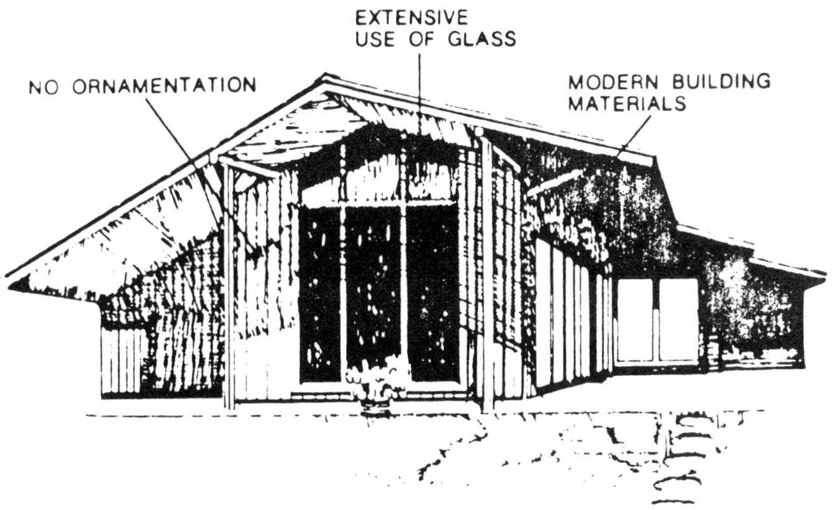

Figure 237 Contemporary house

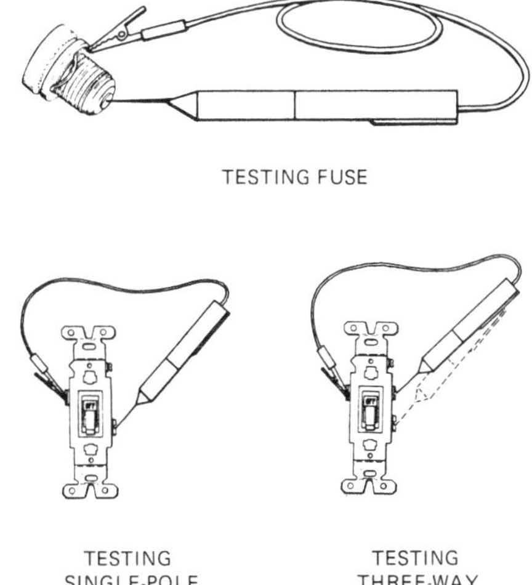

Figure 238 Continuity tester

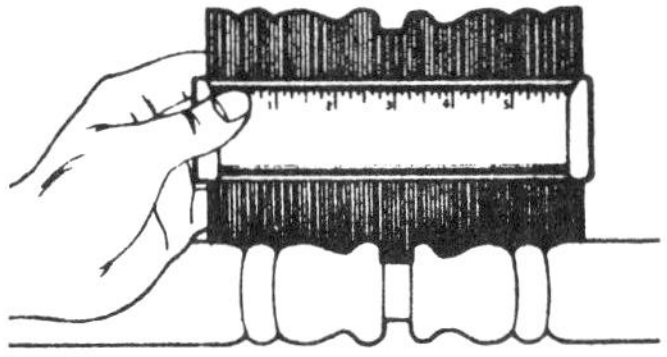

Figure 239 Contour gauge

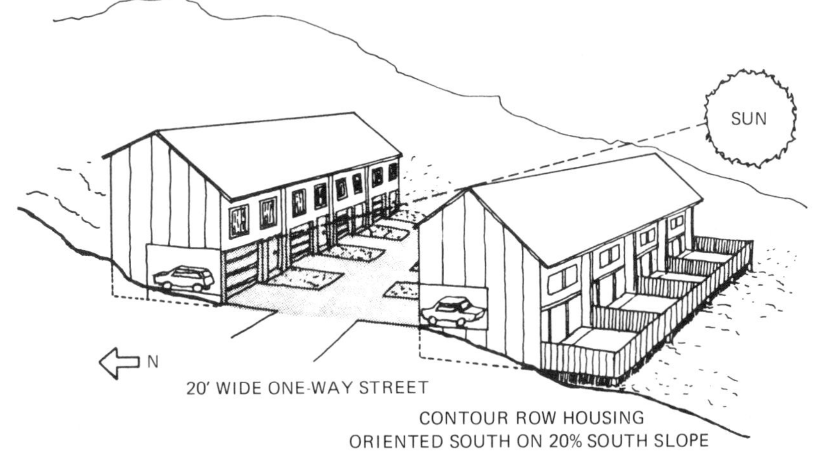

Figure 240 Contour housing

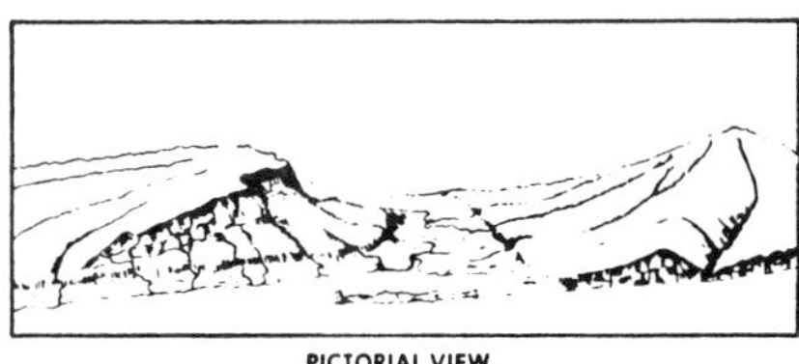

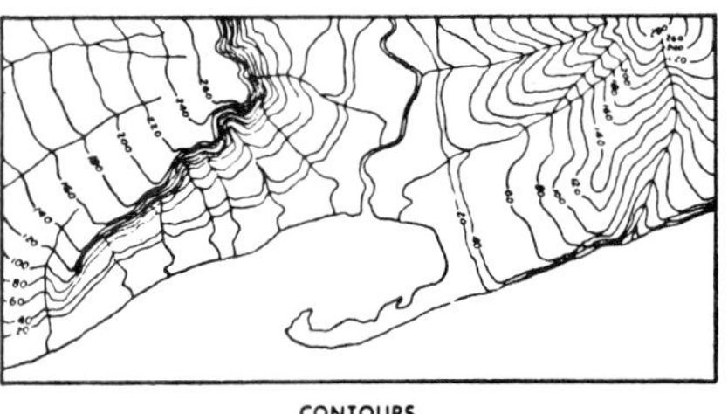

Figure 241 Contour lines

contour gauge Gauge with wires that slide up and down to conform to a contour. (Figure 239.)

contour housing Placement of buildings to follow the contour of a hill. (Figure 240.) On a south slope, following the contour allows solar access.

contour line A line on a plan, as a plat or plot plan, or grading plan, denoting a continuous elevation; all points have the same height (elevation). (Figure 241.) See also *grading plan.*

contraction In cement masonry, a shrinkage due to temperature change; contraction occurs with a temperature drop; expansion occurs with a temperature rise.

contraction joint A *control joint.* Provides for contraction of concrete. Groove in concrete creates a weakness so concrete will break evenly along control joint if there is shrinkage or shifting.

contractor One who agrees to supply materials and perform certain types of work for a specified sum of money, as a building contractor who erects a structure according to a written agreement or contract. Someone who manages a building project.

control joint In concrete flatwork, a shallow groove formed or sawed in a slab to control random cracking caused by *contraction and expansion.* (Figures 242; *954*, p. 422.) Joint can be formed with a groover or sawed with a concrete saw. Also, a vertical joint made in a wall, usually over a major break in the wall, to control cracking caused by expansion and contraction. (Figure 243.) Also called a *contraction joint.* In drywall construction, an open-slot metal connector used between gypsum panels in large wall and ceiling areas to allow for expansion and contraction. (Figure 244.)

controller Device, such as a switch, used to start and stop electric motors or electrical apparatus.

convection Motion of air or a fluid because of heat differential; the colder, denser material flows downward. Being denser, the colder material is heavier and is pulled downward by gravity; the warmer, less dense material flows upward. Forced convection uses a fan or pump to move heated air or liquid.

convector Type of heating outlet; air passing over the fins collects heat and rises. Generally, convectors have a fan to assist the natural convection. (Figure 245.)

convenience outlet Electrical outlet for ordinary 120-V household appliances or lamps.

conveyor belt A wide, continuous belt which, while moving, conveys goods or materials from one area to another.

convex Curving outward, like an inverted bowl. Opposite of *concave.*

coolant Solution used for cooling.

cooperative In a cooperative or co-op all the living units are jointly owned. Each owner owns one share in the co-op and lives in a single unit.

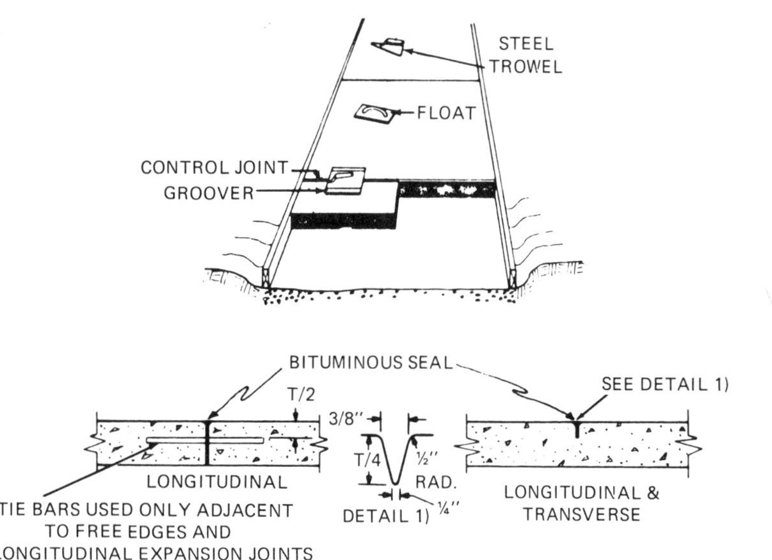

Figure 242 Control joint: concrete flatwork

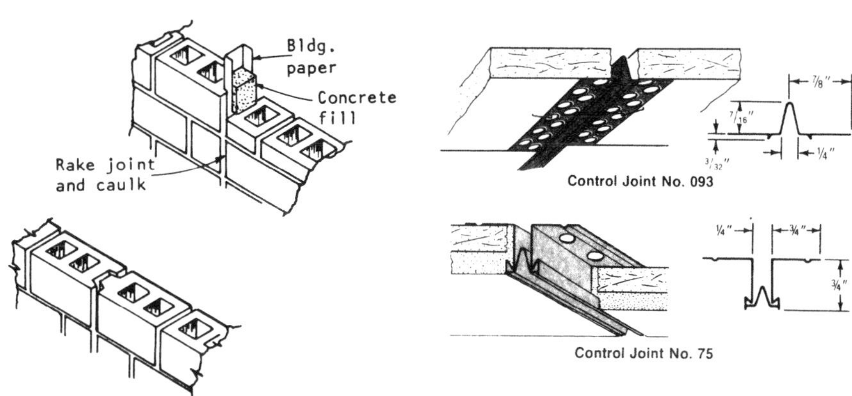

Figure 243 Control joint: vertical concrete block

Figure 244 Control joint: drywall

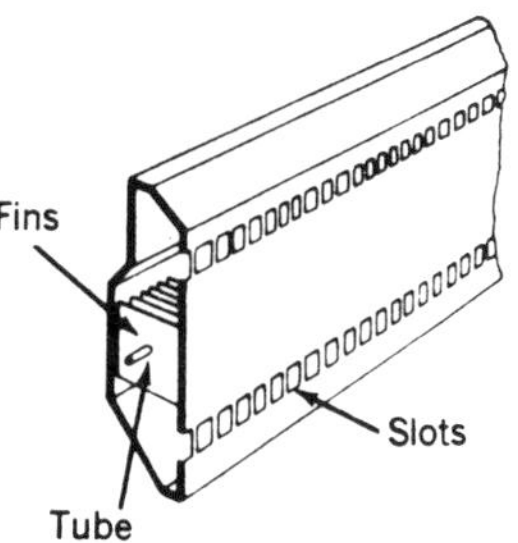

Figure 245 Convector: baseboard type

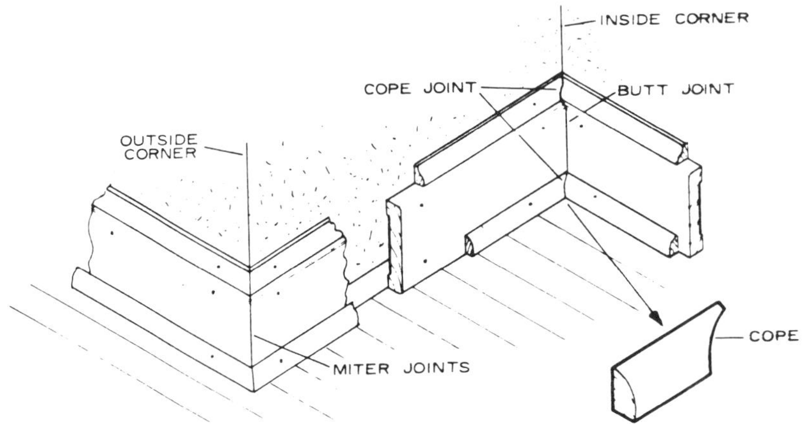

Figure 246 Coped joint

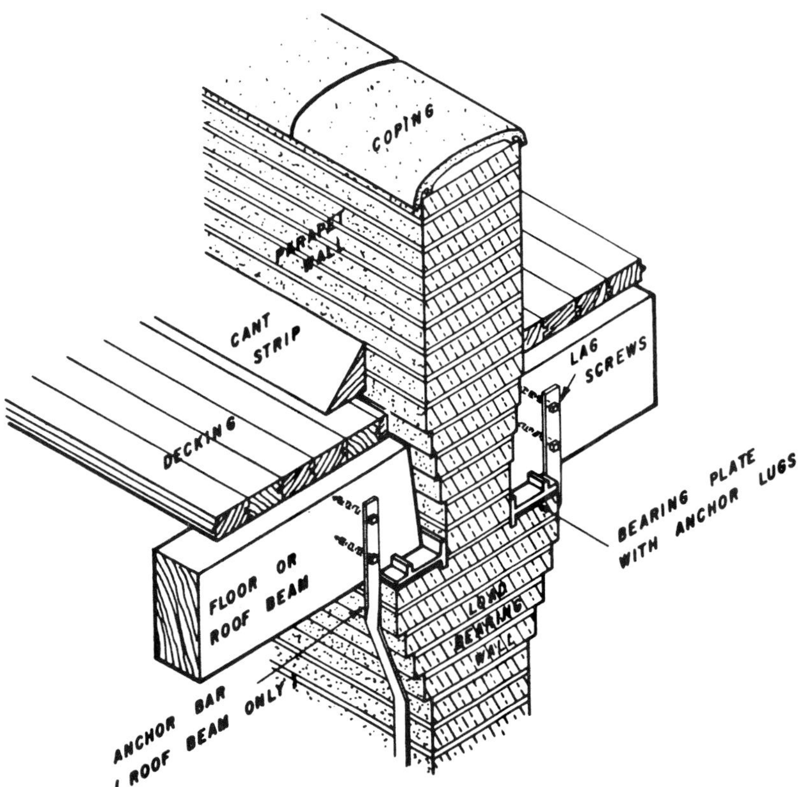

Figure 247 Coping at a parapet roof wall

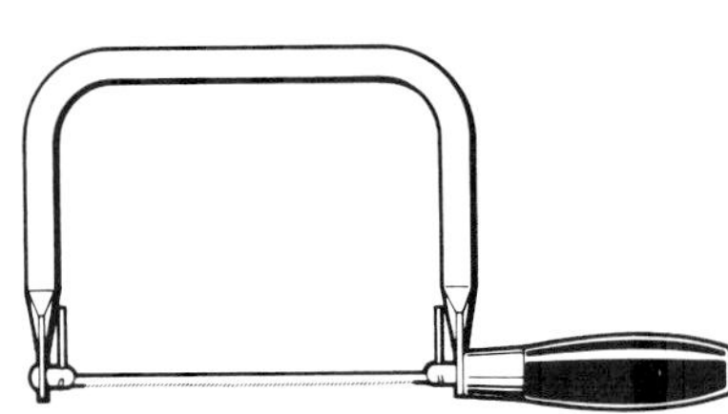

Figure 248 Coping saw

cope To cut away a part to make room for another piece or to allow two lumber pieces to fit together at a joint. Top course of masonry wall, often with units projecting out. A *coping*.

coped joint Cutting of wood to fit an irregular surface; cutting of a wood end to fit against a shaped molding, one end being cut to fit the profile of the molding. (Figure 246.)

coping Wall capping or covering, generally slanted to throw off water; the covering of a masonry wall; the top course. (Figure 247.) A *coped joint*.

coping saw Special saw for cutting close curves, as for coped joints for moldings. Has a fine, thin blade mounted in a frame. The blade is easily detached for threading through a hole for cutting out center stock. (Figure 248.)

copper A *soldering copper*.

copper tubing Tubing used for plumbing water supply, drainage, and venting. Drawn (hard copper) and annealed (soft copper) tubings are used. Sizes vary from $\frac{1}{4}''$ to 12″ in diameter. (Figures 249, 250.)

copying attachment Pattern attached to a wood lathe for copying a turned piece. Used when making a number of identical turned pieces.

corbel Built-out masonry wall or part of a wall where the masonry units project out further as they go higher. Often used to form a support shelf, as for a fireplace hearth. Masonry courses that project out from the wall. (Figure 247.) Bracket or timber end masonry unit that projects out from a wall, used as a support base. Also, a steel bracket or reinforced concrete shelf on the side of a reinforced concrete column or panel to provide a bearing surface for another member.

corbelling A projecting masonry course, commonly a projecting brick course. (Figure 247.)

corbel out To project out masonry courses from a wall. (Figure 247.)

cord Cut tree wood measure, a pile 4′ × 4′ × 8′ long.

core The center; the center piece. In plywood construction, the center ply, center lumber piece, or center pressed-wood board. See *plywood*. The center of a hollow-core flush door. Wood material removed from *cored hole;* materials removed from a mortise. In heavy construction, a center wall assemblage in a multistory building. The core normally forms a square, open tube in the center of the building. In masonry, the hole in a concrete block or brick. In prestressed concrete, the long holes in the slab. Also, a cylindrical rock sample cut out of bedrock with a core drill.

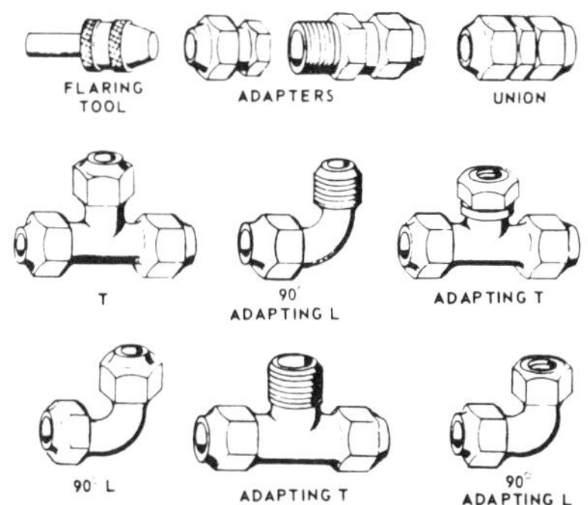

Figure 249 Copper tubing: flared joint fittings

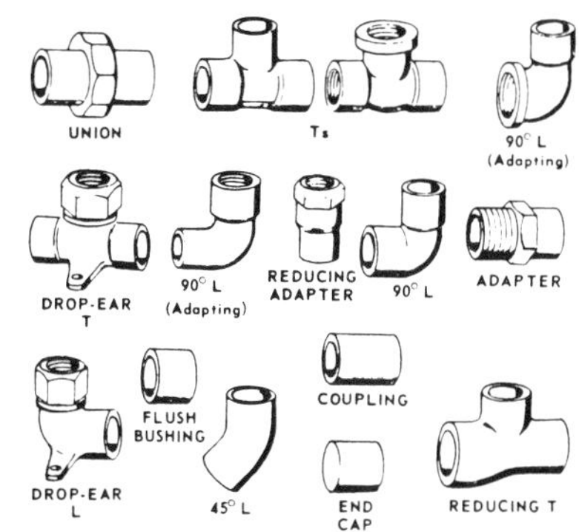

Figure 250 Copper tubing: solder-type fittings

cored brick Regular-sized brick that has two rows of five holes running down the length. They are considered to be as strong and as moisture-resistant as regular solid bricks.

cored hole Hole bored so that a central wood piece is removed; a *hollow bit* is used.

cored solder Solder with a core of flux.

core wall A hollow-core wall.

Corinthian order See *order*.

corner angles Angle-iron fasteners used at corners to strengthen the corner; either an inside or outside corner angle can be used.

corner bead Metal molding used at exterior corners of plastered and gypsum paneled walls. Used to protect the corner. (Figure 251.)

corner block In woodworking, a wood support piece glued or glued and screwed at a right angle inside a joint. Also, a masonry unit, such as a concrete block, designed to finish a corner.

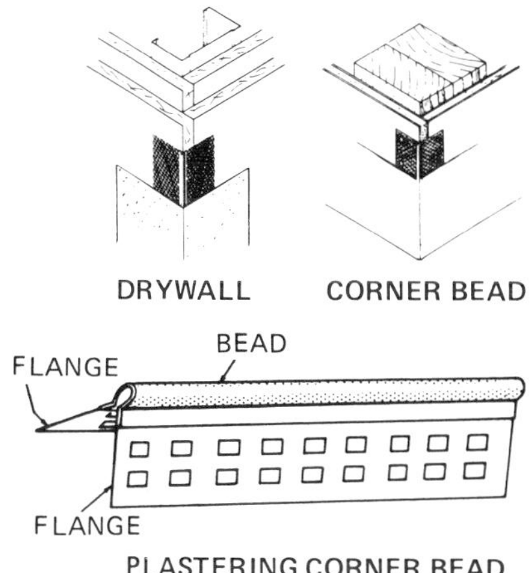

Figure 251 Corner bead

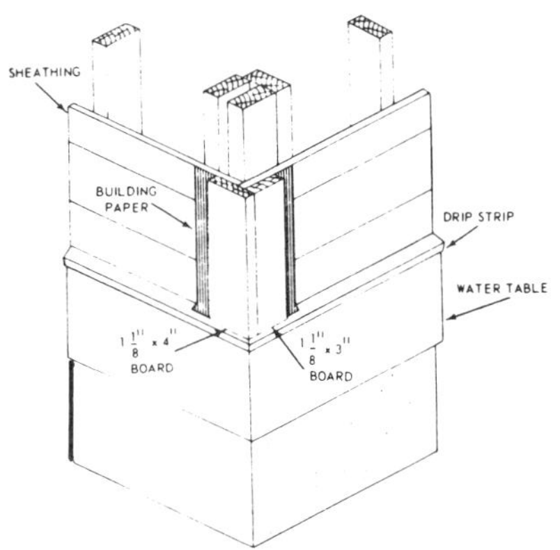

Figure 252 Corner boards

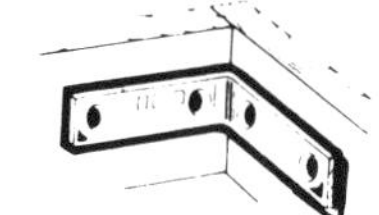

Figure 253 Corner brace

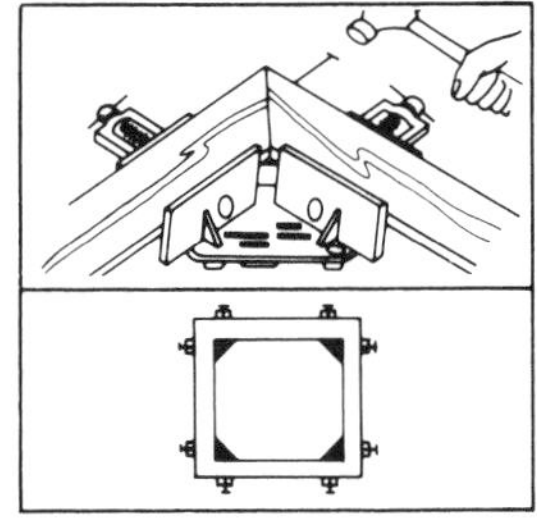

Figure 254 Corner clamps

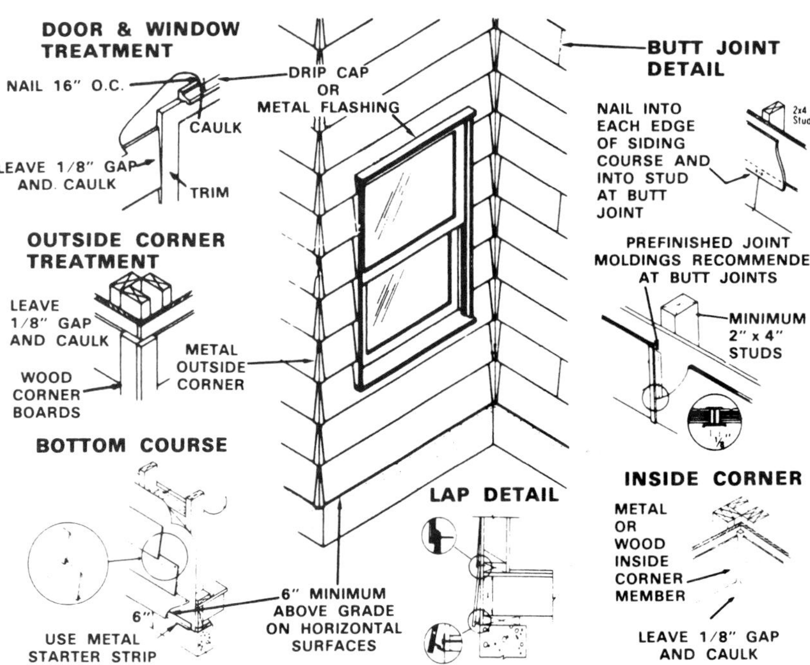

Figure 255 Corner construction

corner boards Boards used to finish the corner of a frame house. (Figure 252.) See also *corner construction.*

corner brace Diagonal brace or rigid sheathing (such as plywood) used at the corners of a frame house to strengthen the corner. Also, for light framing and woodworking, an L-shaped metal piece used for strengthening inside corners. Holes are bored and countersunk in the metal so wood screws may be used. (Figure 253.) Also called a *corner iron.* See *flat corner iron, mending plate, T brace.*

corner chisel A wood chisel with an L-shaped cutting edge, used for cutting out mortises.

corner clamp Clamp used to hold a corner joint together while glue sets. (Figure 254.)

corner construction Structure and finish of a corner, as the exterior and interior corners of a building. Figure 255 shows corner construction. See also *corner boards.*

corner frame clamp A clamp set used for making 90° corners. Holds wood pieces up to 3″ wide. Matching pieces are each cut at 45°. (Figure 256.) See also *corner clamps, miter vise.*

corner iron Metal strip bent in an L shape used for strengthening inside corners. Holes are bored and countersunk so wood screws may be used. Also called a *corner brace.* See also *flat corner iron, mending plate.*

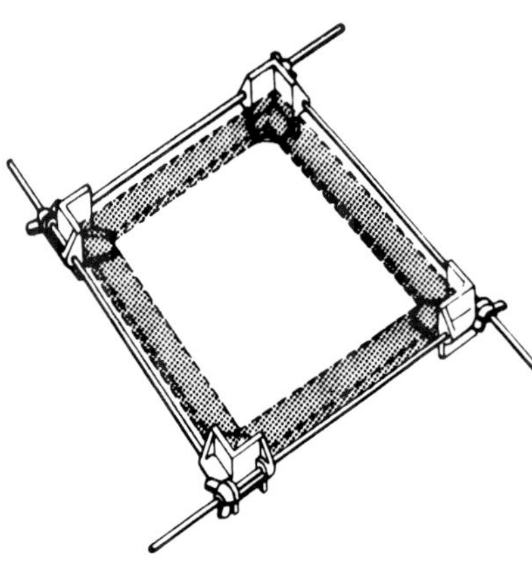

Figure 256 Corner frame clamp set

cornerite In plastering, metal lath strips bent at a right angle and used in interior corners of walls and ceilings. Designed to prevent cracking. See *corner bead.*

corner joint weld Weld of two pieces that meet at a right angle at a corner; a *fillet weld.*

corner lead See *lead, mason's lead.*

corner tape creaser Special tool used for applying tape to inside drywall corners before finish.

corner trowel Right-angle trowels used for finishing the corners of a concrete step. See *step trowel, trowel.* Also, special L-shaped trowels used for plastering and finishing drywall corners, and for finishing plaster wall corners. Both inside and outside corner trowels are used. See *trowel.*

cornice Built-up construction at the rafter ends at the eaves, consisting mainly of the *soffit* and *fascia* in a closed cornice. (Figure 257.) The area where eaves and outside walls meet. See *open cornice.* In masonry, the upper course of a wall.

cornice brake Sheet metal brake used for bending long metal sections. See *brake.*

cornice return A cornice that is finished into the gable end of the house at the corner. (Figure 258.)

cornice trim Exterior finish of a cornice.

corporation cock Valve by the water main that is controlled by the water company to turn on or shut off water to a building. See *water service.*

corrosion Oxidation or rusting of metal.

corrugated Having a surface with ridges and valleys, as corrugated roofing.

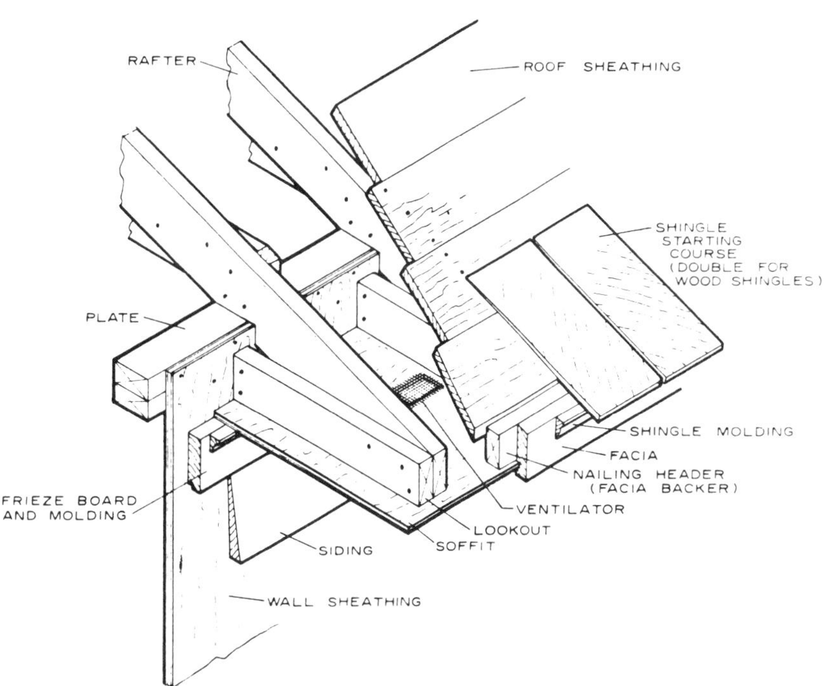

Figure 257 Cornice

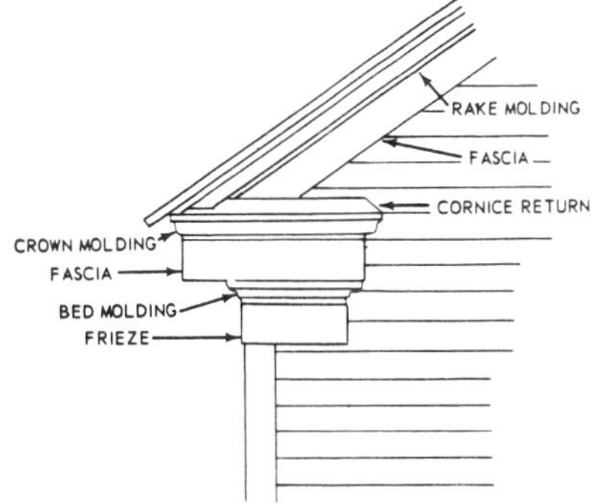

Figure 258 Cornice return

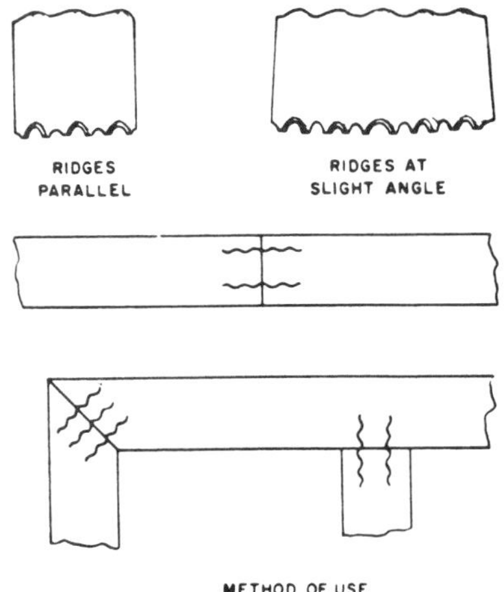

Figure 259 Corrugated fasteners

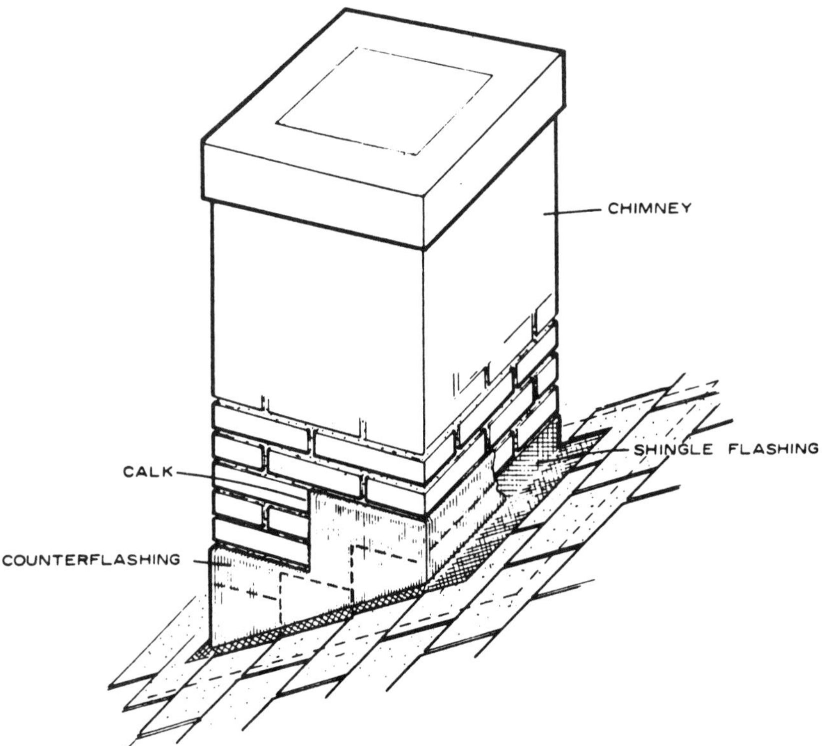

Figure 260 Counterflashing

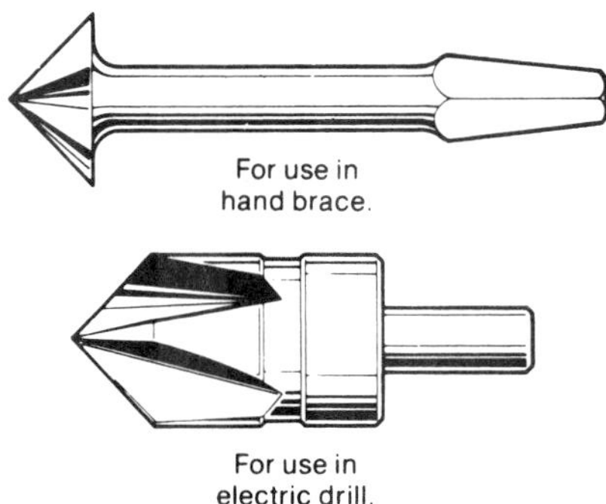

Figure 261 Countersink bits

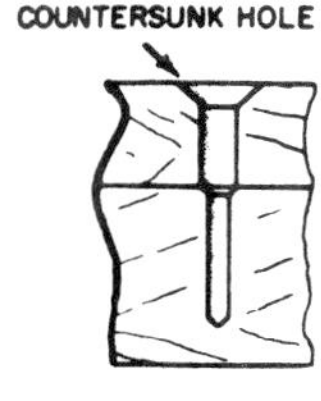

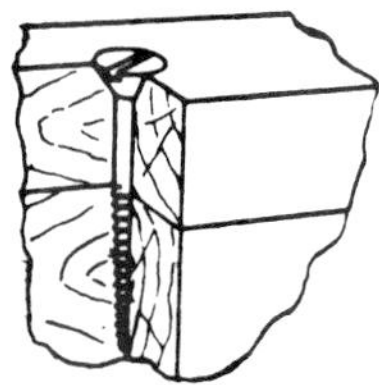

Figure 262 Countersunk hole

corrugated fastener Metal device used for strengthening corner joints. (Figure 259.) Special power staplers are available for setting this fastener.

counterbore To increase the size of a hole for part of its length by boring with a larger drill. (*Figure 841*, p. 370.) See *drill and counterbore bit.*

counterbracing Bracing that supports in an opposite direction to the main bracing.

counterclockwise Rotation opposite to the movement of the hands on a clock.

counterflashing A flashing used to cover over other flashing to protect the top edge, as the flashing used with a chimney to cover shingle flashing. (Figure 260.) Flashing used at masonry wall intersection: top of counterflashing is embedded into the masonry; used at shingle roof intersections.

countersink To increase the size of the top of a hole by boring a cone-shaped opening. A special countersink bit is used. (Figure 261.) The flared opening receives the head of a screw. (Figure 262.)

countersink nail Nail with a small head, designed to be sunk below the wood surface with a nail set. The nail head has a small cup or depression to receive the nail set tip. A *finish nail.*

countersunk Having the head (of a screw or nail) sunk below the surface. (Figure 262.)

counter-top lavatory A *vanity.*

coupling In plumbing, a short open section of pipe with inside threads at both ends, used to join two threaded pipe ends.

course Horizontal row of masonry units, as a brick course. (*Figure 125*, p. 47.) In surveying, the bearing of a line with reference to the true meridian.

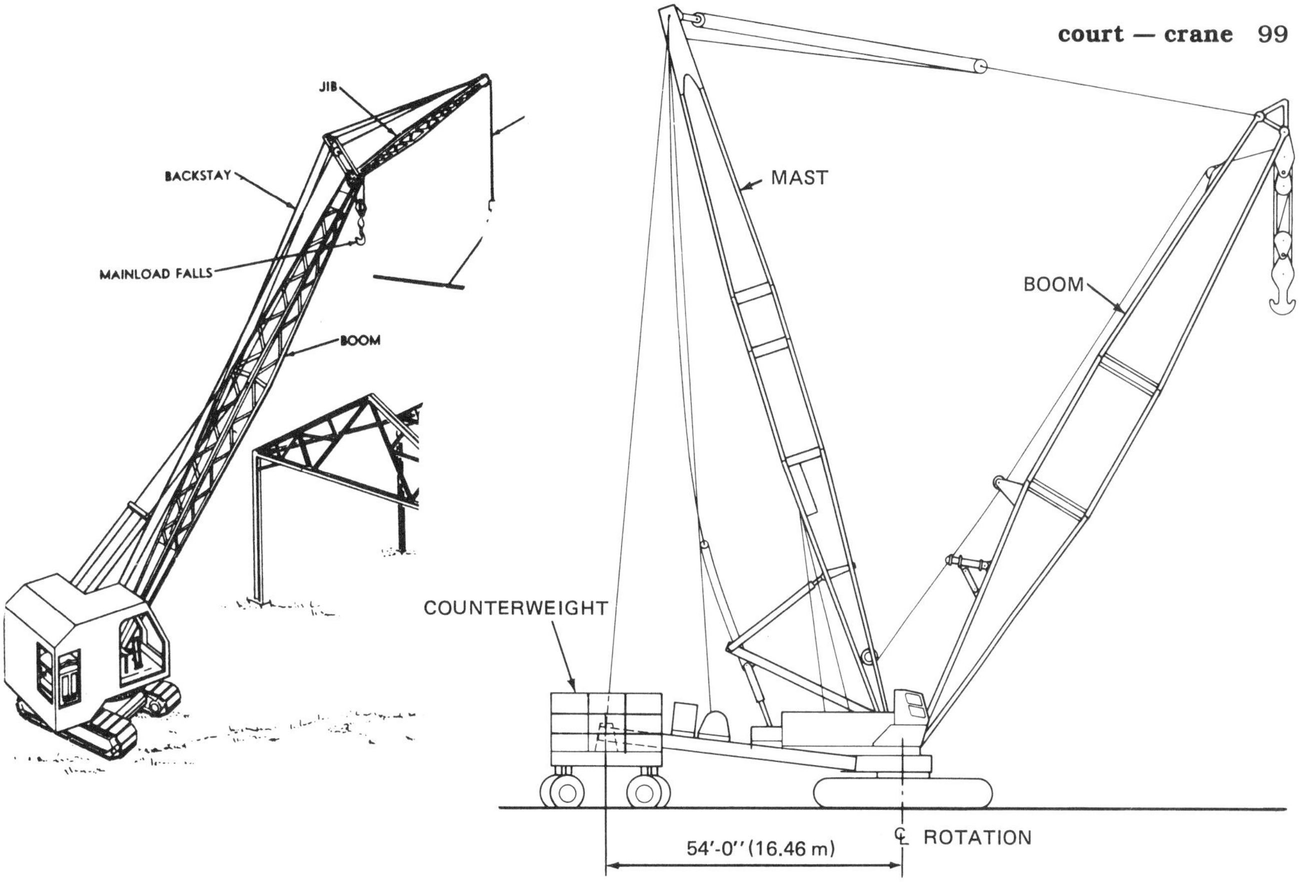

Figure 263 Crane

court Open area enclosed by the sides of a building or several buildings, or enclosed by walls.

cove Concave trim, a *cove molding.* A concave, curving arch where the wall joins the ceiling.

cove lighting Lighting that is shielded so no direct light is seen; light is provided by reflection from walls or ceiling.

cove molding Molding, often vinyl, with a flared-out bottom edge used as counter-top trim or as baseboard; concave molding used where wall and ceiling meet. See *molding.*

covered electrode Welding electrode that is coated with a shielding material. The coating provides a protective slag layer over the weld area.

CPVC A light-colored plastic (chlorinated polyvinyl chloride) pipe used for cold and hot water distribution. See *plastic piping.*

cracking In oxyacetylene welding, a momentary opening of a gas cylinder to blow out any dust or dirt in the outlet nozzle.

craftsman An experienced tradesman or worker; a journeyman. One who can build true and make a sound structure or part.

crane Heavy hoisting machine that uses a boom with cables to lift and move material. Boom may be raised or lowered or swung horizontally. Used for moving earth and to deliver heavy material, steel, and prefabricated or precast parts to where needed on a structure. (Figure 263.) Also called a *derrick crane.* See *climbing crane, derrick, tower crane, truck crane.* See also *boom, clamshell.* Figure 264 shows attachments used with a crane.

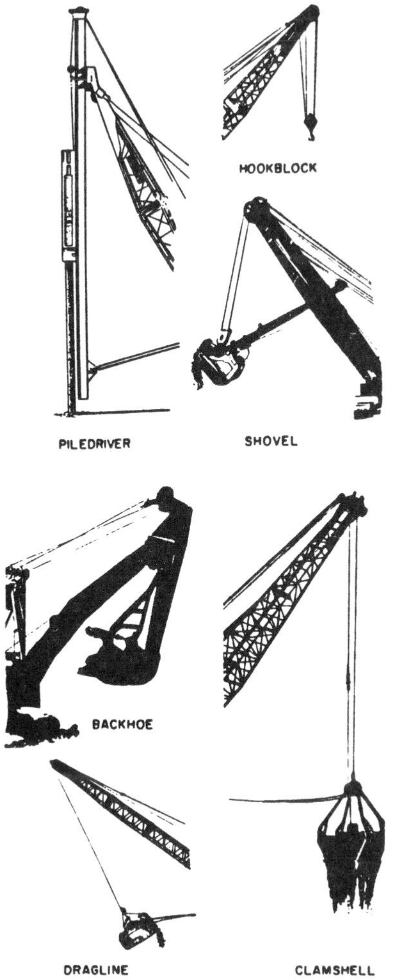

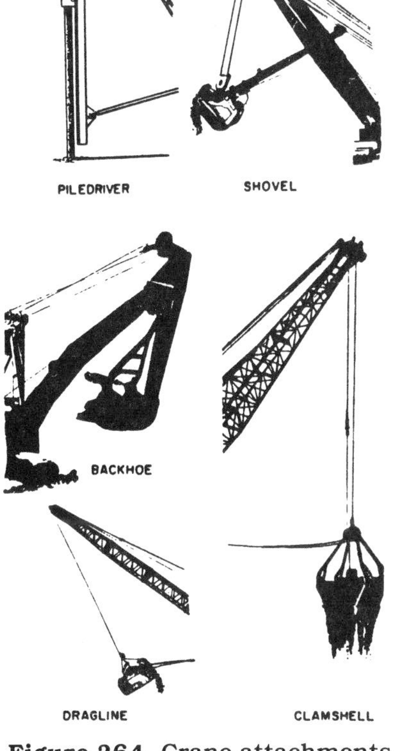

Figure 264 Crane attachments

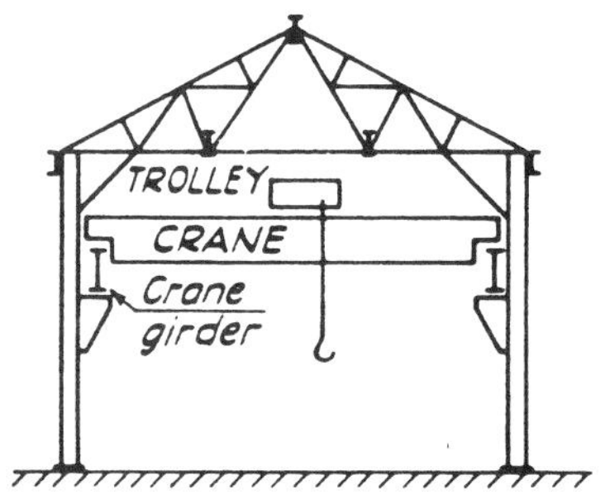

Figure 265 Crane girder

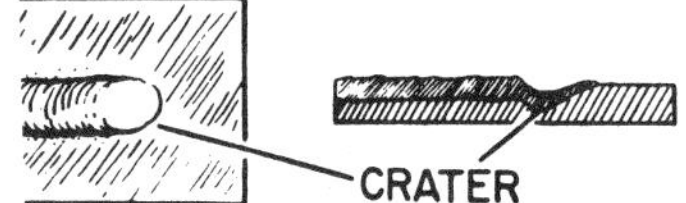

Figure 266 Crater

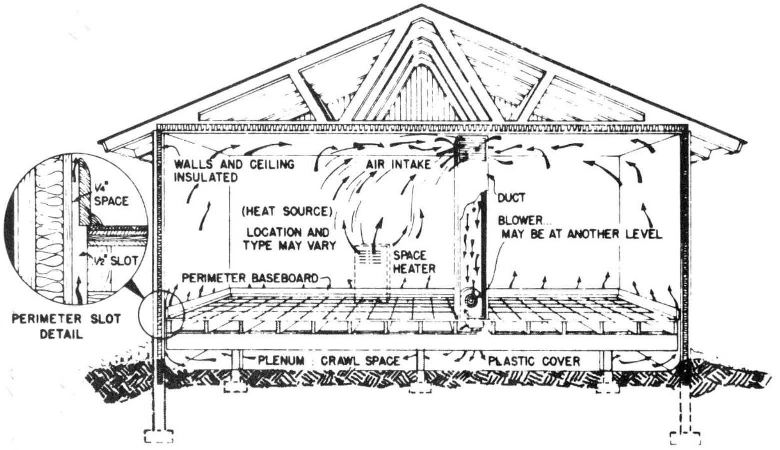

Figure 267 Crawl-space plenum

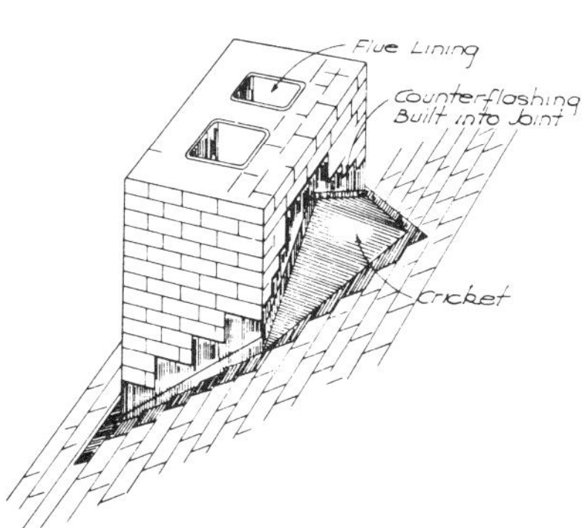

Figure 268 Cricket

crane columns Structural steel columns designed for supporting a rolling crane in a factory.

crane girder Girder in a building that supports a moving crane support. (Figure 265.) Crane trolley moves from side to side on rails supported by crane girders.

crater In arc welding, a hole at the end of the weld or in the molten weld pool. (Figure 266.)

crawler Earth-moving equipment with steel treads, opposed to rubber-wheeled equipment. See *loading shovel* for comparison. See also *bulldozer.*

crawl space In houses with no basements, the space between the first floor and the surface of the ground; space is high enough to crawl through for repairs and installation of utilities. See *perimeter wall foundation.*

crawl-space plenum Forced-air heating system that uses an insulated crawl space as a plenum. Warm air is blown into the crawl space and exits through outlets in the various rooms. (Figure 267.)

crawl-space ventilation Circulation of air into and through the crawl space. Openings are left in the foundation wall to allow air to flow through the crawl space. See *wall vent.* Air ventilation cools the space in summer and prevents moisture build-up in winter.

crazing Fine cracks that appear on a paint or concrete surface.

creep In gluing, a tendency of certain glues to remain elastic after setting so the joint may move or creep under pressure. Common in white glues (polyvinyl resin emulsions). Also, in concrete, a movement after concrete is placed.

crenelated Having evenly spaced square notches (in an edge, wall, molding, or other part).

creosote Tar-based material used as a wood preservative.

creosoting Impregnating wood with creosote.

crescent wrench General-use adjustable-jaw wrench. See *wrench.*

crest Ridge of a roof; top of a hill.

crib A heavy timber framework built as a support or as a retaining wall. A built-up boxlike timber support, often filled with rubble, used as a support or retaining wall in a waterway.

cricket Small, double-sloped structure built on the high side of a brick chimney to divert water. Also called a *saddle.* (Figure 268.)

crimp To crush or indent pipe or tube ends, flanges, or other metal ends together — often designed to hold one part to another. (*Figure 83,* p. 32.) See *crimping tool.*

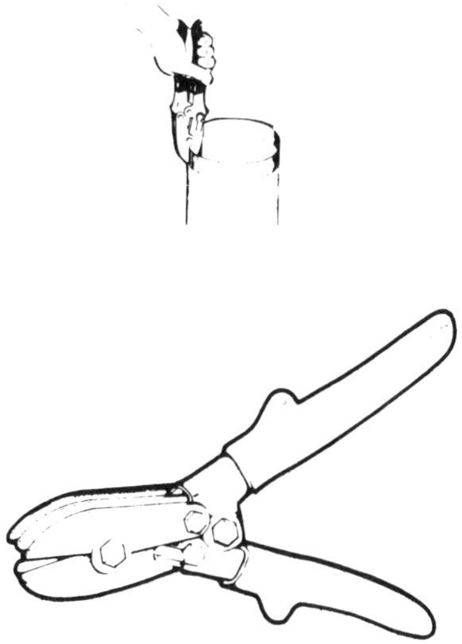

Figure 269 Hand crimper: sheet metal

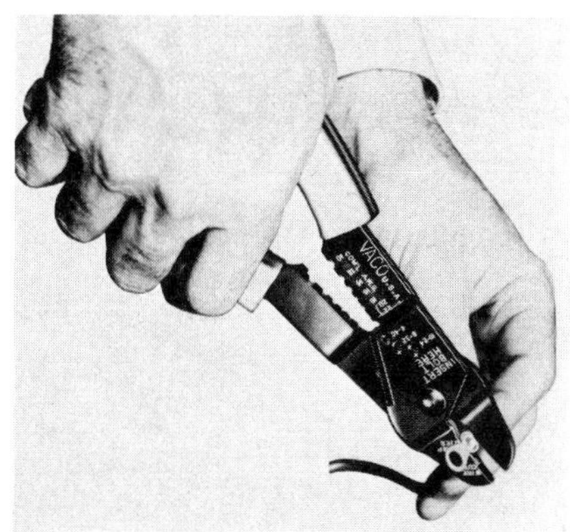

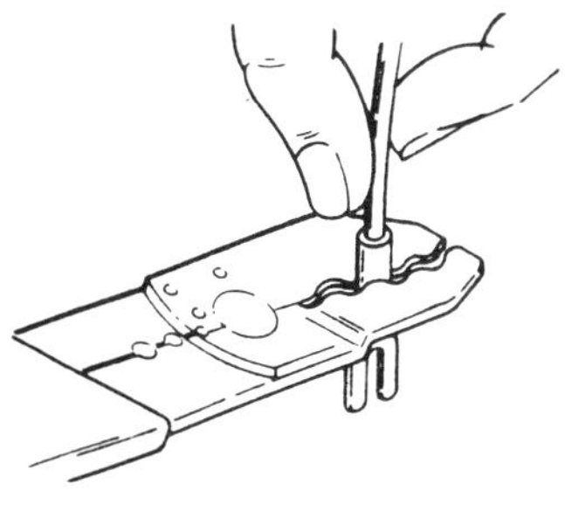

Figure 270 Crimping tool

crimper Sheet metal tool for crimping. (Figure 269.)

crimping tool Special tool used for crimping pipe, tube ends, or solderless terminals. (Figure 270.) See also *multipurpose tool.*

cripple Short framing member, as the short studs above or below a window opening. (Figure 271.) Short joists around an opening; *tail joists.* (*Figure 531,* p. 228.)

cripple jack rafter Rafter that runs between two other rafters, as between a valley rafter and hip rafter or between two valley rafters. See *jack rafter.*

cripple rafter A *cripple jack rafter.*

cripple stud Short studs, as those over or under a window frame. (Figure 271.)

critical angle In stair building, a stair slope over 50°; considered unsafe for general use. See *preferred angle.*

crook Lumber defect — the end of a board warps, bending up or down to cause a sharp curve. (Figure 272.) See *warp.*

crossbands In plywood, veneer plies with grain running at a right angle to the face plies. See *plywood.*

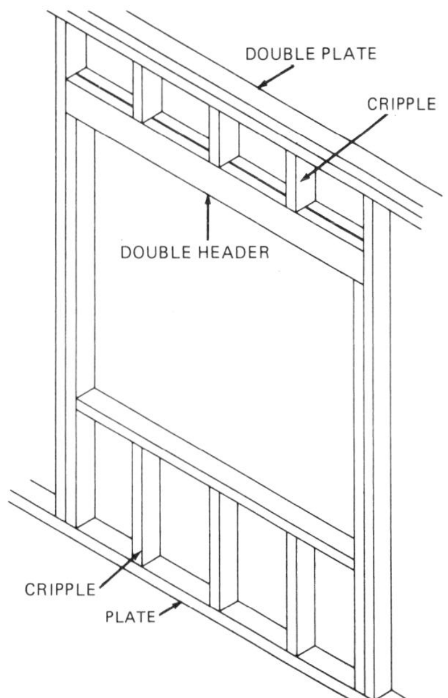

Figure 271 Cripples

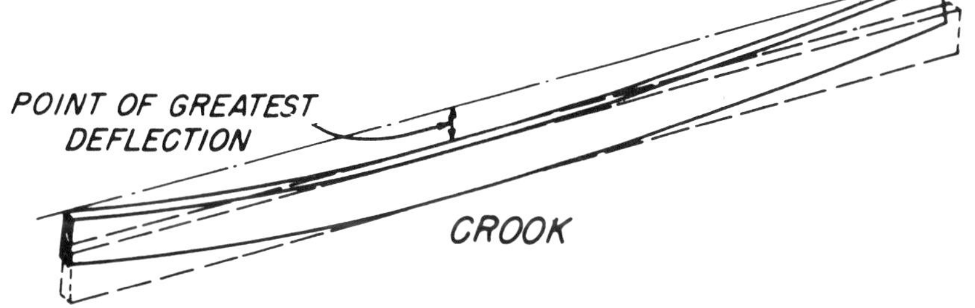

Figure 272 Crook (lumber defect)

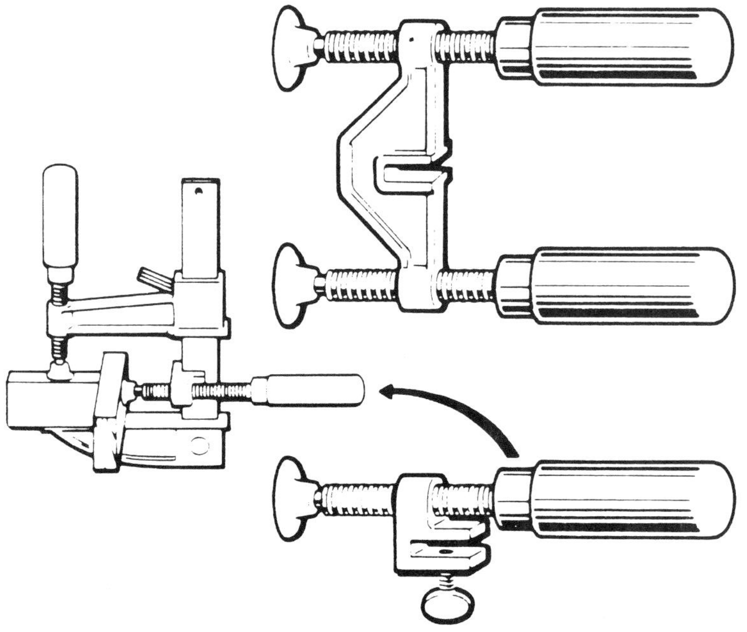

Figure 273 Cross clamp

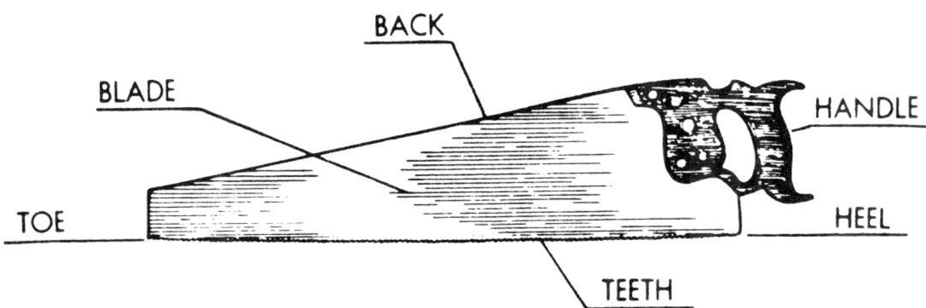

THE SIZE OF A SAW IS DETERMINED BY THE LENGTH OF THE BLADE IN INCHES
SOME POPULAR SIZES ARE 24' AND 26'
THE COARSENESS OR FINENESS OF A SAW IS DETERMINED BY THE NUMBER OF POINTS PER INCH
A COARSE SAW IS BETTER FOR FAST WORK AND FOR GREEN WOOD
A FINE SAW IS BETTER FOR SMOOTH ACCURATE CUTTING AND FOR DRY SEASONED WOOD
5-1/2 AND 6 POINTS ARE IN COMMON USE FOR RIP SAWS
7 AND 8 POINTS ARE IN COMMON USE FOR CROSS CUT SAWS
SAW TEETH ARE SET, EVERY OTHER TOOTH IS BENT TO THE RIGHT AND THOSE BETWEEN TO THE LEFT, TO MAKE THE KERF WIDER THAN THE SAW

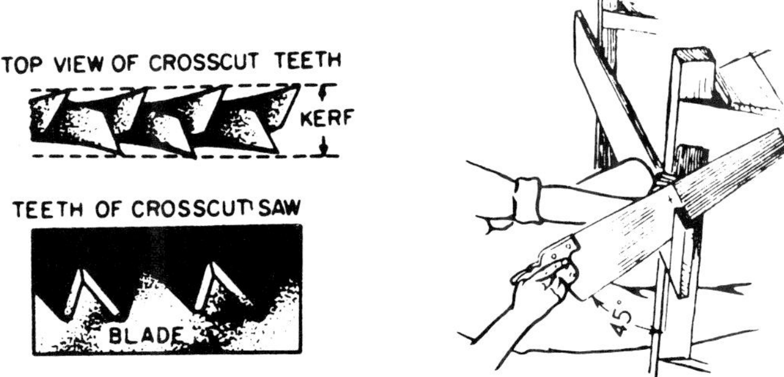

ABOUT 65°
ONE INCH
8 POINTS PER INCH, 7 TEETH
15°
45°
60°
SET
SET

CROSS CUT SAW TEETH ARE LIKE KNIFE POINTS. THEY CUT LIKE TWO ROWS OF KNIFE POINTS AND CRUMBLE OUT THE WOOD BETWEEN THE CUTS

Figure 274 Crosscut saw

cross brace Support that runs between (across) two members to brace them; *cross bridging.*

cross bridging Pairs of diagonal braces running between support members to stiffen and give support to the members, as the cross bridging between floor joists. See *bridging.* (*Figure 130,* p. 49.)

cross clamp Clamp used for corner and edge gluing. (Figure 273.)

crosscut blade Circular power saw blade that has teeth specially cut for crosscut cutting. See *saw blade.* (*Figure 828,* p. 364.)

crosscut saw Hand saw used for cutting wood across the grain (at a right angle). A 26-inch, eight-point saw is commonly used. (Figure 274.) See also *ripsaw, saw.*

crosscutting Sawing of wood across the grain, at a right angle to the direction of the grain. Cutting across the trunk of a tree. Opposed to *ripping.*

cross grain In wood, grain that runs at an angle to the length of a wood piece; grain that runs diagonally on the face of a board. Wood cut at the right angle to the long wood fibers, as cutting through the trunk of a tree.

cross peen hammer Metalworking hammer used for shaping and flatening metal. The peen is at a right angle to or crosses the handle. (Figure 275.) See *ball peen, hammer, straight peen.*

cross section In drafting, a view made by taking an imaginary vertical cut, then showing the exposed structure. See *section.*

cross ventilation Natural flow of air across a room, as from openings on opposite sides.

crotch veneer Veneer wood that is cut from a crotch area in a tree (where the trunk divides into two main parts or limbs). An attractive pattern or figure is obtained. See *feather crotch, moonshine crotch, swirl crotch.*

crowbar Heavy steel bar used for prying and levering. (Figure 276.) See also *nail puller.*

crown In highway construction, the high-curved, center part of the finished road.

crown molding Molding run at the top of a wall at the intersection with the ceiling. See *chair rail, molding.*

CRSI Concrete Reinforcing Steel Institute.

CRT Cathode-ray tube; viewer used with a computer.

CSI specifications format See *Construction Specifications Institute.*

cubage Cubic content; cubic feet in a room or building; the volume.

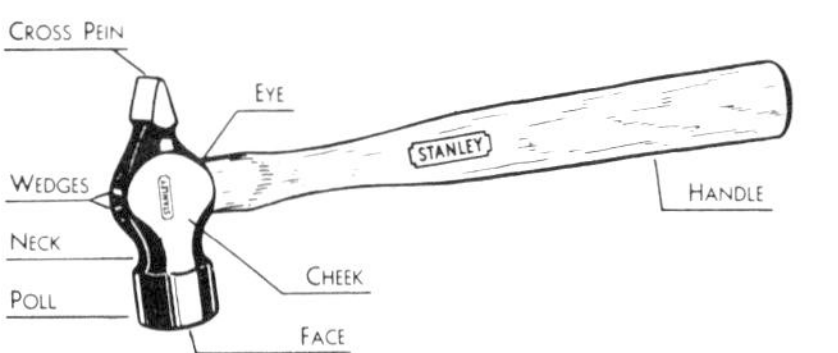

Figure 275 Cross peen

Figure 276 Crowbar

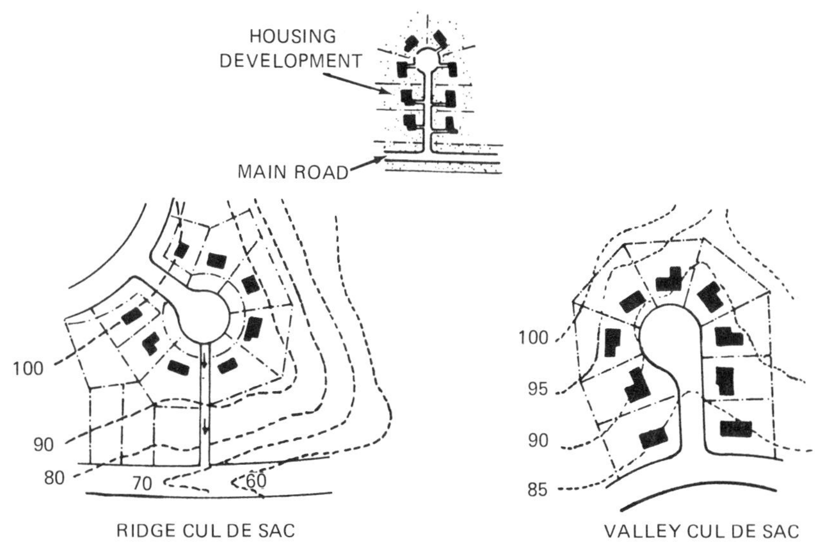

Figure 277 Cul de sac

cubic content The volume; the number of cubic feet (1′ × 1′ × 1′) in a room, house, or structure; used in estimating costs, air conditioning needs, ventilation needs, and so on.

cubic measure A volume measure in cubic inches, feet, or yards.

1728 cubic inches	=	1 cubic foot
27 cubic feet	=	1 cubic yard
231 cubic inches	=	1 gallon

See also *area.*

cul-de-sac Street, road, or alley with one end closed, usually with a widened area for turning around. A dead end. (Figure 277.)

culls Rejects; items unsuitable for use or below standard.

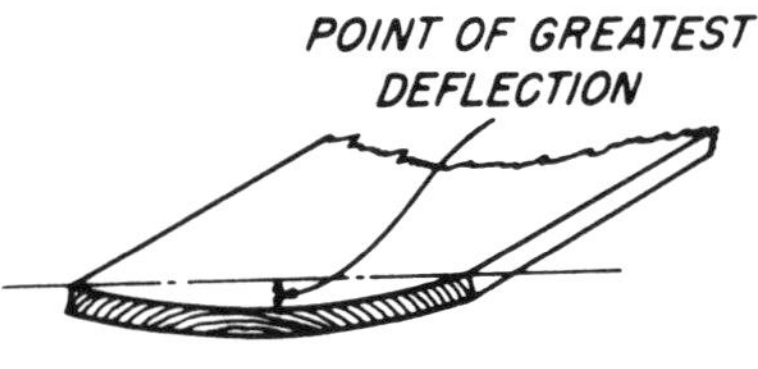

Figure 278 Cup (lumber defect)

Figure 279 Cup hook

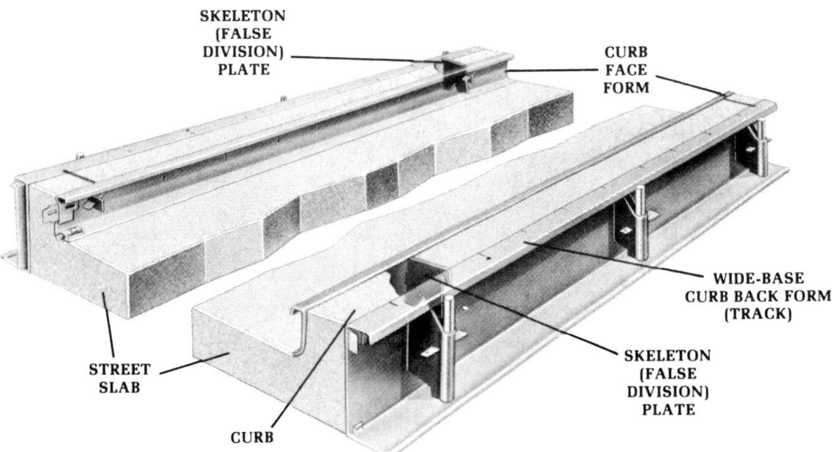

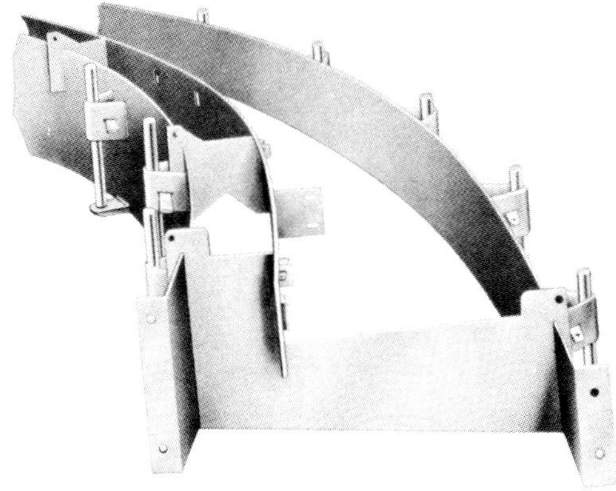

Figure 280 Curb forms

culvert a drainage structure beneath and across a road or other traveled way.

cup Lumber defect, a warp that results in a curve across the width of a board. (Figure 278.) See *warp*.

cup hook Light, screwed hanger with an open hook on the unthreaded end. (Figure 279.) See *L hook screw, screw eye, screw hook*.

cupola vent Boxlike vent that fits on the ridge of a roof.

cupping the mortar Cutting into the mortar with a trowel and rolling over a trowel load.

cup shake Lumber defect, where the annual rings of a tree separate to form curving open areas. See *shake*. Since it is thought that this defect is caused by wind, it is often called *windshake*.

curb cock Valve that controls water supply to a building. Located at the bottom of an access pipe called a *stop box* or *buffalo box*. See *water service*.

curb edger Cement mason trowel used for finishing off the edge of a concrete border, such as a sidewalk or driveway edge. See *edger, trowel*.

curb forms Forms used for making street curbs. Figure 280 shows a special metal form used. See also *sidewalk forms*.

curb joint The joint or angle made by two intersecting *gambrel* roof slopes. Also called a *knuckle joint*.

curb roof Roof side with a double slope, as a gambrel roof.

curb stop A *curb cock*.

curb trowel Right-angle curving trowel similar to a *step trowel*. Used for finishing highway and street curbs.

curing In concrete work, the setting or hardening of concrete by chemical action. Various methods are used to assure that proper curing takes place, such as covering the concrete with plastic sheeting or some other material to prevent evaporation or loss of moisture from the concrete. The hardening process for mortar or concrete. When completely cured, usually in 28 days, the mortar or concrete is said to *set*.

curly grain Wood grain that swirls in a tight "S" pattern.

current Rate of electrical flow; *amperes.*

$$\text{current (amperes)} = \frac{\text{watts}}{\text{voltage}}$$

$$\text{current (amperes)} = \frac{\text{volts}}{\text{ohms}}$$

curtain In reinforced concrete, a vertical mat of reinforcing bars wired together to give support to a wall.

curtain wall An outside non-bearing wall, supported by the building framework.

curved claw hammer Carpenter's hammer with a curved claw, widely used for driving nails. The claw is used for pulling nails or separating nailed boards. Also called a *nail hammer.* See *hammer.*

cut and fill Process of earth movement: earth is removed from a high area and filled into a low area or built up in a separate area. (*Figure 466,* p. 191.) See *filling, top spreading.*

cut brick Bricks that are broken or shaped by the bricklayer. They are used to fill odd-sized spaces, as at a corner. Figure 281 shows several types of cut brick.

cut-in brace Diagonal bracing made by using short sections of 2 × 4 between the studs. The 2 × 4's are cut at an angle at each end to fit between the studs. (Figure 282.) See also *let-in brace.*

cut nail Squared, blunt-pointed nail made in imitation of old-fashioned nails that were made by hand, opposed to the modern *wire nail.* See *nail.*

cut-off saw See *abrasive cut-off saw.*

cut-off wall Wall built around a building foundation to hold back ground water. Also called a *slurry wall.*

cutouts Slots cut in two-tab, three-tab, or lock *asphalt shingles.*

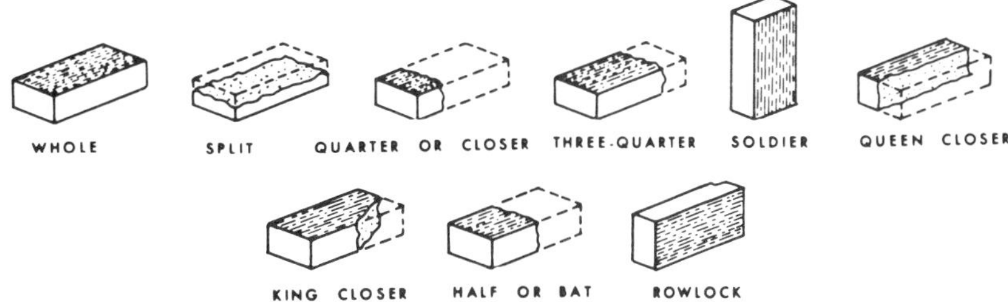

Figure 281 Cut brick shapes

cut stone Squared stone building units made by cutting from solid rock. Sizes vary.

cutting In masonry work, cutting off a section of mixed mortar and rolling it on the hawk.

cutting gauge A gauge very similar to a marking gauge except that, instead of a sharp pin, a flat blade is used for marking. The blade can be used for cutting veneer.

cutting plane Imaginary plane for cutting building to produce floor-plan views and sections.

cutting torch An oxyfuel torch used specifically for cutting. A special lever allows manual control of the oxygen flow. Special cutting tips are used.

C value Number of Btu's that will pass through a square foot of material (regardless of thickness) in an hour (assuming a 1°F differential between the sides of the material), as for a wall. A 0.8 C value would mean $\frac{8}{10}$ of a Btu would pass in an hour. See *K value, R value, U value.*

cylinder Container for the storage of compressed gas, as an oxygen or acetylene cylinder.

cylinder lock Heavy-duty cylindrical lock used on exterior doors. (Figure 283.) See *lock.*

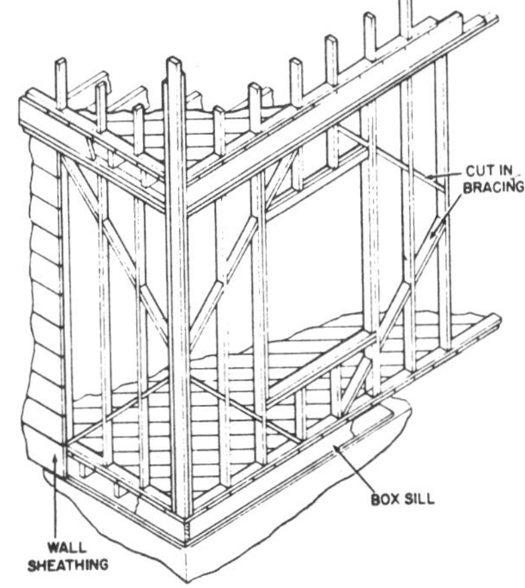

Figure 282 Cut-in braces

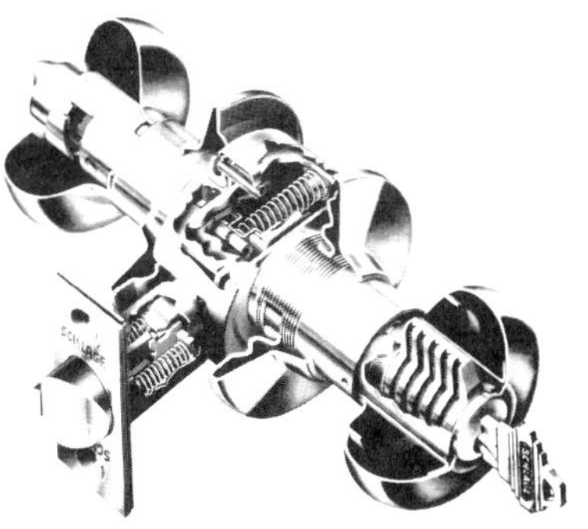

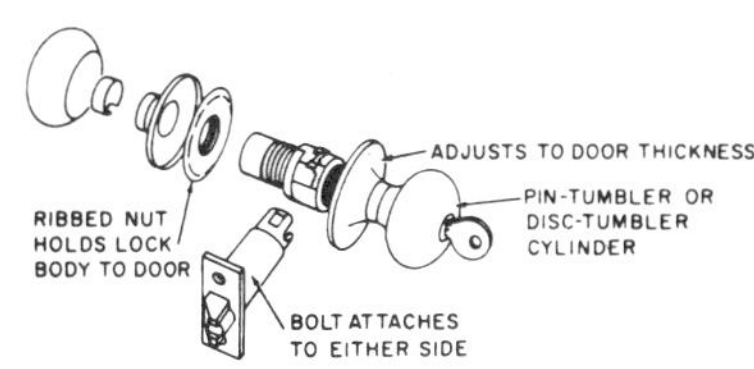

Figure 283 Cylindrical lock

d

d See *penny*.

dado Rectangular groove normally cut at a right angle to the grain of the wood piece. See *groove, wood joints*.

dado head An assemblage of cutting blades used with a stationary power saw to cut a dado. (Figure 284.) Two blades and different-width chippers are assembled together to produce a dado of the correct width.

dado joint Joint made by one board being notched into the side of another. (Figure 285.) See dood joint.

damper Plate or valve that controls the flow of air or gas. Device that regulates the draft allowed into a fire, as at a fireplace or in a gas heater. See *fireplace*. (*Figure 396*, p. 159.)

damp proofing Protection against the penetration of moisture. Treatment of a wall to prevent moisture from seeping through.

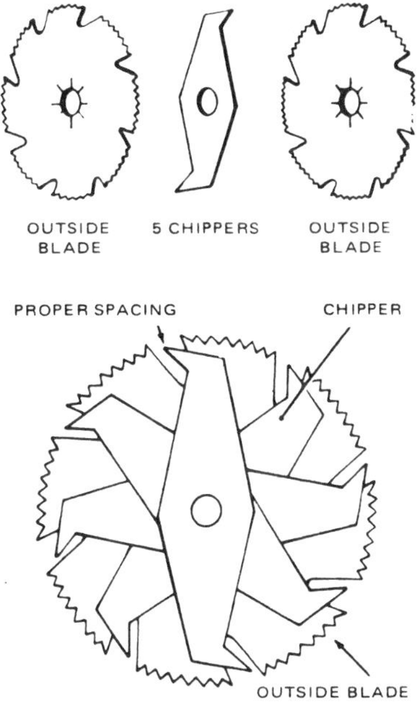

Figure 284 Dado head set

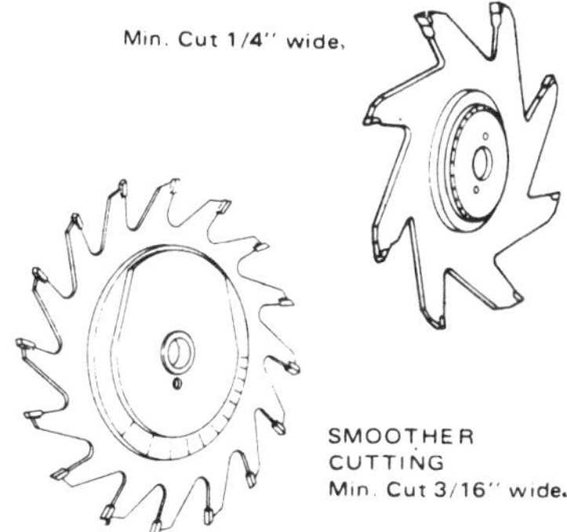

Figure 285 (continued)

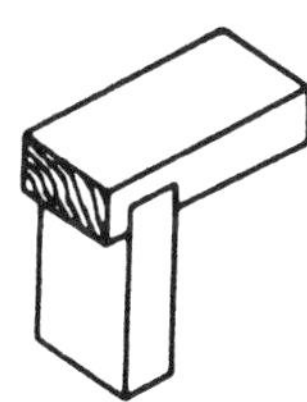

Figure 285 Dado joint

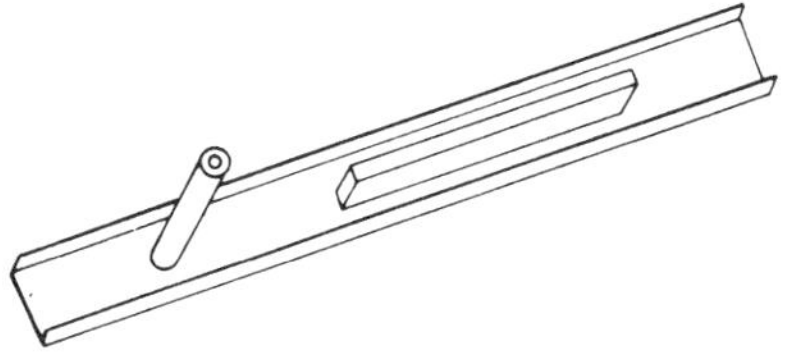

Figure 286 Darby

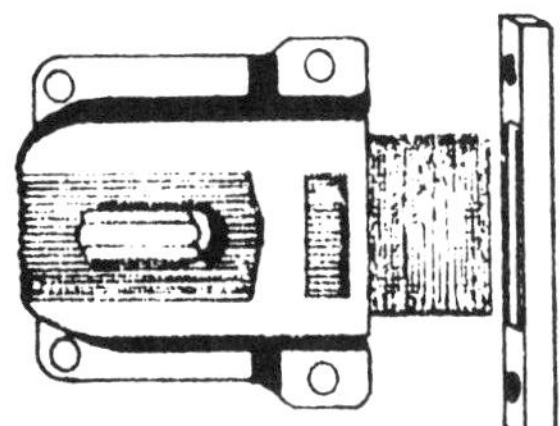

Figure 287 Dead-bolt lock

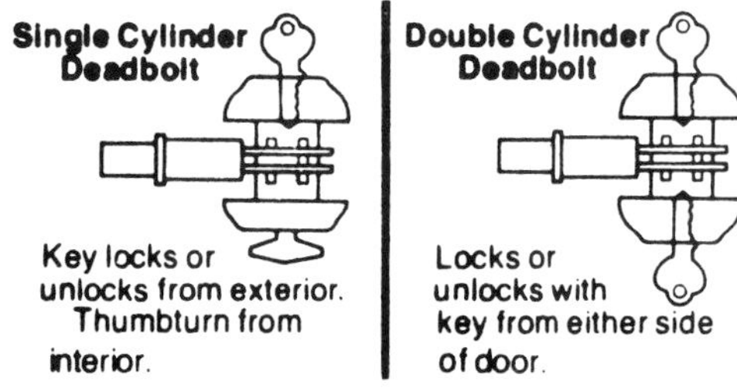

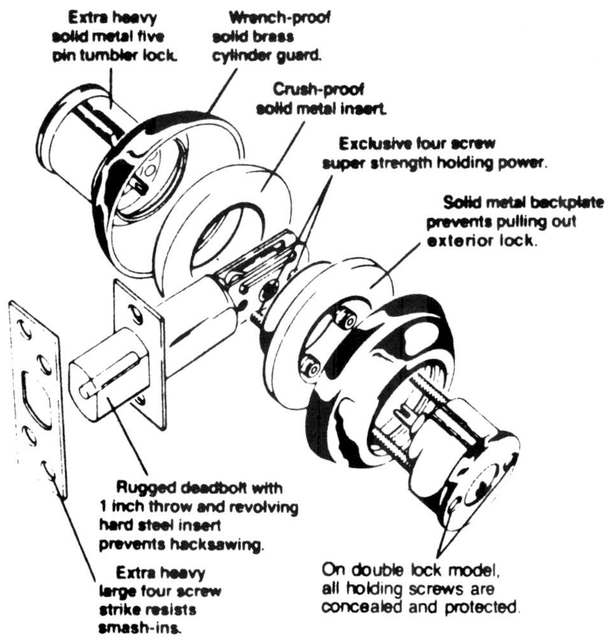

Figure 288 Deadlock

darby Plastering tool used to level a plaster surface. Also, in concrete flatwork, a smoothing tool used after *screeding*. (Figure 286.)

darbying In concrete work, the smoothing out of flatwork after screeding. *See bull floating.*

datum Horizontal plane from which measurements are made to determine elevation; a local, established elevation used as a reference point in building construction and surveying. In general, the mean sea level.

datum point A point with a known elevation, established by local authorities.

DC See *direct current.*

DC rectifier welder See *rectifier, transformer-rectifier, welding machine.*

dead blow hammer A hammer or mallet that has a hollow head filled with lead shot. When hammered, it has very little recoil or bounce.

dead bolt Square-head bolt used in a deadlock. Also used to refer to the whole lock. (Figure 287.) See *deadlock.*

dead end street Public street with one end closed.

dead level Absolutely horizontal, as a roof with zero slope.

dead load A permanent, inert load whose pressure on a building is steady and constant. The structural members and the total empty building give the dead load. See *design load, lateral thrust, live load.*

deadlock Door lock with a dead bolt (square-head bolt). The door is opened by turning the key in the lock to retract the bolt; no knob is turned. (Figure 288.) Some deadlocks are double-keyed; a key must be used on each side to retract the dead bolt.

dead man An anchor made by burying a heavy object, such as a timber, in the ground.

decay Wood rot caused by fungi.

decibel Unit for measuring sound energy or power. The more powerful the sound, the higher the decibel rating.

deciduous *Hardwood* tree with broad leaves that fall off with the cold season.

decimal equivalent Fraction expressed as a decimal, as $\frac{1}{2}$ equals 0.50.

decimal system A system based on tenths. A decimal point determines the order of the tenths. For example, 0.2 reads $\frac{2}{10}$; 0.25 reads $\frac{25}{100}$; 0.255 reads $\frac{255}{1000}$. Figure 289 shows the relationship between common fractions and decimal equivalents.

decking stapler Air-driven stapler used for fastening decking. Staples from $1\frac{1}{2}''$ to $2''$ are used. Also called a *sheathing stapler.* See also *air-driven nailer, nailer, stapler.*

deck An outside platform; an open, second-story porch. The floor of a house, including subfloor and finish floor. A roof surface on which finish roofing is applied. (Figure 290.) Often used to refer to a flat roof only. The base for roof finish. See *timber construction.*

declination of the needle See *magnetic declination.*

decorative blocks Shaped or patterned concrete masonry units used in decorative walls or for special effect.

defect Wood irregularity that decreases the structural value or strength of the wood. See *blemish* (which is a visual defect but does not affect wood strength). See *bow, checks, cup, decay, knot, shake, wane, warp.*

deflection Slight bending of a horizontal support member, as a beam. The slight give of a structure, such as a bridge. Allowable deflection is calculated for star dard uses.

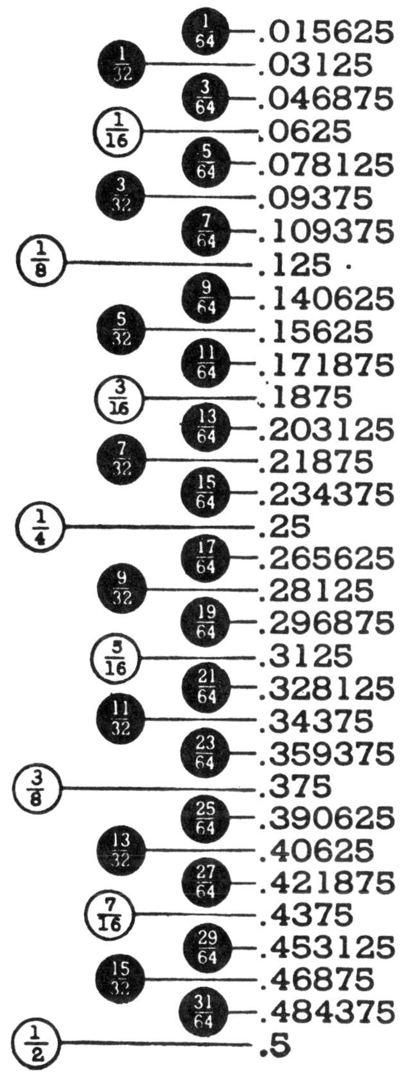

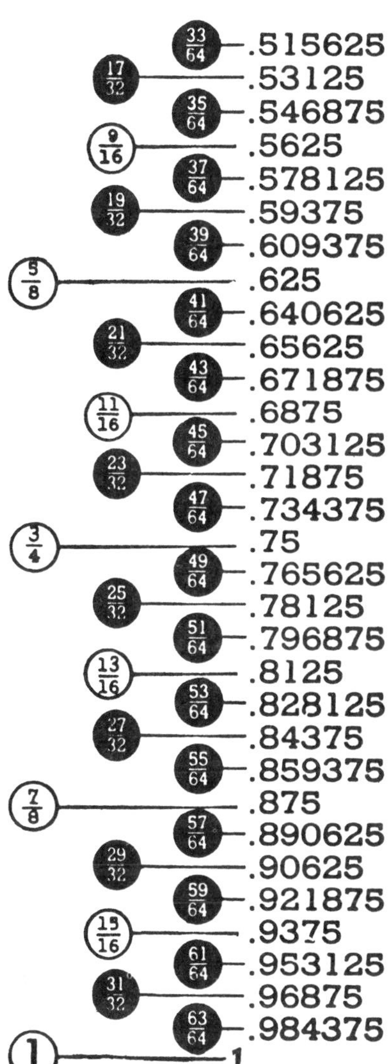

Figure 289 Decimal equivalents

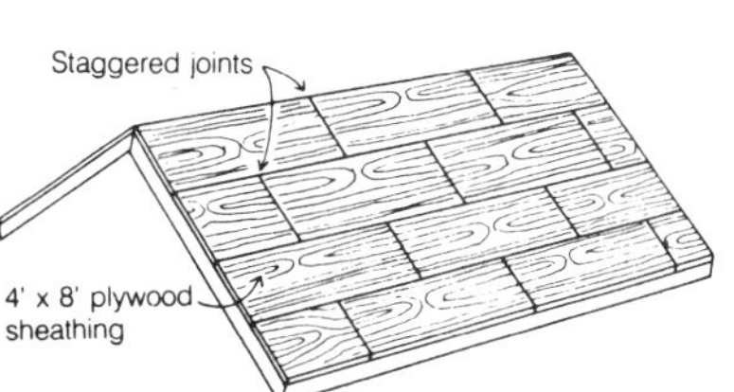

Figure 290 Deck (plywood)

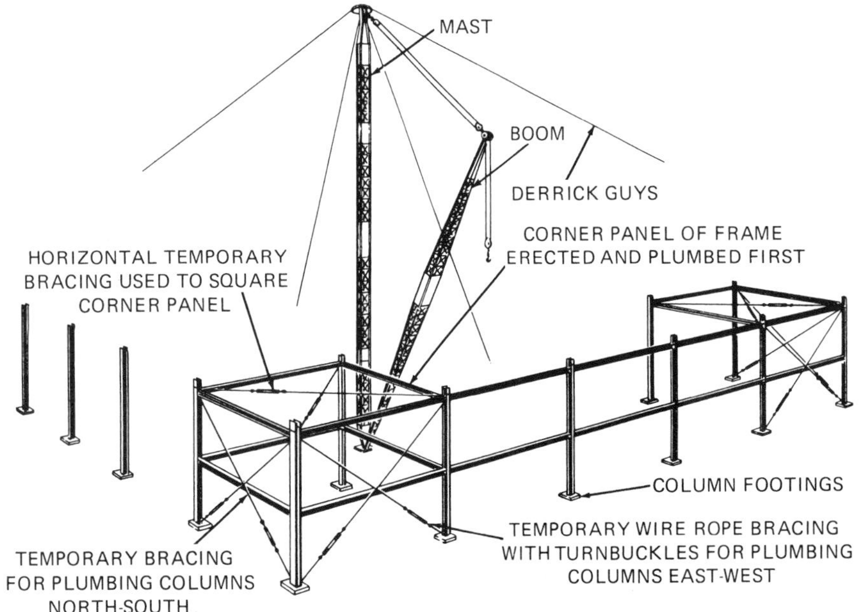

Figure 291 Derrick

deformation A change from original shape due to the load of the structure.

deformed bar Reinforcing bar with ridges. See *bar*. (*Figure 60*, p. 24.)

degree day Unit used in calculating heating and cooling requirements. Unit represents 1°F deviation from some fixed reference point, usually 65°F, in the mean daily outdoor temperature. See *winter degree day*.

dehumidifier In a total air conditioning system, a device that removes moisture from the air to dehumidify.

dehumidify To reduce the amount of moisture in the air.

delamination Separation of plies or a laminate from the base.

delayed-action fuse In electrical wiring, a special fuse that will allow a momentary overload without blowing but will blow with a continuous overload.

demising wall A *party wall*.

demountable partitions Partitions that are designed to be disassembled and reassembled as needed to form new room layouts.

density zoning Ordinances restricting number of living units in an area, generally used in relation to subdivisions.

depth of fusion In welding, the distance of weld penetration into the base metal.

derrick Hoisting machine with a long arm or boom that controls the cable; boom pivots from the base and may be raised or lowered or swung horizontally. (Figure 291.) See also *crane*.

derrick crane See *crane*.

desiccate To make dry by removing moisture, as lumber is dried in a kiln.

design drawings In steel construction, engineering drawings that give the information necessary to fabricate steel members; size and location of members are specified. See *engineering drawings, erection drawings.*

design load The load which structural members are designed to bear or hold. Total design load for a building is calculated by adding the dead loads of the structure and the live loads of occupancy. Figure 292 shows the principle of total design load. (Dead loads are assumed.)

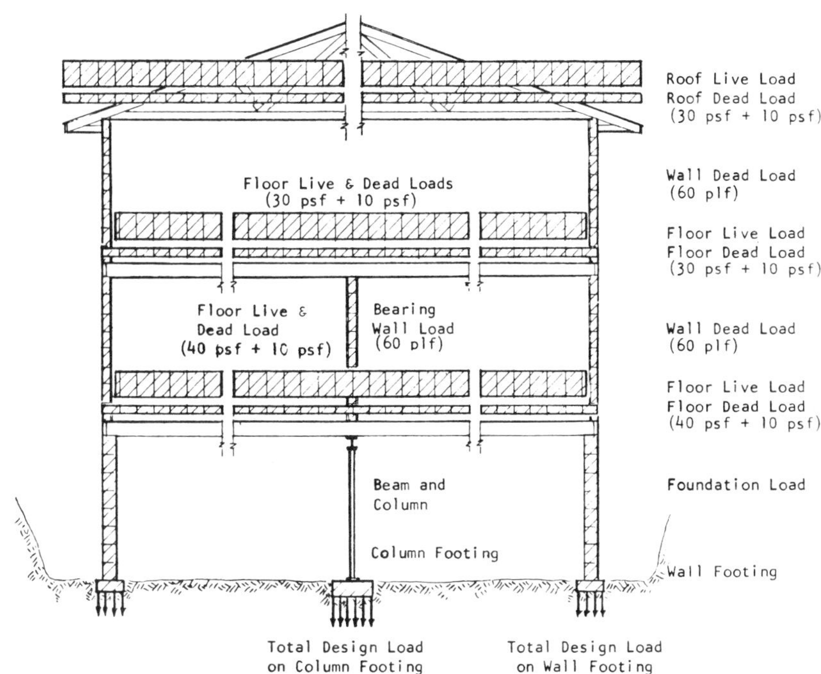

Figure 292 Design load. Total live and dead loads act on wall and column footings. Earthquake and wind loads are specified by local codes

detail In drafting, a large-scale drawing of a small part of a structure. Details are drawn in *isometric* or *orthographic* views. Sometimes an *interior elevation* is called a detail. See also *door section, section, window section.*

developer Someone who prepares land for construction; one who builds.

development Tract of land subdivided for housing; large-scale housing project; a whole community laid out and built from a master plan. Also, sheet metal drawing showing layout of work.

dewatering Removal of water from a construction site, such as by pumping.

diagonal bracing Bracing to resist laterial loading, such as wind and seismic shock. See *cut-in brace, let-in brace, tension straps, wall braces, wind brace.*

diagonal cutting pliers Pliers with side cutters used for cutting wires or other thin metal pieces. (Figure 293.) See also *pliers.*

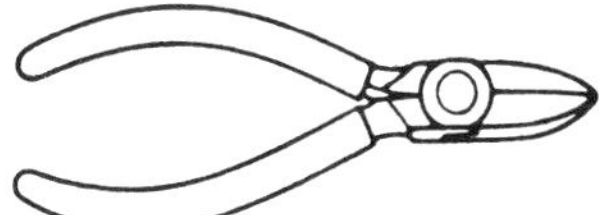

Figure 293 Diagonal cutting pliers

diagonal grain Lumber that is sawed at an angle to the grain of the wood.

diameter Distance across a circle running through the center. One-half the diameter is the radius. The symbol is ϕ.

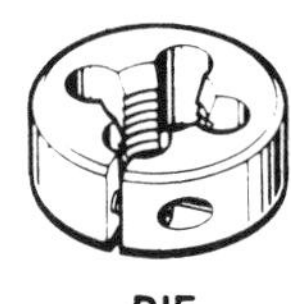

Figure 294 Die held in die stock for turning external threads

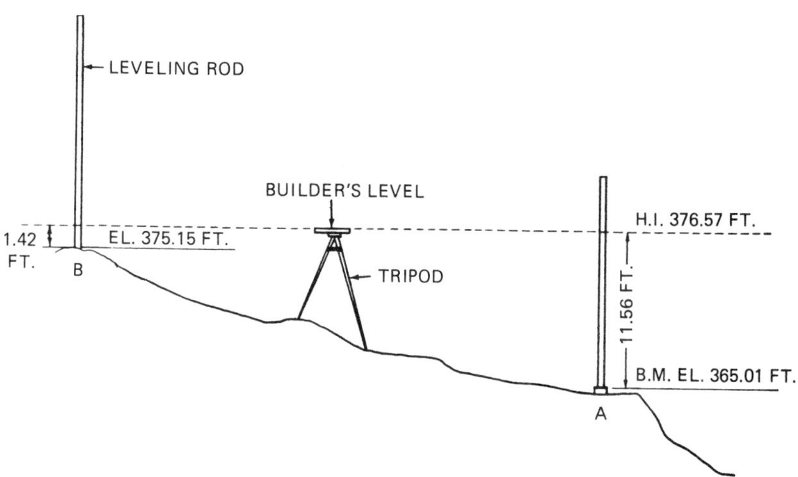

Figure 295 Differential leveling

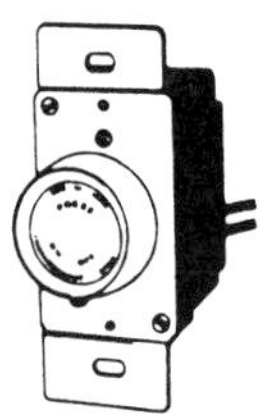

Figure 296 Dimmer switch

diamond-point chisel Wood chisel with a V-shaped cutting edge, used to cut V grooves. See *chisel.*

diazo process Method of reproducing a drawing in a special machine. The working drawing and a sensitized sheet of paper are exposed to light in the diazo machine; then the sensitized paper is developed to form a blueprint (black or blue lines printed on the white sheet). See *drawing reproduction.*

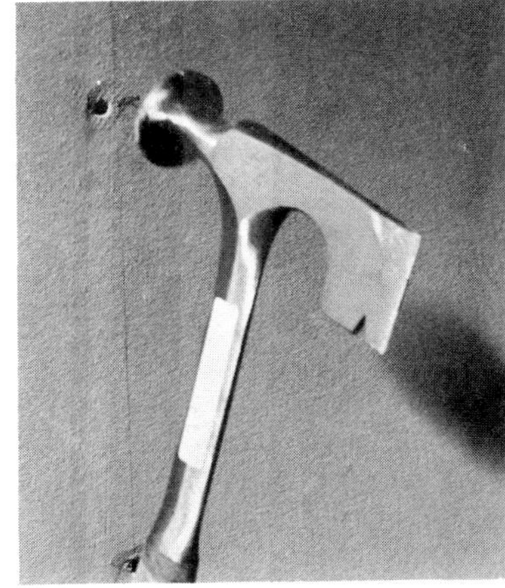

Figure 297 Dimple made with drywall hammer

die A special tool for making external threads on metal conduits or plumbing pipes. (Figure 294.) The die is held in a die stock. See also *tap.*

dielectric fitting An insulating fitting for connecting pipes of dissimilar metals.

differential leveling Finding the difference in elevation of two separate points. (Figure 295.) See *backsight, foresight, height of instrument, level, leveling, turning point.*

diffuser Outlet for air from a forced-air heating system, usually located on the ceiling. The outlet is designed with air vanes which break up the air flow into various directions.

diffusion A spreading out or dispersing, which may lead to a mixing together of different substances, as cold air mixing with warm air.

dimension A distance or measurement of a building part or area; a size marked on a blueprint.

dimension lumber Lumber sized from 2 × 4 up to 4 × 6 inches:

Nominal size	*Actual size*
2 × 4	$1\frac{1}{2} \times 3\frac{1}{2}$
2 × 6	$1\frac{1}{2} \times 5\frac{1}{2}$
2 × 8	$1\frac{1}{2} \times 7\frac{1}{4}$
2 × 10	$1\frac{1}{2} \times 9\frac{1}{4}$
2 × 12	$1\frac{1}{2} \times 11\frac{1}{4}$
4 × 4	$3\frac{1}{2} \times 3\frac{1}{2}$
4 × 6	$3\frac{1}{2} \times 5\frac{1}{2}$

dimmer switch A rotary dimming switch used on lighting circuits. As the switch is revolved clockwise, the lighting brightens. (Figure 296.)

dimple In drywall installation, an impression made in the board by the crowned head of the drywall hammer sinking the nail head. (Figure 297.)

dipping Method of preserving wood by dipping into a wood preservative and water-repellent solution. No pressure treating is involved.

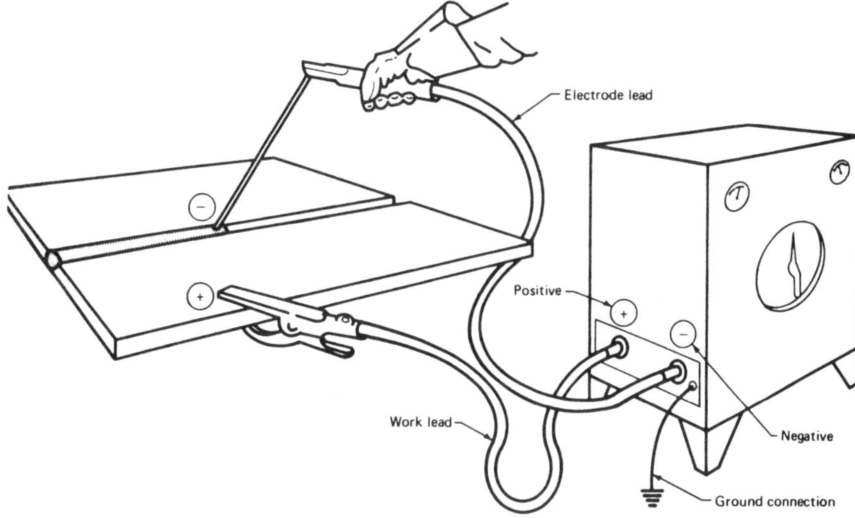

Figure 298 Direct-current electrode negative (straight polarity)

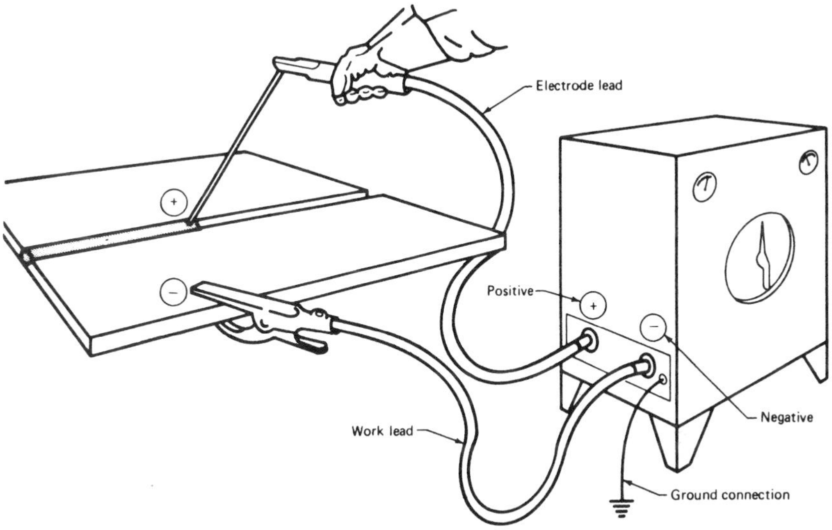

Figure 299 Direct-current electrode positive (reverse polarity)

direct current (DC) Electrical current that supplies a constant flow of electricity in the circuit. See also *alternating current.*

direct-current electrode negative (DCEN) In arc welding, the arrangement of leads from the power source so that the work is the positive pole and the electrode is the negative pole. (Figure 298.) Also called *straight polarity.* Opposed to *direct-current electrode positive.*

direct-current electrode positive (DCEP) In arc welding, the arrangement of leads from the power source so that the work is the negative pole and the electrode is the positive pole. (Figure 299.) Also called *reverse polarity.* Opposed to *direct-current electrode negative.*

disappearing stair See *folding stair.*

Figure 300 Disc sander: portable

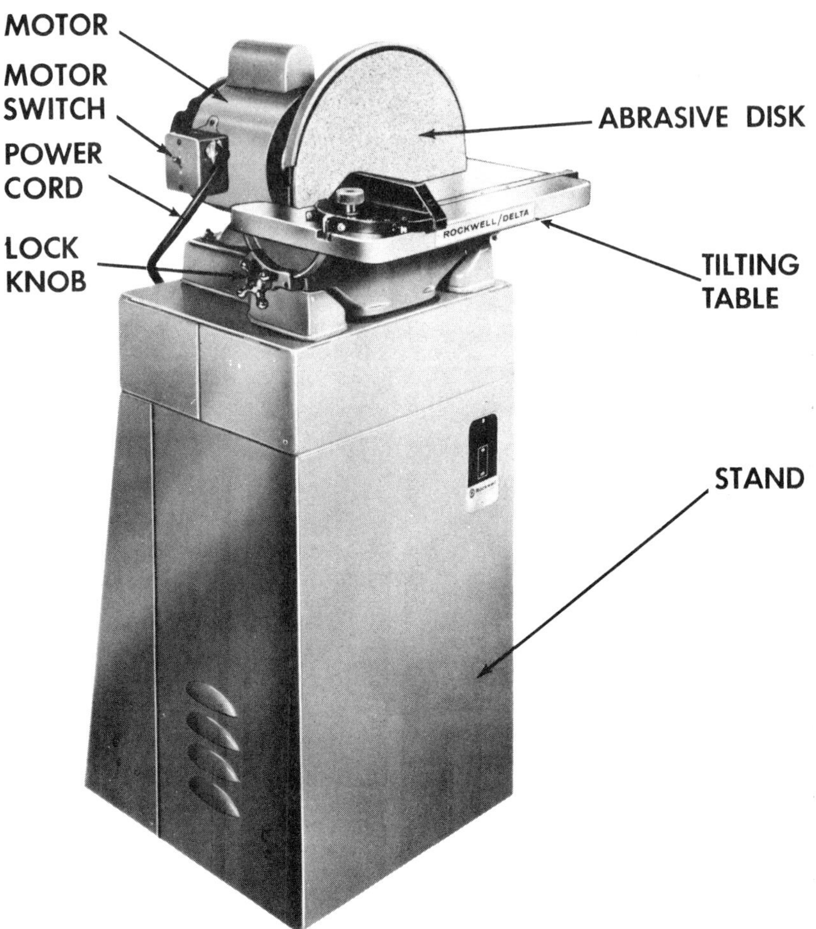

Figure 301 Disc sander: stationary

disc sander Power sander with a circular sanding disc. The portable disc sander is used for finishing around the edges of floors, etc. (Figure 300.) The stationary disc sander is commonly used for metal finishing and for taking down rough wood pieces. (Figure 301.)

display drawings Presentation drawings, often pictorial, used to give the general arrangement and appearance of a structure, showing what it will look like when built.

disposal field Private waste disposal system. See *absorption field, distribution field.*

distributed load Building load spread evenly over the support member or surface, calculated in pounds or tons per square foot.

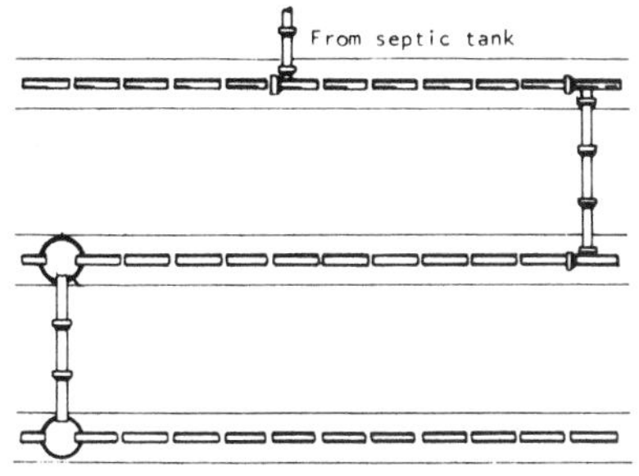

Figure 302 Distribution field

distribution box In plumbing, a concrete collection box that holds the waste before it flows into a private waste disposal system or *absorption field.* In electricity, an electrical box from which connections are made to branch circuits; a *distribution panel.*

distribution field A private waste system that has a pattern of tile laid out so waste can flow out and seep into the earth, where it is purified. Also called *absorption field.*

distribution panel The *panelboard.* In some electrical systems separate distribution panels run from the main panelboard, and each may distribute to several branch circuits.

distribution tile Tile laid out in an absorption or distribution field so wastes can leak through the tile and percolate into the ground for purification. (Figure 302.) See *absorption field.*

diversion Surface drain or depression used to direct flow of water runoff at a construction site; water is directed to a stable area or to a holding area, such as a *sediment pool* or trap. (*Figure 855*, p. 276.)

dividers Scribing or measuring tool with two movable legs with metal points. In construction, the *carpenter's dividers* are used for scribing surfaces. (Figure 303.) In drafting, the *drafter's dividers* are used to step off or transfer distances.

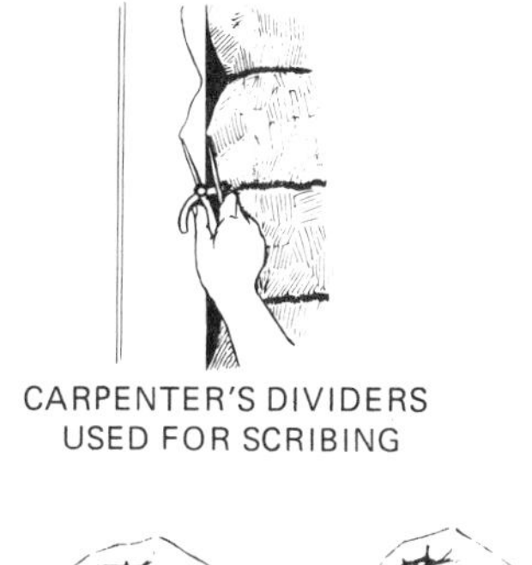

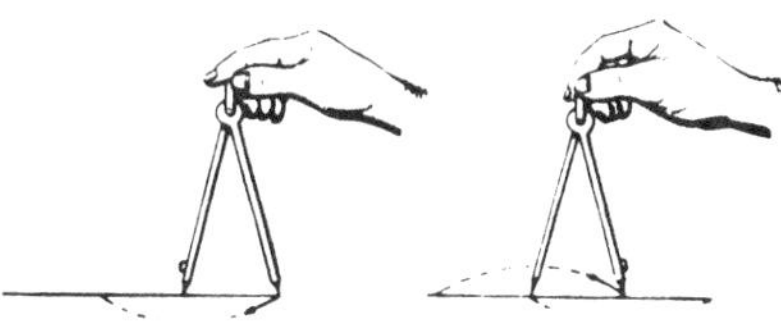

Figure 303 Dividers

dobie *Concrete block bar support.*

dog A *dog anchor.*

dog anchor A heavy steel bar with each pointed end bent at a right angle like a large staple, used to make a temporary connection between two timber pieces; one end of the dog is driven into each timber piece. Also called a *dog* or *dog iron.* See *pinch dog.*

dog leg Anything that changes direction sharply (to create an angle similar to the angle of a dog's leg).

dog's tooth Wall that has brick corners projecting out.

Dolly Varden siding A tapered siding similar to regular bevel siding except that the thick butt end has a shoulder cut to receive another siding edge. See *siding.*

Figure 304 Dome house (reinforced concrete)

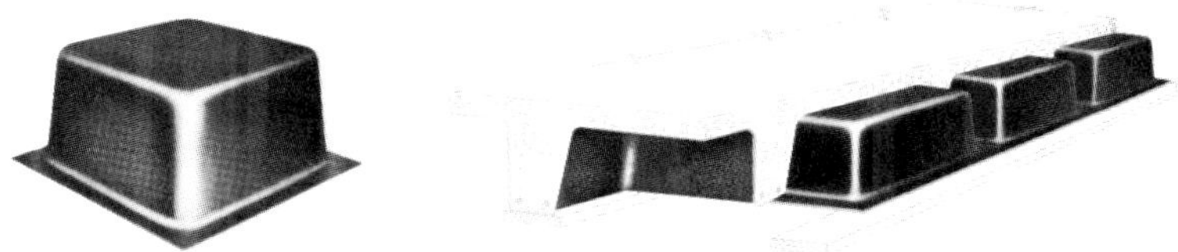

Figure 305 Dome forms used for two-way reinforced concrete joist construction

dome A semispherical roofing: reinforced concrete, wood, glued-laminated wood, or steel is used. Radial ribs are framed from the center to the perimeter. A geodesic, triangular-shaped interlocking of building members is also used. Concrete is placed monolithically. Figure 304 shows a reinforced concrete dome house. See *geodesic dome, lamella.*

dome forms Molded forms used for shaping concrete floor systems. Figure 305 shows dome forms used for two-way concrete joist construction. Also called *pan forms.* See also *flange forms, pan.*

domicile Place of permanent habitation.

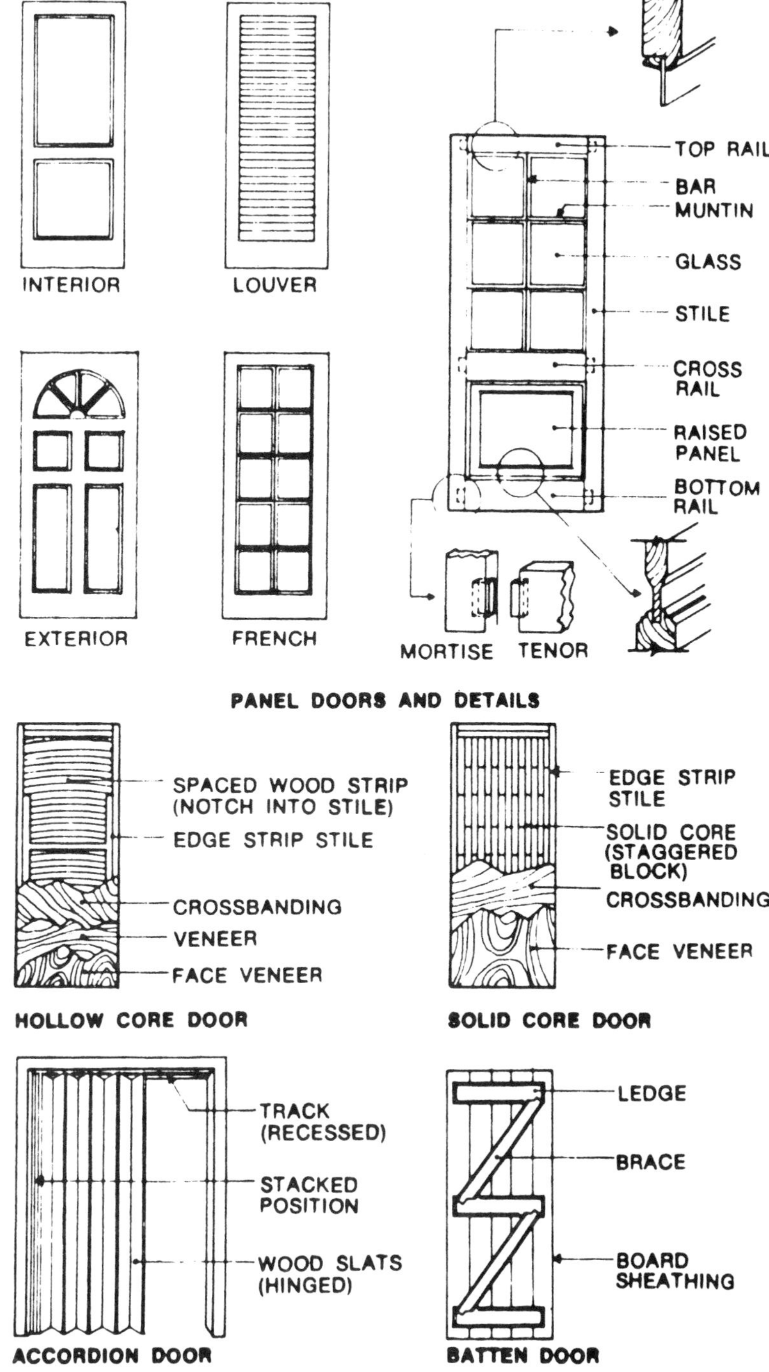

Figure 306 Doors

door The assembly that fits in a door frame. Two types of swinging doors are in common use: flush and panel. Figure 306 shows common door styles. See *door frame.*

doorbell A device inside the house that rings when the doorbell button (located outside by the front or back doors) is pushed. See door *chimes.*

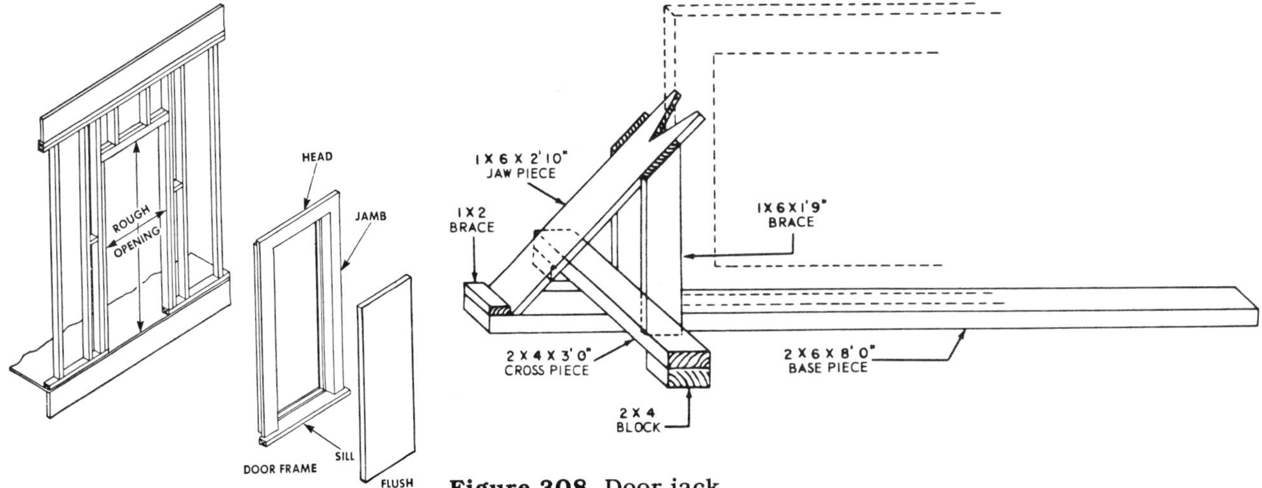

Figure 308 Door jack

Figure 307 Door frame wedged into door opening

door bevel Slight angle made on a door edge to better fit against the jamb.

door buck Metal door frame set into the wall. Bucks are normally used in masonry walls.

door casing Finish trim around the door opening.

door check Mechanism that retards the closing of a door. Also called a *door closer* or *automatic door closer.*

door chimes A device inside the home that chimes when the doorbell button (located outside by the front or back door) is pushed. See *doorbell.*

door details Section views of parts of the door: head, jamb, and sill. See *door sections.*

door frame Supporting framework that holds the door, consisting of the sill (bottom), jambs (sides), and head (top). Figure 307 shows how the door frame relates to the door opening.

door head Top part of the door frame. (Figures 307, 309.)

door holder Device for holding a door open.

door jack Special wood frame for holding a door while it is being worked on. (Figure 308.)

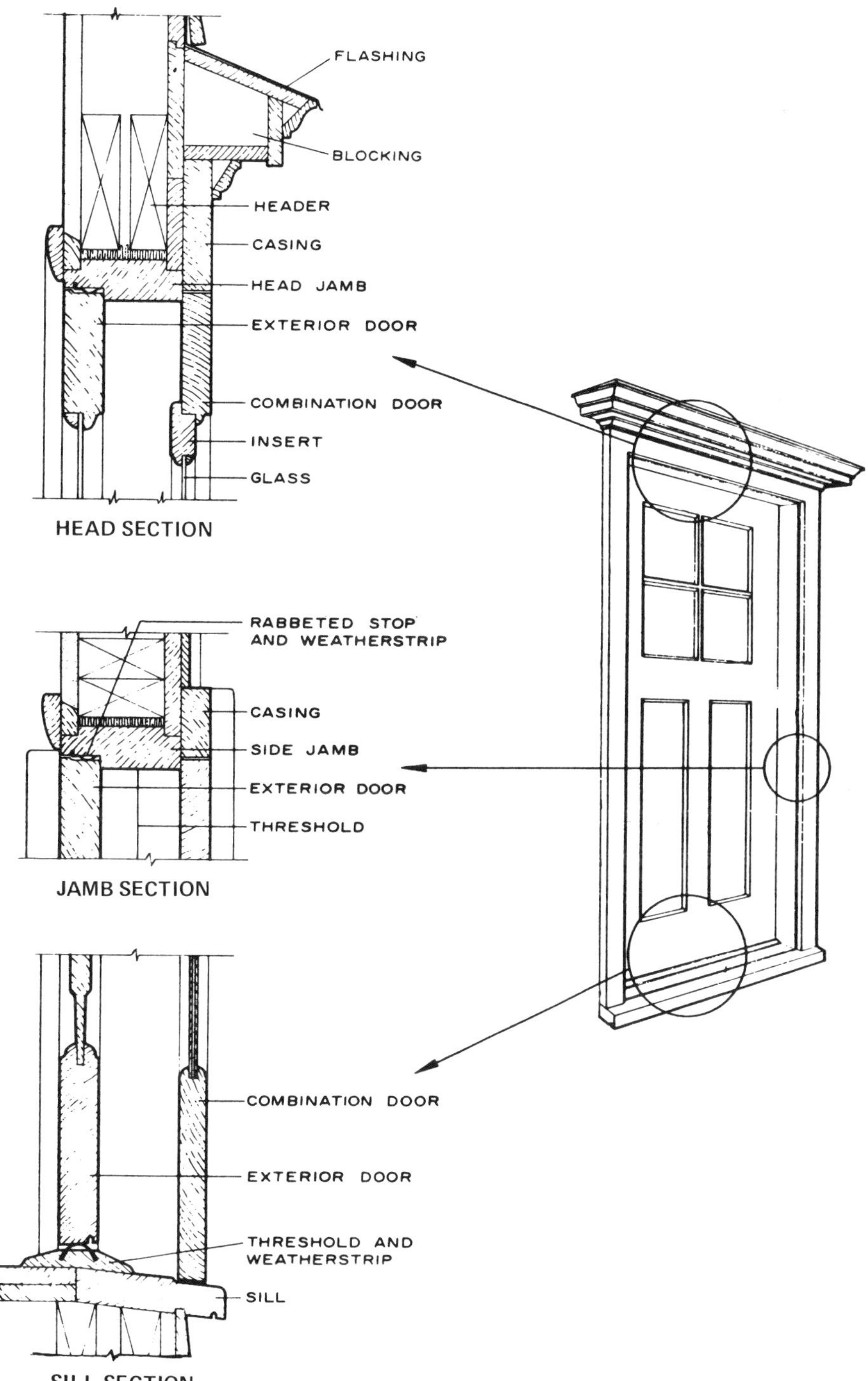

Figure 309 Door sections

door jamb A general term relating to the complete door frame, including the two side jambs and the head jamb. Specifically, the two side jambs. (Figures 307, 309.)

door lock bit Bit for boring holes for cylinder-type locks. Also used for boring pipe and conduit holes. See *lockset bit.*

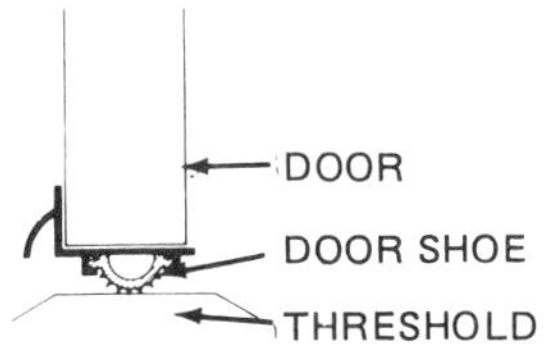

Figure 310 Door shoe

Figure 311 Door stop

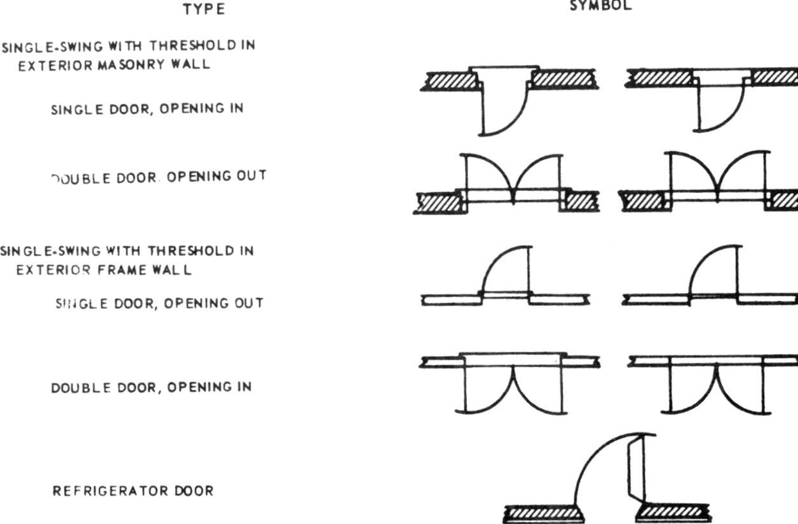

Figure 312 Door symbols (swing doors)

door knob Handle for turning a door latch so door may be opened.

door lock Device used to secure a door in the frame when closed. A key operates the lock to allow the door to open. A latch is opened by turning the door knob. See *dead bolt, lock.*

door post General name for the sides (jambs) of a door frame.

door pull Handle for opening and closing a door. Unlike a *door knob,* it is fixed.

door schedule Table which identifies all the doors by symbol, giving number, size, type, manufacturer, hardware, and glazing (if any).

door section Section made to show the internal parts of a door: the head, jamb, and sill. (Figure 309.) A door section is sometimes referred to as a *door detail.*

door shoe Vinyl seal affixed on the bottom of a door. (Figure 310.)

door sill Bottom part or threshold of a door. (Figures 307, 309.)

door step One or more steps leading up to an outside door.

door stile Vertical support pieces in a panel door.

door stop Device that a swinging door opens against. (Figure 311.) Also, the part of the door frame that the edge of a swinging door closes against. (Figure 307.)

door symbols Symbols used on floor plans to locate and identify the door type. A curving line shows the swing of the swinging door edge. (Figure 312.) See *plan symbols, symbols.*

door trim Interior casing around door frame.

Doric order Greek order. See *orders.*

dormer Framed opening in the side of sloped roof. Figures 313 and 314 show the basic dormers and typical framing.

dormer rafter The main rafters on a dormer roof; rafters are full length from plate to ridge board or plate to roof base. See *dormer.*

dormer window Window in a small dormer constructed to bring light into the space under the roof.

dote Wood decay.

dots Plaster projections made on the wall to show how thick the final plaster coat should be; plaster is built up to the face of the dot. See *grounds.*

double-acting hinge Hinge that allows a swinging door to swing in either direction. See *hinge.*

double floor Floor that also functions as the ceiling of the room below.

double framing Doubling of framing members, such as joists, rafters, or studs, around openings to give added strength. (*Figures 489*, p. 204; *490*, p. 204; *531*, p. 228; *962*, p. 425; *966*, p. 427.) Also, doubling of support under bearing partitions or under heavy fixtures, such as a bathtub. (Figure 315.)

double glazing Window sash with two panes of glass with an air space sealed in, used for the insulative value. (Figure 316.) See *triple glazing.*

double header Double framing of structural members at a right angle to joists or studs, used at openings over doors and windows, or at stairwell openings. (*Figures 489*, p. 204; *490*, p. 204; *966*, p. 427.)

double hem edge In sheet metal work, an edge that is folded back twice. See *edge.*

double-hung window Window with upper and lower sashes that slide up and down. (Figure 317.)

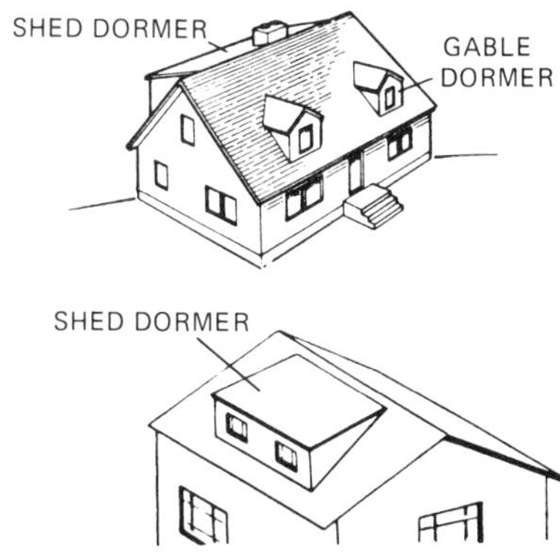

Figure 313 Dormers

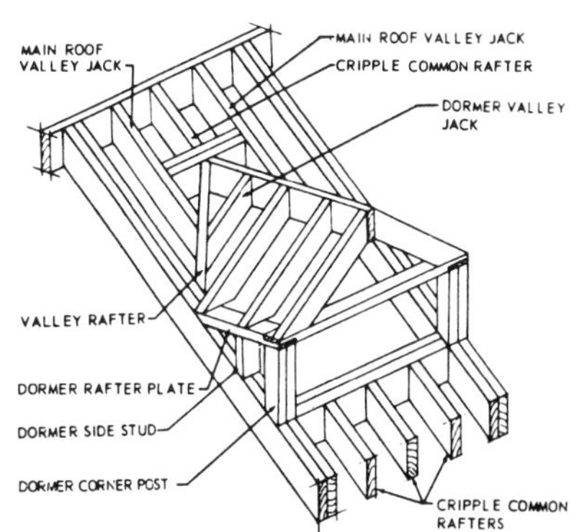

Figure 314 Dormer framing (gable dormer)

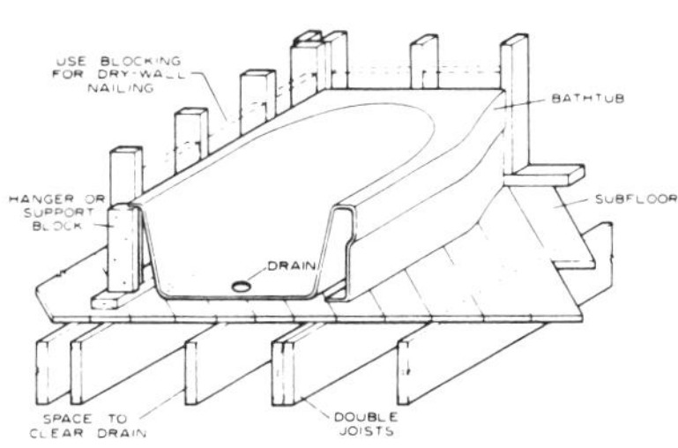

Figure 315 Double joist framing to support bathtub weight

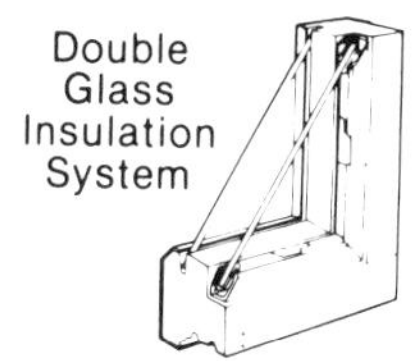

Figure 316 Double glazing

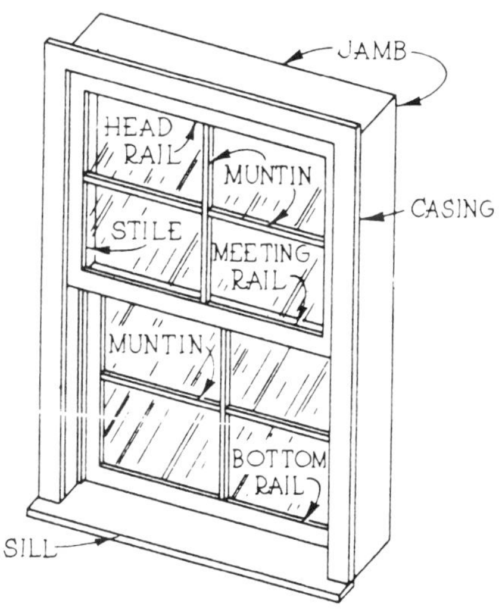

Figure 317 Double-hung window

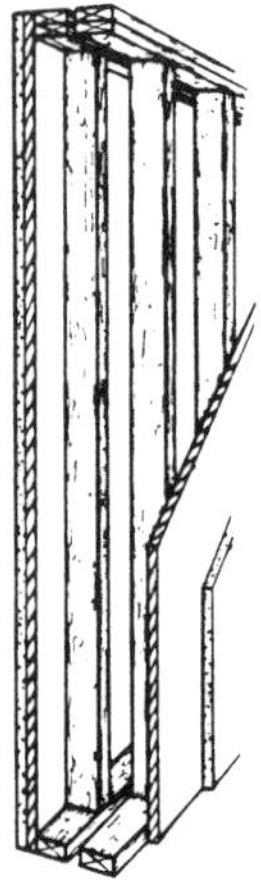

Figure 318 Double-stud wall

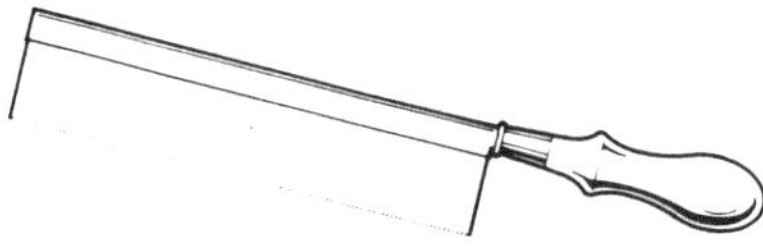

Figure 319 Dovetail saw

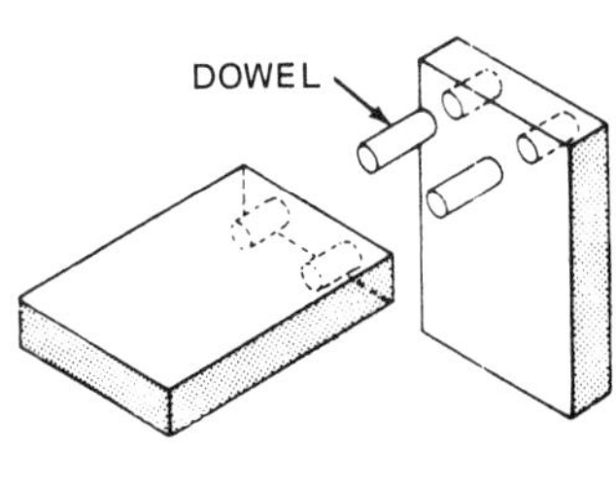

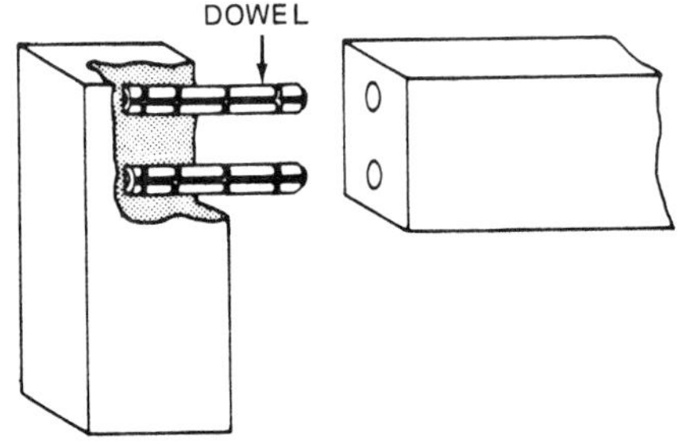

Figure 320 Dowels used to join wood pieces

double-insulated Said of electrical tools that have special insulation against shorting; an internal ground in the electrical tool is designed to prevent electrical fault.

double sighting In surveying, to take a double sight first with the telescope in normal position, then with the telescope plunged and revolved 180° around the vertical axis.

double studding Two studs nailed together, as on the side of a door or window.

double-stud wall Wall having two separate rows of studding, used for sound deadening and to create space for soil pipe. Often used between separate living units. (Figure 318.) See also *staggered-stud partition.*

double window Window frame with two separate sets of sashes.

dovetail Joint where two boards are joined at right angles with interlocking dovetail-shaped or fan-shaped wood pieces. See *wood joints.*

dovetailing Fastening boards together with *dovetails.*

dovetail saw Straight-handled, fine-toothed back saw. (Figure 319.) Used for making fine cuts as for *dovetails.* See also *backsaw.*

dovetail seam In sheet metal work, a method of joining pipes or circular collars to a rectangular surface.

dowel Wood pin used to strengthen wood joints; pin fits into holes in the wood pieces to be joined and is glued in. (Figure 320.) Used widely in furniture construction. Also, a reinforcing bar that connects two separate concrete sections, as a footing doweled into a column. Slab rod ends are bent up into the footing to form dowels.

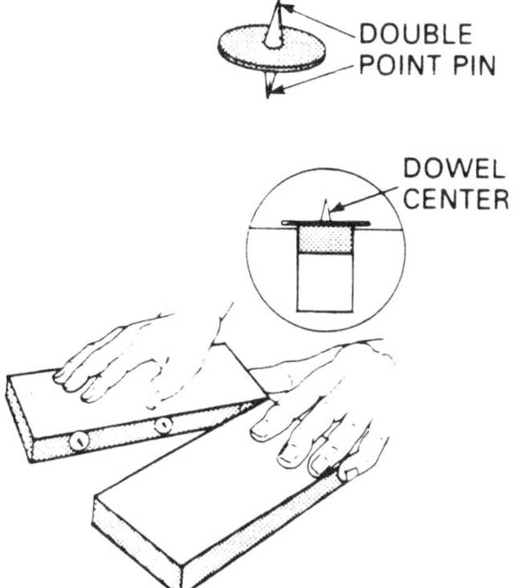

Figure 321 Dowel centers

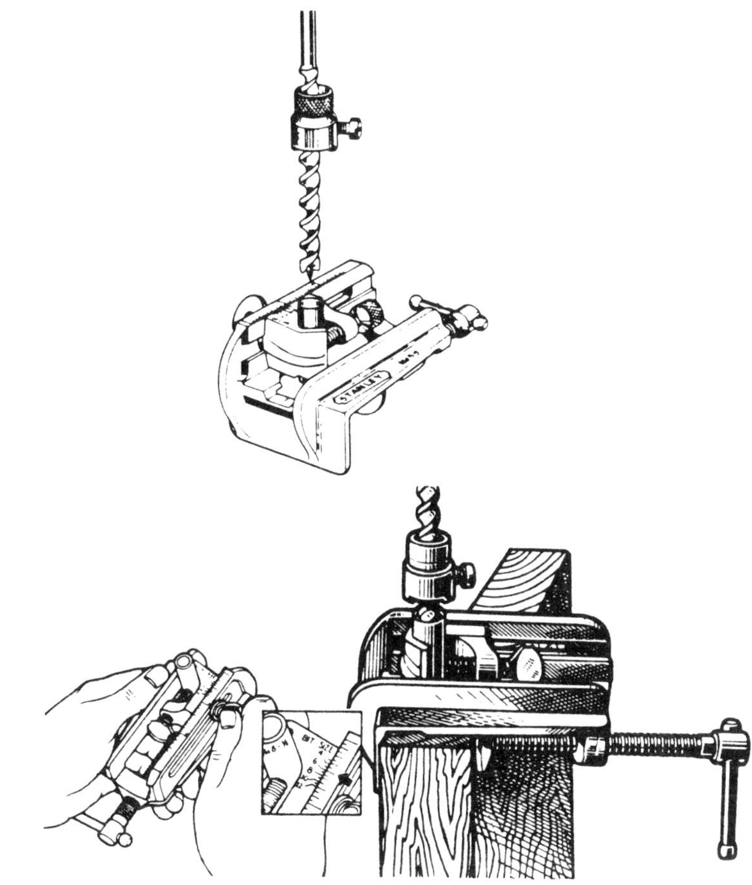

Figure 322 Doweling jig

dowel bit Special bit for drilling dowel holes in end grain. Point is designed to hold the bit true without moving off course to follow the wood grain.

dowel center Metal piece with a centered pin. (Figure 321.) Dowel center fits into one dowel hole and, when the piece is fitted in its final location, the pin marks the center of the opposite dowel hole. This allows the matching hole to be accurately located.

doweling Fastening wood pieces together using *dowels.* See *wood joints.*

doweling jig Guide for boring dowel holes in end, edge, or surface of work piece. (Figure 322.) Controls exact location and depth of bore.

dowel template In reinforced concrete, a pattern that has holes for locating dowels (reinforcing bars) the proper distance apart. Dowel ends run through the holes into the fresh concrete. Allows the dowels to be accurately located.

down draft Downward flow of air, as in a chimney.

downflow furnace Furnace where air flows downward and is discharged at the bottom. See *upflow furnace.*

downspout Pipe for carrying water down from the roof gutters. (Figure 323.) Also called a *leader* or *conductor.* See also *splash block.*

doze Wood decay.

draft An air flow; air needed for a furnace or hot-water heater flame. In precast concrete, splay or slope provided on surfaces of the precast concrete unit to allow ease of removal from a mold or form. Also called *draw* or *batter.*

drafter One who draws working drawings under the direction of the architect. Formerly called a draftsman.

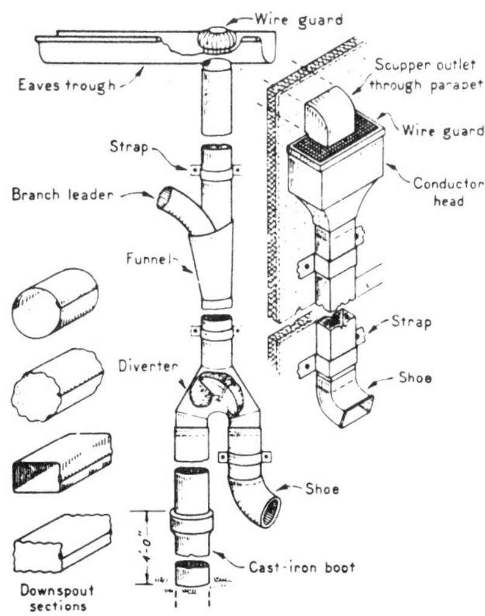

Figure 323 Downspout

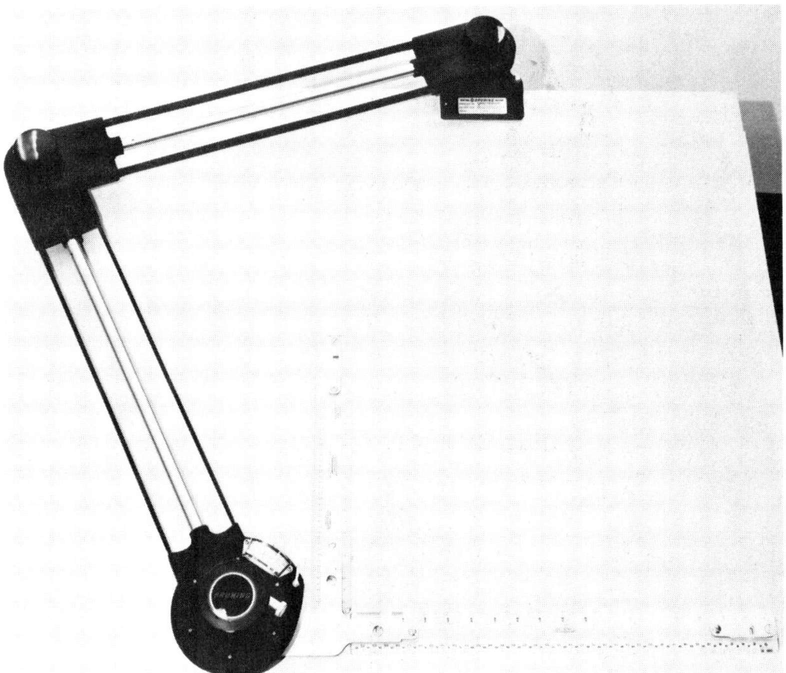

ARM DRAFTING MACHINE

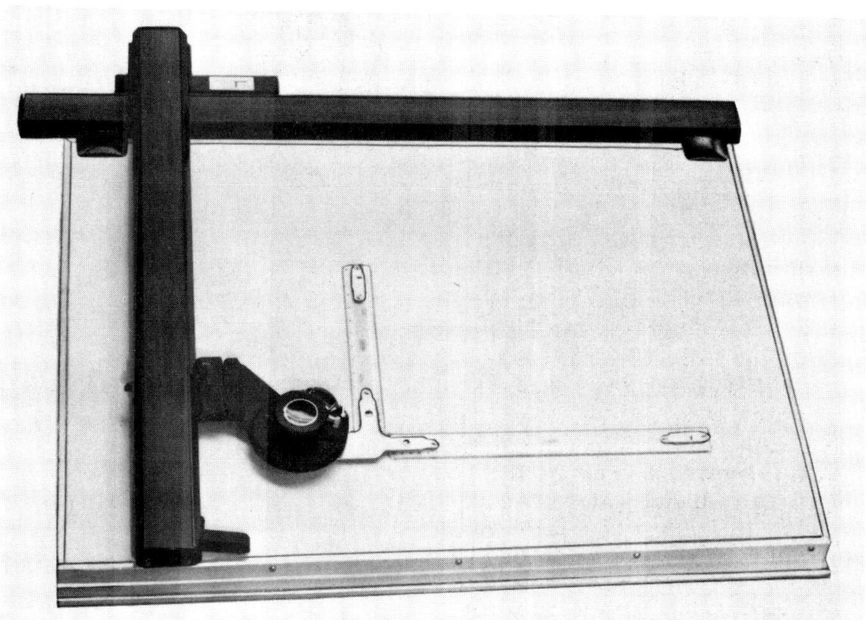

TRACK DRAFTING MACHINE

Figure 324 Drafting machines

drafting Mechanical drawing; use of drafting instruments to make finished drawings. *Alphabet of lines* shows the standard lines used in drafting.

drafting machine Mechanical device used for drafting, replaces T square, triangles, and protractor. (Figure 324.)

Figure 325 Drain augers: hand and power

drafting sheets Three kinds of sheets are used by the drafter; paper (including vellum), cloth, and film. Sheets come in standard inch and millimeter sizes. Different-sized sheets are available. See *drawing sizes.*

draftsman A *drafter.*

draft stop See *firestops.*

drag Masonry tool with steel teeth used for dressing a steel surface; a *comb.*

drain Pipe for taking a fluid discharge.

drainage The flow of water as on a building site; system for carrying off waste water through piping.

drainage easement Right to use property to direct flow of water, normally used for highway development. See *planting easement.*

drain auger A flexible coil-spring auger or snake that is run into a plumbing drain to clean out obstructions. Both hand and power drain augers are used. (Figure 325.) The flexible cable is stored on a revolving drum. See also *sewer rod.*

drain tile Clay tile or plastic tubing used to drain water away from a structure. See *absorption field, clay tile, perimeter drain tile.*

drain trap Water trap in the drainage system to prevent sewer gas from escaping into the living area. See *P trap, S trap.*

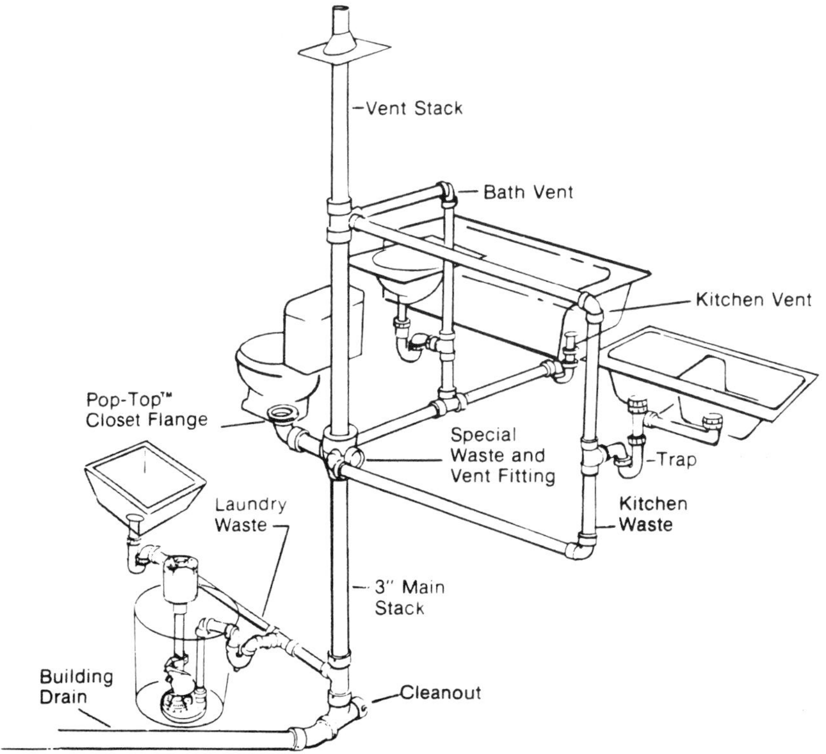

Figure 326 Drain-waste-vent system

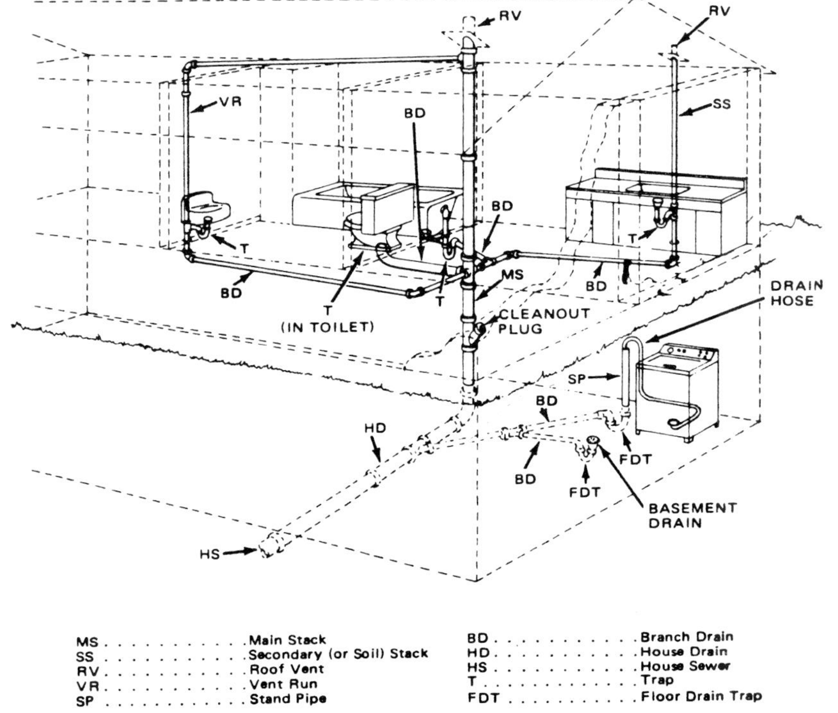

Figure 327 Drain-waste-vent system

drain-waste-vent system (DWV) In plumbing, the system that includes the drain and waste piping and the venting through the roof to maintain air circulation. (Figures 326, 327.) See also *plumbing system.*

draw *Draft.*

drawer pull Handle for pulling open a drawer.

drawer slip A guide on which a drawer slides when it is opened.

drawfile To file by moving a file sideways, as opposed to lengthwise, across the workpiece. (Figure 328.)

drawing board Wood board to which drawing paper is attached.

drawing reproduction Blueprints or copies of working drawing are made on a reproduction machine using the diazo process. The original working drawing and a sensitized sheet of paper are run through the machine. (*Figure 103*, p. 40.) An exact copy with blue or black lines (a "blueprint") is made.

drawing sizes Standard paper sizes used by the drafter when drawing. Both inch and metric sizes are available. Inch sizes are identified by letter. Figure 329 shows standard inch sizes for individual sheets.

drawknife Woodworking tool used for smoothing wood, has a blade with handles at both ends. The drawknife is pulled over the wood by the handles. (Figure 330.)

dress To smooth or finish wood or stone.

dressed and matched (D & M) Of planks, having edges that have been machined to create a groove in one edge and a tongue in the other plank edge. They fit together to provide a firm joint. See *tongue and groove.*

dressed lumber Seasoned lumber that is surfaced (dressed) in a planing machine to smooth out the surface and establish a uniform size.

drier Substance in a varnish or paint that causes the oil to oxidize quickly and harden.

Figure 328 Drawfiling

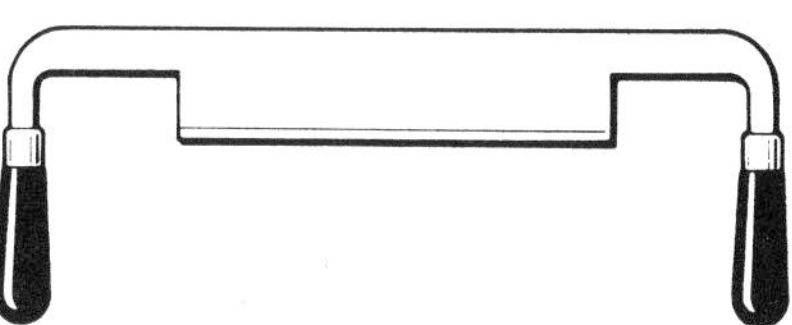

Figure 330 Drawknife

Size Designation	Width (Vertical)	Length (Horizontal)	Margin
			Horizontal
A (Horiz)	8.5	11.0	0.38
A (Vert)	11.0	8.5	0.25
B	11.0	17.0	0.38
C	17.0	22.0	0.75
D	22.0	34.0	0.50
E	34.0	44.0	1.00
F	28.0	40.0	0.50

Figure 329 Drawing-paper sizes

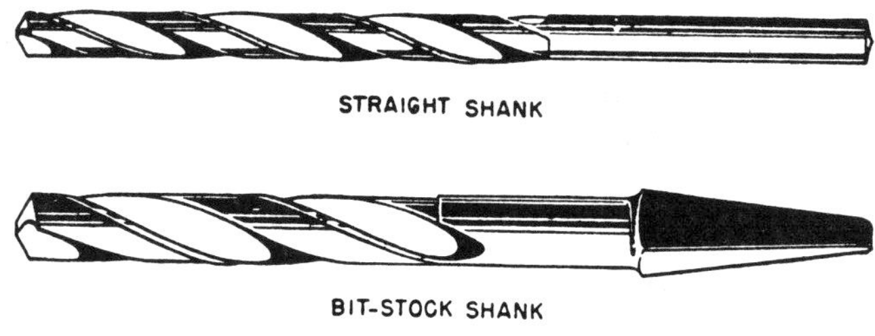

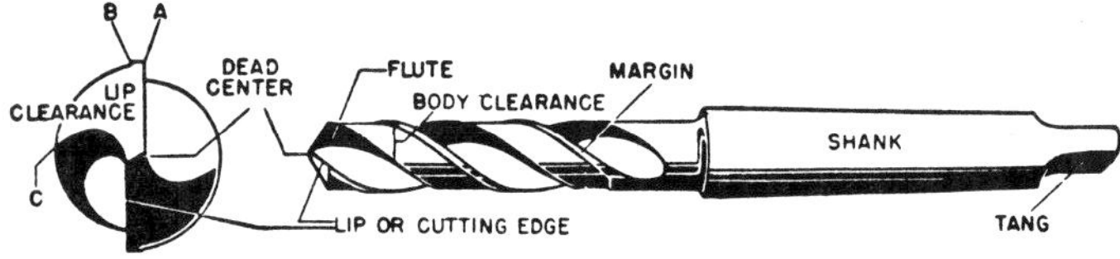

Figure 331 Drills

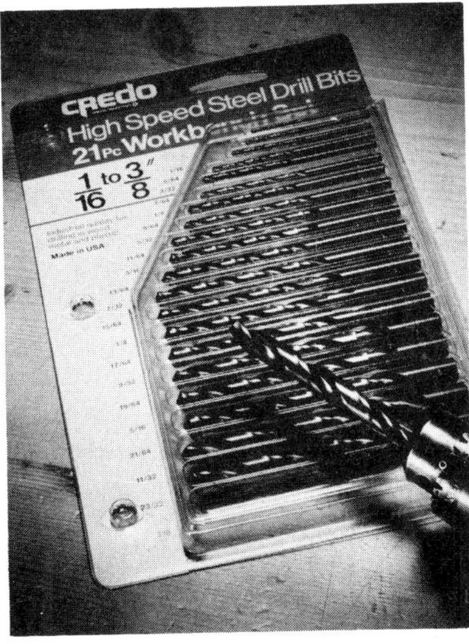

Figure 332 Drill set

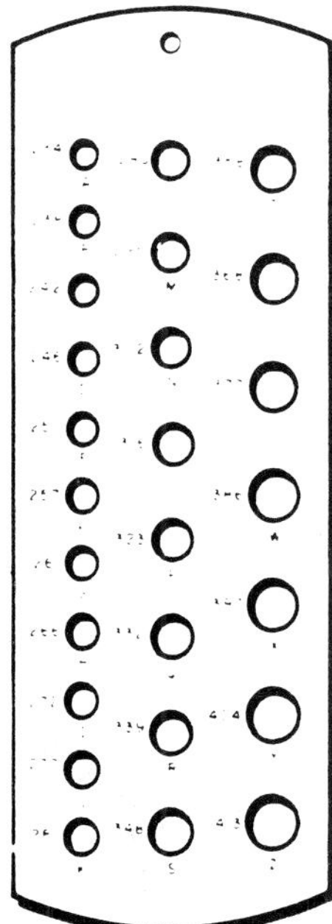

Figure 333 Drill gauge

drift bolt Bolt used in fastening heavy timbers together.

drift down In roofing, a shingle course that is not straight and angles down at one end.

drift pin Steel, double-tapered pin used to align holes between structural steel members so they can be bolted together. See also *bull pin, erection wrench.*

drill Small, straight-shank boring bit used in hand drills and electric drills to drill into wood or metal. (Figure 331.) Also called a *drill bit, drill point,* or *twist drill.* Drills are sized by number from 1 to 80; the larger numbers representing a smaller drill. Large drills, from 0.234″ to 0.413″ diameter, are represented by letters from A to Z. The higher letters are for the larger sizes. See *drill gauge.* Additional sizes are specified by fraction. Figure 332 shows a drill set marked in fractional inches. Size is stamped on the drill shank. See also *bit, drill gauge, drill press, electric drill, hand drill, masonry bit, push drill, right-angle drill, star drill, tap drill sizes, Unified Threads.*

drill and counterbore bit Special bit that simultaneously drills a hole and a counterbore. Also called a *screw bit.*

drill gauge A metal template with holes made to the various drill sizes. Used for checking drill sizes if the drill is not clearly marked. Sizes are given in decimal equivalents, letter codes, or fractions. (Figure 333.)

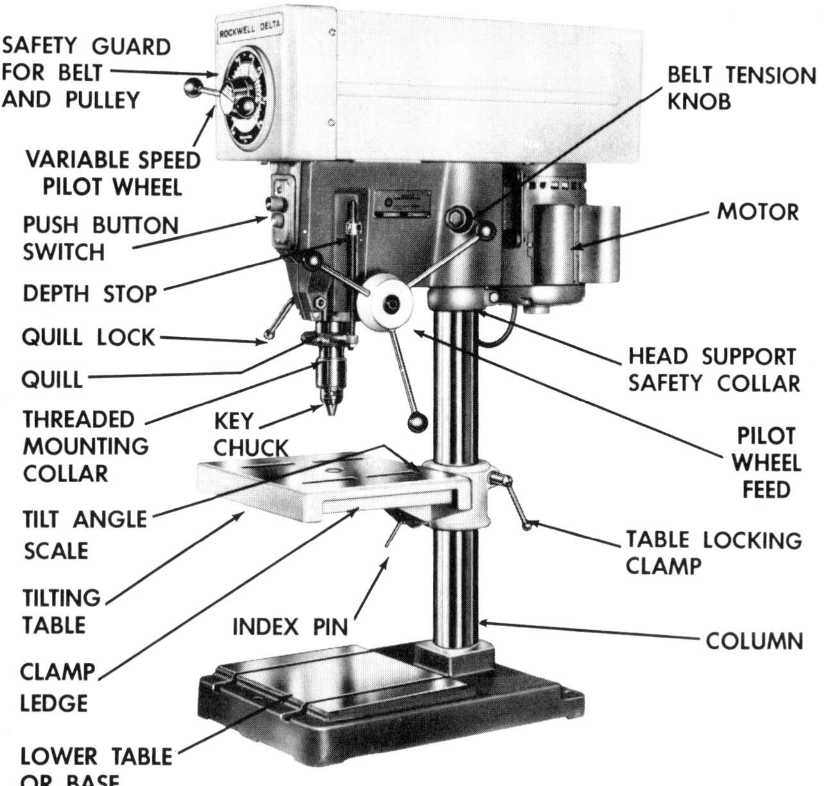

Figure 334 Drill press

Figure 335 Drill saw

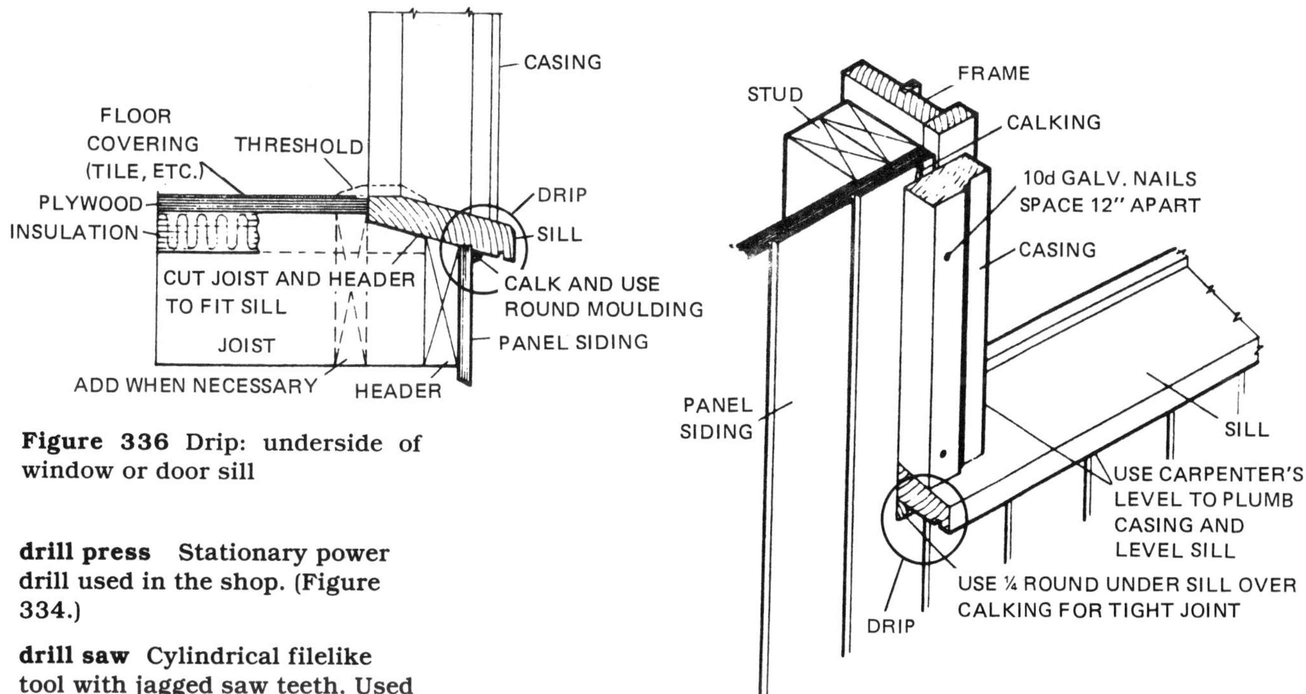

Figure 336 Drip: underside of window or door sill

drill press Stationary power drill used in the shop. (Figure 334.)

drill saw Cylindrical filelike tool with jagged saw teeth. Used for enlarging holes. (Figure 335.) Also called a *round file*.

drip A channel or groove cut on the underside of a sill or drip cap to prevent water from flowing on the bottom side of the member and into the wall. (Figure 336.) See *molding*.

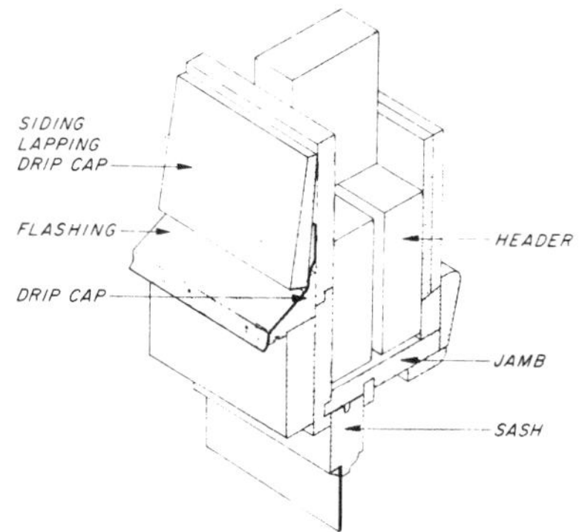

Figure 337 Drip cap

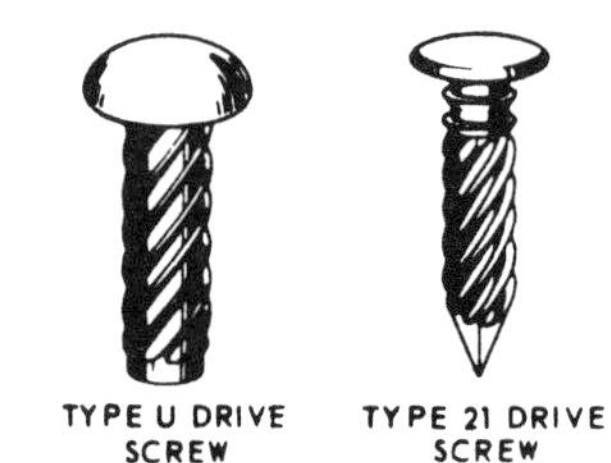

Figure 338 Drive screw nails

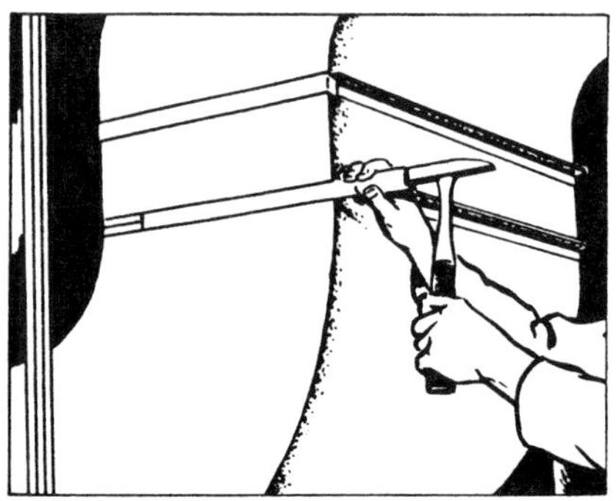

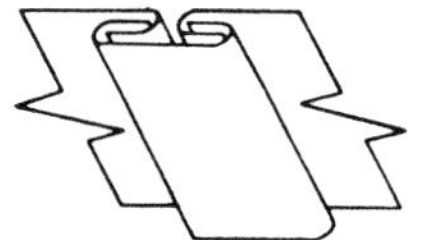

Figure 339 Drive slip connection

drip cap Projecting molding run over the windows and doors, or just below the sheathing, to catch rain water and allow it to drip away from the wall. (Figures 337; *252*, p. 96; *225*, p. 84.) Also called a *drip mold.* See *molding.*

drip edge Metal or vinyl projection along eaves or rakes to allow water to run off or drip away from the underlying construction.

drive screw nail A nail with a swirl cut on the shank for holding power. (Figure 338.)

drive slip Sheet metal connection made by fitting formed piece over bent edges of two sections. (Figure 339.) Also called a *cap strip seam.* See *locked corner seam.*

drop Vertical pipe or duct that allows water, gas, or air to flow downward. See *riser.* Also a decorative pendant used on the bottom part of architectural details; the drop points downward; also called a *drop finial.* The opposite is called a *finial* and points upward from an architecture detail.

drop ceiling Ceiling suspended on wire hangers. A *suspended ceiling.*

drop chute Trough for moving concrete from the mixer or truck to where it is placed in the structure.

drop finial See *finial.*

drop panel In reinforced concrete, a thickness designed in a slab to give additional support for a column. (*Figure 784*, p. 344.) See also *capital, shearhead.*

drop siding A horizontal board siding; top edge to one board fits into the bottom edge of another. *Tongue-and-groove siding.* See *siding.*

drop wire Wires supplying electrical power from the utility pole to the house or structure, connected to the *service drop.*

drum wall In solar heating, a heat-storage system that uses a wall of stacked water-filled oil drums. Heat is absorbed during the day and radiated out during the night.

drying in In roofing, the placement of roofing felt on the roof. Also called *capping in.*

dry lumber Lumber seasoned to a low moisture content, specified as 19% or less moisture.

dry masonry Masonry units laid without mortar.

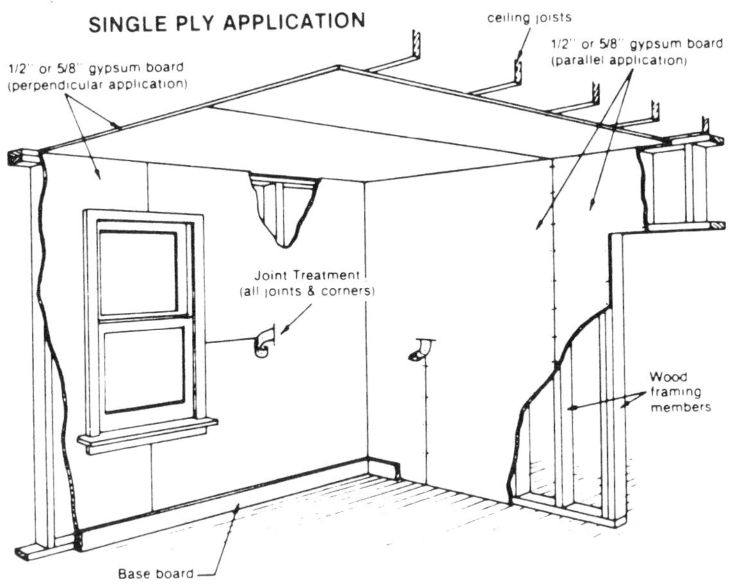

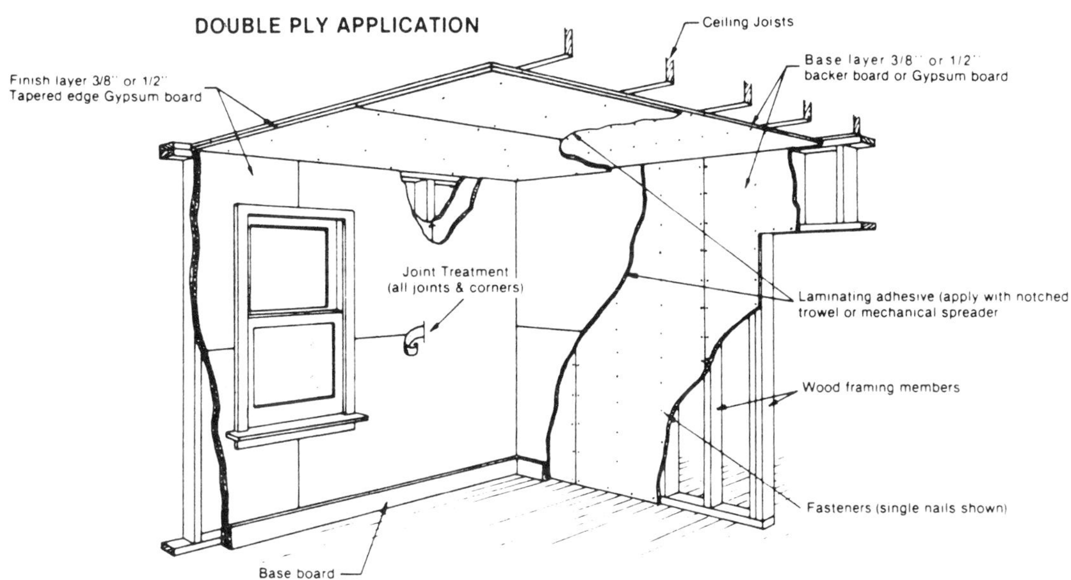

Figure 340 Drywall construction

dryout Condition in wet mortar caused by excessive dry wind; the wind removes moisture before the chemical process of setting can be completed. The mortar turns into powder.

dry-pipe sprinkler system Automatic sprinkler system that uses air pressure in the piping to hold back water. Heat form a fire will melt a locking device on each sprinkler head, releasing the air pressure, which then allows water to flow through the system. See *wet-pipe sprinkler system*.

dry rot General term for wood decay. In the advanced stage the wood becomes crumbly and breaks into a powder.

dry wall A masonry wall laid up without mortar.

drywall General term used to designate *drywall construction*. Also, a specific term for the gypsum wallboard used.

drywall construction Wall covered with gypsum wallboard finished with tape and joint compound. (Figures 340, 341.)

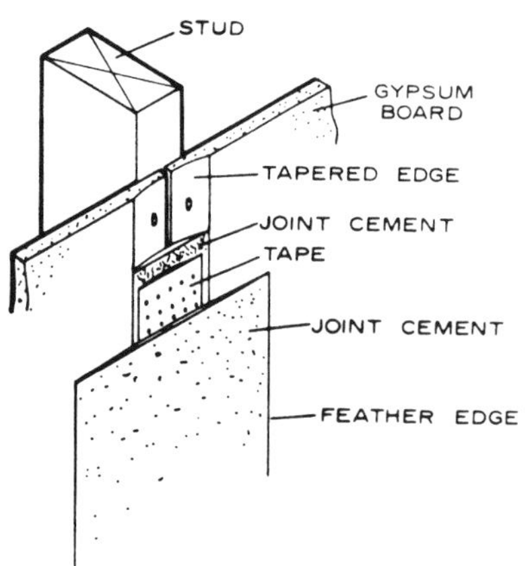

Figure 341 Drywall construction

Figure 343 Drywall knife

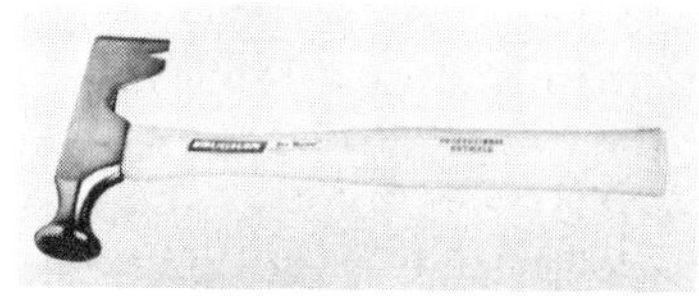

Figure 342 Drywall hammer

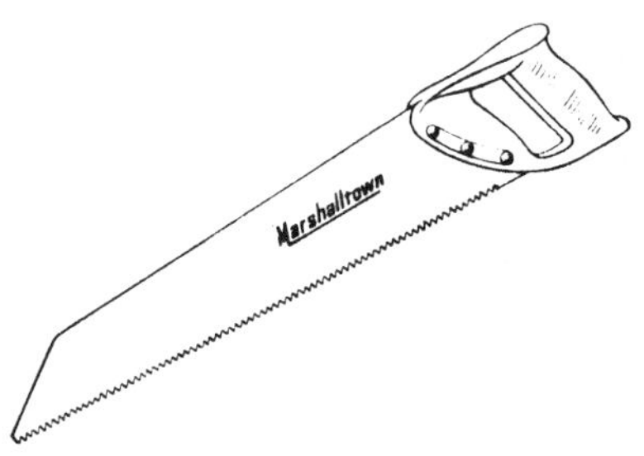

Figure 344 Drywall saw

drywall hammer Hammer designed for nailing gypsum wallboard. Has a convex face for driving nails below the face of the panel, leaving a dimple, which is then filled in with plaster. (Figure 342.)

drywall installer One who installs gypsum board inside a structure on walls, partitions, and ceilings.

drywall joint tool A rectangular trowel with a slight upward curvature used for finishing drywall joints. See *drywall trowel, trowel.*

drywall knife Wide-bladed knife for applying drywall joint compound. (Figure 343.) See *automatic taper, banjo taper, drywall trowel, taping knife, trowel.*

drywall nails Annular-shanked nails used to fasten gypsum panels.

drywall saw A short, blunt-nosed hand saw used for sawing gypsum board. (Figure 344.)

drywall screws Special screws used for installing gypsum wall boards. Figure 345 shows a few of the types available. The head has a bugle shape so it can be easily driven into the gypsum panel slightly below the surface. Some drywall screws are self-tapping and cut their own hole into metal studs. The screw is driven with a power drywall screwdriver.

drywall screwdriver Special power screwdriver used to drive drywall screws. (Figure 346.) The screw head is set slightly below the surface. See also *power screwdriver*.

drywall taper See *banjo drywall taper*.

drywall taping See *taping*.

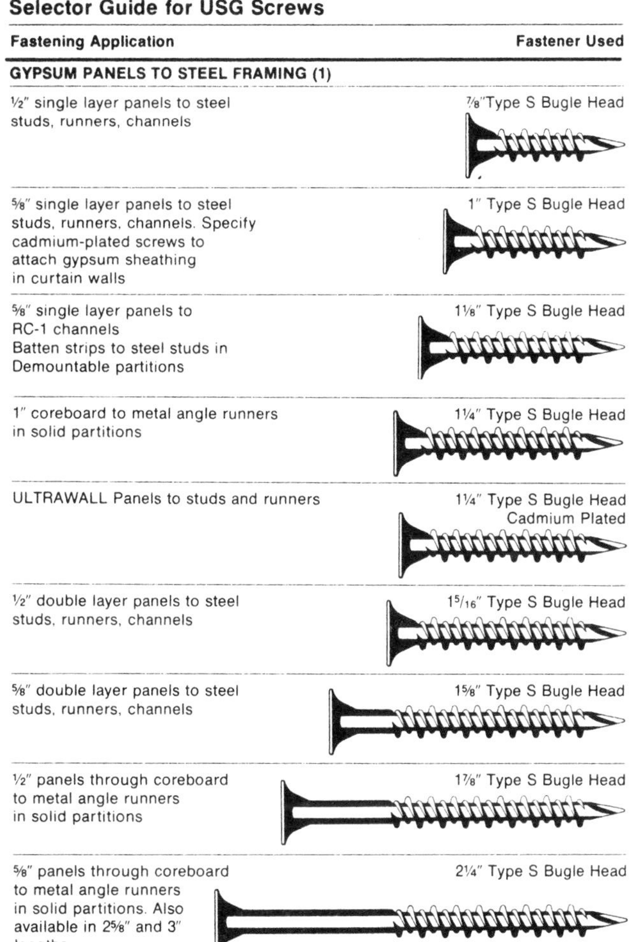

Selector Guide for USG Screws

Fastening Application	Fastener Used
GYPSUM PANELS TO STEEL FRAMING (1)	
½" single layer panels to steel studs, runners, channels	⅞" Type S Bugle Head
⅝" single layer panels to steel studs, runners, channels. Specify cadmium-plated screws to attach gypsum sheathing in curtain walls	1" Type S Bugle Head
⅝" single layer panels to RC-1 channels Batten strips to steel studs in Demountable partitions	1⅛" Type S Bugle Head
1" coreboard to metal angle runners in solid partitions	1¼" Type S Bugle Head
ULTRAWALL Panels to studs and runners	1¼" Type S Bugle Head Cadmium Plated
½" double layer panels to steel studs, runners, channels	1 5/16" Type S Bugle Head
⅝" double layer panels to steel studs, runners, channels	1⅝" Type S Bugle Head
½" panels through coreboard to metal angle runners in solid partitions	1⅞" Type S Bugle Head
⅝" panels through coreboard to metal angle runners in solid partitions. Also available in 2⅝" and 3" lengths	2¼" Type S Bugle Head

Figure 345 Drywall screws

Figure 346 Drywall screwdriver

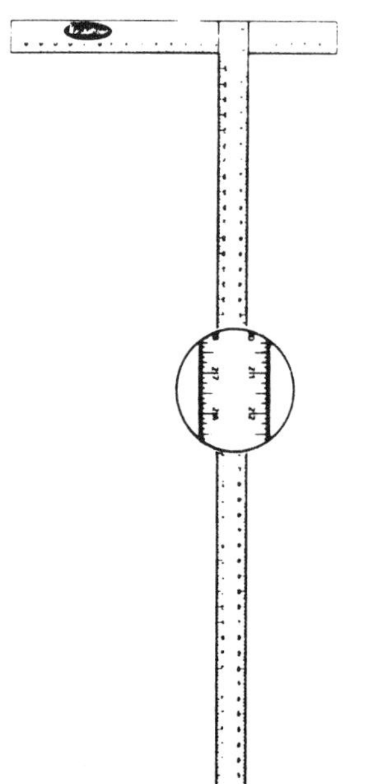

Figure 347 Drywall T square

Figure 348 Duckbill trowel

drywall T square Large quare, 22″ × 48″, used for laying out and cutting gypsum panels and other large panels. (Figure 347.)

drywell Well for the disposal of water from foundation drain or crawl space.

duckbill pliers Long-nose pliers with flat, rounded-end jaws. Very similar to *flat-nose pliers*. Used for wiring and in sheet metal work. See *pliers*.

duckbill trowel Pointing trowel with a duckbill shape. (Figure 348.)

duct fan Fan installed in ductwork to force air through the duct system.

duct furnace Furnace unit that is installed in the ductwork.

ducts Passageways for forced warm air; rectangular sheet metal or vinyl pipes are used. Also called *ductwork*. Also, an enclosed metal assembly that holds electrical power cables; see *raceway*.

ductwork In heating, rectangular or round sheet metal or vinyl piping that carries warm air to room outlets and returns cool air to the furnace. Figure 349 shows ductwork and fittings.

dummy joint In concrete work, a *control joint*.

dumpy level See *builder's level*.

duplex Building with two living units, either side by side or above each other.

duplex nail Double-headed nail used for temporary fastening. Second head makes it easy to withdraw. (Figure 350.) Also called a double-head or a scaffolding nail. See also *nail*.

duplex receptacle Outlet with two receptacles.

dust mask Protective mask worn for protection against high dust concentrations.

Dutch bond See *English cross bond*.

Dutch door Exterior door that is divided horizontally so that either the bottom or top part can be opened independently.

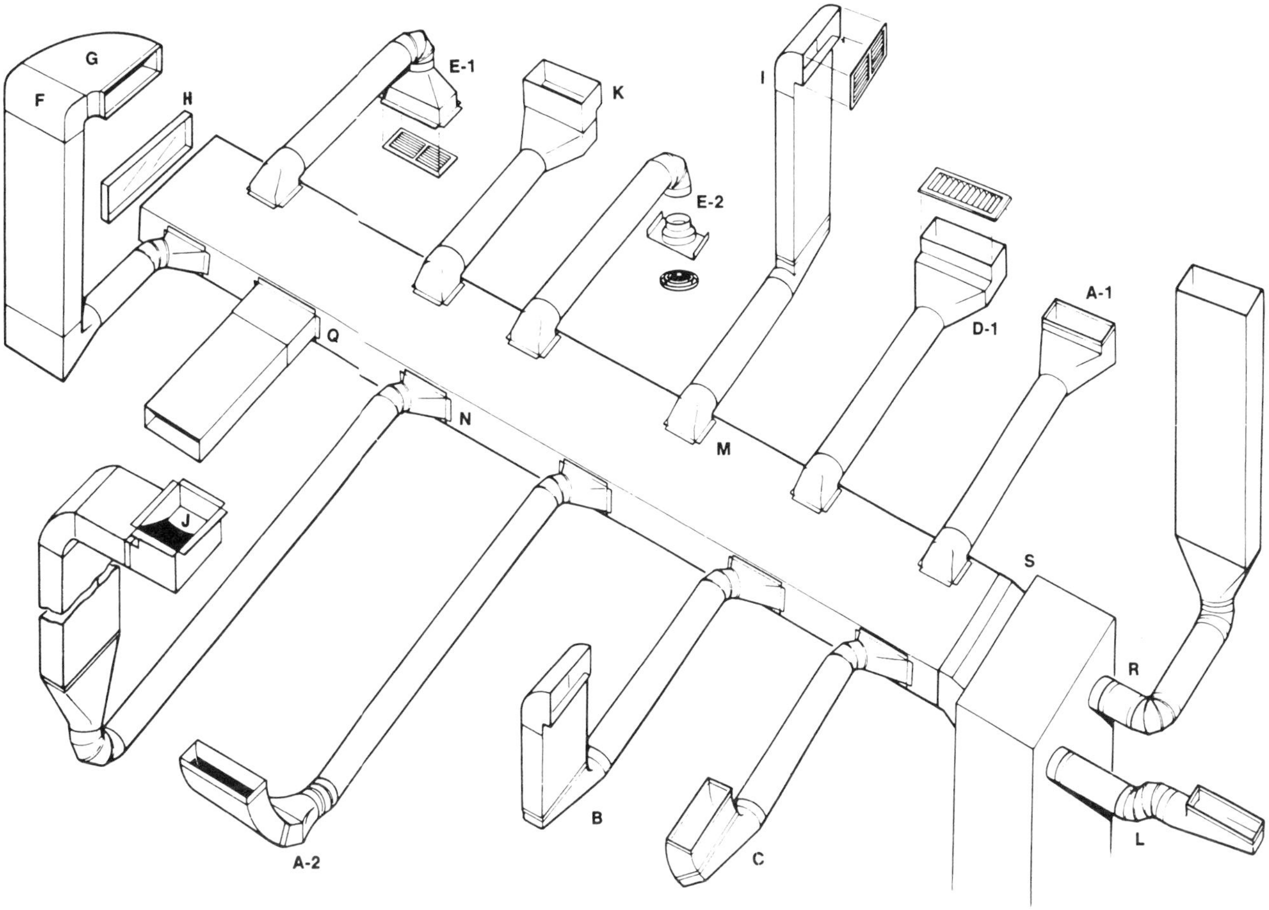

Figure 349 Ductwork

KEY NO.	IDENTIFICATION OF FITTINGS
A-1	90 Angle Boot
A-2	Straight Boot w/45 Angle
B	Center end boot
C	Center end boot w/45 Angle
D-1	Straight Boot w/floor pan
E-1	Floor or ceiling outlet pan
E-2	Ceiling outlet
F	Shortway 90 elbow
G	Longway 90 elbow
H	End Cap
I	Stackhead
J	Floor or ceiling outlet pan
K	Stack Reducer
L	Flex elbow
M	Top Takeoff
N	Side Takeoff
Q	Flat Takeoff
R	Starting Collar
S	Offset Takeoff Collar

duty cycle In welding, the amount of time (during a set period of 10 minutes) a power supply can operate normally at its rated capacity.

dwarf wall A low wall, specifically a retaining wall. A *parapet,* or *toe wall.*

dwelling A building containing one or more dwelling units that are used for living purposes.

dwelling unit Single living unit providing facilities for living, eating, sleeping, cooking, and sanitation; a *living unit.*

DWV The drainage, waste, and vent part of a plumbing system.

DWV copper tubing Thinwall, drain-waste-vent copper tubing used above ground with special DWV solder-joint copper fittings. See *copper tubing.*

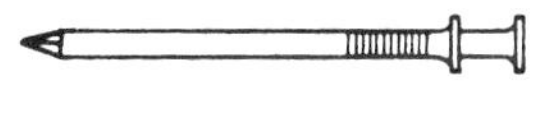

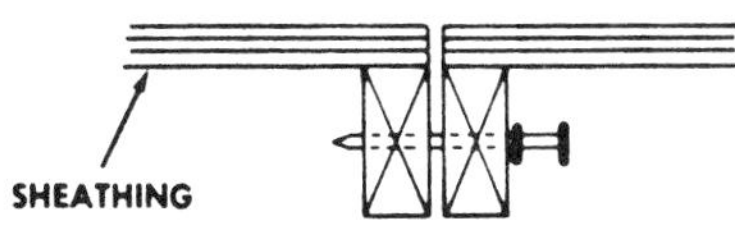

Figure 350 Duplex nail

early wood Springwood.

earth change plan Part of the total site development that covers location of building site with proposed grading and drainage. See also *erosion and sedimentation control plan, grading plan, landscape plan, sediment pool, diversion.*

earthquake load Special loading required in the design of structures in some areas. The load is caused by the ground vibration and the inertia or resistance of the building to movement.

earth insulation Underground or partially underground structures use the earth itself as insulation. Most earth, below the surface level, is a constant low 50°F, only 18°F colder than a termperature required for comfort. Earth also allows very little heat loss.

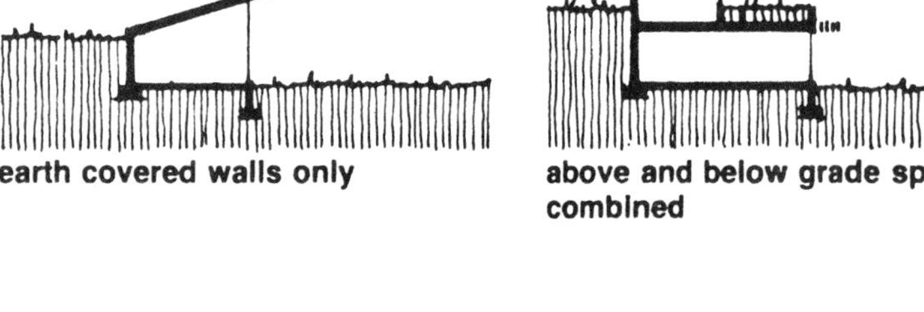

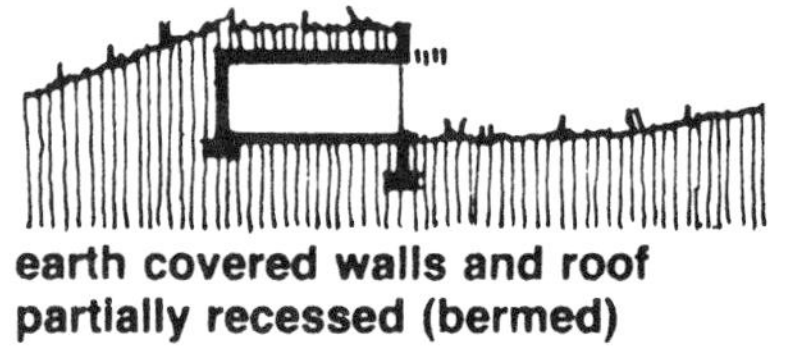

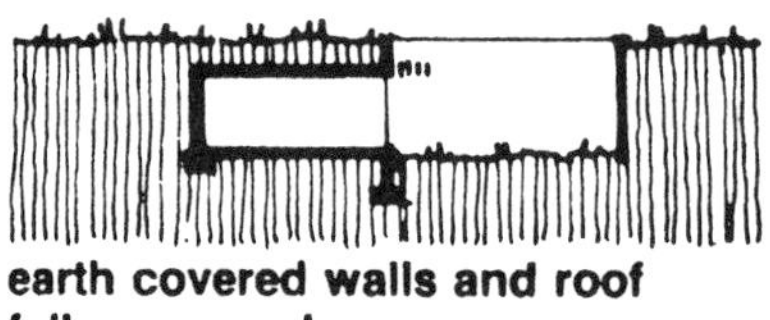

Figure 351 Earth-sheltered construction (cross sections)

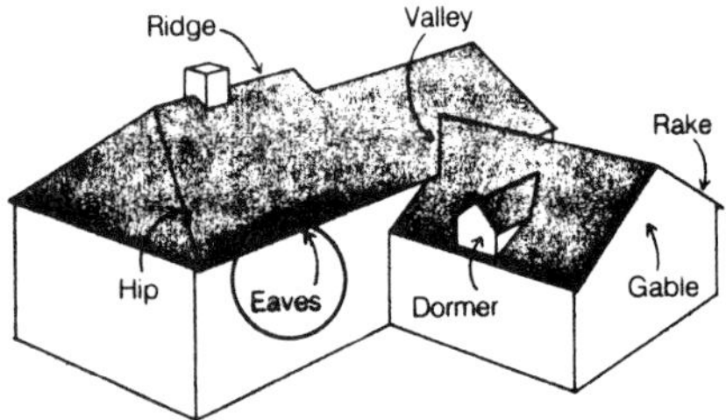

Figure 352 Eave

Figure 353 Edge: sheet metal

earth-sheltered construction Homes and other structures built partly or wholly underground. (Figure 351.) The insulative quality of the earth saves heating and cooling costs. The earth itself remains at a constant low 50°F in most areas. Commonly, a southern-facing glass window wall is used for *solar heating.* See also *berm.*

eased edge Rounded edges on surfaced lumber.

easement A right of access to property; a legal right to use property for a specific purpose, as the right to run electrical power lines. See *drainage easement, light and air easement, planting easement, solar skyspace easement.*

eave The part of a roof that projects over the side walls. (Figures 352, *323*, p. 123.) See *rake* for projection of roof over end walls.

eave course Shingle or tile course by the eave.

eave strut In a rigid frame steel building, the steel framing member running along the end of the roof over the eave. It transmits load from the roof and truss into the wall bracing.

eave trough Guttering along an eave.

eave vent Vent screen in the eave soffit to allow air to flow into the attic area. (*Figure 257,* p. 97.)

eclecticism In architectural design, a borrowing and blending together of styles from different periods or areas.

ecologically fragile area Area that has a sensitive balance of nature, such as marshlands, wetlands, flood plains, and could be upset or destroyed by development or construction. See also *environmental impact statement.*

edge The narrow side of a piece. (*Figure 380,* p. 152.) See *face.* In sheet metal work, the turned-over edge of the metal. (Figure 353.) A simple *hem edge* is made by folding back the metal. If folded over a second time, it is called a *double hem edge.* Edges that are rolled or wrapped over a wire are called a *wired edge.* See *seam.*

edge grain Softwood lumber sawn so the annual growth rings are at an angle between 45° and 90° to the face of the lumber. The same condition for hardwood is called *quartersawn.* Also called *rift grain, vertical grain.*

edge insulation Insulation that runs around the outside edge of a concrete slab between the slab and the outside foundation wall. Serves as an *expansion joint.* Also called *expansion strip.*

edge joint Lumber pieces joined edge to edge. See *joints.* Also, a weld made between the edges of two pieces. See *wood joints.*

edger Special trowel used to round and finish the edge of concrete flatwork, as the edge of a sidewalk or driveway. (Figure 354.) See *trowel.*

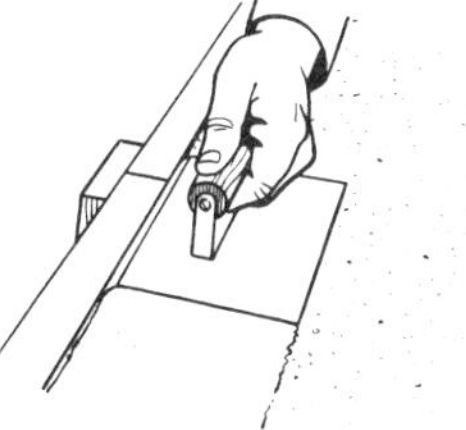

Figure 354 Edger

edge weld Welds made on thin material by welding turned-up edges. (Figure 355.)

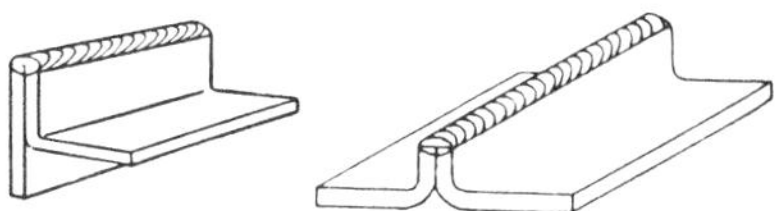

Figure 355 Edge weld

edging Finishing the edge of concrete flatwork such as a sidewalk or driveway. *Jointing* the edge of a board.

edging clamp Three-way adjustable C clamp used for gluing edges. (Figure 356.)

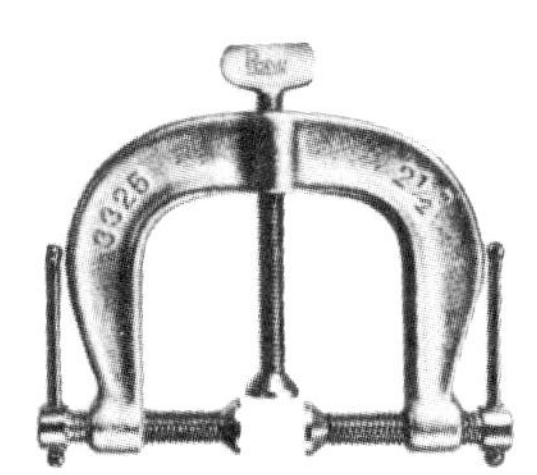

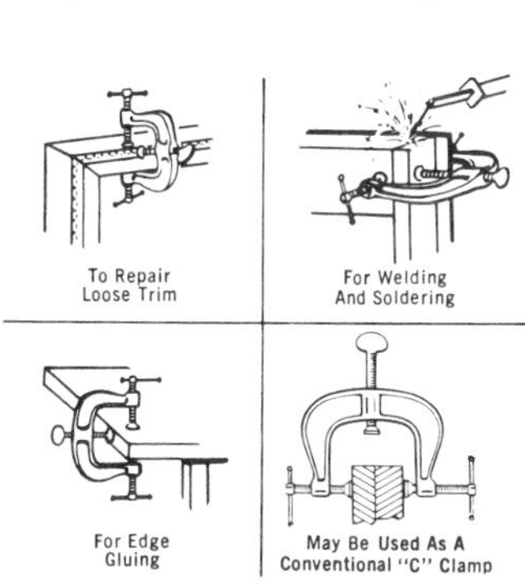

Figure 356 Edging clamp

Edison base The screw base used for standard light bulbs. The older plug fuses have Edison bases.

effective span Horizontal distance measured from the center of one support to the center of the opposite support. Used in calculating strengths. See *clear span, span.*

efflorescence Chalklike stain or crust on concrete, mortar, or plaster surface caused by moisture's bringing out salts (calcium carbide). Also called *bloom.*

effluent In plumbing, the liquid discharge from a waste system, especially sewage from a septic tank or municipal water treatment plant.

elbow "L" shaped, as a *plumbing fitting.*

electrical drawing Blueprints that show electrical layouts. See *electrical plan.*

electrical fitting A mechanical part of the electrical system used to hold, fasten, or support, such as a lock nut or a bushing.

electrical metallic tubing (EMT) A rigid steel tubing used in all locations except where subject to severe physical damage, in corrosive areas, or in cinder fill. Conductors are fished through the installed cable. (Figure 357.) Also called *thinwall conduit.* See *conduit.*

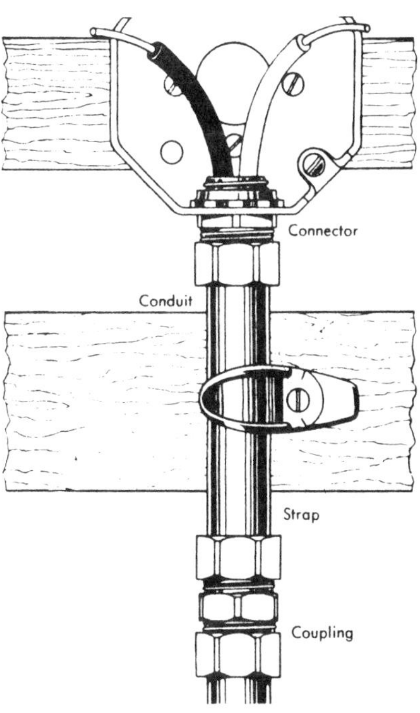

Figure 357 Electrical metallic tubing

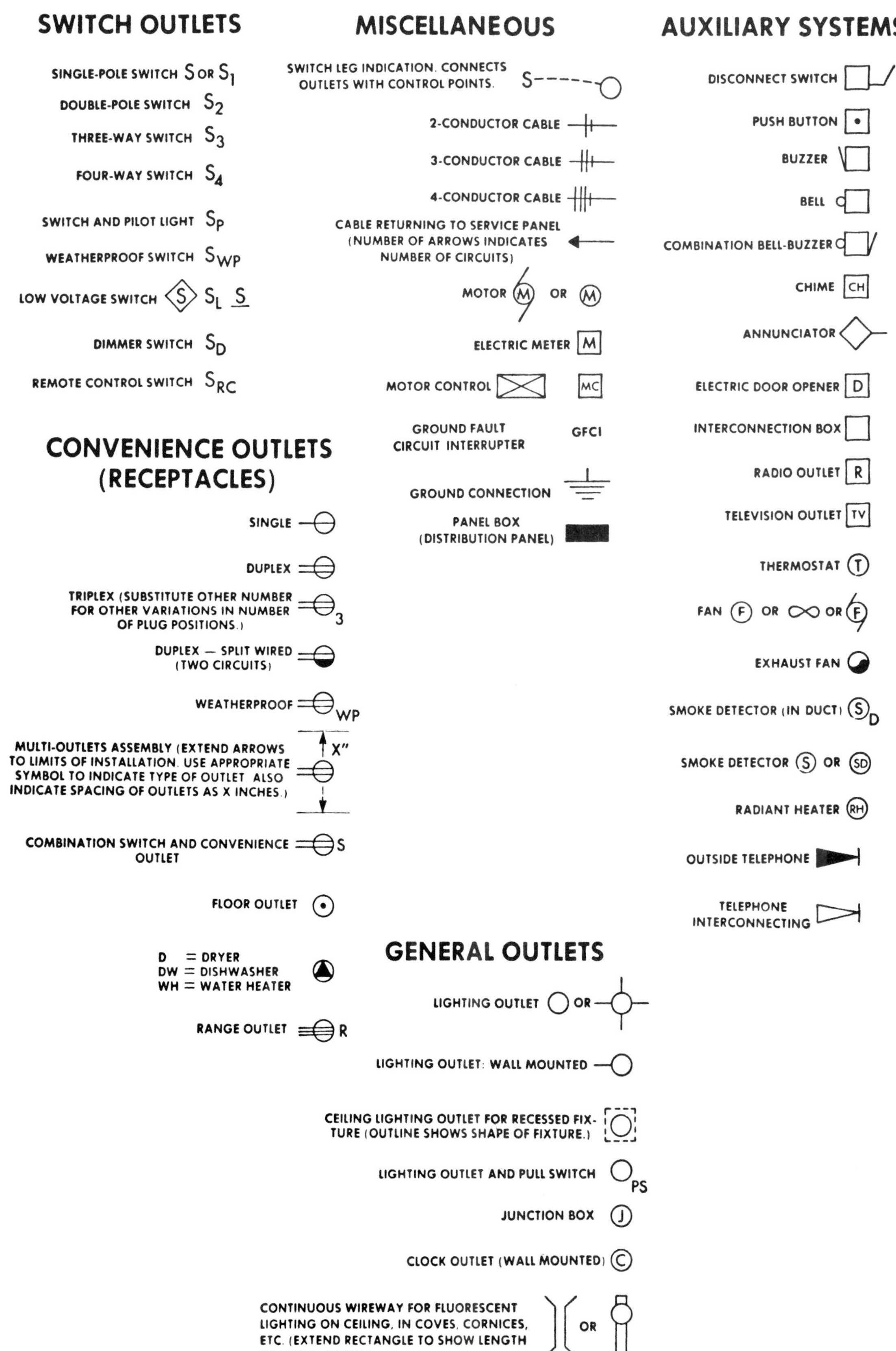

Figure 358 Electrical symbols

electrical plan Drawing that shows a plan view of the location of electrical switches, receptacles, and lights. Normally, in residential construction, electrical plan information goes on the regular floor plans. If a separate electrical plan is prepared, it will be identified with an E in the title box. See *electrical symbols, electrical wiring plan.*

electrical power Electrical wattage.

power (watts) = amperes × volts.

Electrical power is generally expressed in kilowatts. A kilowatt = 1000 watts.

electrical symbols Symbols used to show electrical fixtures, outlets, switches, and so on. (Figure 358.) See also *plan symbols, symbols.*

electrical system The total system that delivers electrical energy to a building. Power is brought in from the transformer at the public utility pole to the service drop. (Figure 359.) Power may also come underground. A *meter* measures the kilowatt-hour (kWh) usage. The *main switch* can shut off all power. The *panelboard* receives the power and distributes it to the various *branch circuits. Fuses* or *circuit breakers* protect each branch circuit in the panelboard. The branch circuits deliver electricity to the various outlets, such as *receptacles* or *lights. Switches* control the various outlets.

electrical wire sizes See *American standard wire gauge.*

electrical wiring color codes The following colors are generally followed in manufacturing and installing wiring:

Black: hot conductors
Red: hot conductors
White or gray: neutral conductors
Green: ground

A bare wire or metal strip in a cable is also ground.

Figure 359 Electrical system

electrical wiring plan Drawing that shows a plan view of the actual wiring runs. See *electrical plan, electrical symbols.*

electric-arc welding *Arc welding.* Welding process where an electric arc is produced between the electrode end and the work piece.

electric circular saw See *circular saw.*

electric drill Power drill used with a boring bit or drill for boring holes. (Figure 360.) See also *power screwdriver, right-angle drill.*

electric generator Arc welder that supplies DC electrical power.

electric heating Heating by use of resistance cable in the ceiling, floor, or walls. See *radiant heating, resistance wiring.*

Figure 360 Electric drill

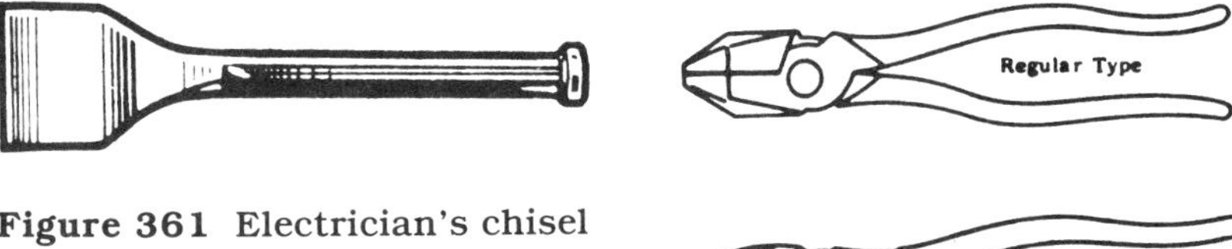

Figure 361 Electrician's chisel

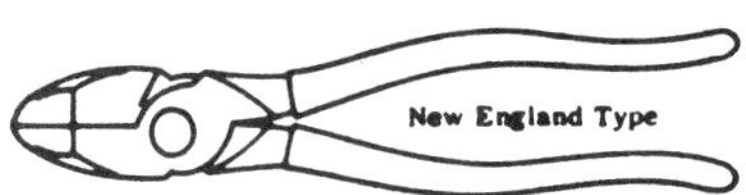

Figure 362 Electrician's pliers

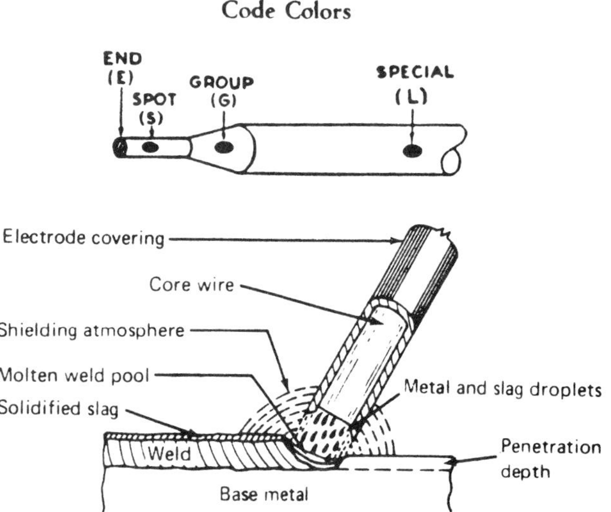

Figure 363 Electrode: shielded metal-arc welding

FOURTH DIGIT	TYPE OF COATING	WELDING CURRENT
0	cellulose sodium	DCEP
1	cellulose potassium	AC or DCEP
2	titania sodium	AC or DCEN
3	titania potassium	AC or DCEN or DCEP
4	iron powder titania	AC or DCEN or DCEP
5	low hydrogen sodium	DCEP
6	low hydrogen potassium	AC or DCEP
7	iron powder iron oxide	AC or DCEN
8	low hydrogen potassium iron powder	AC or DCEP
E6020	iron oxide sodium	AC or DCEN or DCEP
E6022	high iron oxide	AC or DCEN or DCEP
E7048	low hydrogen potassium iron powder	AC or DCEP

DCEP—Direct Current Electrode Positive   DCEN—Direct Current Electrode Negative

EXAMPLE

E-8018-B1

- ELECTRODE
- 80,000 psi minimum tensile strength requirement (stress relieved)
- For AC or DCEP
- Chemical Composition of weld metal deposit
- all position

Figure 364 Electrodes: AWS classification numbers

electrician (construction) One who lays out, assembles, installs, and tests electrical wiring systems and fixtures and electrical equipment and machinery. Connects and tests all electrical equipment, fixtures, and controls.

electrician's chisel Alloy steel chisel for cutting through metal lath and other building materials. (Figure 361.)

electrician's pliers Special, siding-cutting pliers used for cutting, bending, and twisting electrical wires. Some have a crimping die at the hinge. The handles are often covered with insulative vinyl. (Figure 362.) See *ironworker's pliers, side-cutting pliers.*

electric screwdriver See *power screwdriver.*

electric tester See *circuit tester.*

electrode In arc welding, a consumable wire or stick that conducts current from the electrode holder to the base metal. (Figure 363.) An arc occurs between the end of the electrode and the base metal to create the high temperatures needed for welding. Different electrodes are used for different base metals and different thicknesses. AWS codes are used to identify the different electrodes. (Figure 364.) For example, the code E7011 can be read as follows: "E" refers to electric-arc welding, the welding process the electrode is used for. The first two digits, "70", refer to the minimum tensile strength of the weld: 70,000 pounds per square inch. The third digit, "1", relates to the position the electrode can be used in:

1 All positions
2 Horizontal and flat positions
3 Flat position only

See *welding position.* The last (fourth) digit "1" relates to the kind of current used:

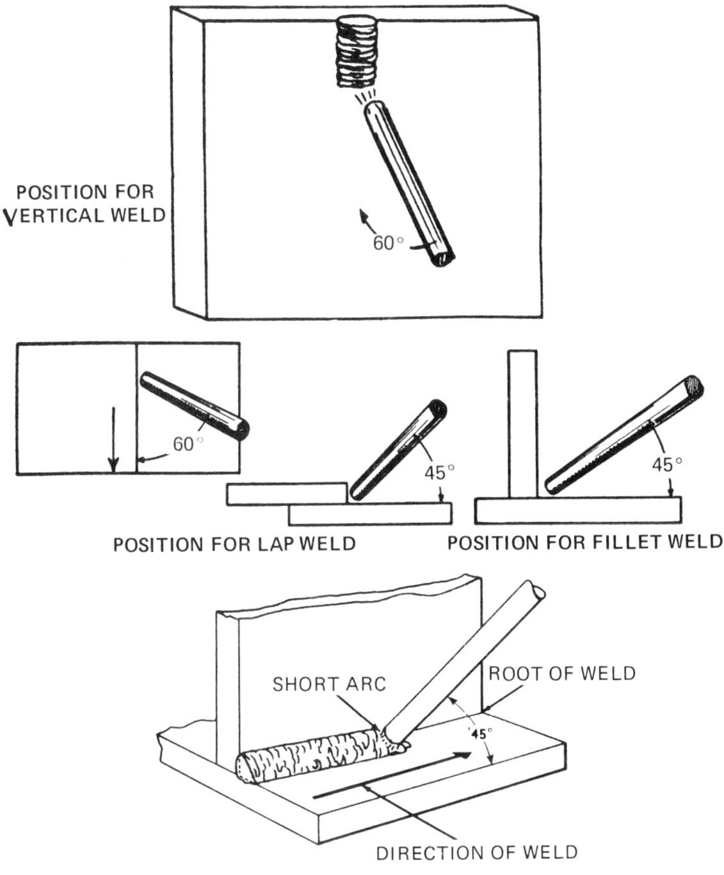

Figure 365 Electrode angle

0 DC with reverse polarity
1 AC; DC with reverse polarity
3 AC; DC with either polarity
4 AC or DC positive
5 AC with reverse polarity
6 AC

See *bare electrode, covered electrode.*

electrode angle In arc welding, the angle at which the electrode is held to the work. Generally the electrode is held at a right angle to the line of the weld and bisects the angle between work pieces. (Figure 365.)

electrode holder In arc welding, the handle (attached to the cable) that holds the electrode. (Figure 366.)

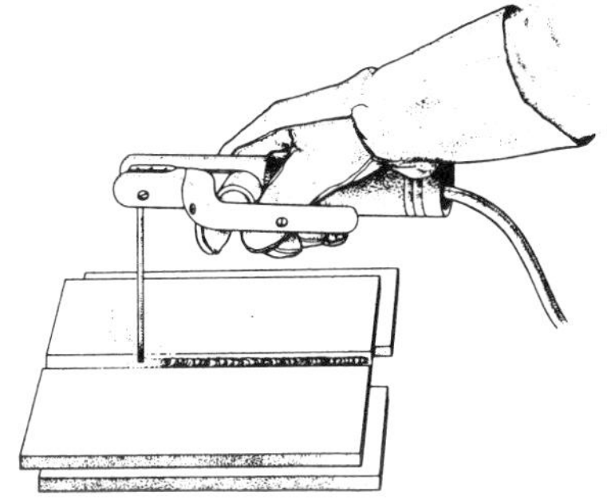

Figure 366 Electrode holder

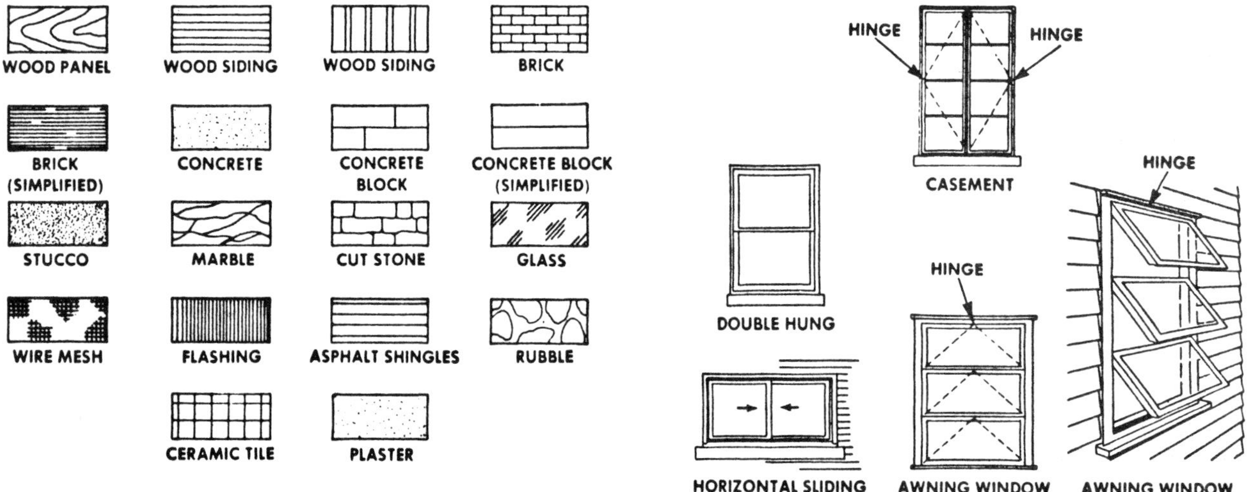

Figure 367 Elevation symbols

electroplated nails Nails that have been electroplated with a thin coating of metal, such as brass, cadmium, copper, nickel, tin, or zinc.

electrostatic filter Filter that removes dust and other airborne particles from the air by giving them an electrical charge and collecting them on an oppositely charged plate.

elevation Height. In surveying, a height (elevation) above sea level or above an established local *datum* or *bench mark*. On a plat or plot plan, land with the same height is joined together by a continuous line called a *contour line*. In the actual building, key levels, such as the floors, are given an elevation based on local datum or on the height above the foundation top. Also, a drawing showing an exterior, orthographic view of a building. See also *interior elevation, plan*.

elevation symbols Symbols used to show materials and features on the exterior of a structure. (Figure 367.) See also *reference line*.

Elizabethan construction A house style common in England in Elizabethan times (late sixteenth century) and used today in period architecture. House is two or two-and-a-half stories with exposed timbers; space between the timbers is filled with stucco. (Figure 368.) See *architectural styles.*

Figure 368 Elizabethan style

ell A part of a building that meets the main structure at a right angle; the ell and main building form an "L" when seen from above. In general, a right-angle shape.

emergency light Automatic lighting required by code in public buildings, such as exit lights over doors. Also, room lights that automatically come on when the building power fails.

emery A natural abrasive mineral made of iron oxide and corundum. Widely used on abrasive papers, cloths, belts, and discs for finishing metal. Also made into wheels for use with a grinder. See *abrasive.*

EMT *Electrical metallic tubing,* used for running electrical wiring. Also called *thinwall conduit.*

enameled nails Nails with baked-on colored enamel coating.

encased steel structure Steel frame structure where all of the steel framing members are encased in concrete.

end-cutting nippers Nippers used for cutting wire or nails close to the end, as for cutting off the head of a nail. (Figure 369.)

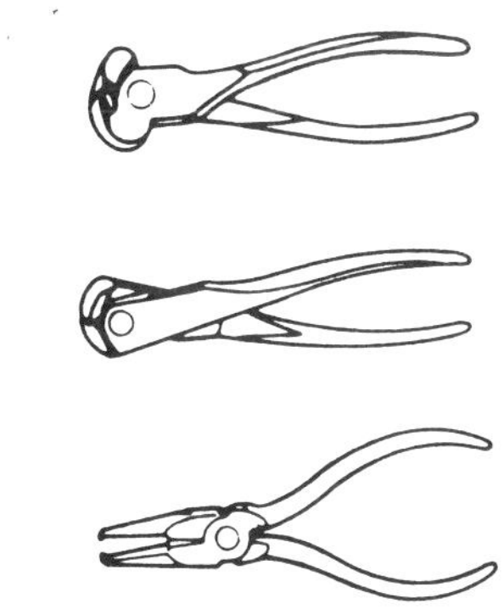

Figure 369 End-cutting nippers

end grain Wood piece showing the end view of the grain; view when the wood is cut at a right angle to the long wood grain; a cut across the tree trunk.

end joint Lumber pieces joined end to end, a *butt joint.* See *joints.*

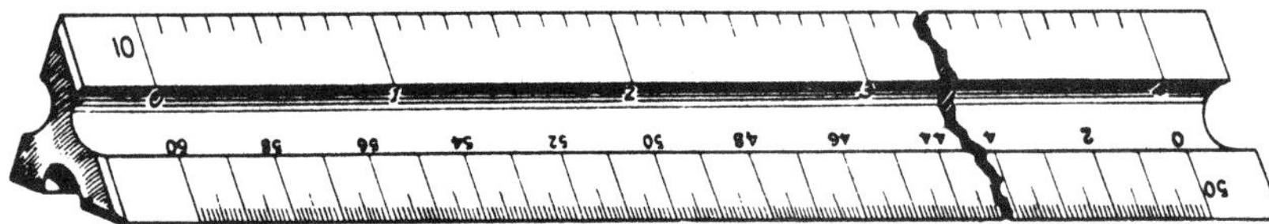

Figure 370 Engineer's scale

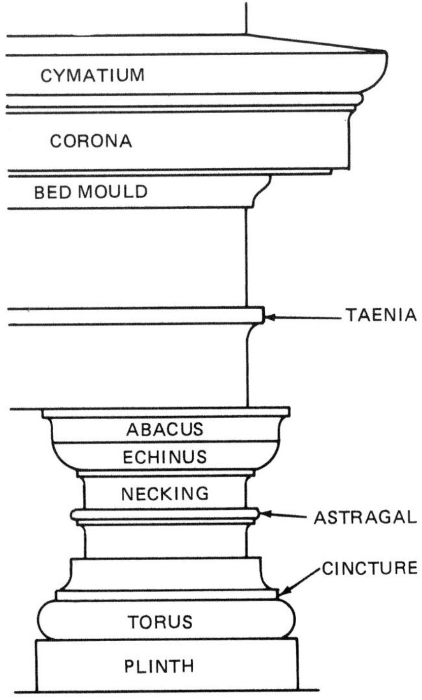

Figure 371 Entablature (Tuscan order)

end joists Joists running on the outside of the floor platform in the same direction (parallel to) the common floor joists. (Outside joists that run at a right angle to the common floor joists are called *header joists.*) See also *band joists, rim joists, sill joists.*

end view In drafting, a view looking straight at the end of a building, an elevation.

energy-control window shades Heavy thermal shades pulled down to cut down heat loss through a window area.

energy loss Heat loss through conduction or infiltration.

engaged column Column partly embedded in a wall.

engineering drawings Design drawings used in steel, and reinforced and prestressed concrete construction, to show general structural layout and spacing of support members; used for the shop fabrication of needed members and reinforcement and sometimes for field erection and placement. Used as the basis for the development of *erection drawings* and *placing drawings.*

engineer's scale Scale divided into decimal graduations used for map making. (Figure 370.)

English architecture Architectural styles characteristic of periods of English history. See *Elizabethan, Georgian, Tudor.* See *architectural styles.*

English bond Masonry bond pattern that alternates a course of all stretchers with a course of all headers. The headers are centered on the stretchers. Joints between stretchers in alternate rows line up. See *brick bond.*

English brick Kiln-baked brick slightly larger than the standard building brick. The English brick is 3″ × $4\frac{1}{2}$″ × 9″.

English cross bond Masonry bond pattern that is a variety of the *English bond.* Joints between stretchers in alternate rows do not line up. See *brick bond.* Also called *Dutch bond.*

entablature In classical architecture, the horizontal assembly above the capital of the column. (Figure 371.) See *orders.*

entry Entrance to a building.

environmental impact The effect construction will have on the surrounding area.

environmental impact statement (EIS) Evaluation required by HUD (and other agencies and local governments) detailing how any proposed construction will affect the total environment, including human environment, vegetation, flora, fauna, water, air, traffic, and waste disposal. The EIS details the proposed construction or change and evaluates the effect of the change on the total environment; alternatives to the proposed change are analyzed.

Environmental Protection Agency (EPA) Federal agency charged with the protection of the environment; monitors state pollution control plans.

epoxy cement A two-part cement which, when mixed together, sets very hard and strong.

epoxy glue A two-part glue with a white liquid epoxy resin and a dark liquid catalyst which are blended together. Dries to an extremely rigid and strong bond.

epoxy paint A two-part paint which, when mixed together, leaves a hard and very tough coating.

erasing shield Thin metal shield used to erase through to avoid erasing adjacent areas.

erection drawings In steel or prestressed concrete construction, drawings that show how a steel or prestressed concrete frame is to be erected. Each part or piece is identified with a number and located. See also *anchor bolt plan, beam number, column number, design drawings, engineering drawings, field work drawings.*

erection mark In steel construction or in reinforced prestressed concrete or precast concrete, an identifying mark, letter, or number placed on members so they can be easily identified and correctly placed in the structure. See *marking.*

erection wrench An open-jaw wrench with a pointed or tapering handle. The handle is used for aligning bolt holes in structural steel members; the wrench is used for tightening bolts to hold steel beams. Figure 372. Both open-end and adjustable wrenches are available. Also called a *spud wrench.* See also *bull pin, power wrench, turn-of-the-nut method.*

erosion Wearing or washing away of soil from an area, such as a construction site. See *erosion and sedimentation control plan, sediment pool, diversion, scarification.*

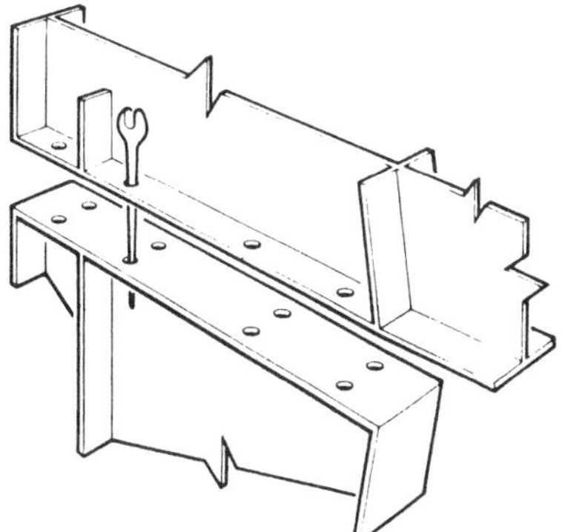

ALIGNING HOLES

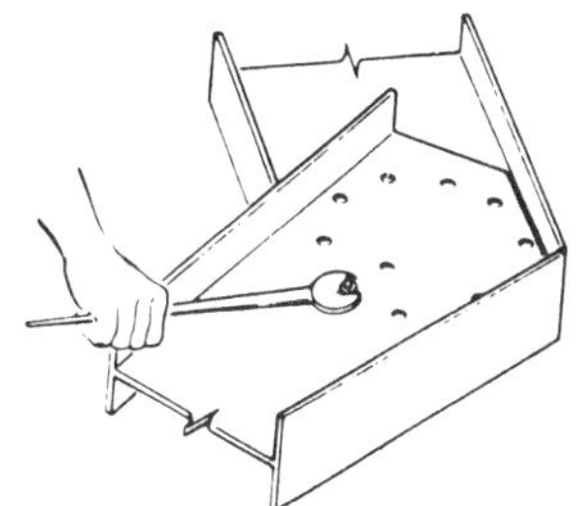

TIGHTENING NUT ON BOLT

Figure 372 Erection wrench

erosion and sedimentation control plan Part of the total site development that covers temporary and permanent erosion control at a construction site or development; covers earth change, on-site drainage, and all erosion controls. See *diversion, earth change plan, environmental impact statement, plot plan, sediment pool, vegetative filter.*

escutcheon Plate that covers the locking mechanism of a door lock; metal plate surrounding a keyhole. Decorative metal cover that fits around a pipe where it exits from the wall. Decorative piece that fits over a faucet assembly.

Figure 373 Excavator

escutcheon pin Decorative nail with a rounded head used to fasten ornamental parts, such as a hinge, to wood.

essex board measure A table on a *rafter square* used for calculating board feet.

estimating Determination, before starting construction, of a cost or an amount of material needed in a project.

estimator One who calculates labor and materials for a building job by a close study of blueprints and specifications.

estimator's criteria sheet Outline giving information and financial conditions on a project, used in preparing job estimates.

etching Eating away of a surface with an acid or with abrasive action to form a design.

evaporator In an air conditioning system, a device that allows the refrigerant to expand from a fluid to a vapor, absorbing heat from the surrounding atmosphere. See *air conditioning.*

evergreen *Softwood* tree with needlelike leaves that do not fall off with the cold season. A *conifer.*

excavation Hole made by removing earth; the removing of soil from a site. See *cut and fill, trench, earth change plan.*

excavation line Line made by a string stretched between *batter boards* to show the location of the edge of the excavation, the foundation, and building components.

excavator Heavy-duty earthmoving machine with a backhoe-type trenching bucket. (Figure 373.) See also *backhoe.*

exit A way out; in commercial structures the exit will be marked with a lighted sign.

expanding cement A cement that expands after mixing with water, used for patching and sealing cracks and holes. See *expansive cement.*

expansion bit A bit having a cutting spur that can be expanded, or extended out, to cut a larger hole. (Figure 374.) Bit is adjustable from $\frac{1}{2}''$ up to $3''$. See *bit.*

expansion bolt Bolt with a sleeve whose size can be increased once in the hole, used to create a tight fit in masonry. (Figure 375.) See *stud anchor.* Also called an *expansion anchor bolt.*

expansion joint In concrete flatwork, an *isolation joint*; a separation between concrete materials with different expansion rates. A resilient filler or strip is placed or poured in the joint. Also, in roofing, separation of different building elements to prevent stresses from splitting the roof membrane.

expansion shield See *lag shield.*

expansion strip Strip of resilient material that fits into an expansion joint.

expansion tank Tank in a heating system that allows heated water to expand. See *hot-water heating.*

expansive bit See *expansion bit.*

expansive cement A cement that, when mixed with water, expands; used to fill cracks and in areas where any shrinkage might cause a problem. Also called *expanding cement.*

explosive-driven fasteners Various fasteners or studs driven with a powder gun. Fastener is driven mostly into concrete or steel to attach wood members or metal track. Powder cartridge is selected to fit the type and thickness of material driven into. See *powder gun.*

exposed aggregate Concrete finish which has crushed stone or gravel exposed on the surface.

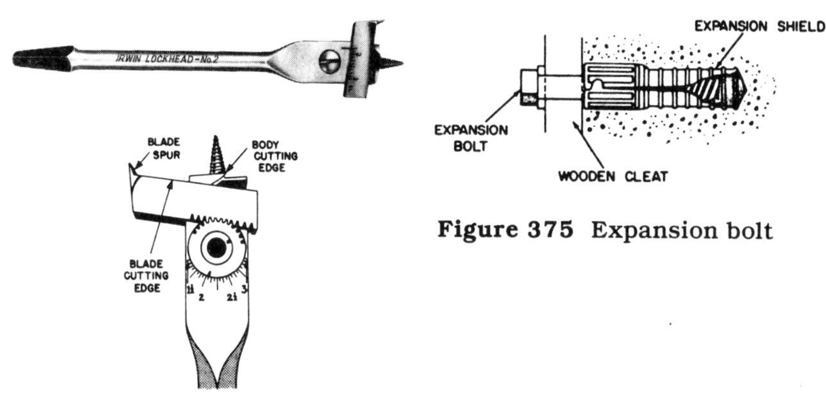

Figure 375 Expansion bolt

Figure 374 Expansion bit

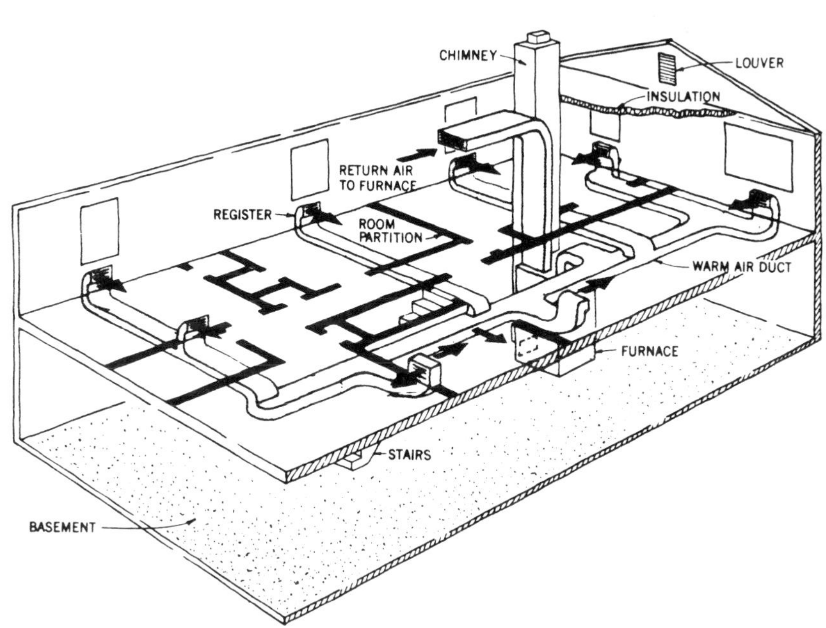

Figure 376 Extended plenum: forced warm-air heating system

exposure Distance exposed to eather, as shingles. (*Figure 1125*, p. 497.)

extended plenum heating A type of *forced warm-air heating* that uses a central large duct area or *plenum.* Warm air is forced with a blower through the warm-air outlets. Cool air is returned through cool-air returns. (Figure 376.) See also *perimeter loop heating, radial perimeter heating.*

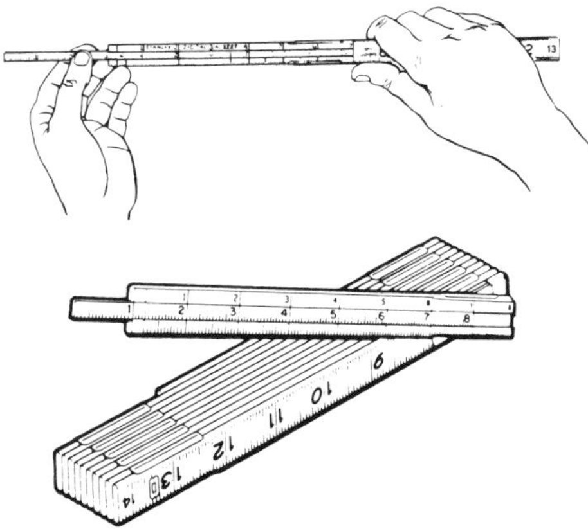

Figure 377 Extension rule

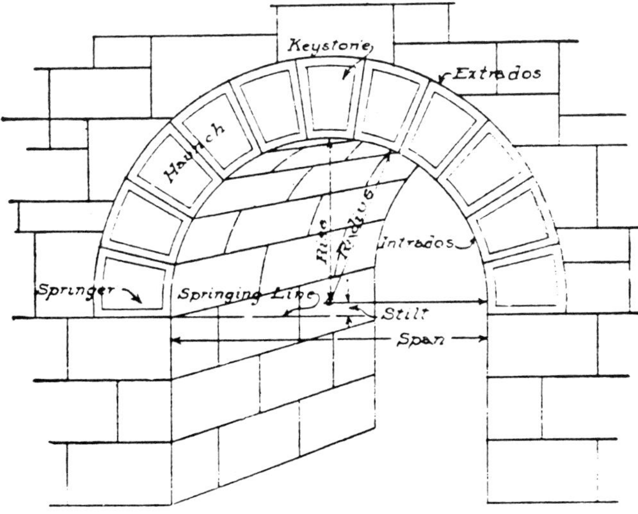

Figure 378 Extrados and intrados

extension ladder Two-part ladder that can be extended up to longer lengths. The bottom part is called the *base section*, the top part that extends out is called the *fly section*. When extended, the fly section is locked in place.

extension rule Rule having a section that can be slid out, used for making accurate measurements between framing members or between existing parts of the building. (Figure 377.) See also *folding rule*.

exterior finish Exterior covering of the building, including roof and wall covering, window and door trim, cornices, and any trim.

externally operated disconnect Main control switch for disconnecting power to a building. A handle is located on the outside of the service box. Abbreviated E-X-O.

extrados On a masonry arch, the curve bounding the far extremity of the arch. (Figure 378.) See *intrados*.

extreme fiber stress Allowable bending stress in a member, as a joist, before rupture, or failure. See *extreme fiber stress in bending*.

extreme fiber stress in bending (F_b) The stress on a horizontal member with a direct load applied. Tension is produced in the wood fibers farthest from the applied load and compression in the fibers closest to the applied load. (*Figure 219*, p. 82.) Design tables are available for different woods and grades for this and other stresses in pounds per square inch. See *compression parallel to grain, compression perpendicular to grain, modulus of elasticity*.

eye Opening on the head of a tool to receive a handle, as for a carpenter's hammer or a hatchet.

eye bolt Screw bolt with an eye (circular) head. See *turnbuckle*.

eye protection Protective glasses, goggles, or shield worn over the eyes. See *face shield, goggles, welding helmet*.

f

fabricating To put something together in the shop, as to assemble a reinforced steel part or section.

facade The front exterior of a building; the face.

facade forms Reusable formwork that goes on the outside of the building. As the building goes up, the forms are stripped and hoisted to the next level. (Figure 379.) See *wall forms*.

face The main side or the front of something. The front of a building, wall, or arch; the wide flat side of a board or a timber (Figure 380); the surfaced side of a board; the wide, front surface of a plywood or drywall sheet; the exposed surface of steel beam. The striking surface of a tool, such as a *hammer*. The working surface of a tool, as the cutting surface on a saw tooth. See *brick* (*Figure 125*, p. 47) for the face side.

Figure 379 Facade forms

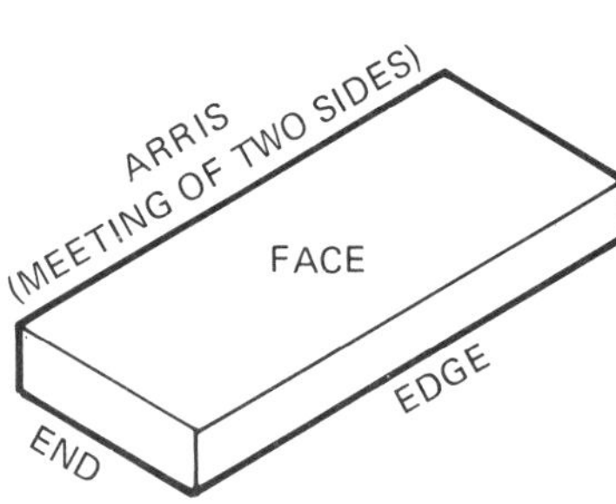

Figure 380 Face of lumber

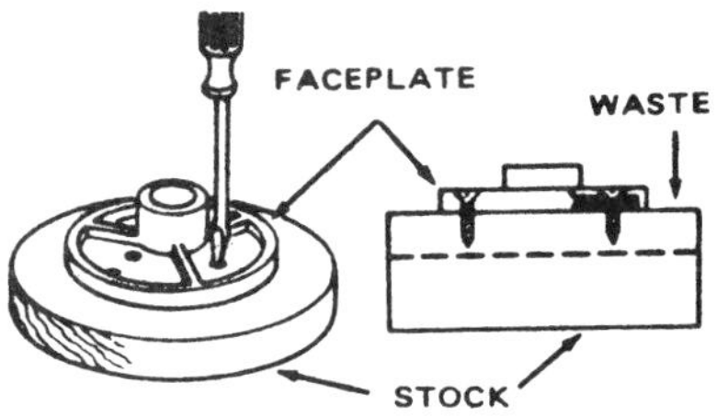

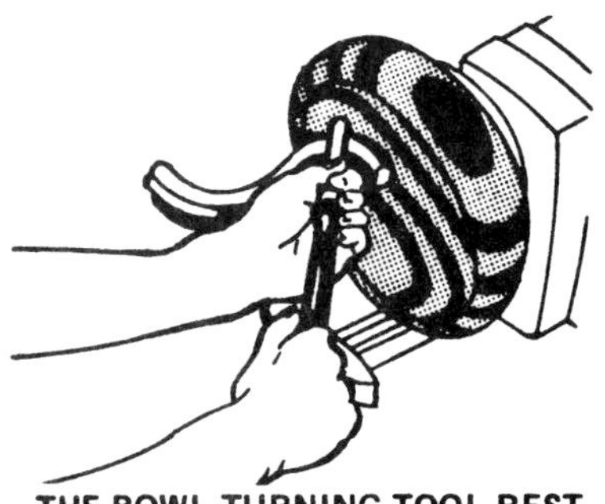

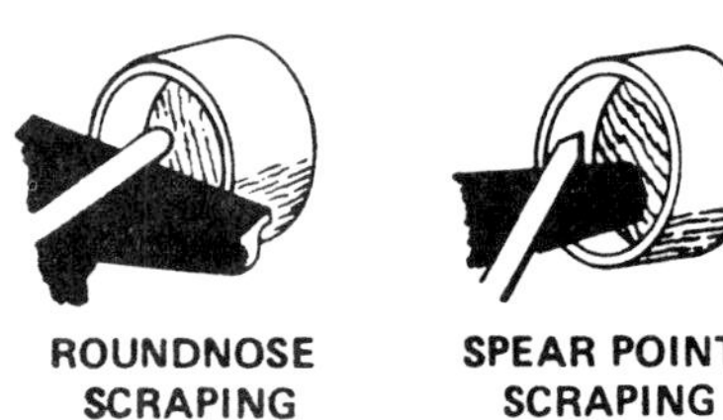

Figure 381 Faceplate turning

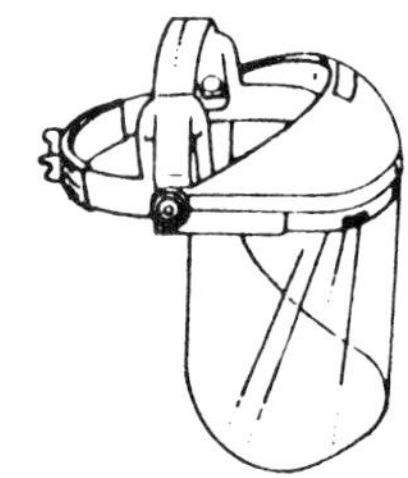

Figure 382 Face shield

face brick A better brick that is used on the exposed face of a wall.

faced wall A veneer wall that is integrated with the support wall of the building so as to become one with it and become a load-bearing wall.

face joint Exposed masonry joint, as in the wall face. A face joint is always jointed or tooled. See *masonry joints.*

face measure Measure across the face, as across the width of a board (from edge to edge, not including any tongue).

face nailing Nails driven perpendicular to a surface. In roofing, nails that are driven into the exposed part of a shingle.

face of a square Front of the framing square with the manufacturer's name.

face of weld The surface on which the weld was made.

faceplate turning Turning a wood piece on a lathe with only one end attached. The wood piece is glued to a faceplate, which is attached to the head stock. (Figure 381.) The rotating wood piece is shaped with woodturning tools.

faceplate work In wood turning, working on the open face of a turning wood piece. The wood is held glued to a faceplate.

face puttying Spreading putty (with a putty knife) into the rabbet holding a window light in the sash.

face shield A full plastic shield that is worn over the face for protection. (Figure 382.) See *eye protection.*

face string Exposed string or carriage of a stair.

face veneer Face ply on a plywood panel.

facia See *fascia.*

facing Straightening or smoothing the face of a board with a jointer or planer. The outside surface of a masonry wall; the *face.*

facing hammer Hammer for dressing stone or precast slabs.

factor of safety The margin of safety. See *safety factor.*

factory edge In roofing, the original, uncut shingle edge.

factory lumber Shop lumber.

Fahrenheit Temperature measure. Abbreviated F. On this scale the freezing point of water is 32°F and the boiling point is 212°F. (*Figure 172*, p. 64, compares the Fahrenheit scale to the metric Celsius scale.)

faience Type of glazed tile or brick used for ornamental facing or for a floor.

failure Breaking or giving away; collapse of a support or a structure.

fair housing laws Guarantees that no person will be discriminated against in regard to housing on the basis of race, color, religion, sex, or national origin.

fall rope Free end of a rope in a *block and tackle.* You pull on the fall rope to lift the load using the block.

false ceiling A ceiling hung or suspended from a higher ceiling. A *suspended ceiling.*

false set In concrete or mortar, a rapid apparent hardening that may occur before the actual initial set (which involves a chemical change with the release of heat). See *flash set.*

falsework Temporary support structure erected during the process of construction. The falsework is removed after completion of the structure.

fan light A window shaped like a fan, used in front doors.

fascia Outside, horizontal board on a cornice. (Figures 383; *201*, p. 73.)

fastener Device for holding something together, as a nail, screw, staple, or bolt. A locking device for latching windows, cabinet doors, house doors.

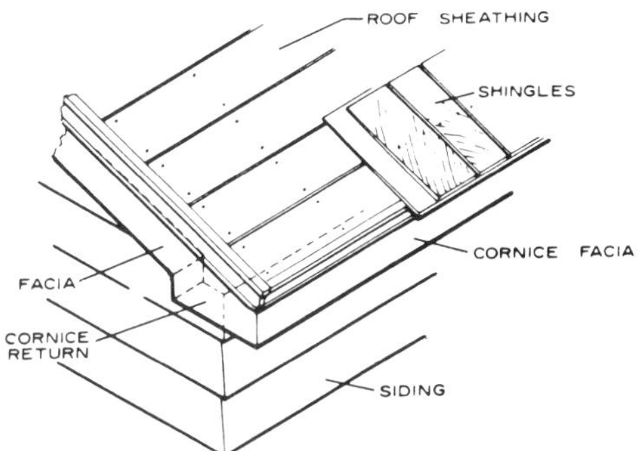

Figure 383 Fascia

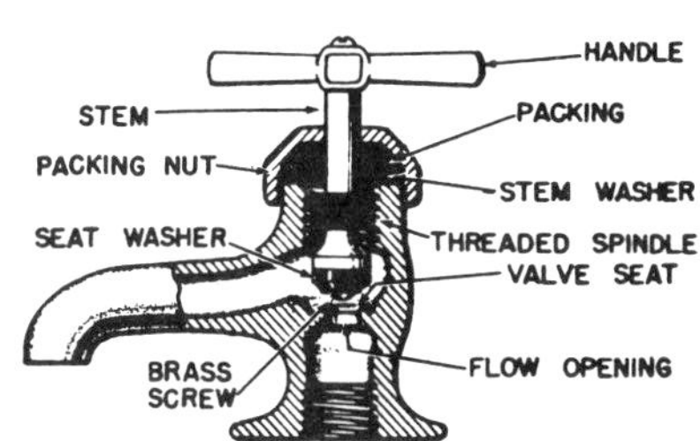

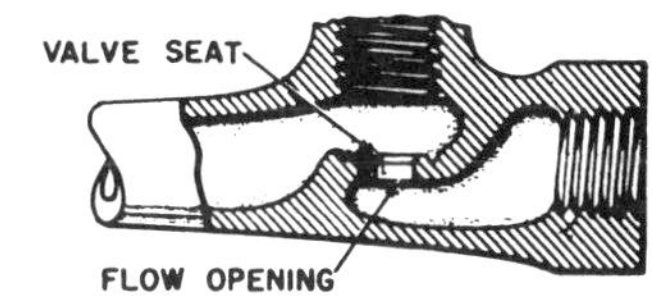

Figure 384 Faucet

fast-track construction Starting on-site construction before contract documents are completed; foundation is started before all drawings and specifications are complete.

fat Thicker than normal; referring to a liquid or plastic mix, as a mortar, concrete, or paint.

fat mortar Mortar with a high percentage of cementitious material; mix will be strong and may stick to the trowel. Opposite to *lean mortar.*

fattening In painting, a thickening of the paint during storage.

faucet Plumbing fixture for the outlet of water. (Figure 384.)

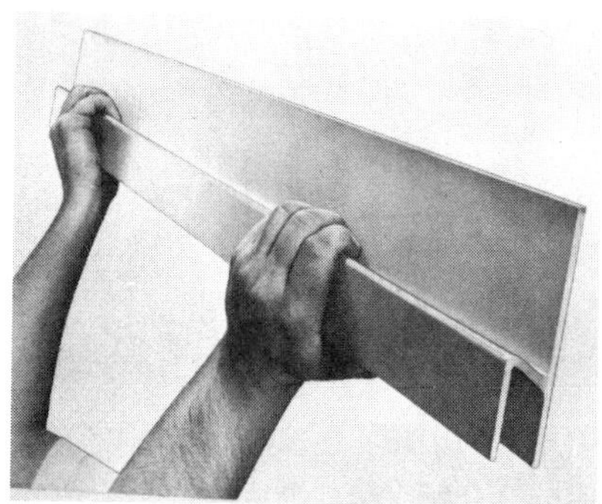

Figure 385 Featheredge

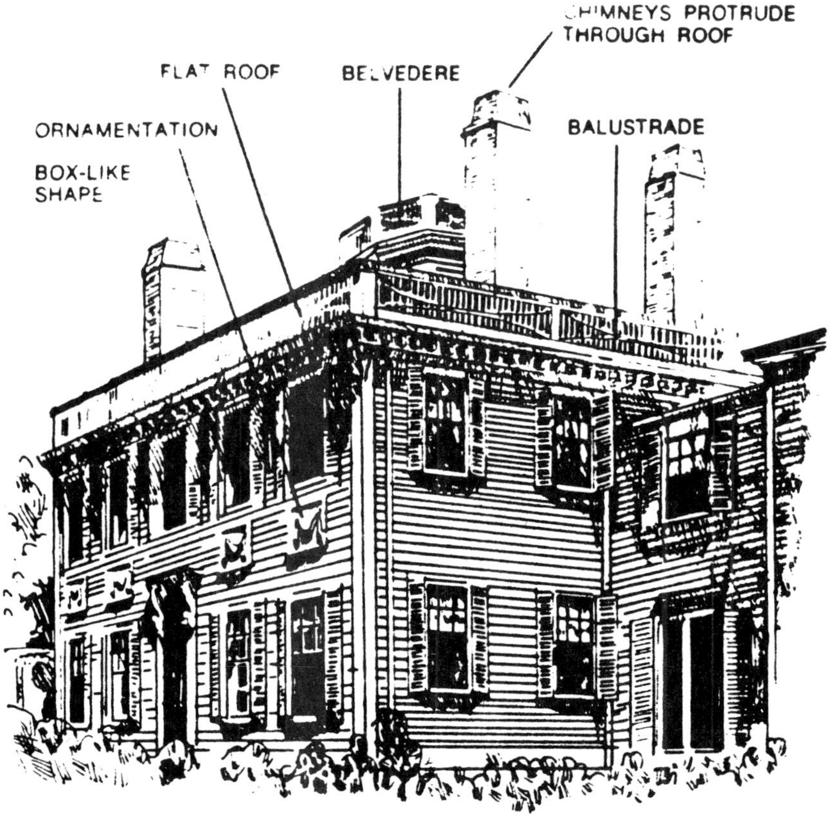

Figure 386 Federal style

feather In woodworking, a thin section of wood inserted flush in the outside corner of a miter joint. Designed to strengthen the joint.

feather crotch Attractive veneer figure cut from the crotch area of a tree. The design or figure resembles a feather.

featheredge Tapering to a thin edge. Also, a finishing tool used in plastering. (Figure 385.) See also *darby, plasterer's shingle, slicker.*

feathering In sheet metal work, edges that are flattened out together. Applied plaster or drywall joint compound that is thinned out at the edges.

Federal Flat-roofed, multistory Colonial house. (Figure 386.) See also *architectural styles.*

Federal Housing Authority (FHA) Agency operated under the Department of Housing and Urban Development (HUD) to insure housing purchase; guarantees loans made for purchase, but the real estate must be approved by FHA and loans must meet interest requirements.

feeder A main source that feeds into a secondary. In electrical wiring, conductors running between the control center and the final branch circuit breaker or fuse. In commercial or industrial electric wiring, a power connection from a switchboard (which receives the main power supply from the service company) to panelboards (each of which in turn supplies power to branch circuits). In plumbing, the main pipe that supplies water.

feed pipe Main plumbing pipe carrying water supply; a *feeder.*

felt Asphalt-impregnated fibrous sheet used in built-up roofing and waterproofing.

female thread Threads on the inside of a pipe or fitting.

fence A guide or guard; used on stationary power saws to guide the material being cut and determine the depth of cut. See *table saw.* A fence can also be attached to a portable circular saw.

fenestration The design or arrangement of windows in a structure.

ferrule A sleeve or loop on the end of a tool handle where it joins the metal tool, designed to prevent the handle from splitting.

festoon In woodworking, carving, and art metalwork: a fanciful, decorative garland of flowers and leaves worked into, raised out of, or applied to a surface.

FHA Federal Housing Authority.

fiberboard Panel made of pressed wood and vegetable fibers, used for insulating outside walls and roofs over living areas.

fiberglass Glass spun out in fine hairs, used in thick blankets or batts for insulating walls, floors, and ceilings.

fiberplug A plastic holding device that expands when a screw is screwed in. A single, hollow plastic section is used. The fiberplug is inserted into a drilled hole, then the screw is screwed in to fasten whatever is to be held. The fiberplug expands to a tight fit in the hole. See also *anchor, plastic anchor, rawlplug.*

fiber saturation point In lumber seasoning, the point at which the wood cells have lost their water from the cell opening but still have water in the cell walls, around 30% moisture content.

field book In surveying, a notebook in which the survey record is kept.

field connection A connection made at the job site, as by bolting together steel beams. Opposed to *shop connections.*

field stone Irregular stone as found natural in a field.

field tile Tile used around the house to carry water away from the foundation.

field weld Weld made in the field (on the job), as opposed to a *shop weld.*

field work drawing Drawings showing field fabrication and attachment in steel construction. Part of the *erection drawings.*

figure Pattern or design developed in wood by growth rings, irregular wood fiber, knots, rays, and the like. Important in veneering and in furniture making.

file Very hard metal tool cut so as to form an abrading surface. Used for abrading wood or metal. Single- and double-cut files are used. Figure 387 shows the types of file cut. The fineness of the cut determines the use. Single-cut files have one set of parallel cuts running diagonally across the face. The bastard cut is rough, the second cut is medium, and the smooth cut is fine. Double-cut files have two sets of criss-crossing parallel cuts on the face. Files are also defined by shape: flat or rectangular, square, triangular, round, or half round. A file is always used with a handle. See also *auger bit file, cabinet rasp, drill saw, file card, flat file, half-round cabinet file, half-round file, mill file, needle file, pillar file, rattail file, saw file, wood file.*

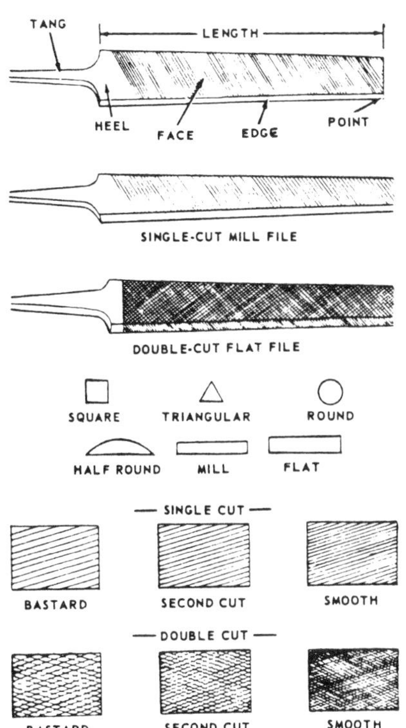

Figure 387 File terminology

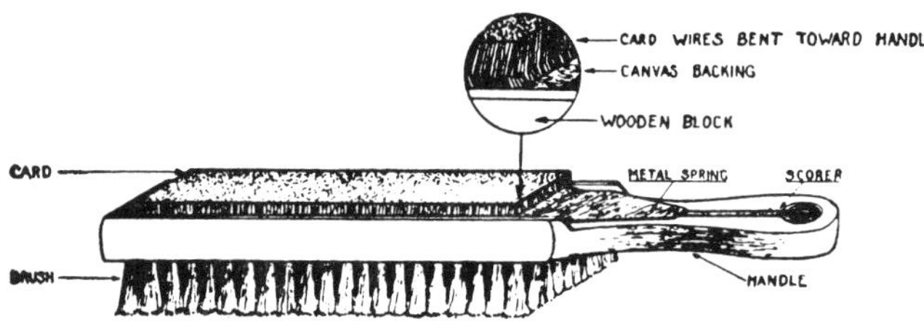

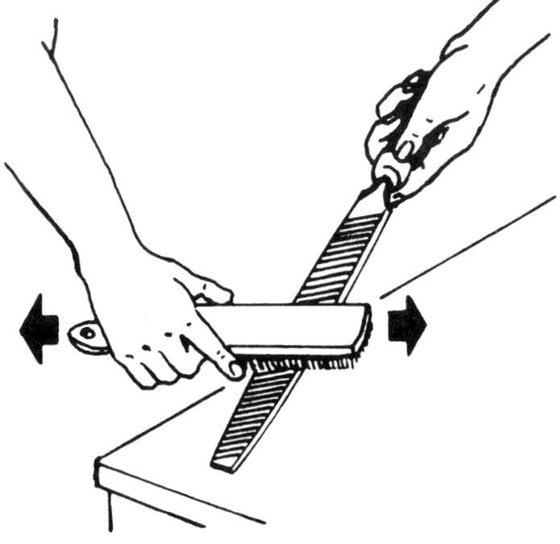

Figure 388 File card

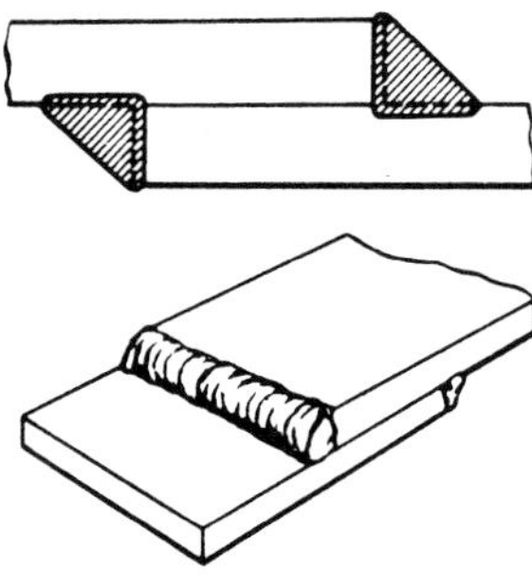

Figure 389 Fillet weld

file card Handled brush with stiff bristles used for cleaning files. (Figure 388.)

fill Earth and rocks used to fill in around a foundation, replacing material removed during the excavation. Any material used to fill in a low spot in the ground. See *cut and fill, filling.*

filler A wood putty used to fill in holes and cracks in the wood before the finish.

filler coat In painting, the base coat generally designed to fill in the wood pores, the *primer coat.*

filler metal In welding, metal that is added to the weld. See *welding rod.* Filler metal can also be solder or brazing filler metal.

filler rod In oxyfuel welding, a welding rod used to provide extra metal. The $\frac{3}{32}''$ iron powder general-purpose rod is probably most commonly used.

filler stud An expression sometimes used to describe a short stud, as a trimmer at a door opening.

fillet weld Weld made between two pieces that are at right angles to each other, as in a lap joint, tee joint, or corner joint. (Figure 389.) The fillet weld is approximately triangular in cross section.

fill-in specifications Short technical specifications that use a standard blank form. Materials and sizes are filled in for a specific structure. The Federal Housing Authority (FHA) supplies a four-page fill-in specification form. See *specification.*

filling In site development, the leveling or filling in of low ground. See *cut and fill, top dressing.*

fill insulation Loose insulation that is poured or blown in place between the framing members.

fillister A groove.

filter In a forced-air heating system, a device in the duct system that collects dust and airborne particles.

final set In concrete, the complete hardening that reaches a specified strength. Specifications often state the period of full-strength set is 28 days. Final set is determined by compressive resistance in pounds per square inch; the psi depends on the type of concrete.

fine aggregate Sand used in a concrete mix.

fines In concrete flatwork, a fine cement mixture worked over the surface to fill in the holes. Also, small aggregates; sand.

fine-textured wood Wood with very close fibers or grain.

finger joint End-to-end gluing of lumber using a series of V-shaped teeth in each end. (Figure 390.)

finial A decorative point that finishes off the top of an architectural detail, as on a spire or a stair newel. Its opposite is called a *drop* or *drop finial* and is used on the bottom of an architectural feature, pointing downward.

finish Completion of the interior of a structure, including installation of trim, carpeting, painting. In electrical work, the covering of outlet boxes and service equipment. In plumbing, the installation of plumbing fixtures; see *finish plumbing.*

finish carpentry Installation of trim and finish. See also *rough carpentry.*

finish coat Final finish coat of plaster. Figure 391 shows some of the plaster finishes.

finish floor Final finished floor, such as hardwood, in a room (*Figure 970*, p. 428.)

finish grade (FG) The final elevation and contour of the building site, after all grading and filling is completed. Opposed to the original contour or *natural grade* (NG).

finish hardware Visible hardware used to finish off the interior of a building, such as drawer pulls, doorknobs, and brackets. Opposed to *rough hardware.*

finishing trowel Rectangular, flat cement trowel used for smoothing concrete. See *trowel.*

finish plumbing The installation of plumbing fixtures, such as a water closet or bathtub. Also called *finishing.* See also *rough plumbing.*

finish stringer Outside decorated or finished stringer or carriage of a stair.

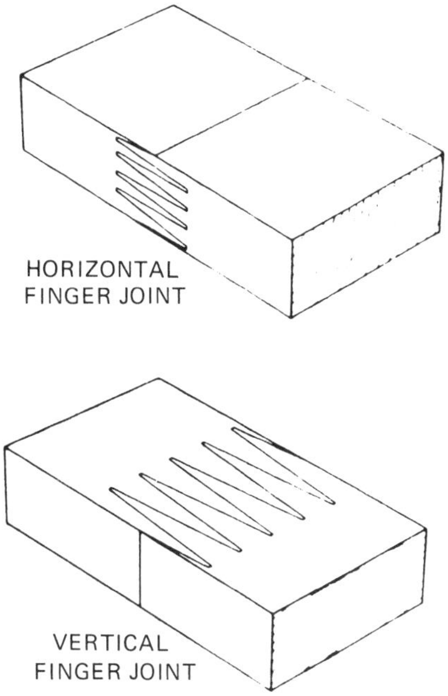

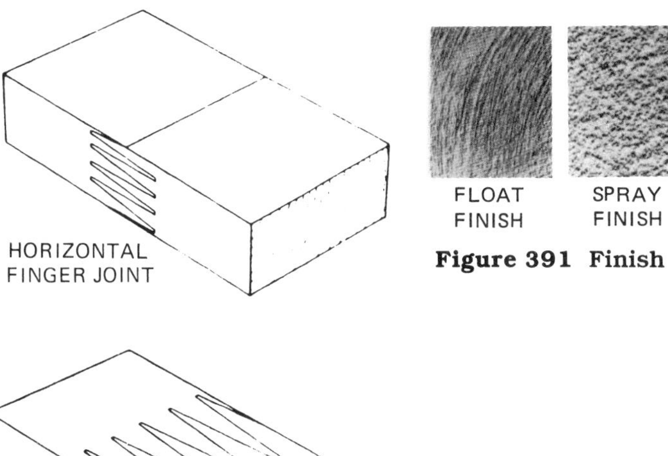

VERTICAL
FINGER JOINT

Figure 390 Finger joints

FLOAT FINISH
SPRAY FINISH
TEXTURE FINISH

Figure 391 Finish coats

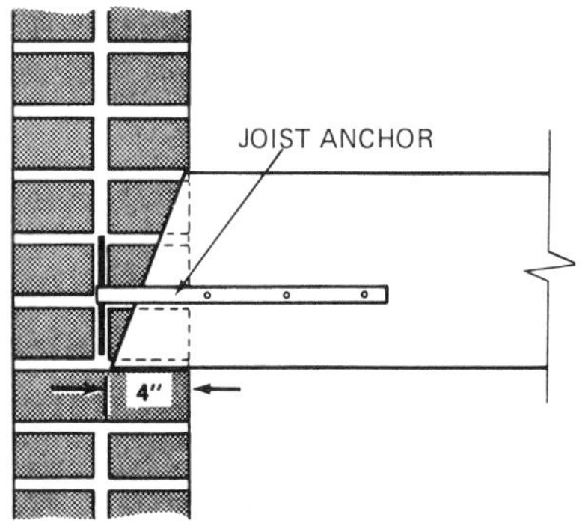

Figure 392 Firecut

Class of Fire	Typical Fuel Involved	Suitable Types of Extinguishers	Approximate Discharge Range	Approximate Discharge Time
A	Wood	Water types[1]	30-40 ft.	30 sec.-1 min.
	Paper			(Pump-tank may be
	Cloth	Foam[1] (very small		somewhat longer)
	Rubber	fires only)	30-40 ft.	40 sec.-2 min.
	Plastics			
	Rubbish	Multipurpose	5-20 ft.	8-25 sec.
	Upholstery	dry chemical		
B	Gasoline	Foam[1]	30-40 ft.	40 sec.-2 min.
	Oil	Carbon dioxide[2]	3- 8 ft.	8-30 sec.
	Grease	Halogenated agent[3]	4-15 ft.	8-15 sec.
	Paint	Regular or ordinary		
	Lighter fluid	dry chemical	5-20 ft.	8-25 sec.
		Multipurpose dry		
		chemical[4]	5-20 ft.	8-25 sec.
C	Motors	Carbon dioxide[2]	3- 8 ft.	8-30 sec.
	Appliances	Halogenated agent[3]	4-15 ft.	8-15 sec.
	Wiring	Regular or ordinary		
	Fuse boxes	dry chemical	5-20 ft.	8-25 sec.
	Switchboards	Multipurpose dry		
		chemical	5-20 ft.	8-25 sec.
D	Aluminum	Dry powder extin-		
	Magnesium	guishers and dry		
	Potassium	powder agents		
	Sodium	*only.*		
	Titanium			
	Zirconium		5-20 ft.	25-30 sec.

Footnotes:

[1]Cartridge operated water, foam and soda acid type extinguishers are no longer manufactured. These extinguishers should be removed from service when they become due for their next hydrostatic pressure test.

[2]Caution should be exercised when using in an unventilated, confined space.

[3]Avoid breathing the extinguishing agent (a gas) and the gases produced when the agent is applied to a fire, as they may injure the operator.

[4]Should not be used on fires in liquefied fat or oil of appreciable depth.

Figure 393 Fire extinguishers

finishing Completing the final operation, as the final house finish, or the smoothing of concrete flatwork.

fink truss A *W truss.* The term "fink truss" is generally used in heavy construction to refer to a steel W-shaped truss. See *truss.*

fire blocks Short wood blocks nailed between studs or joists to prevent the spread of fire. See *firestops.*

firebox Combustion chamber in a furnace.

firebrick Special refractory brick made of fire clay, used around combustion areas as in the back, sides, and bottom of a fireplace.

firecut Angular cut made in the ends of joists or beams that frame into a masonry wall. (Figure 392.) The cut allows the members to fall (as in the case of fire) without prying the wall apart.

fire door Door made of fire-resistant material.

fire escape Outside steel stairway used to escape a building in the event of fire.

fire extinguishers Four types of fire extinguishers are used (Figure 393): Classes A, B, C, and D.

Class A For ordinary combustibles, such as wood, paper, or cloth. Water or CO_2 is used in Class A extinguishers. *Symbol:* Green triangle with letter A.

Class B For flammable liquids such as oil or gasoline. Foam and CO_2 is used in Class B extinguishers (never water!). *Symbol:* Red square with letter B.

Class C For fires in electrical equipment. Nonconducting foam and CO_2 is used in Class C fire extinguishers (never water!). *Symbol:* Blue circle with the letter C.

Class D Fires in combustible metals such as magnesium, potassium, and powdered aluminum. A dry powder is used to smother the fire. *Symbol:* Yellow star with the letter D.

Fire extinguishers have a label noting the type of extinguisher (Class A, B, C, or D). Many extinguishers are suitable for several types of fire and are labeled with the appropriate symbols and letters. (Figure 394.)

Figure 394 Fire extinguisher: general-purpose

fireplace A masonry or metal opening with a flue, used for burning wood; a hearth. Figure 395 shows a masonry fireplace.

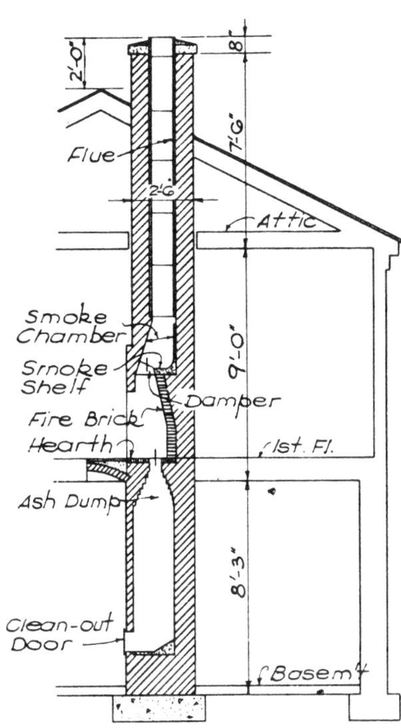

Figure 395 Fireplace section

fireplace, prefabricated Standing metal unit with metal flue for burning wood. See *zero-clearance fireplace.*

fireplace section Section or vertical cutaway made through the fireplace, chimney, and footing to show interior fireplace detail. (Figure 396.)

fireproof A term often used to designate *fire-resistant* material or construction.

fireproofing Any process that encloses structural members with fire-resistant material, as spraying steel with insulating material. Use of fire-resistant material in floors and walls.

fireproofing machine Air gun used for spraying insulating material on a surface.

fireproof structure Building that uses *fire-resistant* building materials. The structure will resist fire for a certain time. See *construction classifications.*

fire resistance Property of construction materials that prevents or retards passage of excessive heat, hot gases, or flames. See *construction classifications.*

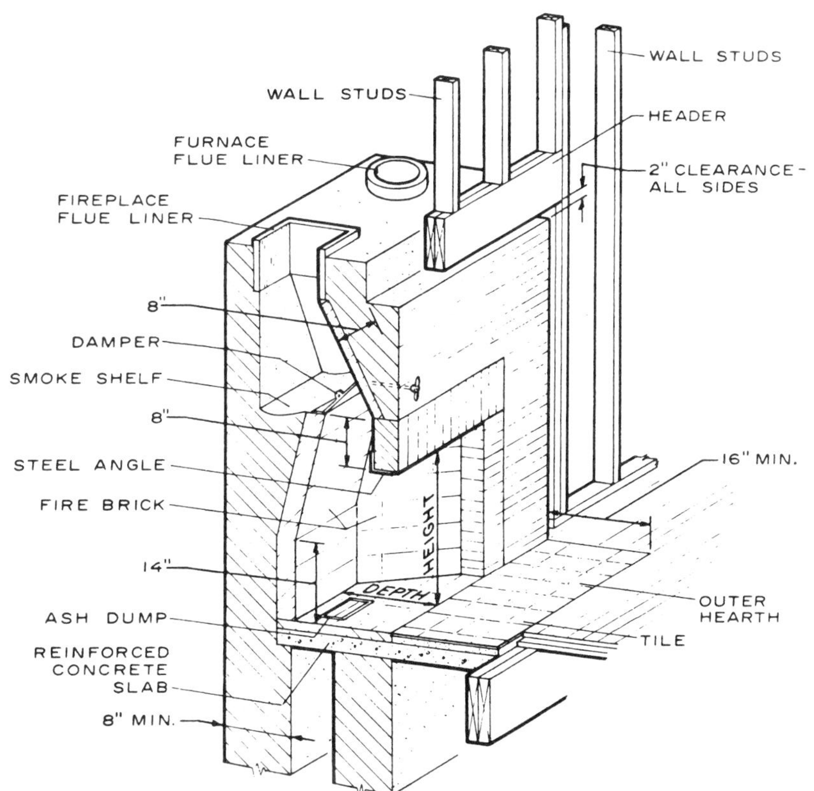

Figure 396 Fireplace: masonry

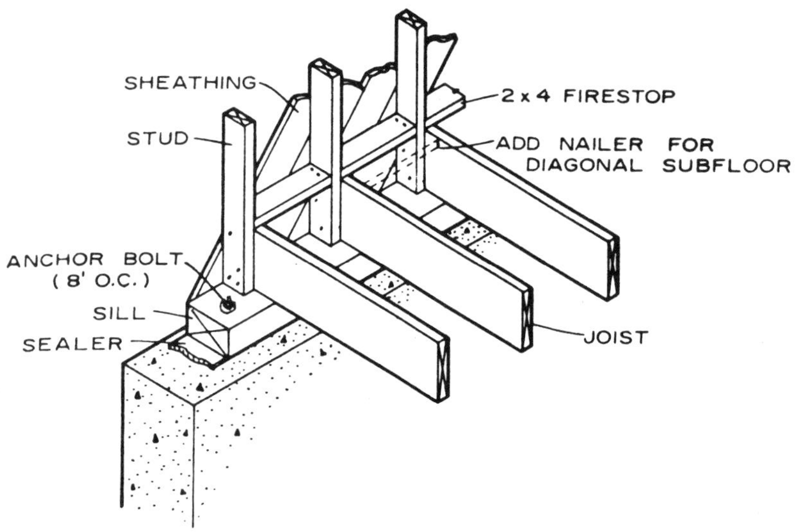

Figure 397 Firestops used between studs (balloon framing)

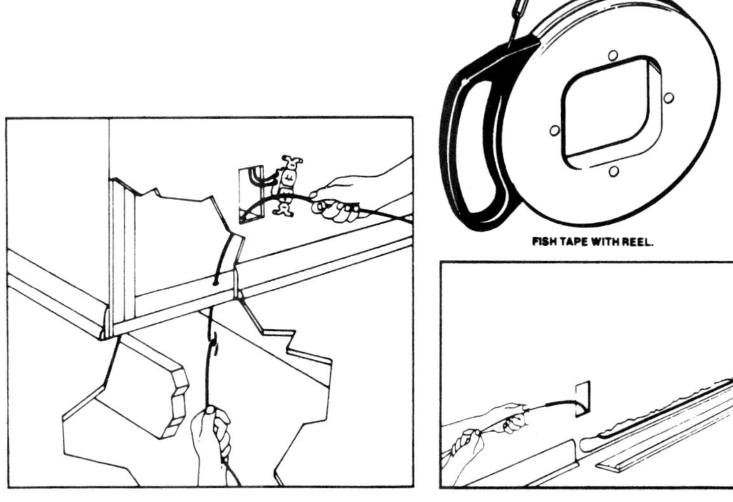

Figure 398 Fish tape

fire-resistance rating Ability of a material to resist the spread of fire, excessive heat, or hot gases for a set time, as for example $2\frac{1}{2}$ hours (determined by exposure of material in a furnace). See *construction classifications.*

fire-resistive material See *incombustible building material.*

firestops Wood blocks nailed between studding or joists to retard the spread of fire, heat, or hot gases. Also called a *draft stop.* (Figure 397.)

fire wall Wall designed to restrict the spread of fire, used between living units in multiunit construction. Wall may extend from foundation to roof. Also used in commercial and industrial construction.

firmer tools Ordinary bench woodworking tools.

firming chisel All-purpose wood chisel with a straight, flat blade; often used to clean and trim joints. It may be operated by using both hands or by striking with a mallet. See *wood chisels.*

first floor plan Plan view of the first floor. An orthographic view that shows location of walls, partitions, doors, windows, stairs. Symbols show the location of heating, plumbing, and electrical fixtures. See *floor plan.*

fishing Drawing electrical conductors through conduit.

fish mouth In roofing, an asphalt shingle that does not lie flat and bubbles or humps up.

fishplate Wood or plywood strips used to fasten two wood members together end-to-end at a butt joint. Pieces or joining plates are used on both sides of the joint to form a sandwich. The pieces are glued and screwed or nailed. See *splice* (*wood*). (*Figure 924*, p. 408.)

fish tape Flexible steel tape that is run through walls or electrical conduit and then used to pull electrical wires through. (Figure 398.)

fittings In sheet metal, manufactured ductwork pieces that fit onto the straight ductwork run. See *ductwork.* In electrical work, mechanical connectors screwed onto the end of conduit, such as a lock nut or bushing. In plumbing, a connector used to fasten piping or tubing ends together. (*Figures 167*, p. 62; *249*, p. 95; *250*, p. 95; *444*, p. 182; *691*, p. 297; *708*, p. 306.) Also *plumbing trim:* hardware, fittings, and supplies, such as faucets, shower heads; or a waste outlet, such as a sink drain strainer and fixture traps.

fixed arch Unhinged arch. See *arch.*

fixed beam Beam with fixed ends.

fixed light Window that does not open, a fixed window.

fixed-sash window Window whose sash is permanently part of the window frame and cannot be opened, such as a picture window or bay window.

fixture Heating, plumbing, and electrical units added to a building, as hot-water heaters, furnaces, bathtubs, shower stalls, sinks, light outlets, and range hoods. In a plumbing system, a receptacle that discharges or receives water or waste, such as a faucet, shower head, water closet, or floor drain. (*Figures 76*, p. 30; *384*, 153.) In woodworking or metal working, a device that holds the work during a manufacturing or fabricating process. See *jig.*

flag A flat stone, *flagstone.* In drafting, a symbol calling attention to or pointing to a drawing change or a special condition.

flagging Walkway made of *flagstone* or *flags.* The natural stone itself.

flagstone Natural flat slabs of stone used for making walkways or flooring. One stone is called a *flag.*

flake board A manufactured pressed board made up of wood flakes and a resin binder.

flame-cut plate Steel plate that has been cut from a larger plate with an oxyfuel (oxyacetylene) torch.

Figure 399 Flanged pipe valve

flame cutting Cutting metal using an oxyfuel (oxyacetylene) cutting torch. A flow of pure oxygen is used to rapidly oxidize the metal and burn a cut in the metal. See *cutting torch.*

flame spread Rate at which combustion will spread in a material or structure.

flange The flat top or bottom surface of a *steel beam* or *channel iron.* In cross section a flange is at a right angle to the *web.* See *channel iron, steel beam.* In reinforced precast concrete, the horizontal part of a T slab or beam. See *tee.* In plumbing, a fitting which has a rim or edge on the end. The flanges bolt to each other. On batt insulation, a flange may be provided for stapling to joists or studs.

flanged pipe Piping with flanges on the end; the flanges bolt together. (Figure 399.)

flange forms Molded forms used for shaping concrete floor systems. Figure 400 shows a flange form used for one-way concrete joist construction. See also *dome forms.*

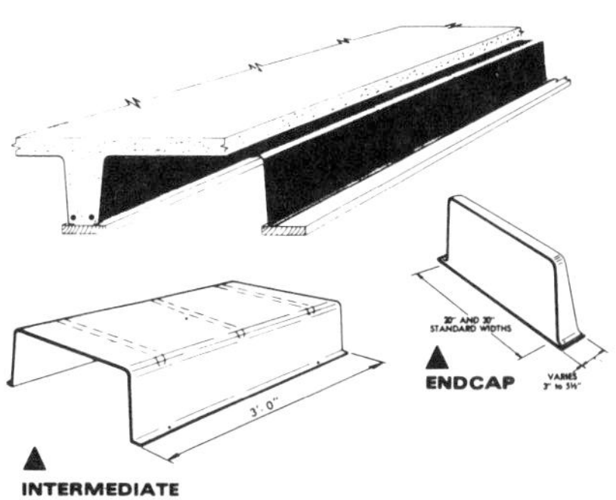

Figure 400 Flange form used for one-way reinforced concrete joist construction

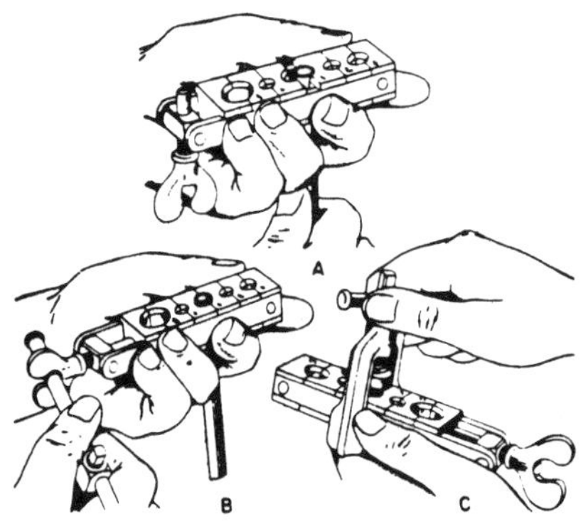

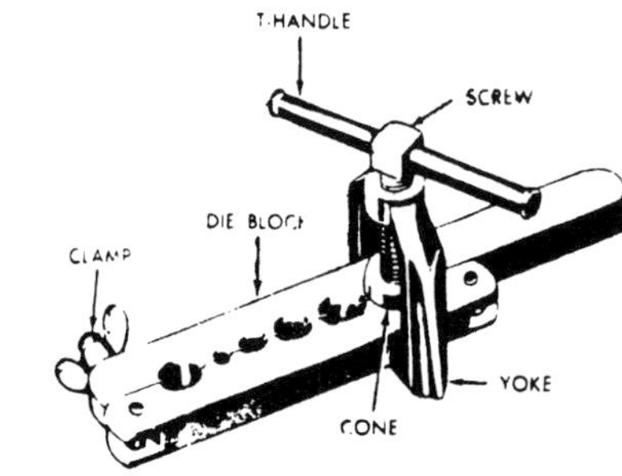

Figure 401 Flaring tool

flap wheel sander See *abrasive flap wheel.*

flare Widening the end of copper tubing so another piece of copper can fit in the mouth or so the widened mouth can fit over a nipple. See *flaring tool.* In general, to spread outward.

flared-joint fittings Cast-bronze fittings used to connect soft-drawn copper tubing for water supply piping. See also *copper tubing.*

flaring tool Tool used to flare the end of copper tubing. (Figure 401.)

flashback In oxyfuel (oxyacetylene) welding, a condition where a flame starts burning inside the torch, indicating a malfunction or damaged torch. The torch should be turned off and fixed before any further welding is done.

flashing Sheet metal or plastic sheathing used at major breaks in a roof, around or over exterior openings, or wherever water might leak in. Examples: at intersecting roof; roof valleys; where a roof abuts against the side of a house, as a dormer or shed roof; around chimneys or skylights; over windows and outside doors. (Figure 402.) The flashing creates a seal and is covered by the roofing or siding. See *brick veneer, cap flashing, counterflashing.*

flash point Temperature at which a volatile liquid gives off enough fumes to form an ignitable mixture.

flash set In concrete or mortar, a very rapid undesirable setting that does not develop a specified strength. A *quick set.*

flat In apartment buildings, sometimes used in reference to an apartment or living unit. In painting, a paint that has a dull finish when dry. Also, a horizontal, level surface.

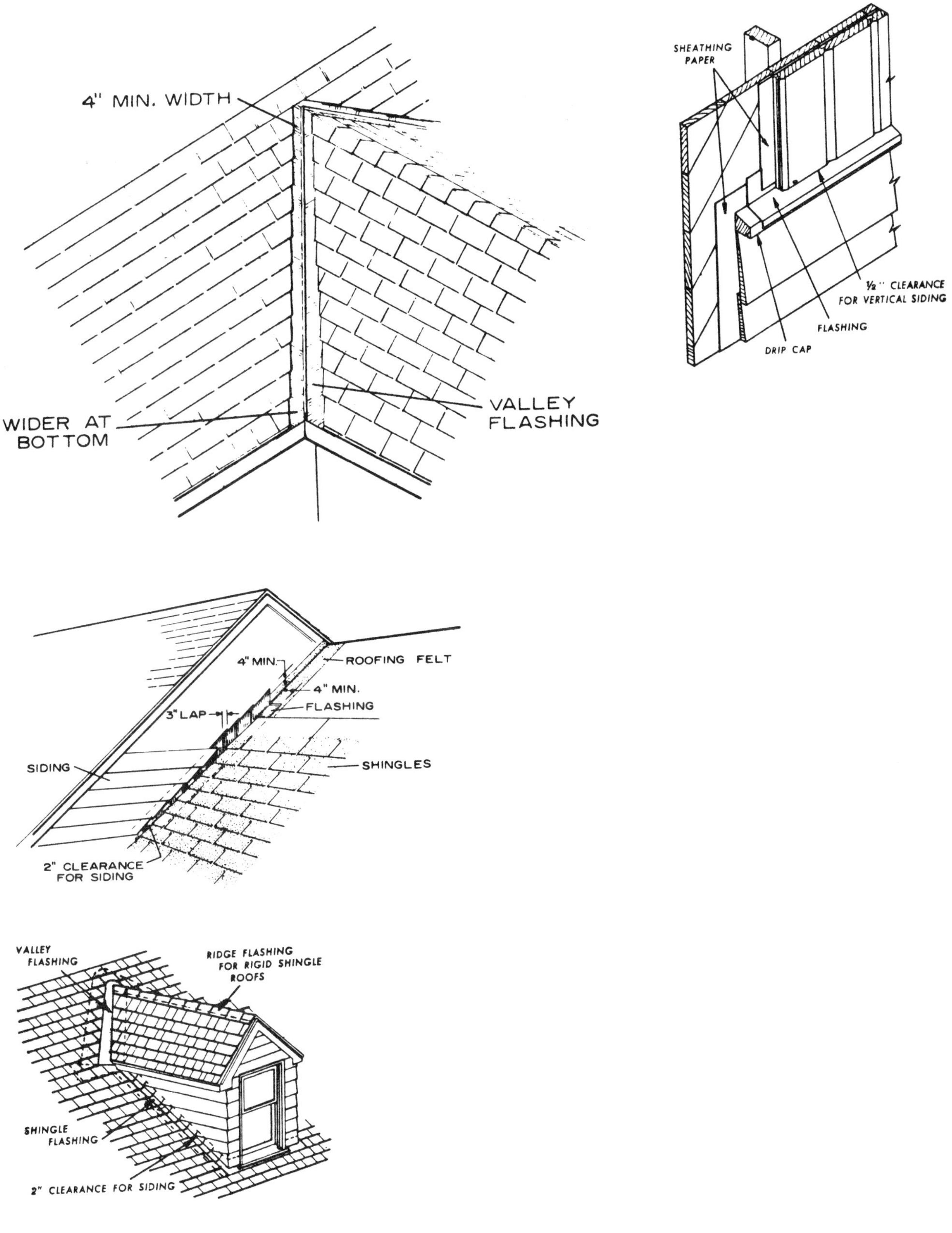

Figure 402 Flashing

Figure 403 Flat file

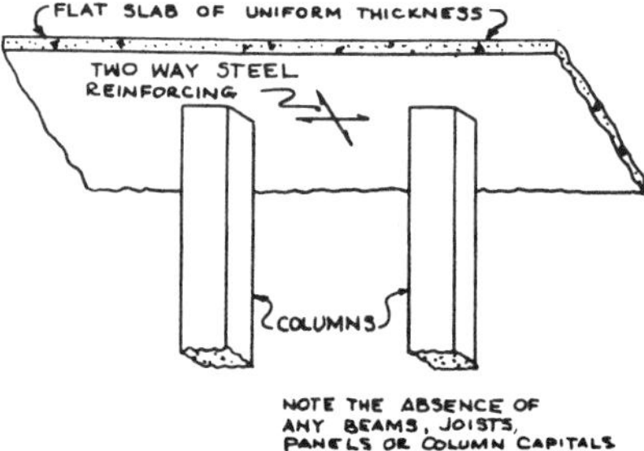

Figure 405 Flat-plate concrete floor

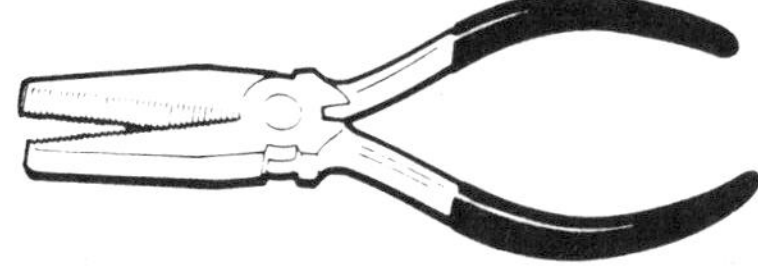

Figure 404 Flat-nose pliers

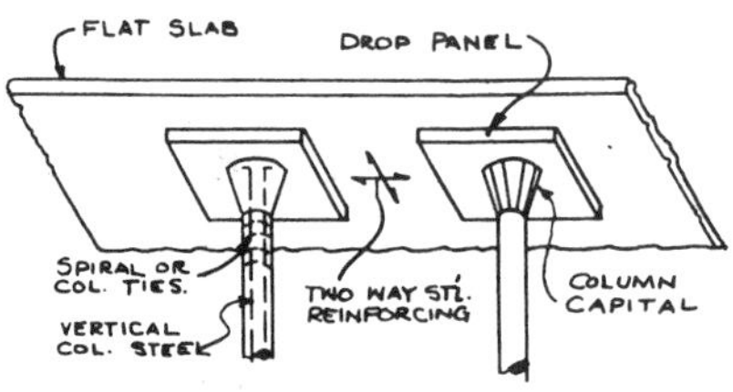

Figure 406 Flat-slab concrete floor

flat corner iron A flat L-shaped metal piece used for strengthening corner joints. Holes are bored and countersunk so wood screws may be used. See *L brace.* See also *corner brace, mending plate, T brace.*

flat file Flat file used for filing metal surfaces. (Figure 403.) See *file.*

flat grain Softwood lumber with the grain (annual growth rings) running at less than a 45° angle on the face of the board. Called *plainsawed* in hardwood lumber. *Flatsawn.*

flat-head wire nail Thin-shank nail used for general purpose.

flat-nose pliers Long-nose pliers with flat, square-end jaws. (Figure 404.) Used for wiring, bending, and handling sheet metal.

flat-plate concrete floor Reinforced concrete slab floor with bars running in two directions supported by simple columns. (Figure 405; *784,* p. 344.) See *reinforced concrete*

flat-plate solar collector Solar-heat collector that has a flat surface oriented (slanted) to the south to catch the sun's rays. The sun's rays strike a flat black surface. The amount of solar radiation striking the collector is called *insolation.* See *concentrating solar collector.*

flat-position welding Weld made from the upper side of a joint when the work is laid flat and horizontal. See *welding position.*

flat roof Roof that is almost flat but still has a slight slope for water runoff. See *roof.*

flatsawn Softwood lumber with the grain (annual growth rings) running at less than 45° to the face of the board. (*Figure 699,* p. 301.) Called *plainsawed* in hardwood lumber.

flat-slab concrete floor Reinforced concrete slab floor with bars running in two directions and supported by columns with enlarged tops (capitals). Two-way reinforcement forms a criss-cross pattern, and the floor is often called a two-way flat slab. (Figure 406.) See *reinforced concrete floors.*

flat wood rasp Rough flat rasp for working wood. See *rasp.*

flatwork A concrete slab, as a sidewalk, driveway, or floor.

flaw A defect.

Flemish bond Masonry bond pattern where the course is made up of alternate stretchers and headers; headers in alternate courses are centered over stretchers above and below. See *brick bond.* See also *Flemish common bond.*

Flemish common bond Masonry bond pattern that is a variety of the *Flemish bond.* Several courses of *running bond* are alternated with a course of alternate stretchers and headers. See *brick bond.*

flex handle Steel handle with a hinged square drive at the end. Used for turning sockets.

flexible armored cable See *armored cable.*

flexible curve In drafting, a flexible but stiff guide used to draw long or complicated curves.

flexible metal conduit A flexible conduit used in dry nonhazardous, above-ground locations where not exposed to oil or gasoline. Conductors are fished through the installed conduit. See *conduit.*

flexible metallic tubing A flexible tubing used in dry, concealed, above-ground locations where not subject to physical damage, as in ceilings or ductwork. Conductors are fished through the installed tubing. Only short runs are normally made.

flexible paving Surfacing with aggregate and bituminous material. (Figure 407.) See also *paver, rigid paving.*

flexural strength Resistance to a bending force.

flight Stair run unbroken by a landing.

flight of stairs A straight, continuous run of stairs without a break (as for a landing). See *stair run.*

flint A natural abrasive mineral (white quartz) widely used on abrasive papers, belts, and discs. It was the first widely available abrasive. Its widespread use and popularity led to the development of the designation *sandpaper,* since sand is mainly quartz particles. See *abrasive.*

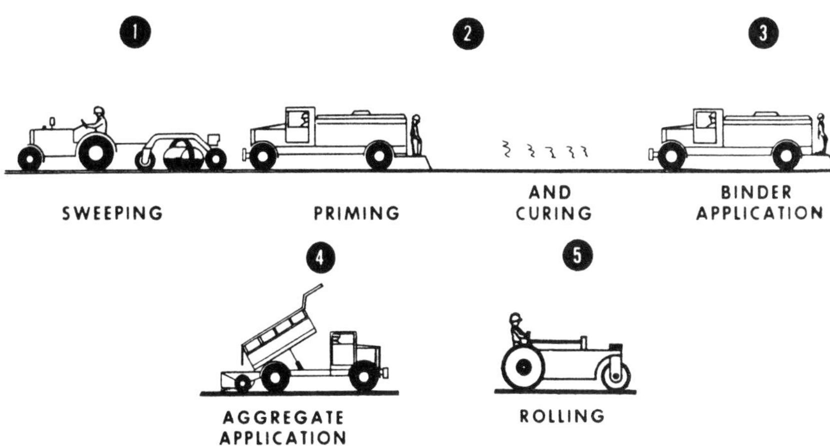

Figure 407 Flexible paving: single coat

flitch Lumber with bark on it — cut from the outside of a tree. In veneer cutting, a log or bolt from which the veneer is cut. The process of cutting veneer plies off the bolt. Also a *flitch beam.*

flitch beam A beam built up of lumber pieces with steel plates; the assemblage is bolted together to form a strong unit.

float In concrete flatwork and plastering, a kind of flat trowel used for final finish to assure a smooth surface. Magnesium, wood, or rubber-based floats are used. (Figure 408.) See also *bull-float, trowel.*

float coat In cement work, a final finish put on with a float.

floated coat In plastering, the coat applied over the base or scratch coat.

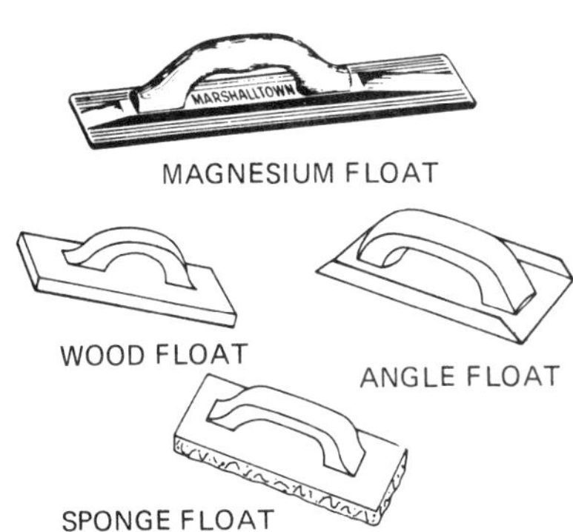

Figure 408 Floats

Figure 409 Floating

floating The spreading and smoothing of plaster or concrete with a float. (Figure 409.)

floating screed A strip of plaster laid down as a guide for the plaster thickness. See *screed.*

floating slab A reinforced concrete slab that spans from outside walls to outside walls.

float time Slack time in a work schedule.

floodlighting Highlighting a structure at night by carefully placed lights, usually hidden or recessed near the ground. Reflectors or shields are used so that light is directed only toward the structure.

floor Bottom surface of a room; the level in a building, as basement floor, first floor, second floor, and so on.

floor chisel All-steel chisel used in cutting floor boards.

floor-covering installer One who installs carpeting or composition floor tile.

floor drain Waste fitting located in basements, garages, and the like that receives any water leakage, overflow, or spillage. It normally has its own *P trap.*

floor framing Structured members, such as joists, headers, sills, that are framed together to form a floor. (Figure 410.) See also *platform framing.*

floor framing plan Plan view showing the layout, size, and length of the structural members of a floor.

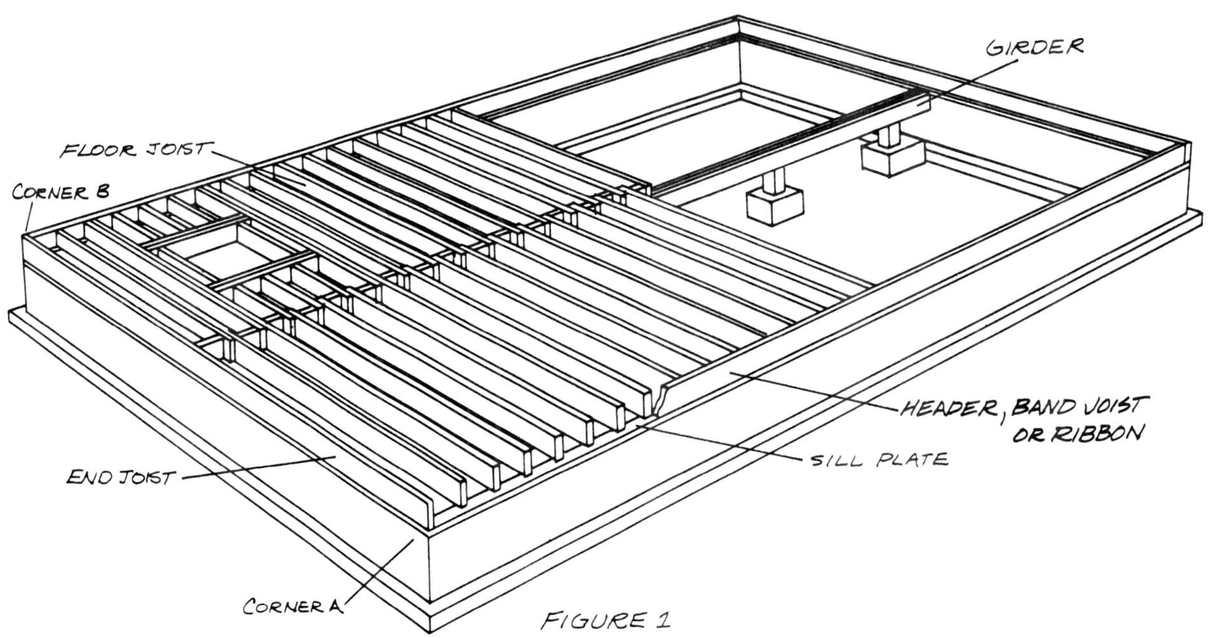

Figure 410 Floor framing

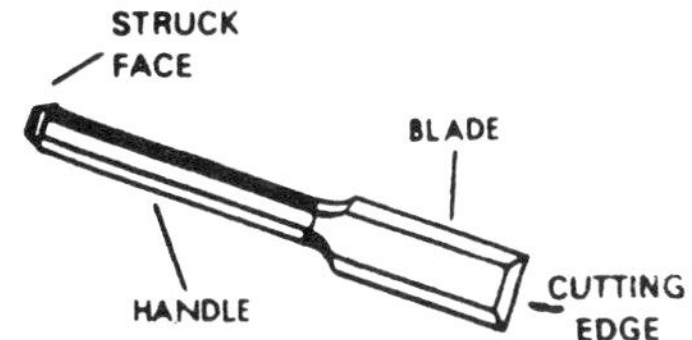

Figure 411 Flooring chisel

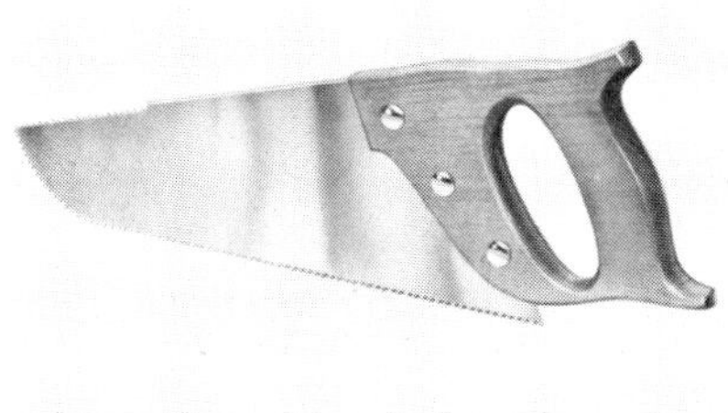

Figure 412 Flooring saw

flooring Building materials used to construct a floor.

flooring blade Circular power-saw blade for cutting through flooring, including nails. See *saw blades.*

flooring chisel All-metal chisel used for rough cutting work where metal, such as nails, may be struck. (Figure 411.)

flooring saw Special saw designed for cutting through flooring. It has a curved blade with teeth on both sides at the top. (Figure 412.)

floor jack A *jack post.*

floor joist Main framing member in a floor.

floor plan Plan view (orthographic view looking straight down) of a floor. An imaginary cut is made across the building 2 or 3 feet off the floor to expose the floor. Walls, partitions, doors, windows, fixtures, stairs, and dimensions are shown. Figure 413 shows a simplified floor plan and indicates how the drawing is made. Floor plans are identified by their location in a structure: a *first-floor plan, second-floor plan, basement plan, foundation plan.* See also *plans, plan symbols.*

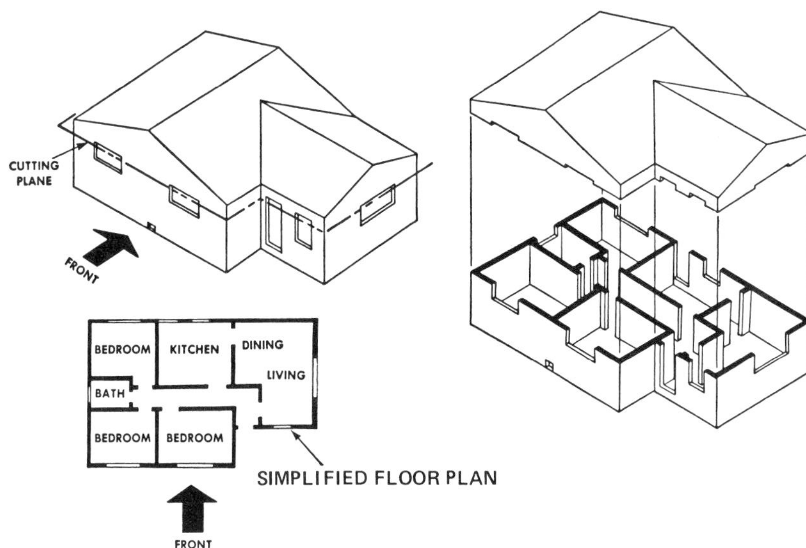

Figure 413 Floor plan

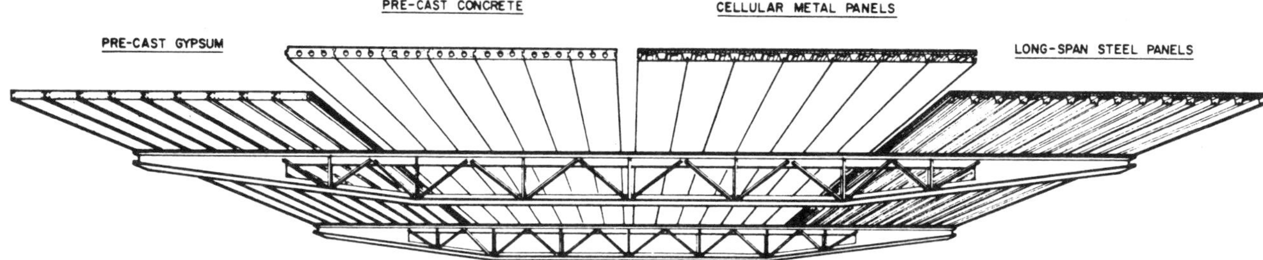

Figure 414 Floor systems: industrial

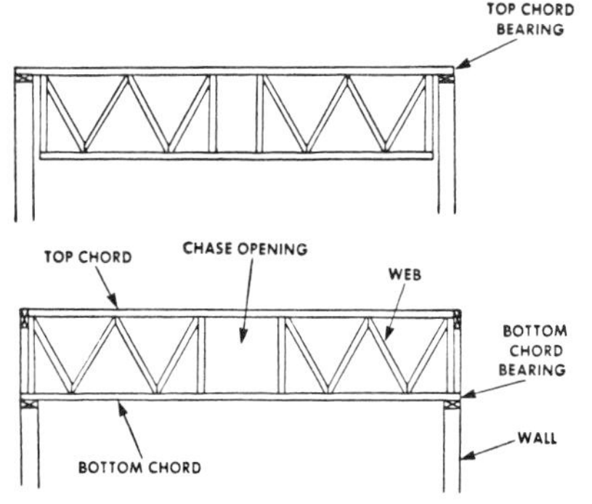

Figure 415 Floor trusses

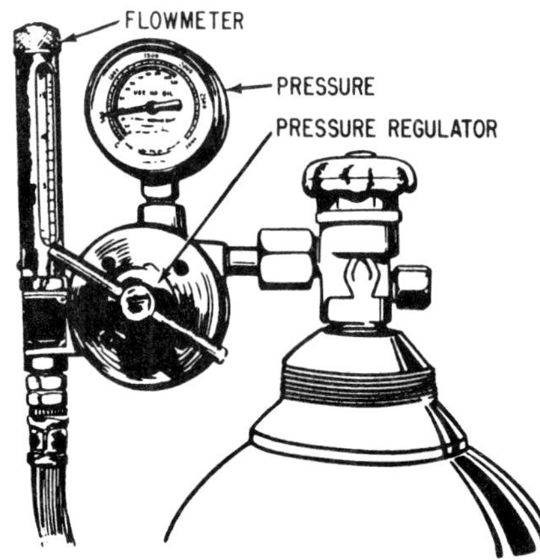

Figure 416 Flowmeter

floor sander Portable power sander used to finish floors. See *belt sander, orbital sander.*

floor system Complete makeup of a floor, including subfloor and the finish floor. A residential frame floor would commonly have a plywood subfloor with a hardwood or carpeting finish. Figure 414 shows flooring systems used in industrial building.

floor truss A flat, manufactured truss used for floor joists. (Figure 415.) See *truss.*

flowmeter Measure used in gas shielded arc welding to indicate gas flow rate. (Figure 416.)

flow rate In plumbing, the volume of water in gallons per minute that goes through a fixture.

flue Vertical passage that carries off smoke and gas fumes from a fireplace or furnace. See *fireplace.*

fluid-type solar system An active solar heating system that uses a liquid (water, antifreeze, oil) to collect and carry solar heat. A heat exchanger is used to transfer heat to the house heating system. See *air-type solar system.*

fluorescent lamp A lamp having with a long glass tube with a fluorescent coating and filled with mercury vapor.

fluorescent lighting Lighting using tube fluorescent lamps.

flush Two or more surfaces that are even or level with each other.

flush door Door that has flat surfaces on each side; may have a light. See *door.* Flush doors have a hollow core or a solid center with plywood faces.

flush valve Valve for flushing, as for a water-closet tank. See *water closet.* (*Figure 1075*, p. 475.)

flush weld Weld that is level (flush) to the pieces joined.

flute reamer Tool for cleaning out burrs on the inside of a pipe. (*Figure 787*, p. 345.)

flutes Long decorative grooves cut in a member, as in a small wood column.

flux In soldering or welding, a material or paste used to free the metal parts from oxides so they can be joined.

flux-cored arc welding (FCAW) Arc-welding process. Arc is created between a continuous electrode and the work piece. Electrode provides filler metal, shielding is provided from the vaporization of the flux creating a cloud over the weld area. The shielding protects the weld from the oxygen in the air, which could cause contamination. (Figures 417, 418.)

fly ash Soot used as an aggregate in making lightweight concrete.

flying bond A *common bond.*

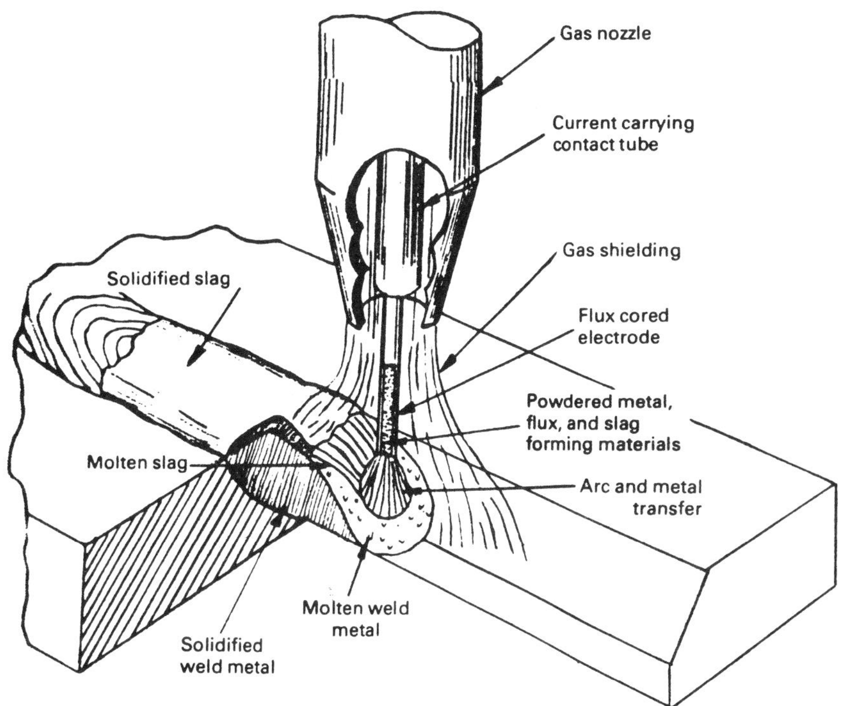

Figure 417 Flux-cored arc-welding process

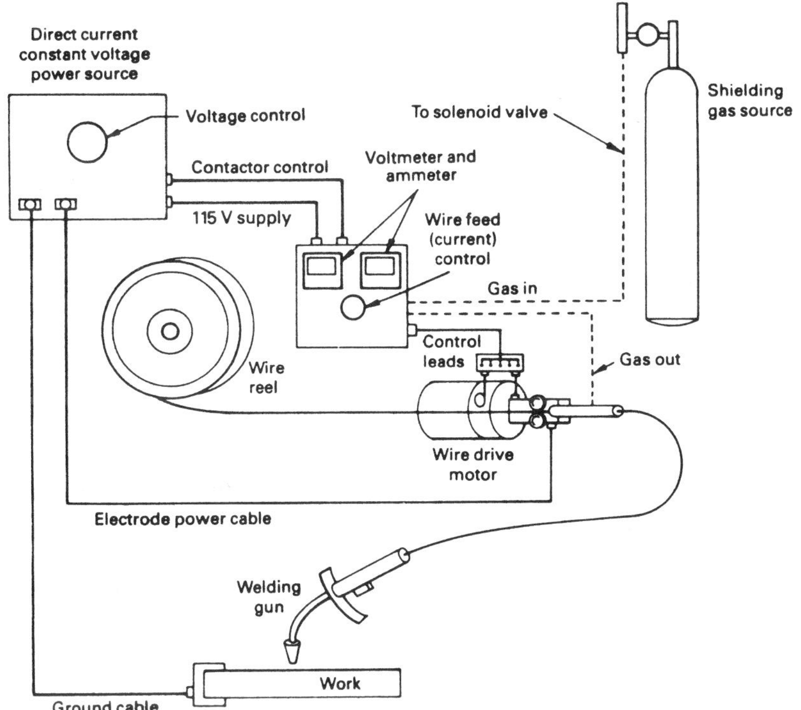

NOTE: Gas shielding is used only with flux cored electrodes that require it.

Figure 418 Flux-cored arc-welding equipment

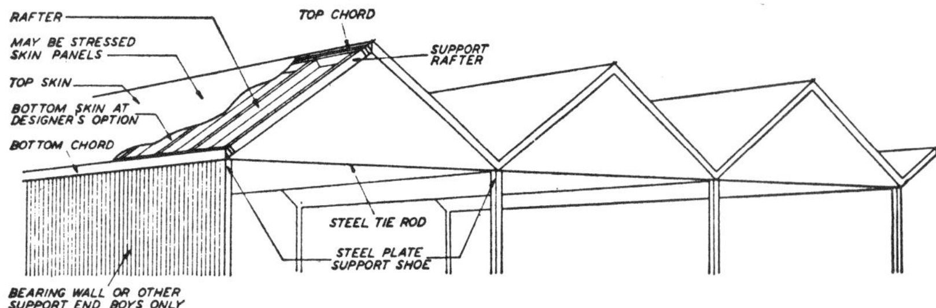

Figure 419 Folded-plate roof system

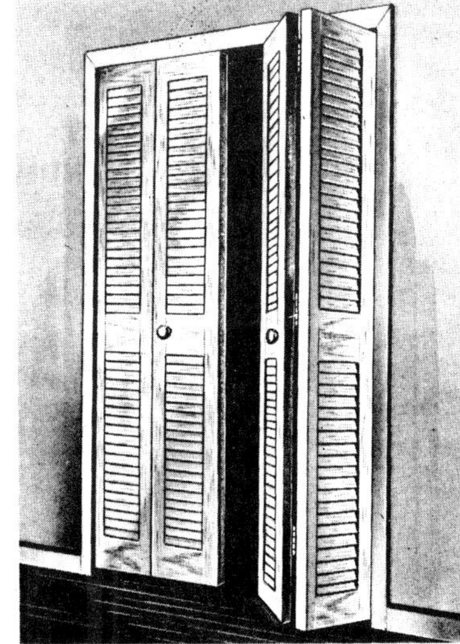

Figure 420 Folding door

Figure 421 Folding rule

flying buttress A masonry support (like part of an arch) that arcs out from a wall to give support to the wall.

flying form Large, prefabricated form normally with all support work in place; form is shifted with a crane.

fly rafter Rafter at a gable end overhang that is supported out from the gable by *lookouts* or a *ladder.* The fly rafter is framed against the extended ridgeboard. See *barge rafter.*

fogging Spraying concrete with a fine mist as an aid in curing.

foil back Insulation with a metal foil on one side to serve as a heat reflector.

foil-back gypsum board Gypsum-base wallboard with an aluminum foil coverage on the back. The foil works as a vapor barrier, heat reflector, and provides insulation.

foil insulation Insulation with metallic (aluminum) foil backing. The foil backing always faces the inside of the structure.

folded-plate roof system Series of inverted-V shaped panels that work as giant beams. Used to span large spaces without internal support. (Figure 419.)

folding door Door with sections that fold back together when open. (Figure 420.)

folding rule Foot rule that folds together. (Figure 421.) A ***zigzag rule.*** See also ***extension rule.***

folding stair Stair that folds up into the ceiling. Used as a stair into an attic area.

foot A measurement of 12″; the base of support, as a column or pier base; bottom support of a furniture piece, as a chair or table foot.

foot cut Cut near the end of a rafter to create a seat against the plate; a *seat cut.*

footing Support base for a foundation wall, column, pier, or chimney. (Figures 422; *529*, p. 227; *784*, p. 344.) A widened concrete base that spreads the weight over a greater area. Footings are normally concrete, but other material, including treated wood, is used.

footing drainage Drain tile run around the outside and/or inside of the building footing to carry off ground water. See *foundation drainage.*

footing forms Retaining walls (forms) to hold the footing concrete until it sets. (Figure 423.)

footing plate In a wood foundation, the base or footing upon which the wood wall rests.

force The loads or forces that act on the structural parts of a building or structure: *compression, shear, tension, torsion.* See *stress.*

force cup A rubber suction device and handle used for clearing clogged drains. The cup is fitted over the drain opening and pushed down, creating a surge of pressure, then pulled up to create a suction. Also called a *plumber's friend* or *plumber's helper.*

forced hot-water heating See *hot-water heating.*

forced warm-air heating Heating by distributing warm air thrugh a ductwork system; a blower is used to "force" the warm air through the system. Three types of ductwork systems are used to distribute the air: *extended plenum* (*Figure 376*, p. 149), *perimeter loop* (*Figure 673*, p. 291), and *radial perimeter* (*Figure 760*, p. 334). See also *ductwork, gravity warm-air heating, furnace, plenum, under-floor plenum.*

forehand welding A welding technique where the welding torch slants in the same direction as the process of the weld. (Figure 424.) Opposite to *backhand welding.*

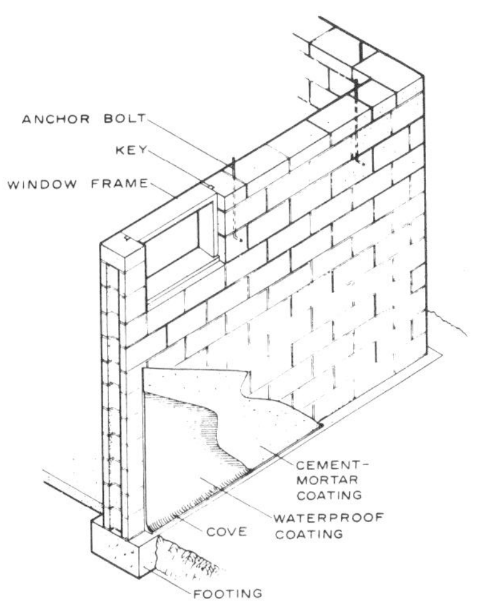

Figure 422 Footing supporting the foundation wall

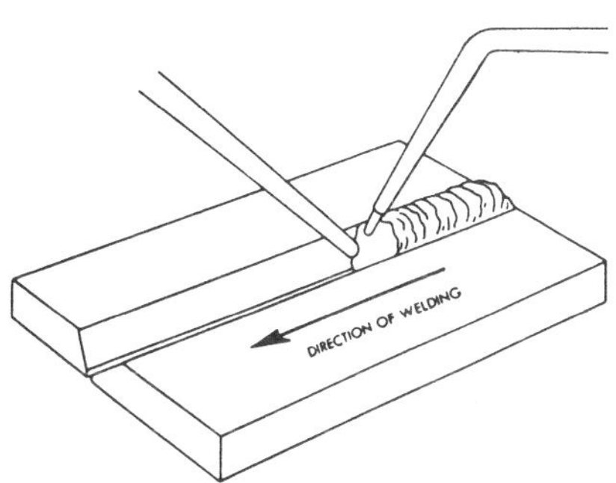

Figure 424 Forehand welding

fore plane A long bench plane, commonly 18″ with a wide $2\frac{3}{8}$″ blade, used on long work pieces. See also *plane.*

foresight In surveying, a rod reading taken at an unknown location, as opposed to *backsight,* which is a sighting taken to a previous, known point. Also called a *minus sight.* Abbreviated as F.S. or −S. (*Figure 54,* p. 22.)

form clamp Clamp that holds the forms in place while the concrete sets. See *forms.*

form hanger Bent hanger used to hold forms around steel beams. (Figure 425.)

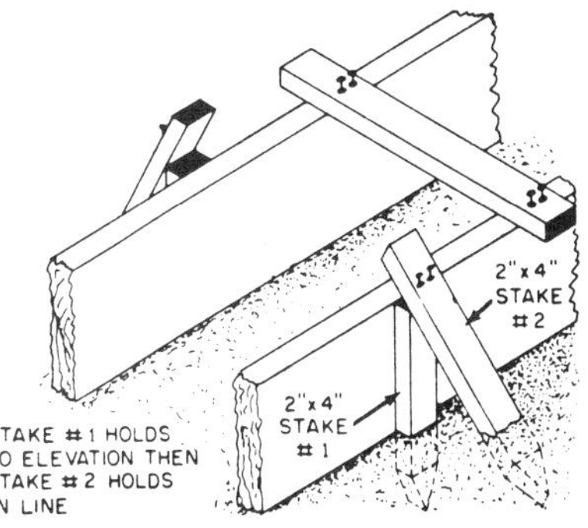

Figure 423 Footing form

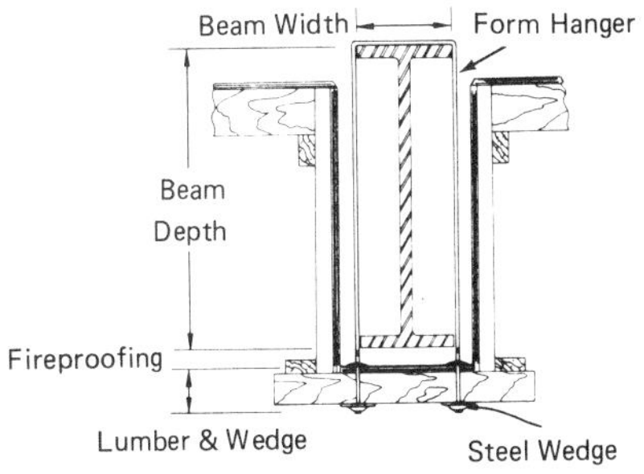

Figure 425 Form hanger

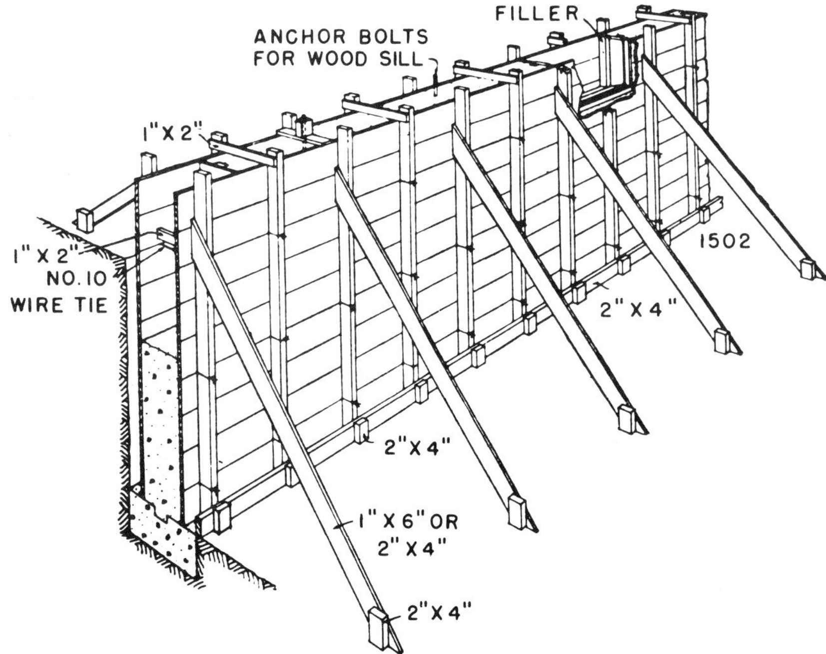

Figure 426 Forms holding the plastic concrete until it sets

form oil Oil spread on the inside surface of concrete forms to allow the forms to be easily separated (stripped) when the concrete is set.

form-panel layout Plan view of the location of the various form panels for a concrete wall; size of each panel is specified.

forms Retaining walls built to hold concrete until it sets. Figure 426 shows forms used to hold foundation walls. See also *bulkhead, column forms, curb forms, dome forms, facade forms, flange forms, flying form, form hanger, ganged forms, she bolt, sidewalk forms, slab formwork, wale, wall forms.*

form stop A braced plank placed across the width of a form, running from top to bottom inside the form, to hold the concrete at the end of a pour. Also called a *bulkhead.*

form tie A metal bar, strap, or wire used to hold concrete forms together. The metal is broken off at the end after the concrete sets, so most of the form tie remains in the concrete. See *bar tie, coil tie, snaptie, tie rod, tie wire.*

formwork The support constructed to retain concrete until it sets. See *forms.*

foundation The lowest division of a wall for a structure intended for permanent use; that part of a wall on which the building is erected. The part of a building which is below the surface of the ground and on which the superstructure rests. See *basement wall foundation, perimeter wall foundation, pier foundation, slab foundation, wood foundation.*

foundation bolt Bolt set in the foundation and bolted to the sill. Normal size is $\frac{1}{2}''$ diameter by 6″, 8″ 10″, or 12″ long. See *anchor bolt.* (*Figure 21*, p. 11.)

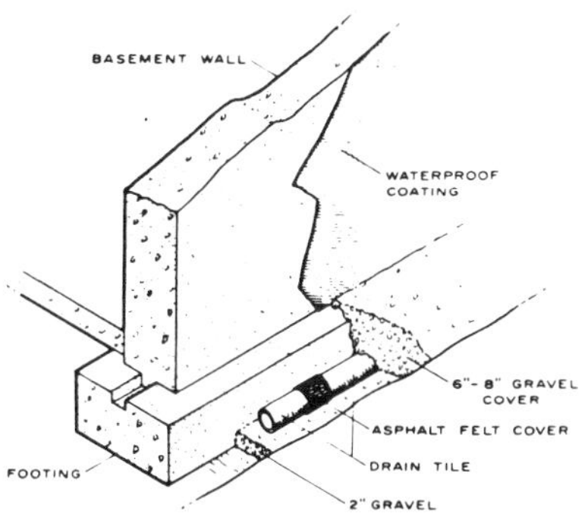

Figure 427 Foundation drainage

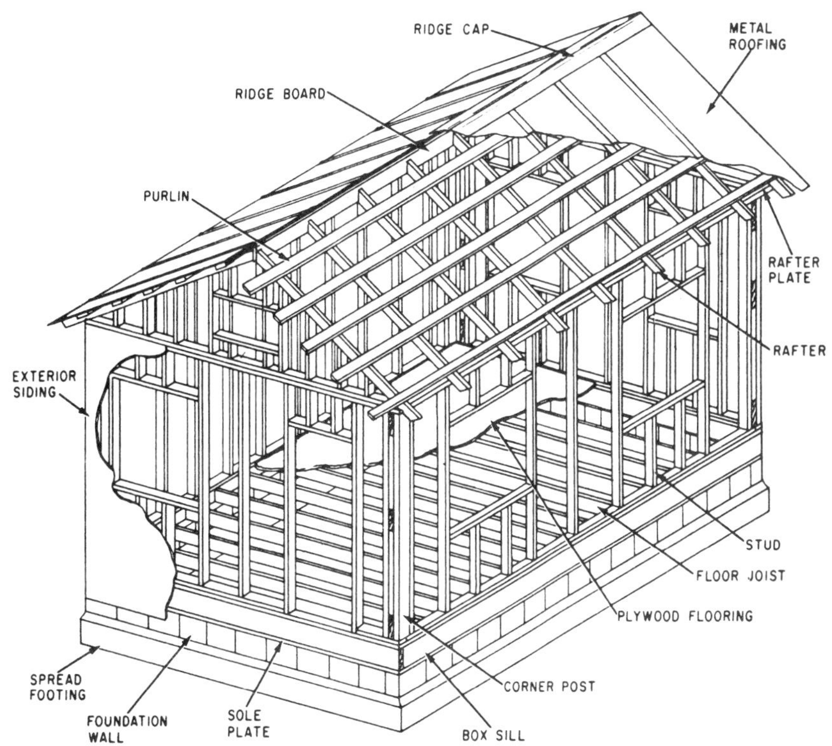

Figure 428 Frame construction

foundation drainage Drain around the outside edge of a building foundation, designed to carry away ground water. (Figure 427.) Water seeps through the loose joints between each pipe section, or a perforated tile or plastic tubing is used.

foundation plan Plan view that shows the footings and foundation wall. See also *basement plan.*

foundation vent Ventilation openings in foundation walls. See *wall vent.*

foundation wall Bearing wall that supports a structure; the wall runs around the outside of the house and supports the joists and any beams. Foundation walls are used with a crawl space or with a structure with basement. (*Figures 267*, p. 100, 72, p. 28.)

frame Wood structure; the wood framework of a house.

frame construction A type of construction in which the individual structural parts are mainly of wood (or wood substitute); a structure that depends upon a wood frame or skeleton for support. Figure 428 shows basic frame construction. See also *platform framing.*

framed beam connection In steel construction, a steel beam that is connected by fittings (angles or tees) to the web of another beam or the flange of a column. The load is transferred from the beam to the fittings to the web of the support. Figure 429 shows a beam framed into the web of another beam. See also *seated beam connection, spread.*

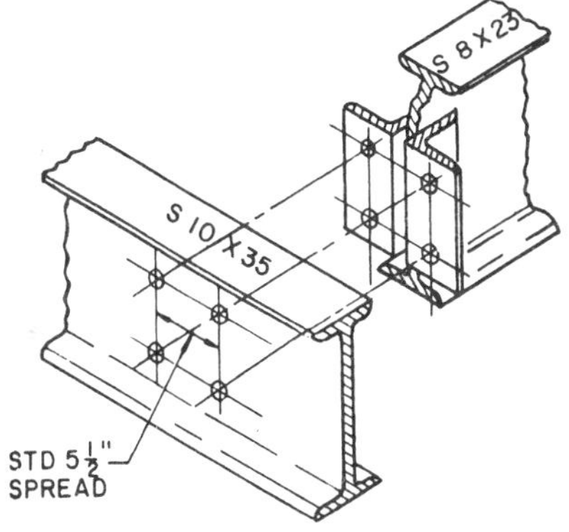

Figure 429 Framed beam connection

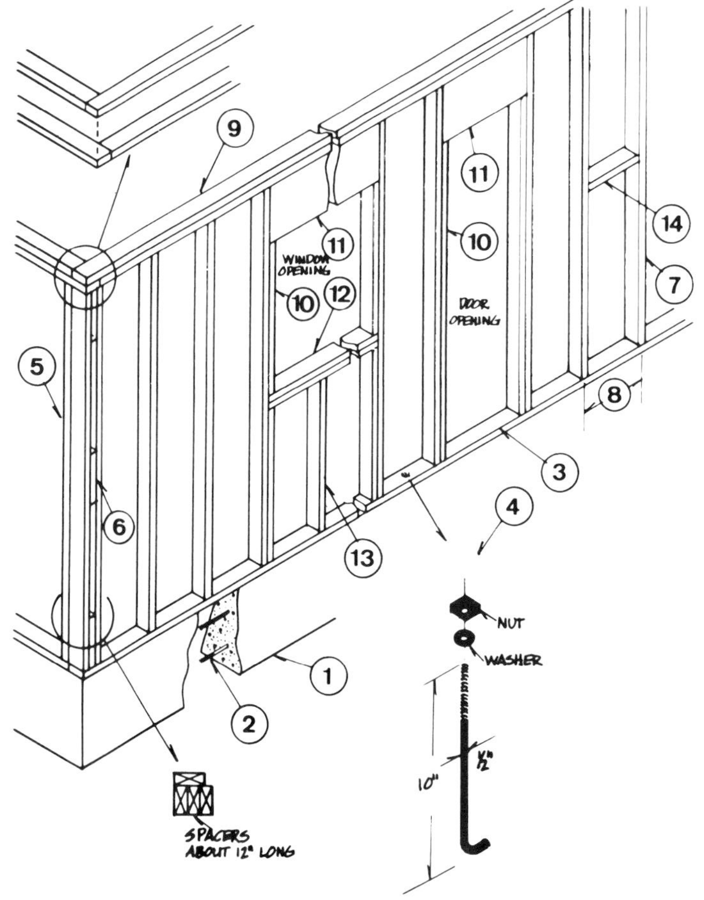

COMMON WALL FRAMING

(Sole plate attachment is typical for a concrete slab.)

1 foundation
2 reinforcement rods
3 2x4 sole or base plate (treated)
4 anchor bolts
5 corner post assembly (all 2x4s)
6 3 spacers
7 2x4 studs
8 stud spacing — 16″ on centers typical
9 doubled 2x4 top or rafter plate
10 2x4 trimmers
11 4x12 headers
12 doubled 2x4 rough sill
13 2x4 cripples
14 fire stops — optional (check local codes)

Figure 430 Frame wall (platform)

frame wall Wall constructed with individual support members, as 2″ × 4″ wood studs or metal studs. Figure 430 shows a frame wall (platform).

framework Structural members that support a structure.

framing Putting together or erecting the structural support of a building.

framing anchor Shaped metal fastener used for holding timbers or lumber pieces together, especially rafters or trusses to top plates. Figure 431 shows typical framing anchors. Also called a *framing clip.*

framing chisel Heavy wood chisel used to clean and trim a joint or for rough cutting. It is normally struck with a mallet or soft-head hammer; the bevel is turned toward the work. See *wood chisels.*

framing clamp A four-sided adjustable framing clamp used for making 90° corners, as for picture frames. See *corner clamp.*

framing clip See *framing anchor.*

framing in Closing up a building with roof, outside sheathing, and all doors and windows in place. Allows interior to be finished away from the weather and secures the building.

framing plan Plan that shows location of structural elements in a reinforced concrete or steel structure. See *erection drawing, placing drawing.*

framing square Carpenter's measuring tool. See *steel square.*

framing square gauge Two marking gauges that can be attached to a steel square to mark a particular measurement on each arm of the square, useful in stair and rafter layouts. See *stair gauges.* (*Figure 938,* p. 414.)

freezeback In roofing, a condition at the eaves where thawing ice refreezes, causing water to be forced back under the shingles.

freeze-thaw cycle A cycle or sequence of freezing and thawing, expanding and contracting. Unprotected concrete, when exposed to a sequence of freeze-thaw, may crack or separate.

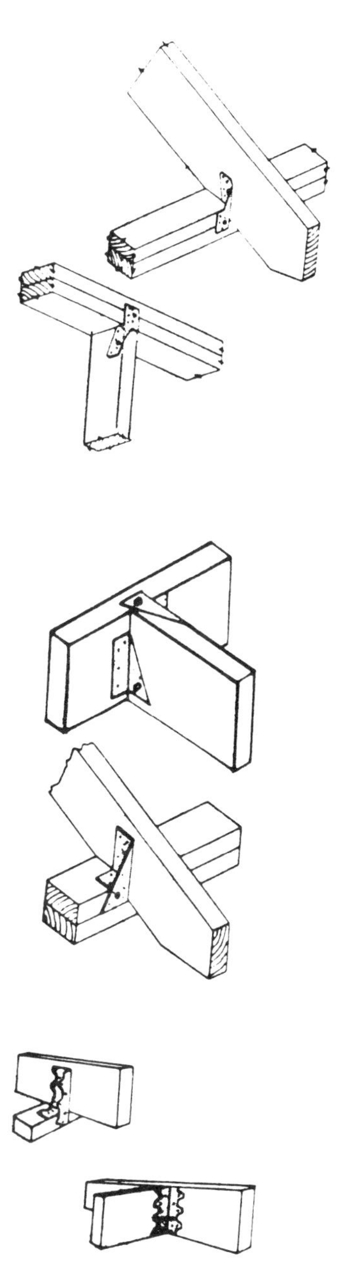

Figure 431 Framing anchors

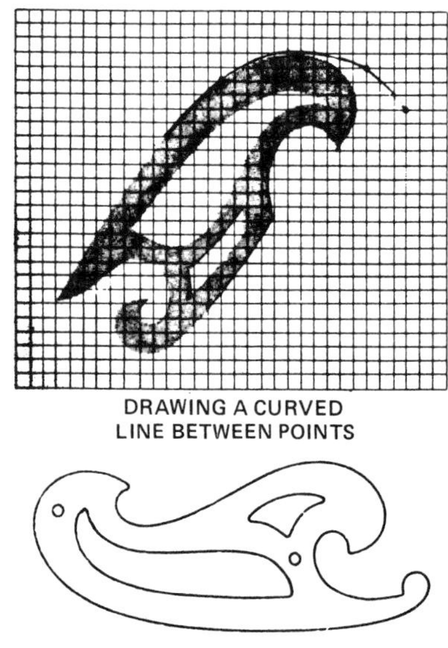

Figure 432 French curve

Figure 433 French door

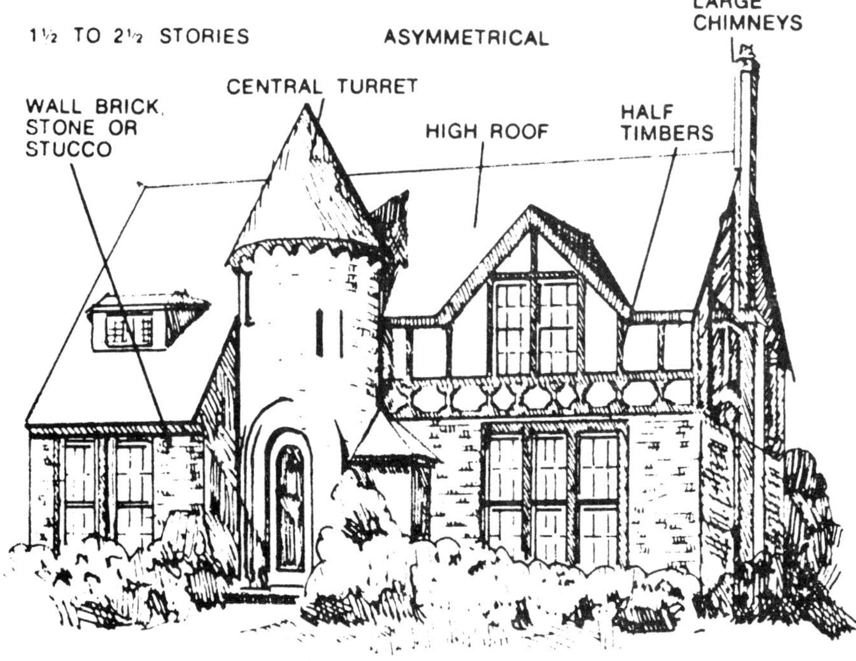

Figure 434 French Normandy style

French architecture Architectural styles characteristic of periods of French history. See *French Normandy, French Provincial.* See *architectural styles, American mansard.*

French curve In drafting, an irregularly shaped curving figure used for drawing curves. (Figure 432.)

French door Pair of swinging doors with a large top-to-bottom glass area. (Figure 433.) See *door.*

French Normandy architecture Asymmetrical, one-and-a-half to two-and-a-half story house with central turret. (Figure 434.) See also *architectural styles.*

French polish A fine finish made of shellac thinned with alcohol and boiled linseed oil. The mixture is rubbed on with a pad. It is used on small pieces or to repair a damaged or scratched finish.

Figure 435 French Provincial style

French Provincial architecture Balanced one-and-a-half to two-and-a-half story brick house with high hip roof. (Figure 435.) See also *architectural styles.*

freon A gas used in refrigeration systems.

fresco Art form of painting into wet plaster.

friction Resistance to any movement.

friction catch Mechanical device for holding a cabinet door closed. A stud on the door is caught in a small two-sided holding device mounted on the carcass. See *catches.*

friction tape Electrical tape used for securing wiring connections.

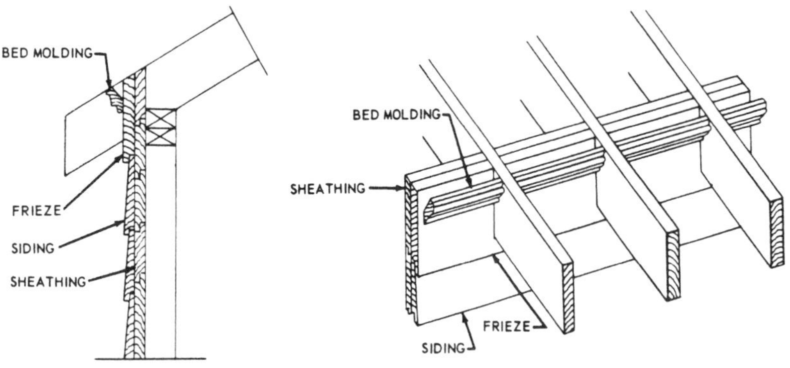

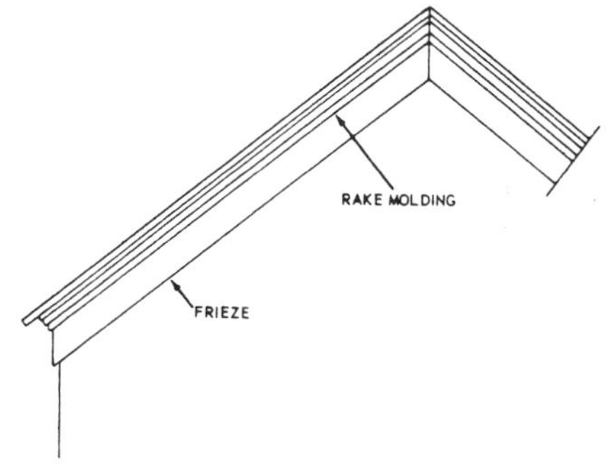

Figure 436 Frieze

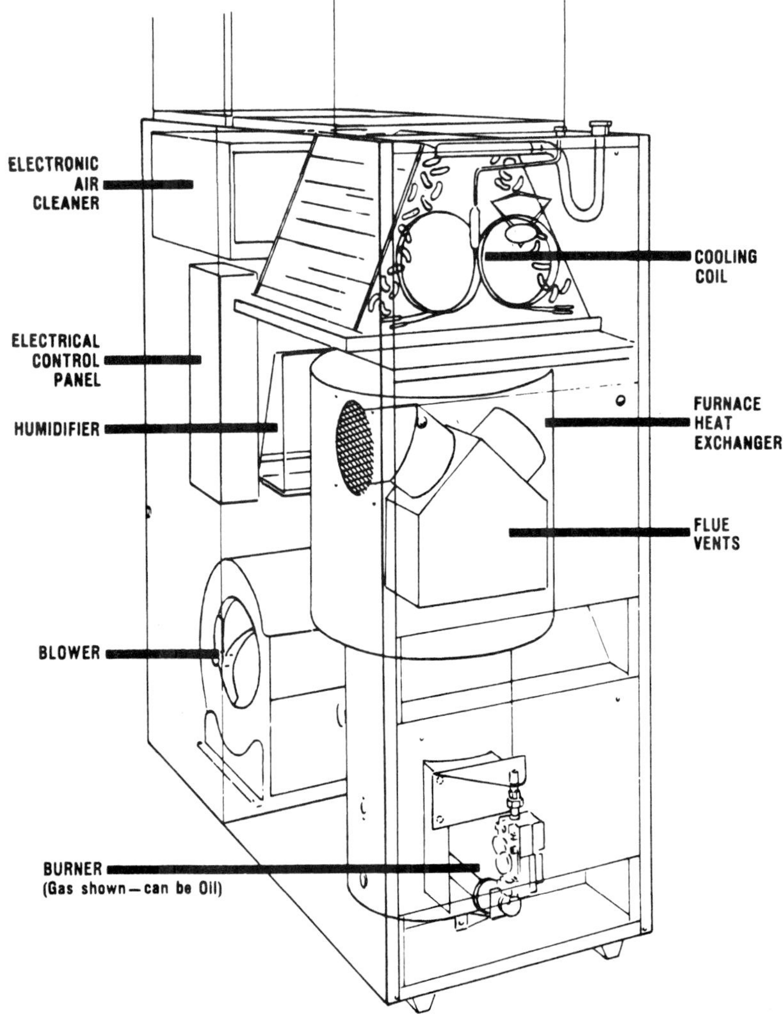

Figure 437 Furnace: gas-fired, forced warm air

friction-type connection In steel construction, a bolted connection where the frictional resistance between contact surfaces transfers the load. See also *bearing-type connection.*

frieze Board or molding at the top of siding where it butts against the cornice or rake. (Figure 436.)

front Side of a building with the main entrance.

frontage The part of a lot or structure running along a public street or highway, waterway, or other thoroughfare.

front of lot The part of a lot that borders on a street or highway; in case of uncertainty, one side is designated as the front.

front view Orthographic view of the front or main side of a building.

frost heave Lifting or heaving action caused by the freezing of soil.

frostline Depth of frost penetration in any area. Bottom of footings must be placed below the local frostline to avoid freezing and heaving action under the footing. Frostline may be specified by local code.

functional modern architecture See *contemporary architecture.*

fungi Microscopic plants that feed on wood, causing decay and stains.

furnace Heating unit in a heating system; may use gas or oil or may be electrical. (Figure 437.) See *air conditioning, heat pump.*

furring Wood or metal strips fastened to a wall or other surface to form an even base for application of structural or finish material, such as wallboard or flooring. Often used on a concrete wall to provide a nailing base. (Figure 438.) Also used to provide an air space.

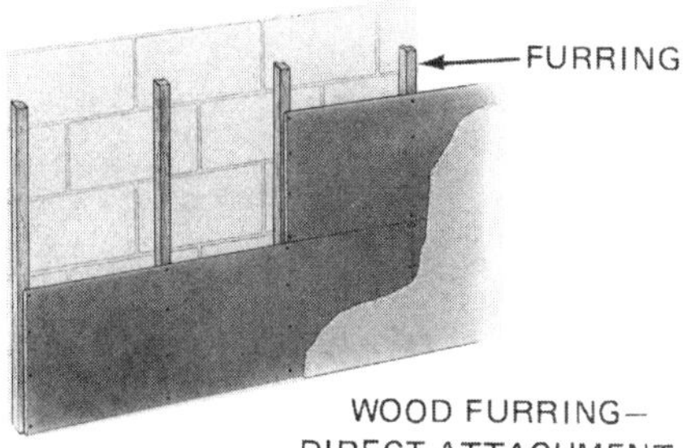

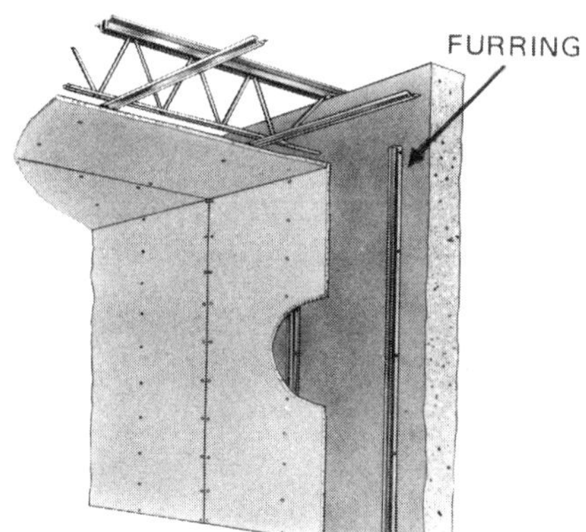

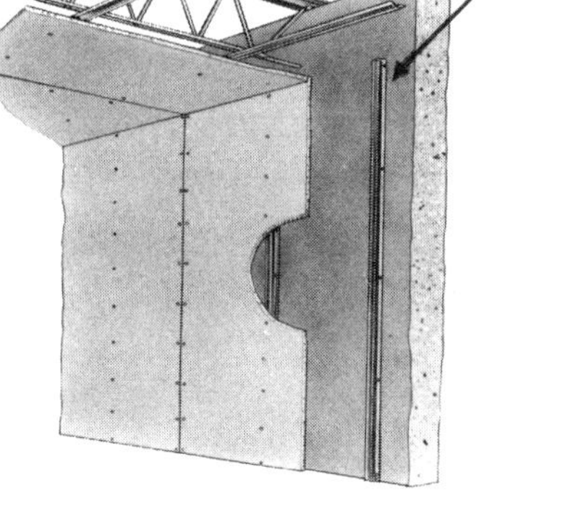

Figure 438 Furring

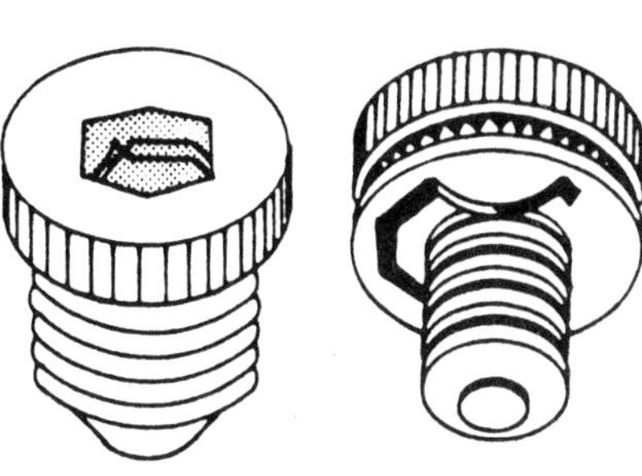

Figure 439 Fuses

fuse Protective device used in an electrical system. (Figure 439.) The fuse burns out (the metal filament is heated and severed) with an overload and must be replaced to reconnect the circuit. Two basic types of fuses are used: plug fuses and cartridge fuses. *Plug fuses* screw in and should not be used in circuits over 125 volts. Figure 440 shows typical plug fuses. There are three main types: standard, time-delay, and Type S (time-delay, nontamperable). Note: Because fuses of different amperage can be switched, the older standard and time-delay (Edison-base) fuses are no longer permitted by NEC in new construction. Type S fuses are permitted because a special adapter is used so that only the proper amp size can be screwed in. Type S is designated as a nontamperable fuse. (Figure 441.) *Cartridge fuses* fit into special fuse clips and are used in circuits up to 600 volts. Cartridge fuse holders are designed to make it difficult to put in cartridges designed for a lower or higher voltage. *Figure 439*, top, shows typical cartridge fuses. A special time-delay *fusetron*, Class R, is available. See also *circuit breakers, fusestat, fusetron.*

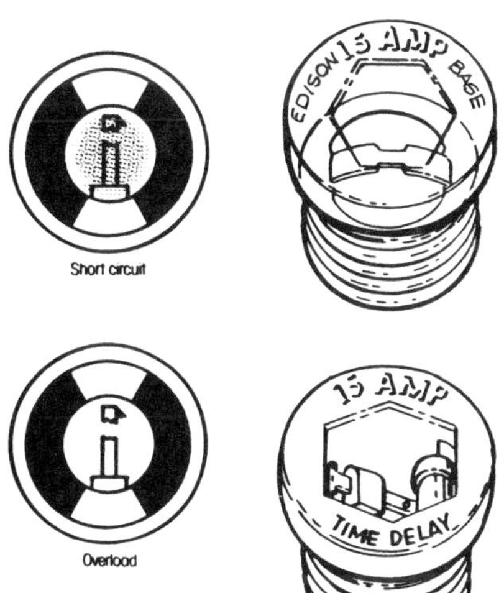

Figure 440 Fuses: plug type

Figure 441 Fuse: Type S with adapter

fuse box Electrical box containing fuses, one for each branch circuit. (*Figure 656*, p. 284.)

fuse panel Panelboard with plug fuses.

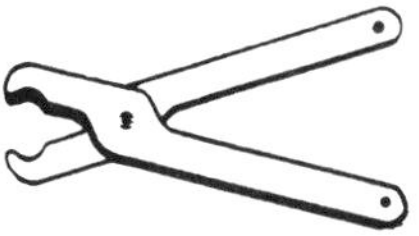

Figure 442 Fuse puller: used to remove cartridge fuses

fuse puller Special tool used to remove or pull out cartridge fuses. (Figure 442.) Different-sized fuse pullers are available for different-sized fuses. Fuses are pulled only when the power is off.

fusestat In electrical work, a time-delay plug fuse used with motor circuits.

fusetron In electrical work, a special time-delay fuse, both plug and cartridge. Specifically, a time-delay Class R cartridge fuse used with motor circuits where there is a brief power surge when the motor starts up. See *fuse.*

fusion welding Welding metal pieces together by melting the parts and allowing them to flow together and fuse. A melting together of two pieces to be joined, a melting together of separate pieces and a filler metal.

g

gable The triangular ends of a building under a gable roof. (*Figure 805*, p. 354.)

gable dormer Small dormer with a gable roof See *dormer.* (*Figure 805*, p. 354.)

gable roof Roof with an equal slope on each side. See *roof.* (*Figure 805*, p. 354.)

gable ventilator Louvered ventilators at the peak of the gable ends. See *attic ventilation.*

gage See gauge.

gage line In steel construction, a line of holes on a steel plate, beam, angle, or channel, parallel to a working edge or the back of the angle or channel. (Figure 443.) Holes are punched for bolts or rivets. See also *gages, GOL.*

gages In steel construction the distance between two *gage lines* or the distance from a gage line to the back of an angle or channel. See also *pitch, spread.*

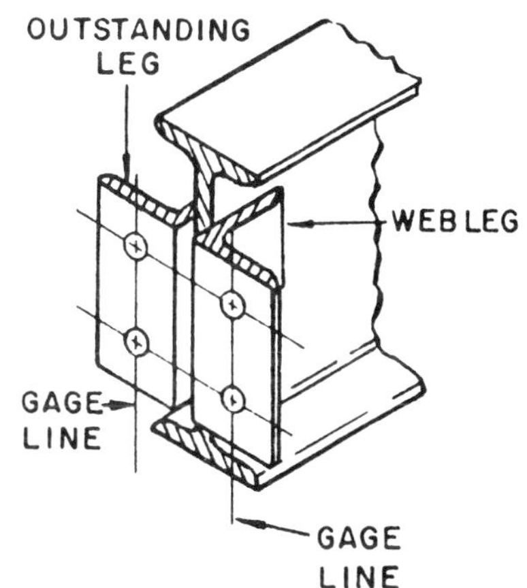

Figure 443 Gage line

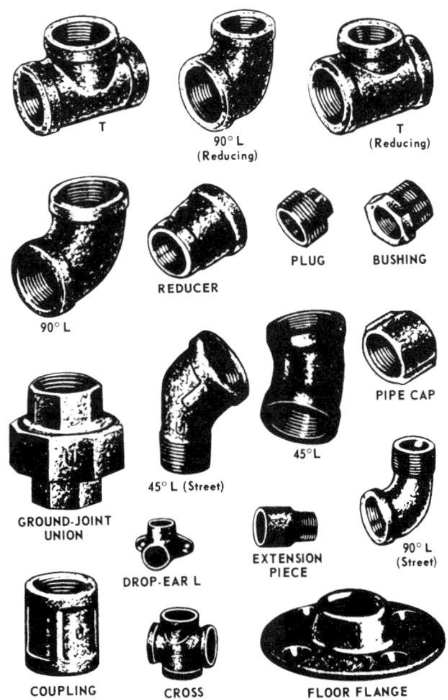

Figure 444 Galvanized malleable iron pipe fittings

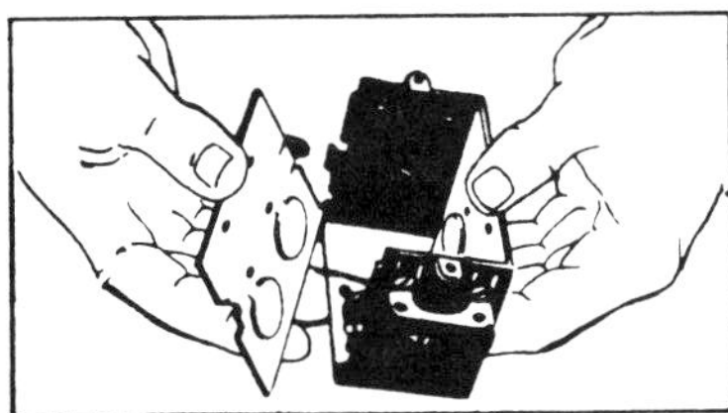

Figure 445 Gang box

gain Notch cut in a lumber piece to receive another board or piece. Cut recess made to receive a butt hinge. See *wood joints.*

galleting Mixing of small stone splinters (gallets) into fresh mortar used in coarse rubble joints. Also called *garreting.*

galvanized Iron that is coated with zinc to prevent rusting, as galvanized roofing. Abbreviated G.I.

galvanized malleable iron fittings Cast-iron fittings that are heat treated to make them more durable. They are galvanized to prevent rusting. (Figure 444.) Used for water supply piping. See *cast-iron fittings.*

galvanized steel pipe Galvanized steel plumbing pipe used above ground for water supply, drainage, and venting. Sizes vary from $\frac{1}{8}''$ to 12″ in diameter. Available as standard size (called schedule 40) or extrastrong (called schedule 80); the extrastrong has a thicker wall, although the outside diameter is the same. Three types of fittings are used: standard cast-iron threaded fitting (used for vent piping), cast-iron recessed drainage fitting (used for sewer and drainage piping), and galvanized malleable iron fitting (used for water-supply piping). See *cast-iron fittings, galvanized malleable iron fittings.*

gambrel roof Roof with two slopes on each side, the bottom slope has a sharper angle. See *roof.* (*Figure 804*, p. 354.)

gang box Electrical outlet box made by joining two or more boxes together, forming a larger enclosure, so several wiring devices can be ganged (put together) in the same outlet. (Figure 445.)

ganged Said of several similar parts when assembled together, as when concrete forms are ganged to make a larger whole.

ganged forms Prefabricated forms that are joined together to make a larger unit.

gangway Temporary board walkway used on the job site.

gardenwall bond A *common bond.*

garnet A natural abrasive mineral used on abrasive papers. See *abrasive.*

garret Living space under a roof; an *attic.*

garreting Mixing of small stone splinters into fresh mortar used in coarse rubble joints. Also called *galleting.*

gas cock Valve that controls gas flow in a gas line.

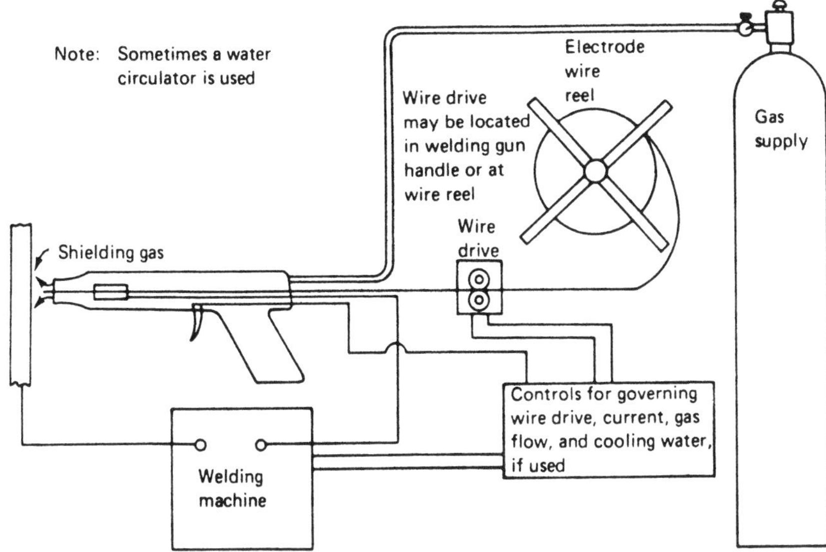

Figure 446 Gas metal-arc welding equipment

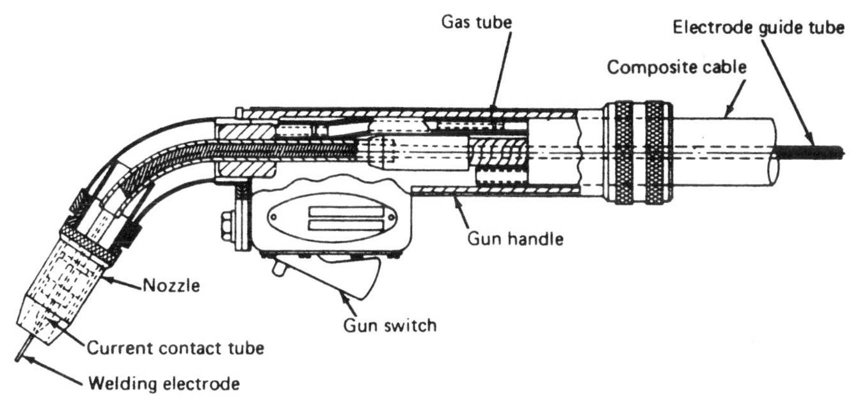

Figure 447 Gas metal-arc welding gun

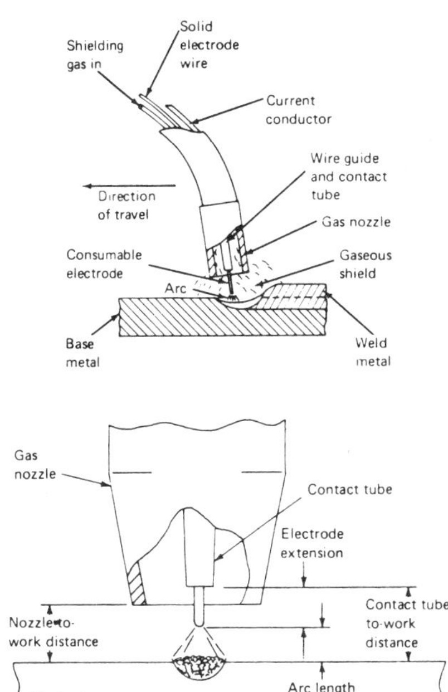

Figure 448 Gas metal-arc welding process and terminology

gas cylinders In *oxyacetylene welding*, the oxygen and acetylene cylinders that hold the compressed gases.

gasket A material placed between two surfaces to create a leakproof joint, as a head gasket on an automotive engine.

gas metal arc welding (GMAW) Arc-welding process that uses an inert-gas shield (argon or helium) over the weld puddle to prevent contamination and oxidation. Continuous wire electrode is fed into the puddle at a controlled speed. An electric arc between the electrode and the work piece creates the heat to cause fusion. (Figures 446, 447, 448.) Also called *mig*.

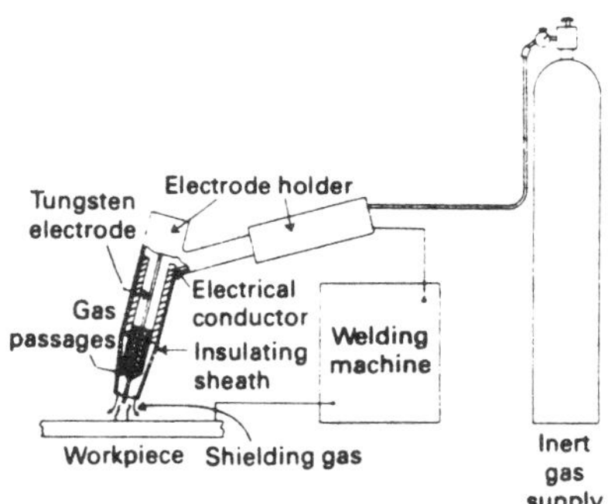

Figure 449 Gas tungsten-arc welding equipment

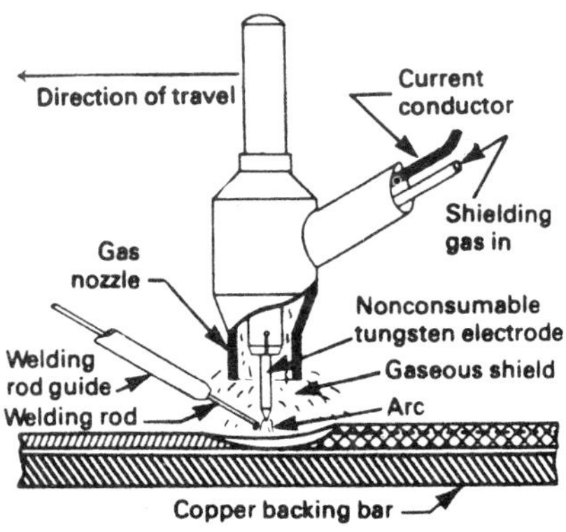

Figure 450 Gas tungsten-arc welding process and terminology

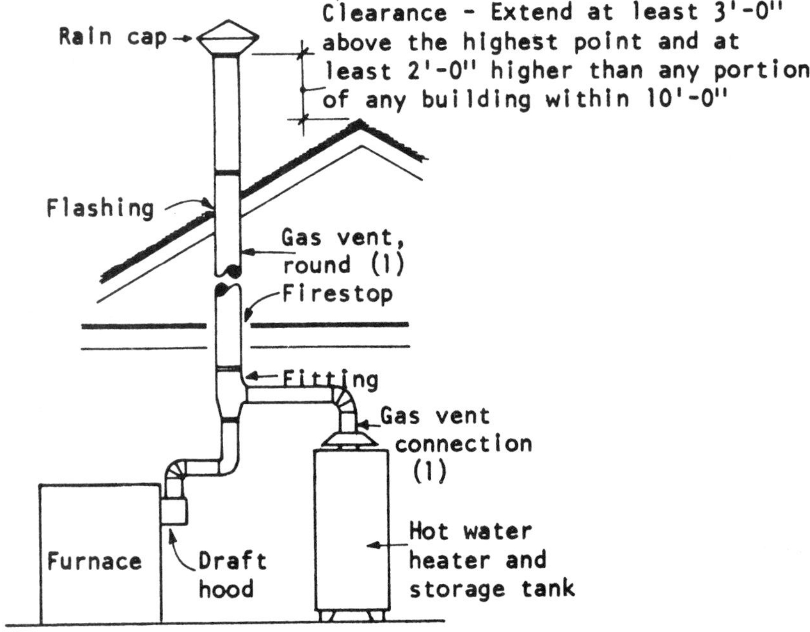

Figure 451 Gas venting

gas-pipe pliers Special pliers with jaws shaped for holding a pipe. One handle has a pipe reamer at the end.

gas pocket A void in a weld. See *porosity*.

gas-shielded arc welding Arc-welding process using a shielded puddle and an electric arc. See *gas metal arc welding* and *gas tungsten arc welding*. See also *flowmeter*.

gas tungsten arc welding (GTAW) Arc-welding process that uses an inert-gas shield (argon or helium) over the weld puddle to prevent contamination and oxidation. A nonconsumable tungsten electrode is used to produce an electric arc to heat the weld area and cause fusion. A filler rod may be used. (Figures 449, 450.) Also called *tig*.

gas vent Vent pipe to the outside that allows fumes to be expelled. (Figure 451.) See also *vent pipe*.

gas welding Welding with the heat of burning gases, such as oxygen and acetylene. See *oxyacetylene welding, oxyfuel welding*.

gate valve A hand-wheel operated plumbing valve that regulates the flow of water by lowering a wedge disk or gate into the valve opening. (Figure 452.) The merit of the gate valve is that when it is completely open, the water flows through without restriction. The valve is designed to work fully opened or fully closed. Compare to the *globe valve.*

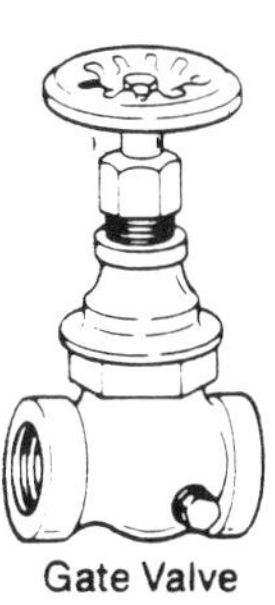

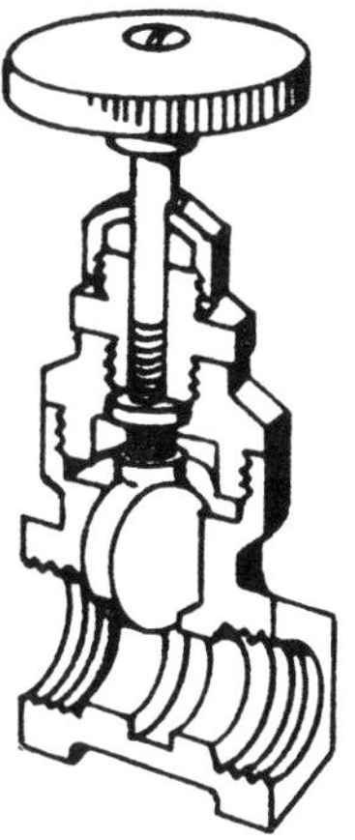

Figure 452 Gate valve

Figure 453 Gauging trowel

Figure 454 Generator

gauge Thickness of wire or sheet metal. A wire gauge used for determining thickness of a wire. See *American Standard Wire Gauge.* In roofing, the amount of shingle laid to weather. Also, the device on a roofing hatchet for measuring shingle exposure. The process itself of measuring shingle exposure. In construction, the process of marking a line parallel to the edge of a board with a marking gauge. In steel construction, the distance between two *gage lines* or the distance from a gage line to the edge of the steel member.

gauge board Template used in stair building to determine tread width and riser height on a stringer or carriage.

gauged mortar Mortar with plaster of Paris added for faster setting.

gauging Cutting of brick or other masonry units to a designed size. Adding plaster of Paris to a mortar for quick setting.

gauging trowel Trowel used to mix gauging plaster. (Figure 453.)

general contractor Overall contractor who is responsible for the construction of the total project; works with subcontractors, such as the plumbing, HVAC, and electrical subcontractors, to coordinate and complete the building.

generator Gasoline- or diesel-operated power supply (120/240 V) that delivers electricity to power electrical tools at building site. (Figure 454.)

geodesic Structure made using triangular supports joined together as a *geodesic dome.*

geodesic dome An engineered dome based on the triangle as the basic building unit. Short sections are joined together to create an overall, very strong dome frame. See *dome.*

geodetic survey Survey of large land masses where the curvature of the earth must play a part. See *aerial survey, plane survey.*

geometry A math that deals with the relationship of lines, surfaces, and solids.

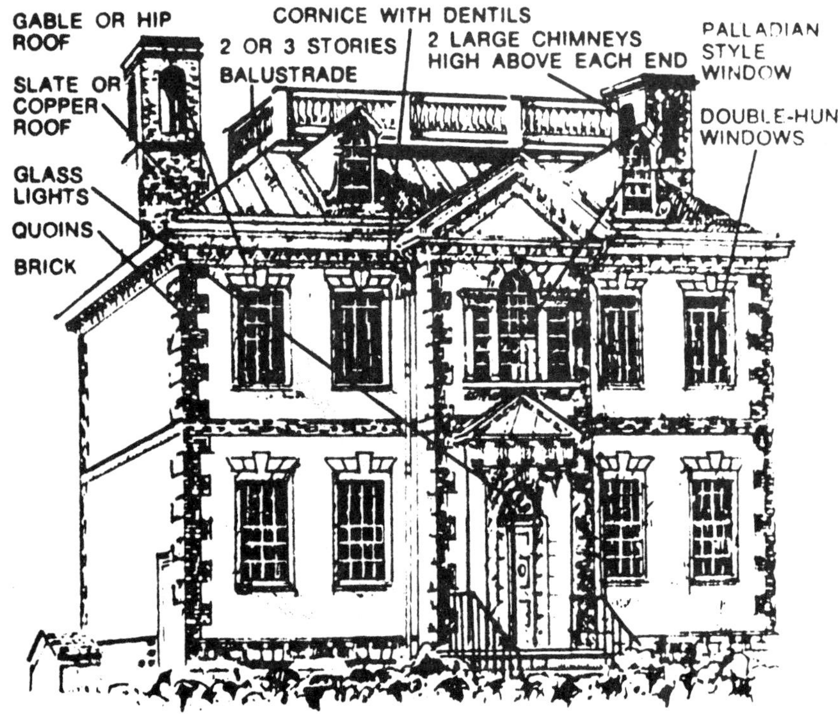

Figure 455 Georgian style

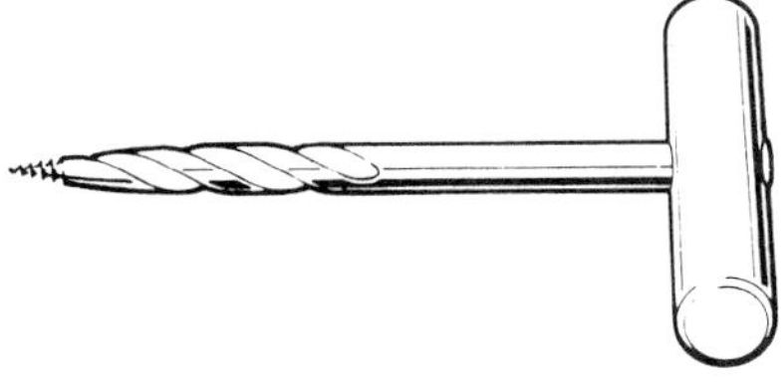

Figure 456 Gimlet

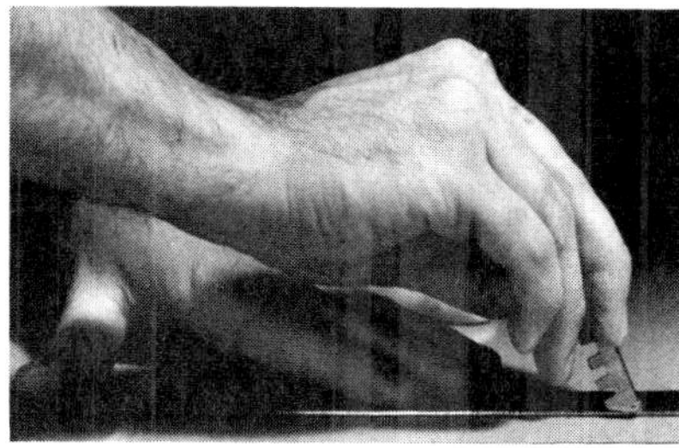

Figure 457 Glass cutter

Georgian architecture Brick two- or three-story rectangular house developed in England under the four King Georges. Popular in the United States in the 1700s and 1800s. (Figure 455.) See also *architectural styles.*

geothermal energy Energy extracted from the heat of the earth's interior, as from hot water or steam.

gesso Hard plasterlike material made by mixing together plaster of Paris and a glue. Used for casting.

GFCI *Ground-fault circuit interrupter.* Also called a ground-fault interrupter (GFI).

gimlet Small tool for starting or boring a hole in wood. (Figure 456.)

girder Heavy steel, reinforced concrete, timber, or laminated horizontal support member that supports beams. Sometimes used to refer to a *beam.*

girt Horizontal bracing that runs around the outside of rigid frame steel or timber construction. See *rigid frame construction.* (*Figure 792*, p. 349.)

girth Distance around a cylindrical object, as a tree trunk, pole, or lally column.

girt strip A cut-in wood member or ledger board used in wood balloon framing to support floor joists. See *balloon framing.*

glass A manufactured clear silicate used for glazing. Various types are used: quartz (very pure glass, transmits ultraviolet rays), obscure (translucent), shatterproof, stained (pigmented; different pieces are held together with lead strips), structural (thick glass used as tiling), and wire glass (wire mesh is embedded). See *glass block.*

glass block Opaque or semi-opaque glass blocks used for constructing light-admitting, but private, window areas.

glass cutter Cutter for scratching glass so it can be broken. (Figure 457.)

glass holders Vacuum cups with handles used for lifting glass.

glass pliers Special wide-nose pliers used for holding and breaking off narrow strips of glass. See *glazier's pliers.*

glaze In sharpening or honing a metal edge, the glassy metal deposit that forms on the stone or leather strop. Also called *swarf.* To install glass (lights). A glasslike surface.

glazed Equipped with glass (lights).

glazed brick A brick that has one surface glazed with a colored ceramic glazing. The glaze forms a glasslike coating and is especially suitable in areas where clean, impermeable surfaces are needed, such as in a hospital.

glazed tile Tile with a vitreous or glassy surface.

glazier One who installs all ornamental glass, plate glass, and mirrors.

glazier's chisel Wide-blade chisel used for removing hardened putty.

glazier's pliers Special pliers for grasping and breaking off a narrow glass piece after it has been scratched with a glass cutter. (Figure 458.)

glazier's points Metal points for holding glass in place in the window frame. Points are pressed into the wood frame next to the glass 2″ to 6″ apart. (Figure 459.) Putty or a glazing compound is used as a seal and for additional holding power after the points are in.

glazing The glass or plastic used for windows. The process of installing glass (lights) in windows or doors; installing mirrors. Also, a polished surface that appears on an overused hone, whetstone, or grindstone, consisting of embedded metallic particles. See *double glazing, triple glazing.*

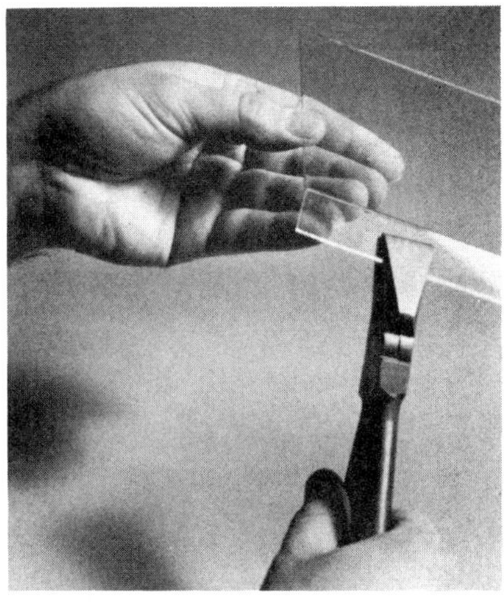

Figure 458 Glazier's pliers

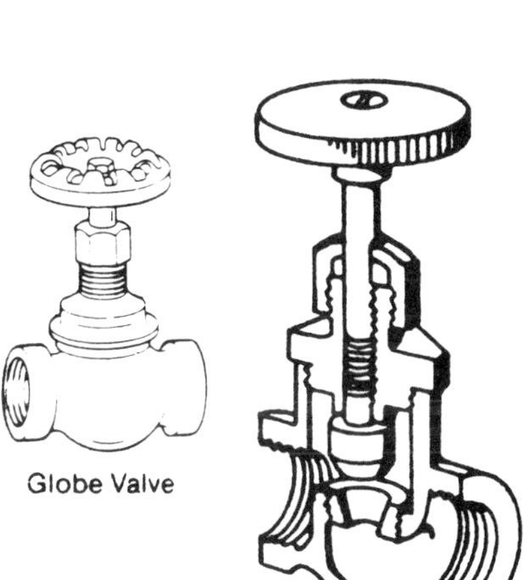

Figure 460 Globe valve

glitter finish Plaster finish produced by throwing shiny flakes into the wet plaster.

glitter gun Gun for blowing glitter (colored bits) on wet plaster.

globe valve A hand-wheel operated plumbing valve that regulates the flow of water by lowering or raising a disk into an annular ring seat. (Figure 460.) When the valve is fully open, the water flows up into and through the annular seat area; this restricted flow causes turbulence and pressure drop. See *angle valve.* Compare to the *gate valve.*

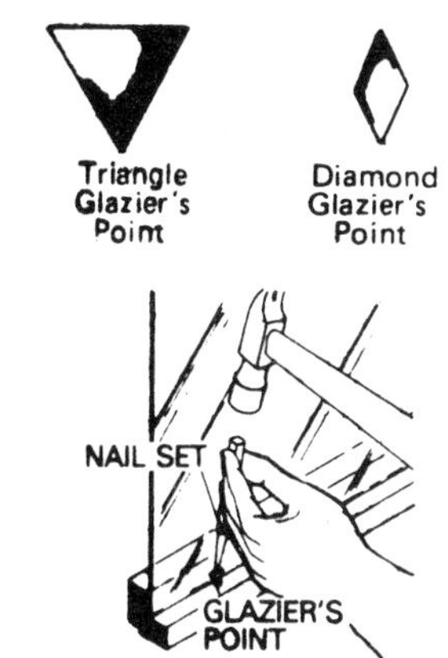

Figure 459 Glazier's points

EYE CUP GOGGLE

EYE CUP GOGGLE

PROTECTIVE GLASSES (FLASH GOGGLE)

FACE GOGGLE

Figure 461 Goggles

globular transfer In arc welding, a process where the filler metal falls into the weld in large drops.

gloss Enamel paint that dries to a high shine.

glue Type of adhesive for holding two surfaces together in a bond. Various glues are commonly available. *Polyvinyl resin emulsion* (PVA): The common white glue that comes as a liquid in plastic bottles. The white glue is a widely used household glue. Used for bonding wood pieces together. Sets in $1\frac{1}{2}$ hours. It is water soluble and remains slightly plastic after setting. Also, it is heat sensitive and will soften if exposed to heat for long periods. *Aliphatic resin* (AR): A widely used yellow glue that comes in a plastic bottle or in large containers. Used in construction. Similar to white glue but is more water resistant and sets more quickly. *Urea-formaldehyde resin* (UF): Comes in a powder or in a liquid to which a powdered catalyst is added. The powder is mixed with water. It sets in 4 to 6 hours, leaving a hard brown joint. With higher temperature the setting time can be lowered to as little as 1 to $1\frac{1}{2}$ hours. Used for wood joints that may be subject to moisture. *Resorcinal-formaldehyde resin* (RF): A two-part glue with a liquid resin and a powdered hardener or catalyst. Sets in 6 to 10 hours with a hard dark joint. Used for wood joints in high-moisture areas. *Epoxy resin:* A two-part glue with a liquid white epoxy resin and a dark liquid catalyst which are blended together in equal amounts. Dries to an extremely rigid and strong joint. Used for bonding metal to wood or metal to metal. *Contact cement:* A liquid cement with a neoprene rubber base. Cement is applied to the two separate surfaces and allowed to dry. When the two prepared surfaces come in contact with each other, they make an instant and permanent bond. Used for applying veneers. *Animal glues:* A glue made from animal hides and bones. Comes in a dry or liquid form. The dry form is soaked and melted in water, then heated. It is applied hot to wood joints. No longer widely used in woodworking today. *Casein glue:* A milk-base glue. Comes in a powder which is mixed in with water. Sets in 6 to 8 hours. *Plastic cement:* A tube glue used for minor repairs. It sets quickly and is used for model building. *Rubber cement:* Rubber-base cement. Since it does not contain water, it is useful in bonding paper. See *adhesive.*

glue injector Syringe for injecting glue into hard-to-get places, such as a hole.

glue-nailed Glue joint or connection that is also nailed.

glue nailing Fastening of panels by setting into an adhesive and nailing, as for plywood onto joists, or underlayment onto the subfloor.

glulam Glued-laminated structural timber; large beam or arch made by bonding lumber pieces together with adhesive. See *wood beam.*

goggles Eye protection worn when working in hazardous area. (Figure 461.) See *eye protection.*

GOL In steel construction, the *gage* on the *outside leg* of an angle iron. GOL = gage, outside leg.

gouge A type of hand-held cutting chisel that has a curved or V-shaped cutting edge. Various shapes and sizes are available for cutting different grooves. A gouge is worked by hand or struck with a mallet. Basically there are two types: firmer and paring. The *firmer gouge* has a bevel ground either on the outside or inside and is used for cutting hollows and grooves. The *paring gouge* has a bevel ground on the inside and is used to cut pieces and ends with irregular surfaces. (Figure 462.) The curve of the gouge is called the *sweep.* The sweep or curve varies from a slight curve, called flat; to a medium; and finally to a sharp V-shaped curve, called quick. See also *woodturning tools,* which are a kind of gouge with a long handle used for cutting coves and curves in wood pieces being turned in a *lathe.*

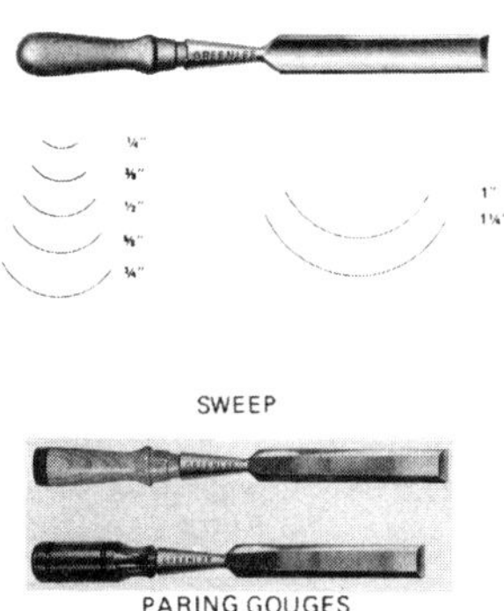

Figure 462 Gouges

gouging A groove formed by cutting metal away with a gas flame or an electric arc. Cutting with a wood gouge.

government survey system *Rectangular survey system.*

grade In general, an elevation or the slope of the ground. The ground level at building site. See *finish grade, natural grade.* In plumbing, the slope or pitch of the horizontal drainage piping, usually $\frac{1}{4}''$ per foot. In lumber and plywood, the quality of the material. See *lumber grades, plywood grades.*

grade beam A reinforced concrete beam that rests on caissons or piers placed at regular intervals around the edge of the foundation. (Figure 463.)

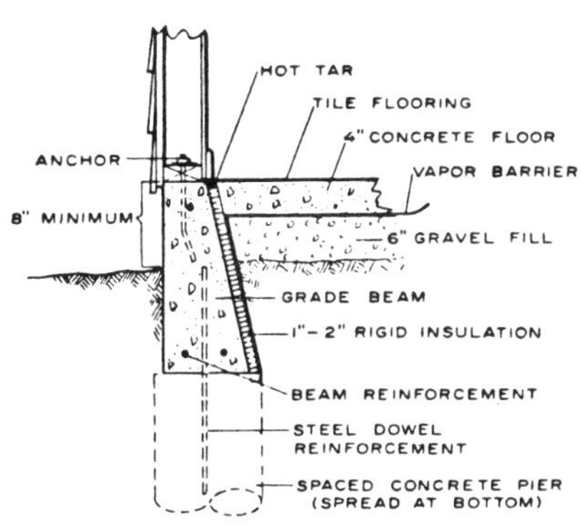

Figure 463 Grade beam

grade level Elevation of the ground at a building site or around a building. Also called grade line.

grade line See *grade level.*

grade mark A stamp or mark in lumber, plywood, or steel bars denoting quality.

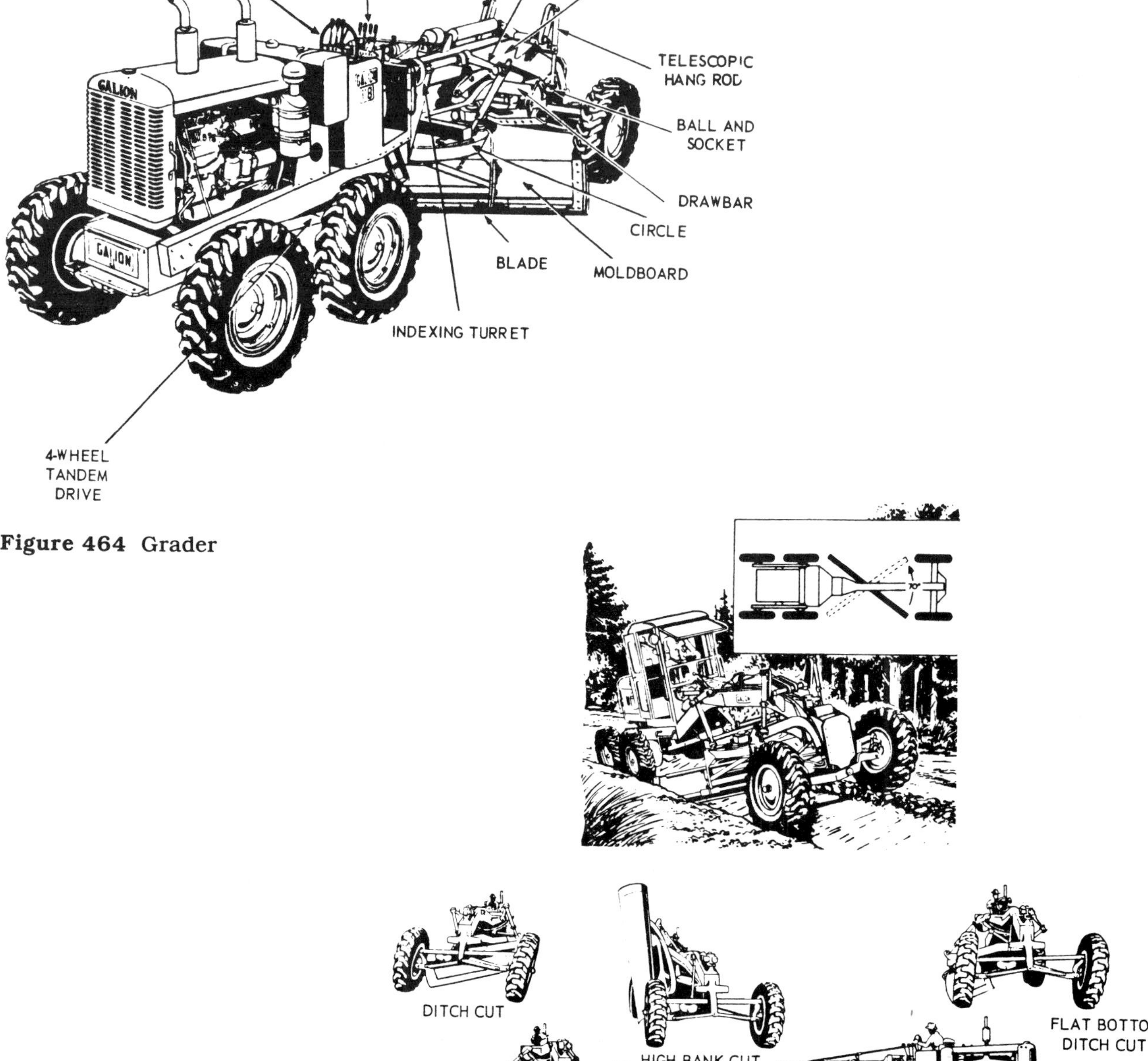

Figure 464 Grader

Figure 465 Grader blade positions

grader Heavy earth-moving machine with a scraping blade, used for leveling a site and filling in low places. Widely used in developing road beds for highways. Figure 464 shows a grader and identifies the main parts. Figure 465 shows the normal grader blade positions. See also *bulldozer, scraper.*

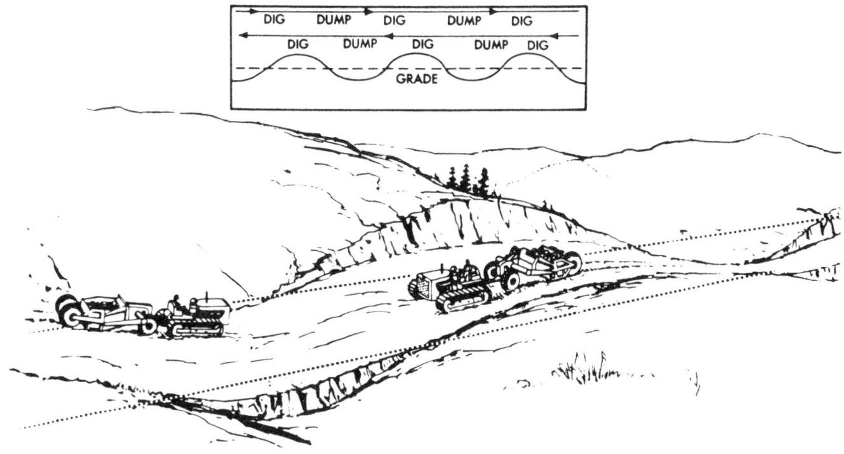

Figure 466 Grading: low ground is filled in with earth from high ground

gradient Incline or slope of the ground.

grading In site development, the leveling of site; moving of earth to fill in low spots and to lower high spots. (Figure 466.) See *filling, finish grade, natural grade, cut and fill, compacting.*

grading plan Plan of a building site that shows existing contours and the final, finish contours after grading. (Figure 467.) See also *earth change plan, contour line.*

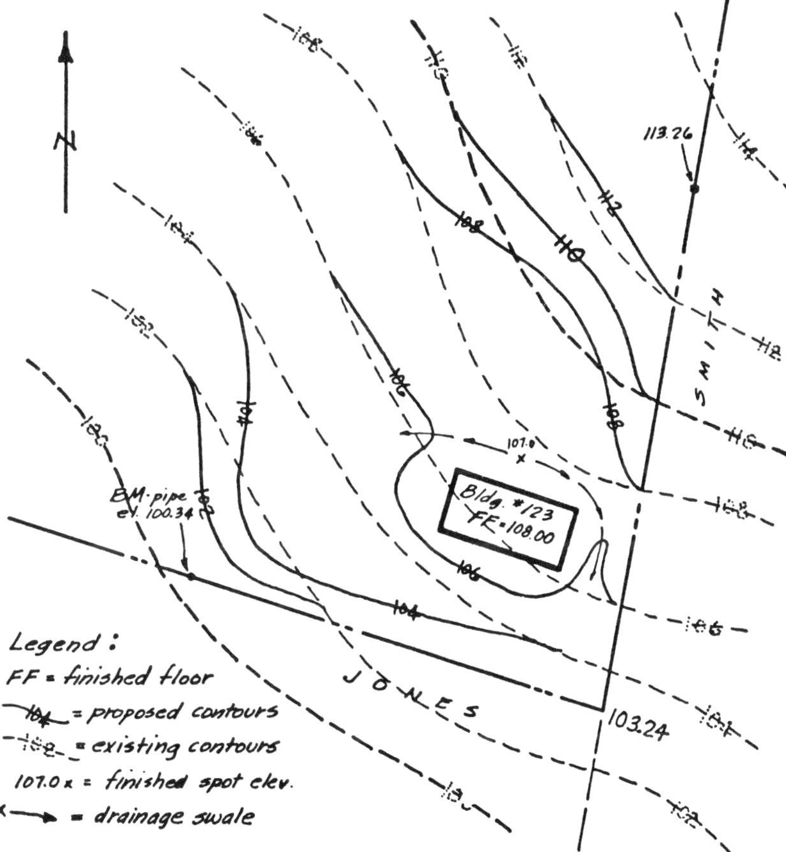

Figure 467 Grading plan

graduate To mark off in equal parts, as a foot rule is graduated in one-inch increments.

grain In wood, the arrangement, size, direction, and appearance of the wood fibers.

grain direction The direction in which the long wood fibers run.

granular Composed of or containing many small grains, as a fine aggregate or beach sand.

gravel Fairly coarse, naturally formed aggregate, as beach pebbles, sized from pea and marble size up to fist size.

gravel envelope Gravel that surrounds a drain or well to facilitate waterflow.

gravel fill Placement of gravel, usually a bed 4″ thick, as the base for a concrete slab or concrete work; allows water to drain away from underneath the concrete.

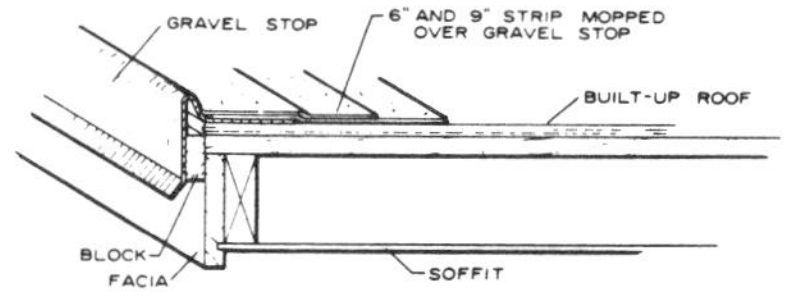

Figure 468 Gravel stop

graveled roof Built-up roof finished off with exposed gravel.

gravel stop Raised metal bead nailed around the edge of a graveled roof to catch any gravel that washes down. (Figure 468.)

gravity hot-water heating Central heating system that allows heated water to be distributed to the various outlets by a natural process of rising or flowing upward (because hot water is lighter than cool water). The cool water flows back to the boiler through a separate set of piping. Has been generally replaced by *forced hot-water heating.*

gravity warm-air heating Central heating system that allows heated air to be distributed by rising through a ductwork (because heated air is lighter than cool air). The cool air falls through gravity back to the furnace for reheating. Has been replaced by *forced warm-air heating.*

grease trap Catch basin in a plumbing drain system designed to retain grease in the waste water.

green Lumber not completely seasoned (dried). Moisture content is over 19%. Also, fresh mortar or cement.

greenhouse effect Solar heating of an area by sunlight passing through glass or plastic. The short-wave radiation of sunlight passes through the glass (or clear vinyl sheet), striking objects in the room and transferring heat. The radiation is changed to long-wave radiation which does not pass back out through the glass (or vinyl), thus remaining in the area and heating it. A solar heating caused by covering something with clear plastic, as covering a green concrete slab to allow the heating to aid in curing.

green lumber Fresh, unseasoned lumber, specified as having a moisture content over 19%.

grid Criss-cross pattern. A small criss-cross pattern on paper (grid paper) used as an aid in drafting or sketching. A modular layout (4") used to scale on blueprints or full-scale on a structure. A large electrical power network. Also, a 40-acre parcel; one-sixteenth of a section.

grid lines In drafting, reference lines printed on the drafting paper representing set distances, normally based on a 4" module.

grill Criss-cross grating used over air-duct opening.

grinder Power-operated abrasive wheel used for sharpening or abrading away of metal. (Figure 469.)

grit Size of abrasive used on sanding papers, disks, and belts. See *abrasive.*

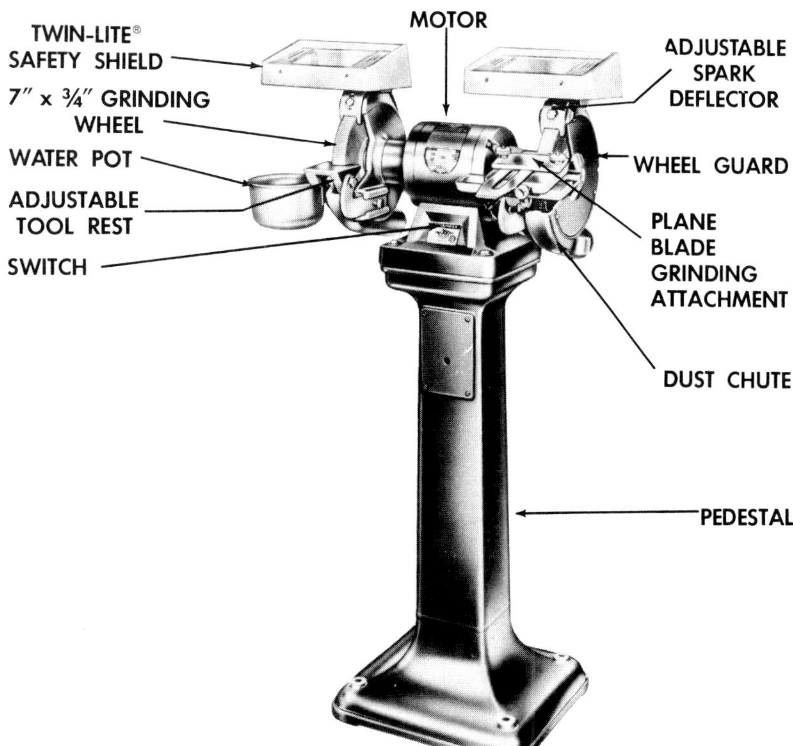

Figure 469 Grinder

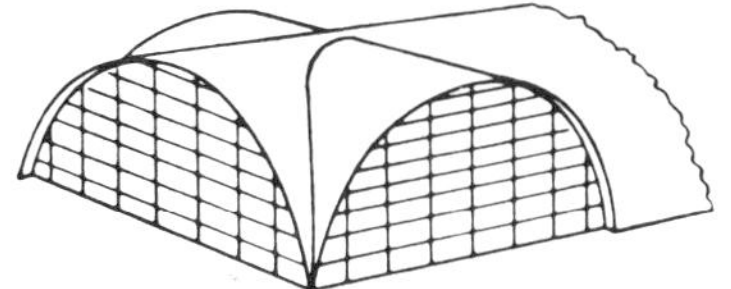

Figure 470 Groined vault roof

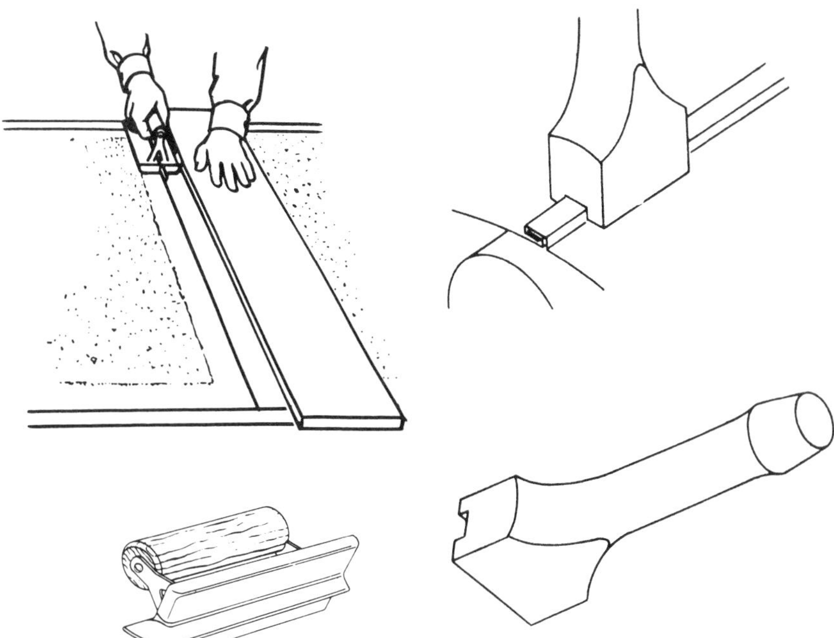

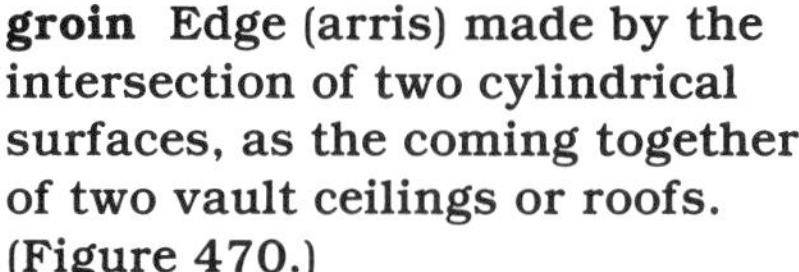

Figure 471 Groover (cement)

Figure 472 Groover (sheet metal)

groin Edge (arris) made by the intersection of two cylindrical surfaces, as the coming together of two vault ceilings or roofs. (Figure 470.)

grommet Metal eyelet, mounted near the edge of canvas coverings, through which a cord is run for lashing down.

groove Small rectangular channel cut in wood with the grain, as a groove cut in a plywood or board edge to receive a tongue from another member. Channel pressed into sheet metal. Channel troweled into concrete flatwork.

groove-joint pliers See *adjustable-joint pliers.*

groover Special trowel used for making grooves in fresh concrete flatwork. (Figure 471.) Sometimes referred to as a *jointer* or *jointing tool.* See *trowel.* Sheet metal tool used for setting or locking seams (Figure 472.)

groove weld Weld made between two pieces that come together in a groove. (Figure 473.) Standard groove welds are: square, single bevel, single, flare bevel, single flare V, single J, single V, single U, double bevel, double flare bevel, double flare V, double J, double V, double U. See *backing weld.*

grooving machine Sheet metal machine for running grooves. Used to make a joint seam.

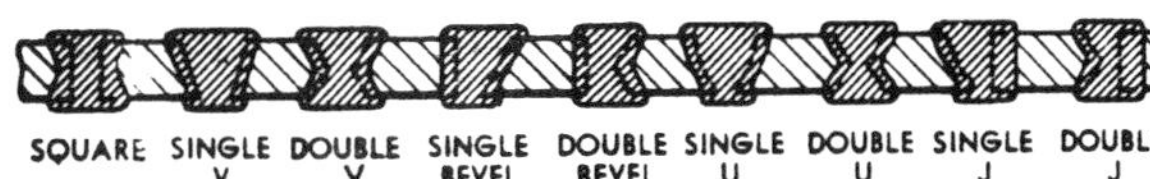

Figure 473 Groove welds

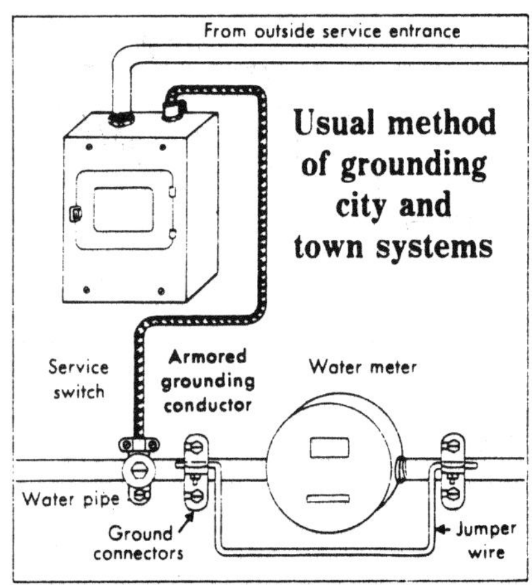

Figure 474 Grounding to cold-water pipe

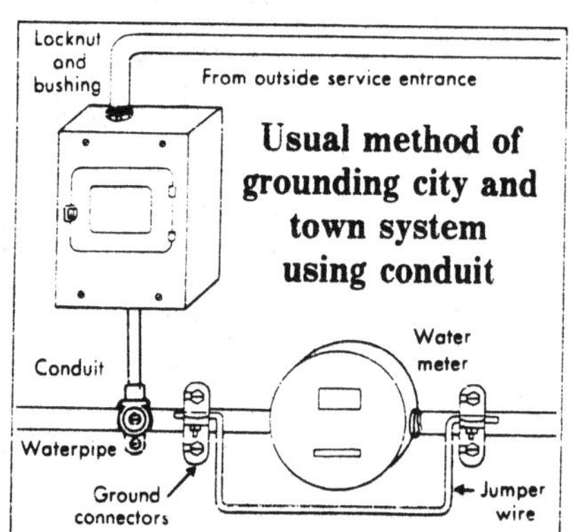

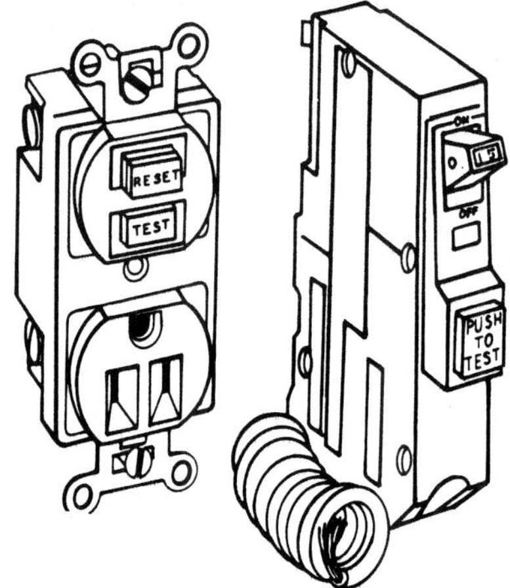

Figure 475 Ground-fault circuit interrupter

ground A strip of wood or metal fastened on a plaster wall flush with the finish plaster surface, used as a nailing base and as a thickness gauge to guide the final wall finish depth. See *dots.* In an electrical system, a connection to earth (Figure 474); also, an accidental connection of a lead or wire to a conducting material outside the electrical system.

ground clamp Clamp used to connect conduit, armored cable, or wire to a ground, such as a cold-water pipe.

ground course First or base masonry course.

grounded Electrical circuit connected to earth or other conductor.

grounded outlet Electrical receptacle that is grounded. In new construction, according to the National Electrical Code, all electrical outlets are grounded.

ground-fault circuit interrupter (GFCI) Special device to interrupt electrical circuit where a weak electrical loss or ground occurs. (Figure 475.) Required on receptacles in bathrooms, garages, and swimming-pool areas. The GFCI provides protection against accidental grounding in the circuit that is not sufficient to trip the circuit breaker. The GFCI senses the unbalanced current and opens the circuit to prevent a possible shock to someone using the circuit. GFCI's are also required on temporary receptacle outlets on construction sites.

ground floor First or main floor of a structure.

grounding conductor In an electrical wiring system, a conductor used to make ground to a conducting body or to earth. The bond or connection is made by a pressure connector clip, clamp, screw, metal strap, lug, or bushing. (Solder is not used.) (Figure 474.) Conductors used for grounding an electrical system are normally white or gray colored. Outside grounding is made using a copper ground rod (Figure 476) or to an underground metallic water-pipe system.

grounding wire Wire used in a flexible cord for making a ground. (Figure 477.) Wire is green and connects to the grounding screw on a receptacle. See *bond wire.*

ground lead In arc welding, a *work lead.*

ground line *Grade line;* elevation of the ground. A reference line in perspective drawing. In *perspective drawing* the line of intersection of the *picture plane* and the *ground plane.*

ground plane In *perspective drawing,* the plane upon which the object rests and upon which the viewer stands.

ground rod Metal rod driven into the earth and used as a ground for an electrical system.

grounds Plastering *ground.*

ground water Water that naturally occurs underground; reached by drilling a well.

grout A thin cement mortar that is used to fill cracks and small, hard-to-get-at places. Also used as a sealing wash over a masonry surface. Epoxy resin and vinyl grouts are also available.

grouting Filling cracks or other small areas with grout. Also, specifically, filling the cores of concrete block with grout; laying the mortar base on the foundation top to receive the sill; applying grout for ceramic tile installation.

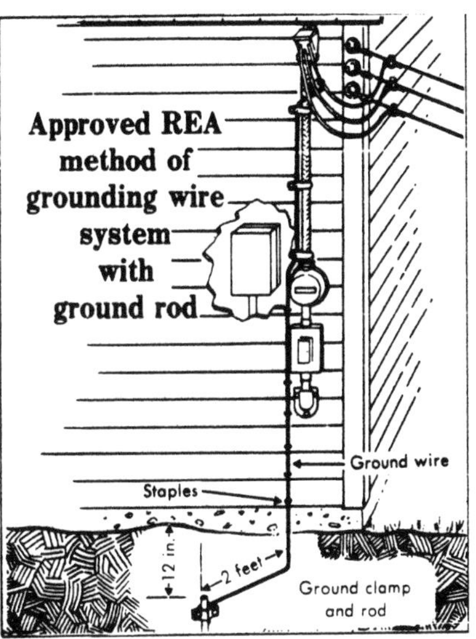

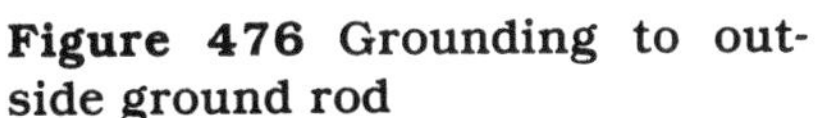
Figure 476 Grounding to outside ground rod

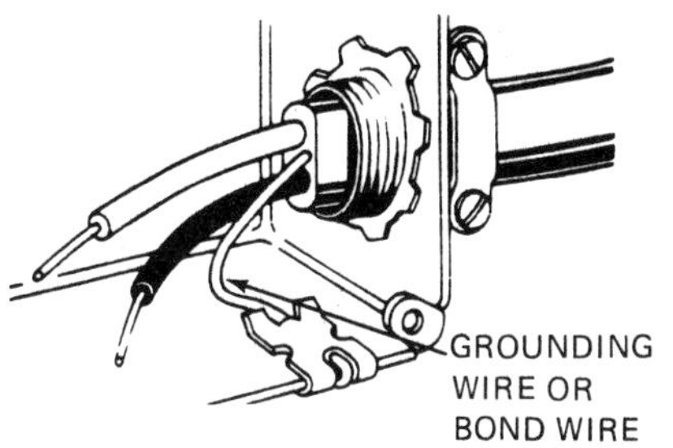

Figure 477 Grounding wire

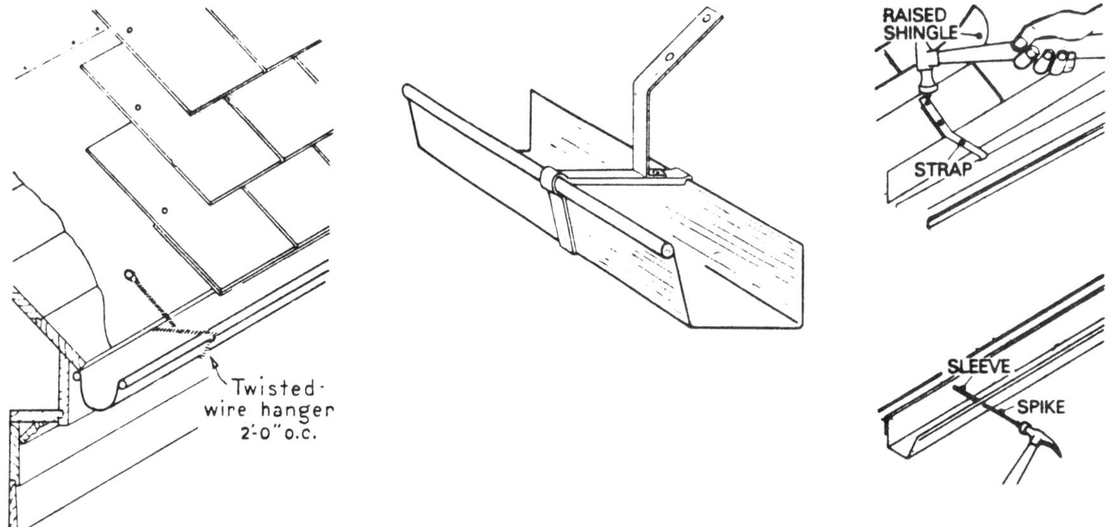

Figure 478 Gutter

growth ring Annual growth ring on a tree.

grubbing Removing trees, roots, brush from a construction site.

guide bar In cabinetmaking, the support on which a drawer slides.

gullet The notch between two saw teeth.

gumming Grinding and shaping the gullets (bottoms between teeth) on a circular saw blade. A special grinding wheel is used.

gunite Pneumatically sprayed concrete. Dry cement and sand are blown through a gun, where water is added to make concrete which is sprayed on a surface. See also *shotcrete.*

gunning Spraying of cementitious material, such as shotcrete, out of a nozzle (gun) onto a surface.

Gunter's chain A 66-foot chain with 100 links each 7.92" long. Used for land surveying. See *chain.*

gusset Thin wood or plywood piece used as a connector for wood joints. Nails or screws are fastened through the gusset into the wood joint; glue may also be used. (*Figure 539*, p. 231.) A *gusset plate.* Also, a brace or bracket used to strengthen a structure at a corner.

gusset plate Metal, wood, or plywood plate fastened to wood or steel members that come together to form a joint, used to support the joint. See also *fishplate, splice plate.*

gutter Channel attached at the eave to catch and divert rain water or snow melt. Gutter has a slight slope so that water runs to one end, where it is carried down by a *downspout* (leader). (Figure 478.)

gutter trowel Right-angled curving trowel similar to a *corner trowel.* Used for finishing highway gutters.

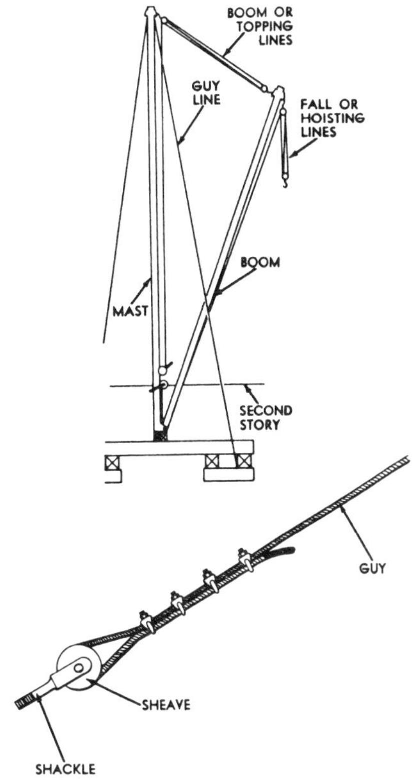

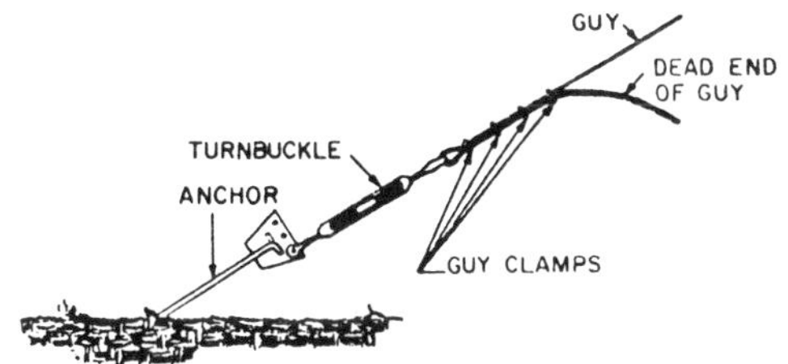

Figure 479 Guy line

guy Steel rope or wire used to support a structure, runs from the top of the structure at an angle to ground where it is secured. (Figure 479.)

gypsum Mineral used to make drywall sheets. When heated and part of the water is removed, it becomes plaster of Paris.

gypsum concrete Quick-setting concrete made using a dehydrated gypsum (plaster of Paris) and aggregates.

gypsum lath A sheet with a gypsum core covered on each side with fibrous paper. Used as a base for plastering. See also *perforated gypsum lath.*

gypsum plaster Plaster of Paris.

gypsum wallboard Board with a gypsum core covered on both sides with heavy paper. The edges are also covered with paper. Used in *drywall construction.*

hacksaw Fine-toothed saw used for cutting metal. (Figure 480.) Blades are available in 8″, 10″, or 12″ lengths. See also *stab saw.*

haft Handle of a cutting tool, as a knife. See also *helve.*

hairline A fine crack, especially in concrete, plaster, or where woodworking joints are made.

hairpin In steel and reinforced concrete construction, long hairpin-shaped or U-shaped steel rods used for reinforcement. Often used at the base of a steel column, where the column footing is tied into the slab.

half bat Half a building brick. See *cut brick.*

half bath Bathroom with a lavatory and water closet only.

half hatchet Heavy hatchet with a hammer face and a cutting blade. See *hatchet.*

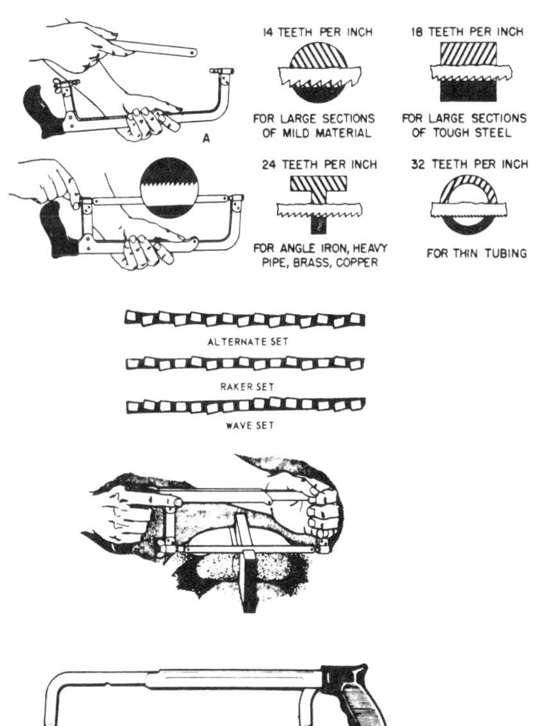

Figure 480 Hacksaw

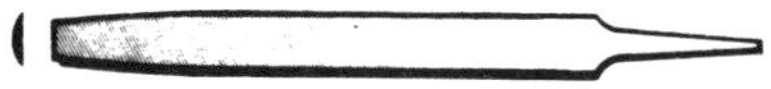

Figure 481 Half-round file

half lap joint Lap joint made by cutting away half the thickness of each piece where they meet. When joined, the surfaces are flush with each other. See *wood joints.*

half round A molding that is semicircular in cross section. Normally two half rounds are cut from a full circular or cylindrical molding. See *molding.*

half-round cabinet file A file or rasp used for smoothing wood. See *file.*

half-round file File with one flat and one curved face, used for filing wood or metal. (Figure 481.) See *file, round file.*

half-round rasp Rasp with one flat and one curved face, used for smoothing wood. See *rasp.*

half section In a symmetrical piece, only half the section (vertical cross-cut view) may be detailed; detail runs to the centerline; the other half is assumed to be identical to the half shown.

half story Living area under a roof area; interior side walls will be partly slanted to follow the roof line.

half-timbering House construction that uses exposed timbers with the space between them filled with stucco, as in *English Elizabethan* houses. See *architectural styles.*

hammer Hand tool for driving nails or for shaping metal. Figure 482 shows common hammers. See also *air-driven nailer, ball peen hammer, brick hammer, bush hammer, chipping hammer, claw hammer, curved claw hammer, dead-blow hammer, drywall hammer, mash hammer, sledge, soft-face hammer, straight claw hammer, tack hammer.*

hammer fastener Special hammer and fastener holder for setting pins and studs into concrete or steel.

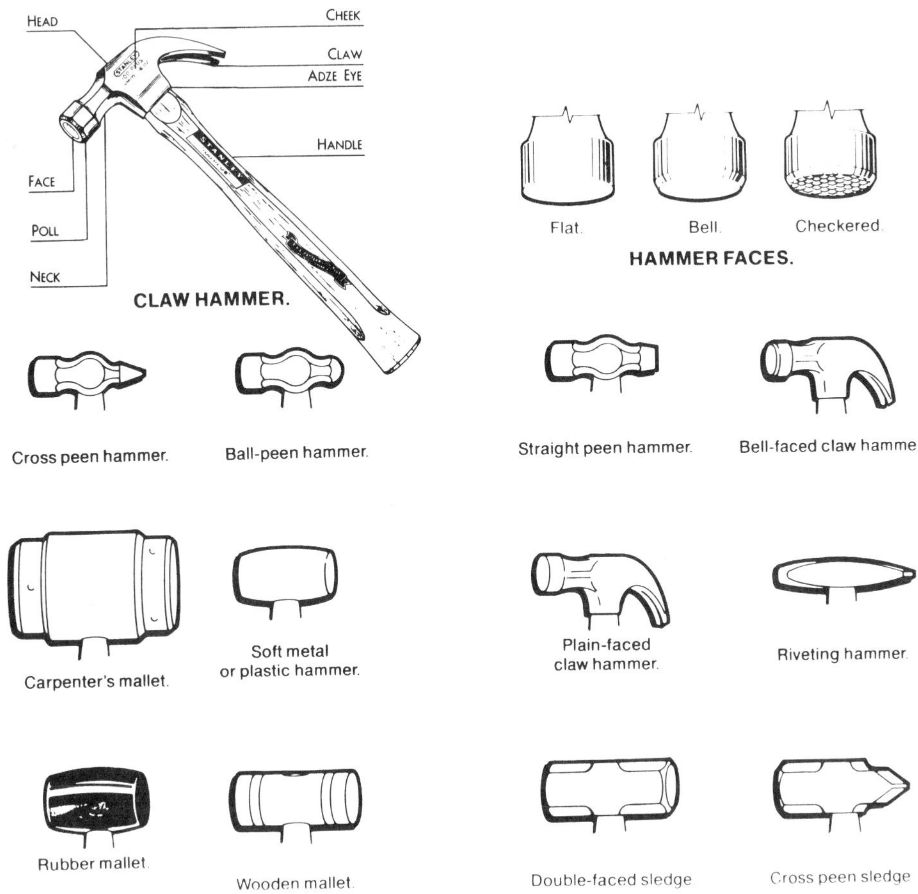

Figure 482 Hammers

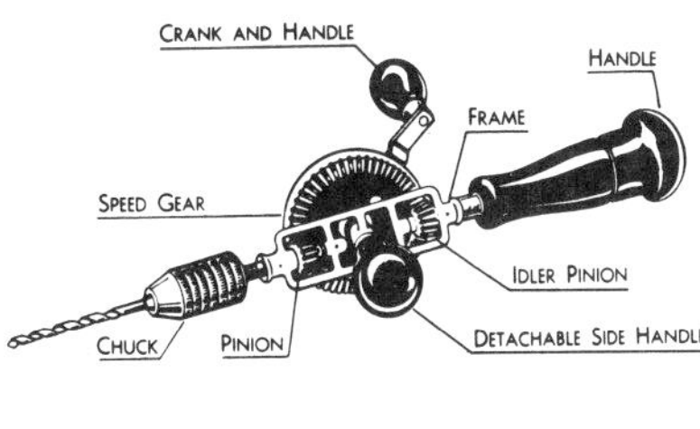

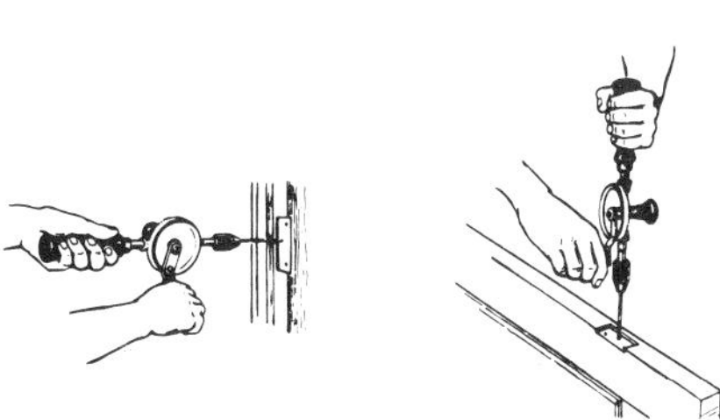

Figure 483 Hand drill

hand brace Hand tool for boring holes. See *brace*.

hand drill Crank-turned tool used for driving a drill for boring holes. (Figure 483.) Drill is secured in a three-jaw chuck.

hand file Flat abrading tool for smoothing metal surfaces. See *file*.

hand groover Tool for cutting grooves in sheet metal.

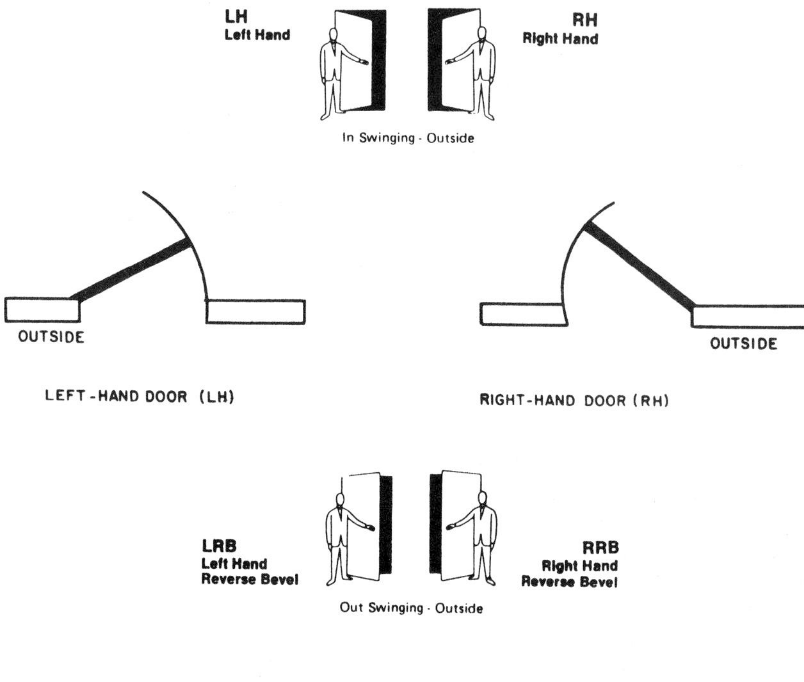

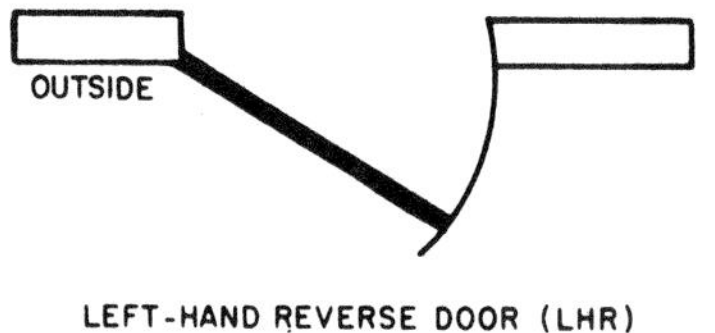

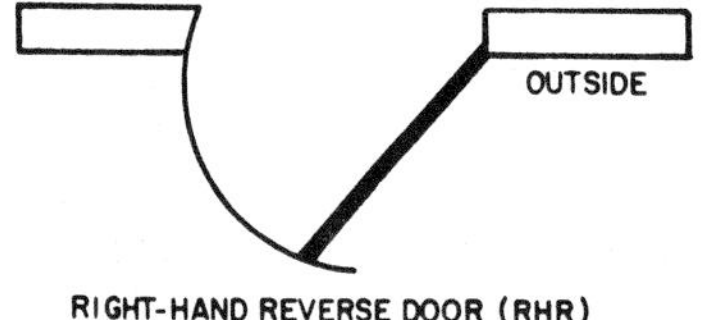

Figure 484 Hand of doors

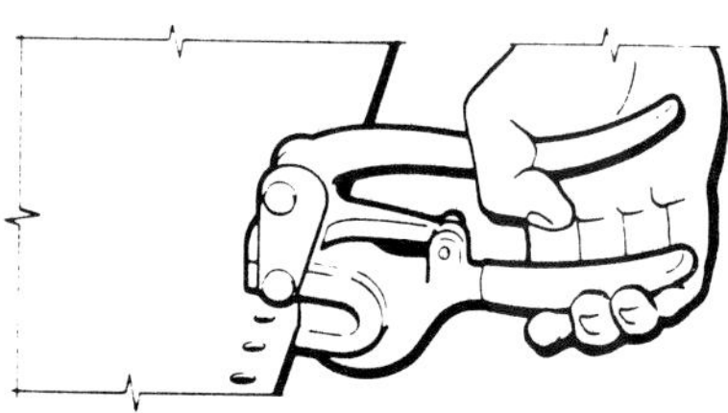

Figure 485 Hand punch

Figure 486 Hand shield

hand of doors Description of how a door swings. Swinging doors are installed to open in or out. The hinges may be either on the left or right. How a door opens, in or out, left or right, is called the hand of the door. Left-hand doors have hinges on the left; righthand doors have hinges on the right. Inward swing is standard; outward swing is reverse. (Figure 484.)

LH = lefthand (hinge on left; opens inward)
RH = righthand (hinge on right; opens inward)
LHR = lefthand reverse (hinge on left; opens outward)
RHR = righthand reverse (hinge on right; opens outward)

hand punch Hand tool for punching holes in sheet metal, as for rivets. (Figure 485.)

handrail Support used along the side of a stairs, walkway, or scaffolding walk. The handrail is supported by balusters. See also *molding.*

handrail bolt Special bolt used in stair construction which is threaded at both ends. Used to hold handrail to *balusters* or *newels.*

hand saw Hand-operated saw. See *back saw, coping saw, cross-cut saw, dovetail saw, drywall saw, flooring saw, hacksaw, keyhole saw, ripsaw, stab saw.*

hand scraper Flat metal piece with a turned-back edge. The tool is pulled over the wood edge to produce a smooth finish. See *cabinet scraper.*

handscrew clamps *Parallel clamps.*

hand shield In arc welding, a protective face cover held in the hand before the face (rather than worn over the head). Used for making a brief inspection of a weld process, as in a teaching situation. (Figure 486.)

hanger Metal strap used for supporting or holding ductwork, piping, guttering, joists, beams, electrical conduit, and the like. A *stirrup.* Wires used to support a *drop* or *suspended ceiling.*

hanging scaffold Scaffold suspended by cables; a *swing-stage scaffold.*

hardboard Manufactured, pressed-wood panels used for wall coverings.

hardened steel nail A very hard nail similar to common nails but with a smaller-diameter shank, used for its extra strength.

hardener In glues, the catalyst or accelerator that is mixed with the resin to make the glue, as an epoxy glue.

hard facing *Hard surfacing.*

hardpan Very hard, compacted layer of earth or clay sometimes encountered in excavations.

hard surfacing A welding process where a worn surface, such as an edge or point, is built up with weld beads. Also called *hard facing.*

hardware Metal parts used in finish construction, such as hinges, catches, pulls, and locks. Sometimes called *builder's hardware.*

hard water Water that has a high amount of calcium and magnesium minerals. Hard water is objectionable because it is hard to wash with, leaves a scum, and allows a build-up of calcium scale.

hardwood A broadleaf tree that loses its leaves with the cold season. A *deciduous* tree such as oak, maple, or walnut. The wood of such trees often takes a high finish, although it may not always be hard.

hardy A tanged-bottom cutting chisel that is inserted into the tail of an anvil. Used as a cutting base. See *anvil.*

hatch Small access door, as to an attic.

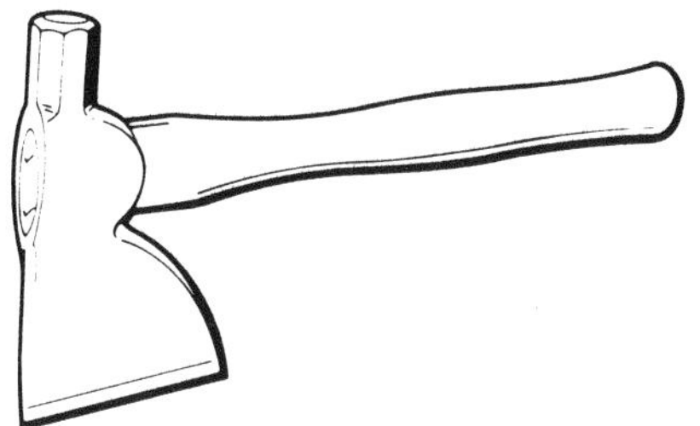

Figure 487 Hatchet

hatchet Hand tool with a cutting blade and a striking face, used for rough cutting in forming and framing. A notch is generally provided in the blade for pulling nails. Figure 487 shows a half hatchet with a striking face similar to a regular hammer face. A full hatchet has a very wide face similar to an axehead face.

haul In roadway construction, the amount of excavated earth that must be transported to fills or waste areas to create the roadway.

haul road Temporary road used to transport construction material to and from site.

haunch Thickening of a support member to give added strength, as the added thickening of a reinforced beam over a column or a widened knee at the juncture of the column and rafter in a rigid frame building. Bracket that projects from a column or panel to provide support for another member.

hawk Small square board with a handle attached to the bottom used for carrying mortar or plaster. (Figure 488.) See also *hod, mud holder.*

Figure 488 Hawk

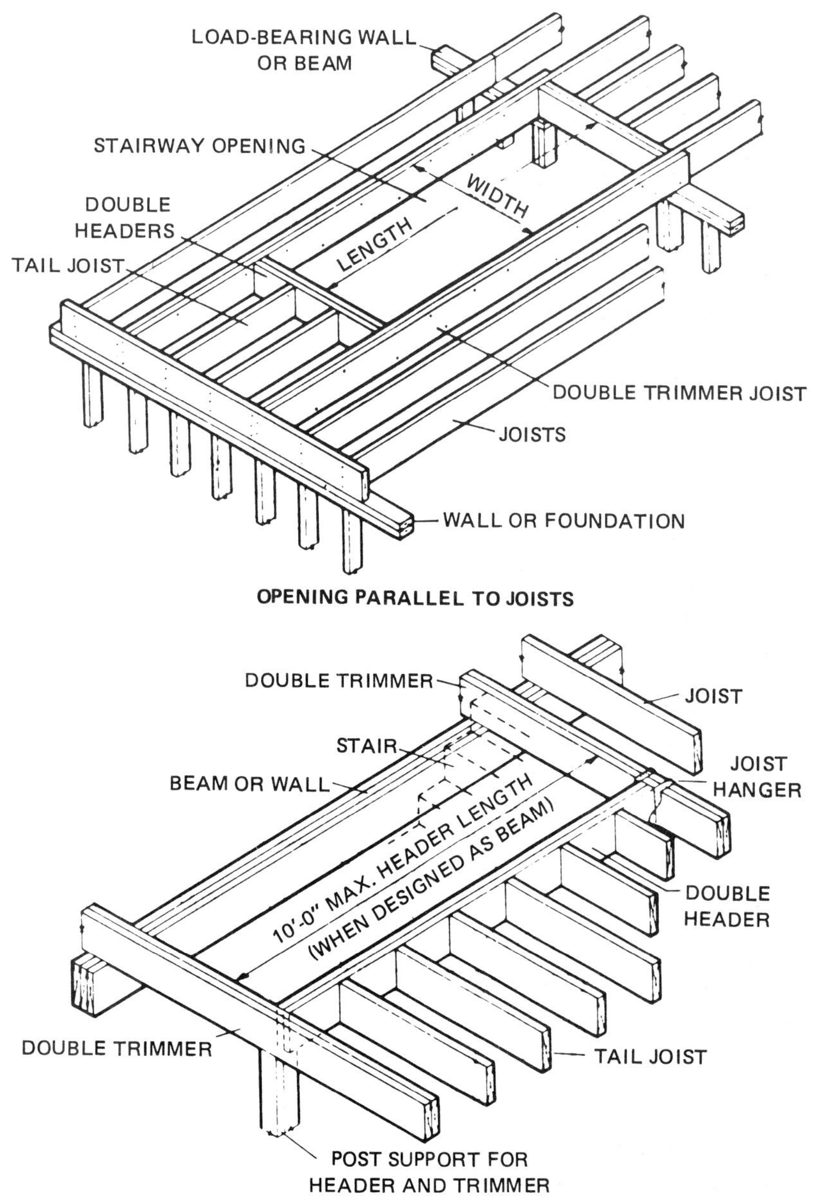

Figure 489 Headers used at floor openings

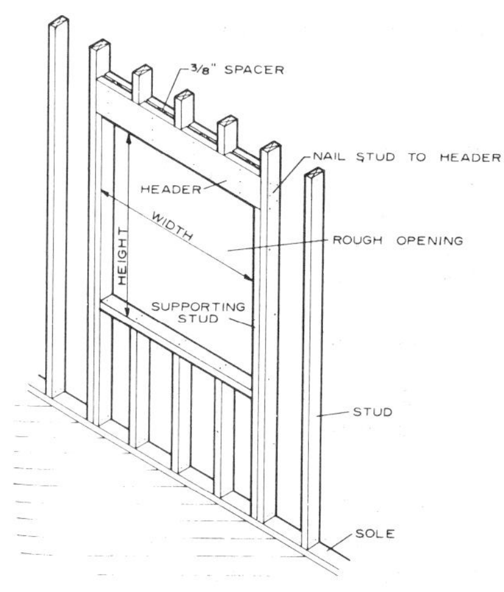

Figure 490 Headers used over window opening

H beam Steel beam with an H-shape.

head Top member of a window or door frame. See *door section, window section.*

head casing Outside trim over a window frame, includes the *drip cap.*

header Lumber piece fastened at a right angle to joists, rafters, or studs framed around an opening. Figure 489. Framing members at the top of window or door framing. (Figures 490; *966*, p. 427). Brick or concrete block laid with the end facing out. (*Figure 125*, p. 47.) In electrical wiring, a metal raceway leading from a distribution center to a cell (enclosed tubular space) in a metal floor or precast concrete floor. Wiring and cables run horizontally in the cell. In plumbing, a waste pipe that receives the discharge from two or more pipes.

header bond Masonry bond pattern in which all courses are laid as headers. It forms a wall 8″ thick.

header course Course of bricks or concrete block laid as headers (ends facing out).

header joists Joists running on the outside of the floor platform at a right angle to the common floor joists. (Outside joists that run the same direction as the common floor joists are called *end joists*). See also *band joist, rim joist, sill joist.*

heading bond A *header bond.*

head jamb The *head* of a window or door frame.

head joint Vertical masonry joint, opposed to a horizontal masonry joint (*bed joint*). Also called *builds.*

headroom In a stairs, the clear height from the treads to the ceiling above, normally not less than 6½ feet. (*Figure 937,* p. 413.) In a doorway, the clear distance from the threshold to the bottom of the head jamb.

head shield A welding helmet with a filter lens designed to protect against arc-welding rays.

heart The center or pith of a tree; the *heartwood.*

heart center Pith of a tree.

hearth Fireplace floor; the area just in front of the fireplace opening. See *fireplace.*

heart rot Rotting of heartwood, often found in a living tree; results in a hollow tree.

heartshake Lumber defect consisting of cracks opening around the center (or heart) of a tree. When the tree is cut across through the trunk, the cracks sometimes appear to radiate away from the center, giving a starlike appearance. Also called *star shake.* (Figure 491.)

heartwood Wood near the center of a tree as opposed to *sapwood,* which is nearer the outside of a tree. (*Figure 1021,* p. 451.) Heartwood is harder and stronger than sapwood.

heat ducts Ductwork that carries warm air to the various outlets. See *ductwork.*

Figure 491 Heartshake (lumber defect)

heat exchanger In a heating system, a chamber where heat from the furnace is transferred to the air being circulated to heat the building. In solar heating, a device for exchanging heat from an indirect solar fluid system to the household water. A simple heat exchanger would be a copper coil (containing solar-heated fluid) immersed in a tank of household water; heat is transferred from the coil of solar-heated water to the household water.

heat gain Increase in the amount of heat in a given space from transmission from outside the area; heat gain is caused by solar radiation, cooking, electrical equipment, lights, body heat, and so on.

heating and ventilating plan Drawing that shows a plan view of heating and cooling ductwork or piping. Runs and sizes are shown. Normally, in residential construction HVAC information goes on the regular floor plans.

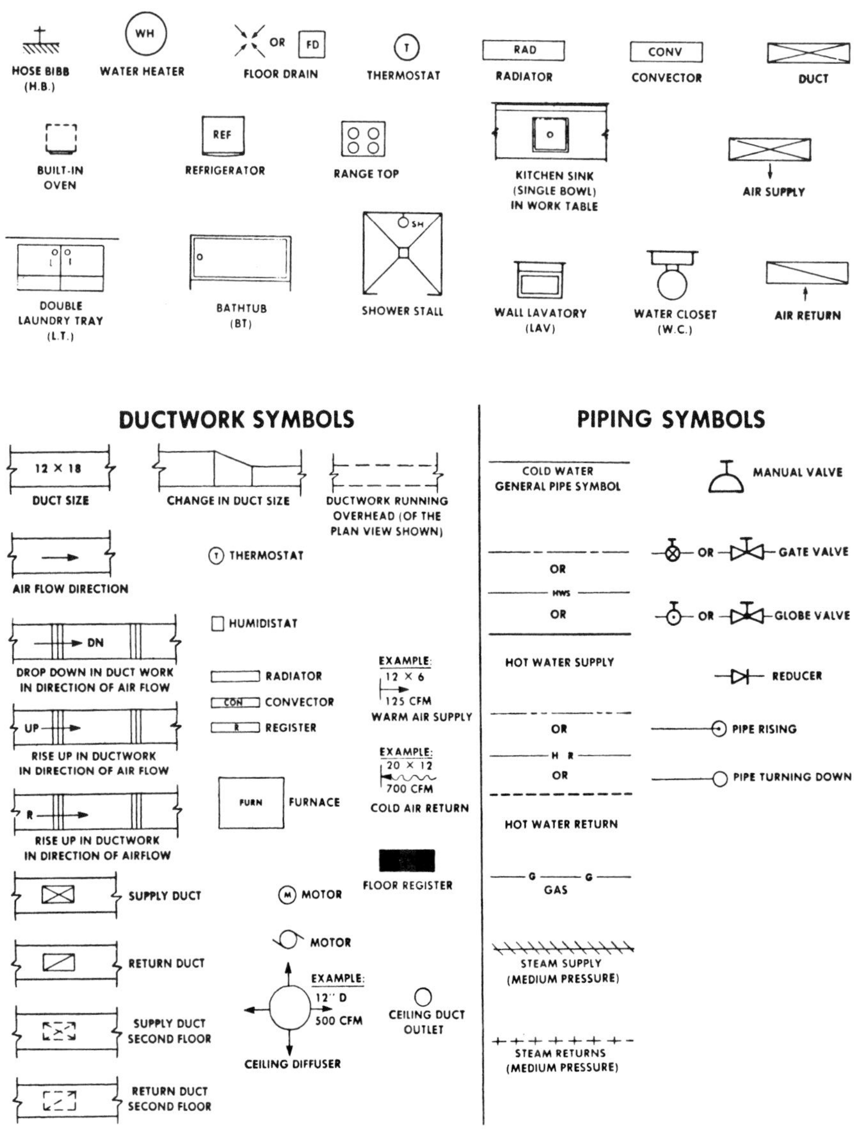

Figure 492 Heating and ventilating symbols

heating and ventilating symbols Symbols used on blueprints to show heating and cooling fixtures. (Figure 492.) See also *plan symbols, symbols.*

heating cable Electrical resistance cable used for heating. Often placed in a ceiling or floor to create a radiant heating panel. Also used in gutters and on roof eaves to melt ice and snow.

heating plan Drawing that shows a plan view of the heating ductwork or piping, with sizes and outlets noted.

heating plant The total system for heating a building, from the furnace, to outlets, to return, and venting.

heating system The total system for controlling temperature in the building. The most common systems are: *forced warm air, hor water,* and *steam.* See also *baseboard heating, electric heating, heat pump, radiant heating, solar heating, unit heater.*

heat loss Decrease in the amount of heat in a given space by transmission of heat inside a structure to the outside. Heat is lost through conductance or infiltration through walls, floors, ceilings, windows, or doors. See *heat gain.*

heat of hydration Heat released by chemical change in concrete, caused by reaction between water and cement.

heat pump Device for extracting heat from the ground or the outside air and using it to warm a house or structure in the winter. A kind of reverse refrigeration. (Figure 493.) In summer, heat is extracted from the inside and removed to the outside. See *air conditioning.*

heat sink Material or structure designed to absorb and store heat. A thick stone wall may function as a heat sink if exposed to direct sunlight.

heat transfer Flow or movement of heat from one place to another by *conduction, convection,* or *radiation.*

heat treated Refers to metal that is heated, then quenched in water or oil to give strength and hardness.

heavy timber construction Construction using large solid, built-up, or laminated timbers. See *post and beam construction, timber construction.*

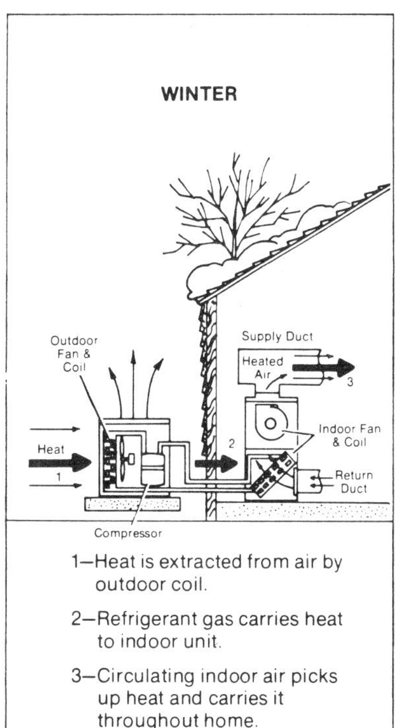

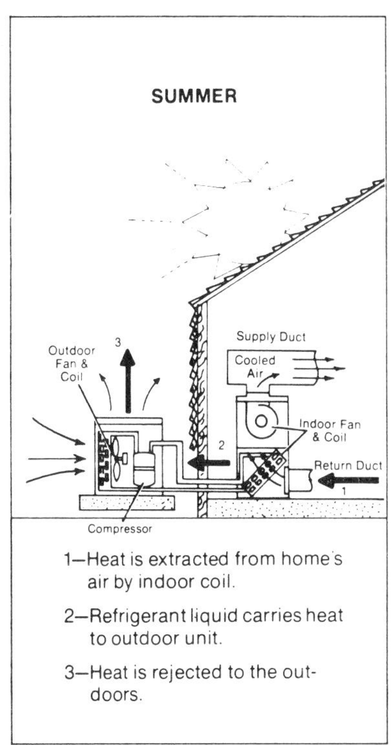

Figure 493 Heat pump

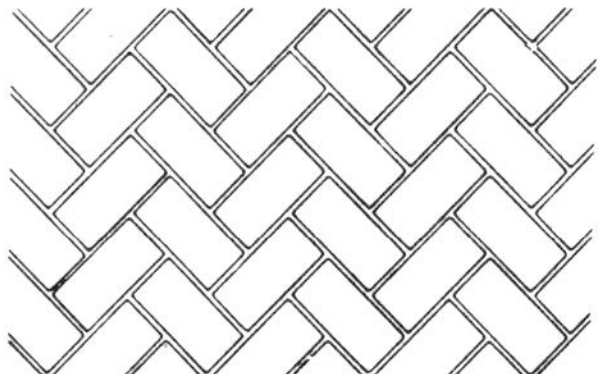

Figure 494 Herringbone bond

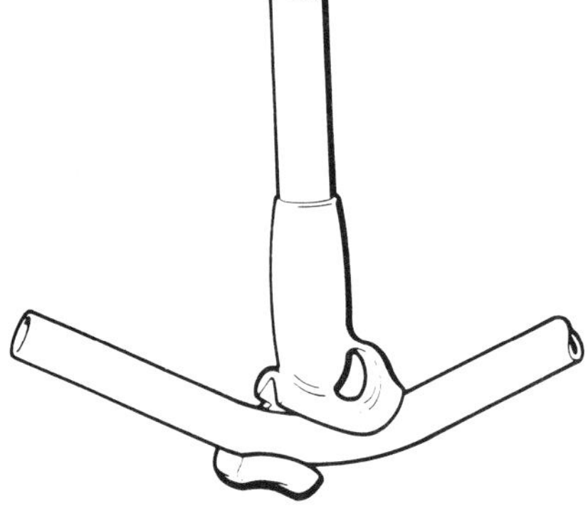

Figure 495 Hickey

heel The part of a rafter or joist that rests on a wall plate. (*Figure 771*, p. 339.) The end bearing surface or seat of a horizontal support member as it rests on a vertical support. Ends of a truss that rest on supports; point where the top and bottom chords intersect. See *rafter cuts.*

height of instrument In surveying, the height or exact elevation of the line of sight in the level or transit telescope. Abbreviated H.I.

helically grooved nail Nail with a spiral groove on the shank.

helve Handle of swinging tool, as a hammer or axe. See also *haft.*

hem edge In sheet metal work, an edge that is simply folded back. See *edge.*

herringbone A criss-cross pattern used in decorative brick walls and floors, and in hardwood floors.

herringbone bond Masonry bond pattern where the brick are laid at an angle in a zigzag pattern. Bricks join each other at a right angle. (Figure 494.)

Hertz (Hz) Electrical unit: one cycle per second. The standard AC house current of 60 Hertz.

hewing Cutting a timber to shape with an axe or aze.

hexagon An even, six-sided figure.

hex tapping screw Self-drilling screw with a hex head, used for attaching light metal.

hickey Hand tool with a side opening used for bending plumbing pipe, rigid conduit, or reinforcing bars. (Figure 495.) See also *conduit bender.*

hidden line In drafting, lines that are not seen directly in a view but are behind the surface of any view. Hidden lines are normally not shown in architectural blueprints unless essential to an understanding of the structure. Foundations are shown in elevation with hidden lines. See *alphabet of lines.*

high chairs Bent rod supports used for holding up reinforcing bars and welded-wire mesh so concrete can flow evenly around the steel. High chairs are used where there are two levels of bars or mesh. They support the upper level. See *bolsters, chairs.*

high-early-strength concrete Concrete that produces a strength faster than normal; a special cement or admixture is used.

high-pressure boiler Boiler furnishing steam at pressures over 15 psi, or hot water at pressures over 160 psi or at temperatures over 250°F.

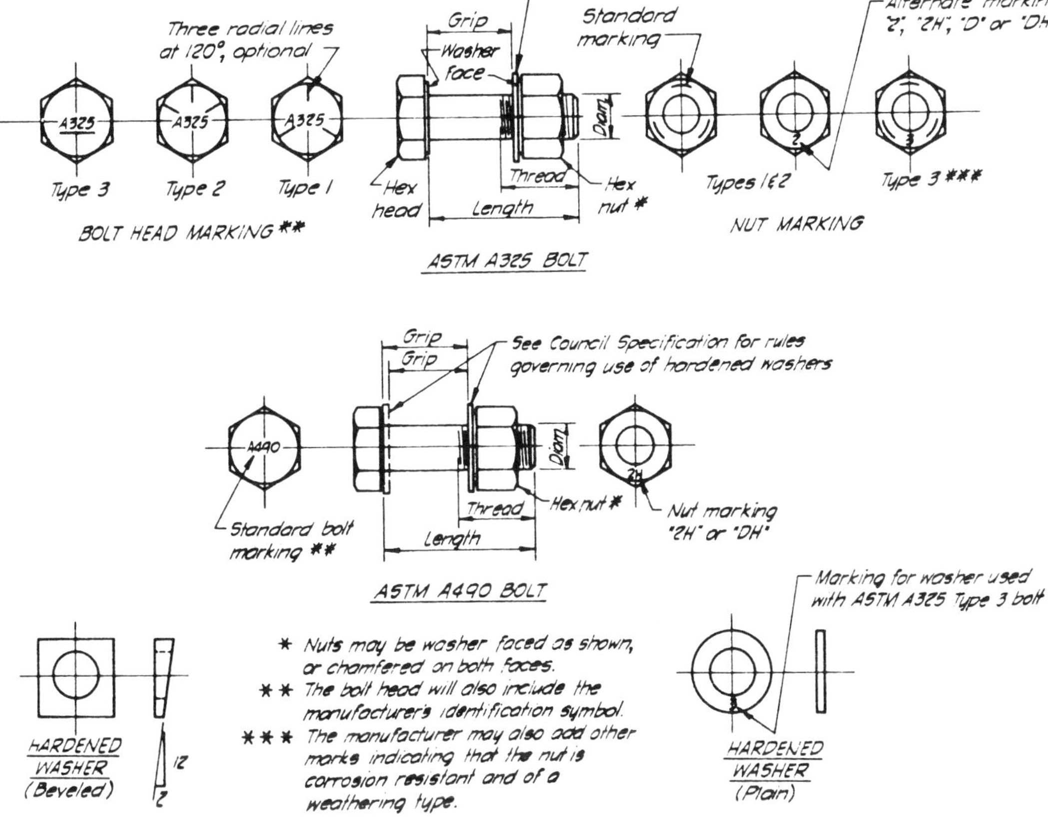

Figure 496 High-strength-bolt terminology

high-rise Tall, many-storied structure that provides living area combined with in-building shopping and convenience areas.

high-strength bolts Special bolts used to fasten steel or precast panels. (Figure 496.) See also *bearing-type connection.*

hi-load nails Special nails with a high resistance to shear or withdrawal.

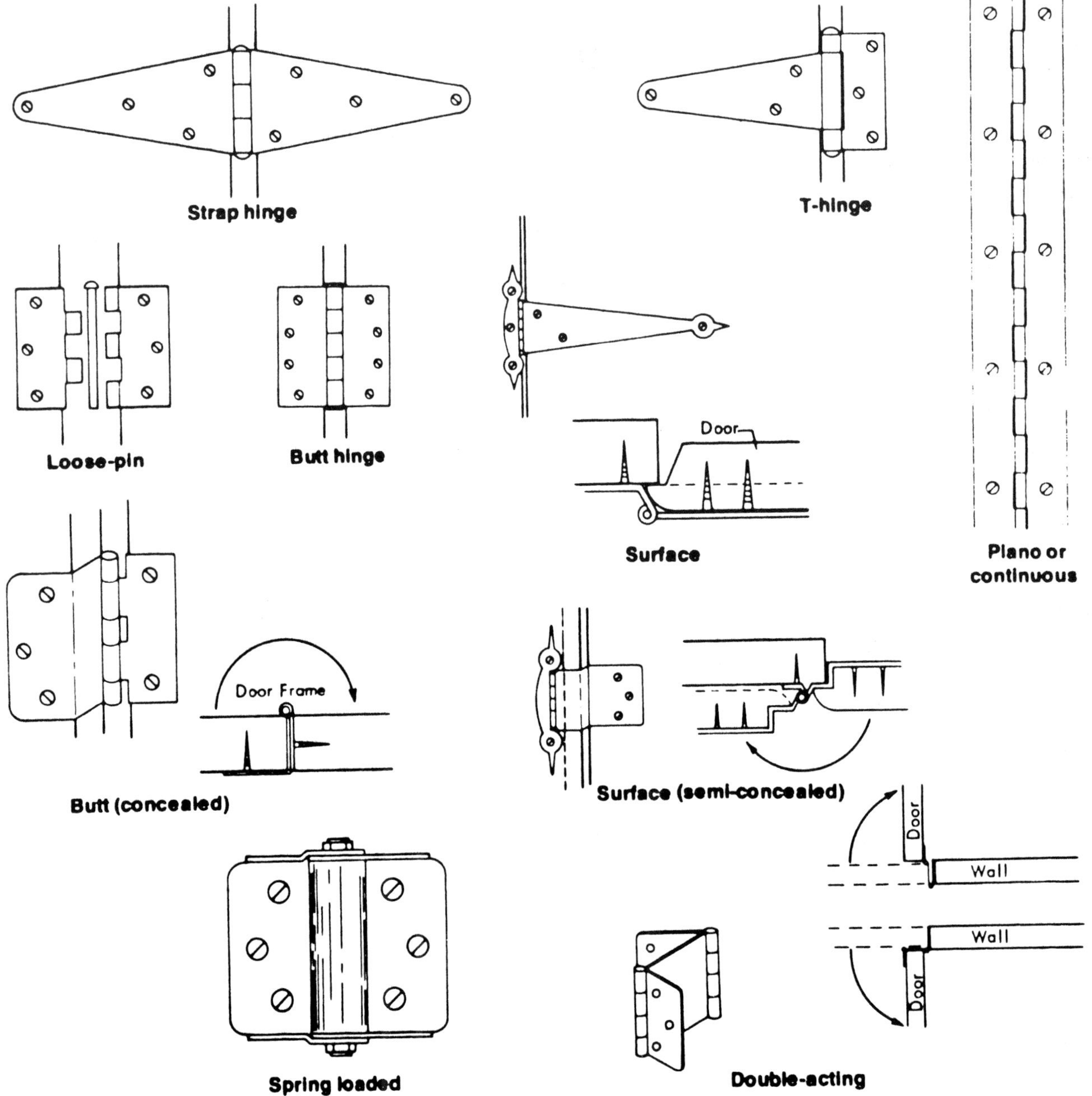

Figure 497 Hinges

hinge A two-part moving joint used to hang doors so they swing. The two movable plates or leaves fit together and are held with a pin. Some pins may be removed. Various hinges are available. (Figure 497.) See *butt hinge*. Also, in arch construction, the main point of support. A two-hinge arch is supported at each end. A three-hinge arch is in two sections, supported at each end and joined at the middle with a pivot or hinge. See *arch*.

hinge hasp Metal strap that passes over a staple; padlock fits through staple on closed hinge to fasten. (Figure 498.)

hip External angle made by hip rafters joining the main roof. See *roof, roof framing.*

hip drop Amount the bird's mouth of a hip rafter is deepened to allow hip to be even with jack and common rafters without beveling the hip. Referred to as dropping the hip.

hip jack rafter Rafter that runs between a hip rafter and the plate. A jack rafter.

hip rafters The two main rafters that join the ridge to form a hip on the end of a roof. See *roof framing.*

hip roof Roof with two gable sides and inclined roof ends (hips). See *roof, roof framing.*

hips Hip rafters. See *roof framing.*

hitch A fastening made to attach a rope or cable to a load. See *cable attachments.* (*Figure 149*, p. 57.)

H molding Aluminum H-shaped channel used to receive ceiling panel edge.

hod Long-handled V-shaped carrier for bricks and mortar.

hog Camber; upward curvature on a prestressed beam.

hogging In woodworking, the rough hand planing of lumber.

hoist Lifting device that uses a chain or rope strung through blocks to obtain a lifting advantage. See *block and tackle, ratchet lever hoist.*

hole cutter Special drill attachment for cutting holes (up to 12″) in sheet metal. (Figure 499.)

hole punch In sheet metal work, a hand-operated punch used for punching holes in metal. (Figure 500.)

Figure 498 Hinge hasp

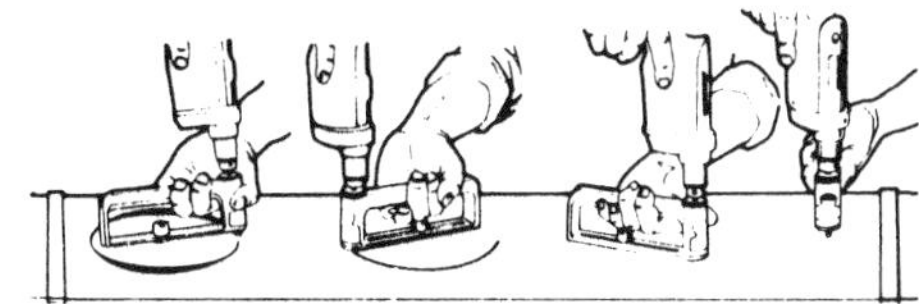

Figure 499 Hole cutter

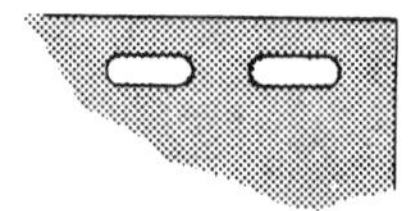

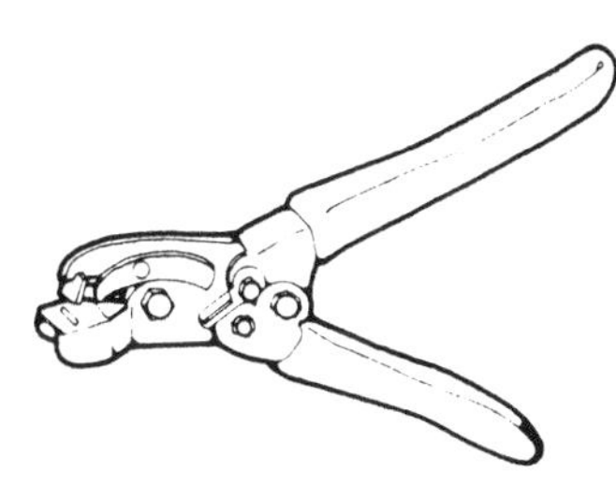

Figure 500 Hole punch: making of nail slot

Figure 501 Hole saw

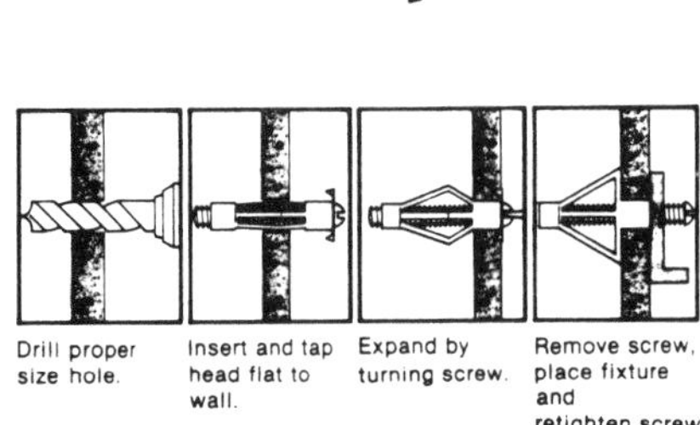

Figure 502 Hollow-wall anchor

hole saw A circular, sawlike cutting device used for cutting holes. (Figure 501.) Sizes vary from $\frac{9}{16}$″ up to 6″ in diameter.

hollow-core door Flush door with a criss-cross internal framework made of pressed paper or polyurethane. The core supports the outer plies. See *door*.

hollow-core slab Reinforced, prestressed floor-panel unit with hollow cores running its length. See *hollow slab*.

hollow knot Otherwise sound knot with a small hole through it.

hollow masonry Masonry unit with voids that exceed 25% of the cross-sectional area, such as concrete block. See *solid masonry*.

hollow slab Prestressed concrete deck or floor member that has hollow cores running its length. Slabs are placed side by side to form the floor or deck. (*Figure 785*, p. 344.)

hollow wall Masonry wythes separated by an air space, usually 2″. Brick and brick, brick and concrete block, or concrete block and concrete block are used. (*Figure 1072*, p. 473.) Also called a *cavity wall*.

hollow-wall anchor Metal anchoring device used to fasten into a hollow wall, such as drywall. The anchor folds together to fit through a hole, then the sides spring open and are tightened back against the inside of the wall. (Figure 502.) See also *anchors, plastic anchor, plastic toggle, toggle bolt*.

homogeneous All the same thing or type.

hone A fine-grained stone used for finishing a cutting edge to a fine sharpness. A further degree of sharpness can be achieved only by *stropping* the edge on a leather strap. See *oilstone, waterstone*.

honeycomb An open, cell-like structure. A concrete that has set with many voids or openings.

honing A final sharpening on a fine sharpening stone. You push the blade edge *into* the stone. See *stropping*.

hood Flared metal canopy designed to collect air and fumes, as a range hood used over the kitchen range to collect fumes for ventilation to the outside.

hook A bent device for holding or pulling something. In concrete reinforcement, a turning back of a reinforcing bar to form a hook, used as anchorage. (Figures 503, *210*, p. 78.)

hopper window Window with a sash that swings inward; hinges are on the bottom. See *window*.

horizon line In perspective drawing, the line in the distance where all the lines of the object converge to form *vanishing points*. See *perspective drawing*. (*Figure 676*, p. 292.) The line that forms the horizon.

horizontal Level to the ground or the horizon. Opposed to *vertical* (straight up and down).

horizontal position welding Fillet weld on the upper side of the horizontal surface. Also, horizontal groove weld on a vertical surface. See *welding position*.

horizontal shear (F_v) Stress calculated in pounds per square inch on a horizontal shear. Values are available in published tables for different woods and are used in design calculations. See also *extreme fiber stress in bending*.

horn The rounded end or beak of an *anvil*.

horse A base used to support work. A *sawhorse*.

horsepower Work unit of 33,000 foot-pounds per minute. A 3-horsepower (hp) motor can do 99,000 foot-pounds of work per minute.

horse rasp Very rough rasp for roughing out a wood piece. See *rasp*.

hose bibb An outside faucet that is threaded to receive a water hose. Also called a *bib*, *bibcock*, or *sill cock*.

hose cock Faucet with a threaded end for a hose. A *hose bibb*.

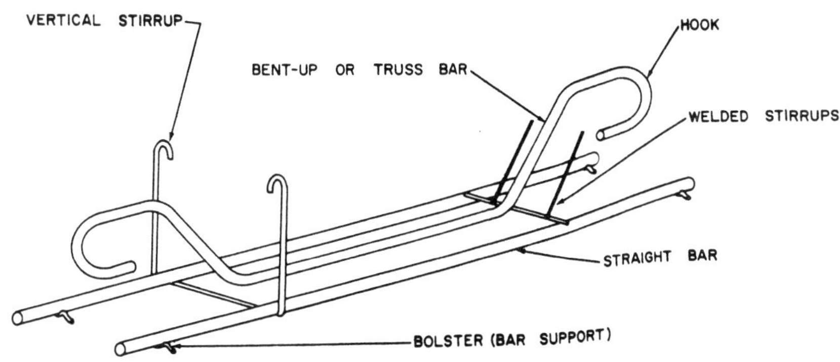

Figure 503 Hook or reinforcing bar

hot In an electrical system, a live wire or circuit. In plastering or cement work, a very dry surface, one that is hungry for moisture.

hot glue Woodworking glues that are mixed from a dry powder, heated, and applied hot.

hot stuff In roofing, hot asphalt or bitumen.

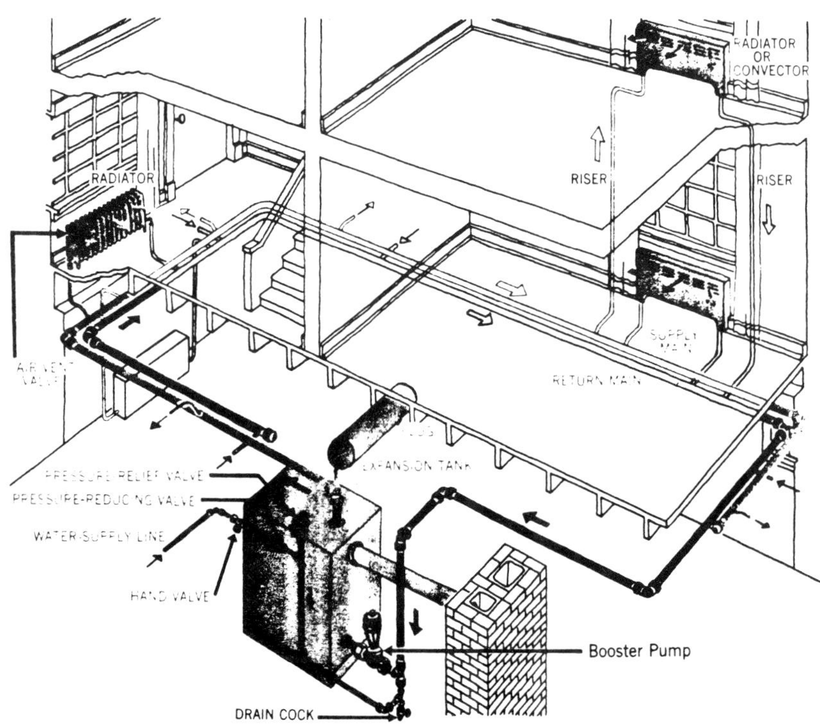

Figure 504 Hot-water heating: two-pipe forced hot water

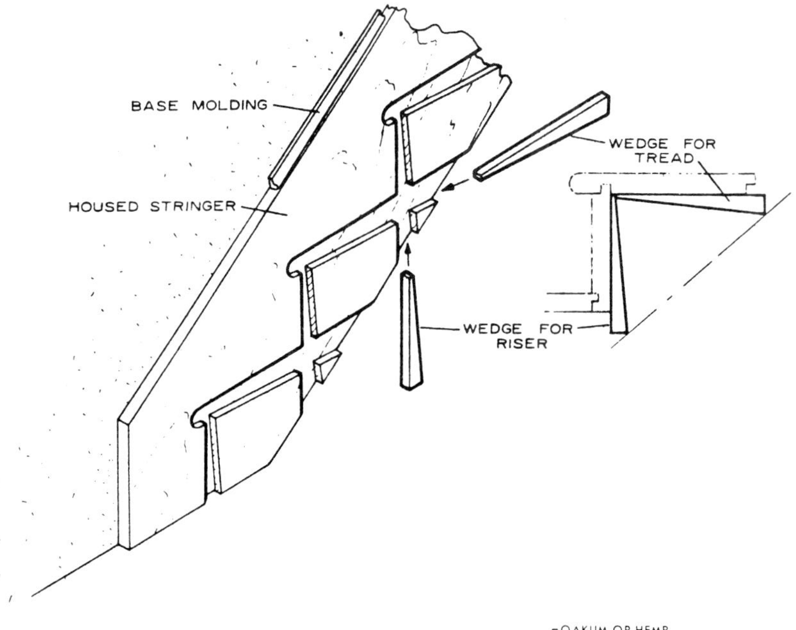

Figure 505 Housed stringer

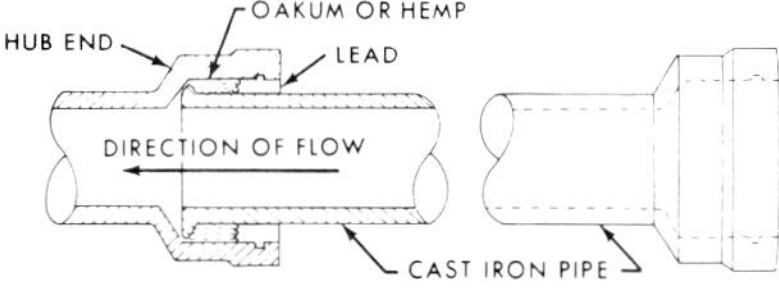

Figure 506 Hub

hot-water heating Heating system that heats by circulating hot water to the various heating outlets. One- or two-pipe systems are used to circulate the hot water. The most widely used is the *two-pipe hot-water system.* (Figure 504.) See also *baseboard heating, gravity hot-water heating.*

hot wires Conductors carrying current.

housed Of a lumber piece, cut out to receive a second piece of lumber, as a *mortise* which receives a *tendon,* or a *housed stair.*

house drain House drainage system that runs to the house sewer or private disposal system. A *building drain.*

housed stair Stair where the outside stringers (carriages) are housed (have grooves made) to receive the tread and riser ends. (Figure 505.)

housed stringer Stringer (carriage) that is housed out (cut with grooves) to receive the tread and riser ends. (Figure 505.)

house sewer Drainage system that carries liquid waste from the house drain to the public sewer or private disposal field. A *building sewer.*

house trap Water seal in the building drain to prevent the entrance of sewer gases.

housing The cut or opening made into a lumber piece to receive the end of another lumber piece.

hub In plumbing, the flared opening or bell on the end of *bell-and-spigot cast-iron soil pipe.* (Figure 506.) Also called a bell or socket. In surveying, a ground marker (usually a stake) used to indicate a key point in the survey: a corner, angle, or turn made in the survey direction. See also *reference points.*

humidifier In a total air conditioning system, a device that adds moisture to dry air, to humidify. Opposite to a *dehumidifier.*

humidity Amount of moisture in the air, expressed as a percent. When there is a 100% humidity, no further moisture can be held in the air; any further moisture will result in precipitation or in moisture's condensing out in the form of water droplets. See *relative humidity.*

HVAC Heating, ventilating and air conditioning.

hydrated lime Quicklime (calcium oxide, CaO) with water added to form calcium hydroxide, $Ca(OH)_2$.

hydration Chemical reaction between cement and water which results in a hard mortar or concrete. Also used in relation to quicklime when slaked (treated with water) to make a hydrated lime. See *hydrated lime.*

hydraulic cement Cement that will harden under water.

hydraulic pressure Pressure from water; the force of water running out of a faucet is caused by hydraulic pressure.

hydronic heating Forced hot-water heating. See *baseboard heating, hot-water heating.*

hydropower Energy developed when moving water turns turbines.

hydrostatic pressure Water pressure.

hygroscopic Of a substance or a wood, losing and gaining moisture easily.

i

I beam In popular usage, a standard (S) steel beam. Technically the term I beam is no longer recognized by standards organizations and has been replaced by the term *S beam* (standard beam). A built-up wood beam or joist; *I joist.*

I joist A built-up joist with a solid plywood web supporting a top and bottom flange. (Figure 507.) Joist depths vary from $9\frac{1}{2}''$ up to 30″; spans run up to 50′. See *open-web joist, wood beam.*

impact A sudden, hitting force.

impact insulation class (IIC) Number rating which gives an estimate of the impact sound-insulating performance of a floor-ceiling assembly.

impact wrench Heavy-duty power wrench used to set and remove bolts and nuts in steel construction. (Figure 508.)

Figure 507 I joist

Figure 508 Impact wrench

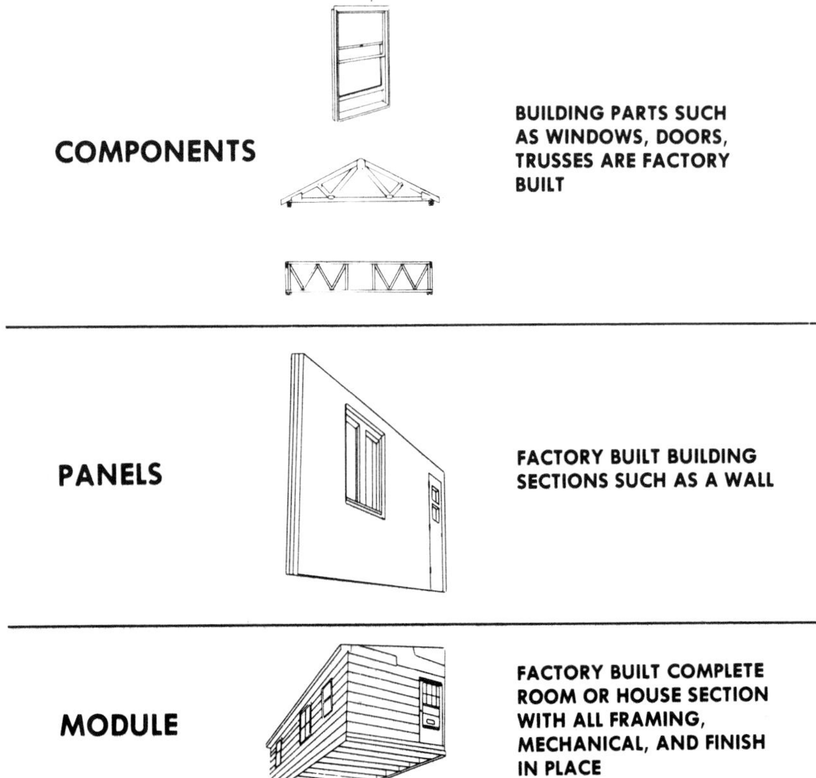

Figure 509 Industrialized building

improvement Change made on a structure or site that increases its value. The change may be adding a building to a site or adding a room to an existing building.

incandescence Light emitted from a heated source, as from a hot weldment.

incandescent lamp Lamp where light is produced by a glowing heated filament in the bulb. Electrical resistance causes the filament to heat up and glow brightly.

incipient decay Early decay, visible but perhaps not yet having softened or destroyed wood fiber.

incise To cut, to engrave, or to gouge a groove.

inclined plane The angle a plane or surface makes in relation to the horizon, referred to as the *angle of inclination.*

incombustible building material Material that will not burn in a standard $2\frac{1}{2}$-hour test in a furnace.

increaser Any pipe section with one end larger than the other.

indirect lighting Lighting accomplished by reflecting the light off a wall or ceiling; no direct lighting is used.

individual vent Vent pipe that vents a single plumbing fixture. The individual vent may run to the roof, to a stack, or to a branch vent.

industrial park A development created especially for commercial and industrial use: structures are built for industry, roads are designed for heavy use, rail spurs are brought in.

industrialized building Construction that uses a high degree of *prefabricated construction* or building *components* or both. (Figure 509.) See *preengineered building.*

inert Something that will not enter into a chemical reaction with another thing, as argon gas.

inert gas A gas, such as argon or helium, that does not readily combine with other substances. Used in arc welding to shield and protect a weld and the base metal during the weld process.

inert-gas metal-arc welding See *shielded metal-arc welding.*

infiltration Flow of air into or out of a structure through openings or cracks.

infrared detector *Smoke detector* that responds to flame.

inhibitor Chemical added to concrete to inhibit (reduce) rust of metal reinforcing bars.

initial set In concrete, a hardening of the concrete which involves a chemical change with a release of heat. The final set takes a longer time. Initial set is sometimes specified as the time needed for the mortar or the concrete to reach a strength of 500 psi penetration resistance.

inlay To decorate by setting woods, mother of pearl, or some metal into a base wood or metal. The base wood is carefully housed out to receive the wood or pearl pieces. The metal is incised to receive metal of a different color, as copper inlay on brass or silver inlay on a base metal.

in-line joint Joint made by butting two lumber pieces end to end; splice piece is used on the sides to give strength.

insert fittings Fitting used to connect PE plastic tubing. Hose fitting clamps the insert fitting in place. (Figure 510.)

inside calipers Calipers with legs that turn outward, used for measuring inside diameters. See *calipers.*

inside corner brace See *corner brace.*

inside diameter (I.D.) Diameter of something measured on the inside opening, as the inside diameter of a pipe. Opposed to *outside diameter.*

inside trim Casing used around the inside of a room, as for windows and doors.

in situ Cast in place.

insolation Total amount of solar radiation striking the cover plate (transparent plastic cover) on a collector, measured in Btu's per square foot per hour per day.

inspection Examination of a structure to see that code restrictions or building requirements have been followed; examination of premises to see that it conforms to health standards, fire safety, electrical code, etc.

insulating board Pressed fiberboard commonly used on the exterior of a frame structure. The 4′ × 8′ board is nailed directly on the studding. Insulative R value is stamped on the board.

insulating concrete Lightweight concrete that has many voids or air spaces that provide insulation; vermiculite may be added to give insulative qualities.

insulating glass Two or three sheets of glass with air spaces sealed between them. See *double glazing, triple glazing.*

insulating wallboard Gypsum base wallboard with an aluminum foil coverage on the back. The foil works as a vapor barrier and provides insulation.

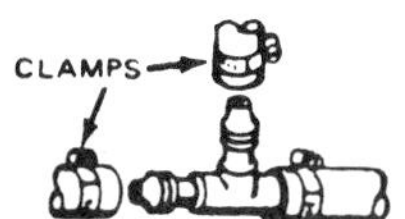

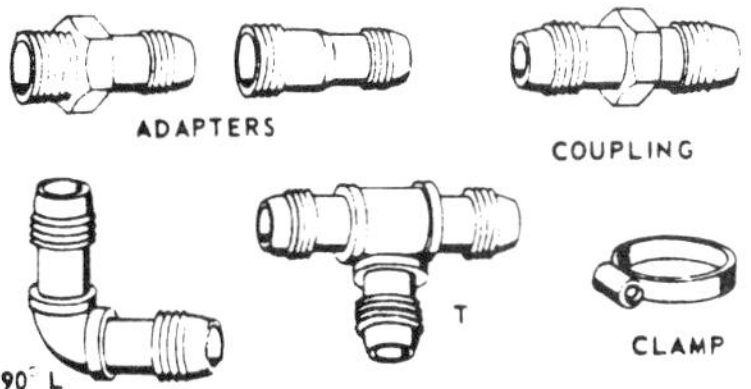

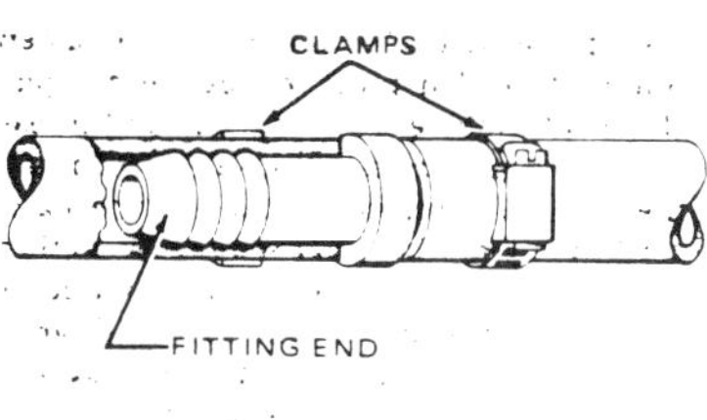

Figure 510 Insert fittings for flexible pipe

Batts — glass fiber, rock wool

Where they're used to insulate:

unfinished attic floor
unfinished attic rafters
underside of floors
open sidewalls

Blankets — glass fiber, rock wool

Where they're used to insulate:

unfinished attic floor
unfinished attic rafters
underside of floors
open sidewalls

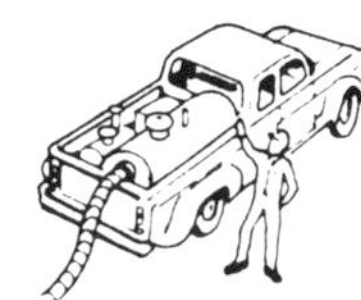

Foamed-in-place — expanded urethane

Where it's used to insulate:

— finished frame walls only

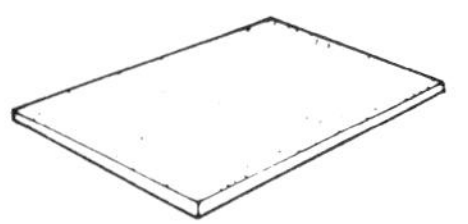

Rigid board—polystyrene (extruded), expanded urethane (preformed), glass fiber, polystrene (molded beads)

Where it's used to insulate:

exterior wall sheathing
floor slab perimeter
basement masonry walls

Loose fill (blown-in) — glass fiber, rock wool, cellulose

Where it's used to insulate:

unfinished attic floor
finished attic floor
finished frame walls
underside of floors

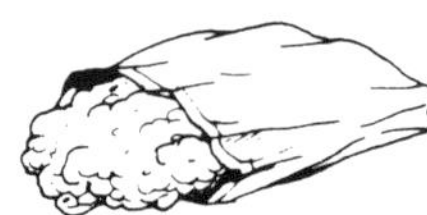

Loose fill (poured-in) — glass fiber, rock wool cellulose, vermiculite, perlite

Where it's used to insulate:

unfinished attic floor

Figure 511 Insulation types

Figure 512 Insulation creates an envelope around the living area. Batt insulation is shown

insulation Any material that resists the passage of heat into or out of a building; used to surround the living or work area. Figure 511 shows common types of insulation used. Insulation is used in floor, walls, and ceilings, and around the foundation. (Figure 512.) Batts, blankets, insulation board, plastic foam sheets, loose insulation, and viscous, blown insulation that sticks on a surface is used. See *batt insulation, R value.* Material that resists heat transmission is called *thermal insulation.* Material that slows down transmission of sound is called *sound insulation.* Also, a nonconducting material, such as rubber, used to cover electrical conductors.

insulation stapler Electric- or air-powdered stapler used for fastening batt insulation and foam panel. A $\frac{7}{8}''$ to $1\frac{1}{2}''$ staple is driven. Also called a *roofing stapler.*

insulator Glass or porcelain material that will not conduct electricity, used in electrical wiring.

intaglio A carving or impression made below the surface.

interactive drafting Computer-assisted drafting where the drafter directly manipulates the drawing project on the screen. An electronic device, such as a light pen, is used to make changes.

interceptor In a plumbing system, a container or receptacle that catches and holds solid matter to prevent its direct entry into the building sewer. A *catch basin.*

interior decorator One who plans color schemes and selects furniture and other interior decoration of the house.

interior elevation A drawing showing an interior, orthographic view of a building wall or part. Often used to show cabinets, carpenter-built details, and fireplaces. See *elevation, plan.*

interior finish All of the final work to complete the interior, including trim, painting, hardware, and so on.

interlocked grain A ribbon pattern in lumber produced by grain running in one direction, then reversing and running in the other, found in quartersawn wood.

intermediate metal conduit A thinwall rigid metal conduit used in all locations except in cinder fill. Threaded fittings are used and conductors are fished through the installed conduit. Type IMC.

intermittent weld Weld bead that is broken and has unwelded sections.

international architecture Simple, unornamented architecture with flat roofs and unbroken wall surfaces. (Figure 513.) Developed by Mies van der Rohe and the Bauhaus architects in the 1920s.

intrados On a masonry arch, the curve on the lower part; a *soffit*. See *extrados*.

invisible hinge Hinge that is completely concealed when in place even when the door is open. See *hinges*.

Ionic Greek order. See *orders*.

ionization detector *Smoke detector* that responds to invisible combustion products.

iron pipe size (IPS) Nominal size galvanized steel pipe.

ironwork Iron or steel used for ornament in a building, as for stair and balcony railings and balustrades. (Figure 514.)

ironworker Ornamental ironworker installs prefabricated ornamental ironwork (other than structural ironwork) such as railings, metal stairways, metal window sashes. Reinforcing ironworker sets steel rods or bars in concrete forms for reinforced concrete.

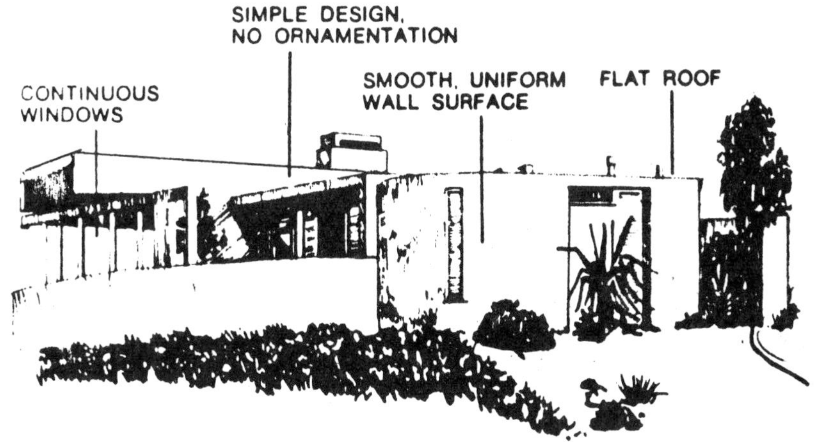

Figure 513 International style

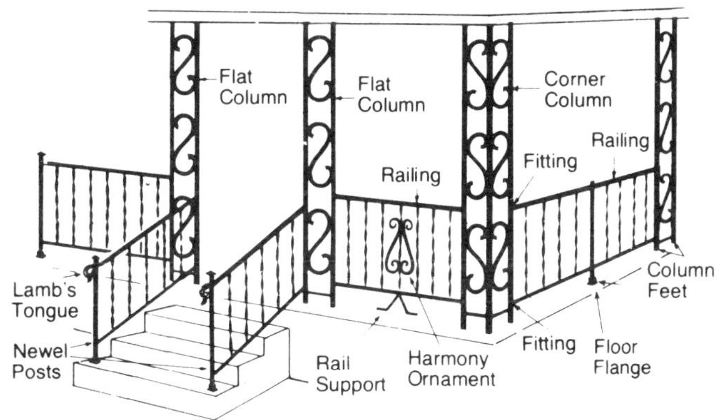

Figure 514 Ironwork

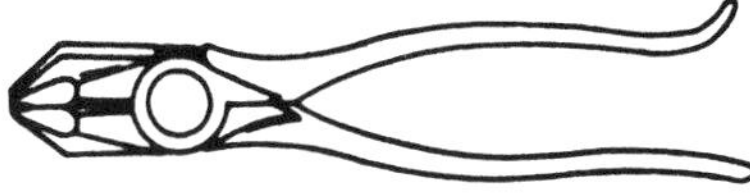

Figure 515 Ironworker's pliers

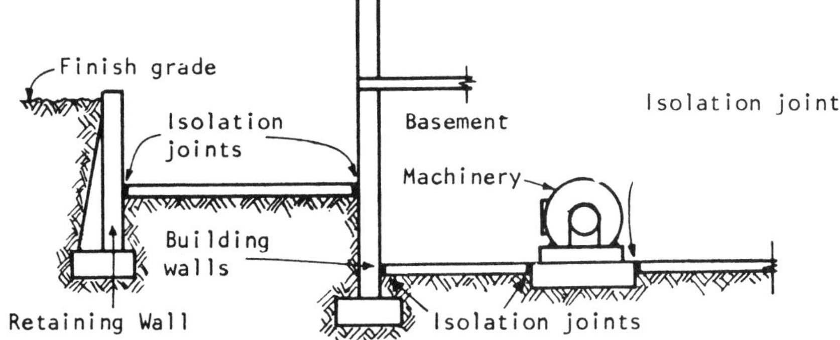

Figure 516 Isolation (expansion) joints

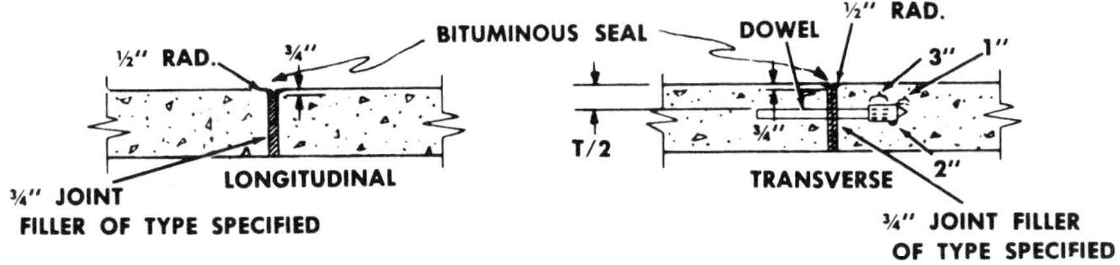

Figure 517 Isolation joint

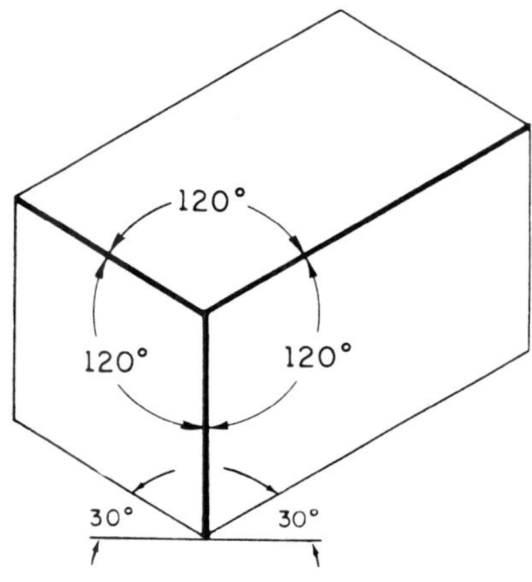

Figure 518 Isometric drawing

ironworker's pliers Side-cutting pliers with a hook bend on one handle and a coil spring to hold the jaws open. (Figure 515.)

irregular curve In drafting, an irregularly shaped curving figure used for drawing curves. Also called a *French curve.*

isogonic chart Chart that gives degrees of *magnetic declination* for different areas of the country. Chart is revised periodically.

isolation joint In concrete work, a joint separation made between concrete materials with different expansion rates, as when a slab butts against a wall. (Figure 516.) A resilient filler is placed between the two materials to allow for expansion or shrinkage and to prevent cracking. (Figure 517.) Also called an *expansion joint.* Also, a joint made in a wall or vertical plane to allow for settlement or movement. Often made over a major break area as over a part of a stepped footing; a *construction joint.*

isometric drawing A special pictorial drawing that has two horizontal lines of the object placed at 30° to the true horizontal. The base vertical line is kept in true vertical proportion. (Figure 518.) See also *cabinet drawing, pictorial drawings* for a comparison of drawing types.

j

jab saw See *stab saw*.

jack Lifting device. Hydraulic jacks or screw jacks are often used to lift or hold small buildings preparatory to moving or constructing a new foundation.

jack hammer Air-powered vibrating hammer with interchangeable tool points, such as chisels, spades, and cutters. Used for breaking concrete or other hard surfaces.

jack plane Widely used general-purpose bench plane. Commonly the bed is 14″ long with a 2″ cutter. Used for smoothing rough stock. (Figure 519.) See also *plane*.

jack post Metal post with another metal tube inside with an adjusting screw at the top. Used in remodeling to correct a sagging floor. Jack is positioned under the sag, raised to the proper length, then the screw is slowly extended over a number of days to raise the floor and allow room for a new column to be set in place. Jack post is sometimes left in place as a permanent column. (Figure 520.) Also called a *floor jack*.

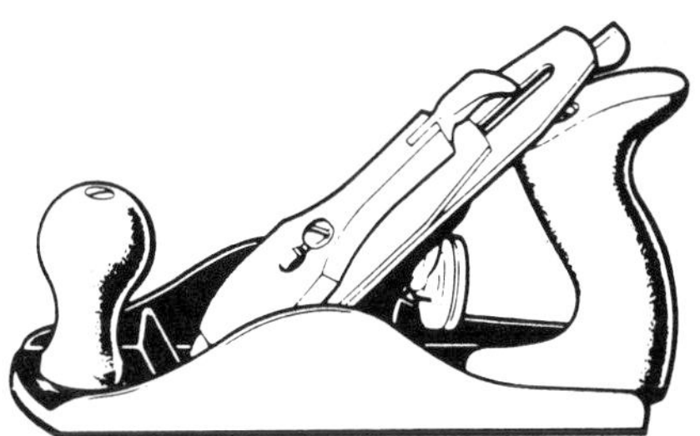

Figure 519 Jack plane

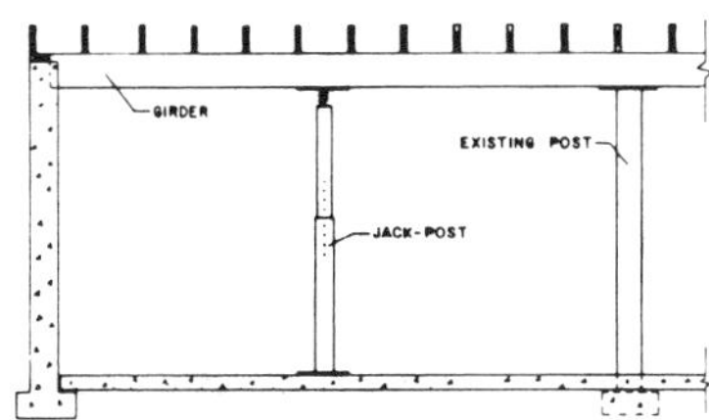

Figure 520 Jack post used to level sagging floor

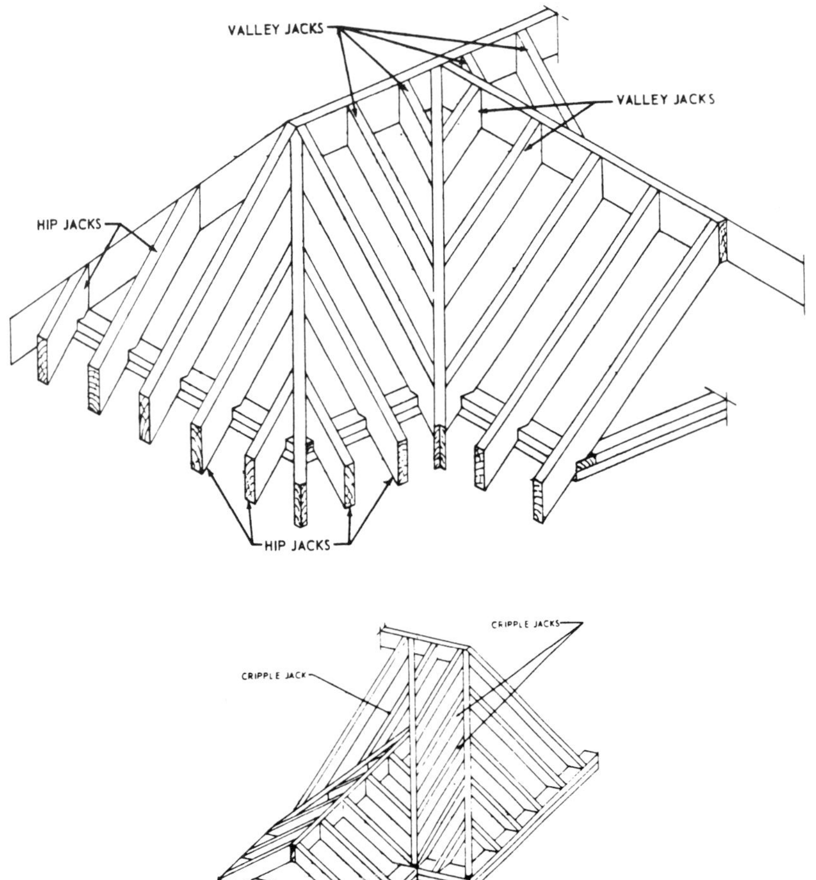

Figure 521 Jack rafters

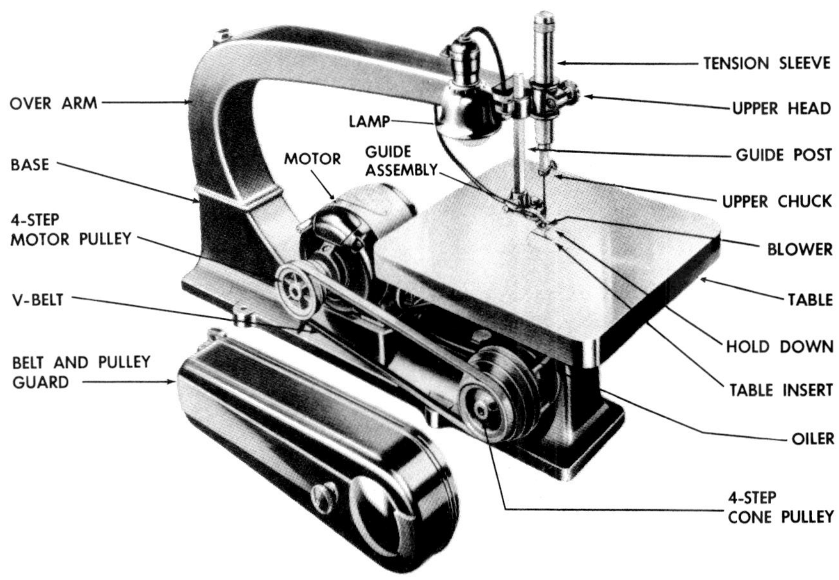

Figure 522 Jig saw

jack rafter Rafter that runs between a hip rafter and the plate (hip jack rafter), between a valley rafter and the ridge (valley jack), or between a valley and a valley or a valley and a hip rafter (cripple jack). (Figure 521.) See also *dormer framing, roof framing.*

jackscrew Jack that is operated by screwing out the center support.

jack stud A short stud, shorter than the standard-length studs around it. The term is sometimes used to refer to a trimmer or cripple stud. Specifically, the short studs at the side of a door or window that frame into the header.

jalousie window Window with movable, horizontal glass panes or slats that crank open to admit air. See *window.*

jamb The vertical sides of a window or door. (*Figures 309,* p. 119; *1111,* p. 489.) (The head or top section is sometimes called the *head jamb.*)

jerry built Something quickly built without plan; an inferior or poorly built structure.

jetting Forcing water under pressure to cut away soil or sand, as under a pile, so member may be set in ground.

jib The projecting arm on a heavy crane that is attached to the *boom.* See *crane.*

jiffy mixer Power mixer used to mix drywall joint cement.

jig In woodworking or metalworking, a tool guiding device; a device that holds a tool or the material being worked on. Also, a template used to align parts. See *fixture.*

jig saw A shop power saw that has a reciprocating (up- and-down) cutting blade. The blade moves up and down, cutting on the downstroke. (Figure 522.) The jig saw is used for cutting curves and following contours. Also called a *scroll saw.* See also *saber saw.*

job site Location of a project or building; work site.

jogged Irregular surface or line; notched lumber.

joggle Projection or shoulder. A metal pin or tie used to reinforce a wood joint or to secure masonry units.

joiner A worker in the shop who assembles basic house parts, such as a window with sash.

joinery Woodworking or construction joints. See *joints.*

joining plates Small lumber or plywood pieces used on the outside of an end-to-end joint. See *fishplate.*

joint Connection of two (or more) wood or steel members. See *wood joints.* See also *weld joints.* In masonry construction, the surface where two masonry units are laid together with mortar. See *masonry joints.* In concrete work, a separation made to control cracking. See *control joint, construction joint, isolation joint.* Also, a connection of two pipe sections. See *fitting.*

Joint Apprentice and Training Committee (JATC) A management-labor group charged with the responsibility of administering apprenticeship programs. Different trades have their own JATC. Sometimes called the Joint Apprenticeship Committee (JAC).

joint bolt A *handrail bolt.*

joint compound Special joint cement used to cover drywall tape at the joints. A taping joint compound is used to embed the tape; a topping joint compound is used for finish coats.

jointed The finishing or tooling of a masonry joint. See *masonry joint.* Also, saw teeth that have been filed even. See *jointing.*

jointer Type of trowel used for finishing masonry joints. Figure 523 shows how a jointer is used. In concrete flatwork, a special jointer is used to make joints between sections; also called a *groover.* See *joint raker; trowel.* Also, a stationary, power-operated woodworking machine used for smoothing rough stock and for planing wood surfaces. Stock is placed on the table and fed by the cutter head. A fence (upright guide table) controls the wood piece. A cutterhead with three knives does the cutting. (Figure 524.) See also *power planer, shaper.*

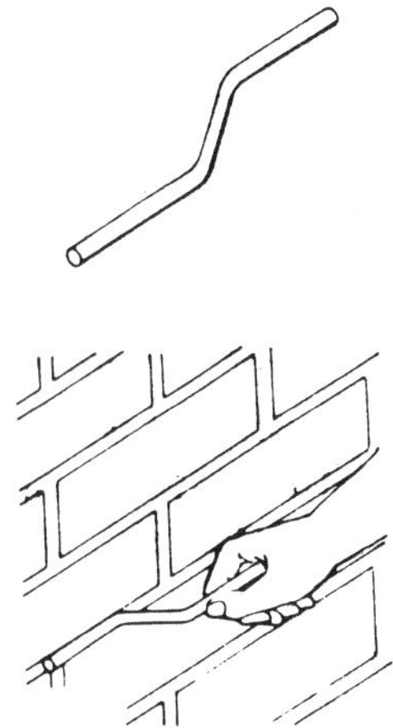

Figure 523 Jointer used to finish brick joints

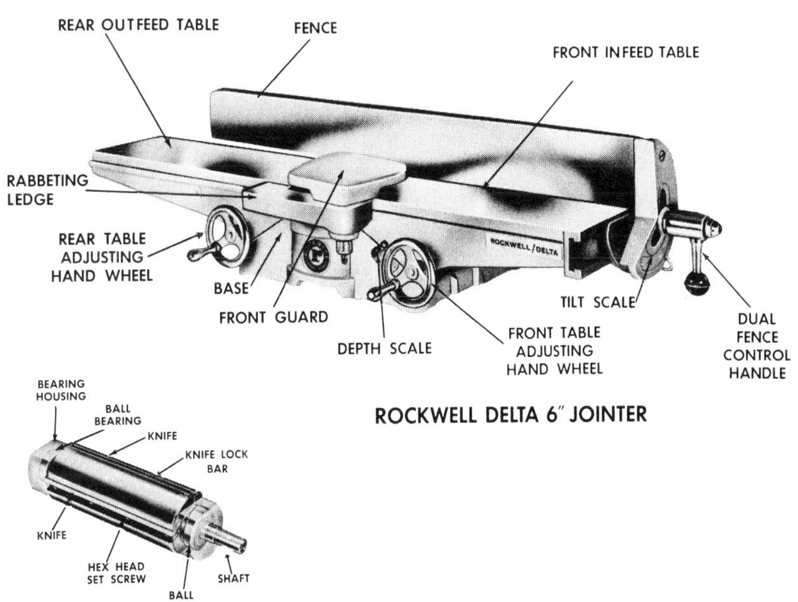

Figure 524 Jointer

Figure 525 Jointer plane

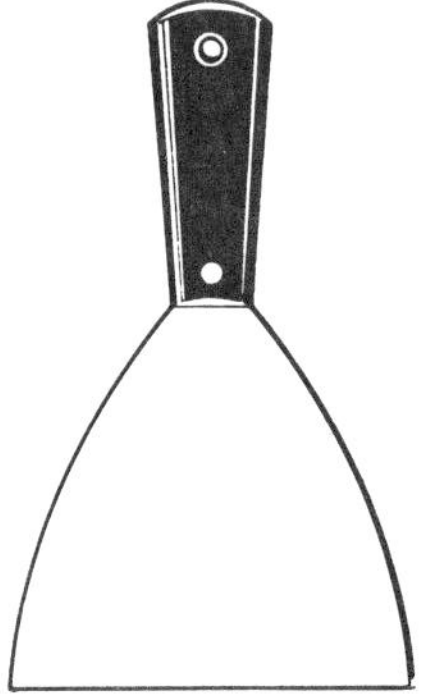

Figure 527 Joint knife for drywall finishing

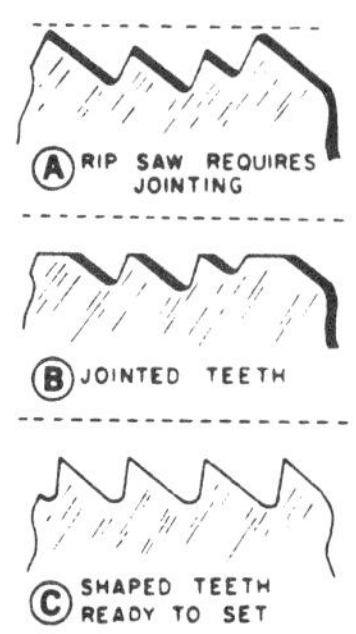

Figure 526 Jointing saw teeth

Figure 528 Joint raker

jointer plane A long bench plane, commonly 22″, used for straightening long pieces and truing edges and surfaces. (Figure 525.) See *plane*.

joint filler Very narrow trowel used for finishing masonry joints. Several sizes are available. The trowel slides over the joint to shape the mortar and push it back into the joint. Also called a *caulking trowel, tuckpointing trowel*. See *trowel*.

jointing The operation of finishing off the joints between masonry units with a jointer. Also called *tooling*. Making joints in concrete flatwork. Operation of a woodworking jointer; straightening or smoothing the edge of a board with a jointer; edge is finished to a right angle to the board face. Also, in saw sharpening, the dressing down of the ends of the saw teeth so the points are all even and in line with each other. (Figure 526.) After jointing, the saw teeth are set (bent to the correct angle) and filed.

joint knife Wide-bladed plastering tool used for finishing drywall joints and for repairing holes. (Figure 527.) See *drywall knife*.

joint raker Mason's tool used for raking out fresh masonry joints. (Figure 528.)

joint rod In plastering, shaped metal piece used for shaping joints, as at a corner. See *miter rod*.

joints A joining or fastening together of two parts. See *masonry joints, weld joints, wood joints*.

joint strike A jointer.

joint tape In drywall construction, a reinforcing strip used to cover gypsum wallboard joints.

joint treatment In drywall construction, method of finishing gypsum wallboard joints by taping and applying layers of joint compound.

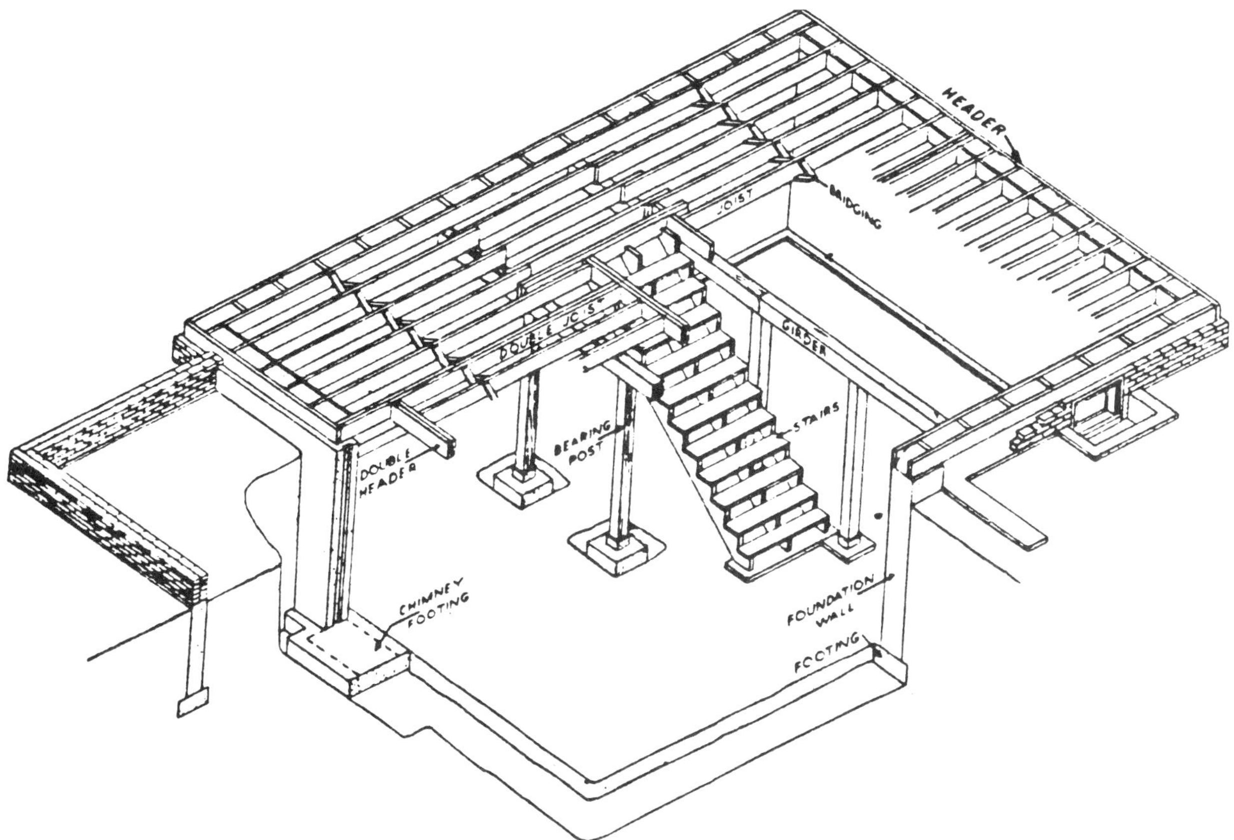

Figure 529 Joists (wood)

joint wiping Wiping a solder joint; wiping a plumbing joint made with hot lead.

joist Horizontal support member used to construct a floor. Joists are laid on edge and parallel to each other 16″ O.C. They are supported at the ends and, if necessary, in the center. (Figures 529, 530; *638*, p. 275.) For outside joists see specifically *end joists, header joists;* also *band joists, rim joists, sill joists.* See also *I joist, open web joist.* Figures *489*, p. 204, and 531 show joist framing around openings. See also *headers.*

joist anchor Metal tie used to hold the end of a joist in a masonry wall. (*Figure 392*, p. 158.) Also called a *wall anchor.* See also *framing anchor, joist hanger, rafter anchor.*

joist chair Wire support that holds the two reinforcing bars in a reinforced concrete joist

Figure 530 Joists (steel)

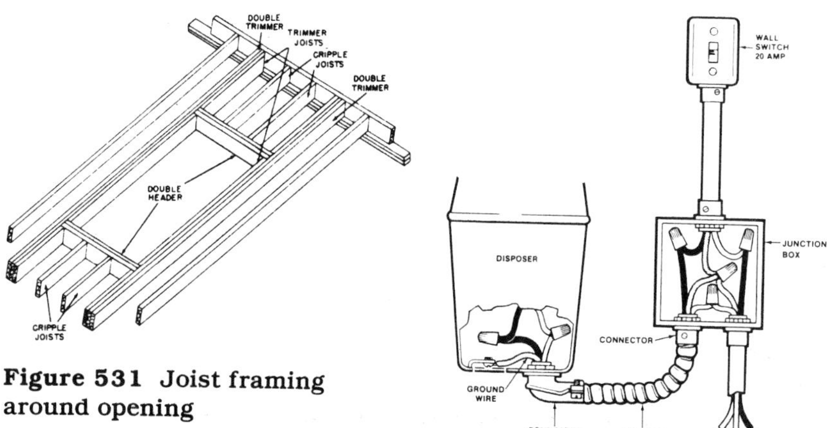

Figure 531 Joist framing around opening

Figure 533 Junction box

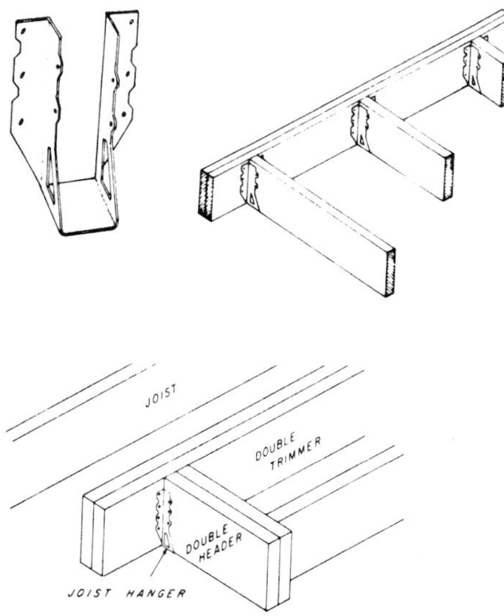

Figure 532 Joist hangers

joist clip See *joist hanger.*

joist hanger Metal stirrup used to support a joist framed into a header or beam. (Figure 532.) Sometimes called a *beam hanger, joist clip.* See also *framing anchor, joist anchor, rafter anchor.*

joist plan Orthographic plan view showing location of joists in floor or ceiling; spacing, length, and size are noted. Not commonly needed unless unusual or complicated framing is involved.

joist plate Plate in which joists rest. See also *top plate.*

joist plates Top plates (horizontal 2 × 4's at wall top) on which joists rest.

journeyman A craftsman who has completed an apprenticeship in a trade.

jumper Temporary connection that bypasses part of an electrical circuit.

junction box Electrical box used for making conductor connections. Used when separate wiring runs come together. The box allows room to make safe connections between conductors. Figure 533 shows connections from a garbage disposal unit into a junction box.

jut To project outward.

k

keeper Strike plate of a door latch.

kerf Shallow slot made by a saw cut. See *set* (*Figure 860*, p. 379).

kerfing Cutting a series of shallow, close parallel cuts in a board. The parallel kerfs allow the board to be bent. (Figure 534.)

key A joint made between separate concrete parts, as between a footing and a wall. Also, a joint between two vertical parts in a wall. (*Figure 134*, p. 51.) Groove or depression made in the top of a foundation footing. (Figure 535.) The key gives a seat to the bottom of the foundation wall. See also *construction joint, control joint*. Also, the various metal or wooden wedges used for securing a wood part, as the wedge sometimes used to tighten a tenon in a mortise. In gluing or plastering, the roughness or tooth of a surface that allows a strong bond. Also, the bond formed by plaster forced through the openings in metal lath.

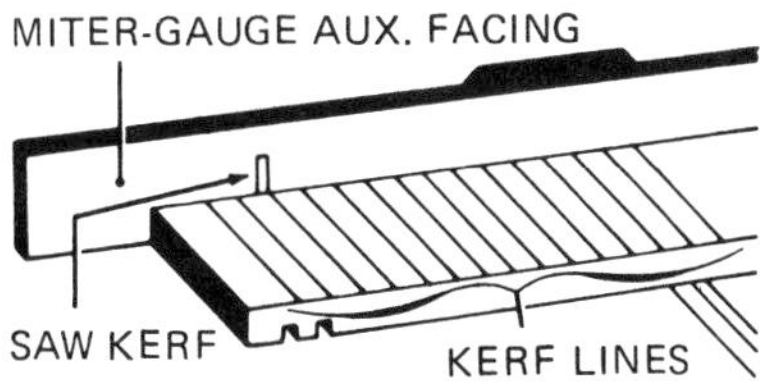

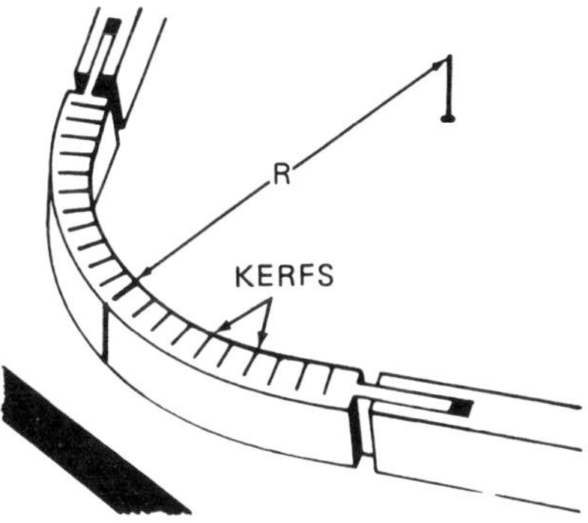

Figure 534 Kerfing

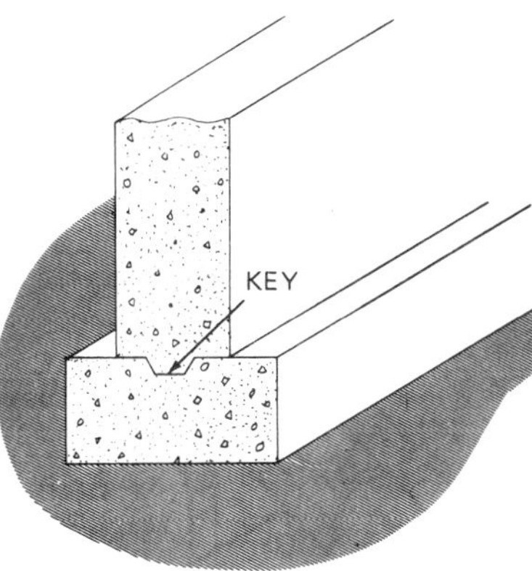

Figure 535 Key

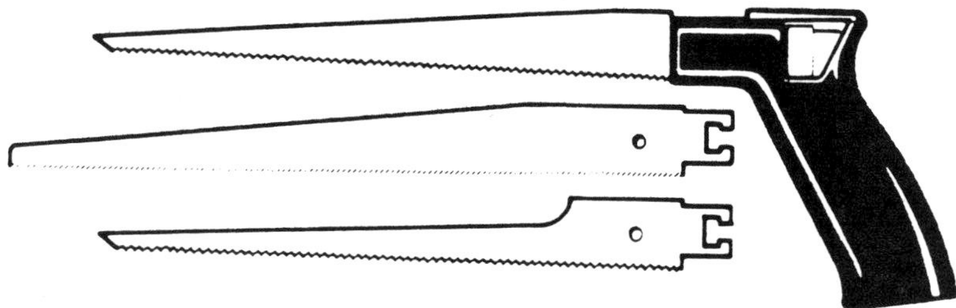

Figure 536 Keyhole saw with set (nest) of blades

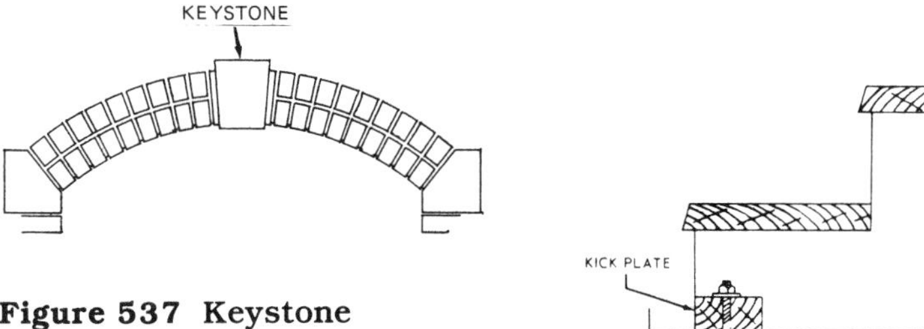

Figure 537 Keystone

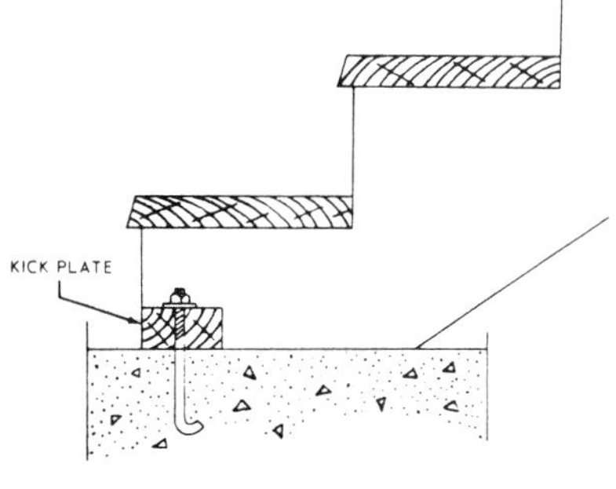

Figure 538 Kick plate

keyhole saw Saw with small tapered blades, used for cutting small openings or notches. Blades are interchangeable. (Figure 536.) Also called a *compass saw*. See also *saw nest*.

keystone Wedge-shaped stone at the top of a masonry arch. (Figures 537; *378*, p. 150.) See *voussoir*.

kicker plate In stair construction, a wood member that receives the stringers at the base of a stairs. The stringer ends are cut to receive the plate. Also called a *kick plate*.

kickout In excavation, an accidental release or failure of a shore or brace.

kick plate Plastic or metal plate attached across the bottom of a door. Also, in stair construction, a horizontal support piece at the base of a stairs to secure the stringers to the floor. (Figure 538.)

kiln Large oven for drying (seasoning) lumber.

kiln-dried Of lumber, having been dried by heating in a kiln, a sort of large oven. As opposed to *air-dried*. The wood is stamped K.D.

kilo Metric unit for 1000. Normally used as a short term for kilogram (1000 grams). Abbreviated: k.

kilowatt (kW) One thousand watts, about $1\frac{1}{3}$ horsepower.

kilowatt-hour (kWh) One thousand watt-hours. A 100-watt bulb lit for 10 hours will use 1 kilowatt-hour of energy.

king closer A brick that is approximately three-fourths the regular brick length, used to close up or finish a course. Larger than a *queen closer*, which is approximately a half brick. See *cut brick*.

king common rafter First two full-length common rafters where a hip roof runs into a gable roof. Also the full-length common rafter that runs down the center of the hip. There are three king common rafters framed in a plain hip roof.

king post Vertical center support in a truss. See *king post truss.*

king post truss Truss with a vertical center (king post) support, often used singly at gable ends. (Figure 539.) See *trusses.*

king stud Full-length stud by a jack stud (trimmer) at the side of door or window.

kip One thousand pounds.

knee brace A cross brace from near the top of a column over to the horizontal support member. (Figure 540.) Used for bracing and to stiffen the structure against wind stress.

knee pads Pads worn on the knees when working, as in cement finishing.

kneewall Short wall normally used in finishing attic areas. Walls run from floor to the angle of the roof. Wall is normally at least 4 feet high. (Figure 541.)

knob and tube wiring An outdated method for running house wiring. Porcelain knobs hold the wires away from the support members. A porcelain tube is used to run wires through the joints or studs. Rarely used today because of the expense, although still recognized by the National Electrical Code.

knocked down Said of a prefabricated assemblage, such as steel scaffolding, when it is completely disassembled.

knockout Punched metal disc in a metal electrical enclosure that is designed to be punched or knocked out to receive conduit or tubing. In some cases different sizes may be removed. (Figure 542.)

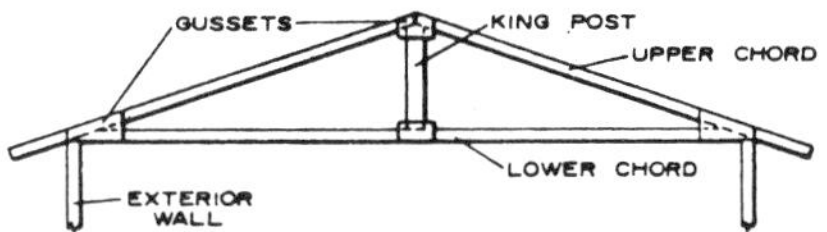

Figure 539 King post truss

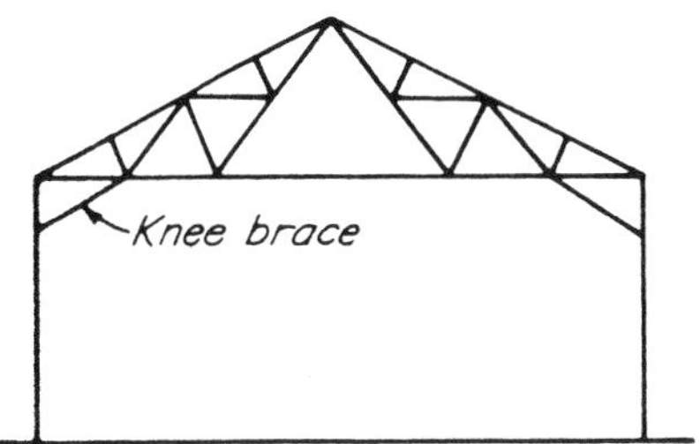

Figure 540 Knee brace

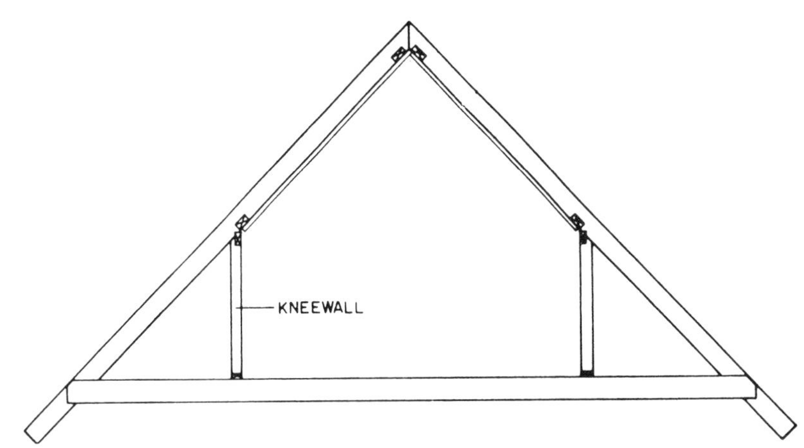

Figure 541 Kneewall

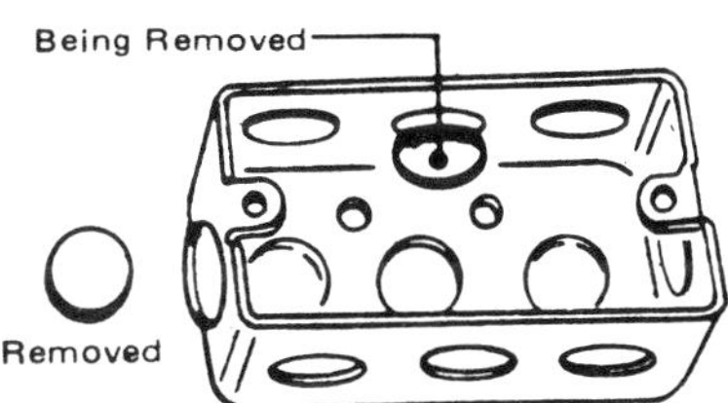

Figure 542 Knockout

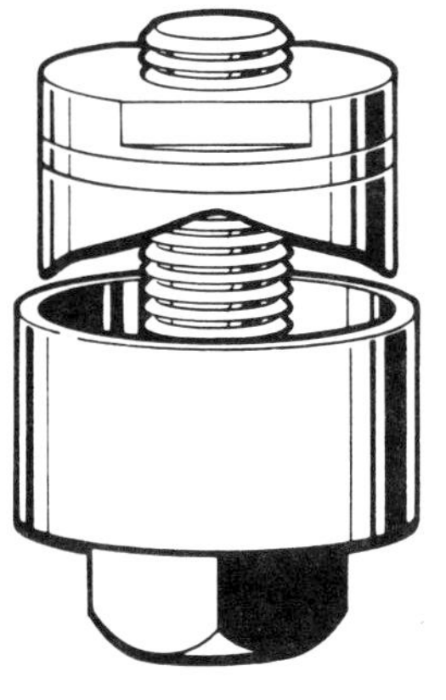

Figure 543 Knockout punch

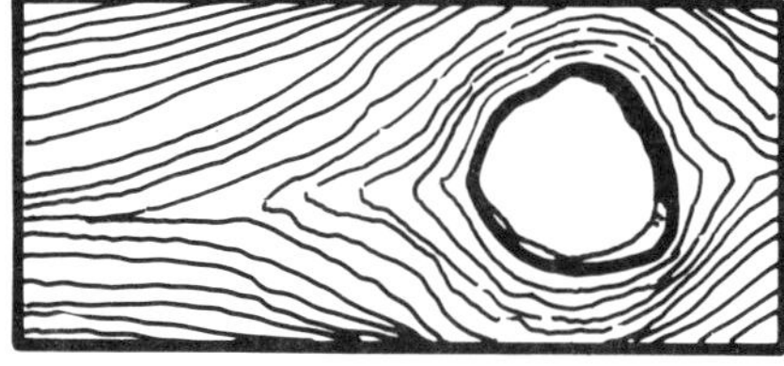

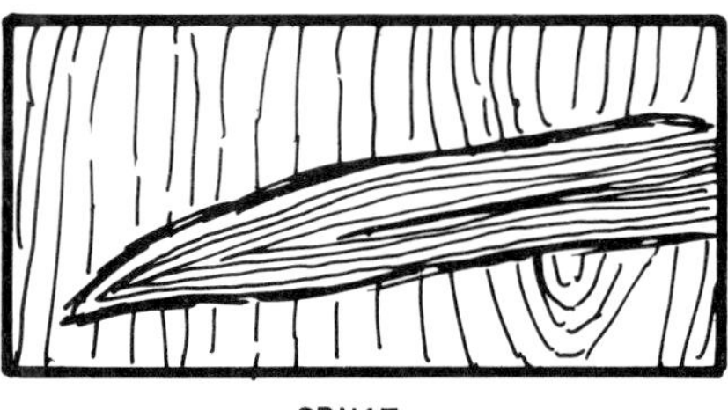

Figure 544 Knots in lumber

knockout punch Punch for making holes in electrical enclosures for conduit. (Figure 543.)

knot Circular area in lumber caused by embedded tree limbs or branches. Figure 544 shows common knots. If tight, the knot is not considered a defect. If loose or fallen out, it is considered a lumber defect. Small tight knots should not affect the structural integrity of the lumber. *Pin knot*: Less than $\frac{1}{2}''$ diameter. *Small knot*: From $\frac{1}{2}''$ to $\frac{3}{4}''$ diameter. *Medium knot:* From $\frac{3}{4}''$ to $1\frac{1}{2}''$ in diameter. *Large knot*: Over $1\frac{1}{2}''$ in diameter. See *decayed knot, enclosed knot, firm knot, fixed knot, hollow knot, intergrown knot, loose knot, sound knot, star-checked knot, tight knot, water-tight knot.* Also, a means of tying or securing rope: see *rope knots.*

knothole Hole left in lumber where a knot has dropped out.

knuckle joint The joint or angle made by the two intersecting *gambrel roof* slopes. Also called a *curb joint.*

knurl Crosshatching milled into metal tool handles to allow for a better grip.

kraft paper A brown paper used as building paper.

kumalong Long-handled, wide-bladed tool for spreading and smoothing concrete after placement. A *concrete placer.* See also *concrete rake.*

K value Thermal conductivity of a material. Number of Btu's that can pass through a material in an hour. (The K value is calculated in Btu's that can go through material 1 foot square and 1 inch thick when temperature differential is 1°F.) See *C value, R value, U value.*

l

laborer One who assists a carpenter to build a structure. A laborer selects and saws lumber, holds for nailing or nails, nails sheathing onto rough frame, removes forms, cleans used lumber and forms, and moves tools and materials as needed.

labor union An association of workers banded together for the purpose of setting standards, upholding the tradition of the craft, and bargaining for their rights.

lacing course Brick course introduced into a stone wall to give strength and to even out the wall.

lacquer Protective wood coating made of natural or synthetic resin dissolved in a solvent. Clear lacquer is commonly used, but pigmented types are also available.

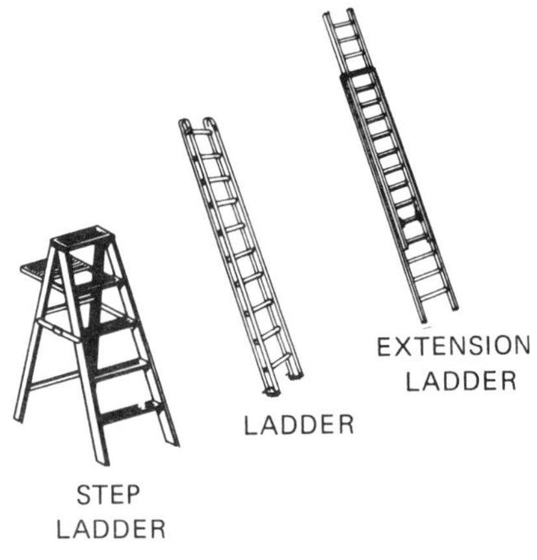

Figure 545 Ladders

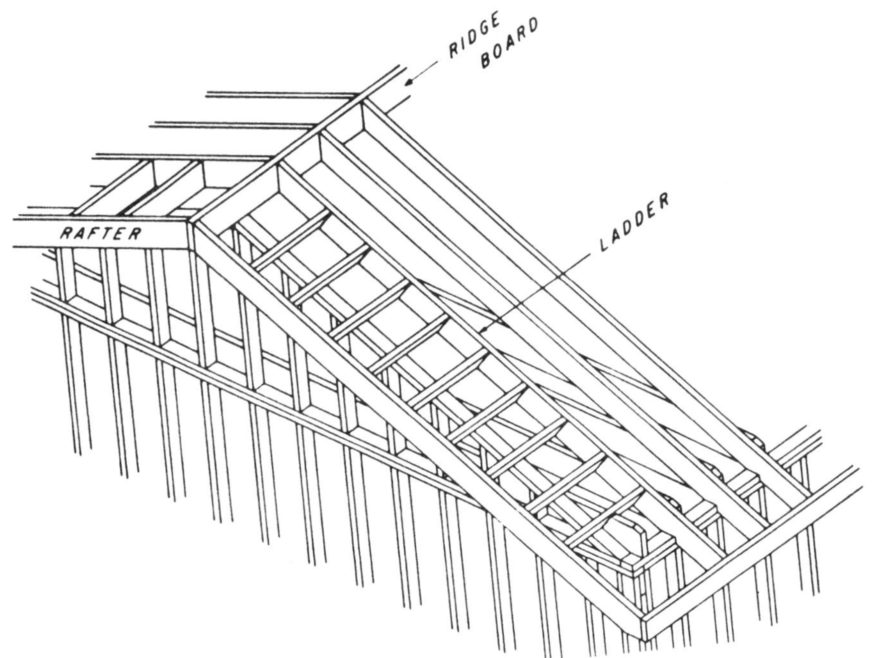

Figure 546 Ladder

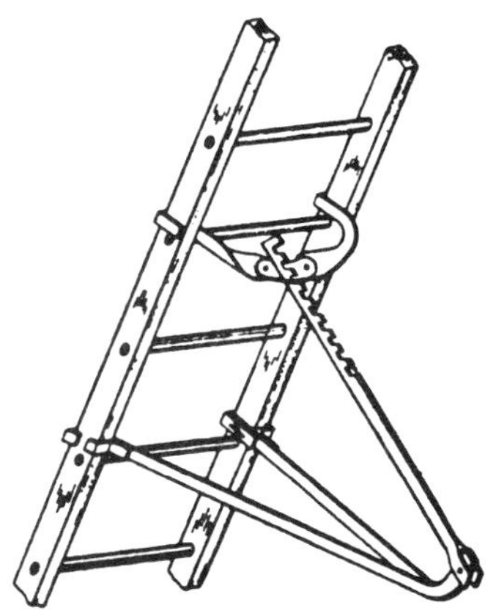

Figure 547 Ladder jack

Figure 548 Ladder safety shoes

ladder Series of rungs or steps supported by two stiles or side rails; used for climbing up on a structure. (Figure 545.) Both metal and wooden ladders are used, although wood is often required on a building site because wood is a nonconductor of electricity and is considered safer. Some ladders may be extended. See *extension ladder*. See also *step ladder*. In roof construction, the series of lookouts or short lumber pieces used on the end of a gable roof to extend the roof out. (Figure 546.)

ladder jack Type of bracket scaffolding. (Figure 547.) See *bracket scaffold*.

ladder safety shoes Supports at the base of a ladder to prevent slippage. (Figure 548.)

lag bolt Large screw with a bolt head, a *lag screw*.

laggings Vertical, braced planks used to support a trench wall; also called *sheet pilings*. See *trench support*. (*Figure 1025*, p. 452.)

lag screw Large screw-threaded bolt used for fastening where a nut could not easily be attached to a regular bolt. It is screwed into a wood member with a wrench. Also called *lag bolt*. (Figure 549.)

lag-screw expansion shield See *lag shield*.

lag-screw shield See *lag shield*.

lag shield A two-part lead tube split in two parts along its length. (Figure 550.) It is inserted in a drilled hole and expanded with a lag screw. Also called *lag-screw shield, lag-screw expansion shield*. See *anchor, lead anchor, machine screw anchor*.

laid to weather Said of a length of shingle or lap siding exposed to the weather.

laitance Soft or weak layer in concrete or mortar; a surface layer sometimes brought to the surface by bleeding caused by excess water in a mix.

lally column Cylindrical metal pipe, sometimes filled with concrete, used as a column.

lamella Type of roof construction where short wood pieces are fastened together in a crisscross diamond pattern.

laminated Built up of glued plies or wood pieces, as plywood or laminated arches.

laminated arches Arches built up of glued lumber pieces. Laminated arches commonly span wide spaces, often over a hundred feet.

laminated beams Beams built up of glued lumber pieces.

lanai Covered walkway (from the Hawaiian).

landing Flat area or platform between two flights of stairs or at the bottom of a stairs. See *stair*.

Figure 549 Lag screw or bolt

landscape plan The total development of an area or site; land-use plan, includes traditional site planning and environmental planning for the total site area. Not to be confused with the *landscaping plan*. See also *land-use plan, soil survey, environmental impact statement*.

landscaping Planned finish on a building site, including sodding, placement of shrubs and trees, erection of privacy screens, and so on. Planned use of natural and man-made materials to create an attractive exterior. See *landscaping plan*.

landscaping plan Part of the total site development that covers location and sequence of plantings on the site. See also *landscape plan, vegetative filter, landscaping*.

land-use plan Plan that covers the total development of a site; *landscape plan*. See *land-use planning*.

land-use planning Long-range planning prescribing, usually through zoning, how land is to be developed and used. See *landscape plan*.

langley Solar radiation measure: equal to one calorie per square centimeter.

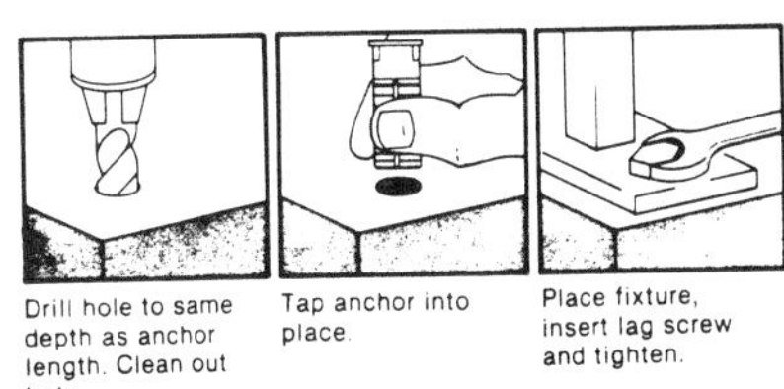

Figure 550 Lag shield

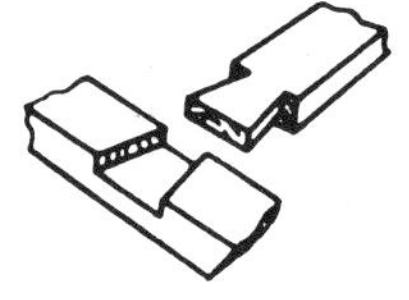

Figure 551 Lap dovetail joint

Figure 552 Latch

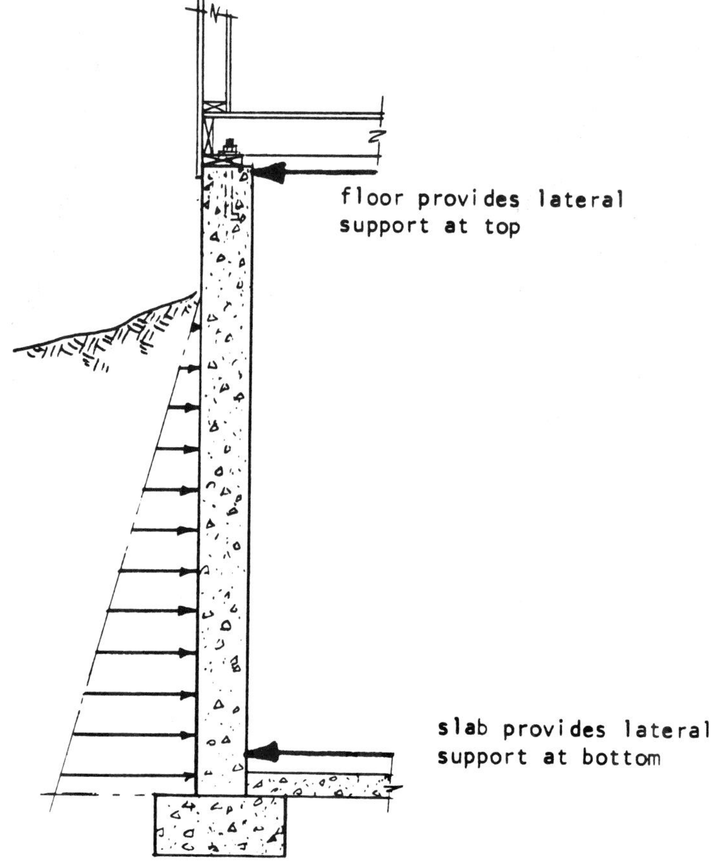

Full Basement

Note: Earth pressure at any given point is equal to equivalent fluid pressure of the soil multiplied by the height of backfill above that point

Figure 553 Lateral thrust

lap To overlap or run framing members side by side, as joists that cross over a support beam. To lay shingles or siding over each other so that part is covered (lapped). To extend one piece over another, as a siding board overlaps another board. Also, the distance two overlapping members overrun each other.

lap dovetail A half dovetail made so that the face of a piece will not show the joint, as in drawer construction. (Figure 551.) See *wood joints*.

lap joint Wood joint made by lapping two wood pieces so the ends ride over each other or cross. See *wood joints*.

lap-joint weld In welding, a weld between two overlapping pieces; a *lap weld*.

lap siding Siding used for exterior house finish. Also called ***bevel siding***. See *siding*.

lap weld Welds made on two pieces that overlap. (*Figure 1099*, p. 485.)

large knot Knot over $1\frac{1}{2}$ inch in diameter.

laser Light amplification by stimulated emission of radiation.

laser level A surveying level that uses a laser beam to establish a level line or exact height. See *transit laser*.

latch The metal tongue or bolt in a lock that is operated by the door knob. (Figure 552.) See *lock*.

lateral To the side.

lateral thrust Pressure or load from the side, as the pressure on the side of a basement wall. (Figure 553.)

late wood Summer wood.

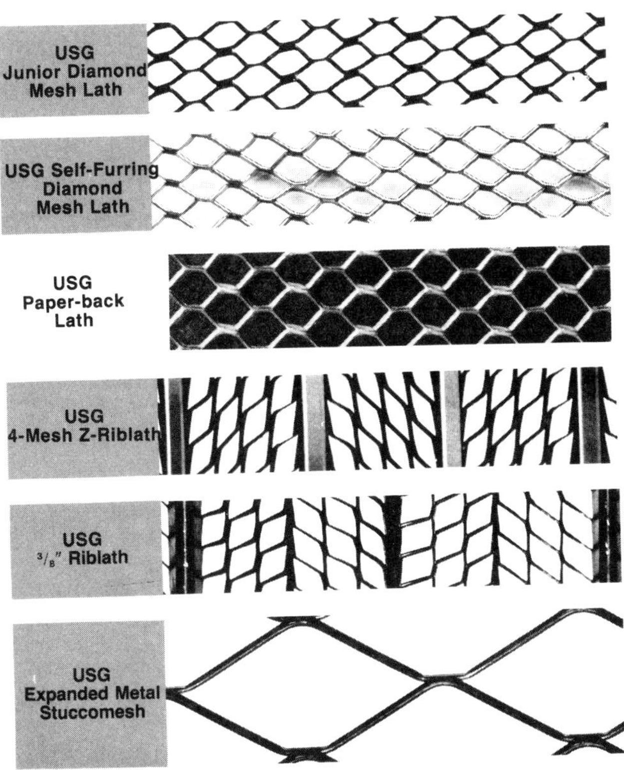

Figure 554 Lath (metal)

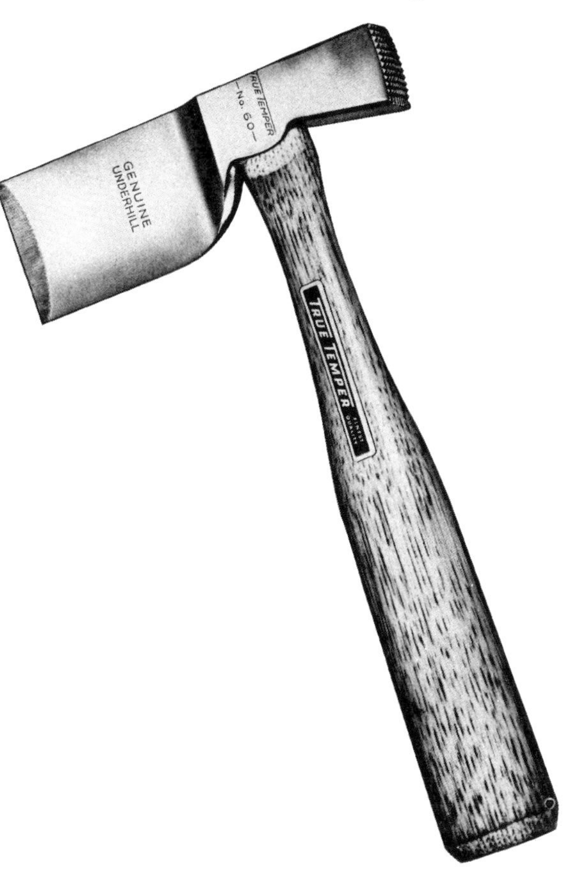

Figure 555 Lathing hatchet or hammer

lath Metal with open spaces that form a plaster base. Lath is stapled or nailed to wall studs; the plaster when applied is forced through the holes to form a permanent key. Also called lathing. (Figures 554; *704*, p. 303.) Also, gypsum wallboard used as a plaster base. Also, a small, thin wood board.

lathe In woodworking, a wood-turning machine, used for making spindles, legs, bowls, and the like. See *wood lathe.*

lathing Metal lath or gypsum wallboard used as a plaster base. See *lath*. The operation of installing metal lath.

lathing hatchet Special hatchet with a cutting blade, used for cutting and nailing lath. (Figure 555.)

lattice bar In steel construction, short, diagonal cross supports between framing members or between parts of a built-up beam or column.

lattice girder See *open-web joist.*

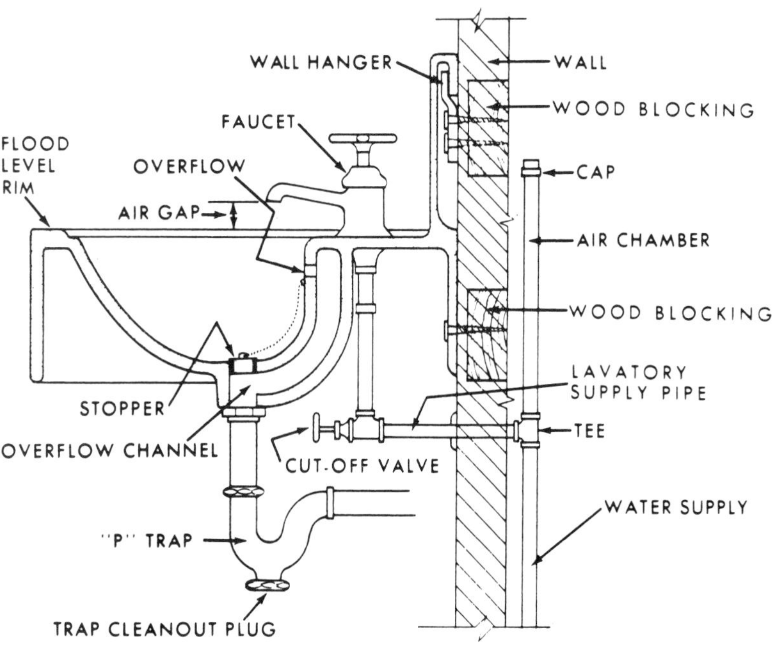

Figure 556 Lavatory

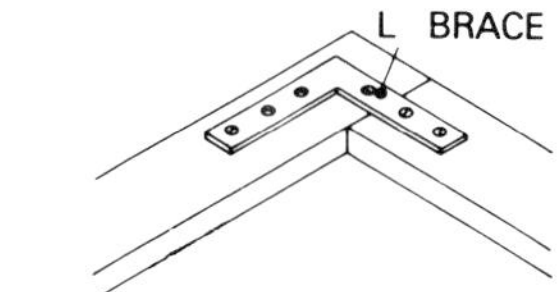

Figure 557 L brace

lattice work Thin metal or wood strips running in a criss-cross pattern.

laundry tray A tray or tub used for washing clothes or for receiving the discharge from an automatic washer. Both single and double units are used.

lavatory In plumbing, a fixture for washing the hands and face in; a wash basin. (Figure 556). In general usage, a bathroom with a lavatory and water closet.

lavatory trim Hot- and cold-water faucet unit and the drain fitting.

lay To spread, as plaster or paint. To put down, as laying tile or bricks. Also, the twist of a rope or wire cable.

laying out Process of locating a structure on the site. Process of measuring and marking material needed for construction. If the same measurements are repeated often, sometimes the layout distances are marked on a *story pole*. The length of the story pole will be cut to fit a major dimension (such as the height of a story); smaller dimensions are marked on the pole.

layout tee A template cut to fit needed rafter angles and cuts, such as side cuts, ridge cuts, and bird's mouth.

lazy susan Storage cabinet shelf that revolves on an axis. Often used in the corner of a kitchen below countertop level in a row of cabinets.

L brace L-shaped brace used to strengthen corners in light framing. (Figure 557.) See *corner brace, T brace.*

leach field See *absorption field, distribution field.*

leaching A dissolving out of materials by the action of water.

lead (pronounced "led"): A heavy, soft metal sometimes used for making plumbing joints.

lead (pronounced "leed"): A built-up masonry corner, used to establish the corner, walls, and the course pattern or bond. (*Figure 223*, p. 83.) See also *mason's lead, rake back lead.* Also: an electrical conductor.

lead anchor A molded lead tube that is inserted in a drilled hole and expanded by a screw. (Figure 558.) See also *anchor, lag shield, machine-screw anchor.*

leaded joint Plumbing joint made with oakum followed by molten lead which is caulked, used in a *bell-and-spigot* pipe. (Figure 559.)

leader A conductor for carrying rain water or snow melt down from the roof gutters; a *downspout* or *conductor*. Ductwork that carries hot air from the main duct to the outlet. In drafting, an arrow on a blueprint that points to something.

lead ladle Metal dipper used for carrying molten lead to the cast-iron pipe hub.

lead pot Metal pot for melting lead in. A small furnace is used to heat the pot.

leads Electrical wires.

lead shield Lead expansion device used for bolting machinery or fastening other equipment to concrete. The shield expands as the bolt is tightened. See *lag shield, lead anchor.*

lead wall anchor Expandable lead anchor used for fastening into masonry walls. See *lead anchor.*

lead washer Washer used with nails to fasten metal roofing or siding, designed to prevent leaking.

lean mortar Mortar with a low percentage of cementitious material; mix will be weak and spread poorly. Opposite of *fat mortar.*

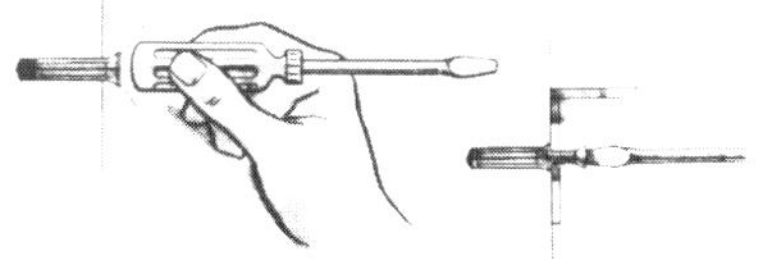

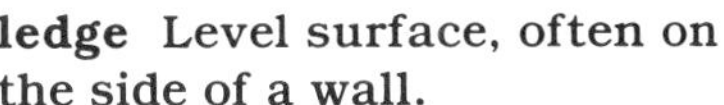

Figure 558 Lead anchor

ledge Level surface, often on the side of a wall.

ledger Wood or metal piece attached to a beam, studding, or a wall to support joist or rafter ends. In *balloon framing* a board that nails across the studs and gives support to the joists. Also called a ledger board or *ribbon*. (*Figure 55*, p. 23.) In *scaffolding*, horizontal boards fastened to vertical support members.

lefthand door Hinges are on the left when you face the door from the outside and the door swings towards you. (If it swings away, it is a reverse lefthand door.) See *hand of doors.*

lefthand stairs Stairs with a rail to the left as you ascend.

leg A support; a side of angle iron; limb of a compass or divider. Also, in furniture construction, vertical support members. See also *post.*

length The long dimension of something, as the length of a stud.

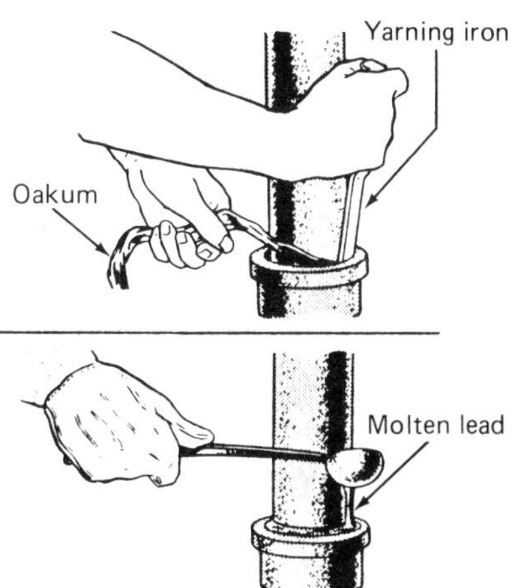

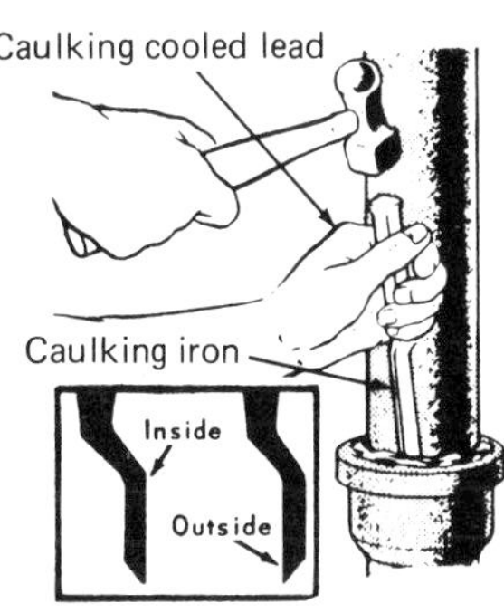

Figure 559 Leaded joint: packing joint with oakum (top), pouring lead into the joint (middle), caulking the lead (bottom)

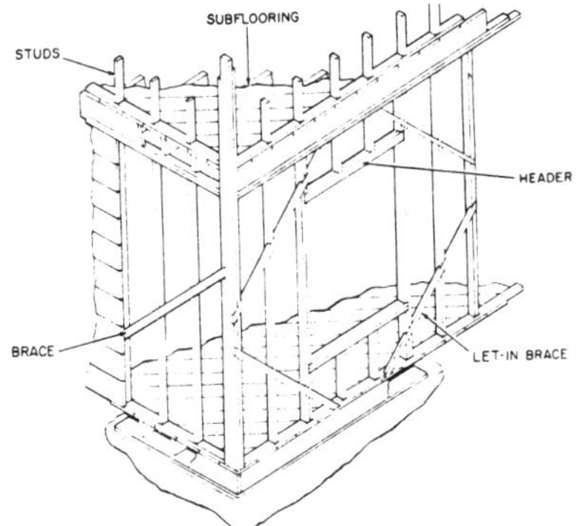

Figure 560 Let-in braces

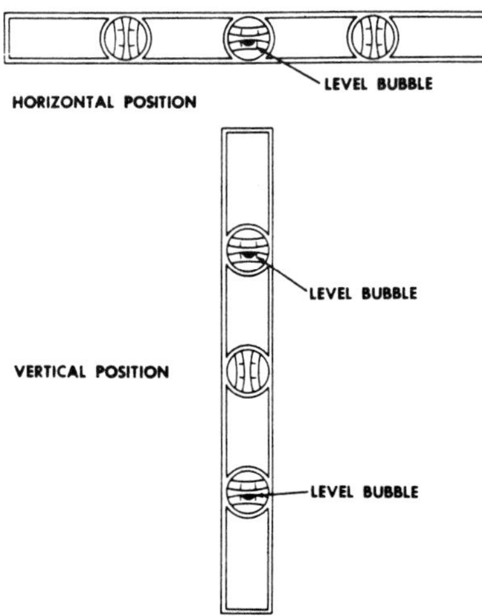

Figure 561 Level

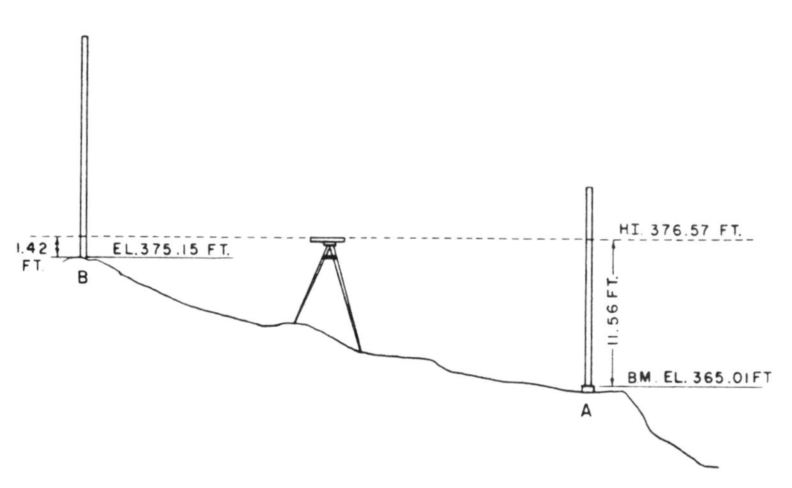

Figure 562 Leveling

let-in brace Diagonal bracing that is let in (notched) into the studs. (Figure 560.) See also *cut-in brace, tension straps, wall brace.*

let into Notched.

lettering Hand lettering of information or notes on drawings. A simple letter printing, in pencil or ink, is normally used; capital lettering is the common practice. Computer-generated drawings have a lettering similar to the traditional simple lettering, but oval letters tend to be squared. In architectural lettering, individual styles are often developed.

lettering guide Drafting device used by drafters to draw even and straight letters.

level A device for establishing exact horizontal or vertical surfaces. (Figure 561.) A bubble in the glass is centered when the surface is true. Two bubbles are used: one for the horizontal and one for vertical measurements. Also called a *carpenter's level, spirit level.* Also, a level and sight mounted on a tripod for establishing elevations and angles on a building site. Commonly called a *builder's level.*

leveling In surveying, transferring of height (elevation) from one point to another with a *builder's level* and *leveling rod.* (Figure 562.) By a series of readings the elevation of any point can be determined by measuring from a known point. Also called *differential leveling.* See also *backsight, foresight, height of instrument, turning point.*

leveling rod Graduated rod used with a builder's level or transit level to establish elevations and distances. Figure 563 shows the rod. Rod is marked off in feet, tenths, and hundredths of a foot. A sliding target may be used to focus on the exact area. See *target rod.*

L hook screw Light screwed hanger with an L-shaped hook on the unthreaded end. (Figure 564.) See also *cup hook, screw eye, screw hook.*

lift Layer of concrete placed in a wall form from one pour, measured in inches.

lift-slab construction Reinforced concrete construction where roof and floor slabs are cast in place one on top of another at ground level. One after another the slabs are jacked up into place to form the roof and floors. See also *tilt-up construction.*

light A glass pane; specifically a window glass.

light and air easement Easement guaranteeing continued use of light and air around a structure; prevents other developers from constructing a building that will interfere with existing light and air use, such as a tall building that will block sun and cut off air flow. See *solar skyspace easement.*

light-box cutter Tool for cutting openings for electrical boxes in gypsum panels. Hole is made in one operation.

lighting fixture Device for holding one or more lamps.

lighting outlet Outlet used for lamps or lighting fixture.

lighting panelboard Electrical panelboard which has more than 10% of the overcurrent devices rated at 30 amperes or less for use with lighting and appliance branch circuits.

lightning rod Metal conductor placed on the high point of a building to receive lightning and thus protect the building. The conductor runs to ground.

lignin Wood ingredient that cements wood cells together. See *cellulose.*

lime Quicklime; calcium oxide (CaO).

Figure 563 Leveling rod

Figure 564 L hook screw

lime mixer Mixer used on the job to mix up plaster.

linear measure Length measurement

12 inches (in.)	=	1 foot (ft)
3 feet	=	1 yard (yd)
$16\frac{1}{2}$ feet	=	1 rod (rd)
320 rods	=	1 mile (mi)
5280 feet	=	1 mile

line blocks and pins In masonry work, factory-made blocks or pins that are attached to the masonry wall to run a line from one corner to another. (Figure 565.)

Figure 565 Line blocks

Figure 566 Line level

line conventions Conventions used by draftsmen to depict different line usages on floor plans, sections, elevations. See *alphabet of lines.*

line drawing Drawing made with pencil or ink lines

line drop Loss of voltage because of the resistance in the conductors.

line level Very small level with a glass and bubble; attaches to a string to establish the true horizontal. (Figure 566.)

lineman's pliers See *side-cutting pliers.* Also called *electrician's pliers.*

line of sight In perspective drawing, the imaginary lines which extend from the viewer's eyes to the object. *Visual rays.* See *perspective drawing.*

line side In electrical wiring, the side of an electrical device that is closest to the current source, opposed to *load side.*

link Measurement used in surveying, a 1/100 part of a chain. A distance of 7.92 inches; one hundred links make a chain (66 feet). On a 100-foot engineer's chain a link is 1 foot.

linoleum knife Wooden-handled knife with curving blade used for cutting linoleum.

linseed oil Oil made from linseed (flax seed) used in paints and for treating unpainted, raw wood. Linseed oil may be raw or boiled; almost all linseed oil used is boiled, since it has a much shorter drying time.

lintel Horizontal support member over an opening, as a window or door opening. Wood, steel, stone, or reinforced concrete is used.

liquidtight flexible metal conduit A special flexible metal conduit that provides protection from liquids or vapors.

liquidus In welding, the lowest temperature at which a metal will still remain molten. Opposed to *solidus.*

listed Equipment that is included on a list published by an approved testing agency.

listing See *listed.*

list of materials In woodworking or building, a list of materials needed to build a project or structure.

live Of an electrical circuit, having a current.

live load The moving load or variable weight to which a building is subjected due to the weight of the people who occupy it, the furnishings, and other movable objects; distinct from the dead load or weight of the structural members and other fixed loads. See also *dead load, design load, earthquake load, snow load, wind load.*

living unit Complete residential unit that includes facilities for one family, with provisions for living, sleeping, eating, cooking, and sanitation. A *dwelling unit.*

L joint Two lumber pieces joined so as to form an "L" shape.

load Building weight. See *dead load, live load.* Also, amount of electrical current flowing in a circuit. See also *concentrated load, design load, uniform load.*

load-bearing wall Wall that supports part of the building's weight; a *bearing wall.* Opposed to a *nonbearing wall.*

load center The *panelboard.*

Figure 567 Loaders: crawler loader (top) and wheel loader (bottom)

Figure 568 Loader-backhoe

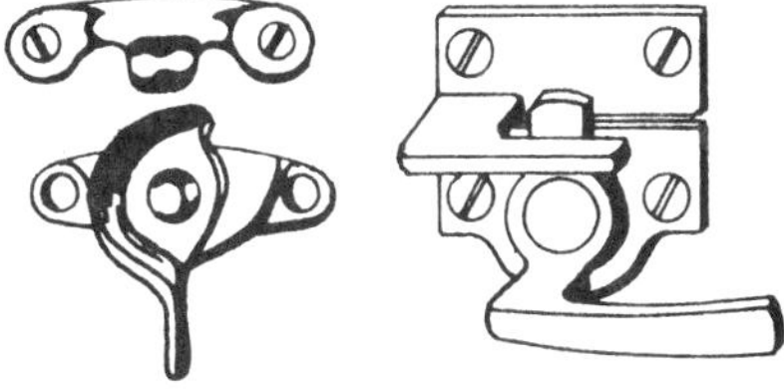

Figure 570 Locks for windows

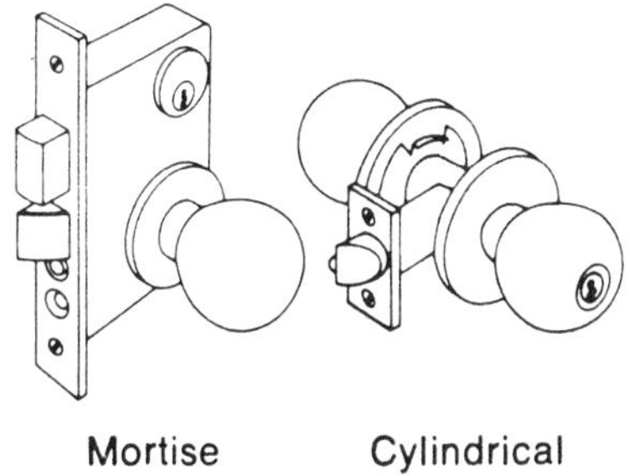

Figure 569 Locks for doors

Figure 571 Locked corner seam

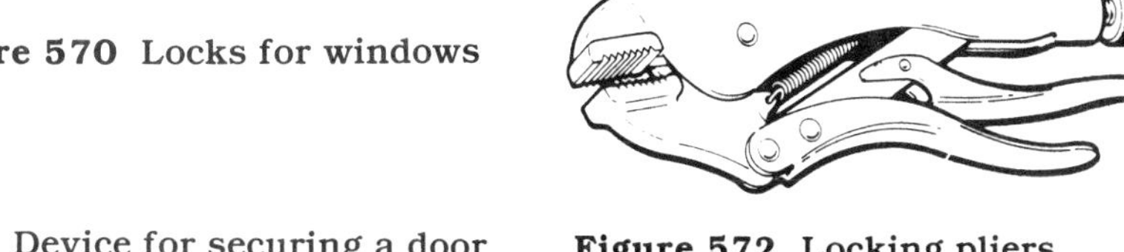

Figure 572 Locking pliers

load distribution center The *panelboard.*

loader Front-end loading excavator used for all types of digging and earth moving. (Figure 567.) Also called a loading shovel.

loader-backhoe Versatile earthmoving machine with a shovel in the front and a backhoe in the rear. (Figure 568.)

load side In electrical wiring, the side of an electrical device that is furthest from the current source and closest to the load where the current is used; opposed to *line side.*

lobby Entrance hall, vestibule.

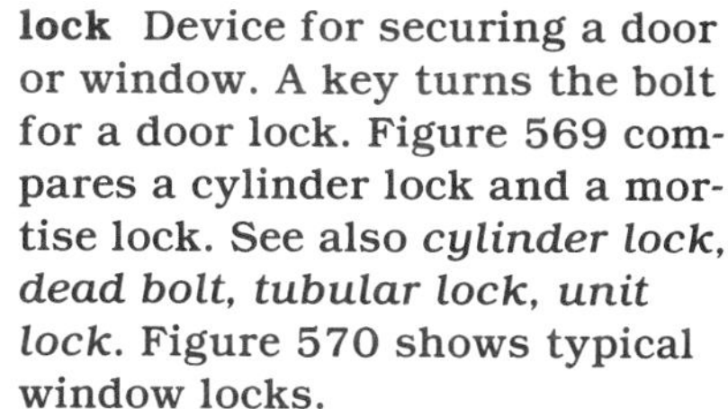

lock Device for securing a door or window. A key turns the bolt for a door lock. Figure 569 compares a cylinder lock and a mortise lock. See also *cylinder lock, dead bolt, tubular lock, unit lock.* Figure 570 shows typical window locks.

lock bevel Bevel angle on the latch of a lock.

lock case Enclosure that holds the lock.

locked corner seam Sheet metal seam at corner; formed piece fits over bent edges. (Figure 571.) See *drive slip.*

locking pliers Adjustable pliers that can be tightened or loosened with a screw in one of the handles. It clamps on the object being held. (Figure 572.) See also *pliers.*

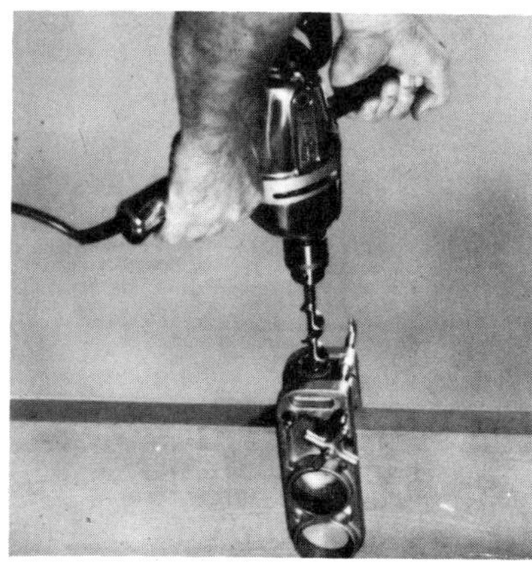

Figure 573 Lock mortise

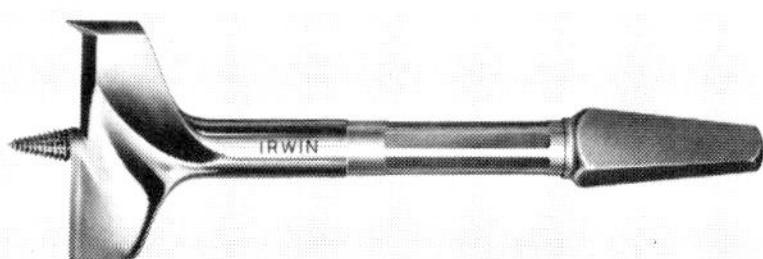

Figure 574 Lockset bit

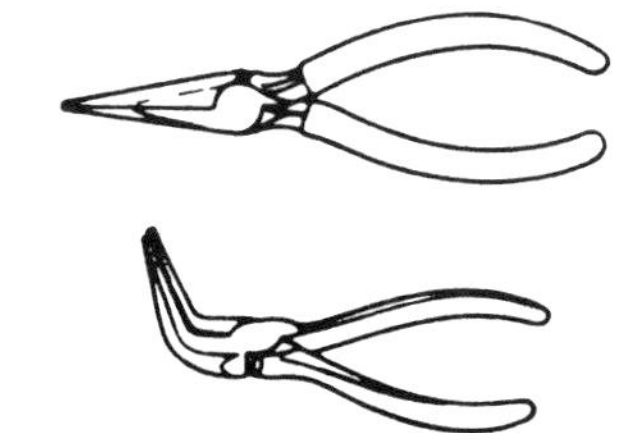

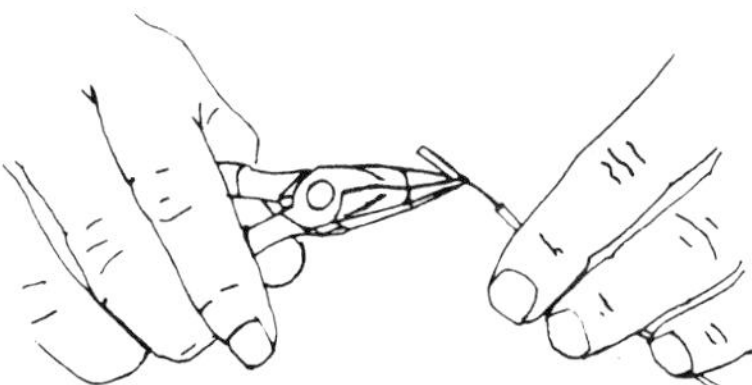

Figure 575 Long-nose pliers

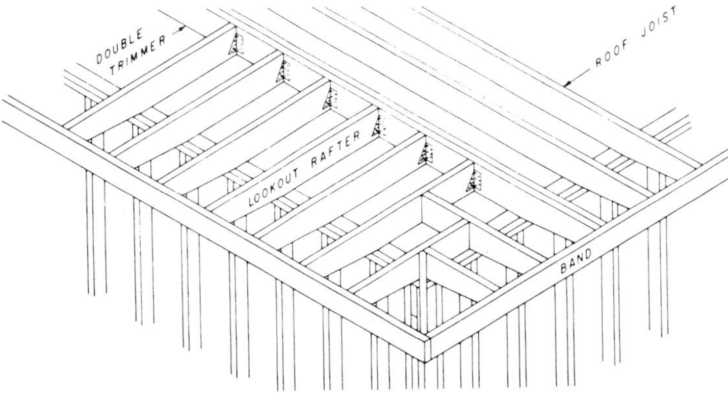

Figure 576 Lookouts (rafter)

lock mortiser A device for cutting a lock mortise in a door edge. The cutter is automatically fed into the cut. (Figure 573.)

lock nut Type of nut used to lock on the primary nut on a bolt, preventing the first nut from turning.

lock rail Horizontal piece on a panel door that receives the lock. See *door*.

lockset bit A large bit used to make holes for the installation of cylindrical locks. (Figure 574.) See also *boring jig*.

log iron A *dog anchor*.

longitudinal section Cross-cut view of a structure made end to end. See *transverse section* (side to side).

long-nose pliers Pliers with a long, often needlelike, nose. (Figure 575.) Used for wiring and small work, such as handling fasteners. Also called *needle-nose pliers*. See also *bent-nose pliers, chain-nose pliers, duck-bill pliers, flat-nose pliers, pliers*.

long oil varnish Varnish with a high percentage of oil. It is used with exterior work where a tough surface is needed. Examples are spar varnish and marine varnish. See *short oil varnish, varnish*.

lookout Extension of roof by using short supports, often cantilevered. (Figure 576.) Short support members are extended out from the roof rake. (*Figure 63*, p. 25.) Also, short support member under the eaves running between the rafter tail and the wall, used to support the plancier or soffit. (Figure 577.)

loose-fill insulation Loose insulative particles that are poured or blown in place.

loose knot Lumber defect in which a knot is loose in the board.

lot Plot of ground, as in a subdivision. Building site.

lot and block description Description of property by reference to lot and block numbers, usually within a subdivision.

lot line Outside perimeter of a lot: established by survey and constituting a legal description of land owned. Often confused with a *building line*, a line established by the municipality regulating how close a structure may be located to a public road, sidewalk, or other public or private features.

lot-line wall Wall adjoining and parallel to the lot line.

lousy In painting, dirt particles in the paint and brush.

louver Series of angled boards designed to admit air but exclude rain. (*Figure 42*, p. 19.)

low- energy power circuit Circuit with a low voltage, commonly 24 V.

low-pressure boiler Boiler furnishing steam at pressures less than 15 psi, or hot water at pressures less than 160 psi or at temperatures less than 250°F. Some low-pressure boilers are limited to a maximum of 15 psi for steam and 30 psi for hot water.

low relief A slight projection out from a background; in sculpture, the figures project out only slightly; *bas relief*.

lug Projection on a building piece. In electricity and electronics, a small metal strip or projection used to make a soldered connection.

lumber Sawn wood used in construction:

- **boards** Lumber 1″ thick and up to 12″ width.
- **plank** Lumber 2″ thick or thicker, 6″ width or more.
- **dimension lumber** Lumber 2″ to 4″ thick up to 6″ width.
- **timber** Lumber 5″ × 5″ or larger.
- **structural lumber** Lumber larger than 2″ × 4″.
- **yard lumber** General rough construction lumber.
- **shop lumber** Lumber that is to be remanufactured. Also called *factory lumber*.
- **rough lumber** Lumber sawed but not dressed.
- **dressed lumber** Lumber that has been surfaced on a planing machine to smooth out the surface and establish a uniform size.
- **worked lumber** Dressed lumber that has been further machined, as tongued and grooved, shiplapped, or patterned lumber.
- **nominal size lumber** Rough overall lumber size before dressing or shrinkage, as a 2″ × 4″. Figure 578 shows common nominal and actual (minimum dressed) sizes for boards, dimension lumber, and timber.
- **actual size lumber** Finish size of lumber after all surfacing and shrinkage, as the actual size of a nominal 2″ × 4″ would be $1\frac{1}{2} \times 3\frac{1}{2}$. (Figure 578.)

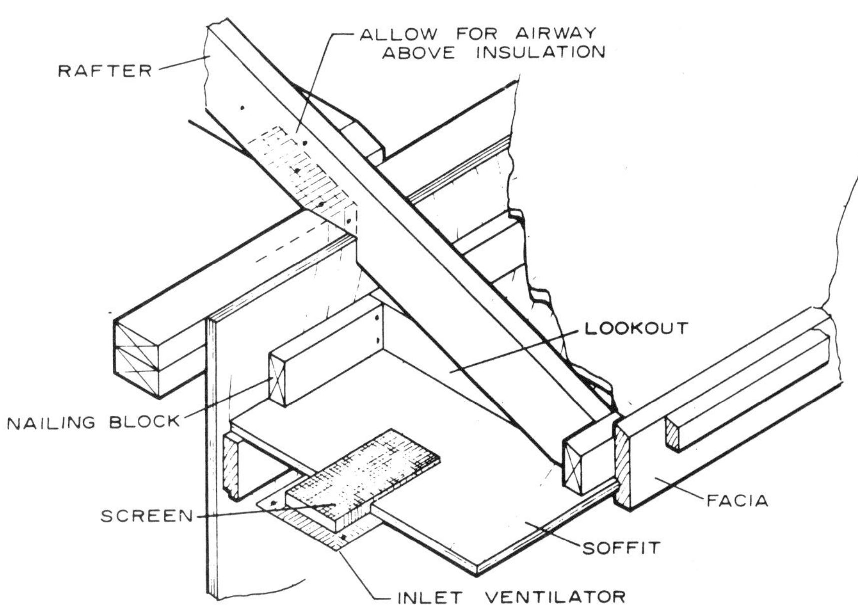

Figure 577 Lookout (eave)

ITEM	THICKNESSES			FACE WIDTHS		
		Minimum Dressed			Minimum Dressed	
	NOMINAL	Dry[1] Inches	Green[1] Inches	NOMINAL	Dry[1] Inches	Green[1] Inches
				2	1-1/2	1-9/16
				3	2-1/2	2-9/16
				4	3-1/2	3-9/16
				5	4-1/2	4-5/8
	1	3/4	25/32	6	5-1/2	5-5/8
				7	6-1/2	6-5/8
Boards[2]	1-1/4	1	1-1/32	8	7-1/4	7-1/2
				9	8-1/4	8-1/2
	1-1/2	1-1/4	1-9/32	10	9-1/4	9-1/2
				11	10-1/4	10-1/2
				12	11-1/4	11-1/2
				14	13-1/4	13-1/2
				16	15-1/4	15-1/2
				2	1-1/2	1-9/16
				3	2-1/2	2-9/16
				4	3-1/2	3-9/16
	2	1-1/2	1-9/16	5	4-1/2	4-5/8
Dimension	2-1/2	2	2-1/16	6	5-1/2	5-5/8
	3	2-1/2	2-9/16	8	7-1/4	7-1/2
	3-1/2	3	3-1/16	10	9-1/4	9-1/2
				12	11-1/4	11-1/2
				14	13-1/4	13-1/2
				16	15-1/4	15-1/2
				2	1-1/2	1-9/16
				3	2-1/2	2-9/16
				4	3-1/2	3-9/16
				5	4-1/2	4-5/8
Dimension	4	3-1/2	3-9/16	6	5-1/2	5-5/8
	4-1/2	4	4-1/16	8	7-1/4	7-1/2
				10	9-1/4	9-1/2
				12	11-1/4	11-1/2
				14		13-1/2
				16		15-1/2
Timbers	5 & Thicker		1/2 Off	5 & Wider		1/2 Off

* **Dry Lumber**—For the purposes of this Standard, dry lumber is defined as lumber which has been seasoned or dried to a moisture content of 19 percent or less.

Green Lumber—For the purpose of this Standard, green lumber is defined as lumber having a moisture content in excess of 19 percent.

* [2] Boards less than the minimum thickness for 1 inch nominal but 5/8 inch or greater thickness dry (11/16 inch green) may be regarded as American Standard Lumber, but such boards shall be marked to show the size and condition of seasoning at the time of dressing. They shall also be distinguished from 1-inch boards on invoices and certificates.

Figure 578 **Lumber: Nominal and minimum-dressed sizes of boards, dimension, and timbers**

lumber abbreviations Various terms are traditionally abbreviated in lumber usage. (Figure 579.) These terms are used in describing softwood lumber.

lumber defect See *defect.*

lumber grades Both softwood and hardwood are graded as to quality and structural soundness. Much of the grading is based on physical appearance. See also *American Softwood Lumber Standards.*

AD	Air-dried
ADF	After deducting freight
ALS	American Lumber Standards
AV or AVG	Average
Bd	Board
Bd. ft.	Board foot or feet
Bdl	Bundle
Bev	Beveled
B/L	Bill of lading
BM	Board Measure
Btr	Better
B&B or B& Btr	B and better
B&S	Beams and stringers
CB1S	Center bead one side
CB2S	Center bead two sides
CF	Cost and freight
CG2E	Center groove two edges
CIF	Cost, insurance, and freight
CIFE	Cost, insurance, freight, and exchange
Clg	Ceiling
Clr	Clear
CM	Center matched
Com	Common
CS	Caulking seam
Csg	Casing
Cu. Ft.	Cubic foot or feet
CV1S	Center Vee one side
CV2S	Center Vee two sides
D&H	Dressed and headed
D&M	Dressed and matched
DB. Clg.	Double-beaded ceiling (E&CB1S)
DB. Part	Double-beaded partition (E&CB2S)
DET	Double end trimmed
Dim	Dimension
Dkg	Decking
D/S or D/Sdg	Drop siding
EB1S	Edge bead one side
EB2S	Edge bead two sides
E&CB1S	Edge and center bead one side
E&CB2S	Edge and center bead two sides
E&CV1S	Edge and center Vee one side
E&CV2S	Edge and center Vee two sides
EE	Eased edges
EG	Edge (vertical) grain
EM	End matched
EV1S	Edge Vee one side
EV2S	Edge Vee two sides
Fac	Factory
FAS	Free alongside (named vessel)
FBM	Foot or feet board measure
FG	Flat (slash) grain
Flg	Flooring
FOB	Free on board (named point)
FOHC	Free of heart center or centers
FOK	Free of knots
Frt	Freight
Ft	Foot or feet
GM	Grade marked
G/R or G/Rfg	Grooved roofing
HB	Hollow back
H&M	Hit-and-miss
H or M	Hit-or-miss
Hrt	Heart
Hrt CC	Heart cubical content
Hrt FA	Heart facial area
Hrt G	Heart girth
IN.	Inch or inches
J&P	Joists and planks
KD	Kiln-dried
Lbr	Lumber
LCL	Less than carload
LFT or Lin. Ft	Linear foot or feet
Lgr	Longer
Lgth	Length
Lin	Linear
Lng	Lining
M	Thousand
MBM	Thousand (feet) board measure
MC	Moisture content
Merch	Merchantable
Mldg	Moulding
No.	Number
N1E	Nosed one edge
N2E	Nosed two edges

Og	Ogee
Ord	Order
Par	Paragraph
Part	Partition
Pat	Pattern
Pc	Piece
Pcs	Pieces
PE	Plain end
PO	Purchase order
P&T	Post and timbers
Reg	Regular
Res	Resawed or resawn
Rfg	Roofing
Rgh	Rough
R/L	Random lengths
R/W	Random widths
R/W&L	Random widths and lengths
Sdg	Siding
Sel	Select
S&E	Side and Edge (surfaced on)
SE Sdg	Square edge siding
SE & S	Square edge and sound
S/L or S/LAP	Shiplap
SL&C	Shipper's load and count
SM. or Std. M	Standard matched
Specs	Specifications
Std	Standard
Stpg	Stepping
Str. or Struc	Structural
S1E	Surfaced one edge
S1S	Surfaced one side
S1S1E	Surfaced one side and one edge
S1S2E	Surfaced one side and two edges
S2E	Surfaced two edges
S2S	Surfaced two sides
S2S1E	Surfaced two sides and one edge
S2S&CM	Surfaced two sides and center matched
S2S&SM	Surfaced two sides and standard matched
S4S	Surfaced four sides
S4S&CS	Surfaced four sides and caulking seam
T&G	Tongued and grooved
VG	Vertical grain
Wdr.	Wider
Wt	Weight

Figure 579 Lumber abbreviations

machine bolt General-use bolt with hexagonal nut. See *bolt.*

machine burn On lumber, a dark blemish resulting from friction caused during machining when the wood jams or binds.

machine expansion shield Heavy-duty fastener often used to bolt equipment to a concrete floor. The base or shield expands when the bolt is tightened to form a permanent base, although the bolt may be removed. See *lag shield, lead anchor*.

machine screw A threaded fastener very similar to a small bolt. The head normally has a slot for a screwdriver. No nut is used with a machine screw; it is designed to screw into a threaded hole. Figure 580 shows machine-screw head types and the various sizes available. Machine screws are identified by the head shape: round, flat, oval, fillister, and so on. Lengths vary from $\frac{1}{8}''$ up to $2''$. See also *bolt, thumb screw*.

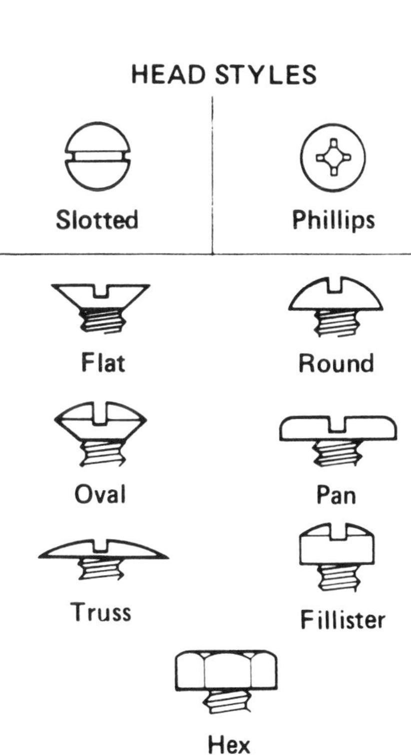

Diameter (Refer to by number)	Threads per inch
2	56
3	48
4	36
4	40
5	40
6	32
8	32
10	24
10	32
12	24
12	32
1/4	20
1/4	28
5/16	18
5/16	24
3/8	16
3/8	24

To determine size of screw or bolt, lay flat within parallel lines. Use circle to identify appropriate size nut.

Figure 580 Machine screws

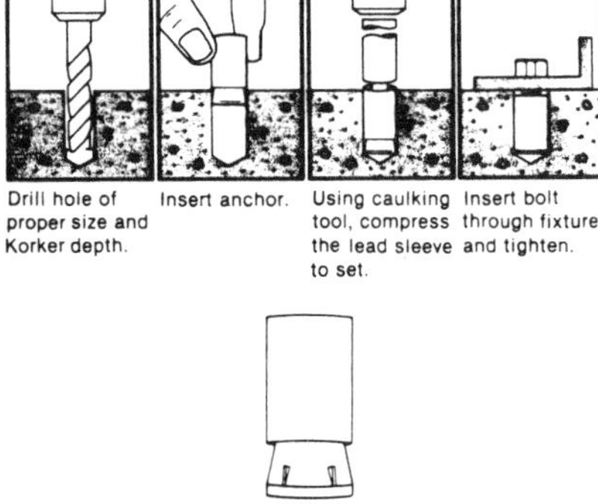

Figure 581 Machine-screw anchor

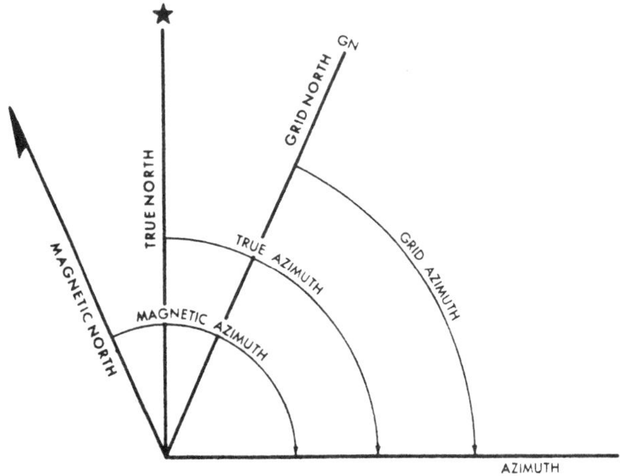

Figure 582 Magnetic meridian (magnetic north)

machine-screw anchor A two-part solid lead anchor. (Figure 581.) The anchor is inserted in a drilled hole in a concrete floor. A screw attaches to expand the anchor for a tight fit. Often used to attach the feet of industrial machines to the floor. Also called a caulking anchor. See also *anchor, lag shield, lead anchor, stud anchor.*

machine welding Welding performed by machines under the control and supervision of operators. See *manual, semiautomatic,* or *automatic welding.* See *welding procedures* for a comparison of the processes.

magnetic bearing Horizontal angle between the *magnetic meridian* and another line or compass reading.

magnetic declination Angle between *true meridian* and *magnetic meridian.* Also called declination of the compass needle. See also *isogonic chart.*

magnetic meridian The direction of the magnetic north pole; the direction a compass needle points. The direction is different for different areas of the world. The magnetic meridian is shown on a map by a line with a half arrow head. (Figure 582.) The angle between magnetic meridian and true meridian is indicated. See also *magnetic declination, true meridian.*

main Utility power line. Major electrical circuit which feeds regular or convenience circuits. In heating or air conditioning, the central heating pipe or duct that feeds into the small branches. The central water supply that feeds into the building water system. See *mains.*

main control center The *panelboard.*

main disconnect The disconnect for power to a structure. Located between the meter and the panelboard. A pull-out block with cartridge fuses is commonly used.

main rafter Rafter running from the ridge to the plate; a *common rafter.*

mains Horizontal pipes or ducts leading away from the heating unit. Mains carry hot water, steam, or hot air. Mains may also carry cooling air in the summer. See *main.*

male thread Threads on the outside of a pipe or fitting.

malleable Metal that can be readily worked, as by hammering.

mallet A hammer-type striking device with a large wood or rubber head. Used for striking woodworking chisels or other tools. (Figure 583.) See *hammer, rubber mallet, wood mallet.*

manhole Access hole to underground water pipes, sewer conduit, or electrical wiring. The manhole is flush to the ground and normally has a round cover.

mansard roof Four-sided roof with a double slope on each side. See *roof.*

mantel The facing around an open fireplace, especially the flat shelf or projection over the fireplace. See *fireplace.*

manual welding Welding completely controlled by the operator. Opposed to *semiautomatic, machine,* or *automatic welding.* See *welding processes* for a comparison.

manufactured building A finished structure or building (not a mobile home) assembled in a manufacturing facility.

marble setters, tile setters, and terrazzo workers Those who cover interior and exterior walls and floors with marble, tile, and/or terrazzo (a manufactured flooring that is poured in and polished).

margin Outside edge of any work. In roof work, the exposed portion of laid slate.

margin of safety The *factor of safety.*

margin trowel Cement finishing trowel used to work up to the edge of the flatwork. (Figure 584.) See also *trowel.*

Figure 583 Mallets

marking In steel, precast concrete, or reinforced prestressed concrete construction, the marking or identification of a building unit so it may be identified and assembled at the correct place in the correct sequence. The unit may be identified as "C" for column, "B" for beam, "G" for girder. "L" is used for left, "R" for right. N, S, E, and W may be used to designate which side of the building a unit is to go in, depending on what direction the side faces (North, South, East, or West). Each marked unit will correspond to a marked unit on the blueprints. Also called an *erection mark.* Also, scribing or drawing cutting lines on lumber or sheet metal.

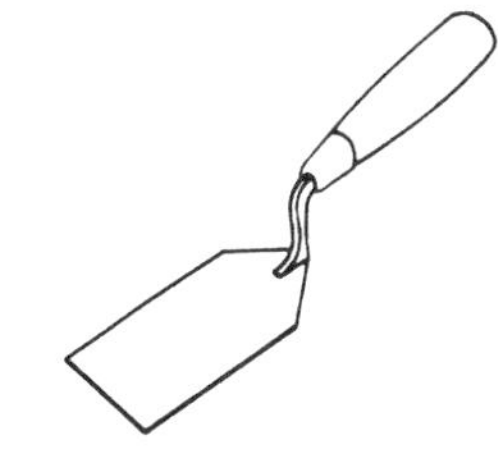

Figure 584 Margin trowel

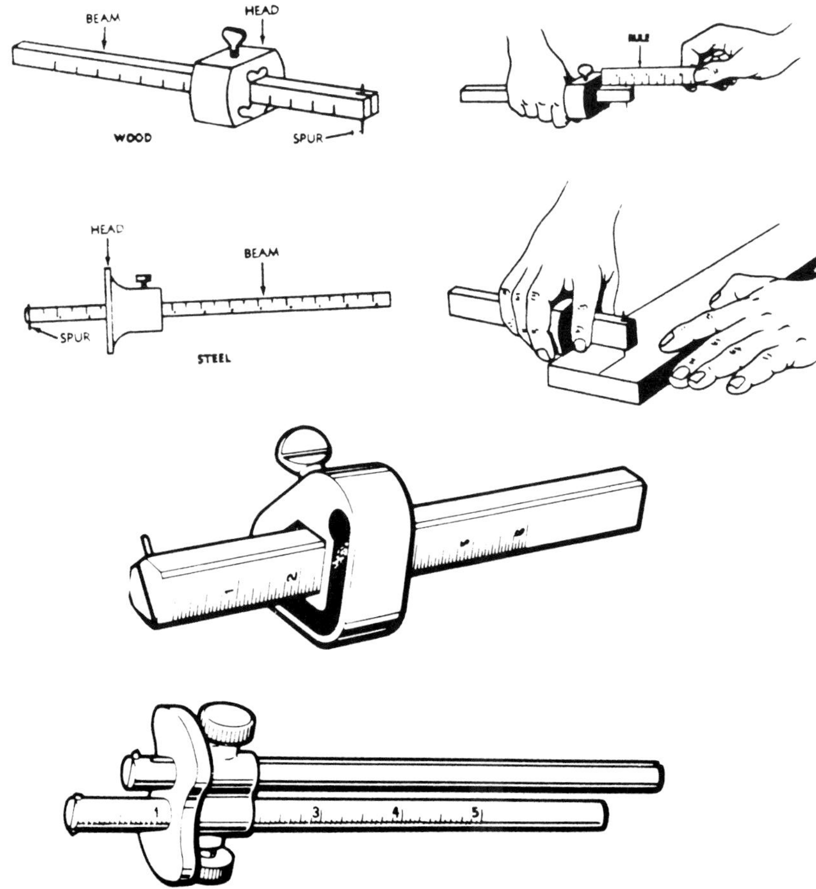

Figure 585 Marking gauges

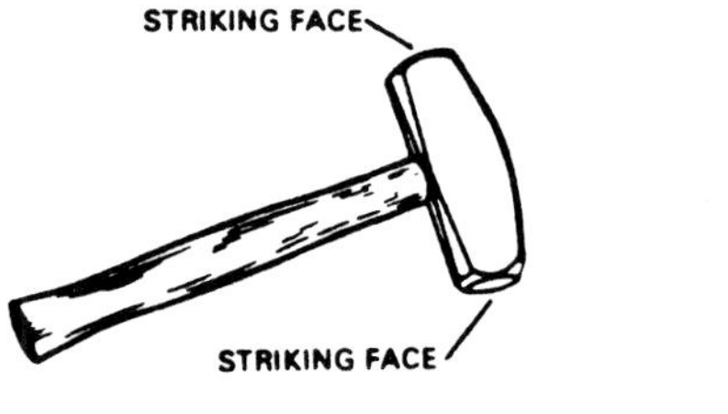

Figure 586 Mash hammer

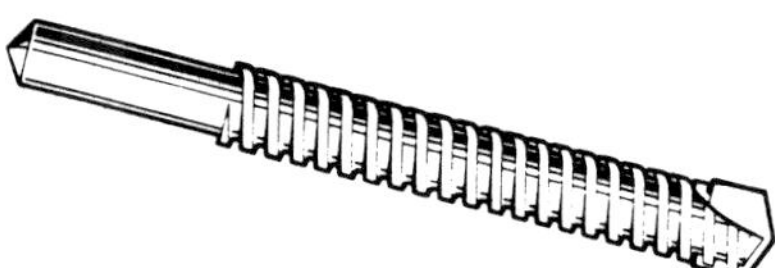

Figure 587 Masonry bit

marking awl Sharp, steel-pointed tool used in layout work to scribe (scratch) wood or metal with cutting lines.

marking gauge Tool used for scribing lines parallel to an edge, as the edge or end of a board, to mark where to make a saw cut. The mark is made with a steel point. (Figure 585.)

mark number Erection mark. See *marking*.

marquetry Wood inlay; colored fine hardwoods are cut in thin strips, then cut to specific shapes and laid into a wood surface in some pattern. Used in fine furniture.

mash hammer Short, heavy-duty hammer used for striking chisels, punches, nails. (Figure 586.)

masking In acoustics, the apparent loss of loudness of a sound because of another noise (which masks the first sound).

masonry General term covering all construction using masonry units, such as brick, concrete block, stone, tile, and glass block, that are set in mortar. The type of masonry unit being worked with may specify the kind of masonry, such as *brick masonry*. The laying of concrete blocks is called *concrete masonry*. *Note*: Sometimes the term *masonry* is used generally to include both masonry units and concrete construction. See also *hollow masonry, solid masonry*.

masonry anchor A metal fastener used to hold masonry parts together or to hold wood structural members to masonry. See *machine screw anchor*.

masonry bit High-strength steel drill or bit with a hard tungsten carbide tip used to drill into concrete or masonry units. (Figure 587.) See *bit*.

masonry cement Cement made mostly of Portland cement, mixed with sand and water to form a mortar.

masonry joints The finished mortar joint made between masonry units, such as brick or concrete block. Figure 588 shows the common masonry joints: concave, flush, struck, weather, raked, and bead. The bead joint is the unfinished joint.

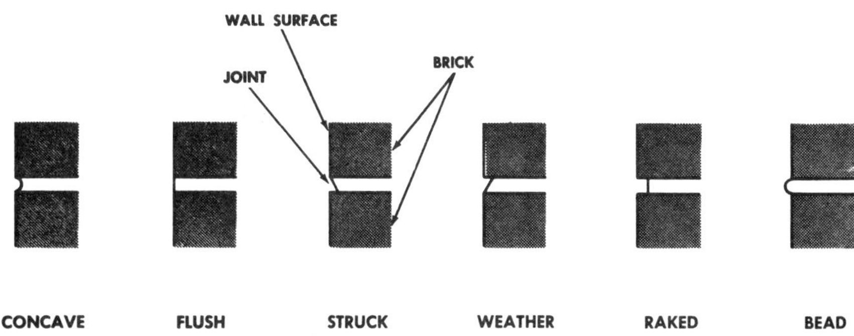

Figure 588 Masonry joints

masonry nails Heavy, hardened steel nails used for driving into masonry; a concrete nail. See *nails*.

masonry saw Portable electrical saw with special cutting blades, including diamond-tipped blades. Commonly used in cutting concrete flatwork. (*Figure 229*, p. 85.) See *abrasive cut-off saw, concrete saw*.

masonry tie Metal fastener used to hold masonry walls together or to hold a veneer wall to the house frame. (*Figures 179*, p. 65; *1072*, p. 473.)

mason's hammer Heavy hammer used for breaking stone. See also *brick hammer*.

mason's lead In masonry work, a squared, built-up brick or concrete block corner; six or seven courses are laid out at a corner stair-step fashion to serve as a guide for the rest of the two walls. (*Figure 223*, p. 83.) Also called a *corner lead*. See *lead, rake back lead*.

mason's scaffold Temporary framing on exterior of wall that supports the mason, bricks, and mortar. (Figure 589.)

Figure 589 Mason's scaffold

mast Pole supporting an antenna or other wiring or cables. The support tower of a tower crane from which the boom projects. Also, on a heavy crawler crane, the upright support pole or framework that acts as a counterbalance to the boom (lifting spar). See *boom*.

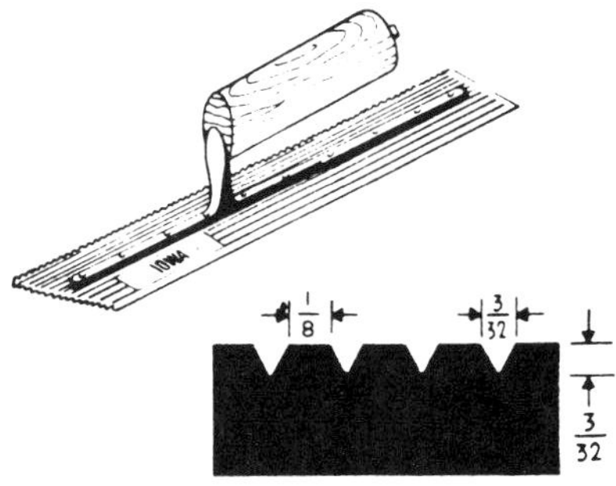

Figure 590 Mastic trowel

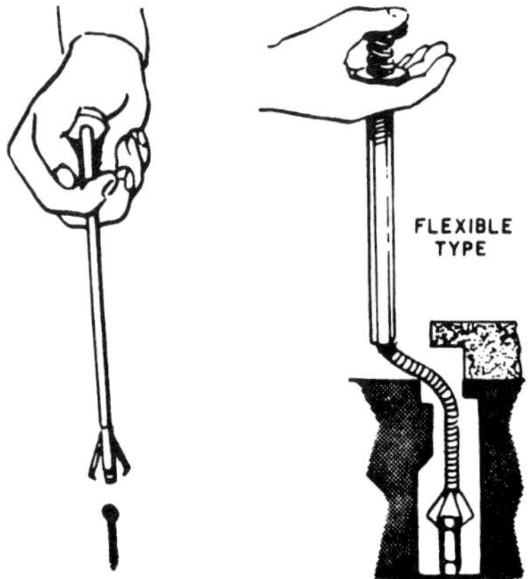

Figure 591 Mechanical fingers

mastic A thick waterproof adhesive widely used in construction for installing plywood panels and flooring. Available in easily used canisters that fit into an adhesive gun. A wide variety of mastics are available, and the type should be selected to fit the specific need. See *adhesive, adhesive gun.*

mastic trowel Toothed, flat trowel used for spreading and smoothing mastic for tile. (Figure 590.) Available with teeth of different shapes.

mat Grid of reinforcing bars laid out and wired together. Also, a large slab used to support a structure; also called a *raft foundation.*

matched joint Tongue-and-groove joint.

matched lumber Edge-finished lumber that is cut for a tongue-and-groove joint.

match line In drafting, a special break line used to separate two parts of a large drawing onto separate sheets; a pair of match lines is used. Part or member ends are normally tied together with identical letters or numbers.

material symbols In drafting, symbols used to designate the various materials used to construct a building. See *plan symbols, symbols.*

matrix Cement mix to which sand is added to make mortar, or to which sand and coarse aggregates are added to make concrete.

matte finish A dull finish.

maul Heavy hammer or sledge used for driving wedges or stakes.

meandered water Lake or stream owned by the government, generally open to the public. Adjacent property owners pay taxes only on the land, not the water area.

meander line Survey line running along a lake, stream, or other natural feature.

mechanic A skilled craftsman or worker.

mechanical drawings Blueprints that show the HVAC or plumbing systems. Identified by an M on the title box. Opposed to *architectural drawings, electrical drawings,* and *structural drawings.* See also *electrical plan, electrical wiring plan, heating and ventilating plan, piping schematic, plumbing plan.*

mechanical fingers Grasping device with a long arm and mechanical fingers that open up to grasp small, inaccessible objects. (Figure 591.)

mechanical pencil A pencil that holds a lead. A *pencil holder.*

median In highway construction, the divided highway strip separating the ways for traffic in opposite directions.

medullary rays In wood, flat bands that extend out from the center of the tree toward the bark. They serve to store and transport food horizontally in the living tree. They appear as flakes in quartersawn wood, are especially prominent in oak. See *tree, wood.*

meeting rail Horizontal wood members in double-hung windows that separate the lower and upper sashes. See *double-hung window.*

melt-thru Complete joint penetration by a weld.

member Any building support unit, as a column or beam.

membrane A continuous, unbroken surface, as from a plastic sheet, asphalt coating, or rubber-based covering. A paint or waterproofing coating forms a membrane. A vinyl sheet laid down on the earth to prevent moisture penetration into a building. A continuous, unbroken-surface roof cover designed to prevent passage of water or moisture, as a *built-up roof membrane.*

membrane curing Covering of fresh concrete with a sheet of vinyl to protect the concrete, inhibit evaporation, and allow curing.

membrane waterproofing Sheet of material impervious to the transfer of moisture, commonly used under slabs and on some foundation walls. Polyethylene, two-ply hot-mopped felts, or 55-lb roll roofing is used.

mending plate A straight, drilled metal strip used for reinforcement in light framing. (Figure 592.) See also *corner brace, L brace, T brace.*

mensuration Measuring.

mesh Criss-cross network of wires or steel rods. See *welded wire mesh.* Also, abrasive grading; see *mesh system.*

mesh system System of grading abrasives. A mesh cloth screen with sized openings is used. Openings range from very coarse #12 (12 openings to the square inch) up to #600 (600 openings per square inch). See *abrasive.*

metal-arc welding Arc-welding process. A metal electrode is used to produce an arc between the electrode end and the work piece. See *gas metal-arc welding* (GMAC), *gas tungsten-arc welding* (GTAW), *shielded metal-arc welding* (SMAW).

Figure 592 Mending plate

metal bridging Metal cross supports used to hold and strengthen joists. See *bridging.*

metal conduit Pipe for running electrical wires and cables.

metal-cutting blade Circular power-saw blade for cutting metal. See *saw blades.*

metal deck Sheet metal roofing.

metal identification See *spark test.*

metal lath Expanded metal sheets which are used as a base for plaster. See *lath.*

metallic clad cable Factory- assembled cable, sheathed in an inflexible metallic tube, used in all locations except where exposed to destructive corrosive conditions. Type MC cable.

metallic inert-gas welding See *gas metal-arc welding.*

metal lock fastener Tool designed to attach metal studs to the flooring channel. (Figure 593.) Fastens pieces and folds metal together. See also *punch-lok riveter.*

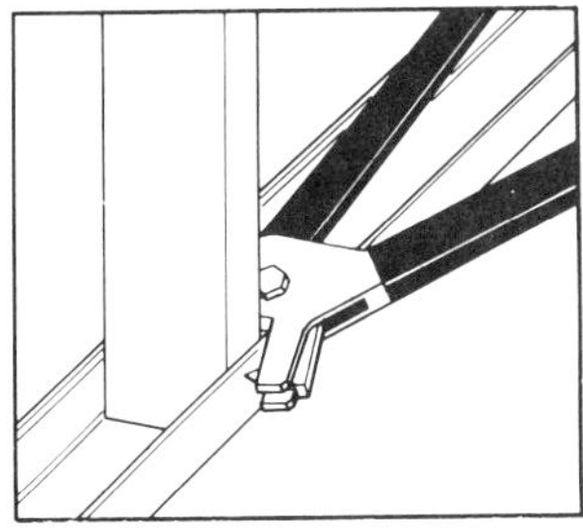

Figure 593 Metal lock fastener

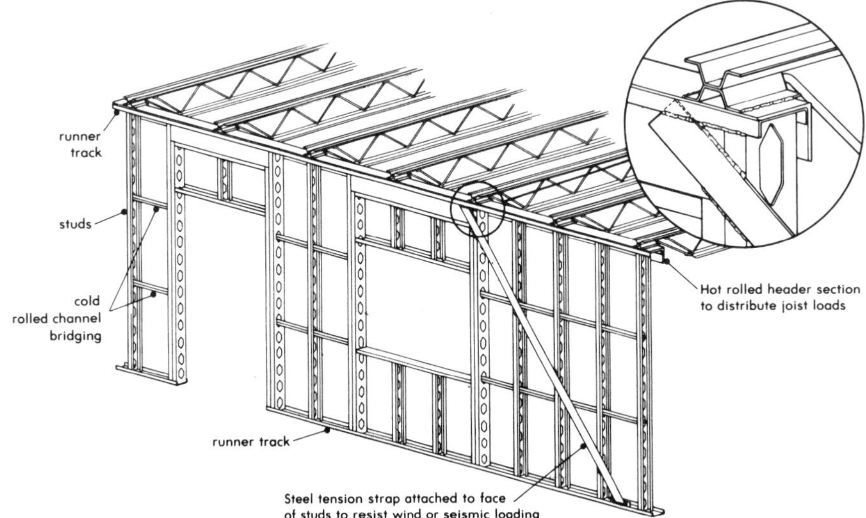

Figure 594 Metal studs

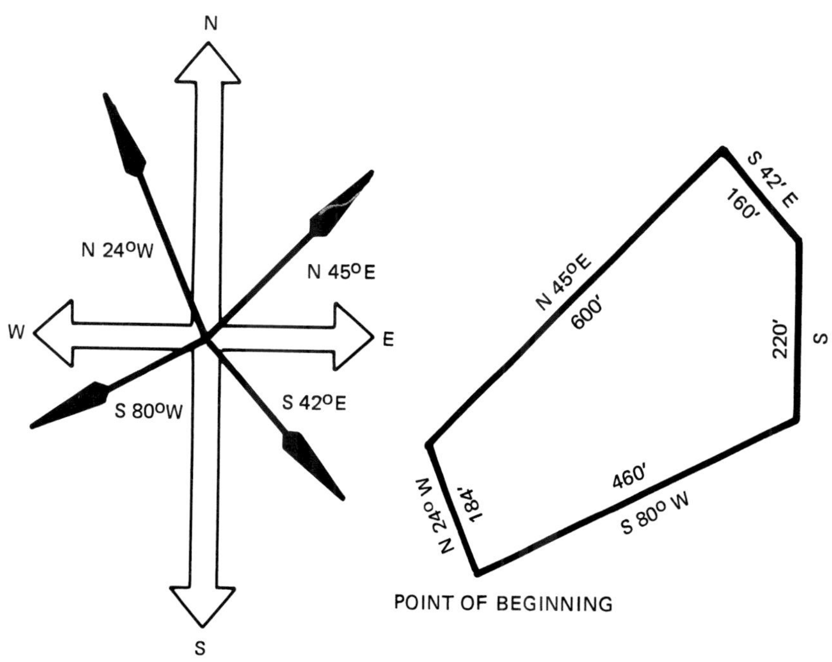

Begin at the beginning point, thence N 24 degrees west 184 feet, thence N 45°E 600 feet, thence 542'E 160', thence S 220' thence S 80° W 460' to point of beginning.

Figure 595 Metes and bounds description of a small tract

metal studs Webbed or hollow steel studding used in light construction. (Figure 594.)

metal ties Steel ties used in masonry construction, used to secure two wythes together in hollow-wall construction or to hold brick veneer to the wood frame. (*Figure 1072*, p. 473.)

metal trim Trim used on the end of gypsum panels at door and window jambs and at other intersections where panels meet other framing materials. In general, any trim made out of metal.

meter In an electrical system, a device for measuring electrical power or current use in kilowatt-hours. One thousand watt-hours is called a kilowatt-hour (kWh). A water meter measures flow of water into the house in gallons. A gas meter measures flow of natural gas into house in cubic feet. Also, a metric measurement of 1000 millimeters or about 39.77 inches. See *metric system*.

meter shunt Grounding device used to bypass the water meter so that electrical ground (to water pipes) is assured. (*Figure 475*, p. 194.)

metes and bounds Description of property by reference to a known point and by a description using directions and distances from this point; description follows all sides of the land parcel and returns to the beginning. (Figure 595.)

METRIC UNITS

1 KILOMETER = 1000 METERS
1 METER = 100 CENTIMETERS
1 CENTIMETER = 100 MILIMETERS
1 LITER = 100 CUBIC CENTIMETERS
1 KILOGRAM = 1000 GRAMS

METRIC TO ENGLISH CONVERSIONS

1 KILOMETER = .62 MILES
1 METER = 39.37 INCHES
1 CENTIMETER = .394 INCHES
1 MILLIMETER = .0394 INCHES
1 LITER = .2642 GALLONS
1 LITER = 1.057 QUARTS
1 KILOGRAM = 2.2046 LBS.
1 DEGREE CENTIGRADE = 1.8 DEGREES FAHRENHEIT

ENGLISH TO METRIC CONVERSION

1 MILE = 1.6093 KILOMETERS
1 YARD = .9144 METERS
1 FOOT = .3048 METERS
1 FOOT = 30.48 CENTIMETERS
1 INCH = 2.54 CN 2.54 CENTIMETERS
1 INCH = 25.4 MILLIMETERS
1 GALLON = 3.785 LITERS
1 QUART = .946 LITERS
1 POUND = .453 KILOGRAMS
1 POUND = 454 GRAMS
1 DEGREE FAHRENHEIT = .555 DEGREES CENTIGRADE

Figure 596 Metric units and conversions

metric conversion Relationship of metric units to the conventional measuring system. Figure 596 shows the conversion of conventional inch-feet units to metric units *and* conversion of metric units to conventional inch-feet units.

metric scale Drawing scale based on millimeter relationships. (Figure 597.)

metric system A universal measuring system used in most parts of the world. Figure 598 shows some of the basic metric units. Units are made larger or smaller by multiplying or dividing by 10 or a multiple of 10. In construction, the units commonly used are the millimeter and meter. Blueprints and all building measurements are based on the millimeter. There are 25.4 millimeters per inch. Longer distances, such as found on plot plans and site plans, are measured in meters. There are 1000 millimeters in one meter.

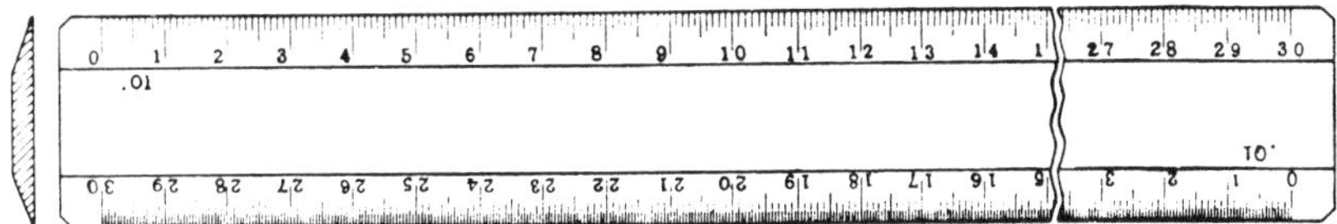

Figure 597 Metric scale

PHYSICAL QUANTITY	UNIT	SYMBOL
LENGTH	meter	m
MASS	kilogram	kg
TIME	second	s
ELECTRIC CURRENT	ampere	A
TEMPERATURE	celsius	°C
LUMINOUS INTENSITY	candela	cd

Basic metric units. These base units may be multiplied or divided by TEN.

MULTIPLE OR SUBMULTIPLE	PREFIX	SYMBOL	MEANING
1000	kilo	k	ONE THOUSAND TIMES
100	hecto	h	ONE HUNDRED TIMES
10	deka	da	TEN TIMES
0.1	deci	d	ONE TENTH OF
0.01	centi	c	ONE HUNDREDTH OF
0.001	milli	m	ONE THOUSANDTH OF

Basic metric terms (prefixes). These prefixes are used to specify base TEN multiples or division. A "prefix" goes *in front of* a metric unit. For example, a *milli-meter* or *millimeter* is 1/1000 of a meter. That is, 1000 millimeters = 1 meter.

Figure 598 Metric units and basic metric prefixes

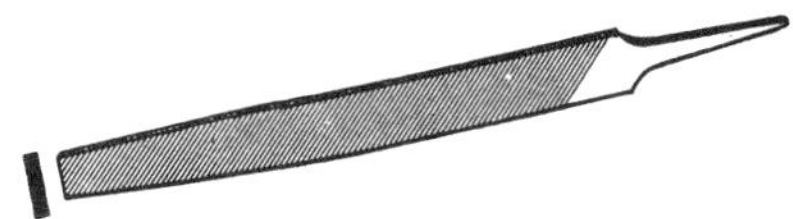

Figure 599 Mill file

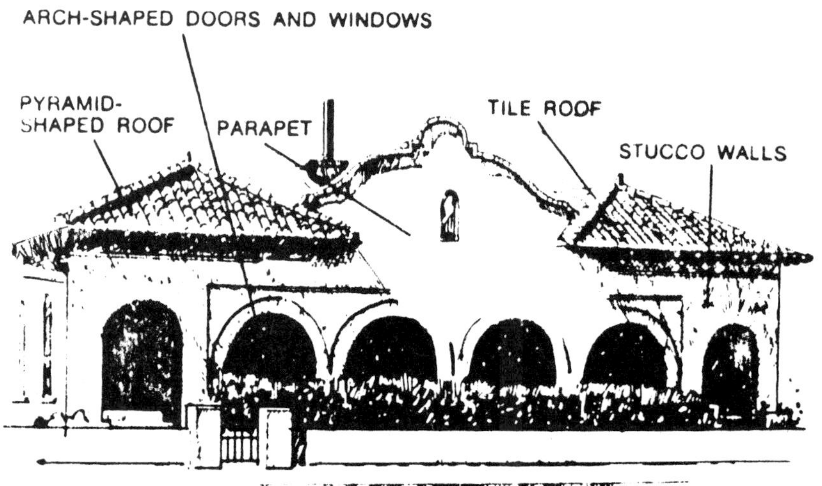

Figure 600 Mission style

metric threads An international thread standard based on millimeters. Similar to the standard *Unified Thread* system but not interchangeable. Two general series are used: coarse (general purpose) and fine (precision).The thread is specified as M with the nominal size. For example: M 4.5 (metric 4.5 = millimeter nominal diameter).

metric tools Tools with sizes based on the metric system. Sizes are in millimeters (mm) rather than inches. Wrenches and sockets are widely available in millimeter sizes.

metrification Use of the metric system (in construction or on drawings).

mezzanine A floor located between the main floor and second floor.

midget trowel Small rectangular trowel used in narrow areas such as window sills or closets.

mig *Gas metal-arc welding (GMAW).*

mil One mil is $\frac{1}{1000}$ inch. In electrical wiring, a unit used for describing a cross-sectional diameter of a wire. See *circular mil* for the area of a wire section.

mildew Airborne fungus that grows wherever moisture and food are available. Often appears on areas of moist paint or wood siding.

mil-foot In electrical wiring, a foot length of wire that is one mil ($\frac{1}{1000}$ inch) in diameter. Used as a base for making electrical wiring calculations.

mill file General use flat file, used for sharpening and other fine work. (Figure 599.) See *file*.

millimeter Basic metric measure. There are approximately 25.4 millimeters to the inch. Abbreviated as mm. Sometimes specified as *Mil*. See *metric system*.

mill material In steel construction, a product ordered for a specific project.

millwork Finished wood materials or parts, such as a window frame, completed in a mill or manufacturing plant.

millwright One who is responsible for installing, aligning, and connecting heavy industrial machinery.

mineral-insulated metal-sheathed cable Copper-sheathed factory-assembled cable used in all locations except in areas exposed to destructive corrosive conditions. Cable Type MI.

minus sight In surveying, a *foresight*.

mission architecture Architecture based on early church (mission) architecture, especially from the California area. (Figure 600.) See also *architectural styles*.

miter A corner that is made by two pieces joining at an evenly divided angle, usually 45°. (Figures 601, *246*, p. 94.) See *wood joints*.

miter box Metal or wood box with slots cut for holding a saw at 45° to cut miter ends; 90° cuts are also made. Some miter boxes have an indexing system to allow any angle from 45° left to 45° right. (Figure 602.) A back saw is used in a miter box. See *power miter box*.

miter clamp Clamp for holding miter joints. (Figure 603.) See also *miter vise*.

mitered stringer Open stair where the joint between risers and stringers is mitered at a 45° angle.

mitering Jointing of equal angle boards to make a corner or miter joint. See *wood joints*.

miter rod In plastering, shaped metal piece used for shaping joints as at corners. Also called *joint rod*.

miter saw Fine-toothed saw with a straight, reinforced back, used in miter boxes for cutting exact angles.

miter try square Small *try square* used for establishing right angles; can also be used for 45° work. (Figure 604.)

miter vise Clamp used for making a 90° corner. (Figure 605.) See also *corner frame clamp*, *miter clamp*.

mix design In concrete work, the specific mix of ingredients to make the specified concrete.

mixed construction In heavy construction, a building that uses two different types of structural building materials, specifically steel and reinforced concrete. Also called *composite construction*.

mixing valve In solar heating, a valve that allows cold water to be added to the solar-heated hot water to maintain a specific temperature.

Figure 601 Miter joint

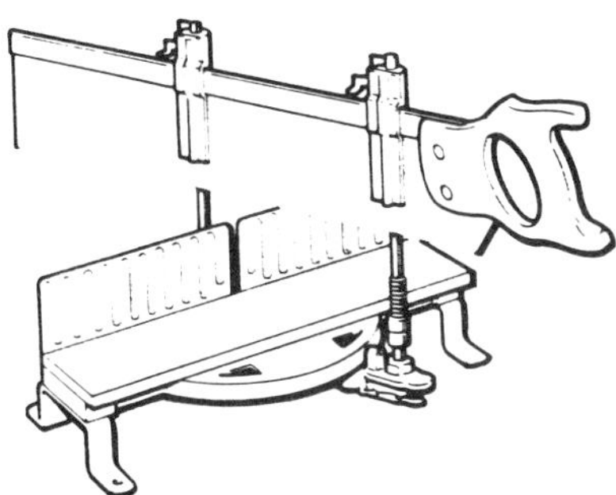

Figure 602 Miter box with backsaw

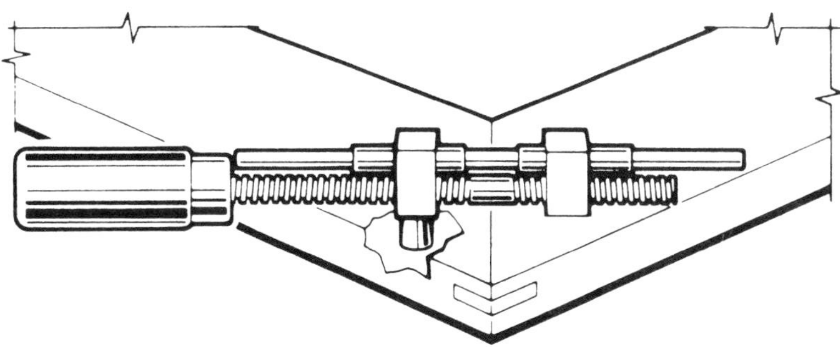

Figure 603 Miter clamp

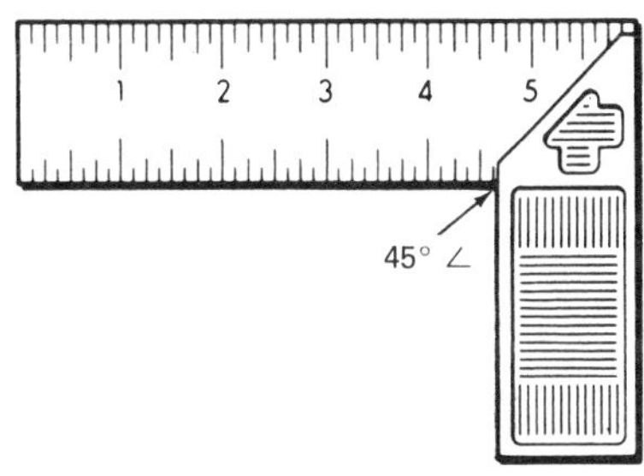

Figure 604 Miter try square

Figure 605 Miter vise

Figure 606 Mobile concrete placer

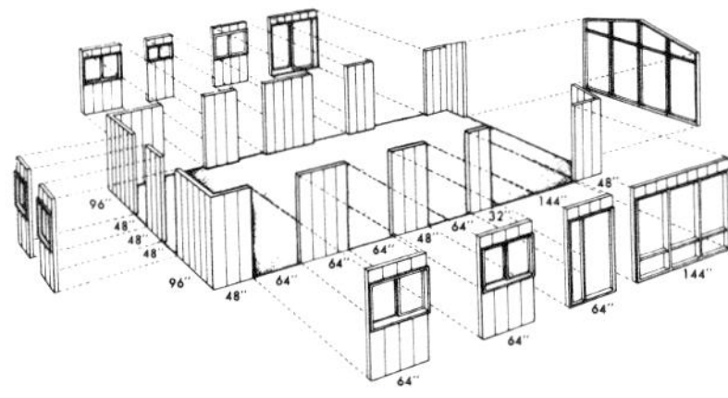

Figure 607 Modular dimensioning: basic 4-inch module

mobile concrete placer Mechanical conveyor used to move concrete from the truck to where needed. (Figure 606).

mobile home Complete house that is designed to be towed or moved. See also *architectural styles.*

mock up Construction of a building part or element at full scale to test structural stability and design.

model In architectural planning, small-scale reproduction of a building or building development. Allows a view of the finished appearance. Often used to show a client what the completed project will look like.

model home Completed and furnished home used for showing prospective buyers what a completed unit looks like; used by developers to sell, rent, or lease homes (often before they are built).

modified wood Processed wood; wood that has been modified to improve its properties; wood changed by chemical, resin, or heat treatment.

modular System built around a base unit which is used throughout. In construction the basic module is 4″.

modular construction Building designed and constructed using a modular unit or dimension, commonly 4″. Large three-dimensional prefinished building components based on a 4-inch or 4-foot module. A module will have one or more rooms. A complete house may be constructed with one, two, or three separate modules. See also *preengineered building, industrialized building.*

modular dimensions Design and construction based on a module; building materials and units are based on a 4-inch dimension or module. Modular dimensions are a multiple of 4″. (Figure 607.)

modular masonry Use of masonry units which, when in place with mortar, form a wall based on a modular measure of 4″.

module A basic construction measure, generally accepted as 4″. A module is thought of as a cube 4″ × 4″ × 4″. Also, a complete building or part of a building, such as a bathroom, constructed in a factory. The building modules are complete and are delivered to the site and installed in place.

modulus of elasticity (*E*) Ratio expressing the amount a material will bend (deflect) with a set load. See also *extreme fiber stress in bending*.

modulus of rupture Ultimate tensile strength at point of rupture, calculated at *extreme fiber stress*.

moist cure Curing of concrete at a constant temperature at a high humidity.

moisture barrier Material, such as vinyl, used in a wall to prevent the passage of water. Also called a *vapor barrier*.

moisture content (**MC**) Amount of moisture in wood. Kiln-dried lumber normally has an MC of 19% (percent of moisture in relation to dry weight of wood).

moisture proofing Applying of moisture-resistant coating to a surface.

moisture test on aggregate In concrete work, a test made to determine the moisture content of sand.

mold Shaped template used in plastering to contour surfaces, such as a plaster cornice molding. Also, in reinforced concrete work, a permanent fiberglass or metal form used for casting concrete or for forming joists or ribs on the bottom side of slabs.

molding Wood, metal, or plaster strip used for finish, as at the top and bottom of walls. Figures 608 and 609 show some of the many wood moldings available. See also *chair rail, panel molding*. Figure 610 shows vinyl molding used on panel edges.

molding cap Decorative wood trim used over a window or door.

molding plaster Finely ground calcined gypsum, used for detail trim, decorative and ornamental work, and casting.

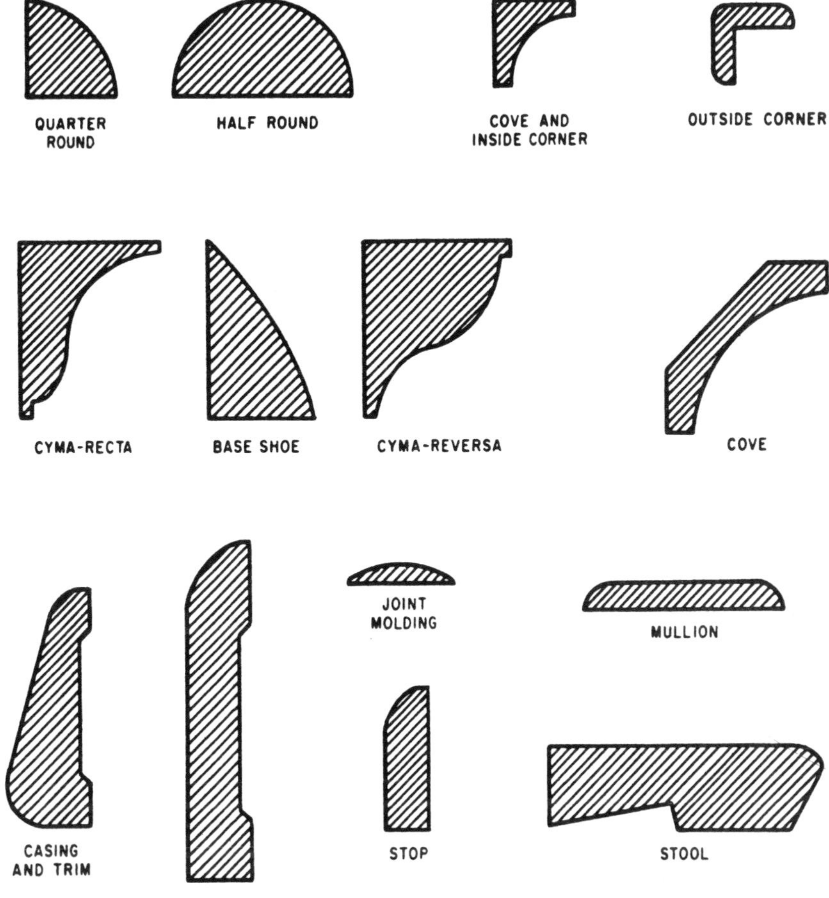

Figure 608 Wood moldings

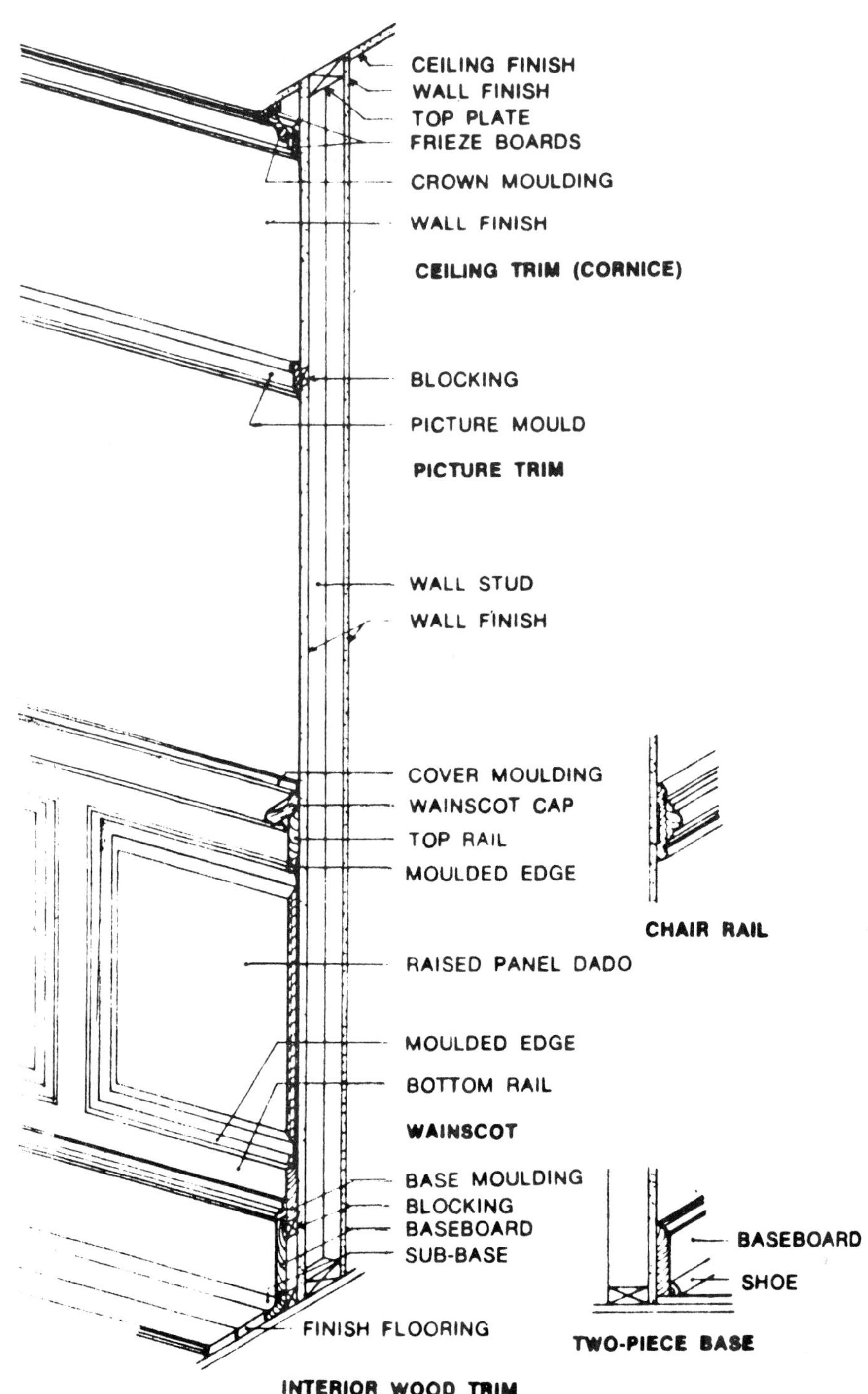

Figure 609 Wood moldings in place

molding, ornamental Decorated plaster molding. Traditional styles are cast in sections, then applied to make a continuous molding.

molly anchor Expansion anchor used in plastered walls and drywalls. (Figure 611.) See *hollow-wall anchor, nylon expansion anchor.*

molten weld pool Liquid state of the weld.

moment Force acting around a point so as to cause a rotational stress. Abbreviated: M.

monial A *mullion.*

monkey wrench Adjustable wrench with flat, smooth jaws, approximately 14″ long, used for tightening nuts or holding or turning plumbing pipe. (Figure 612). See *wrench.*

monolithic Concrete structure poured as all one piece.

moonshine crotch Attractive veneer figure cut from the crotch area of a tree; the wood fibers swirl around.

mop board *Baseboard.*

mopping In built-up roofing, the application of hot asphalt or bitumen to the roof felts to form a bond; a mop is used to spread the hot material.

mortar Bonding mixture of lime cement mixed with sand and water. Used to cement masonry units together. Mortars of various strengths are available for different purposes, based on the type of cement used.

mortar bed Layer of fresh mortar used as a seat for masonry units or any structural member.

mortar board Square board with a handle used to hold mortar when laying bricks, concrete blocks, and the like. See *hawk, hod.*

mortar box Box in which mortar or plaster is mixed.

mortar color Color added to mortar for decorative effect.

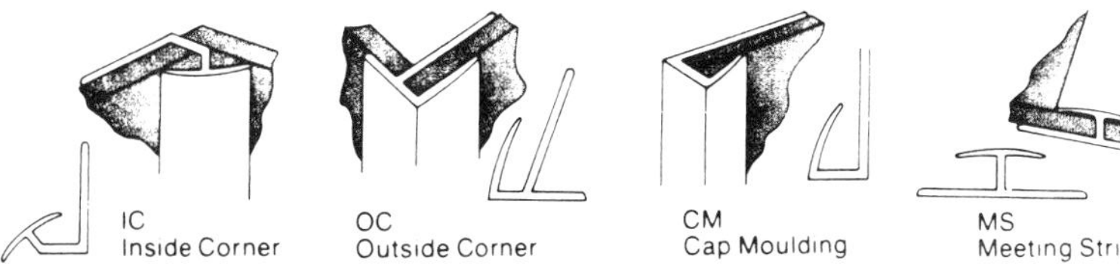

Figure 610 Vinyl moldings used on panel edges

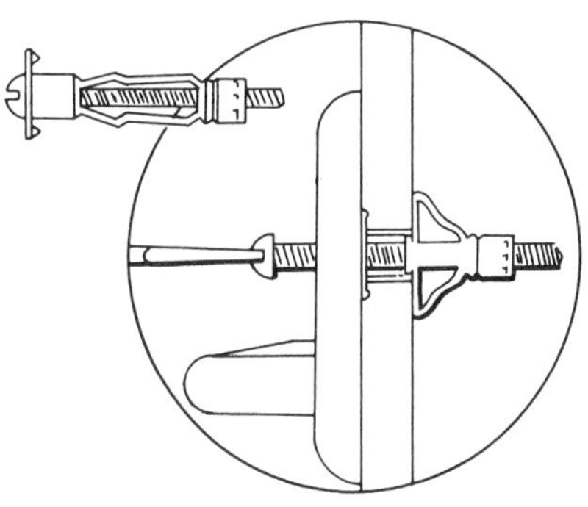

Figure 611 Molly anchor

mortar hoe Hoe with two holes in the blade, used for mixing mortar.

mortar joints Joints made to bond masonry units together. Different external finish is made on the joints so they will not weather. See *masonry joints.*

mortar mixer Mechanical mixer used to blend the mortar ingredients; a rotary drum with paddles.

mortise Deep, rectangular cut made into a wood piece; the cut made to receive a *tenon* and create a *mortise and tenon* joint. See *wood joint.*

mortise and tenon Joint made by fitting a tenon into a mortise. (Figure 613.) See *wood joint.*

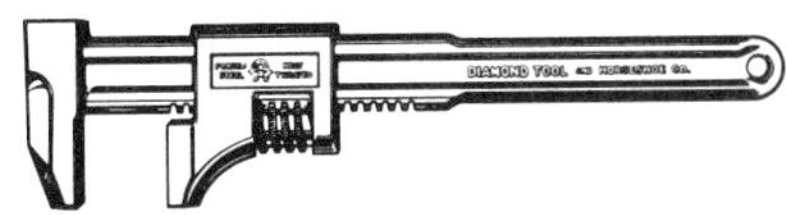

Figure 612 Monkey wrench

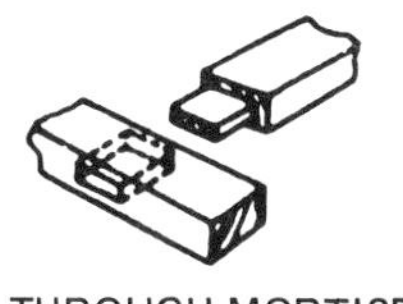

Figure 613 Mortise and tenon

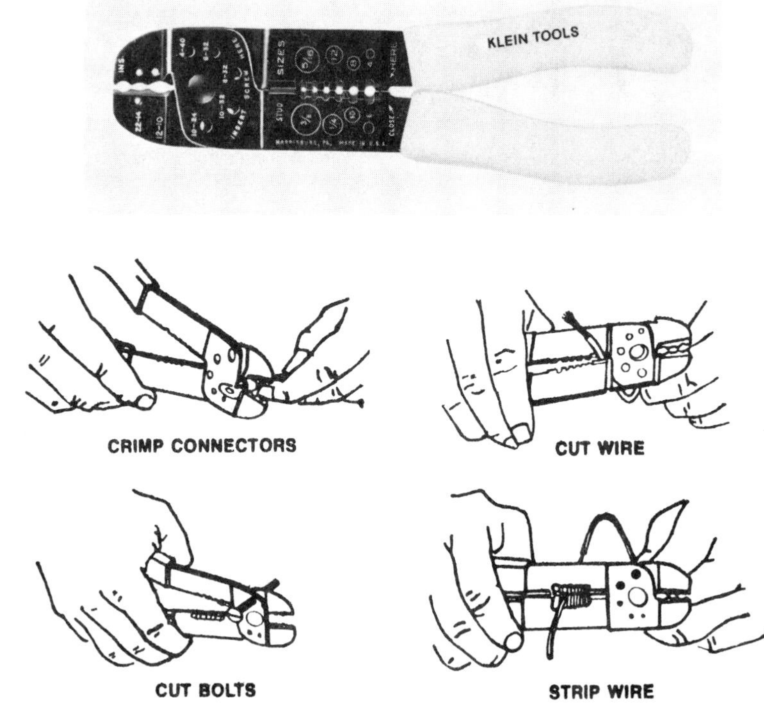

Figure 614 Multipurpose tool

mortise chisel Heavy-bodied wood chisel used for cutting mortises. See *wood chisel.*

mortise gauge A marking gauge with two points. One point is adjustable so that points can be set to mortise width. See also *butt gauge.*

mortise lock Door lock that fits into a mortise in the door edge. See *lock.*

mosaic Decorative design inlaid on a floor or wall using many broken pieces of colored stone or glass. The stone or glass is inlaid in a mortar base.

mosaic tiles Small tiles, about an inch in width, used for covering floors and countertops. Tiles come affixed to 1′ × 1′ or 1′ × 2′ sheets. The tiles are laid onto a mastic base, then the paper backing is soaked off.

motor control General term for electrical control of any type of electrical motor.

M roof Roof formed by joining two gable roofs to form a double roof shaped like an M.

mucking Adjusting steel reinforcing bars during the concrete pour.

mud Quick-setting plaster, used in drywalling to cover holes and tape. Any quick-drying filler material used for patching. Also, wet concrete, mortar.

mud holder A hawk used to carry mud.

mudsill A *sill.* Lumber piece directly atop the foundation.

mullion Vertical framing member that separates window frames in a multiple window. (Do not confuse with *muntin,* which divides lights in a window sash.) Also, in a panel door, the vertical wood piece (also called a muntin) that separates the panels. Also, vertical support members between curtainwall panels. See also *carcass.* (*Figure 164,* p. 61.)

multipurpose tool In electrical work, a plierlike tool used for cutting and stripping wire and for crimping. The nose is used for gripping and bending. Special holes are used for cutting screws commonly found on the job. (Figure 614.) See also *crimping tool, wire cutter, and stripper.*

muntin Small wood bars that divide lights (glass pieces) in a window sash or door. (*Figures 306,* 117; *317,* p. 121.) (Do not confuse with *mullion,* which is a vertical piece separating window frames.) Also, a vertical piece in a door that separates panels.

mushroom construction In reinforced concrete, reinforced concrete columns that flare out at the top where they meet the slab. Used in *flat-slab concrete floor* construction.

n

nail Long-shanked fastener widely used for holding wood members to each other. Nails today are *wire nails.* In the past *cut nails* were exclusively used and they are still available for special uses. Most nails are made of steel, although case-hardened steel is used for concrete and masonry nails. Copper, brass, and aluminum nails are also available. Various head shapes, shank shapes, and points are used. Figure 615 shows some of the different heads and shapes. (See also *nail heads, nail points).* All nails today are still sold by the *penny (d)* system. (*Figure 670,* p. 290.) *Common nails* (2*d* to 60*d*) are generally used for all rough carpentry work. *Box nails* (2*d* to 40*d*) are also widely used. *Casing nails* (2*d* to 40*d*) and *finish nails* (2*d* to 20*d*) are used for finish work and trim. The finish nail is commonly countersunk. *Ratchet nails* are used to apply gypsum panels (drywall).

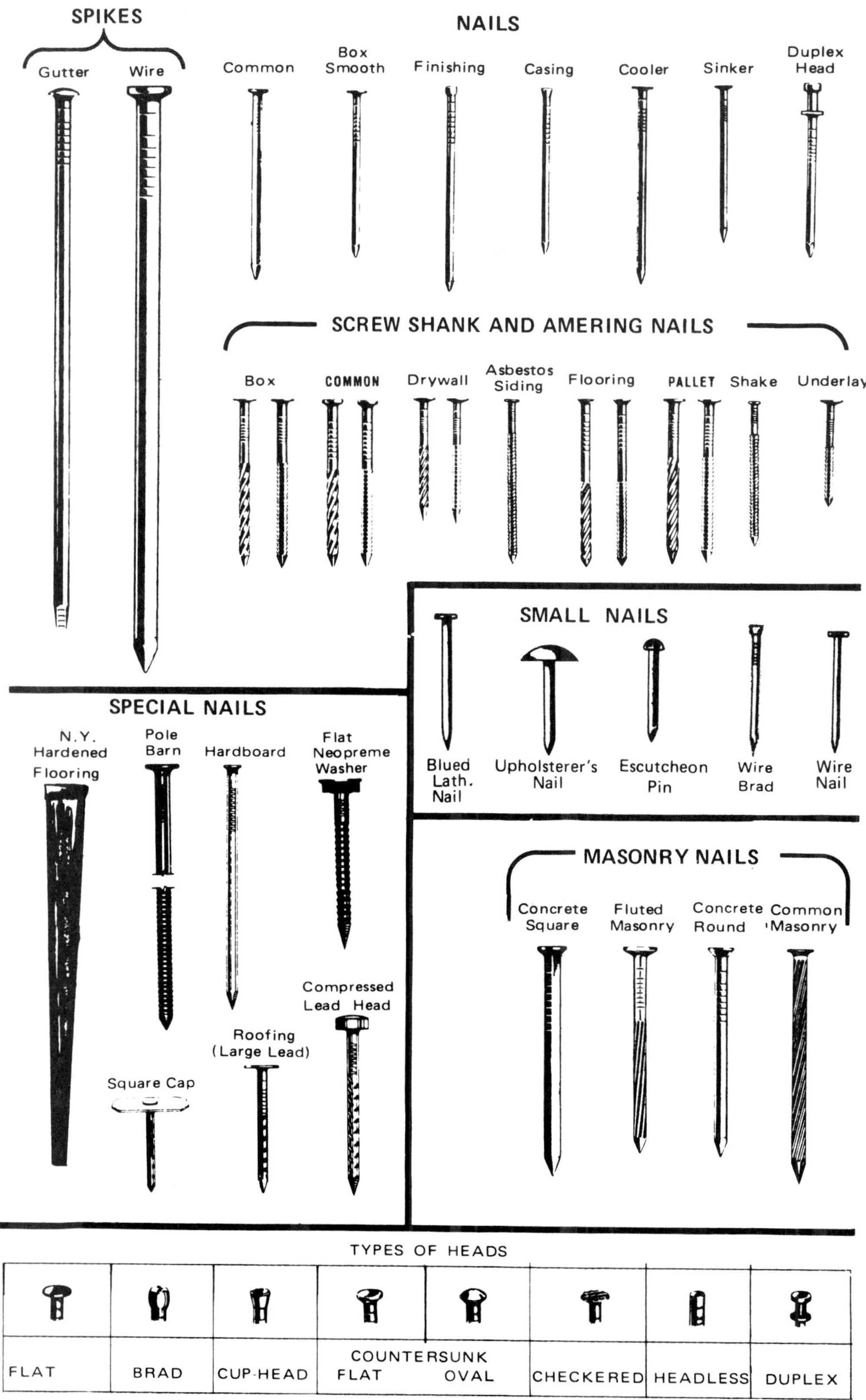

Figure 615 Nails

T nails are used in nailing guns for rough carpenter work. *Shingle nails* are used to apply cedar shingles. *Roofing nails* are used on asphalt roofing. *Duplex* or double-headed nails are used for temporary work where the nail will be removed later, as in wood scaffolding. *Brads* are small nails used for finish work. *Tacks* are very short fasteners used for surface holding. See also the following nail types: *aluminum, annular, blued, box, casing, cement coated, clasp, common, concrete, countersink, cut, drive screw, duplex, enameled, escutcheon, flat-head wire, helically grooved, hi-load, masonry, ratchet, roofing, scaffold, screw, shingle, T nail, vinyl, wire.* See also *staple, tack, wire brad.*

nail claw See *nail puller.*

nailer Air-operated nailing machine. Nails come in clips or coils and are driven at a high rate of speed. Figure 616 shows a coil-fed nailer. The coil holds 300 nails from 2″ to $2\frac{1}{2}$″ and can drive nails at rates up to 10 per second. See also *air-driven nailer.*

nail finishes Finish or coating on the nail. A bright finish is used for nails not exposed to weather. Galvanized or cadmium-plated finishes are used on nails exposed to weather. A painted finish is used when the nails must match the color of the material being nailed. Cement-coated nails are used for extra holding power. Aluminum nails are used for corrosion resistance.

nail-glued See *glued-nailed.*

nail heads Various heads are used on nails: flat, round, double, curved, button, oval, checkered, sinker, projection, cubbed oval, numeral head, countersink, headless, lettered, casing, cut head, oval-countersink, slotted, duplex. (Figure 617.) See also *vinyl-headed nails.*

nailing block Small wood piece used as a base for nailing into.

Figure 616 Nailer (air-operated)

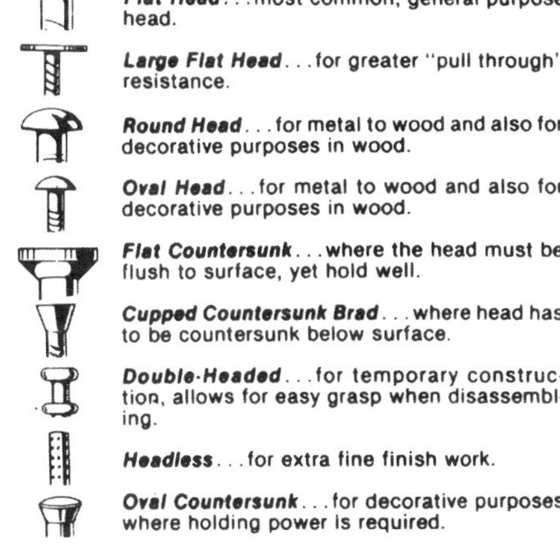

Figure 617 Nail heads

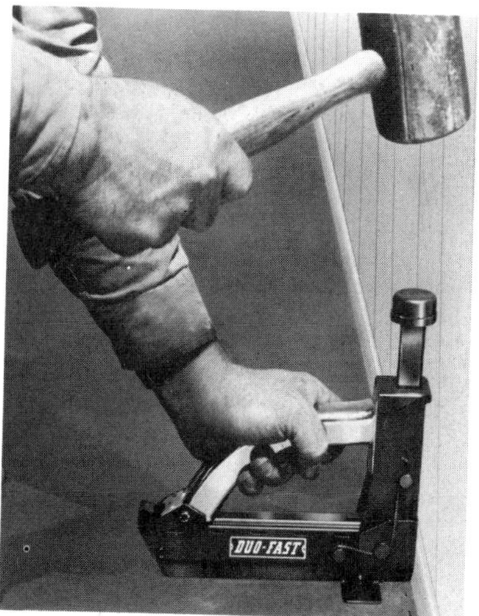

Figure 618 Nailing machine

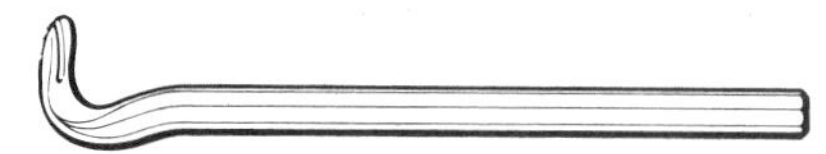

Figure 620 Nail puller

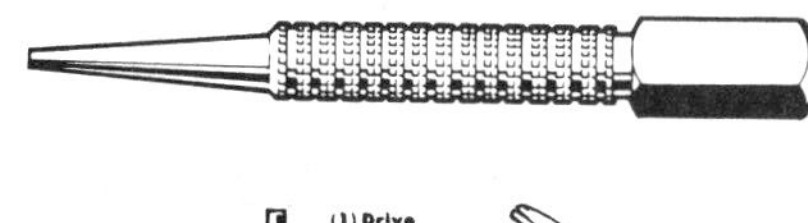

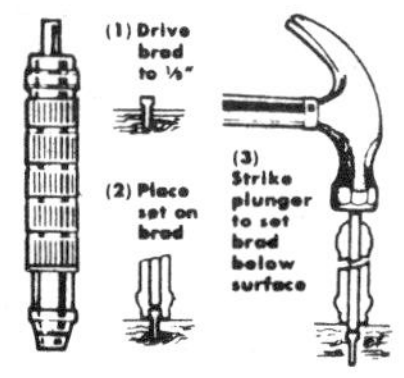

Figure 621 Nail set

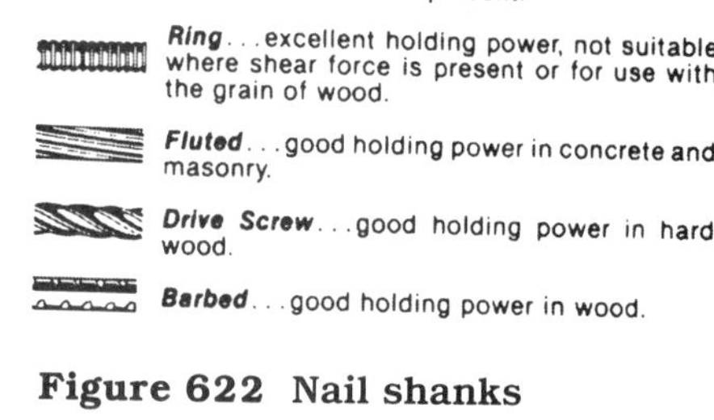

Figure 622 Nail shanks

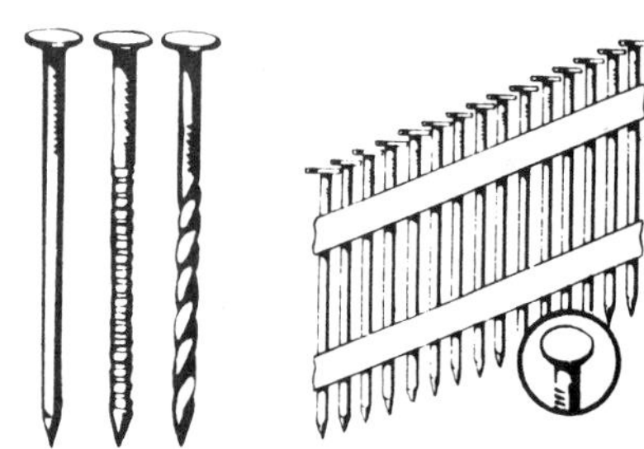

Figure 623 Nail strips

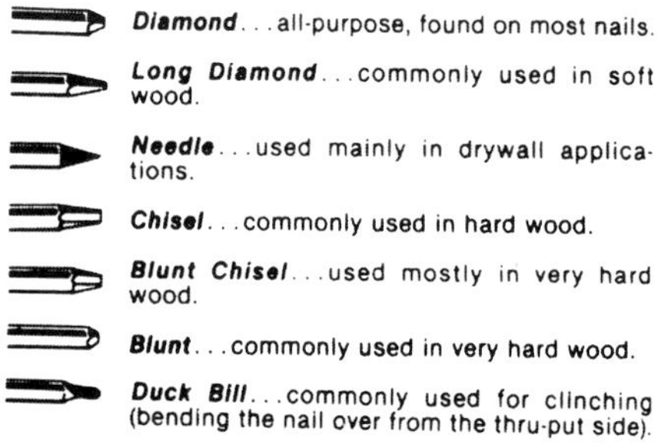

Diamond...all-purpose, found on most nails.

Long Diamond...commonly used in soft wood.

Needle...used mainly in drywall applications.

Chisel...commonly used in hard wood.

Blunt Chisel...used mostly in very hard wood.

Blunt...commonly used in very hard wood.

Duck Bill...commonly used for clinching (bending the nail over from the thru-put side).

Figure 619 Nail points

nailing machine Manual device for nailing flooring underlayments. (Figure 618.) Plunger is struck with a mallet to drive the nail. Long staples are normally used.

nailing pattern The pattern or location of nails used to fasten wood boards or panels together or to a base. Specific patterns are specified, as for the installation of gypsum panels.

nailing strip Piece of wood used as a nailing ground. See *furring strip.*

nail points Various points are used on nails: regular diamond, long diamond, blunt, round, needle, and sheared bevel. (Figure 619.)

nail pop Protrusion of the nail head above the surface. In drywall, outward motion of a nail that causes the plaster covering the nail to fall off.

nail puller Lever-type bar with a forked jaw for grasping and pulling (levering up) nails. (Figure 620.) See also *crowbar.* Also, a levering device with a movable jaw for digging into wood to reach flush nail heads.

nail punch Small punch for driving a nail head below the surface. A *nail set.*

nail set Small punch for driving a nail head below the surface. (Figure 621.) Also called a *nail punch.*

nail shank The body of the nail. Different shank shapes are used: smooth, ring (annular), fluted, drive screw, barbed. (Figure 622.)

nail strip A strip of nails used in a nailing gun. (Figure 623.) Various types and sizes are available. See also *T nail.*

National Electrical Code (NEC) A standard that covers electrical wiring and installation. Revised every three years.

natural aggregates Sand and gravel used in plastering and making concrete.

natural finish Wood finish where the actual wood itself is exposed.

natural grade The original contour and surface before grading and filling. Opposed to *finish grade* (FG), the final graded surface.

naval stores Term for tree products used in a wooden-ship, sailing navy: resins, tars, pitches, oils, and the like — products mostly derived from pine trees.

nave Main part of a church.

neat cement Pure cement and water mixture with no sand or coarse aggregate.

neat plaster Pure plaster and water with no sand.

NEC National Electrical Code; revised every three years.

needle In surveying, a compass needle; a magnetic bearing.

needle file A small, fine file used for very fine work. Needle files are usually sold in a set. See also *file.*

needle-nose pliers Pliers with a long, very thin nose. See *long-nose pliers.*

neoprene Synthetic rubber material used to separate structural members or parts — for example, as an expansion strip between a concrete slab and a concrete wall. Neoprene washers are sometimes used with roofing and paneling nails.

neutral conductor The grounded conductor in a circuit, usually having a white or gray covering.

new wood Woodworking term: wood that has never been worked before.

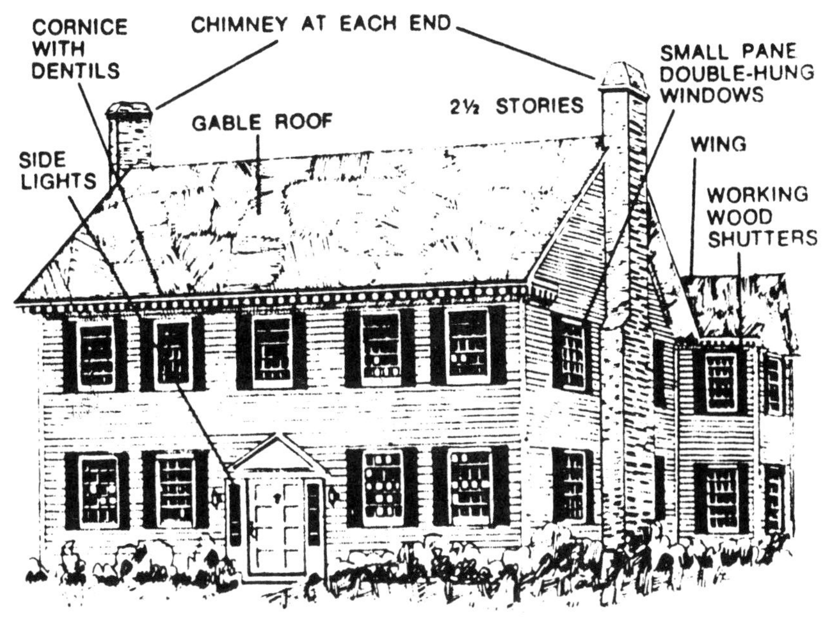

Figure 624 New England Colonial

newel Post which supports a stair handrail. See *stairs.*

newel cap Ornament on top of a newel.

newel post A *newel.*

New England Colonial Symmetrical square or rectangular two-and-a-half story house, usually with a gable roof. (Figure 624.) See also *architectural styles.*

NFPA National Fire Protection Association; publishes the *National Electrical Code* (NEC).

nibbler A power tool used to cut sheet metal. (Figure 625.)

niche A recess or hollow; a small cove.

nineteenth-century American architecture Architecture characteristic of nineteenth-century America. See *American mansard, Mission, Western row house.* See also *architectural styles.*

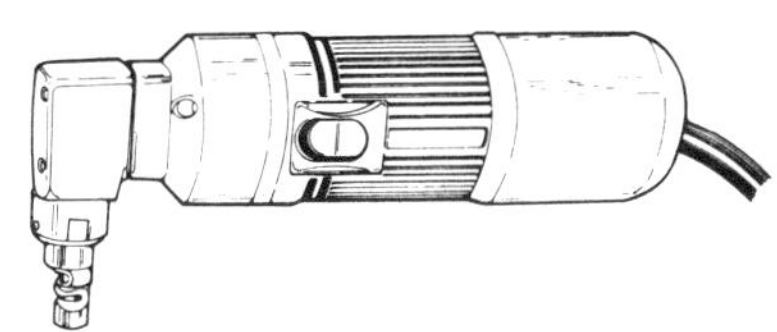

Figure 625 Nibbler

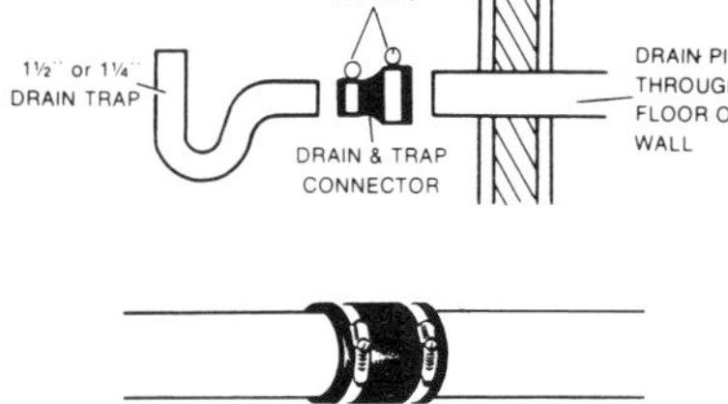

Figure 626 No-hub soil pipe

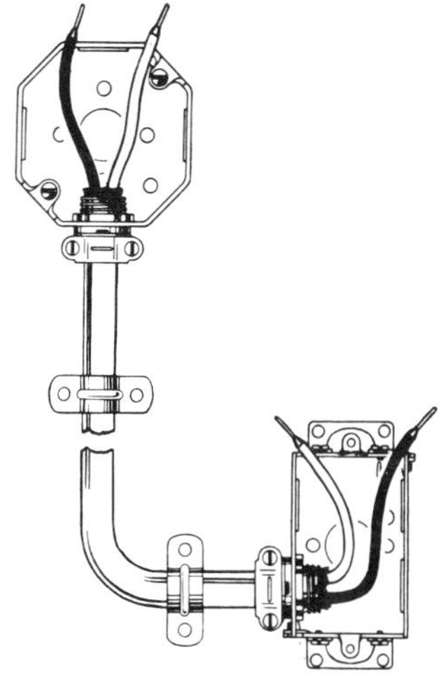

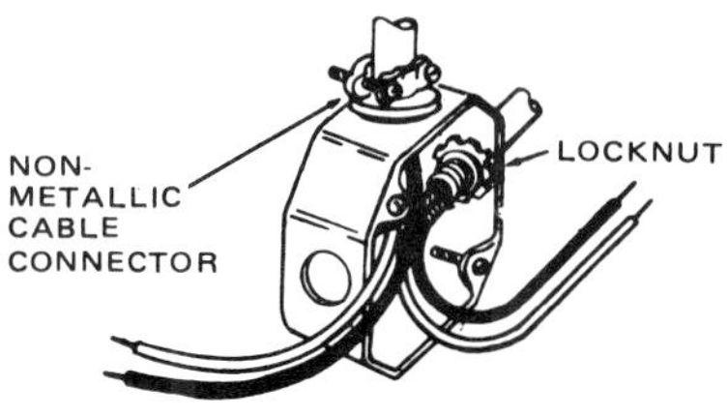

Figure 627 Nonmetallic sheathed cable

nippers Hand tool for cutting small metal pieces such as nails or wires. See *end-cutting nippers.*

nipple Short plumbing pipe externally threaded at both ends.

no-hub soil pipe A cast iron pipe used for sewer and drainage pipe and for vent piping. Sizes vary from $1\frac{1}{2}''$ to 8″ in diameter. Figure 626 shows typical pipe and fittings. The sections are joined together with a neoprene sleeve gasket and stainless steel band with screw clamps.

noiseproof Of a room, insulated to cut down sound transmission; *soundproof.*

nominal Assumed, not actual.

nominal lumber Rough overall lumber size before dressing or shrinkage. Overall nominal sizes are established by tradition and usage — for example, a 2″ × 4″ (nominal size). (*Figure 578,* p. 246.) See *actual size lumber.*

nominal size Assumed size; approximate size of a lumber piece before finishing and surfacing, as opposed to actual size. A nominal size 2″ × 4″ is actually $1\frac{1}{2}''$ × $3\frac{1}{2}''$. Nominal size is used for ordering and specifying lumber. (*Figure 478,* p. 196.)

nonbearing Term describing a support that carries only its own weight; it does not support any of the live or dead load of the structure.

nonbearing wall Wall or partition that only carries its own weight. Opposed to *bearing wall.*

noncombustible Unable to burn or be burned. Noncombustibility is established by a testing authority such as the ASTM (American Society for Testing and Materials).

nonmetallic sheathed cable A two- or three-wire insulated conductor cable covered with vinyl. (Figures *627, 232,* p. 88.) Two types are used: NM and NMC. Both types may be used in exposed or concealed work in dry locations. Type NMC may also be used in damp or corrosive locations. Also called Romex.

nometallic waterproof cable Rubber-insulated conductor enclosed in a rubber sheath.

Norman brick Kiln-baked brick that is larger than the standard building brick. The Norman brick is a $2\frac{3}{4}''$ × 4″ × 12″.

nosing The part of a stair tread that extends beyond the face of the riser. See *stairs.*

notch "V" cut made in a wood piece.

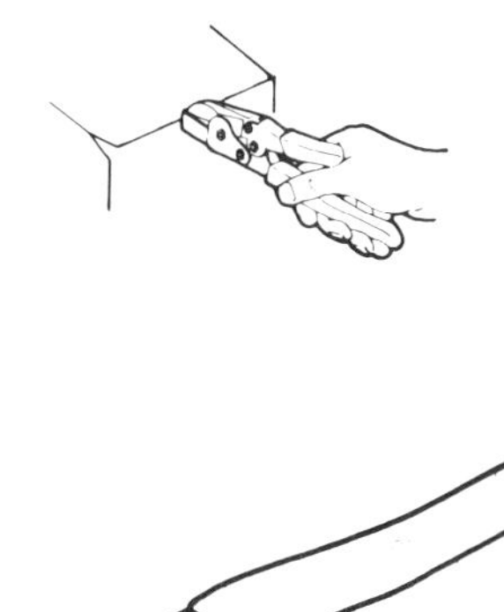

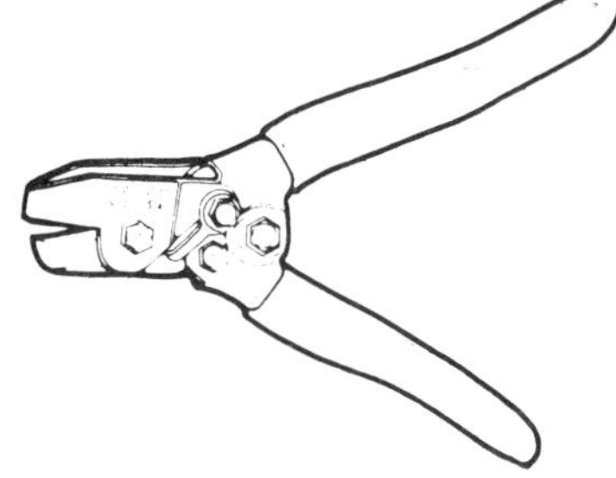

Figure 628 Hand notcher: sheet metal

notchboard A *housed stringer;* board that is notched or housed out to receive the ends of treads and risers.

notcher Hand tool used for cutting notches in sheet metal. (Figure 628.)

note Information printed on a blueprint.

nut Internally threaded fastening device that runs onto the threaded end of a bolt or stud. The nut hole diameter must be the same as that of the bolt end, and the threads per inch must be the same. Figure 629 shows typical nuts. A *washer* may be used underneath the nut to distribute the load and give a more secure friction fit. See *bolt, washer.*

nut driver Screwdriver-shaped tool used for driving hex nuts. (Figure 630.) Nut drivers normally come in a set ranging from $\frac{3}{16}''$ to $\frac{1}{2}''$ or from 5 to 11 mm.

nylon anchor Expandable anchor for holding screws in plaster or drywall construction. (Figure 631.) See *anchor, plastic anchor.*

nylon expansion anchor Plastic-type anchor that expands to form a tight fit when in place, used in drywall construction. See *molly anchor, plastic anchor.*

SIZES: Select nut with same diameter and threads per inch as machine screw or bolt you're using.

TYPES

Hex Square

Hex Cap Wing

Diameter (Refer to by number)	Threads per inch	Outside diameter
2	56	3/16
3	48	3/16
4	36 or 40	1/4
5	40	5/16
6	32	5/16
8	32	11/32
10	24 or 32	3/8
12	24 or 32	7/16
1/4	20 or 28	7/16
5/16	18 or 24	9/16
3/8	16 or 24	5/8

Use this size wrench for hex or square nuts (sizes 3 through 5/16)

Figure 629 Nuts

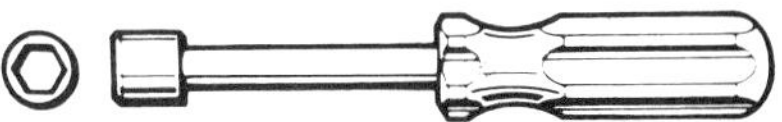

Figure 630 Nut driver

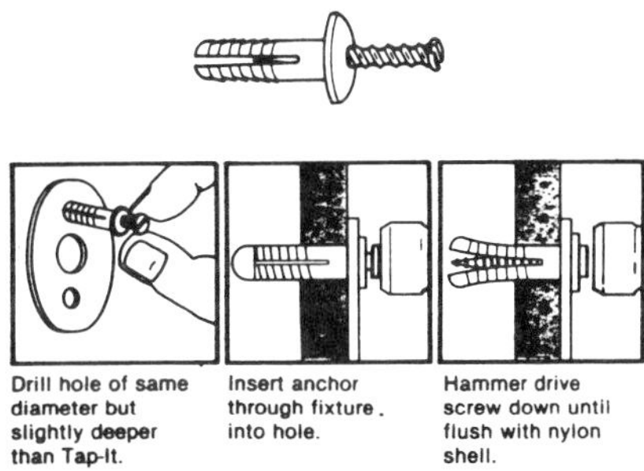

Figure 631 Nylon anchor with screw

O

oakum Untwisted rope fiber used for caulking plumbing joints.

oblique drawing Drawing with lines in the front view in true proportion (a full orthographic view) but with the sides drawn at an angle. (Figure 632.) See *pictorial drawings* for a comparison to other types of drawings.

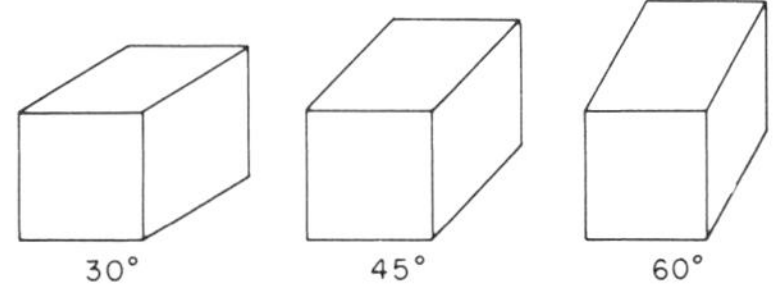

Figure 632 Oblique views of a rectangular block

oblique perspective A three-dimensional drawing that has three vanishing points. Also called *three-point perspective.* See *perspective drawing.*

obtuse angle Angle greater than 90° but less than a straight line. See *angle.*

octagon An even eight-sided figure.

octagon table Rafter framing scale or table on a *rafter square.*

off center. Not on center; not centered.

offset The setting back of one part from another. An L-shaped house has one wall set back (offset) from another.

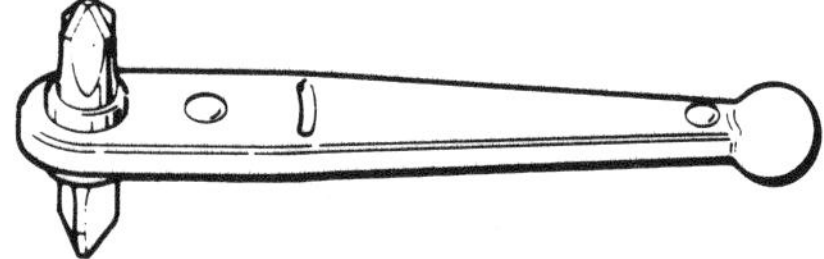

Figure 633 Offset ratchet screwdriver (Phillips-head screws)

Figure 634 Offset screwdriver (slotted screws)

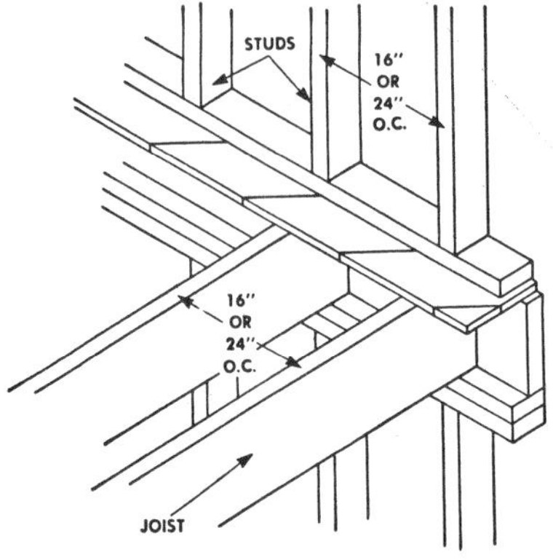

Figure 635 On-center spacings

offset ratchet screwdriver Ratchet screwdriver with end angled at 90°, used to reach tight places. A flat tip and a Phillips tip are used with the ratchet action. Ratchet allows screw to be turned without removing the tip from the screw head. (Figure 633.) See also *offset screwdriver.*

offset screwdriver Screwdriver with the end angled at 90° for reaching awkward areas. (Figure 634.) See also *offset ratchet screwdriver.*

offset wrench Wrench with the head at an angle, used for reaching difficult places. See *wrenches.*

off-site construction Building of house parts off the site (in a factory) to be delivered to the site as units and erected. Opposed to *on-site construction.*

off-site improvements Improvements made near a property that add to its value, such as paved streets, sewer system, etc.

ogee Profile roughly in the form of an S; metal gutters commonly take this shape. Abbreviated O.G. (See *Figure 84*, p. 33, for ogee shape.)

ohm Electrical resistance in a circuit. The symbol for ohm is the Greek capital letter omega (Ω).

$$\text{ohms} = \frac{\text{volts}}{\text{amperes}}$$

Ohm's law Relationship of electrical units of amperes, volts, and ohms as a function of each other:

$$\text{amperes} = \frac{\text{volts}}{\text{ohms}}$$
$$\text{ohms} = \frac{\text{volts}}{\text{amperes}}$$
$$\text{volts} = \text{amperes} \times \text{ohms}$$

oil finish In woodworking, a wood finish made with a penetrating oil, such as boiled linseed oil or tung oil. Oil finishes are wiped on the raw wood and flow into it. See *penetrating oil finish.*

oilstone Fine sharpening stone that is lightly covered with oil when sharpening an edge. To develop a razor-sharp edge three stones are used: coarse, medium, and fine. The oil is used to float off the fine metal bits worn off the edge being sharpened. The most popular oilstone is the Arkansas stone. Silicon carbide and aluminum oxide are also used. See *waterstone.*

oil varnish Varnish with an oil base.

old English bond See *English bond.*

old growth Timber harvested from a natural, mature forest.

old wood Woodworking term: wood that has been worked before.

on center (O.C.) Measurement from the center of one member to the center of a corresponding member; *center to center.* Commonly used in stud, joist, rafter, and truss layout, as 2 × 4 studs 16″ O.C. or 24″ O.C. (Figure 635.)

one-and-a-half story construction House with a full first story and a partial second story developed under the roof. The *Cape Cod* style is typical of this type of construction. See *architectural styles.*

one-pipe heating system In hot-water or steam heating, a system that uses one continuous pipe to carry heat to the outlets and back to the furnace. (Figure 636.) See *two-pipe heating system.*

one-point perspective A three-dimensional drawing that has one vanishing point. Also called *parallel perspective.* See *perspective drawing.*

one-way floor system In reinforced concrete, a slab supported by reinforced concrete joists or girders that run in one direction only. (Figure 637.) Also, floor made with reinforced precast slabs that run in one direction only. Also called a one-way joist system. See *reinforced concrete floors, two-way floor system.*

one-way joist system See *one-way floor system.*

one-way solid slab Solid reinforced concrete slab floor with beams or bearing walls underneath running one way. (Figure 638.) See *two-way solid slab.*

on grade Located at the same height as the grade established as the ground level of a building.

on-site construction A house built on the site from individual building materials. Each piece is cut to fit. Opposed to *off-site construction.*

open circuit In electrical wiring a circuit that is broken and does not have a flow of electricity; a dead circuit. Opposed to a *closed circuit.*

open construction Building that can be inspected at site without damage or disassembly.

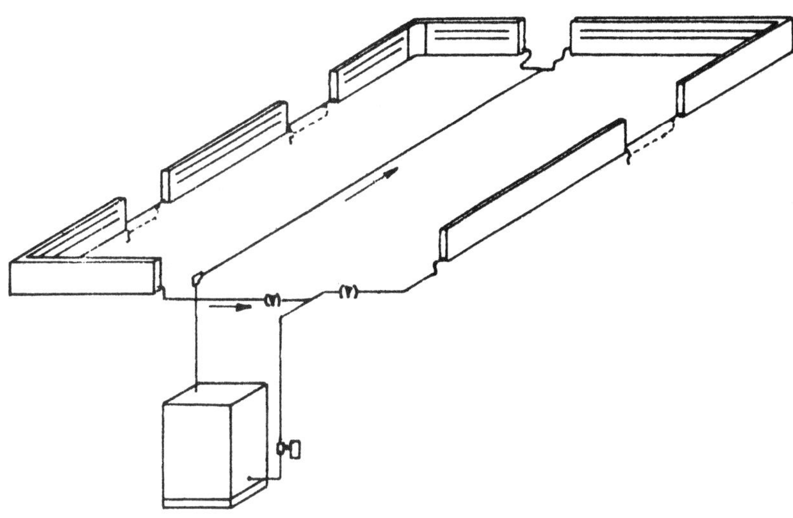

Figure 636 One-pipe heating system

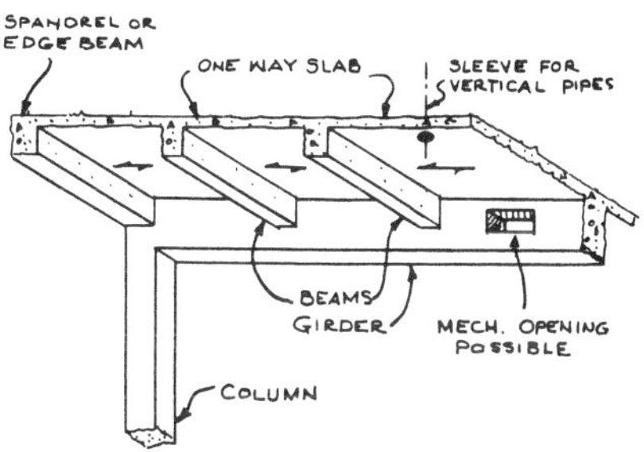

Figure 637 One-way floor system

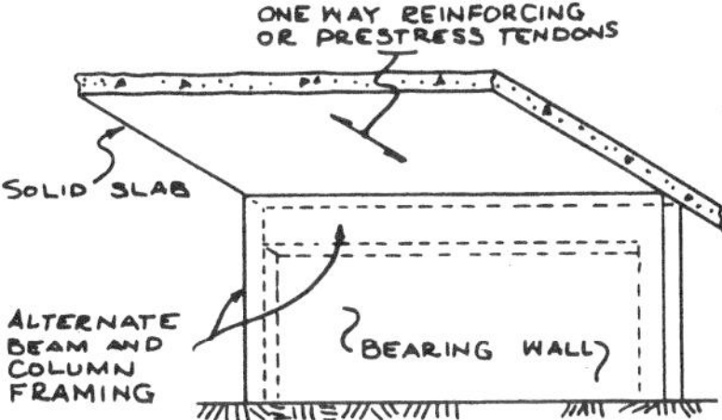

Figure 638 One-way solid slab

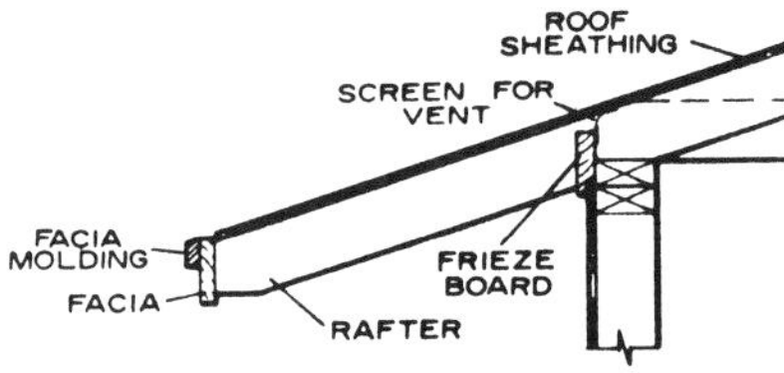

Figure 639 Open cornice

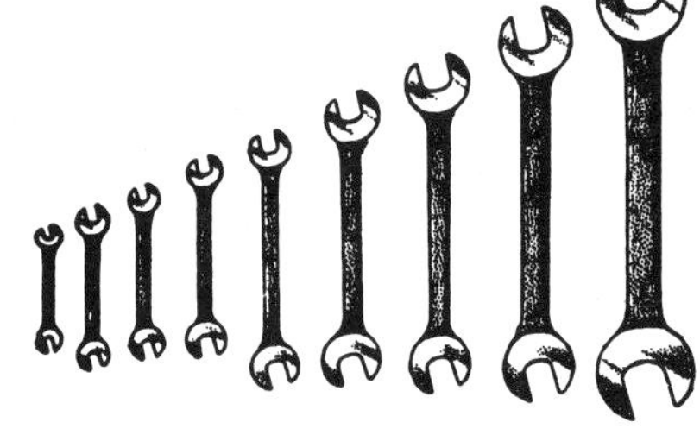

Figure 640 Open-end wrenches

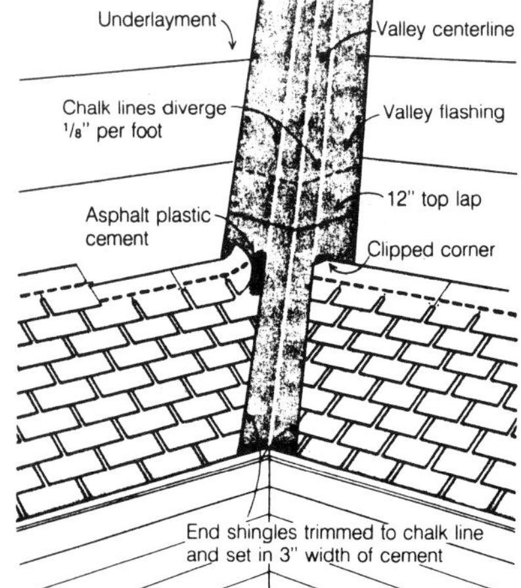

Figure 641 Open valley with flashing

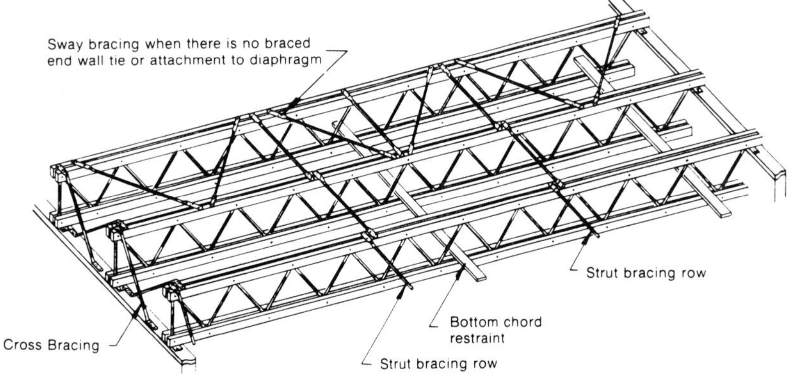

Figure 642 Open-web joist

open-corner fireplace Fireplace at an exterior corner in a room so that it is open on two adjacent sides.

open cornice Cornice where the rafters are exposed. (Figure 639.) Opposed to a *closed cornice.*

open-end wrench Wrench with an open-end jaw. (Figure 640.) See *wrench.*

open grain Wood with a coarse texture.

open-loop solar system Solar-heating active system that circulates household water in a solar collector to heat the water directly. This system is called a direct system. See *closed-loop solar system, thermosiphoning.*

open stairway Stair open on one or both sides. Opposed to a *closed stairway.*

open-string stairs Stair open at the bottom so as to expose the stringers, used as basement stairs.

open time In adhesive work, the length of time an applied adhesive will still form a bond before it dries or sets and material can no longer be applied.

open traverse In surveying, a *traverse* where the connected lines do not come back together to the starting point. See also *closed traverse.*

open valley Open runway made at the intersection of two roof slopes. (Figure 641.) The flashing is exposed. See also *closed valley.*

open-web joist A manufactured joist consisting of top and bottom chords with metal cross braces. Joist depths vary from 14″ to 63″; spans run up to 70′. Figure 642 shows three open-web joists joined together as part of a roof system. Also called a *floor truss, truss joist.* See *I joist.*

Figure 643 Orbital sander

open-web studs All-metal studs constructed with a cross support (web).

operating engineer One who works with construction machinery: backhoes, hoists, bulldozers, power shovels, pile drivers, and so on.

orange peel Paint that raises up to form a rough surface like an orange peel.

orbital sander Portable power sander used for sanding flat surfaces and for finishing. (Figure 643.) A standard abrasive paper is mounted over the orbiting base. Because the sandpaper extends over the edge of the base, it can be used for sanding into corners. Also called a *pad sander.*

orders From the classical period, types of column and entablature. The three types of classical Greek orders are Doric, Ionic, and Corinthian. (Figure 644.) At a later period, in Roman times, two additional orders were added: Tuscan and composite. See also *entablature.*

orientation A direction, as the direction a building faces. The concept of directing the house position to enjoy a good view, a quiet area, the winter sun, or whatever. In solar heating, the number of degrees east or west of south that a solar collector faces.

orifice Small opening, as in a paint gun.

ornament Decorative detail.

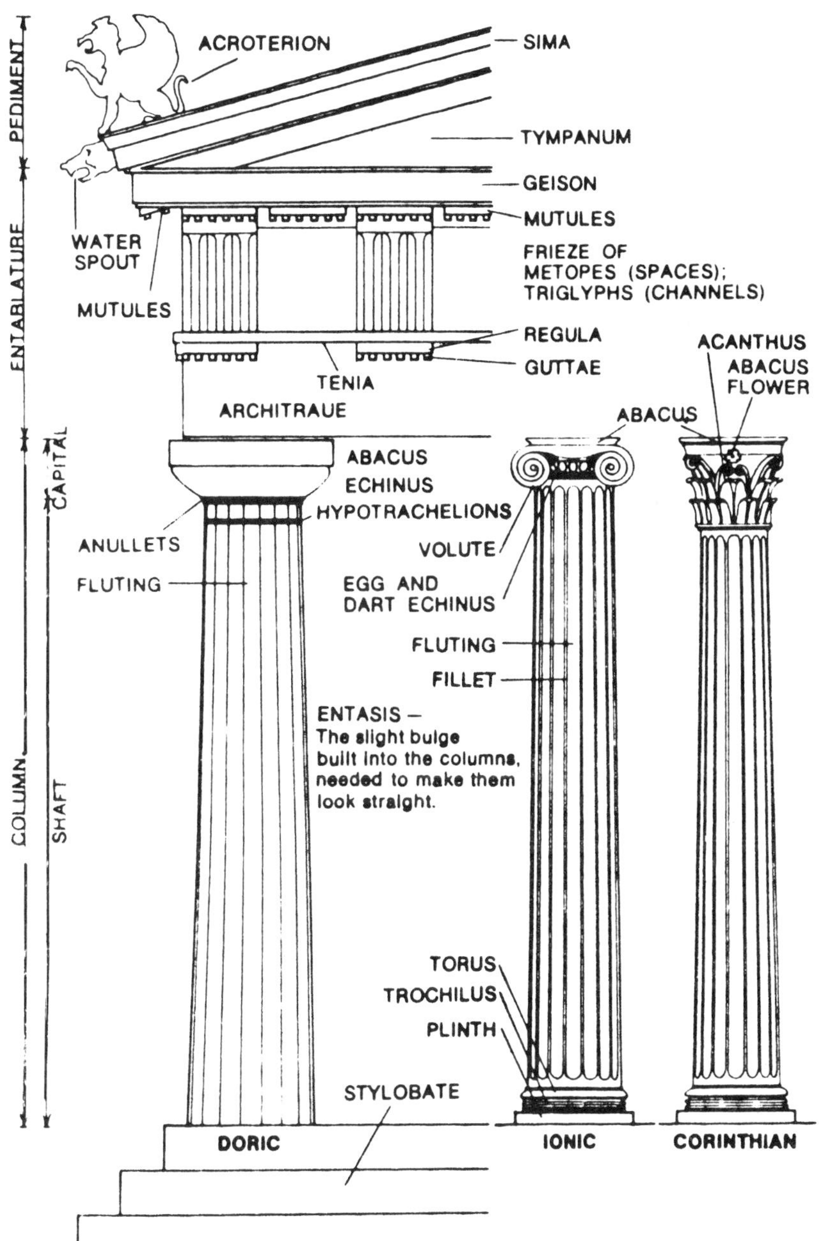

Figure 644 Orders

Figure 645 Ornamental tools

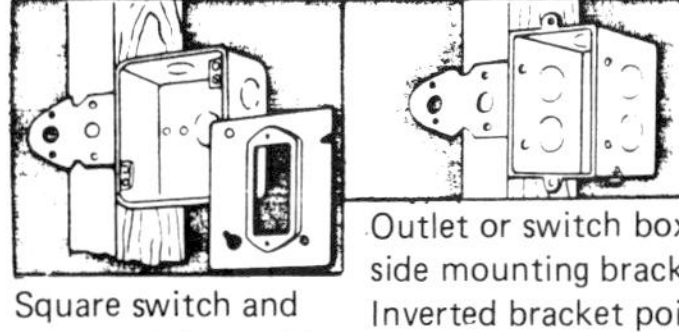

Square switch and receptacle box with side mounting bracket.

Outlet or switch box with side mounting bracket. Inverted bracket points and nails or screws hold the box firmly.

Octagon box. Fixture or junction outlet.

Switch box. Two or more may be ganged together by removing side plate on each box.

Figure 647 Outlet boxes

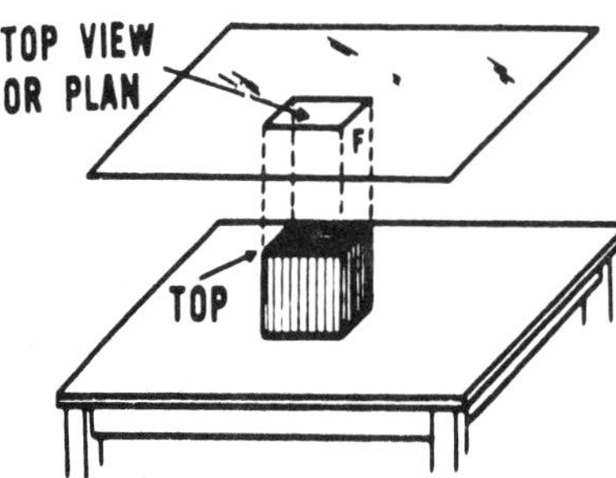

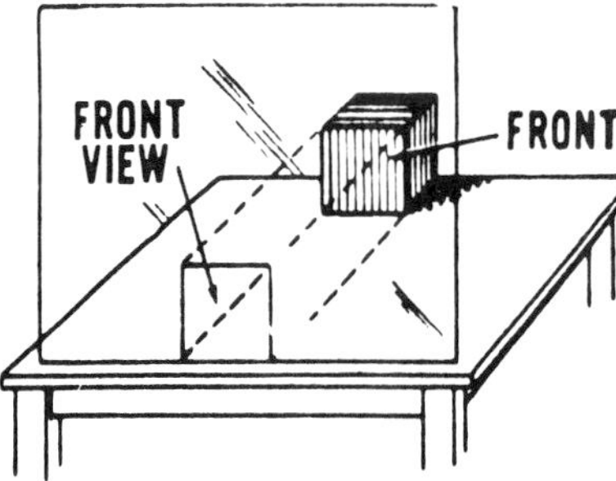

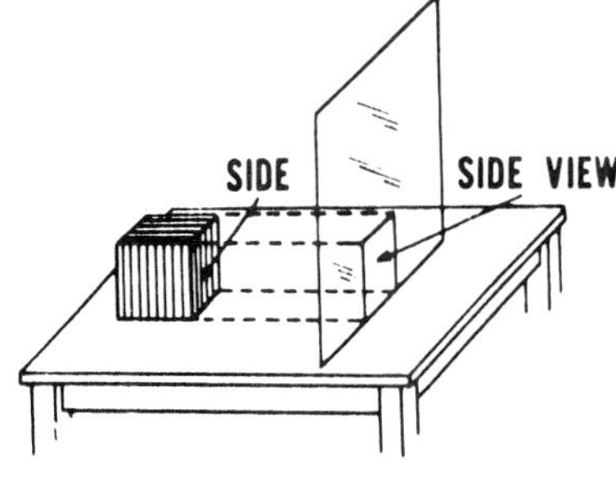

Figure 646 Orthographic projection

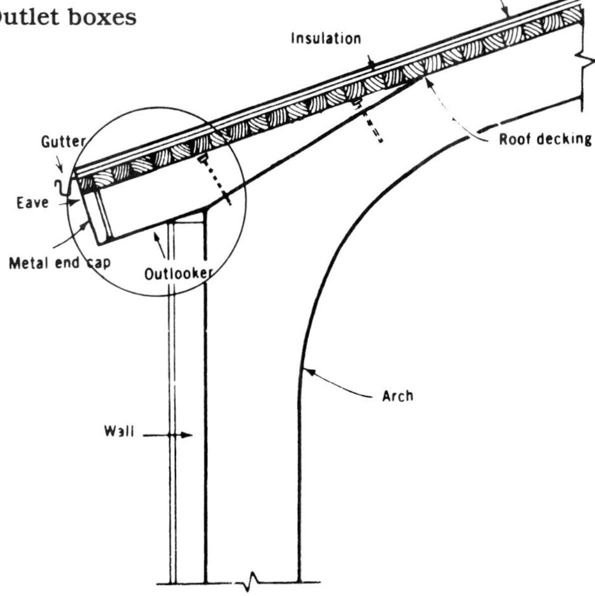

Figure 648 Outlookers

ornamental hinges Exposed, ornately shaped hinges used in cabinetwork. See *hinges.*

ornamental tools Fine tools used for finishing ornamental plaster work. (Figure 645.)

orthographic drawing In drafting, a view made looking straight at the building, as a plan view or an elevation. (Figure 646.) See also *pictorial drawings.*

orthographic projection In drafting, drawing the object looking straight at it, as if imaginary lines were projected out to a viewing plane. (Figure 646.)

orthographic view View or drawing of something taken looking straight at the object. View straight up from an object and parallel to the object. View made at a right angle to a line projected straight up from an object.

outer stringer In stair building, the stringer furthest away from the wall.

outlet Electrical receptacle; the point in an electrical system where current is removed to run electrical lighting, appliances, or equipment. A *power outlet.* See *receptacle.* Also, an opening that directs the discharge of a liquid.

outlet box In electrical wiring, a box used to mount receptacles, switches, and lights. (Figure 647.)

outlet-box connectors Connectors used to join cable or conduit to electrical outlet boxes.

outline In drawing, a line that follows the outside edge of a figure or object.

outline specifications Short technical specifications that give a brief list of materials to be used. See *specifications.*

outlooker Roof extension beyond arch support at wall. (Figure 648.) See also *lookout.*

out of center Not properly centered; off center.

out of plumb Not plumb, not truly vertical.

out of square Not true, not at right angles.

out of true Irregular; not properly aligned or constructed; not exact.

out of wind Masonry work that is true and has no curves or hollows.

outriggers In roof construction, framing members that extend the gable roof end. They are cut into the frame over the gable end rafters back into the side of the next rafter. Compare to *lookout.*

outside caliper Calipers with legs that turn inward, used for measuring outside diameters, such as a pipe. See *calipers.*

outside casing Exterior window or door trim.

outside diameter (O.D.) Diameter of something measured from the outside surface, as the outside diameter of a pipe. Outside diameter does not take into consideration the pipe wall thickness. Opposed to *inside diameter.*

outside leg In steel construction, the leg of an angle pointing away from the structural member. The other leg (web leg) is attached to the web of the member. Often used as a seat for a beam. See *web leg.*

overall Total measurement, as overall width, overall length.

overburden Earth, gravel, or sand on top of solid rock.

overcurrent Electrical current in excess of rated capacity; caused by overload, short circuit, or grounding.

overcurrent device A *circuit breaker* or *fuse.*

overflow pipe Pipe that carries off excess water.

overhand work In bricklaying, the laying of face bricks in a wall from the *inside* of a building; the mason reaches over the face of the wall.

overhang In roof construction, the distance a rafter runs past the wall top plate, measured along the edge of the rafter. Opposed to *projection,* which is measured straight out from the wall to rafter end. (Overhang and projection are sometimes used interchangeably, however.)

overhanging eaves Eaves that project out from the wall two or more feet. Often used to provide shade during the summer and cut down the temperature inside the house.

overhaul To repair by fixing or replacing parts.

overhead door Door that slides up on an overhead track, as a garage door. Also called a *roll-up door.*

overhead-position welding Weld made from the underside of a joint. See *welding positions.*

overlap Lapping of one part over another. In arc welding, a sagging over of the weld bead without fusing to the base metal, caused by too low a current. See *lap.*

overload Operation of equipment or electrical circuits in excess of rated capacity; causes overheating and possible failure. Normally, the circuit breaker will trip or the fuse will blow. In structural design: allowing a load that is greater than the maximum design load of the support.

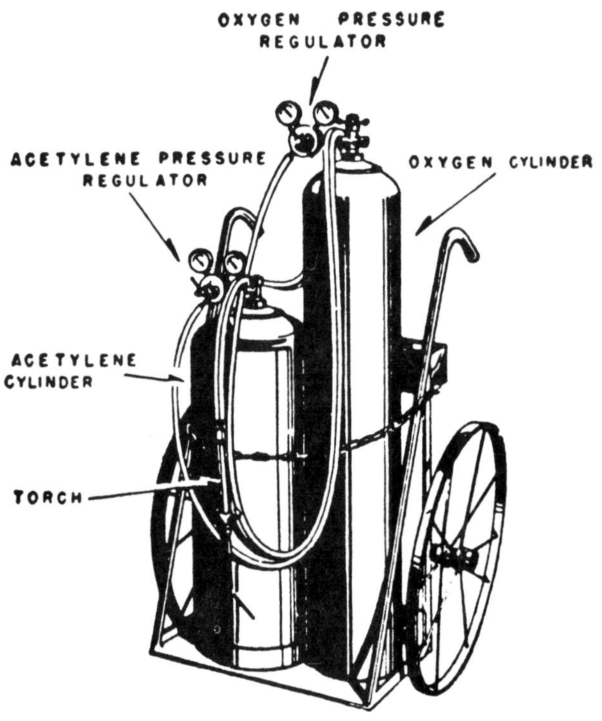

Figure 649 Oxyacetylene welding outfit

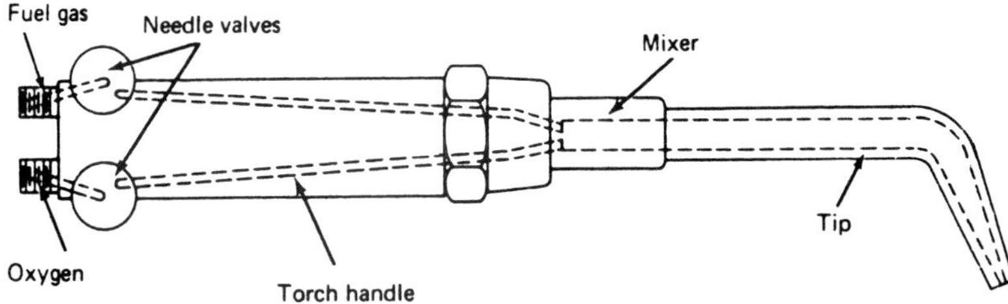

Figure 650 Oxyfuel welding torch

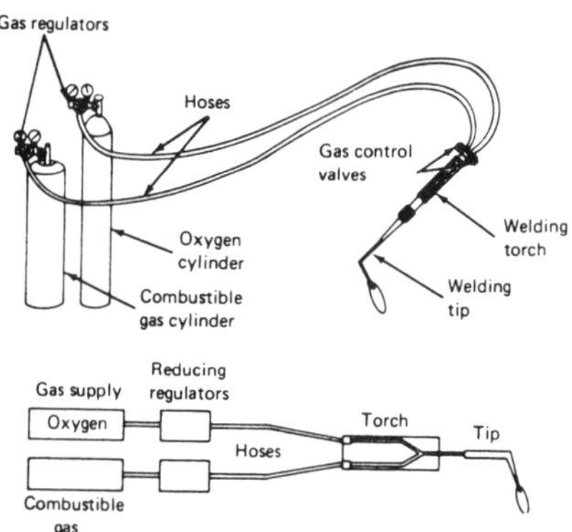

Figure 651 Oxyfuel welding equipment

ovolo Quarter-round molding.

oxidation Chemical change in which oxygen combines to form a new compound; rusting.

oxyacetylene welding Welding with a flame produced by burning an oxygen and acetylene gas mixture. The gases come in pressurized cylinders or bottles. Figure 649 shows a typical portable setup. *Regulators* control the mixture of gases. A *welding torch* directs the flame. (Figure 650.) See *oxyfuel welding.*

oxyfuel welding Welding process where oxygen and a combustible gas, such as acetylene, propane, or natural gas, are mixed together to produce a hot welding flame. (Figure 651.) Commonly referred to as *oxyacetylene welding.*

oxygen A gaseous element that composes 23% of the earth's atmosphere. Used in oxyfuel (oxyacetylene) welding and cutting. It combines readily with other elements to form oxides and therefore is used to oxidize or burn through metal, such as steel.

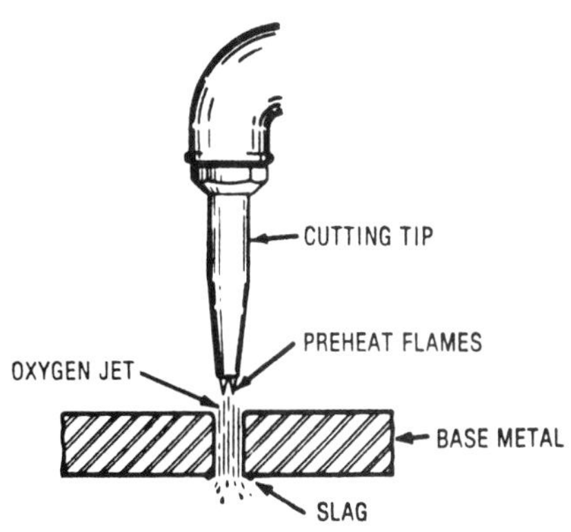

Figure 652 Oxygen cutting

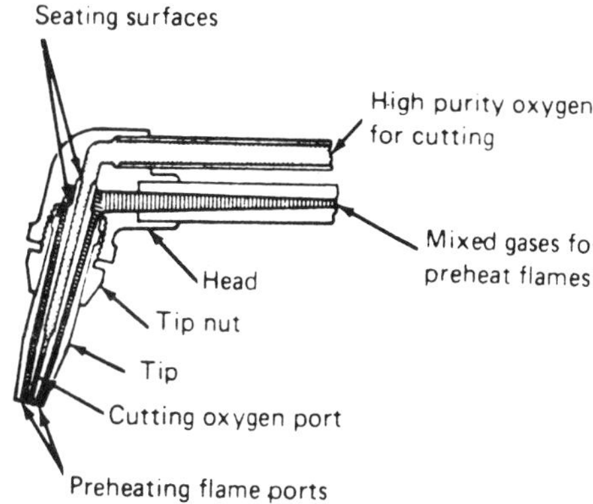

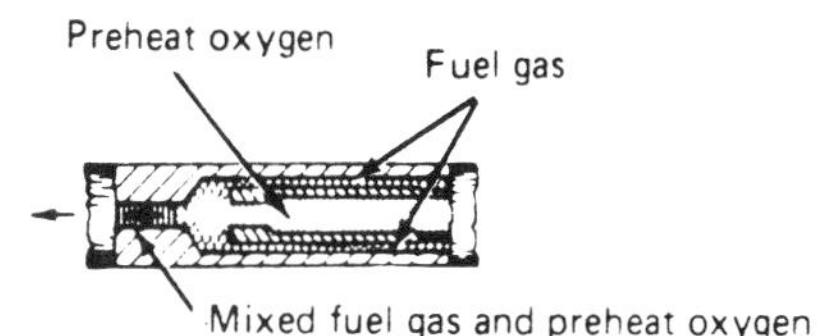

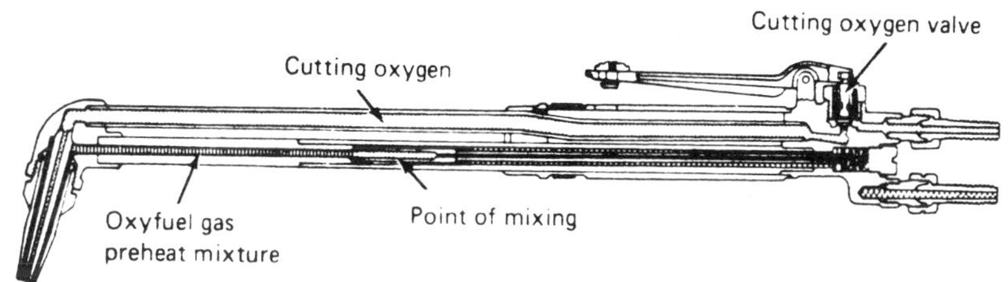

Figure 653 Oxygen cutting torch and tip

oxygen cutting (OC) Cutting metal by heating with a flame to a high temperature and using oxygen to quickly oxidize the metal to produce a cut. (Figure 652.) Also called acetylene cutting or gas cutting. Figure 653 shows the torch and tip.

oxy-MAPP torch A lightweight brazing-welding system that uses oxygen and MAPP gas to develop a flame. An oxygen and a MAPP cylinder are used together to produce the flame.

oxypropane torch A lightweight brazing-welding system that uses oxygen and propane to develop a flame. Two canisters are used and the gases mixed together. (Figure 654.) Widely used for brazing and silver soldering.

oylet Small hole; an *eyelet.*

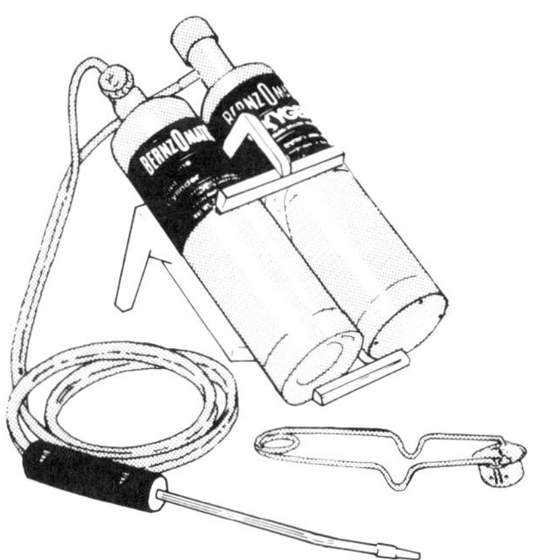

Figure 654 Oxypropane torch

padding In welding, building up an area by depositing several layers of weld beads. Used to build up worn surfaces.

pad sander See *orbital sander.*

pailing Vertical form sheathing.

paint Pigmented finish designed to completely obscure and cover the wood or metal surface. Oil and latex bases are used.

paint base Medium or vehicle which carries the paint pigment.

paint brush Handled painting device with bristles, used for laying and spreading paint, stain, varnish, lacquer, and the like. Brushes are usually flat and come in various sizes and thicknesses. A metal ferrule holds the bristles in place. The most common brushes are the wall brush (normally 4″ wide), flat varnish brush (1″ to 3″ wide), flat sash and trim brush, angular sash and trim brush (bristles have a chisel shape and run at an angle), and the oval sash brush ($\frac{1}{2}$″ to 2″ wide) used for trimming. (Figure 655.) See also *paste brush, smoothing brush.*

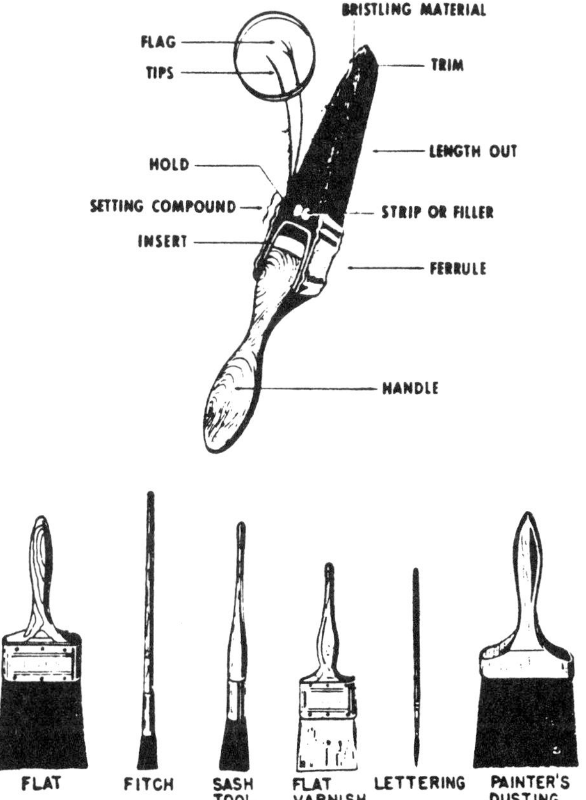

Figure 655 Paint-brush terminology

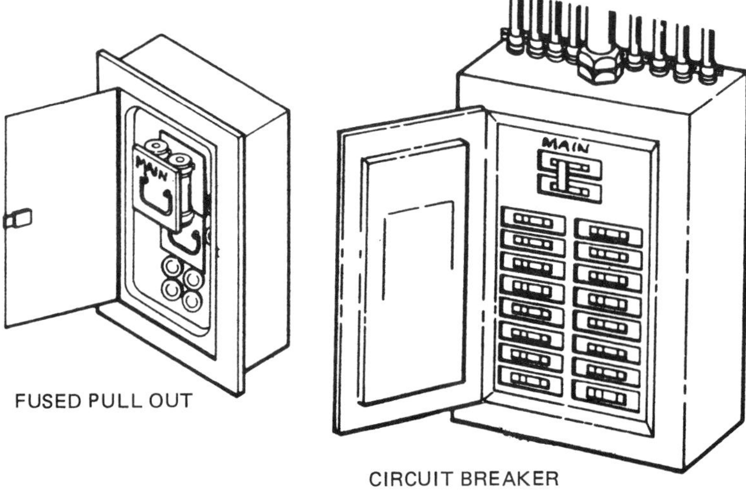

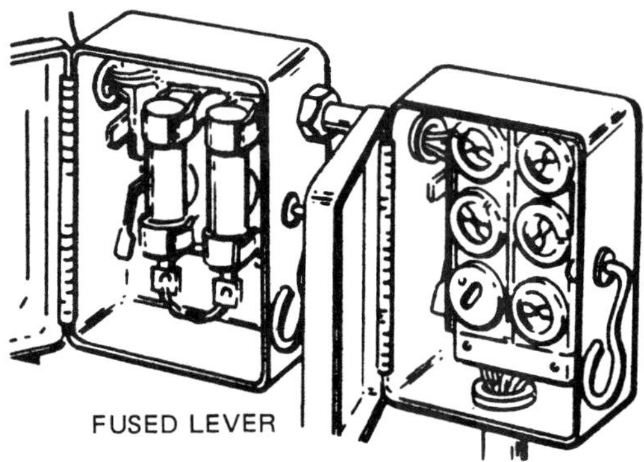

Figure 656 Panelboards

painter One who prepares and applies paint.

paint pad Painting device with a foam rubber pad. The pad takes up the paint, which is then spread on the surface being painted.

paint roller Roller with a handle used for rolling on paint.

paint spraying See *spray can, spray gun.*

paint thinner Additive used to make paint more liquid and easier to apply. Turpentine may be used to thin oil paint. Different types of paint take different thinners.

pan In reinforced concrete, pan-shaped metal or fiberglass forms or molds used to form the underside of a concrete slab floor, as in the *two-way floor system* or *waffle slab.* Also called *dome form.*

pane A window glass; a *light.*

panel A thin, large-surface building part, as a plywood or hardboard panel. Building panels are normally 4′ × 8′. Part of a door between *stiles* and *rails.* Large, sunken area in a wall or ceiling. See *panel molding.* Also, an individual precast concrete wall unit. A complete exterior metal covering or wall system, normally 4′ × 8′, with insulation and exterior finish in place; exterior panels are attached directly to the wall support members. See *cladding.*

panelboard Electrical control center or main distribution panel which receives outside power and distributes it to the various branch circuits. Each branch circuit has an individual circuit breaker or fuse. (Figure 656.) Panelboards are marked with the voltage and current rating and the number of phases (single-phase or three-phase). They are also marked for the type of conductor to be used:

copper (CU) or aluminum (AL) or both. A panelboard, by code, is not allowed to have more than 42 overcurrent devices. Recognized by several names, including load distribution center, load center, main control center. Some panelboards also include the main circuit switch and circuit breaker for the building service. See also *branch circuit, circuit breaker.*

panel box A *panelboard.*

panel building system Construction system that uses precast panels which are erected and fastened together at the site to form the building. See *precast concrete.*

panel clip Metal device for joining two edges of plywood together. (Figure 657.)

panel door Door with *stiles* (vertical members) and *rails* (horizontal members) that hold wood panels. (Figure 658.) See *door.*

panel forms Preformed concrete forms made from plywood sheets (plyforms), fiberglass, or metal.

panel heating Heating by radiation and convection from a large surface area, as a ceiling, wall, or floor. Heating coils are embedded in the panel to heat the total area. Hot-water coils or electrical resistance wire is used.

panelized construction Construction using completed wall sections. Wall section will contain windows and doors. Windows or door sections are constructed on a 4″ module.

panel lifter Foot-powered mechanical device for lifting panels slightly off the floor during installation, as drywall panels. (Figure 659.)

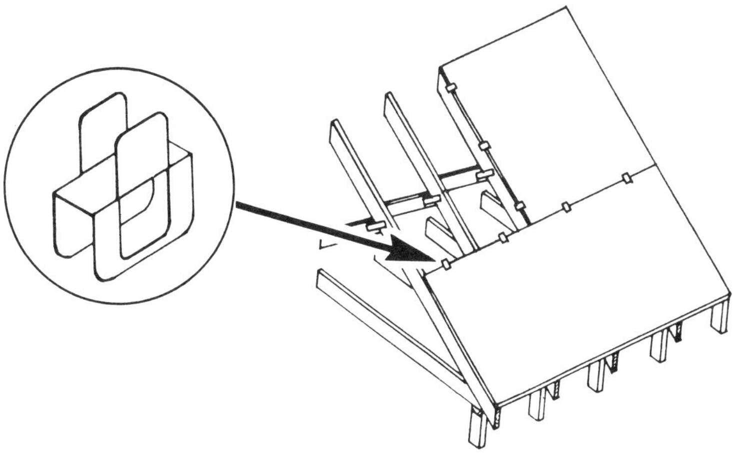

Figure 657 Panel clips

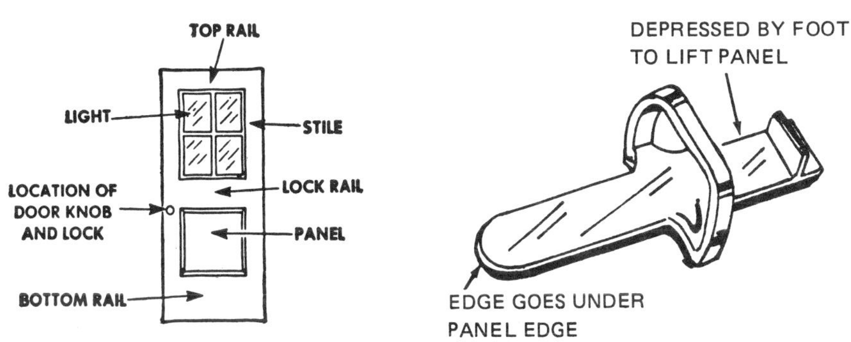

Figure 658 Panel door

Figure 659 Panel lifter

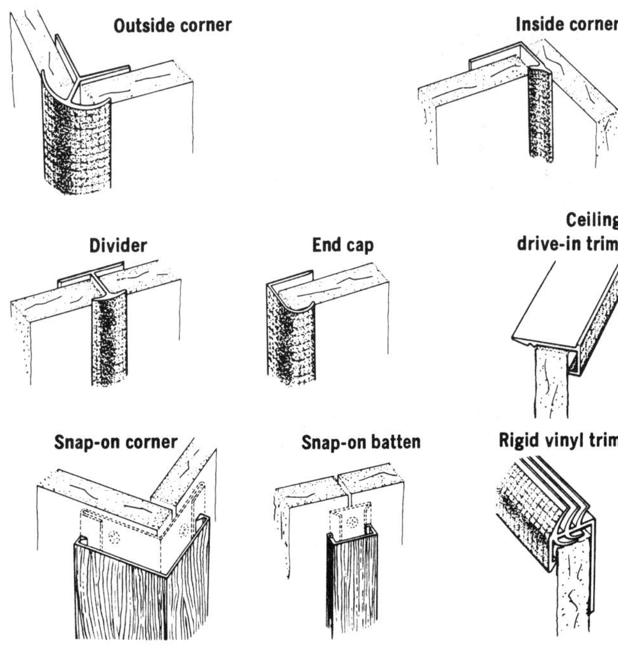

Figure 660 Panel molding

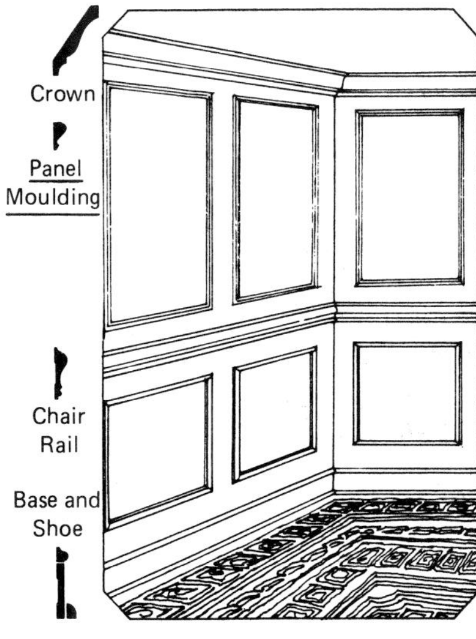

Figure 661 Panel molding

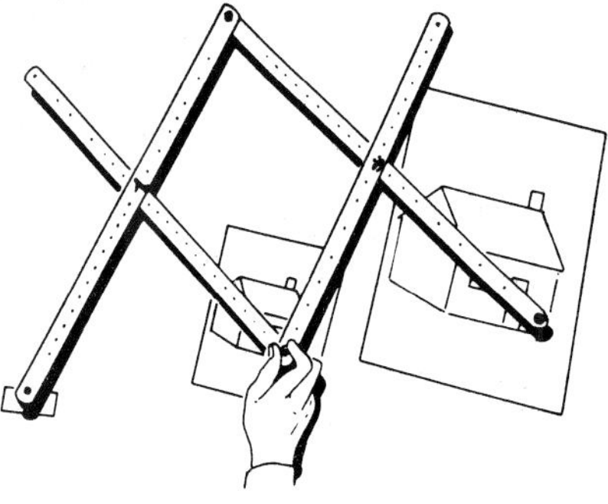

Figure 662 Pantograph

panel moldings Vinyl and metal moldings used to finish interior, prefinished panels. (Figure 660.) Molding around wood panels in an interior finish. (Figure 661.)

panel package house A complete package of materials to build a house, including floors, walls, roof, doors, plumbing, wiring, heating, cabinets, interior trim, and paint.

panel schedule Size, marking, and weights of reinforced prestressed panels.

pan floor A reinforced concrete slab that has pan-shaped depressions on the bottom side. The depressions create a criss-cross of continuous joist support. This type of floor is called a *waffle slab* or *two-way floor system.* See *reinforced concrete floors.*

pan forms See *pan.*

panic exit bar Push bar used on exterior doors in public structures; door opens when the bar is pushed.

pantile Clay roofing tile with a double concave/convex cross section like a flat "S." The tile has a small hole in one end for laying. Tiles are laid in an overlapping pattern. See *shingle.*

pantograph Device used to copy a drawing to a smaller or larger size. As you trace the illustration from the original, it is copied either larger or smaller. (Figure 662.) See also *proportional dividers.*

paperhanger One who installs wallpaper, vinyl, or fabric to cover walls and partitions.

paperhanging Covering interior wall surface with wallpaper. Also, covering a surface with very thin mortar.

papering Applying wallpaper to walls. See *paperhanging, paste brush, smoothing brush, wallpaper hanging.*

parallel clamps Large woodworking clamps used for holding wood pieces together while being glued or worked on. The large wood jaws are adjusted by the two screw rods. (Figure 663.) Also called *handscrew clamps*. See *clamps*.

parallel compression (F_c) See *compression parallel to grain*.

parallel perspective A three-dimensional drawing that has one vanishing point. Also called *one-point perspective*. See *perspective drawing*.

parapet A protective wall.

parapet wall Low wall running around the outside edge of a flat roof; a protective wall; roof fire or party wall. (Figure 664.) See also *coping*.

parent metal In welding, the *base metal* or metal to be welded.

parge A cement mortar used for *parging*.

parge coat Mortar coat used over masonry.

pargeting Decorative plaster work; *parging*.

parging Mortar coat used over masonry. Vertical mortar coat used between brick wythes.

paring chisel Wood chisel with beveled sides used for paring, cutting rabbets, and cutting mortises. See *wood chisel*.

paring gouge Woodworking tool with a curved cutting blade sharpened on the inside. Operated by hand pressure only. See *gouge*.

park electrical wiring system Electrical wiring at all fixtures and equipment in a mobile home park.

parquet Hardwood floor made by inlaying with small pieces of wood, usually square.

parquetry Inlaid wood patterns used on flooring.

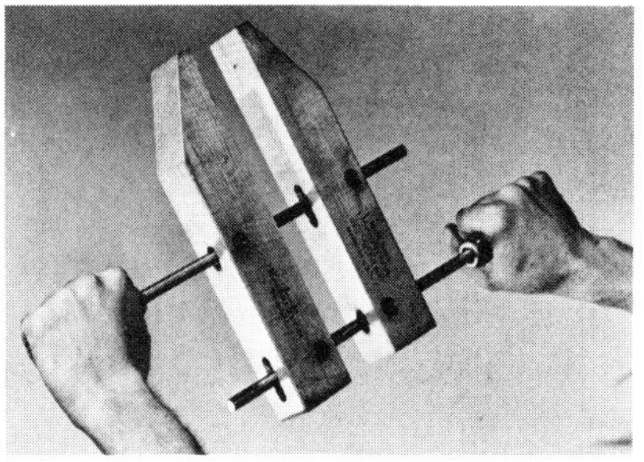

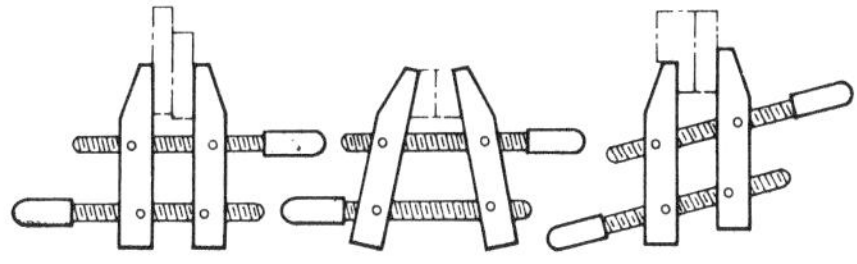

Figure 663 Parallel clamps

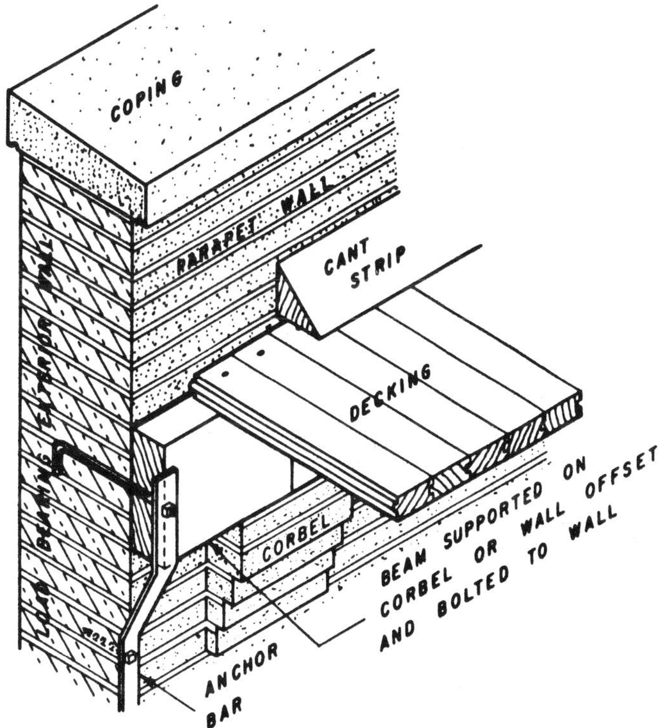

Figure 664 Parapet wall

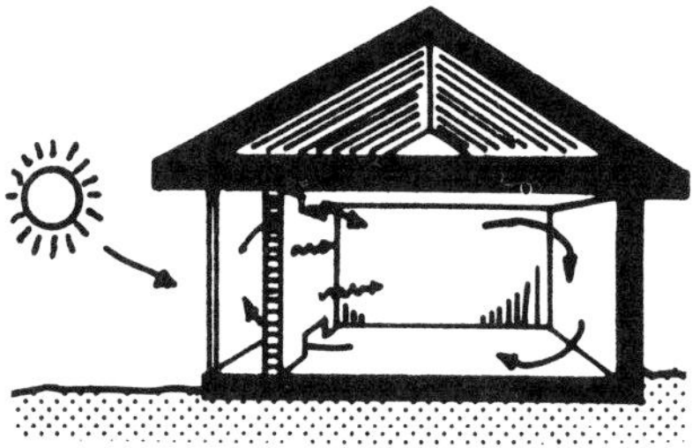

Figure 665 Passive solar heating

particleboard Panels made from reconstituted wood particles bonded with resin under heat and pressure; 4′ × 8′ panels are standard. Commonly used for underlayment; also used for shelving and counter tops (with a laminated vinyl covering).

parting tool Woodworking tool used for cutting grooves on work turning in a lathe. See *woodturning tools.*

partition Interior wall used for dividing rooms from each other, or separating different parts of a building. (*Figure 710,* p. 308.)

partition plate Horizontal framing member over a partition.

party wall A *common wall* between two living units; wall is shared between the two units. Wall between two adjoining parcels of land, used jointly.

pass Weld pass, a completed welding bead.

passive Natural heat movement, as through convection, conduction, direct solar radiation, as opposed to *active.*

passive solar heating Heating a structure by means of direct sunlight, usually through a large window area. (Figure 665.) See *orientation, solar orientation.* Also, circulation of solar heat through natural, nonmechanical means.

pass-through Small, waist-high opening between kitchen and dining area used for passing food from kitchen to dining area.

paste Mixture of cement and water. Also, a mixture used for hanging wallpaper.

paste brush In wallpapering, a wide brush similar to a wall paint brush, used for applying and spreading wallpaper paste on the back of the wallpaper. See *smoothing brush.*

paste wood filler In woodworking, a mixture of silex (fine quartz powder), a solvent, and a drier. The filler is brushed on open-grained woods to fill wood pores and make a smooth surface. The wood is sealed before the filler is brushed on.

patching Repairs and alterations made on a finished plaster job. In general, repairs made on any surface, such as concrete.

patio Open, ground-level paved area just outside the living area, used for recreation; a courtyard.

pattern Model or template used for making copies of something, such as an architectural detail. In roofing, the design formed by the layout of the shingles.

patterned lumber Shaped lumber; lumber shaped to follow a pattern.

pavement structure In highway construction, the subbase, base course, and surface course placed on a *subgrade.* It supports the traffic and distributes the load to the *roadbed.*

paver Heavy highway equipment used for placing pavement. (Figure 666.) A sequence of pavers often work on the road to place and finish concrete. The row of pavers placing and finishing the pavement is called a *paving train*. See also *flexible paving*.

pavilion Open, but roofed, recreational building.

paving construction Surfacing of roads: *rigid paving* uses cement, *flexible paving* uses bituminous material.

PB (polybutylene) A light-colored, flexible piping used in hot- and cold-water supply. See *plastic piping*.

PE (polyethylene) A flexible black plastic tubing used for water supply piping. See *plastic piping*.

peak load Highest predicted use, often used in regard to use of electricity.

pean See *peen*.

peavy A device for rolling timbers or poles. (Figure 667.) Very similar to a *cant hook*, but the peavy has a straight point on the end.

pebble dashing Plaster finish made by throwing small stones onto the wet plaster.

peck Small round imperfections in cedar or cypress. Small decay spots.

pecky Any lumber that has pecks, small decay pockets.

pedestal A support, sometimes used in reference to columns; a short concrete column. (Figure 668.) Support for an art piece, as a statue.

pediment In classical architecture a low gable-type area over the front of a structure, such as a temple. See *entablature, orders*.

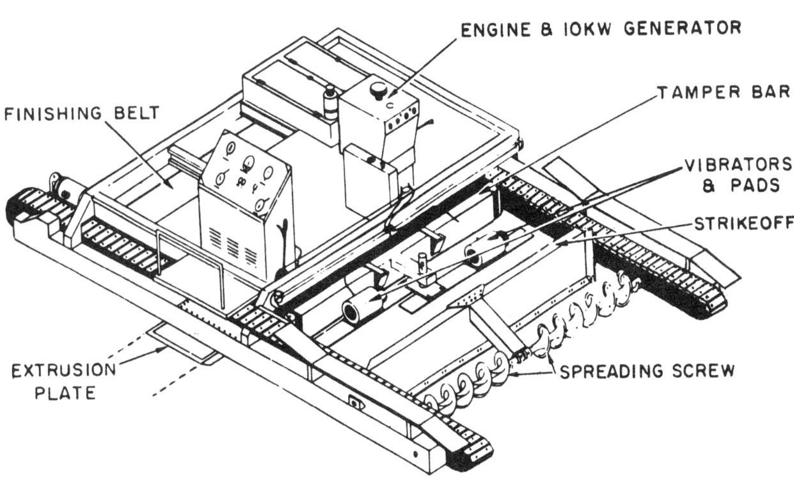

Figure 666 Paver

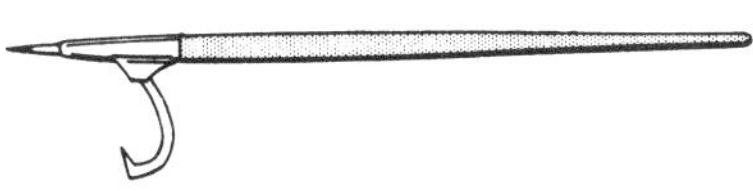

Figure 667 Peavy

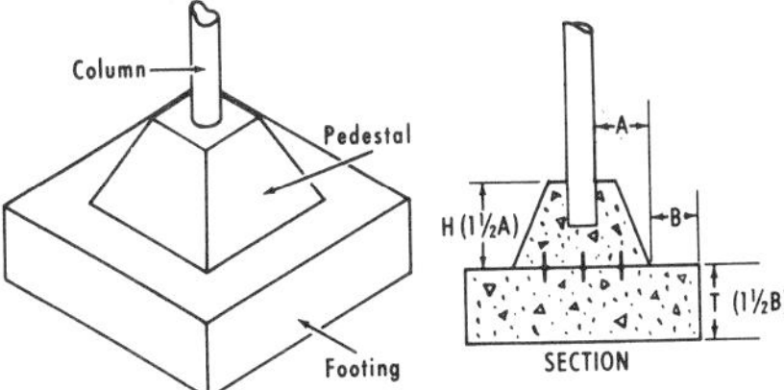

Figure 668 Pedestal and footing support for a column

peen The shaped end of a hammer head opposite the face. The peen may be wedge shaped or rounded and is used for striking, shaping, or bending metal or for chipping or shaping stone. A ball peen is half-round like a ball. See *ball peen hammer*.

peening Working metal by striking, as with a ball peen hammer.

peg Small grooved dowel used for holding furniture parts together. Holes are drilled in the two opposite parts, and the dowel with glue fits into them; the glued parts are then clamped together for drying. See *dowel*.

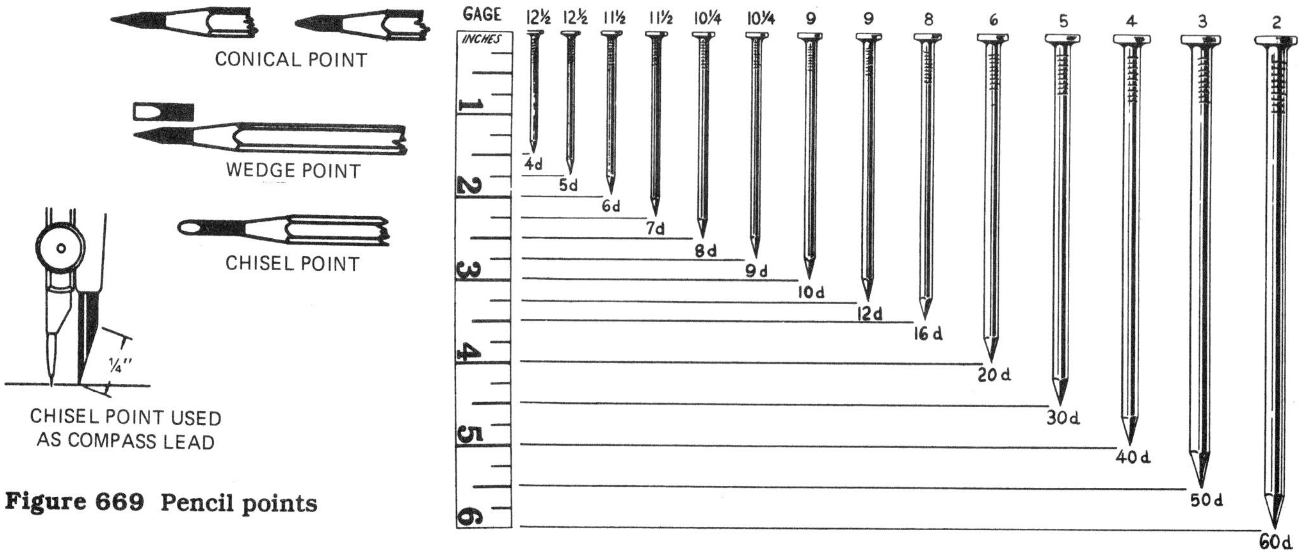

Figure 669 Pencil points

Figure 670 Penny (*d*) system for nails

pegboard Board with many evenly spaced holes, used with hooks to hang various objects.

pencil In drafting, pencils are commonly used to draw the working drawings. Pencils are graded by hardness; the hardness is indicated by number and letter:

Soft 6B, 5B, 4B, 3B, 2B
Medium B, HB, F, H, 2H, 3H
Hard 4H, 5H, 6H, 7H, 8H, 9H

pencil points In drafting, three different pencil points are used: conical, wedge, and chisel. (Figure 669.) The conical point is the standard, common point; the wedge is used for wide lines; the chisel is used with a drafting compass.

pencil holder In drafting, a mechanical pencil holder used to hold a lead; often used in place of standard wood-cased pencils. Very fine leads are often used. See also *technical pen.*

pendant Something that hangs. In architectural design, a decorative ornament that points downward, used in Colonial and Victorian house construction. Some decorative pieces are called *drop pendants.*

penetrating oil finish In woodworking, a wood finish made using an oil and a solvent, such as turpentine or mineral spirits. Common penetrating oils are boiled linseed oil and tung oil. The oil is wiped on and penetrates into the raw wood.

penetration In welding, the fusion depth of the weldment. In wood finishing, the depth of a stain or finish.

penny (*d*) A designation relating to nail length. Historically, the term is based on an old English measure. One explanation is based on cost. For example, the cost of 100 nails of a certain size was said to be 10 pence; thus the nail was called 10 penny. Another explanation is based on weight. For example, the same nail (the 10*d*) would weigh 10 pounds per thousand; the abbreviation *d* was used for pound. Standard penny lengths are shown in Figure 670.

pentagon An even five-sided figure.

perch Stonemason's volume measure equal to $16\frac{1}{2}' \times 1\frac{1}{2}' \times 1'$ or $24\frac{3}{4}$ cubic feet. Length measure: a pole or rod $16\frac{1}{2}$ feet long. See *rod.*

percolation test Test to determine how fast water will be absorbed into the ground.

perforated gypsum lath Gypsum lath with holes through the sheet; designed to give a better key for bonding onto or holding the plaster coat.

perforated strip Strip of flexible metal with holes punched out along its length. Used as a hanger strap, as to support plumbing pipe from a ceiling.

perimeter The distance around the outside edge of something, as the perimeter of a house is the distance measured around the house, the perimeter of a room is the distance around the floor area measured against the wall.

perimeter drain tile Clay tile or plastic tubing around a structure; tile is placed next to the footing. (Figure 671.)

perimeter ductwork Heating ductwork which is normally embedded in concrete to form a loop around the outside wall of the area to be heated. Figure 672 shows a section view of the ductwork in the slab.

perimeter heating Heating system with heating outlets located by the outside walls so heat can rise up in front of the cold windows and cold outside walls. See *perimeter loop heating, radial perimeter heating.*

perimeter loop heating A type of *forced warm-air heating* that uses ductwork running around the perimeter of the area to be heated. (Figure 673.) Warm air is forced with a blower through the system. Cool air returns directly to the centrally located furnace. See also *extended plenum heating, radial perimeter heating.*

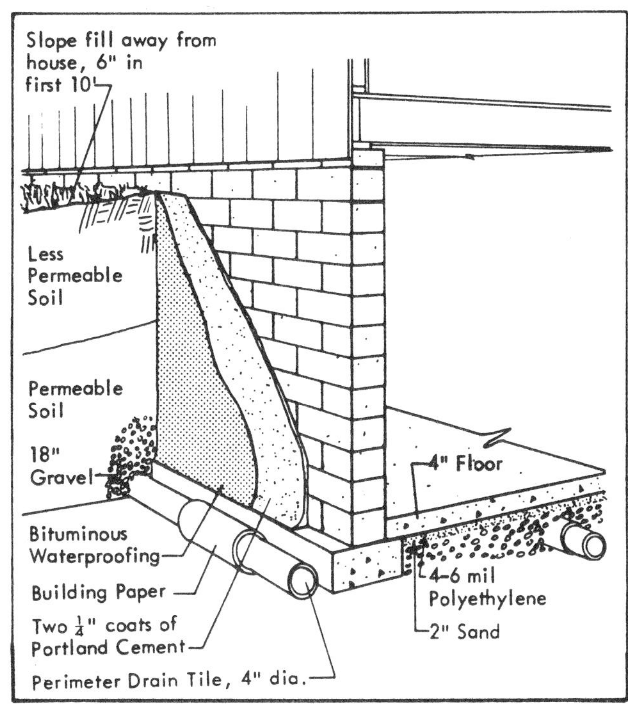

Figure 671 Perimeter drain tile beside footing

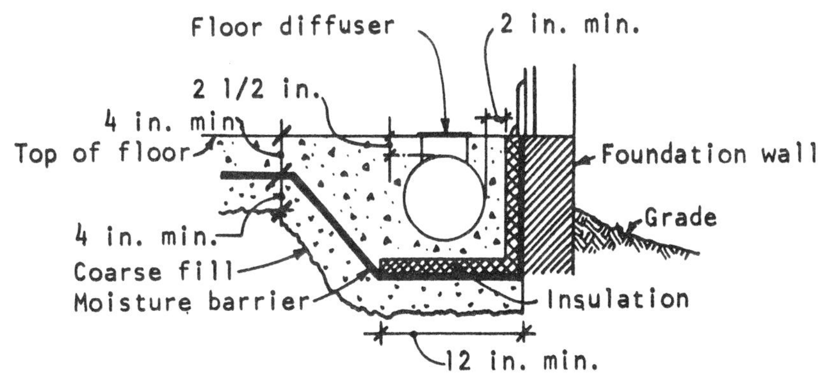

Figure 672 Perimeter ductwork in slab

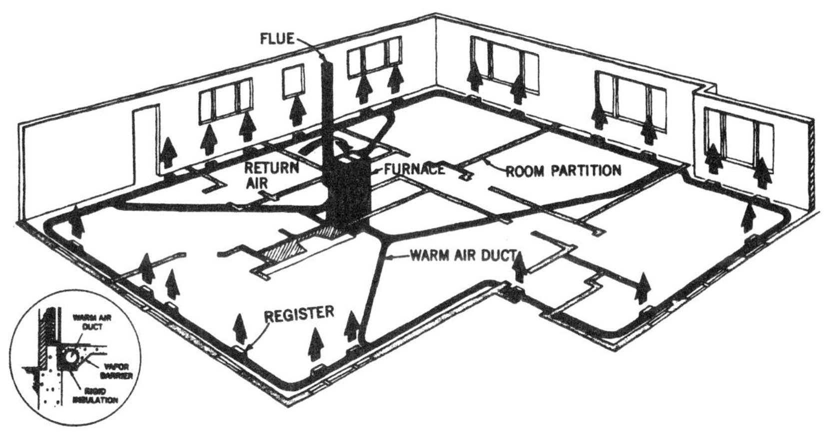

Figure 673 Perimeter loop forced warm-air system

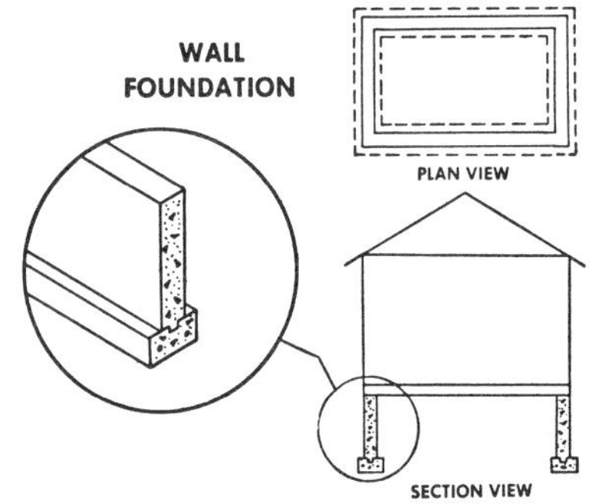

Figure 674 Perimeter wall foundation

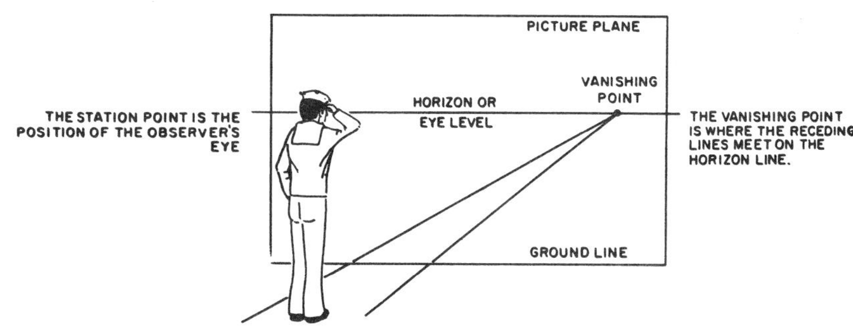

Figure 676 Perspective-drawing terminology

ONE-POINT PERSPECTIVE

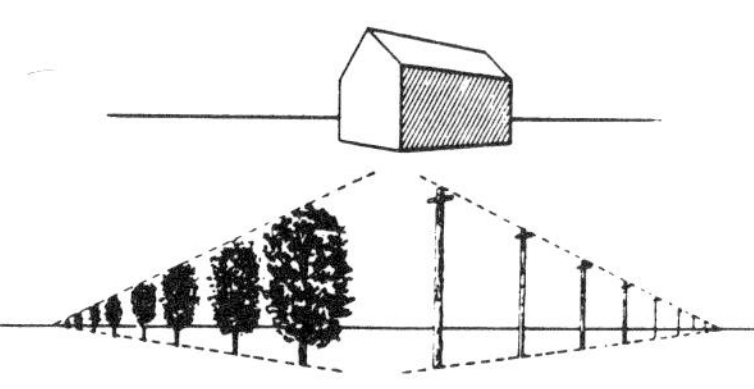

TWO-POINT PERSPECTIVE

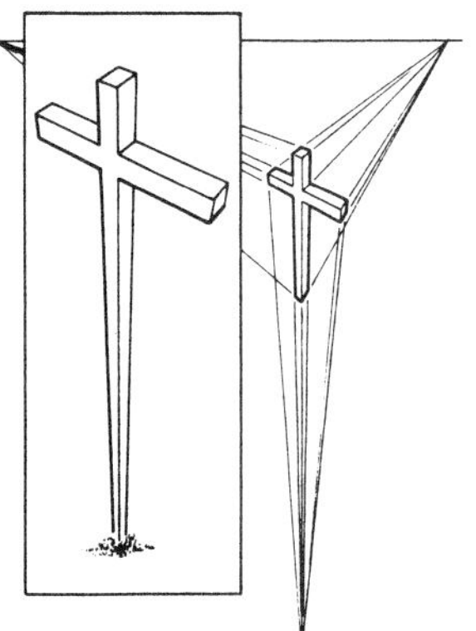

THREE-POINT PERSPECTIVE

Figure 675 Perspective drawing

perimeter wall foundation A continuous outside wall supports the structure. The open space is from 1′–6″ to 2′–0″ deep. (Figure 674.) See *crawl space.*

peripheral Outside edge; outline.

perlite Loose, lightweight insulative particles made by heating a volcanic glass. Used as a lightweight aggregate in concrete or plaster to develop insulative qualities.

perm Unit of measurement of the water vapor permeability of a material. A perm is equal to one grain (7,000 grains equals one pound) of water vapor per square foot per hour per inch of mercury vapor-pressure (.49 psi) difference.

permeability Ability of a liquid to penetrate through a solid material, as water into earth or oil into wood.

perpend Masonry unit that runs through a wall at a right angle, as a header brick.

perpendicular Lines, framing members, walls that meet at a right angle (90°).

perpendicular compression ($F_{c\perp}$) See *compression perpendicular to grain.*

personal property Personal, movable possessions, such as household goods. See *real property.*

perspective A three-dimensional drawing based on the way an object appears to the eye when seen from a specific viewpoint. The sides of the object recede into the distance. (Figure 675.) See *perspective drawing.*

perspective drawing A three-dimensional drawing that presents an object as it would be seen from a specific viewpoint. The sides of the object recede into the distance. Three types of perspective drawing are used: (1) *parallel* (one-point perspective), (2) *angular* (two-point perspective), and (3) *oblique* (three-point perspective). (Figure 675.) In each case one, two, or three *vanishing points* are used. A vanishing point is the point in the drawing where lines come together. In perspective drawing one viewpoint or *station point* is taken from which the object is viewed. The object rests and the viewer stands on a *ground plane,* and the object is viewed by the observer through an imaginary *picture plane.* The picture plane and the ground plane intersect to form a *ground line.* The *horizon line* is the line in the distance where all the lines of the object converge (into the vanishing points). (Figure 676.)

phantom drawing In drafting, a drawing showing a potential position of something, as a machine arm. Dotted lines show the potential position something could move to.

phase Type of electrical alternating current; single-phase (1Ø) is used in ordinary light or residential applications; three-phase (3Ø) is used in heavy or commercial and industrial applications. Equipment specifications will note the electrical phase required. Abbreviated Ø.

phased development In a large building development, such as a subdivision or in heavy construction, the development of different parts of the site at different times to avoid having a large area under development and exposed to erosion. Second phase only starts after first phase is nearly complete and erosion control measures have been taken and the soil stabilized. See *erosion control plan, landscape plan.*

phellem Outer corky tree bark; the cells are dead. See *tree.*

Phillips head Type of screw head having a slot shaped like a deep cross. See *screw.* A *Phillips screwdriver* is used to drive the screw.

Phillips screwdriver A screwdriver with a X-shaped tip used for Phillips screws. (Figure 677.) Comes in four sizes:

1 Phillips screws up to #4
2 #5 to #9
3 #10 to #16
4 #18 and over

See *screw, screwdriver.*

phloem Inner tree bark; the cells are living. See *tree.*

photoelectric control switch A light-sensitive switch used for turning lights on when it gets dark and off when it gets light.

photoelectric detector *Smoke detector* that responds to visible smoke.

Figure 677 Phillips screwdriver and screw

photogrammetry Aerial photographs of land masses taken with cameras at a set height. See *aerial survey.*

piano hinge Special, long butt hinge used in fine cabinetwork. Hinge runs from the top to the bottom of the door. See *hinges.*

pick and dip Bricklaying technique of picking up a brick with one hand while the other hand loads the mortar on the trowel.

picket Board cut to a point at the top, commonly used as a board in a wooden fence.

pictorial drawings Three-dimensional views of an object or structure. Three sides of the object or structure are shown. See *perspective, pictorial view.* Figure 678 compares *cavalier, cabinet, oblique,* and *isometric* drawings.

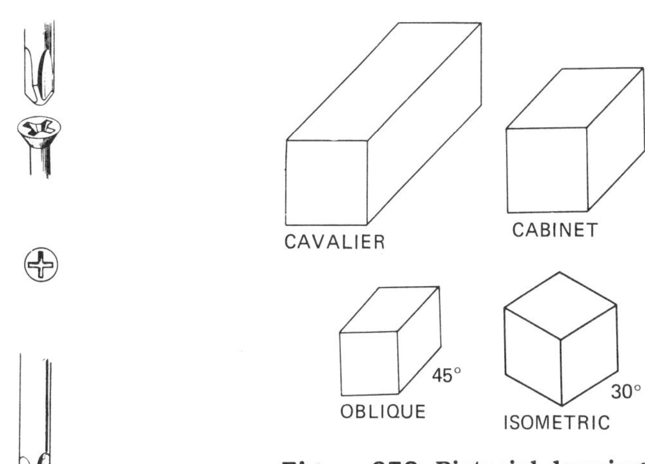

Figure 678 Pictorial drawings

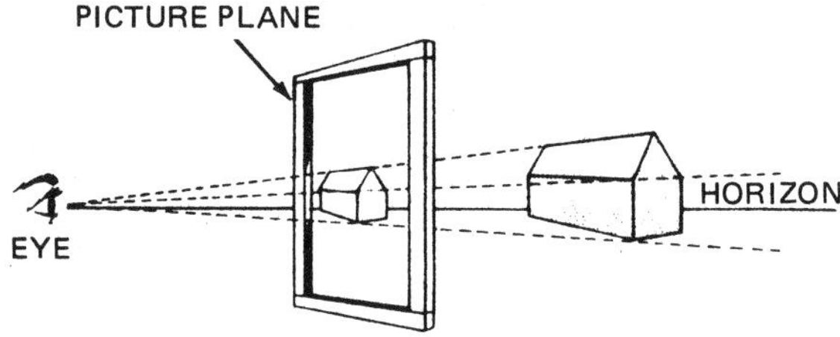

Figure 679 Picture plane

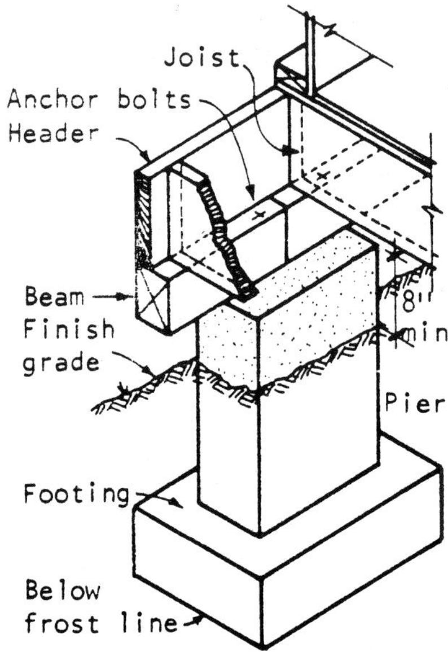

Figure 680 Pier

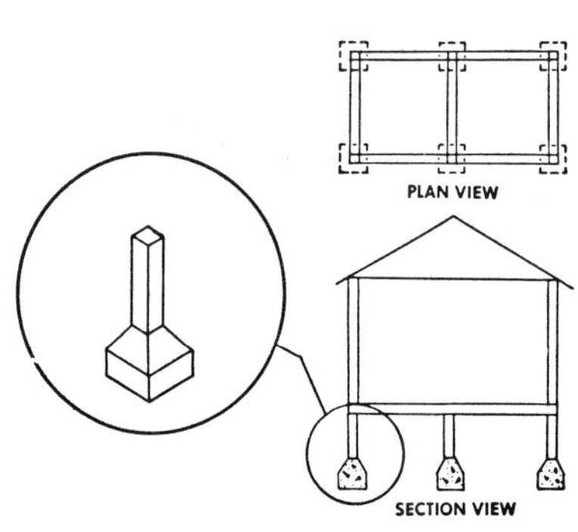

Figure 681 Pier foundation

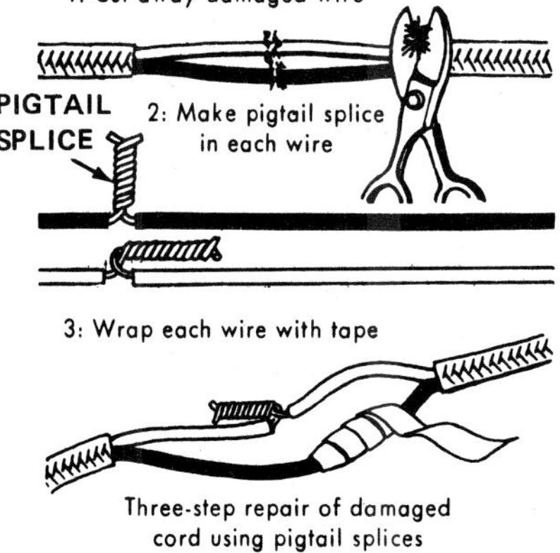

Figure 682 Pigtail splice

pictorial view Drawing showing something in three dimensions, as drawings in *perspective, isometric,* or *oblique.* Perspectives are used in architectural presentations to show the client what the finished structure will look like. Isometrics and sometimes oblique drawings are used to show details on blueprints.

picture-framing vise Tool for holding 90° corners together for nailing and gluing. Holds pieces up to 4″ wide.

picture molding Wall molding used at the ceiling; the molding has a recessed edge that will receive hooks for suspending wires to hang pictures. See *molding.*

picture plane In drawing or drafting, an imaginary viewing plane upon which an object is projected. (Figure 679.) See *orthographic drawing, perspective drawing.*

picture rail A *picture molding.*

picture window Large window area, generally with a fixed sash.

pier Short wood or concrete column used to support a foundation floor beam or decking. (Figure 680.) Arch support; load-bearing brickwork between openings; a built-up landing area by a waterway; a center bridge support.

pier foundation Structure supported on piers. (Figure 681.)

pigment Coloring used in paint.

pigtail splice Simple wire twist used to join electrical wires. Solder or a solderless wire connector is used. (Figure 682.) A soldered splice is taped with an insulating tape. See also *tee splice, Western Union splice.*

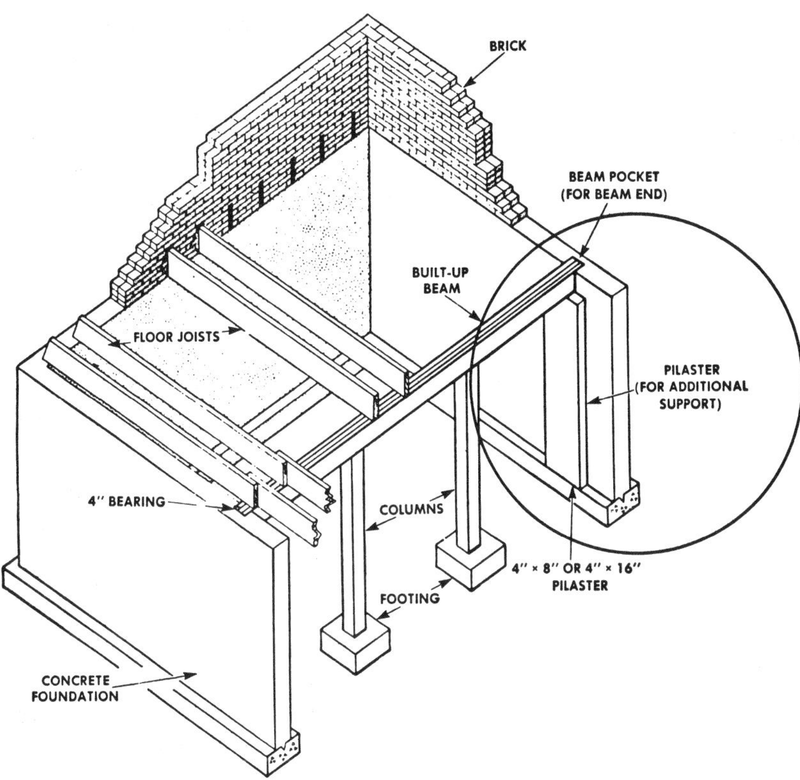

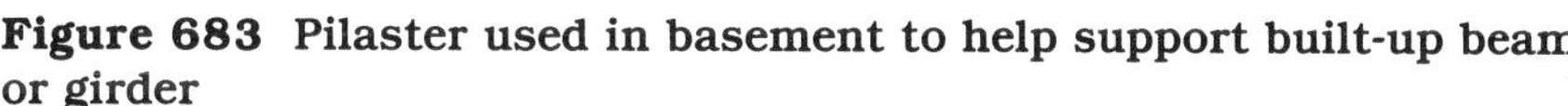
Figure 683 Pilaster used in basement to help support built-up beam or girder

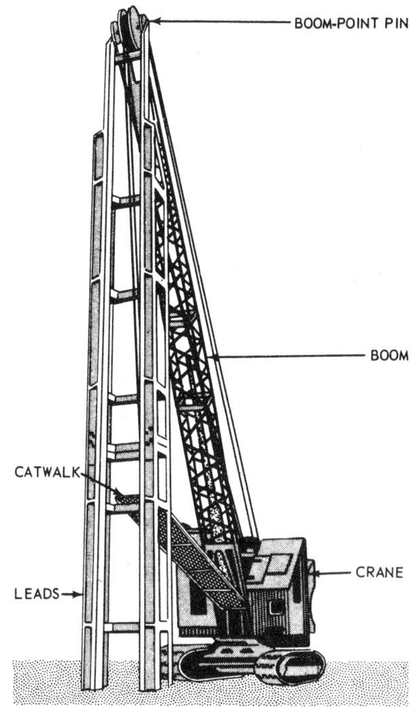

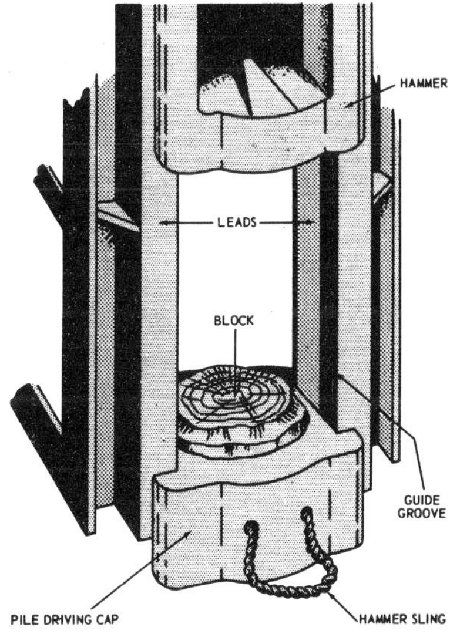

Figure 685 Pile driver

pilaster Support column structurally attached to a masonry or concrete wall. (Figure 683.) In woodworking, a vertical support piece at the side of a carcass frame.

pile Heavy steel, timber, or reinforced concrete member driven into the ground to support a structure, or to resist side pressures of earth or water. (Figure 684.) Piles are also created by drilling and then pouring concrete into the hole.

pile cap Structural member running across the top of several piles to connect the piles and distribute a load. (Figure 684.)

pile driver Heavy machine for driving piles; a heavy weight (hammer) is dropped against the pile head to force it into the earth. (Figure 685.)

piling Structure composed of piles.

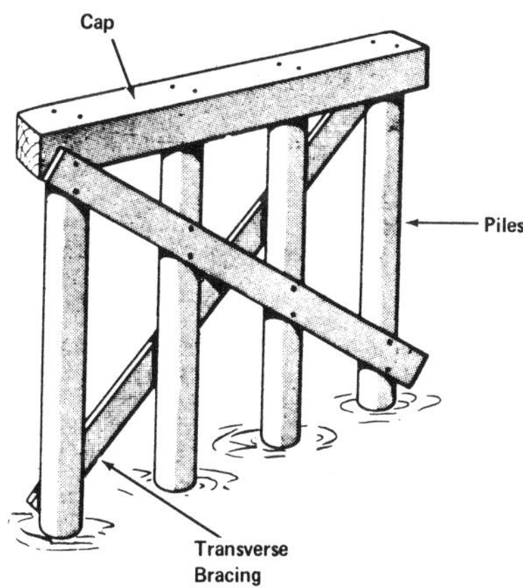

Figure 684 Piles. The braced row of piles is called a bent

Figure 686 Pilot hole

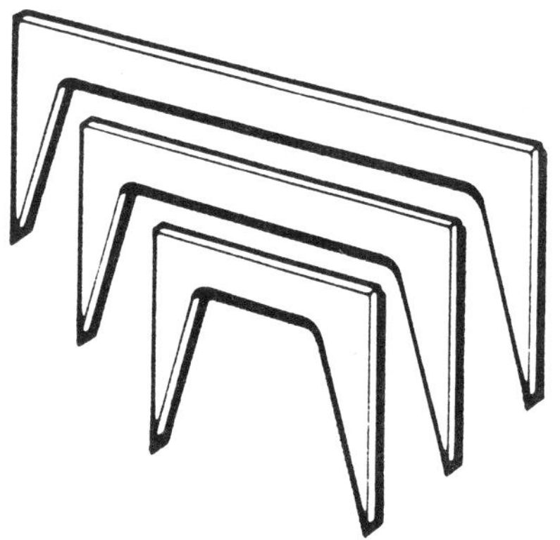

Figure 687 Pinch dogs

pillar Slender stone column.

pillar file A small file used to reach and file in narrow areas or openings.

pilot flame Continuously burning gas flame in a furnace or hot-water heater; used to ignite the burner.

pilot hole Small hole made in wood or metal to serve as a guide for drilling a large hole or for setting a screw or driving a nail. (Figure 686.)

pin Small steel or wood peg used for holding parts together or for stopping movement. A metal shaft in a hinge. A *dowel.*

pinch bar A *crowbar* used for prying and pulling things apart. A *ripping bar, wrecking bar.*

pinch dog Small two-pointed steel joint clamp ($\frac{1}{4}''$ to 3″ wide) used to hold wood pieces together. A pinch dog is shaped like a large staple and is driven into end grain to make a temporary connection. (Figure 687.) See *dog anchor.*

pinchers Hand tool with two jaws used for grasping, holding, bending, or cutting metal pieces; *pliers.*

pin hinge Hinge widely used for cabinet doors. When it is in place and the door is closed, only the pin part of the hinge is visible. See *hinge.*

pinhole Very small hole, less than $\frac{1}{16}''$.

pinholing Small openings in paint caused by bubbles which break, leaving a hole.

pin knot Knot less than $\frac{1}{2}''$ in diameter.

pin members Tenon or pins in a *dovetail joint.* See *tail.*

pin punch A blunt punch with a long neck used for driving out small fasteners, such as a cotter pin. See *punches.*

pintle In general, a heavy pivot pin. A heavy pin used in heavy timber construction to transfer load from a column head through a girder or floor to another column base overhead.

pipe Hollow cylinder used for flowing liquids, gas, or air. Several types of plumbing piping are commonly used. For water service in the building, threaded galvanized steel pipe, plastic pipe, copper tubing, or plastic tubing is used. Cast-iron, no-hub, copper, or plastic piping is used for waste and drainage. Vent piping uses cast-iron, no-hub, galvanized steel, and plastic pipe. Gas piping generally uses $\frac{1}{2}''$ galvanized steel and copper.

Figure 688 Pipe clamps: used here for holding door parts together while gluing

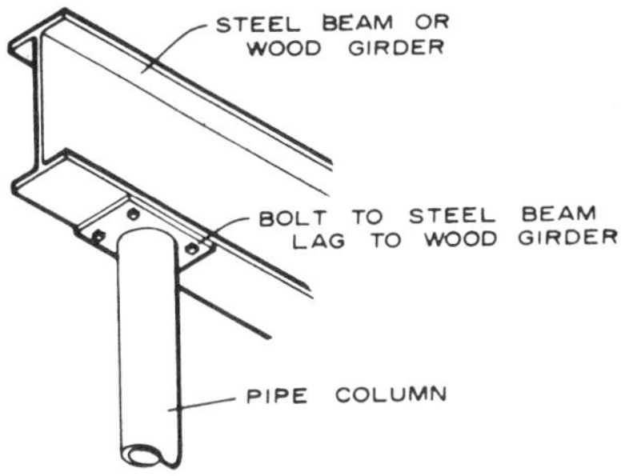

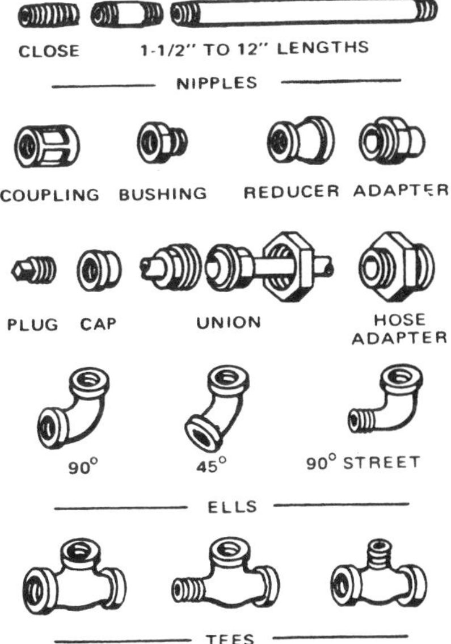

Figure 689 Pipe column

pipe clamp Large woodworking clamp that has jaws that slide on a pipe, used for holding several large pieces together. (Figure 688.) See *bar clamp, clamps.*

pipe column Column made from a steel cylinder; a *lally column.* (Figure 689.)

pipe coupling Short section of pipe threaded at both ends, used for connecting two pipe sections. See *pipe fittings, thread and coupling.*

pipe cutter Plumber's tool for cutting iron or steel pipe. (Figure 690.) As the cutter is turned around the pipe end, the blade cuts into the pipe.

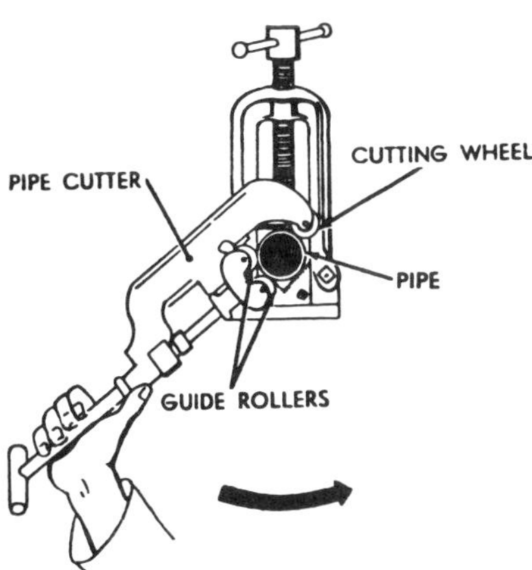

Figure 690 Pipe cutter

pipe die Threaded die for cutting external pipe threads.

pipe dope Lubricant and sealing compound used on threaded piping connections. See *pipe joint compound.*

pipefitter One who lays out, fabricates, assembles, and installs piping for steam and hot-water heating and cooling systems.

pipe fittings Manufactured pipe parts for connecting sections of pipe or tubing together. Figure 691 shows the basic threaded fittings. For the various fittings see: *bell-and-spigot cast-iron pipe, copper tubing, galvanized iron pipe, no-hub soil pipe, plastic piping.*

Figure 691 Pipe fittings (threaded)

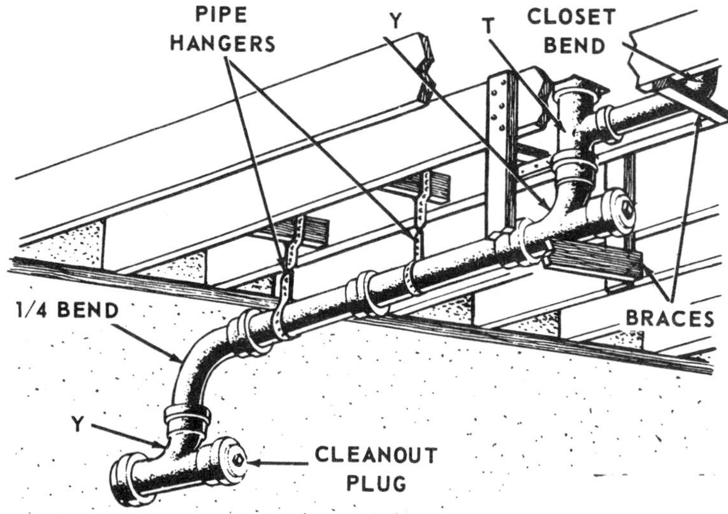

Figure 692 Pipe hangers

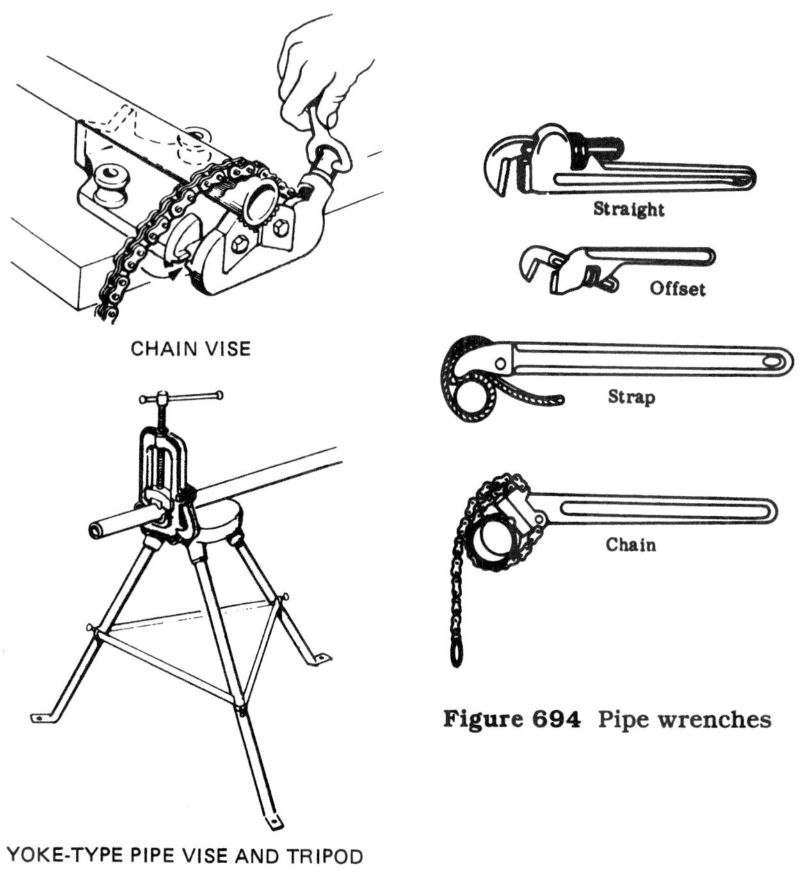

Figure 693 Pipe vises

Figure 694 Pipe wrenches

pipe hangers Clamps and brackets used to support piping. (Figure 692.)

pipe joints Pipes are connected together with threaded fittings (galvanized steel or wrought-iron pipe); fitted together with oakum and lead (cast-iron bell-and-spigot); held together with mechanical fittings (cast-iron no-hub pipe and PE plastic); bolted together (flanged pipe); fitted with flared joints (soft copper tubing and PE plastic); fitted with insert fittings (PE plastic); soldered (copper tubing); welded (steel pipe); and cemented (plastic pipe, plastic tubing). See *bell-and-spigot cast-iron pipe, copper tubing, galvanized steel pipe, no-hub soil pipe,* and *plastic piping* for the various fittings used.

pipe joint compound A plumbing sealing compound used on threaded piping. See *pipe dope.*

pipe size Pipes are specified in nominal size, which refers to the nominal inside diameter. For example, a 1″ galvanized steel pipe may have an actual inside diameter size of 1.049″ or 0.958″ depending on wall thickness (whether a standard wall or extrastrong wall). A nominal 1″ copper tube may vary from 0.995″ to 1.055″ depending on type.

pipe thread A tapered thread used on pipe and pipe fittings.

pipe trowel A long, flat trowel used for working behind pipes and other obstacles. See *trowel.*

pipe vise A portable plumbing tool used for holding pipe while it is being worked on in the field. The vise is used for clamping the pipe down. (Figure 693.)

pipe welding Specialty welding used for welding ends of pipe together. Usually used on larger-diameter pipe.

pipe wrench Heavy wrench used to hold and tighten conduit or plumbing pipe. (Figure 694.)

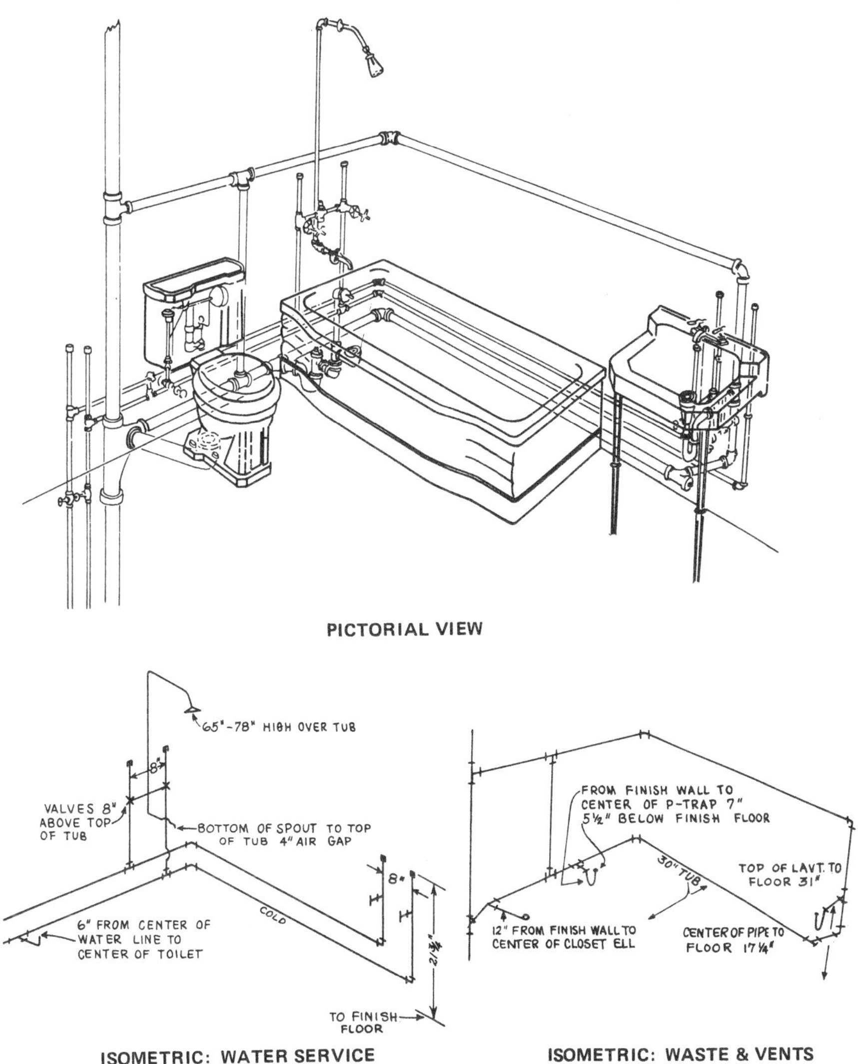

Figure 695 Piping (isometric)

piping plan Drawings that show plan-view layout of concealed piping; horizontal piping runs are shown. See *plumbing plan.*

piping isometric A diagram showing the vertical and horizontal layout and sizes of piping systems. Also called a *piping diagram.* Figure 695 shows typical piping isometrics for a bathroom. A pictorial view is shown at top; separate isometrics are prepared for water service and for waste and vent.

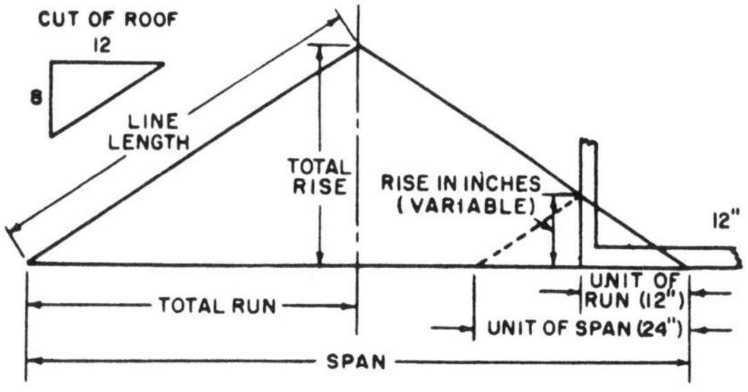

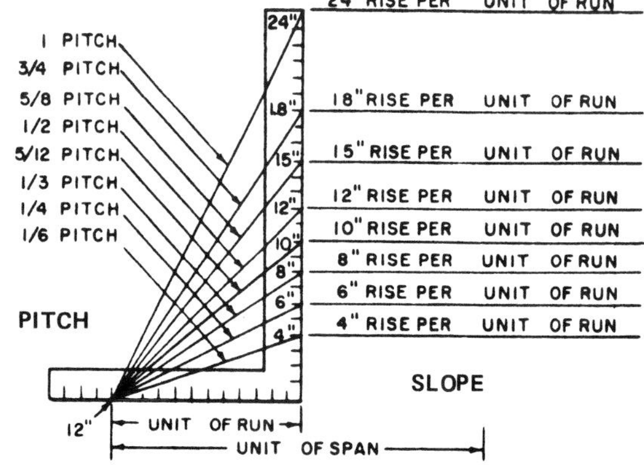

Figure 696 Pitch (roof)

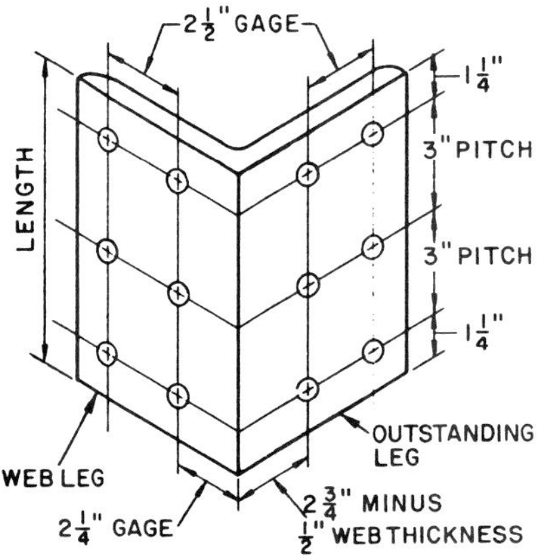

Figure 697 Pitch (steel construction)

piping schematic A diagram showing the vertical and horizontal layout and sizes of piping systems. Almost identical to piping isometrics except that the schematic uses horizontal lines rather than the 30° isometric lines. (*Figure 716,* p. 311.)

pitch Incline or slant. Angle of a roof as a ratio of span to rise. A roof with a span, for example, of 24 feet (clear space between outside wall supports) and a rise of 6 feet would have a pitch of $\frac{6}{24}$ or $\frac{1}{6}$. (Figure 696.) Not to be confused with the more widely used roof *slope,* which is given in inches of rise per 12″ of run. In lumber, a resin excreted by the wood. In concrete flatwork, the slope made for water runoff, as for drainage or a garage floor (given in inches per foot). In steel construction, the distance between holes bored in the steel beam, between holes or connectors in a *gage line;* the distance between rivets or bolts measured along a *gage line.* (Figure 697.) For stairways, pitch refers to ratio of rise to run: an 8–9 pitch would be an 8″ rise with a 9″ run.

pitch board In stair building, a board sawed out to the exact size and shape of the stringer, used as a template to make the stringers.

pitch line In roof construction, a measuring line that runs along the rafter from the top of the bird's mouth to the ridgeboard. The *rafter line.*

pitch of a roof Ratio of the *span* to the *rise.* Mainly used to describe gable roofs. See *pitch.*

pitch pocket In lumber, a small natural opening in the wood which is filled with pitch resin.

pith The core or center of a tree, log, branch, or knot. See *tree.*

pitting Small holes or *pinholes* in the paint coat, caused by small bubbles that break, leaving a hole.

Figure 698 Placing concrete

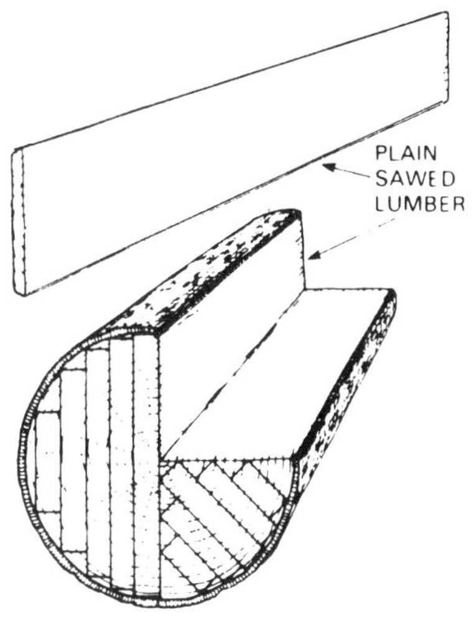

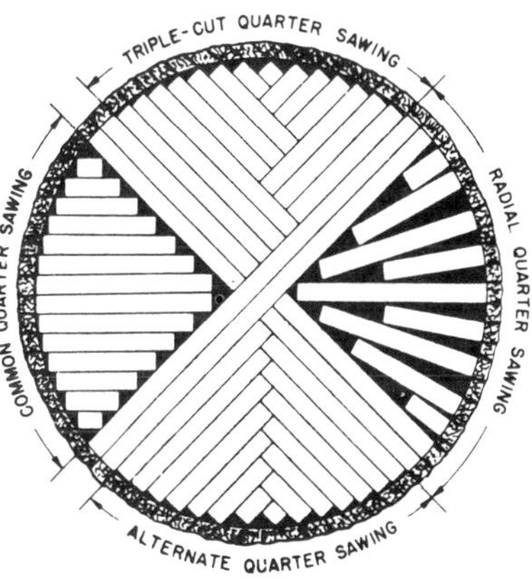

Figure 699 Plainsawn lumber

place To pour fresh concrete.

placing In concrete work, the pouring of the concrete into place. (Figure 698.)

place of beginning See *point of beginning.*

plain concrete Concrete without steel reinforcement.

plainsawn Hardwood lumber with the grain (annual growth rings) running at less than 45° to the face of the board. (Figure 699.) Called *flatsawed* or *flat-grained* in softwood lumber. See also *quartersawn.*

plan In drafting, a drawing made looking straight down (orthographic view) on a structure. A *floor plan* is a plan view of the floor after the top part has been cut away (in imagination) and removed. A plan view is identified by what is shown: *site plan, plot plan, first floor plan, second floor plan, basement plan, foundation plan.* See *floor plan.* Also, when used in the plural, *plans,* a general reference to the complete set of blueprints.

plancher A *plancier.*

plancier Underside of a cornice. (*Figure 117*, p. 44.) A *soffit.*

placing drawings In reinforced concrete construction, drawings that show location and size of concrete members and location, size, and spacing of steel reinforcement. See *engineering drawing.*

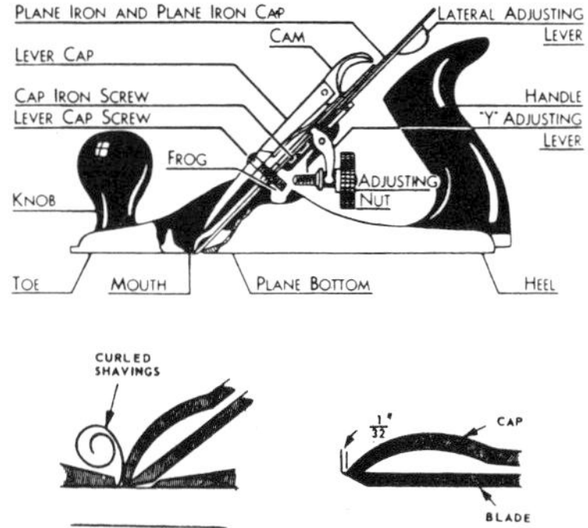

Figure 700 Plane terminology

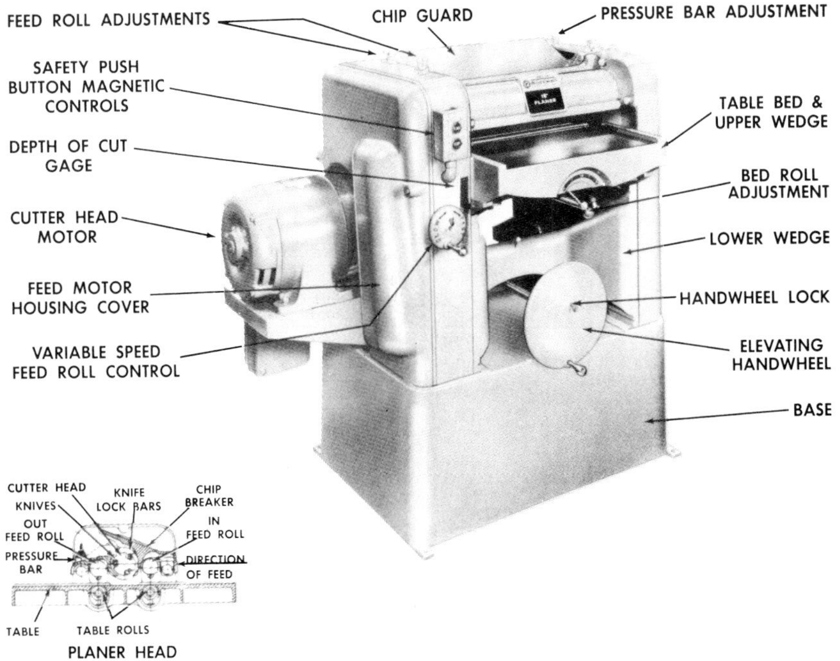

Figure 701 Planer

plane A hand tool that has an adjustable cutting blade, used for planing or smoothing a wood surface. The plane iron in a hand plane protrudes through the mouth slightly to provide a cutting edge. The plane-iron cap directs the shavings up through the mouth. Adjustments are made with the adjusting nut (to move the edge up and down) and the lateral adjusting lever (to move the plane iron left or right). Figure 700 shows the parts of a hand plane. A *jack plane* is a general-purpose 14″ hand plane. A *smooth plane* is a plane commonly 9″ long and is used for planing short pieces. A *block plane* is commonly 6″ long and is used one-handed on very short pieces and for squaring ends. The *fore plane* is a plane commonly 18″ with a wide 2″ blade and is used on long pieces. The *jointer plane* is a hand plane 22″ long or longer and is used for straightening long pieces. The *rabbet plane* is a special hand plane used for cutting rabbets. The *bull-nose rabbet plane,* because of its shape, can work into corners. A *power plane* is an electrically driven, general-use plane widely used for finishing and smoothing. See also *draw knife, power plane, rasp plane, router, spokeshave.*

plane of projection In drafting, an imaginary plane or surface on which points from a figure are projected. See *orthographic.*

planer Power planer used for surfacing lumber to a specified thickness. (Figure 701.)

plane survey Survey of a small land area where the curvature of the earth does not play a part. See *geodetic survey.*

plane table Small table or drawing board mounted on a tripod. An *alidade* is used on the table to take bearings of distant objects. The bearings are transferred directly to a drawing paper on the table. (*Figure 15*, p. 8.)

plank Lumber piece over 2″ thick and 6″ or more wide.

plank and beam construction Construction that uses heavy posts and beams, coupled with heavy planks, to form a rigid load-bearing framework. See *post and beam construction.*

plans A broad term generally referring to the complete set of blueprints for a building or structure. An orthographic view looking straight down on a section of a building. See *floor plan.*

plan shape The shape of the house when viewed directly from above, as on a floor plan. Common shapes are rectangular, square, L-shaped, T-shaped, H-shaped, and U-shaped.

plan symbols Symbols used on floor plans to show wall materials and construction, door swing, windows, partitions. (Figure 702.) See also *electrical symbols, heating and ventilating symbols, plumbing symbols, section symbols, symbols, window symbols.*

planting Sticking ornamental pieces on plaster.

planting easement Right to use land for planting to control erosion, normally used in regard to highway development. See *drainage easement.*

plan view View seen directly from above; *orthographic view.*

plasma-arc welding An electric-arc welding and cutting process where a high-pressure gas is forced into the arc. The gas is superheated into an extremely hot jet which melts or blasts through a metal. (Figure 703.)

plasma welding *Plasma-arc welding.*

plaster A covering troweled on over wire lath or gypsum lath on walls and ceilings. (Figure 704.) A mortar is made of Portland cement combined with sand and water. Plaster is also made of plaster of Paris (calcined gypsum) and is normally used for fine work and casting.

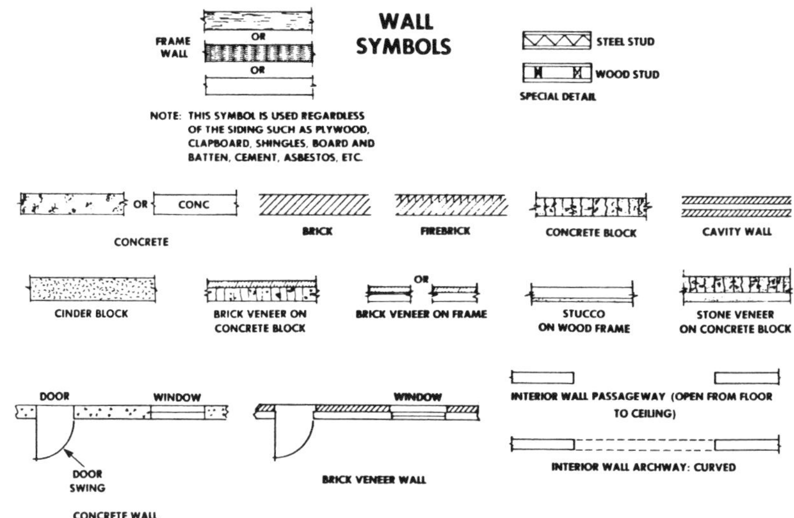

Figure 702 Plan symbols

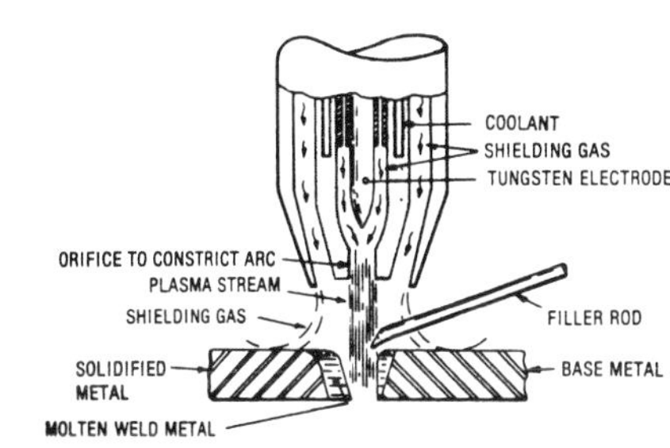

Figure 703 Plasma-arc welding

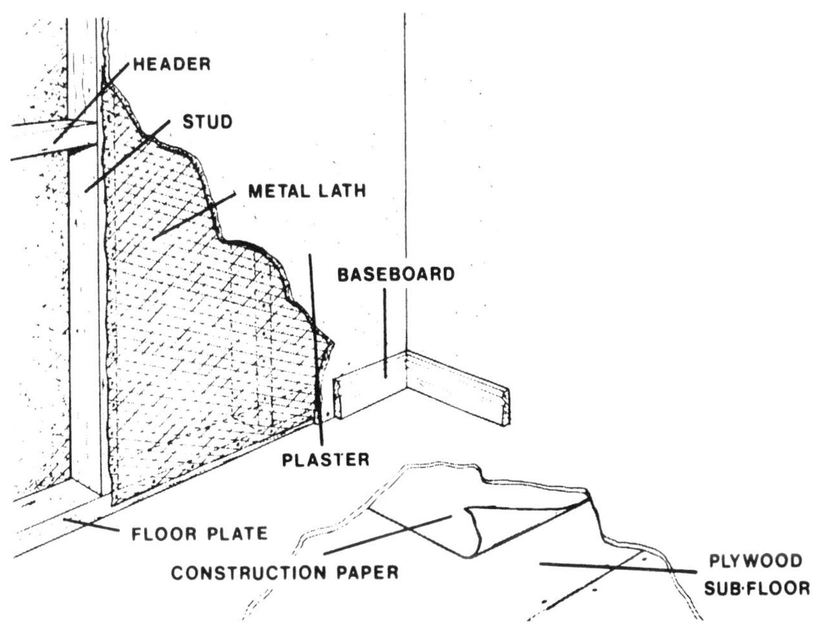

Figure 704 Plaster construction

Figure 705 Plaster mortar mixer

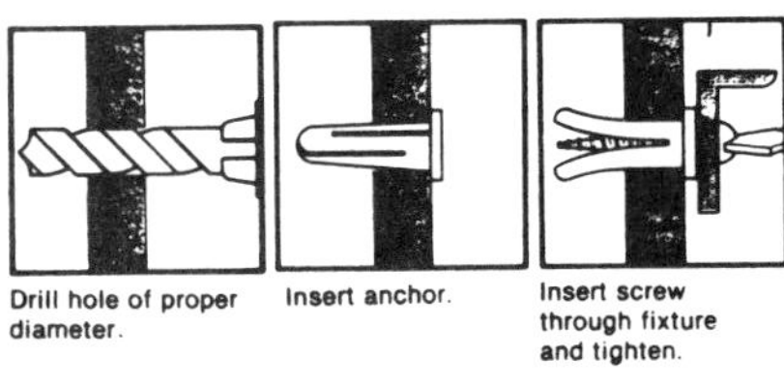

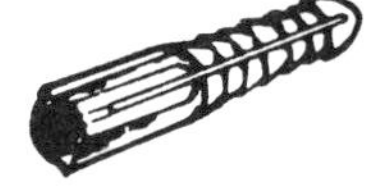

Figure 706 Plastic anchor

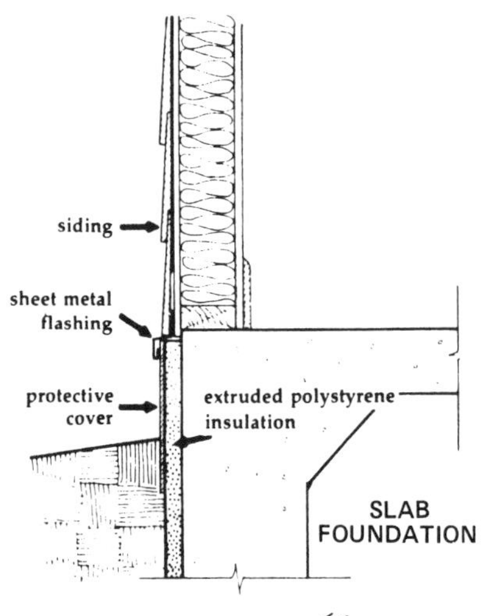

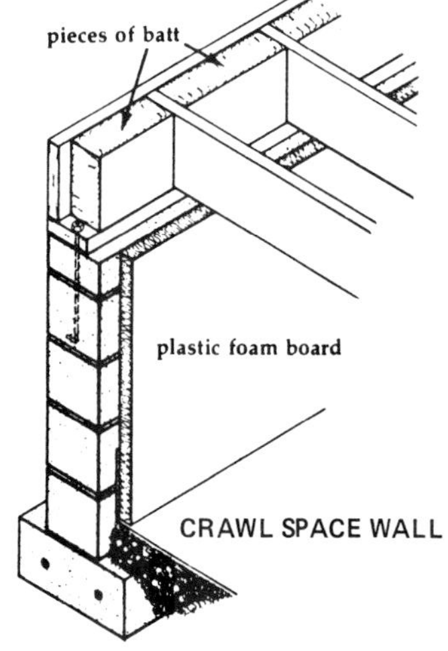

Figure 707 Plastic foam insulation

plasterboard Board with gypsum core covered on both sides with heavy paper. Also called *drywall.* Used in *drywall construction.*

plasterer One who applies plaster (over lath) on interior walls, partitions, and ceilings. Also applies stucco (over lath) to exterior walls and installs ornamental plaster.

plaster grounds Wood strips nailed to wall to serve as a thickness guide for the plasterer.

plasterer's float See *float.*

plasterer's shingle Long, flat tool, used to finish plaster. See also *darby, featheredge, plasterer's shingle.*

plasterer's trowel See *trowel.*

plastering knife Wide-bladed knife used to cover small open spots, nicks, recessed nail heads, and drywall tape with plaster.

plastering machine Air-powered machine with a hose used to blow out premixed plaster onto walls and ceilings. A full coat is applied, then smoothed by hand.

plaster lath Expanded *metal lath* used as a base for plaster. The plaster is forced through the openings in the metal lath to form a permanent bond or key. Formerly, thin wood slats were used as a plaster base; openings were left between the slats or wood laths so that plaster could form a key. See also *gypsum lath.*

plaster mortar mixer Mechanical revolving drum used to mix cement, sand, and water to form plaster. (Figure 705.) Drum tilts so that plaster can be poured out when mixed.

plaster of Paris Calcined gypsum, used for fine plaster detail and for casting.

plasterwork Usually refers to fine work in plaster, as architectural details. In general, any plastering.

plastic anchor Expandable anchor for holding screws in plaster or drywall construction. Anchor is fitted into bored hole, then screw is driven in to expand and wedge the anchor. (Figure 706.) See also *anchor, fiberplug.*

plastic concrete Fresh, wet concrete that can still be worked.

plastic foam An expanded plastic widely used in sheets for insulation. (Figure 707.)

plasticity Ability to be molded, worked, or shaped.

plastic piping Piping used for plumbing, drainage, venting, and water supply. Types include ABS, PVC, CPVC, PB, and PE. Figure 708 shows typical plastic pipe fittings (for ABS). See also *insert fittings*. Figure 709 shows plastic fitting joined with piping to form a total water supply system.

plastic-sheathed cable Cable used as underground feeders and branch-circuit cable. Type UF (underground feeder).

plastic tile Tile used for wall coverings; various colors and shapes are available.

plastic toggle A plastic holding device that tightens back on itself when in place in plaster or drywall. See *nylon expansion anchor*.

plastic veneers A vinyl covering widely used over pressed wood to make counter tops.

plastic welding Bonding of plastic pieces with heat. A plastic filler rod is used to fuse the pieces together.

plastic window lights Unbreakable window panes; widely used in public buildings and in schools; also used in built-in glazed cabinet doors.

plastic wood A premixed combination of wood powder and plastic hardener used for filling cracks or holes in wood. Sometimes called *wood dough*.

plat A survey plan showing the layout of an area, such as a subdivision, including property lines and major geological features and any local bench marks.

plate Horizontal support member in a frame wall, usually a 2 × 4. *Sill plate:* horizontal member on top of the foundation. *Sole plate:* bottom horizontal member of a frame wall. *Top plate:* top horizontal member in a frame wall; normally two top plates are used. (*Figure 967*, p. 427.) Also, a flat strip of iron with bored and countersunk holes for screws; a *mending plate*.

plate anchor See *anchor bolt*.

plate cut A V cut made at the end of a rafter to make a seat for the rafter on the plate. Also called a *bird's mouth, foot cut, seat cut*. (*Figures 763*, p. 335; *769*, p. 337.)

plate glass Large sheet of thick, rolled glass; used in storefronts and in other areas where a large window area is needed. Standard plate glass is available in several thicknesses. A tempered plate glass is also available that is much stronger than the standard plate glass.

platform Stair landing; a raised flat horizontal area, as a raised area for a public speaker or a band.

ABS Plastic Pipe for Drain, Waste and Vent Lines

Easy-to-install ABS (acrylonitrile-butodrine-styrene) plastic pipe for drainage systems has many advantages over copper and cast-iron pipe. It can be cut to length with a saw and requires no threading or soldering at connections. The entire system can be "solvent-welded" into one continuous length letting you accomplish a professional installation quickly and easily.

Tools You'll Need

Here's all you need to install Sears ABS plastic pipe: Measuring Tape, Fine-Tooth Saw, ABS Solvent Cement with Brush in Cap, Clean Rags

Materials

Pipe
Fittings
Adapters

PIPE—Sears ABS Plastic Pipe is available in 3 sizes.
1½" x 10', 2" x 10', 3" x 10'.

FITTINGS—Fittings for use with Sears ABS Plastic pipe come in 1½-in., 2-in. and 3-in. sizes. Couplings, tees, elbows, plugs, bushings, traps, "Y's," increasers, "P" traps and flanges available are shown.

ADAPTERS—Special adapters are available for traps and cleanouts. Adapters are also available for plastic, copper, galvanized steel or soil hub pipe as shown.

Fittings

ABS-DWV Fittings

Pipe
Available in 10 Ft. lengths in 1½", 2" and 3" sizes

Trap Adapters
1½" Pipe to 1¼" Waste and 1½" Pipe to 1½" Waste

Flashings
Available in 1½", 2" and 3" sizes

"Y" Branches
Available in 1½", 2" and 3" sizes

Couplings
Available in 1½", 2" and 3" sizes

Cleanout Adapters
Available in 1½", 2" and 3" sizes

Sanitary Tees
Available in 1½", 2" and 3" sizes

Threaded Plugs
Available in 1½", 2" and 3" sizes

Sanitary Reducing Tees
Available in 2" x 1½" and 3" x 1½" sizes

Coupling Increasers
Available in 1½" x 2", 1½" x 3" and 2" x 3" sizes

90° Elbows
Available in 1½", 2" and 3" sizes

Reducer Bushings
Available in 2" x 1½", 3" x 1½" and 3" x 2" sizes

3" Sanitary Tee
With two 1½" Side Inlets

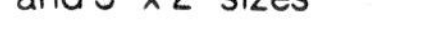

3" Combination "Y" and ⅛" Bend

1½" Slip Plug
For 3" Sanitary Tee with two 1½" Side Inlets

Male Pipe Adapters
Available in 1½", 2" and 3" sizes

Reducing Closet Flange
4" x 3"

Plastic Adapters
Available in 1½" and 2" sizes to 2" Iron Hubs; 3" to 4" Iron Hub and 3" to 4" Plastic Sewer Pipe

45° Elbows
Available in 1½", 2" and 3" sizes

"P" Traps
1½" with double seal union and 2"

ABS Cement
Available in ¼ Pints, to do approx. 25 3" joints or 50 1½" joints

Figure 708 Plastic pipe fittings (ABS)

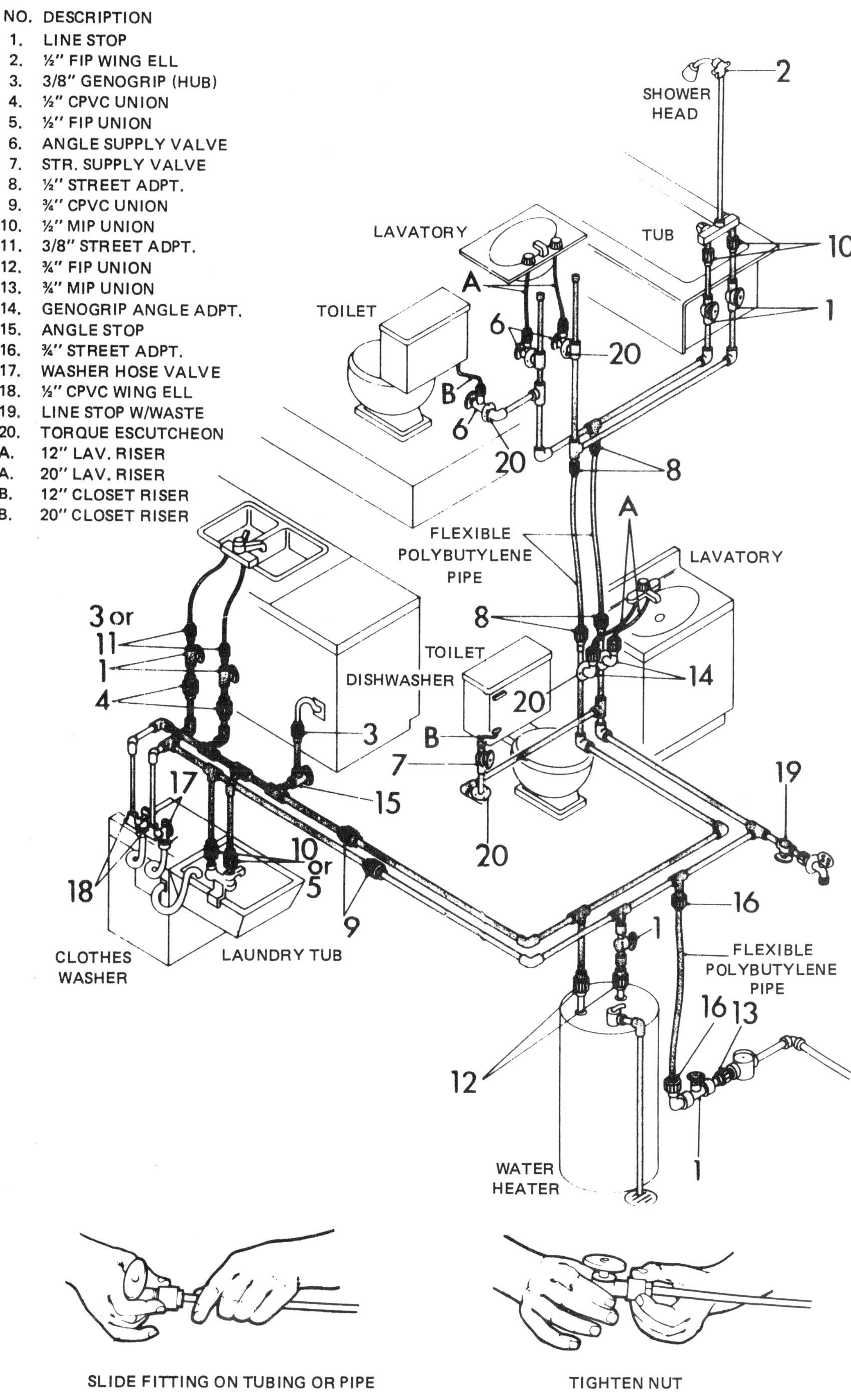

Figure 709 Plastic pipe fittings (CPVC and PB)

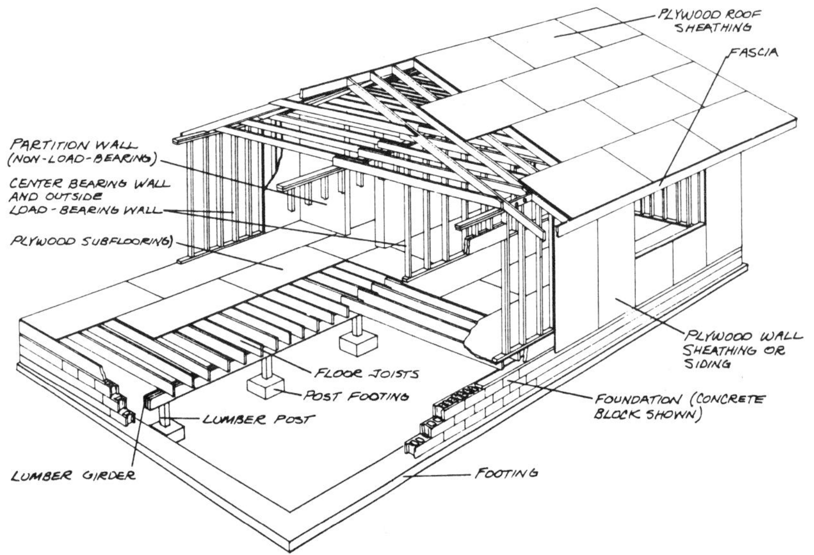

Figure 710 Platform framing

Figure 711 Platform framing: welded steel wall being lifted into place

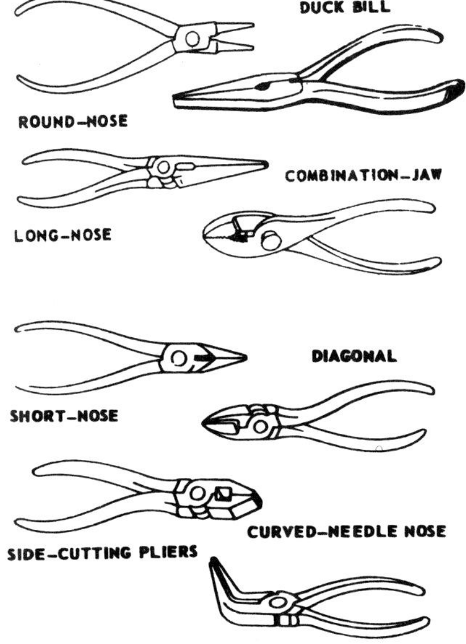

Figure 712 Pliers

platform framing Widely used framing method where a complete platform is constructed at each floor. (Figure 710.) Walls are constructed separately and raised into place on the platform. (Figure 711.) Also called *Western framing.* See *floor framing, frame construction, frame wall.*

platform stairs Stairway with a landing (platform); may have two or even three platforms in a run between floors — one each at the top and bottom and one in mid run to change direction.

plat plan Survey plan of an area. See *plat.*

plenum In forced-air heating, an enlarged area where the air collects under pressure before going through ductwork to the outlets. See also *under-floor plenum.*

plenum system In forced-air heating, the condition of using a pressure in the system higher than the surrounding air. (*Figures 267*, p. 100; *376*, p. 149.)

pliers A type of hand tool having a two-part handle that operates a set of jaws. Used for grasping, holding, bending, twisting, cutting, and the like. A wide variety of pliers are available for different uses. All pliers are used in one hand. Figure 712 shows some of the more common pliers. A wide range of sizes and types are available. See *adjustable-joint pliers, crimping tool, cutting nippers, diagonal cutting pliers, duckbill pliers, electrician's pliers, flat-nose pliers, gas-pipe pliers, glazier's pliers, ironworker's pliers, locking pliers, long-nose pliers, multipurpose tool, side-cutting pliers, slip-joint pliers, wire stripper.*

plinth A square base support for a column, as in neoclassical public buildings. (*Figure 371*, p. 146.) See *orders.* The square base of a statue. The plain baseboard. Base course of rough stonework. In woodworking, the support base for a carcass frame.

plinth block A separate piece at the base of a door trim.

plinth course Base course of a stone rubble wall.

plot A building site or a development site.

plot plan Plan view of a building site with existing features shown and the proposed structure located on the lot. (Figure 713.) Basic lot and building dimensions are noted; north arrow is located. Existing and finish contour lines are shown; elevation of key points is given. See *plan, area plan, grading plan, earth change plan, erosion and sedimentation control plan, landscape plan.*

plot plan symbols Symbols used on a plot plan. (Figure 714.)

plough To cut a groove lengthwise in a wood member.

plow Grooving plane; an adjustable fence allows grooves to be located at different distances to the edge; different cutting blades can be used to form grooves of different shapes. In plastering, a V-shaped trowel used for forming inside corners. See *corner trowel.*

plug Threaded fitting that screws into the end of a plumbing pipe to close it off. In electricity, the pronged end of an appliance or lamp cord that fits into a receptacle to receive power; a screw-in fuse. In woodworking, a cylindrical piece of wood that is inserted in a hole over a screwhead to hide the screw; the plug is finished flush to the wood surface.

plug cutter Special hollow bit used for cutting plugs.

plug fuse Circuit protection used for wiring 125 volts or less. The Edison-base fuse screws into the fuse holder. *Note:* Edison-base plug fuses are no longer permitted by NEC in new construction. See *fuse.*

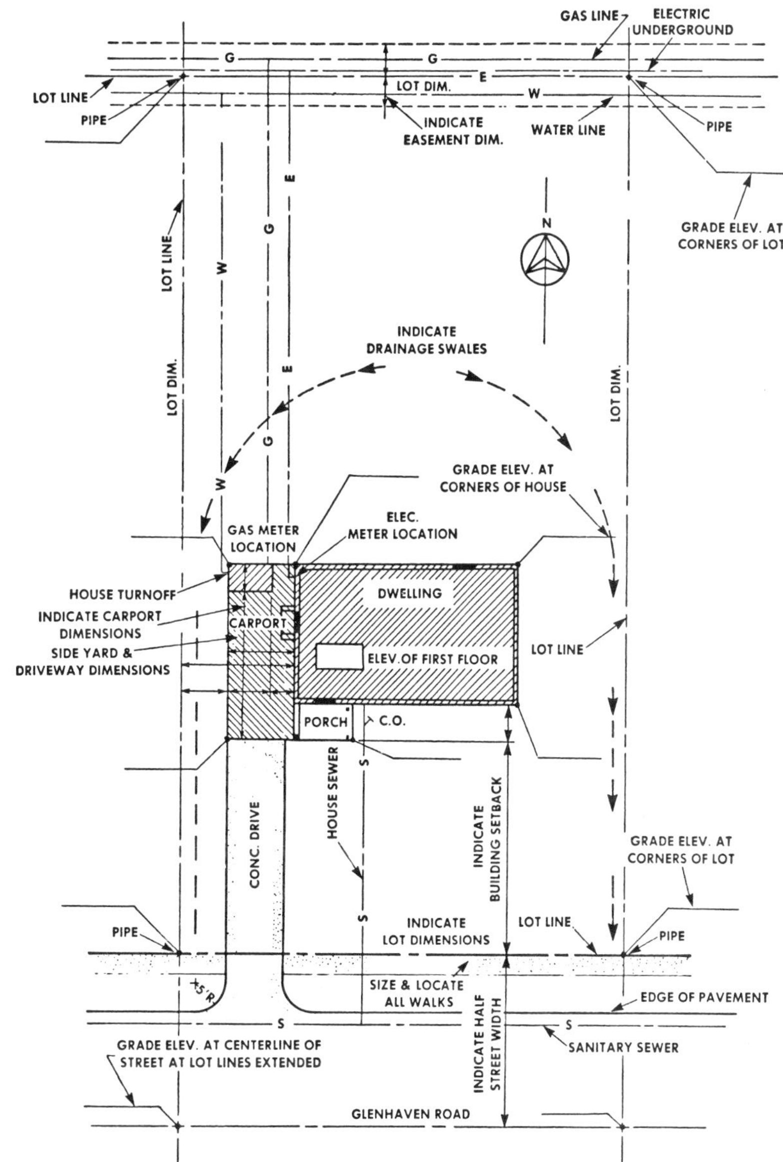

Figure 713 Plot plan: layout and drafting conventions

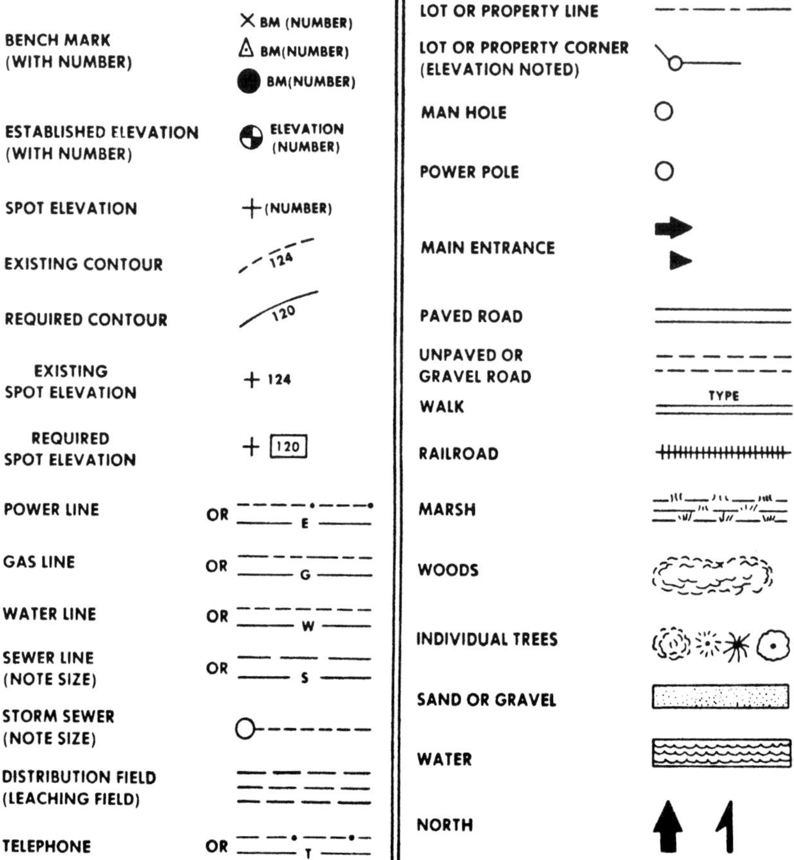

Figure 714 Plot-plan symbols

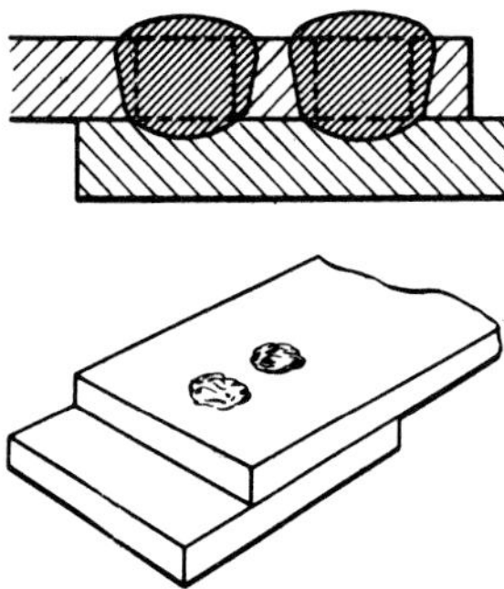

Figure 715 Plug weld

plug weld A weld made through a plug (round opening) in one piece through to another piece below to join the two pieces. (Figure 715.) See *slot weld.*

plumb True vertical; surface or edge trued using a plumb line and bob; a vertical cut or edge.

plumb and level A level used for establishing true horizontal and vertical surfaces.

plumb bob Pointed weight used with a line to establish true vertical. In surveying, the plumb bob is used to locate the center of the level or transit over an exact point. See *plumb-line and bob.*

plumb cut The vertical cut part of a *bird's mouth* at the end of a rafter; cut just at a right angle to the horizontal *seat cut* (also called a *level cut, plate cut*, or *foot cut*). See *bird's mouth, seat cut.* The plumb cut butts against the side of the 2 × 4 plate, while the seat cut bears on top of the plate. The top cut on a rafter where it frames against the ridgeboard. See *rafter cuts.*

plumber One who lays out, fabricates, assembles, and installs cold-water, hot-water, and gas piping and fixtures; installs drain-waste-vent piping; also lays sewage system and distribution field.

plumber's dope Sealing compound used on threaded plumbing joists. Also called *pipe dope, pipe joint compound.*

plumber's friend Rubber suction cup with a wooden handle used to clear blockages in the plumbing system. The rubber cup is fitted over a drain or placed in the water-closet bowl; air is forced into the system by pushing down; suction is created by pulling up. Also called a *plumber's helper* or *plunger* or a *force cup.*

plumber's helper See *force cup, plumber's friend.*

plumber's snake See *drain cleaner.*

plumb heel cut In rafter construction, the vertical cut in a bird's mouth. See *plumb cut.*

plumbing diagram A *plumbing schematic.* See *piping schematic.*

plumbing fittings Connector used to fasten piping or tubing ends together. See *pipe fittings.*

plumbing fixtures A plumbing unit such as a water closet, urinal, lavatory sink, bathtub, or shower that receives waste discharged into the drainage system.

plumbing piping Piping and tubing that brings in a water supply, carries away waste and liquids, and vents the system. See the various types of plumbing pipe: *bell-and-spigot cast-iron pipe, copper tubing, galvanized steel pipe, no-hub soil pipe,* and *plastic piping.* See *pipe.*

plumbing plan Drawing that shows a plan view of the plumbing layout; plumbing fixtures are located. Normally, in residential construction, plumbing plan information goes in the regular floor plans. If a separate plumbing plan is prepared, it will be identified with a P in the title box. See also *piping plan, piping schematic.* Figure 716 compares a plumbing plan (top) with a piping schematic (bottom).

plumbing schematic See *piping schematic.*

plumbing symbols Stylized representation used on blueprints to show the various plumbing fixtures and parts, such as sinks, tubs, showers, water closets, drains, cleanouts, and vents. (Figure 717.) See also *heating and ventilating symbols, plan symbols, symbols.*

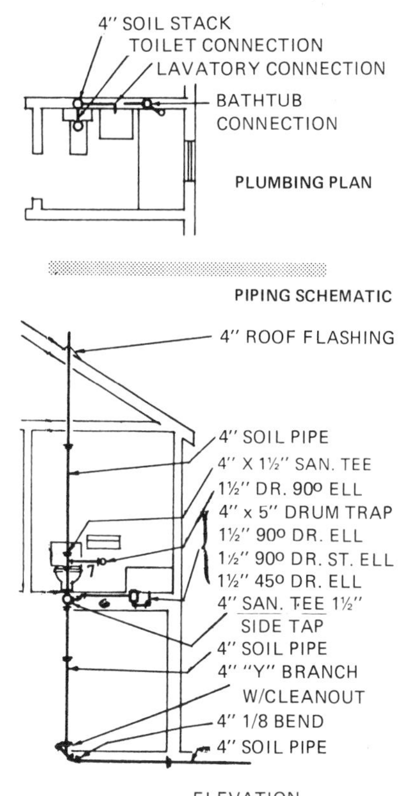

Figure 716 Plumbing plan (top) and schematic (bottom)

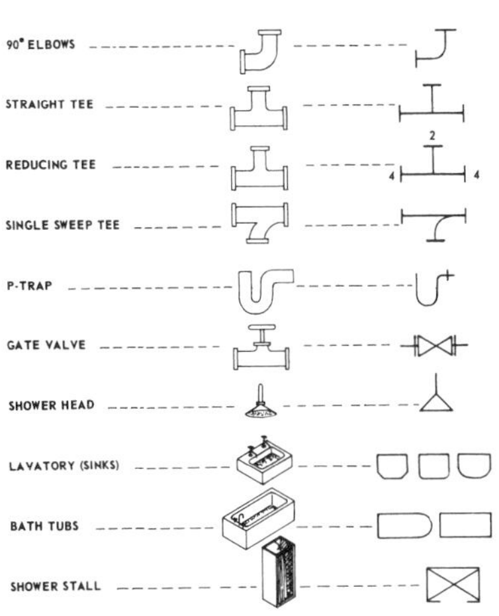

Figure 717 Plumbing symbols

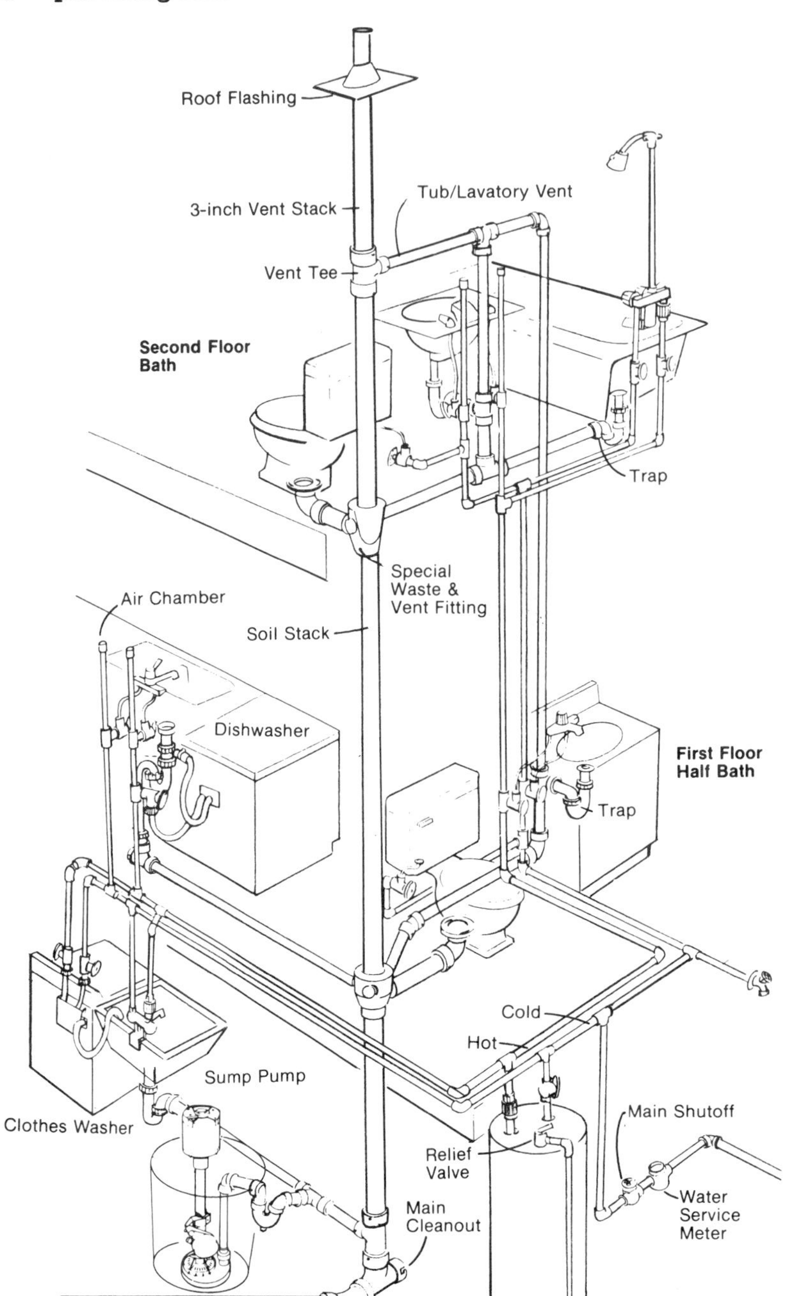

Figure 718 Plumbing system

plumbing system The total plumbing for a building: fresh water supply and drain-waste-vent outlet. (Figure 718.)

plumbing trim Fitting in a plumbing system that controls water outlet (such as faucets or shower heads), or water disposal (such as a sink drain strainer or trap).

plumbline Vertical line cut in a bird's mouth where the end of a rafter butts against the side of the top plate. Also, the line used with a plumb bob. (*Figure 79,* p. 31.)

plumbline and bob Line with a weight hanging on it; for trueness the line runs exactly through the top center of the weight. Used to locate a point (exactly under the plumb bob point) or establish a true vertical surface (exactly parallel to the line). (Figure 719.) See also *plumb bob.*

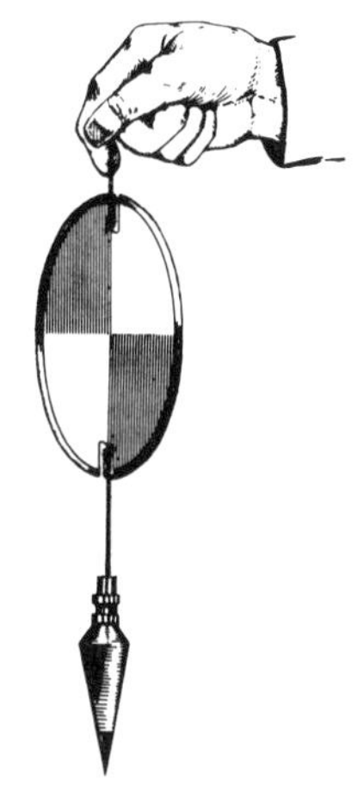

Figure 719 Plumb line and bob (with survey target)

plunging In surveying with a transit, reversing the telescope from its normal position end-for-end around the horizontal axis. Also called *transiting.* See also *double sighting.*

plus sight In surveying, a *backsight.*

ply Thin sheet of material glued against other sheets or against a backing material, as in plywood or veneer construction. In built-up roofing, a sheet of asphalt felt.

plyform Special water-resistant plywood sheets used to make concrete forms; surface is oiled or has a hard resin coating so it will not stick to the set concrete.

plyron Type of plywood that has hardboard faces. Used for counter tops, cabinets, flooring, and the like.

plywood Building material built up by gluing several thin wood sheets (plies) together. There is always an odd number of plies — the center ply and one, two, three plies on *each* side of the core to make three-ply, five-ply, or seven-ply plywood. Opposing plies on each side of the center have wood grain running at right angles to each other to counteract warping in each other. (Figure 720.) A lumber or particleboard center or core is also used. Exterior plywood uses waterproof glues; interior plywood uses water-resistant glues. The plywood sheet is always marked as to whether it is exterior or interior. The standard size is 4′ × 8′, although 4′ × 9′ and 4′ × 10′ sizes are available. The faces may be sanded or unsanded. A special plywood called *plyform* is used to make concrete forms. *Plyron* is a plywood with hardboard faces. See *plywood grades.*

plywood clip See *panel clip.*

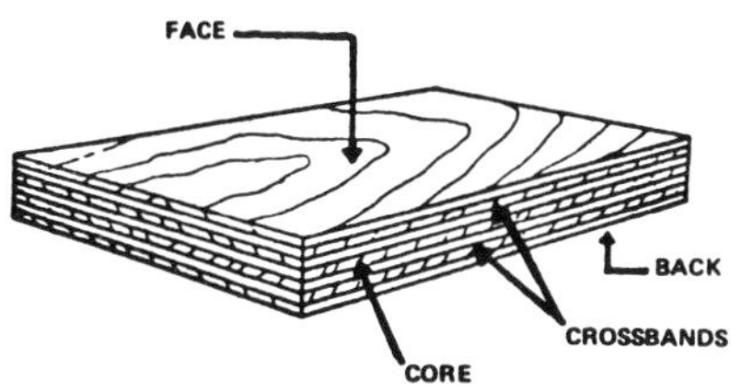

Figure 720 Plywood: seven-ply

plywood grades The structural grade of plywood is identified for the front and back veneer or ply. The American Plywood Association has established the following grades for softwood:

N A special-order "natural finish" veneer; free of open defects; some repairs.
A Smooth with small, neat repair.
B A solid surface veneer with repair plugs and tight knots permitted.
C Knotholes up to 1" with a few knotholes ½" larger permitted. Some splits permitted. This is the minimum grade permitted in exterior-type plywood.
C PLUGGED An improved C veneer with splits limited to ⅛" width and holes to ¼" by ½".
D Allows knots and holes to 2½" width, and ½" larger under certain conditions; limited splits permitted.

Thegrade for front and back veneers in a sheet of plywood is stamped on the sheet. (Figure 721.) A designation of A-C means an A grade on the face, C grade on the back. The standards organization, American Plywood Association, is identified. The use location, interior or exterior, is noted on the grade stamp. The type of wood is identified by group number. Group 2, for example, includes cypress, fir, western hemlock. The product standard governing manufacture is PS1–74.

pneumatic Air-powered; a tool or machine operated by compressed air.

pneumatic impact wrench Air-operated heavy wrench used for tightening nuts in heavy steel construction. The impact wrench is adjusted to a specified torque so that high-strength steel bolts and nuts have the correct, specified value. See *impact wrench.*

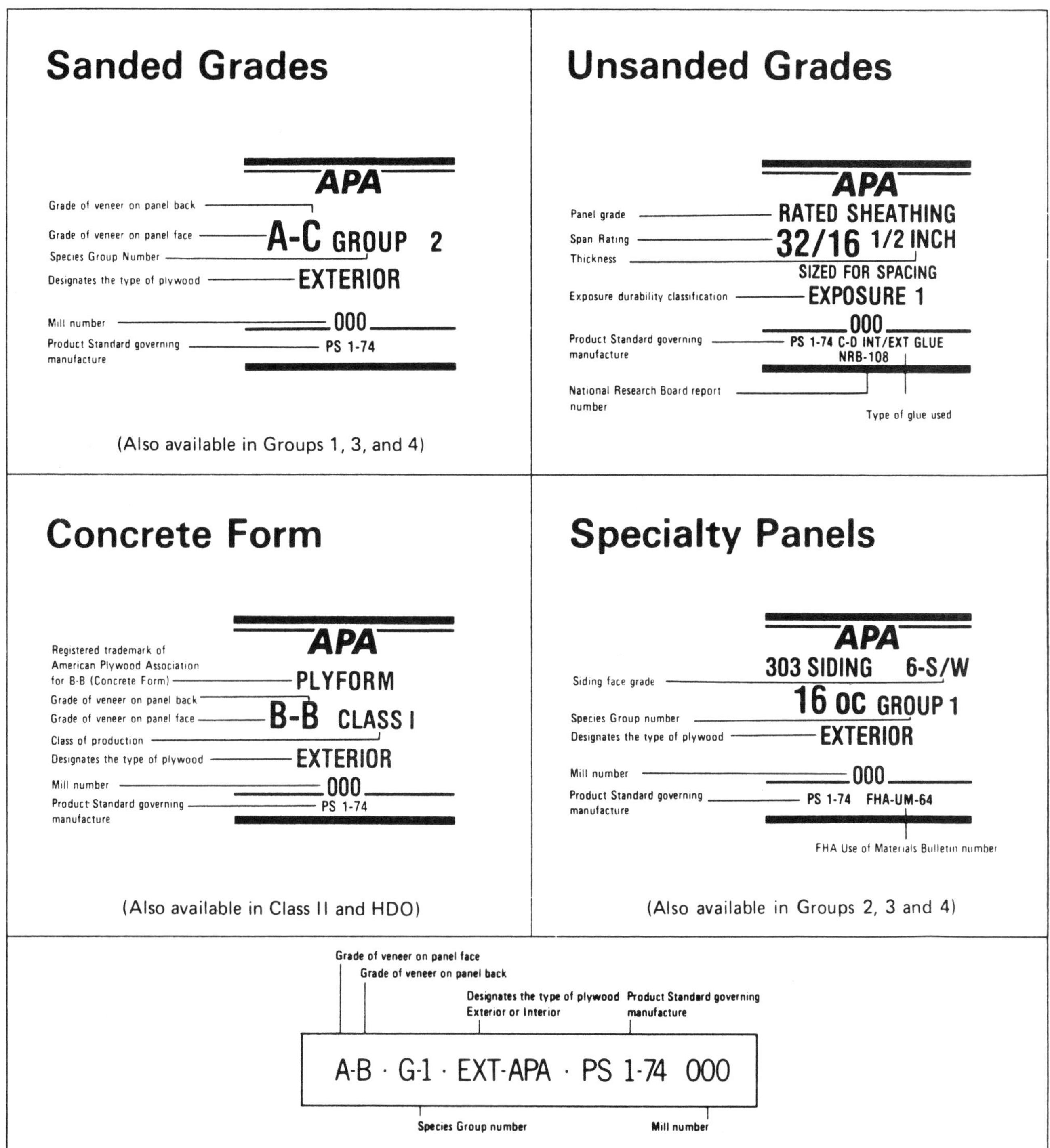

Figure 721 Plywood grade markings

Figure 722 Pocket door

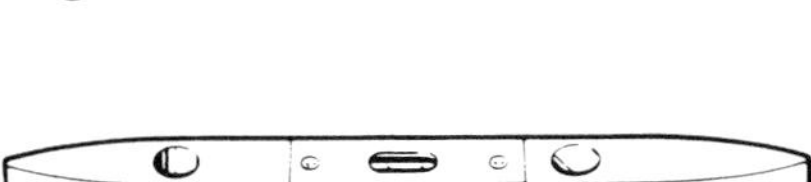

Figure 723 Pocket or torpedo level

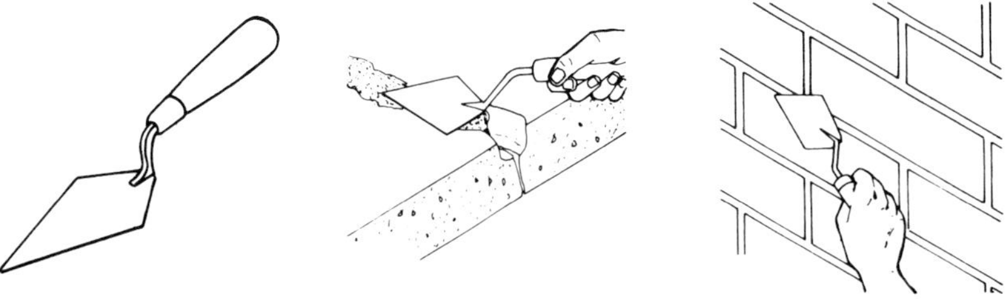

Figure 724 Pointing trowel

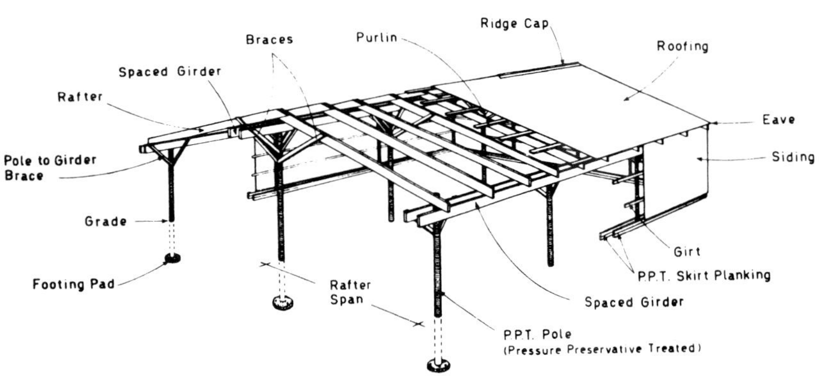

Figure 725 Pole construction (with rafters)

pneumatic structures Flexible, air-inflated or air-supported structures normally used for temporary storage. Air pressure keeps the structure inflated. Double doors are used to hold the air pressure. A clear plastic is sometimes used to create a greenhouse effect for growing plants.

pocket chisel Small wood chisel, about 4″ long, used for fine work.

pocket door Door that slides into an opening in the wall. (Figure 722.)

pocket level Short level used to check trueness of surfaces in tight areas. Sometimes called a *torpedo level.* (Figure 723.)

pocket rot A hole or pocket that has rotted out of the wood.

pointed In masonry, refers to joints that have been raked with a pointing trowel. See *pointing.*

pointing Finishing masonry joints after the masonry units have been laid. This may include applying additional mortar to correct deficiencies or filling in a special mortar after a joint has been raked. Also, filling in and finishing of broken or deteriorated joints; see *tuck pointing.* A pointing trowel is used. See also *jointing, tooling.*

pointing trowel Small triangular trowel used for pointing up masonry joints and for laying mortar for small jobs. In plastering, it is used in places the rectangular trowel will not fit. (Figure 724.) See *trowel.*

point of beginning (POB) In surveying, the beginning of a metes-and-bounds survey. The plot-plan corner that is surveyed directly to the datum point. Also called *place of beginning.*

points The cutting teeth on a saw.

polarity Direction of current flow in a circuit, as in arc welding. See *reverse polarity, straight polarity.*

pole bit A type of ship auger used for drilling into creosoted telephone and power poles and for heavy construction drilling. The pole bit is very similar to the *ship auger* but it has a cutting spur.

pole construction A type of heavy construction that uses poles as the main support members. (Figure 725.) See *post and beam construction.*

polyethylene film In construction, plastic sheets used under concrete slabs, to cover concrete for curing, as a ground cover in crawl spaces, and as vapor barriers in walls and floors.

polygon Irregular closed figure with many uneven sides and angles.

polystyrene A thermoplastic that is made into foam insulation sheets.

poor mortar Mortar short on cement.

pool trowel Special trowel for finishing swimming pools. See *trowels.*

pop out Spalling or small breaks in concrete.

pop riveter Riveting tool that sets *pop rivets.* (Figure 726.) The long end or mandrel of the pop rivet is inserted in the gun. The rivet head is inserted in the hole to be riveted. Squeezing the handle causes the mandrel to draw up, compressing and setting the rivet. The mandrel then breaks off.

pop rivets Rivets used in a *pop riveter.* The stem breaks off, leaving the rivet set in the hole.

porch Covered entryway to a house.

pore Endwise view of a tubelike wood cell; also called a *vessel.* See *wood.*

porosity In welding, gas voids or cavities formed in the weld.

porous wood Wood that has large tubelike cells, often visible as in oaks; general term for hardwoods. The tubelike cell when cut endwise presents a small round opening or *pore.* See *wood.*

portable circular saw See *circular saw.*

post Vertical support member. See *post and beam construction.* In furniture construction, vertical chair supports that form the back legs and back support. See also *rail, splat, stretcher.*

post anchor Shaped metal fastener used for anchoring wood posts to concrete slabs. (Figure 727.) See also *post cap.*

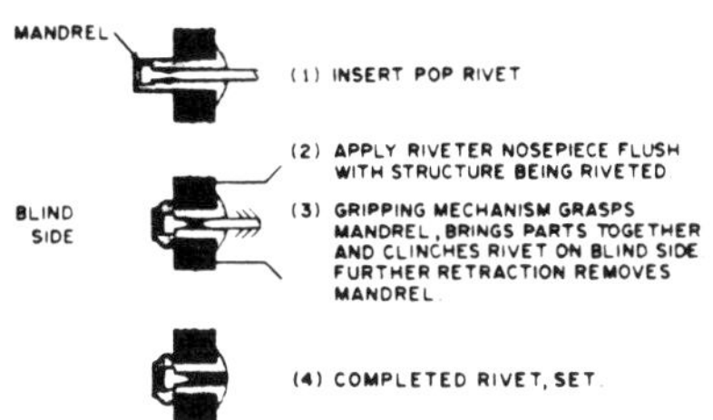

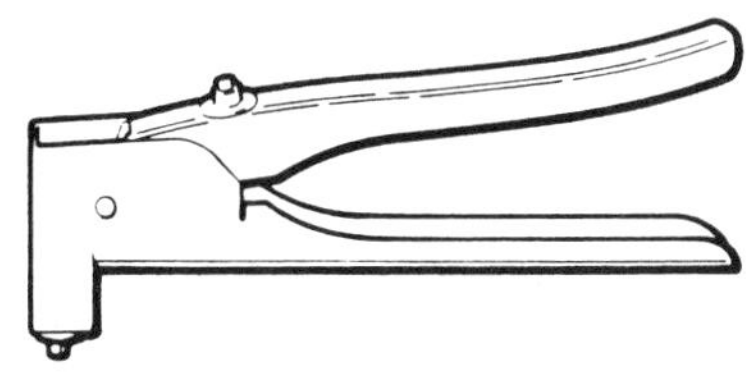

Figure 726 Pop rivets

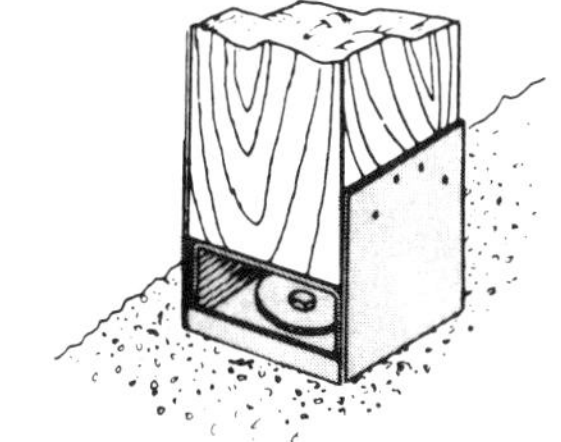

Figure 727 Post anchor

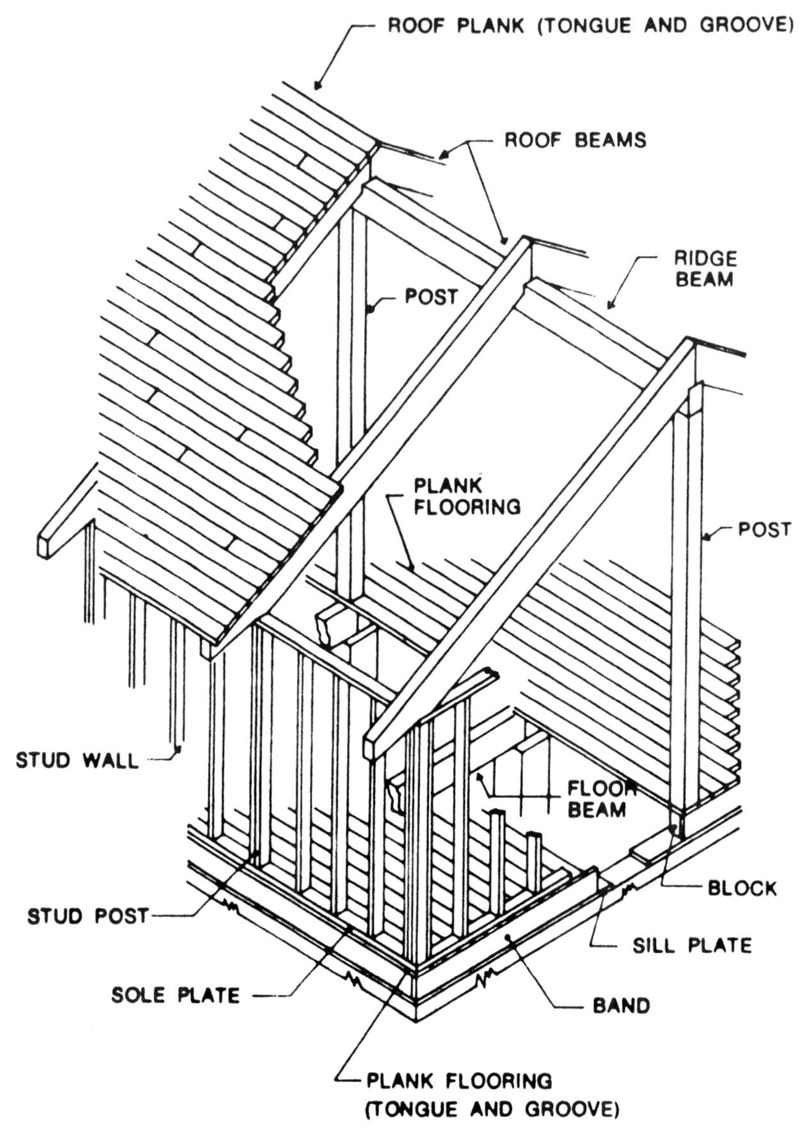

Figure 728 Post and beam construction

post and beam construction Building system that uses heavy posts and beams to form the basic framework. (Figure 728.) The use of timbers creates a load-bearing system that allows an open arrangement; the wall area, except for the posts, is nonbearing. Heavy planks are used on the floor and roof. Also called *plank and beam construction.* Widely used for heavy construction, storage buildings, industrial buildings, and farm buildings. Many farm buildings use poles rather than posts and are called *pole construction.*

post cap Shaped metal fastener used for tying wood posts to beams. (Figure 729.) Also called *beam clip*. See *cap plate, post anchor*.

post construction See *post and beam construction*.

postheating In welding, heating and weldment after completion of the weld process.

post World War II American architecture Architecture developed after the mid 1940s. See *California ranch, contemporary, mobile home*. See also *architectural styles*.

potable Water that is fit for drinking.

pot life In gluing, the working time that a glue is usable.

pouch A nail apron; a carrying compartment worn around the waist or attached to the belt.

poured in place Concrete *cast in place*.

powder-actuated gun Popular gun used for shooting studs or metal pins into concrete or steel. A specially designed blank cartridge drives the pin or stud into the material. (Figure 730.) See also *power hammer*.

powder loads Powder-actuated guns use powder loads of different strengths for different hardnesses of material. Load numbers, cartridge case finish, and colors are used to identify different strengths. The Powder Actuated Tool Manufacturers Institute has established a standard code. The strength used depends on the material. Both .22 and .25 caliber loads are used.

Load number	*Case finish*	*Head or wad color*
1	*Brass*	*Gray*
2	*Brass*	*Brown*
3	*Brass*	*Green*
4	*Brass*	*Yellow*
5	*Brass*	*Red*
6	*Brass*	*Purple*
7	*Nickel*	*Gray*
8	*Nickel*	*Brown*
9	*Nickel*	*Green*
10	*Nickel*	*Yellow*
11	*Nickel*	*Red*
12	*Nickel*	*Purple*

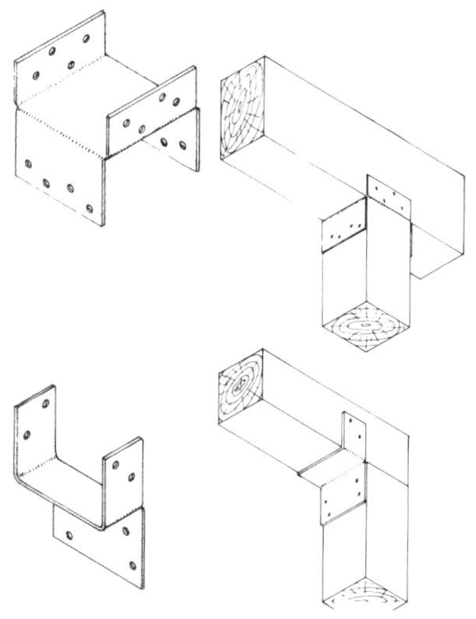

Figure 729 Post caps

power Electrical current used to run electrical equipment. Power is calculated in *watts*.

power block plane See *power plane*.

power bore bit Heavy-duty bit; leaves a cleaner hole than the spade bit. (Figure 731.)

power drill *Electric drill*. See also *power screwdriver*.

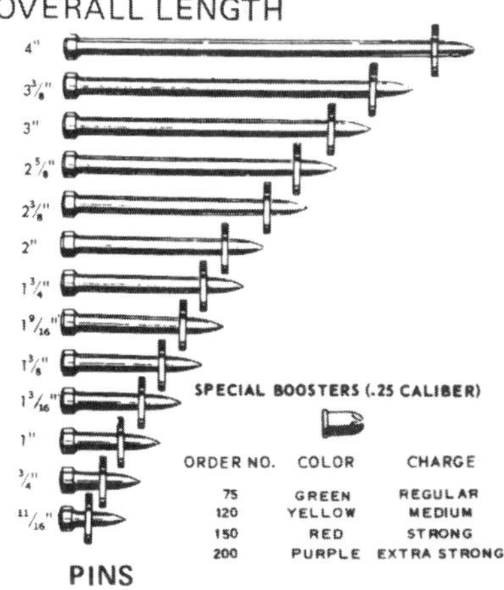

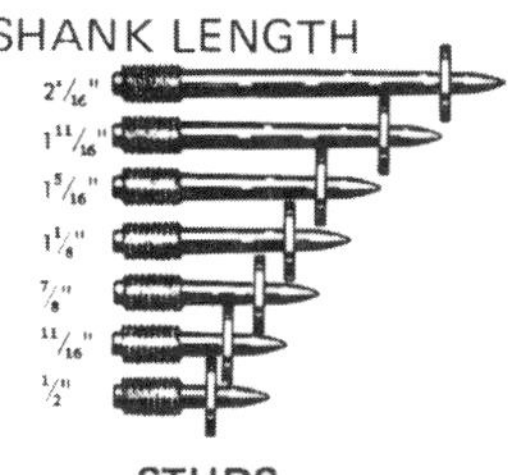

Figure 730 Powder-actuated gun (top). Studs and pins (bottom). Blank cartridges are used to drive the fasteners

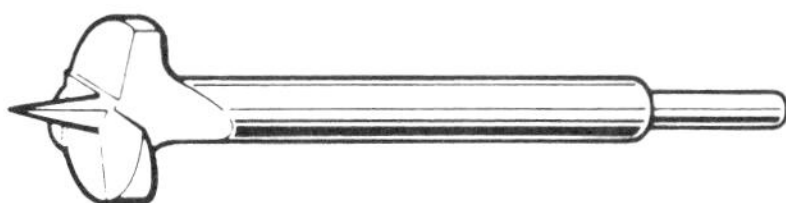

Figure 731 Power bore bit

Figure 732 Power hammer

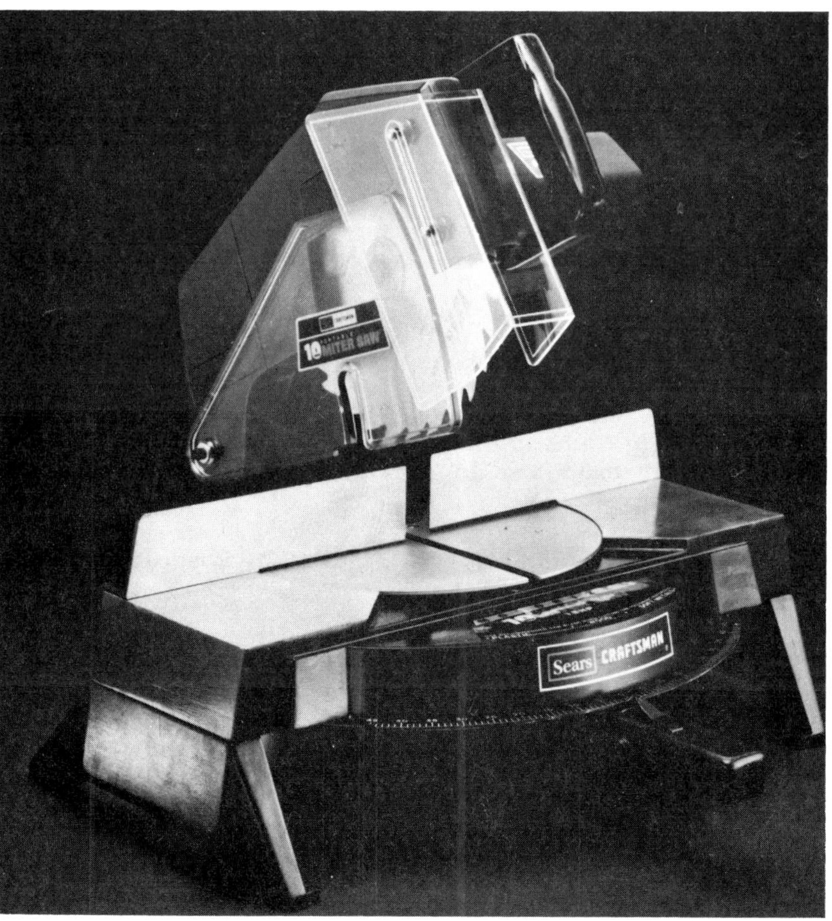

Figure 733 Power miter box

power hammer Powder-actuated gun for driving fasteners into concrete, masonry, or steel. (Figure 732.) Fastener is loaded into end of gun, powder cartridge is loaded into chamber, gun is positioned, and the head is struck with a hammer to explode the cartridge and drive the fastener home. Different loads and fasteners are available for attaching boards or metal parts to concrete, cement block, brick, steel. See also *powder-actuated gun*. Also, an air-powered vibrating hammer with interchangeable tool points, such as chisels, spades, and cutters, used for breaking concrete or other hard surfaces. Also called a *jack hammer*.

power miter box An electric saw and miter box combination used for cutting various angles from 90° to 45°. Cuts wood pieces up to 2″ × 4″ in size. (Figure 733.)

power outlet Any point in an electrical system where current may be used. See *receptacle*. Power-supply outlet for operating temporary facilities or equipment, such as a mobile home, house trailer, recreational vehicle, or boat.

power panel Electrical distribution box. See *panelboard*.

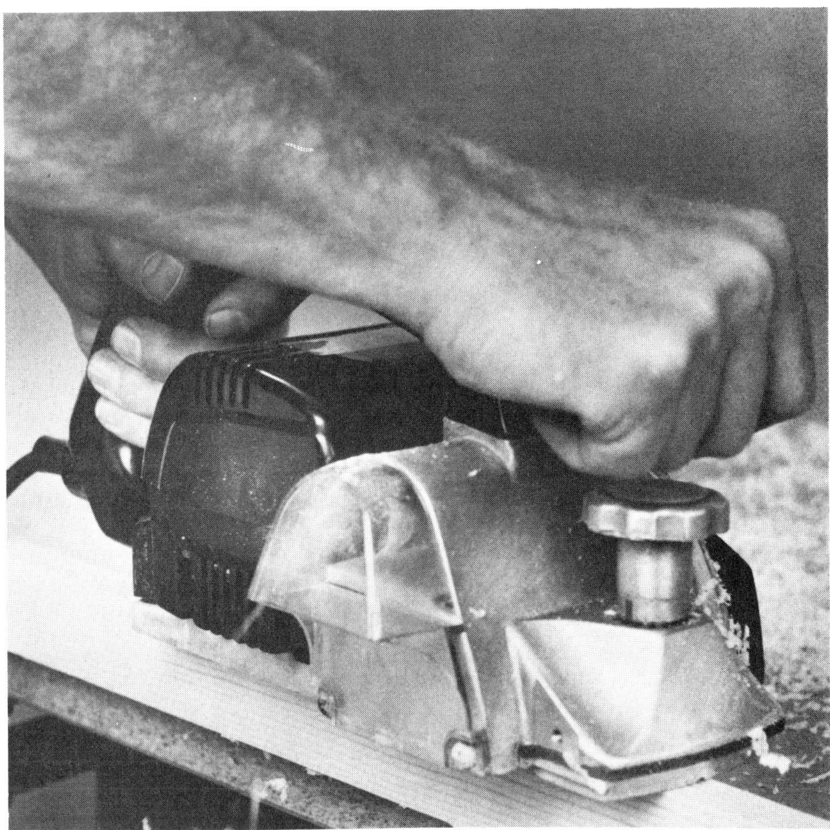

Figure 734 Power plane

power plane Electric plane used for fast work in the shop or on the job. Cutter heads cut the stock away. The larger power plane is similar to the hand jack plane but works the same way as the jointer. The smaller power plane is similar to a block plane and is sometimes called a *power block plane.* (Figure 734.) See also *jointer, router.*

power plant Electrical generating plant, usually coal-fired. Nuclear heat also is used. Heat from coal or a nuclear core creates steam which drives generators to produce electrical power.

power saw Electric or gasoline-operated saw. See *band saw, chain saw, circular saw, concrete saw, jig saw, power miter box, radial saw, reciprocating saw, saber saw, table saw.*

power screed Gasoline-operated screed used for finishing concrete *flatwork.* Two screed boards are mounted under the engine. (Figure 735.)

Figure 735 Power screed

Figure 736 Power screwdriver (cordless)

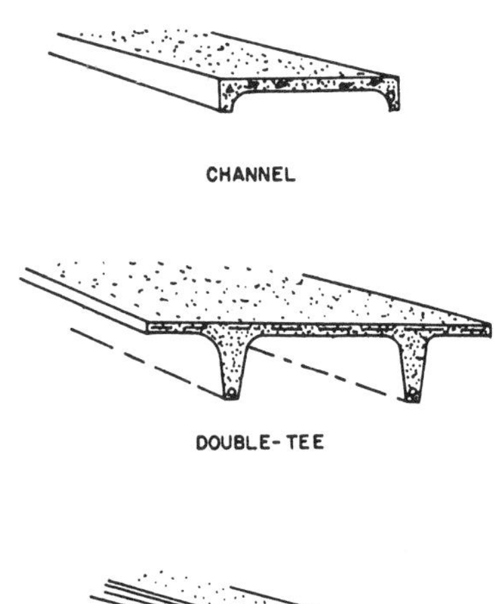

Figure 738 Precast concrete panels

Figure 737 Power trowelers

power screwdriver Electric screwdriver used for driving screws to attach panels and gypsum wallboard and for general use. Most power screwdrivers are also used with drill bits for drilling. On the job a cordless screwdriver/drill powered by rechargeable batteries, is often used. (Figure 736.) See also *drywall screwdriver.*

power troweler Gasoline-powered trowel used for finishing large concrete areas. The trowel blades turn, smoothing out the concrete. (Figure 737.) Hand troweling is still needed in hard-to-reach areas.

power wrench Air-powered driving tool used for installing and removing nuts from bolts or studs. Used in the erection of steel framing members and reinforced prestressed concrete members. The wrench is calibrated for the specified torque.

preapprenticeship training Introductory training to prepare someone for an apprenticeship program.

precast concrete Concrete parts (panels, beams, columns, slabs) that are cast away from the structure, often in factories, and delivered to the site, where they are hoisted into the structure. Some parts are cast at the structure and then tilted into place. (Figure 738.) See *prestressed concrete, tilt-up construction.*

precut Of framing members, cut to needed sizes before delivery to job site or to prefabrication area.

precut lumber construction Construction where framing members, such as studs and joists, are cut at a location away from the building site. Precut pieces are delivered to the site and then nailed into place.

predecorated panels Panels finished in various colors or textures ready for installation. When they are installed, nails with colored heads are used to match the panel colors. Plywood, hardboard, and gypsum wallboard are predecorated.

preengineered buildings Complete steel frame buildings that are manufactured in a plant and erected at a site. Rigid frame and post-and-beam buildings are preengineered. Interior work is done separately. Figure 739 shows a preengineered, modular building using monolithic steel framed sandwich wall panels. See also *industrialized building.*

prefabricate To manufacture a house part, unit, or component in a factory. The component is delivered to the job site, where it is assembled into place. Whole rooms, such as a bath, are prefabricated with all mechanical parts and finish in place. See *module.* Complete houses may be prefabricated. See *industrialized building.*

prefabricated scaffold Scaffold made of tube steel parts. The parts are assembled on-site as needed, then disassembled for later reuse. See *tubular scaffold.*

preferred angle In stair building, a stair slope between 30° and 50° designed to give a safe stair. A slope over 50° is called a *critical angle* and is considered unsafe for general use.

prefinished Any building part or unit that has all the finish in place, as a prefinished door or window, when delivered to the building site.

preglazed Window or door that has lights in place.

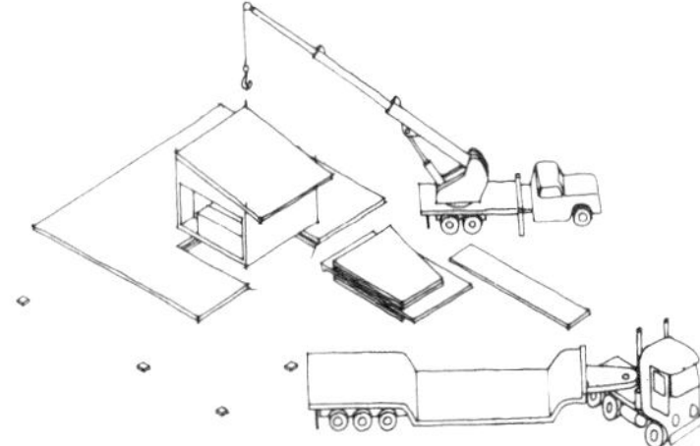

1. Prefab Core Set in Place

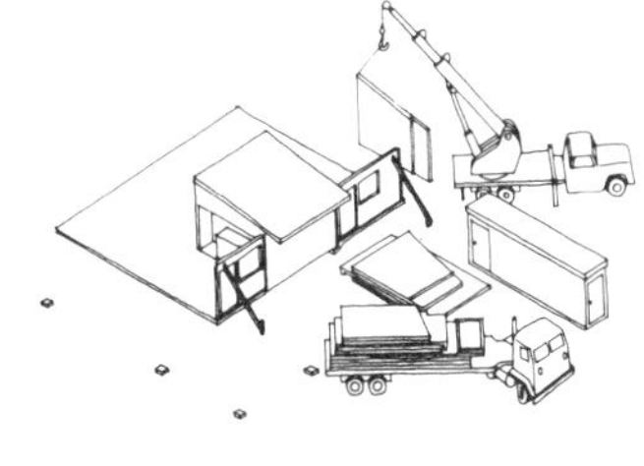

2. Wall Panel Erection Begins

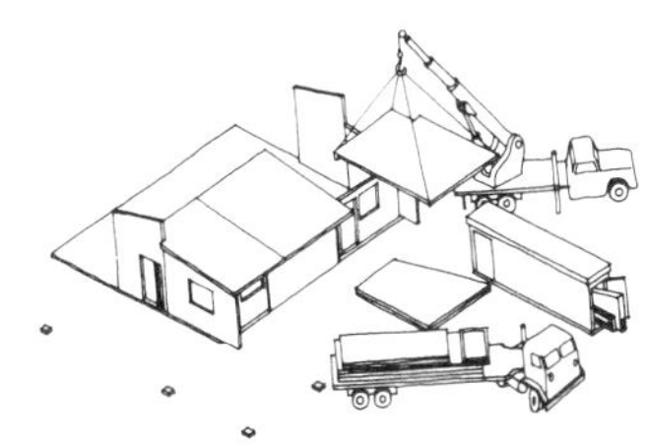

3. Lower Roof Panels Set

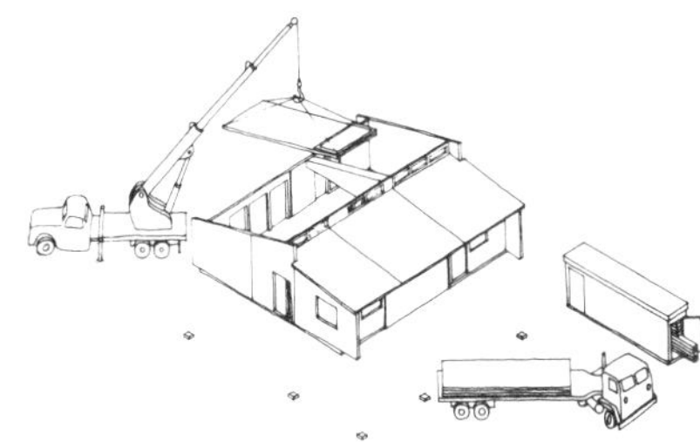

4. Upper Roof Panels Complete Main Structure

Figure 739 Pre-engineered building

preheating In welding, heating the base metal before beginning the weld process. In solar heating, use of solar energy to raise the temperature of household water before it flows into the hot-water heater.

prehung door Door that is hung in the frame.

premix Factory-mixed preparation, such as plasters and cements, that can be mixed with water and used. See *ready mix.*

preservative Chemical that is absorbed into the wood to protect the wood against decay.

pressed wood A manufactured wood product made by pressing wood chips or fibers together with heat. A binding agent, such as a resin, may be used.

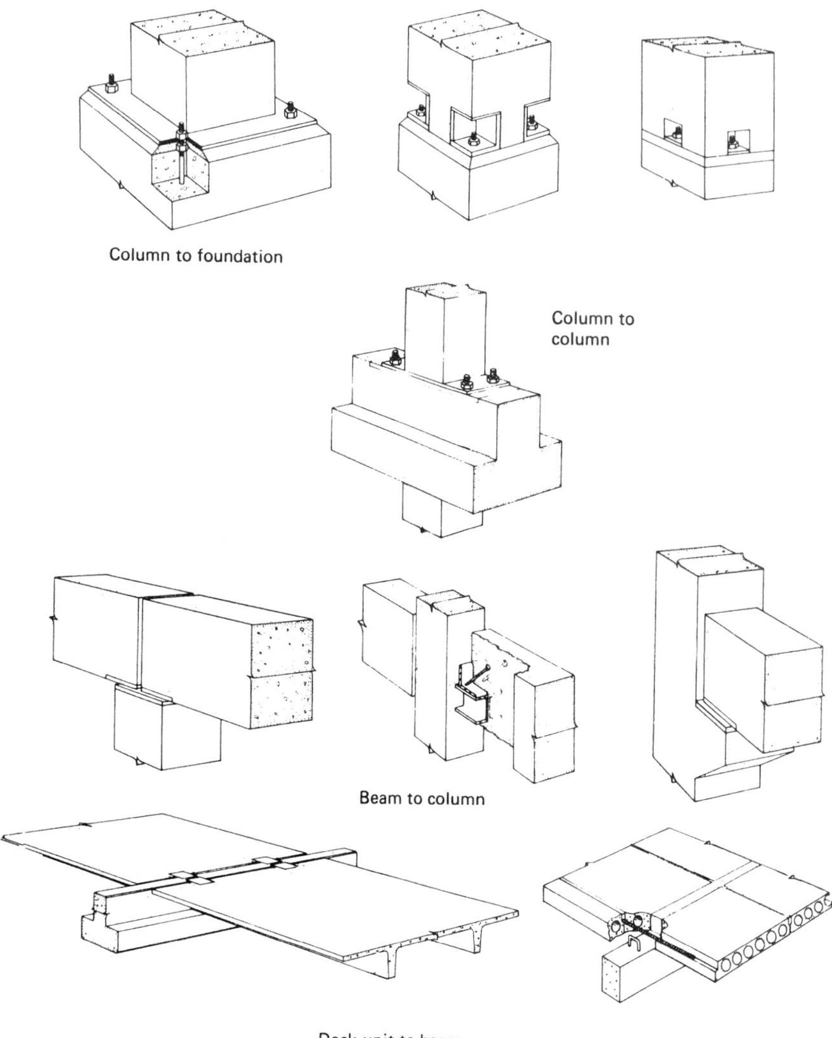

Figure 740 Prestressed concrete connections

press screw See *parallel clamps.*

pressure connector A *solderless wire connector.*

pressure-treated lumber Lumber treated under pressure with chemicals designed to protect against termites, fungi, and the like.

pressure treatment Treatment of wood by exposure to wood preservative under pressure.

prestressed concrete Precast concrete where the steel reinforcement wires are held under tension or stress while the concrete is setting. After setting, the stressed steel is released, causing a permanent stress in the concrete member. The stress in the concrete adds greatly to its strength. Figure 740 shows connection details for prestressed concrete. Also commonly called reinforced prestressed concrete. See panel schedule, precast concrete, pretensioning.

pretensioning In reinforced prestressed concrete, the tensioning of the steel reinforcement wires or tendons, or the strands (six wires in a helix around a seventh, larger wire), before concrete is placed in the mold.

prick punch Punch used to prick holes into sheet metal to locate a cutting line or profile. (*Figure 748*, p. 327.)

primary colors Basic colors used in painting: red, yellow, and blue. See *secondary colors.*

prime contractor Overall contractor with responsibility for the whole project. Normally, he subcontracts parts of the project.

primer coat First paint coat, a base used to fill the wood pores and preserve the wood.

principal meridian In surveying, a north-south longitude line that is used to determine a local line for establishing townships. See *base line, range numbers.*

production welding In-plant welding.

profile Outline; often an outline or contour taken from the side. A section view of a lot area, showing slope. A section.

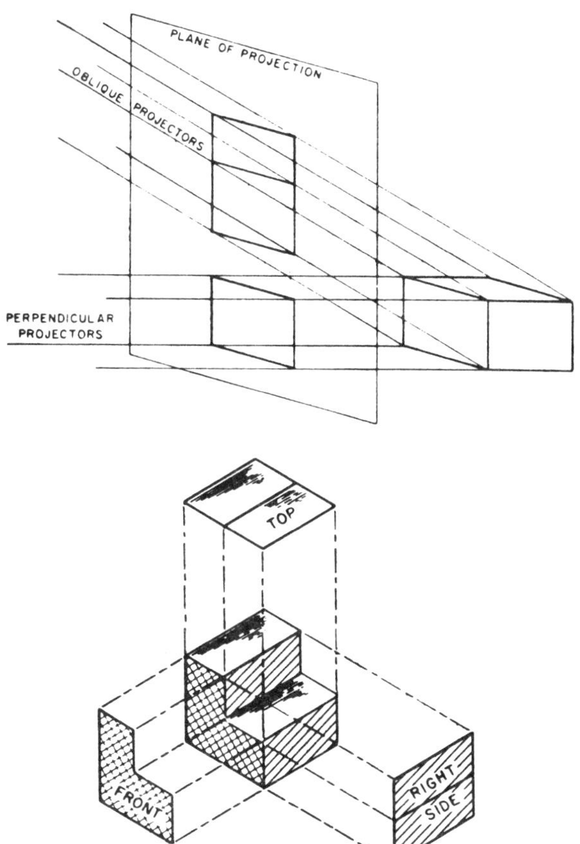

Figure 741 Projection in drafting

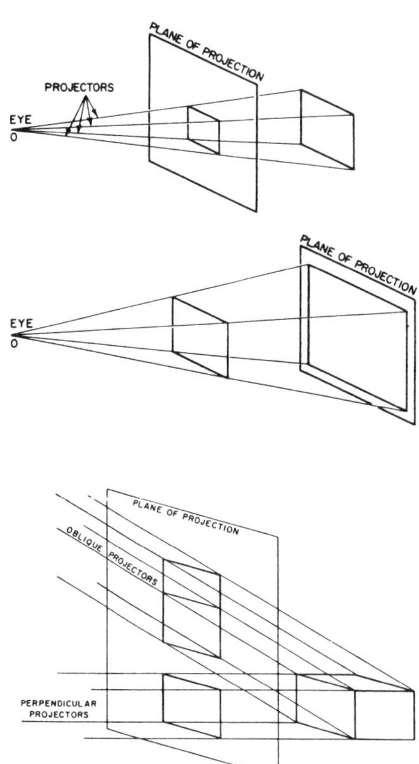

Figure 742 Projectors

project A complete construction job — anything from a house to a development or skyscraper. A design or plan. In woodworking, a finished piece.

projected Jutting out, as a house bay or a masonry course.

projection A horizontal distance out, as rafter extension from the face of a wall. Do not confuse with rafter *overhang*, which is distance measured from the outside edge of the top plate to the rafter end running along the side of the rafter. (However, projection and overhang are sometimes used interchangeably.) In drafting, a drawing made by running lines from a figure or drawing to a drawing. (Figure 741.)

projection plane *Plane of projection.*

project manual Volume which covers the requirements for a building project, including bidding information, contract between owner and contractor, bond and security information, general and supplementary conditions of the contract, addenda to the contract information and documents, and the specifications.

projector The line drawn from a figure to the *plane of projection.* (Figure 742.) See *orthographic projection, projection.*

Figure 743 Propane torch

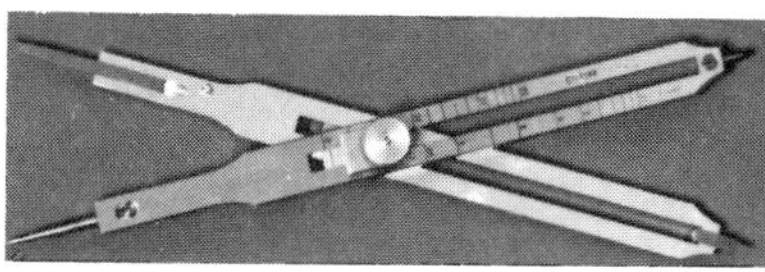

Figure 744 Proportional dividers

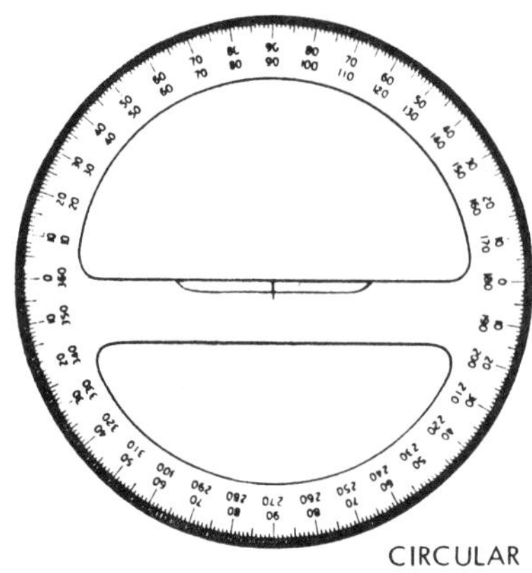

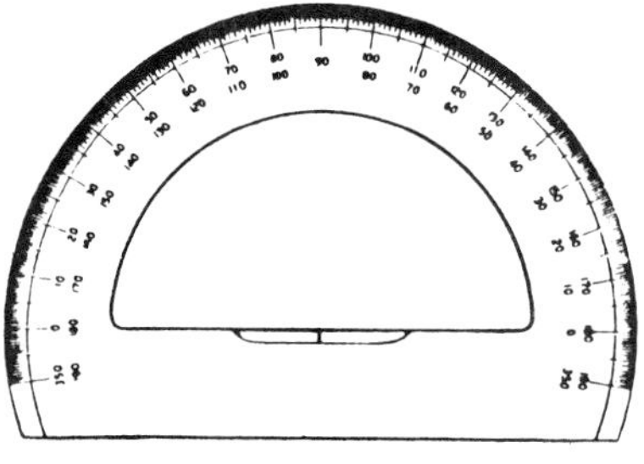

Figure 745 Protractors

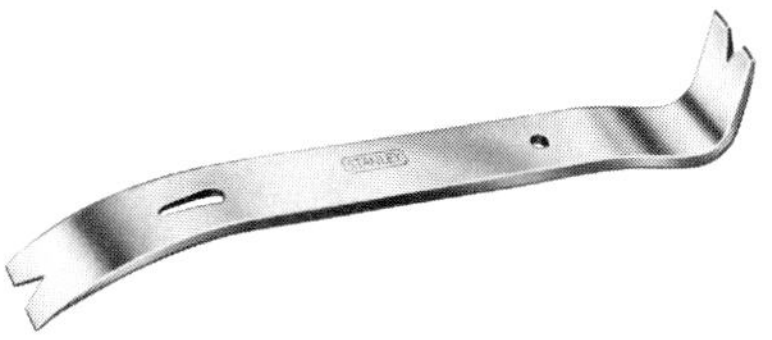

Figure 746 Pry bar

propane torch A small, easily portable torch used for soldering and brazing. The disposable canister is filled with propane gas. (Figure 743.) See also *oxy-propane torch.*

property Building lot and all structures on it.

property line Surveyed lot line of a building site or a property; a *lot line.*

property title Recorded document proving ownership of property.

proportional dividers Special dividers used for reducing or enlarging something or a part of a drawing. (Figure 744.) By adjusting the legs you can obtain a ratio of one part to another. One end of the legs extend over the line or thing to be reduced, the other end gives the desired reduction or enlargement. For example, a 4″ line may be easily reduced to one-fourth the original size by adjusting the scale. Thus the leg opening at one end will be 4″; the leg opening at the other end will be 1″. See also *pantograph.*

proportioning Ratio of one thing to another.

protection shed Covered walkway used over a sidewalk along a construction site. Designed to protect pedestrian traffic.

protective clothing Clothing worn to protect oneself while working on a job. This includes, as appropriate, steel-tipped safety shoes, eye protection, and a hard hat. When welding, additional protective clothing is required, such as goggles, helmet, gloves, apron, leggings, sleeves, and cape.

protractor Semicircular or circular device marked in degrees. Used for measuring and marking angles. (Figure 745.)

pry bar Bent, flat metal levering device used for pulling nails. (Figure 746.) See also *crowbar, nail puller, ripping bar.*

psf Pounds per square foot.

psi Pounds per square inch.

P trap Plumbing trap used under a fixture, such as a sink, to hold water and thus prevent sewer gases from flowing up through the drain. (Figures 747; 556, p. 238.) See also *drain-waste-vent system, S trap.*

public housing Low-rent housing developed with government aid.

public space Legal open space accessible by a public way or street and devoted to public use.

puddle In welding, the molten area being welded.

puddling In concrete work, the consolidating of the freshly placed concrete with a rod or mechanical vibrator.

Pueblo architecture See *adobe.*

pull Drawer handle.

pull chain See *pull switch.*

pull switch Overhead light with a switch operated by pulling a string or chain. Abbreviated ps (pull switch) or pc (pull chain).

pulverize To break into fine parts.

pumice A powder of fine volcanic stone used for polishing. The solid stone itself is sometimes used for smoothing a surface.

pump pliers See *adjustable-joint pliers.*

punch Steel rod-shaped tool with a point; the punch is struck with a hammer to make holes or dents in metal. Figure 748 shows several punches. Center and prick punches are used to make a dent to locate a point. The drift or starting punch is a heavy-duty punch used for removing rivets or pins. The pin punch is used to drive a rivet, pin, or other object from a hole. The aligning punch is used to line up holes in different pieces of metal. See also *hand punch.*

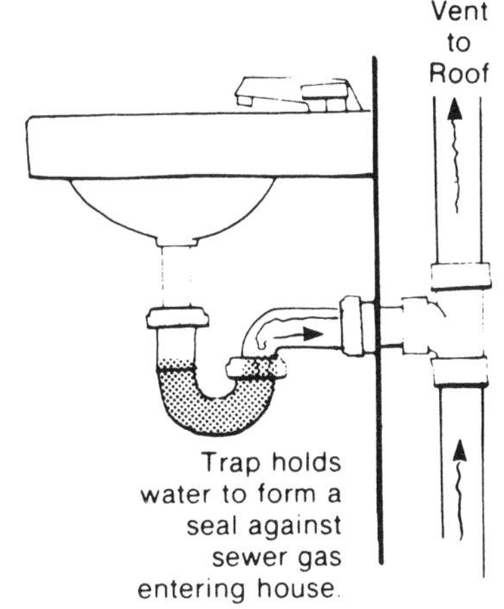

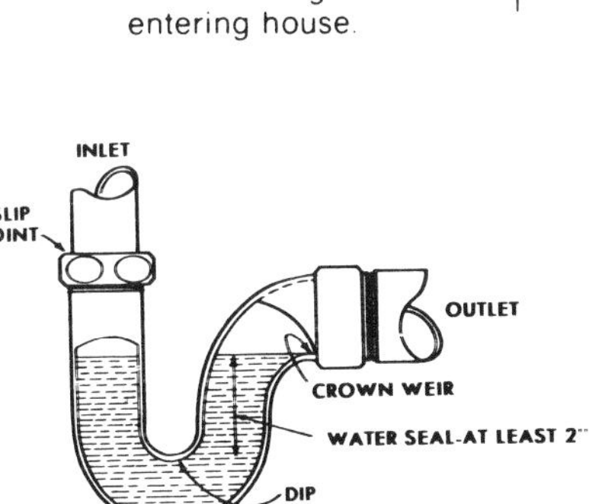
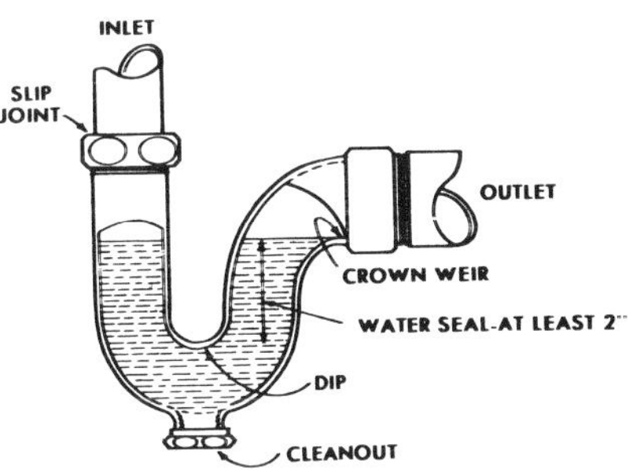

Figure 747 P trap

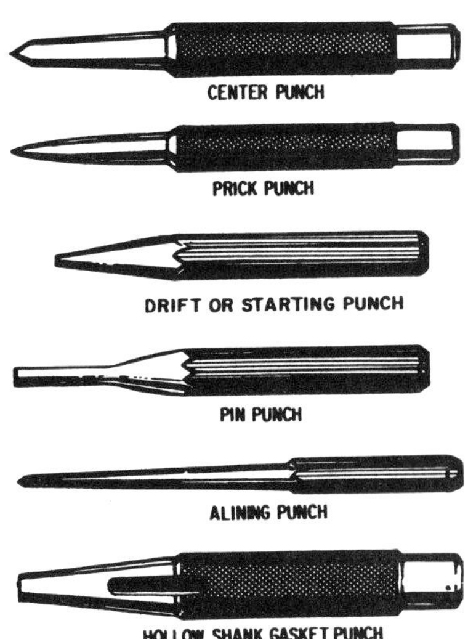

Figure 748 Punches

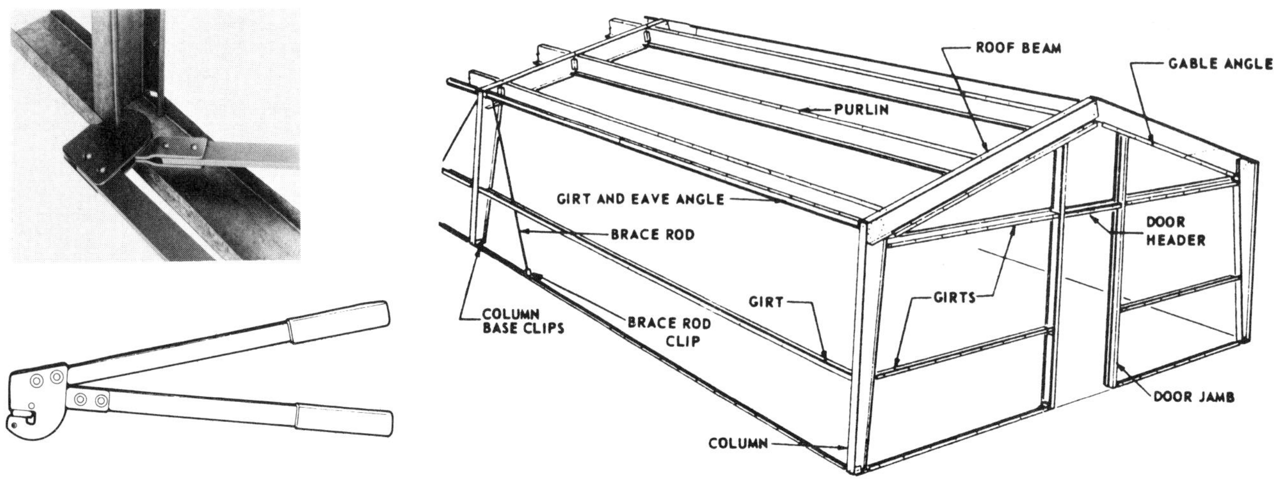

Figure 749 Punch-loc riveter

Figure 750 Purlins (used in rigid frame construction)

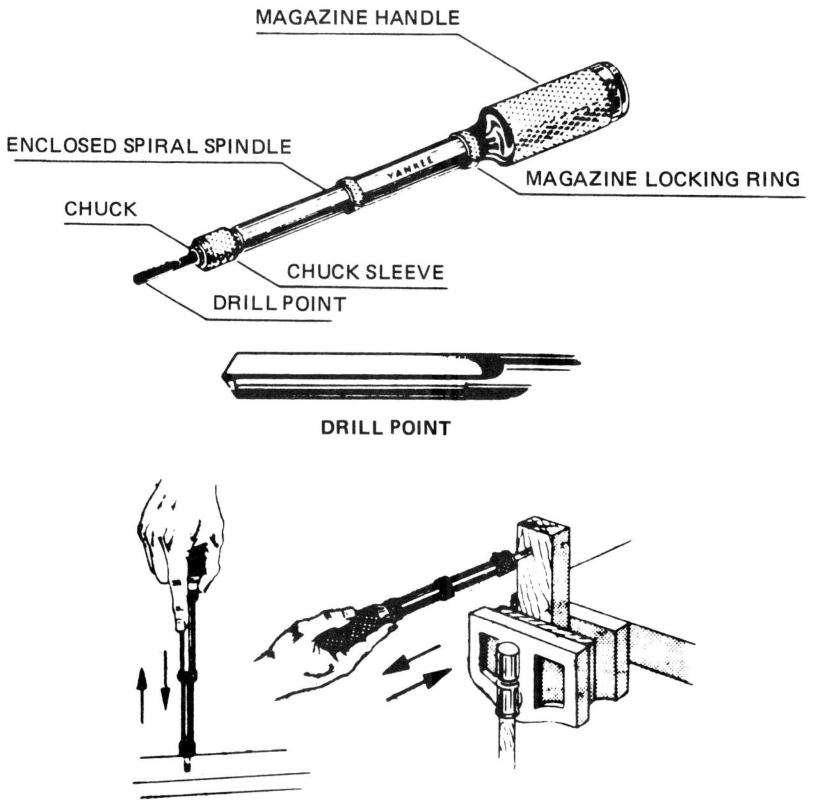

Figure 751 Push drill

punch-lok riveter Lever hand tool for riveting metal studs and runners. (Figure 749.) See also *metal lock fastener.*

purging Removing refrigerant from air conditioning system; removing gas from a gas supply system.

purlin Horizontal support member running over, at a right angle, rafters or trusses. (Figures *750, 428,* p. 173.)

push drill A hand drill that is operated in an up-and-down motion. (Figure 751.) Used for drilling holes through a hinge or other hardware while the hardware is held in place. A straight flute drill point is used. See also *spiral ratchet screwdriver.*

push-pull rule Rule consisting of a steel tape wound up in a case. (Figure 752.) Widely used on the job for layout and measuring. Commonly carried clipped to the belt. The steel tape is marked in feet and inches and commonly with 16-inch modules (for easy layout). Common lengths are 6, 8, 10, 12, and 20 feet.

Figure 752 Push-pull rule

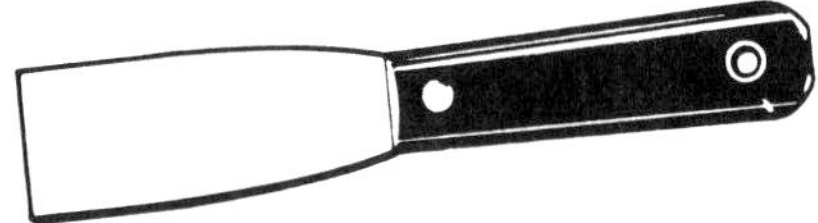

Figure 754 Putty knife

push shoe A hand-made device for pushing a lumber piece through a power finishing machine, such as a saw, jointer, or planer. A push shoe has a flat base with a handle. (Figure 753.) See *push stick.*

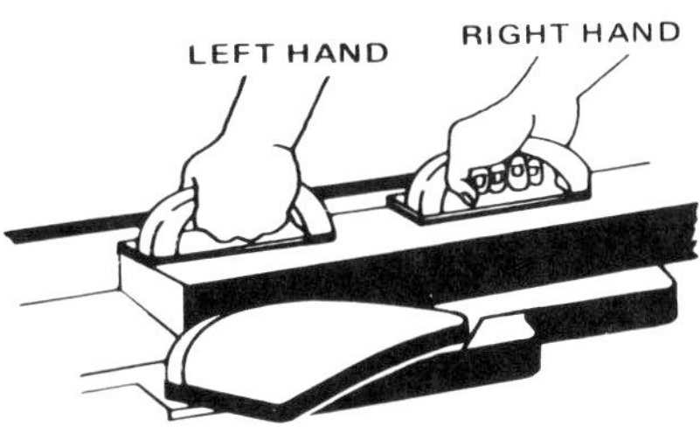

Figure 753 Push shoes: used to push short stock through planer

push stick In woodworking, a hand-made shaped piece of wood used for pushing wood through a power saw, jointer, or planer. Also called a *push shoe,* except that a push shoe has a flat base with a handle.

putlog In scaffolding, a support piece that is attached to or through a wall. (*Figure 834,* p. 366.) Also, base timbers that support the feet of a scaffold.

putty Filler used to fill small holes or to cement window lights in frame. In plastering, quicklime slaked with water, used as a finish coat (mixed with sand or plaster of Paris).

putty coat Plastering finish coat; a *smooth coat.*

putty knife Long-bladed, flat-edged knife used for patching, filling, and seaming with putty or spackle. (Figure 754.)

PVC A light-colored plastic (polyvinyl chloride) piping used for drainage and venting. See *plastic piping.*

q

quadrangle Four-sided open court, often surrounded by buildings.

quadrant A quarter of a compass area, 90°. Often used as a general directional indicator, as the north east quadrant. Measuring tool that has a compass divided into four areas, as a navigational instrument with a 90° dial used for measuring sun or star angles. Figure 755 shows quadrant scales used in surveying; angles are shown marked from the north-south line. See also *compass.*

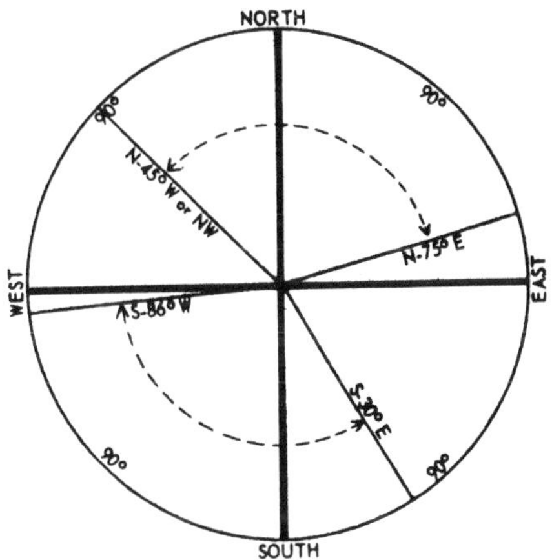

Figure 755 Quadrants. Angles are marked from the north-south line

quadrilateral Four-sided figure.

quadrominium A four-unit or four-plex residence designed for four families but, in appearance, resembling a large, single-family home.

quantity takeoff Detailed listing of materials needed to do a job.

quarry Open pit from which stone is removed. A square stone or tile. Diamond-shaped pane of glass; also called a *quarrel.*

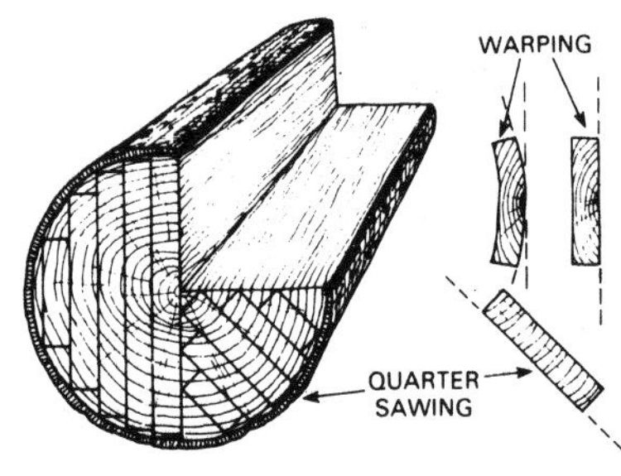

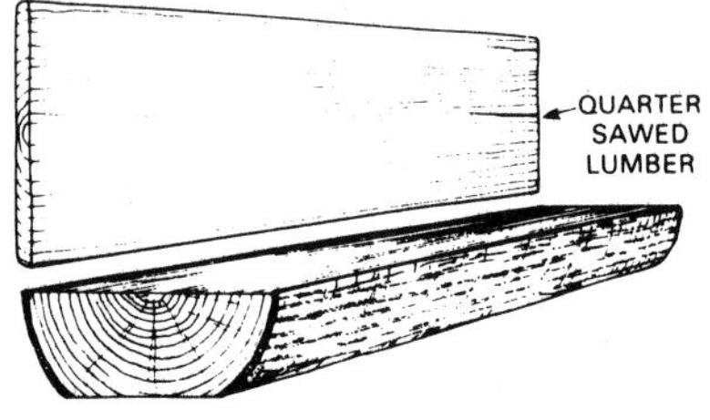

Figure 756 Quartersawn lumber

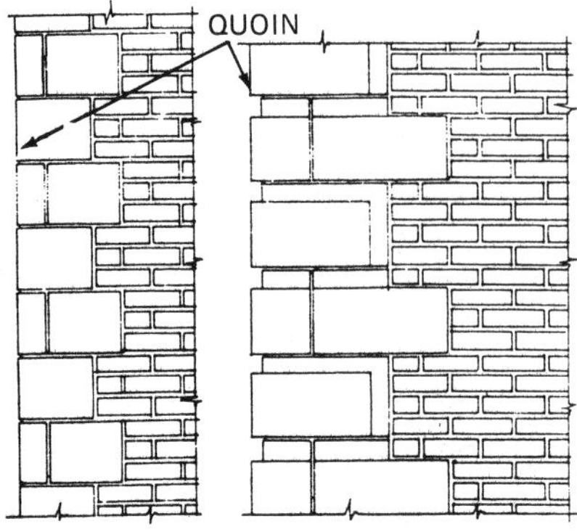

Figure 757 Quoins

quarry masonry Rough-faced, squared quarry stone; a wall built of quarry masonry.

quarry stone bond A bond pattern made using rough stones, as in a rubble wall.

quarter bend Pipe bend of 90°

quarter round Molding which is a quarter of a round (cylinder). See *molding*.

quartersawn Log is sawn lengthwise into four quarters, then each quarter is sawn parallel to the face to produce lumber with the annual growth rings fairly close to a right angle to the face of the lumber. (Figure 756.) Quartersawn lumber has little warpage or checking. Quartersawn is the term used in reference to hardwood. The same cut for softwood is referred to as *edge grain*. See also *plainsawn*.

quarter surface Wood cut parallel to the wood rays; *quartersawn* or *edge grained*.

queen closer Half brick used in a course to offset the vertical joints. Brick is broken in half to make the queen closure. See *cut brick*.

quicklime A caustic lime, calcium oxide (CaO), which, when slaked or hydrated (mixed with water), makes a fine putty used for finish coats. Sometimes referred to as *lime*.

quoin Large squared stones set at the corner of a building. (Figure 757.)

r

rabbet Square or rectangular groove cut in the edge or side of a board or plank. Figure 758 shows a rabbet joint. See *wood joints.*

rabbet plane A low-angle plane used for cabinetwork, especially for planing out rabbets. (Figure 759.) See *plane.*

raceway Metal duct or channel which holds electrical wires or cables or telephone cables.

racking A lateral stress. In masonry work, a procedure followed when courses reach a wall corner — the courses are shortened so vertical joints do not line up at the corner.

radial-arm saw A shop saw having a movable arm that holds the circular saw blade. Blade can be tilted to different angles. See *radial saw.*

radial cut Wood cut made parallel to the wood rays, at a right angle to growth rings; *quartersawn* or *edge grained.*

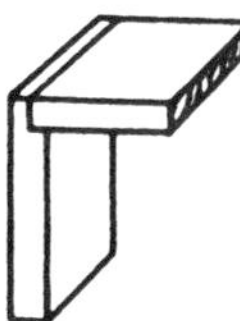

Figure 758 Rabbet joint

Figure 759 Rabbet plane

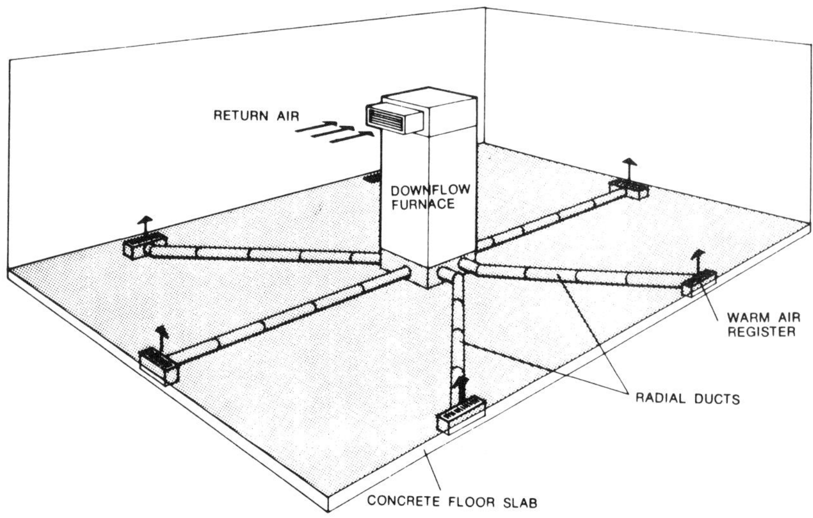

Figure 760 Radial perimeter heating

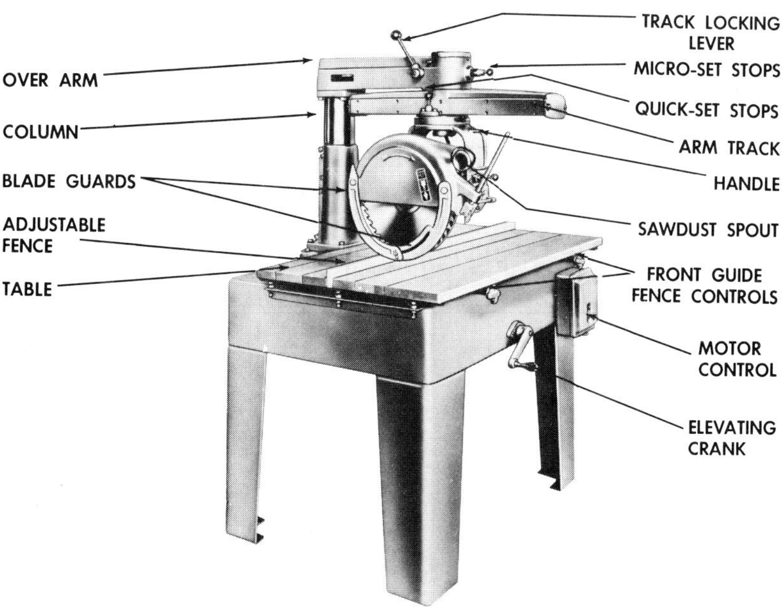

Figure 761 Radial saw

radial perimeter heating A type of *forced warm-air heating* that has ductwork running to individual outlets located around the perimeter of the area to be heated. (Figure 760.) Cool air returns directly to the centrally located furnace. See also *extended plenum heating, perimeter loop heating.*

radial saw Circular saw that moves back and forth on an overhead arm. The saw arm and the saw blade are adjustable for cuts at various angles. The saw is pulled into the work. (Figure 761.)

radiant heating Electrical or hot-water heating by the use of *resistance-wiring* coils or pipe coils in the floor, walls, or ceiling. This creates a heat panel that heats the room by radiation. (Figure 762.)

radiation Heat transfer through means of heat waves. Sunlight is an example of heat through radiation; when the waves strike an object or body, the surface increases in temperature.

radiator Heating outlet for steam and hot-water heating. Heat radiates out from the cast-iron unit.

radius Line from the center of a circle or curve to the outside circumference. In a complete circle the radius is one-half the diameter (distance across the circle through the center).

rafter Wood or metal support member on a sloping roof. (Figure 763.) The basic rafters are:

- **common rafter** Runs from ridge to top plate.
- **valley rafter** Runs at the intersection of two downward-sloping roofs.
- **hip rafter** Runs from ridge to top plate at roof ends to form the hip.
- **jack rafter** Runs from the ridge to a valley rafter (valley jack rafter) or from the hip rafter to top plate (hip jack rafter).
- **cripple rafter** Does *not* touch either the ridge or top plate. Runs from hip to valley (hip-valley cripple) or valley to valley (valley cripple).

In rigid frame construction the horizontal support part of the rigid frame is called a rafter. (Figure 764.) See also *pole construction.*

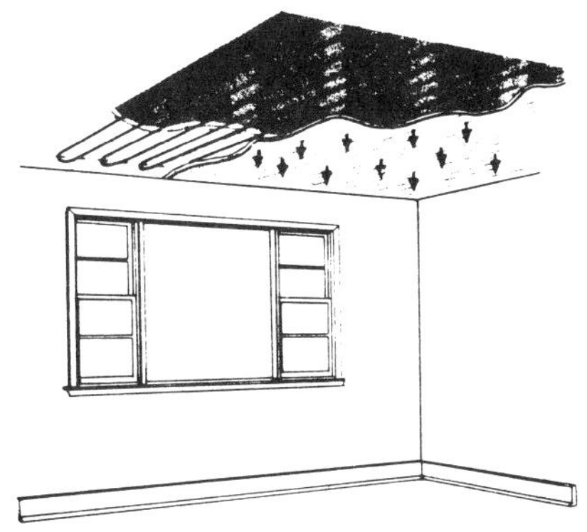

Figure 762 Radiant heating. Coils of electrical heating (resistance) cable are buried in the ceiling to form a heat panel

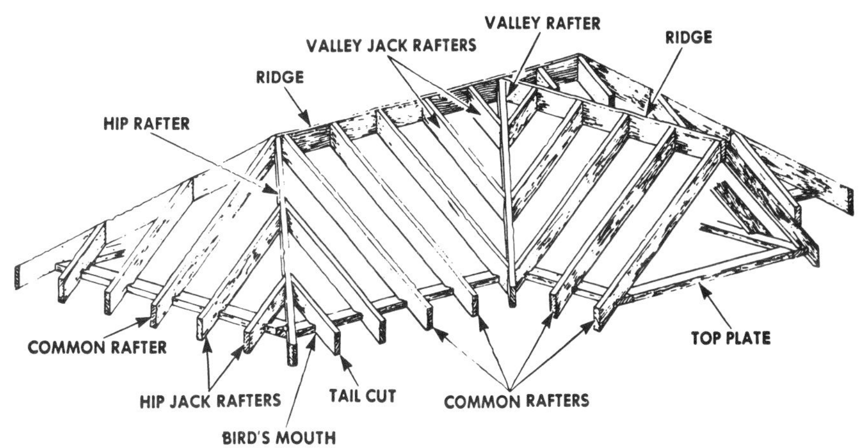

Figure 763 Rafters (wood)

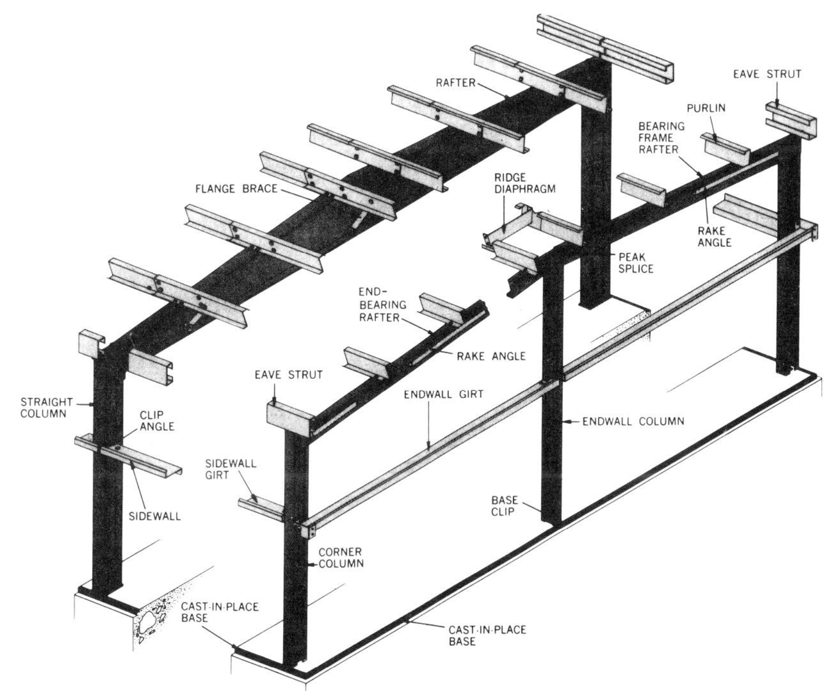

Figure 764 Rafters (steel) used in rigid frame construction

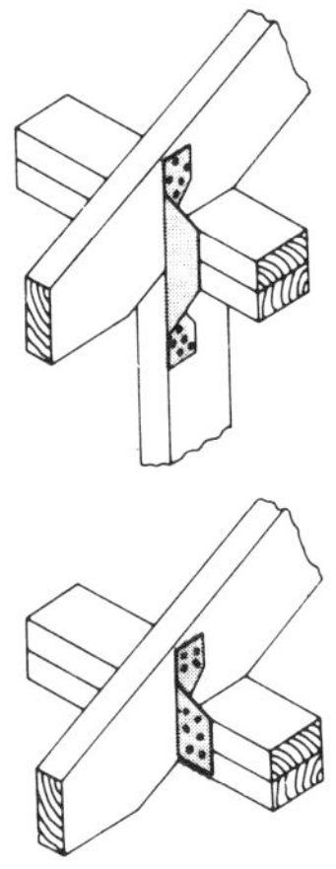

Figure 765 Rafter anchors

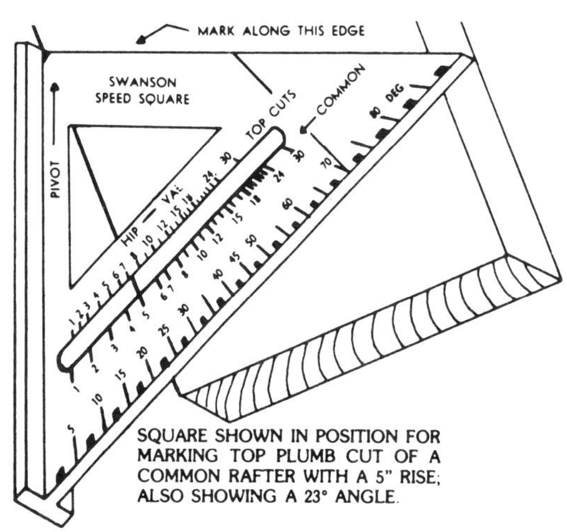

Figure 766 Rafter angle square

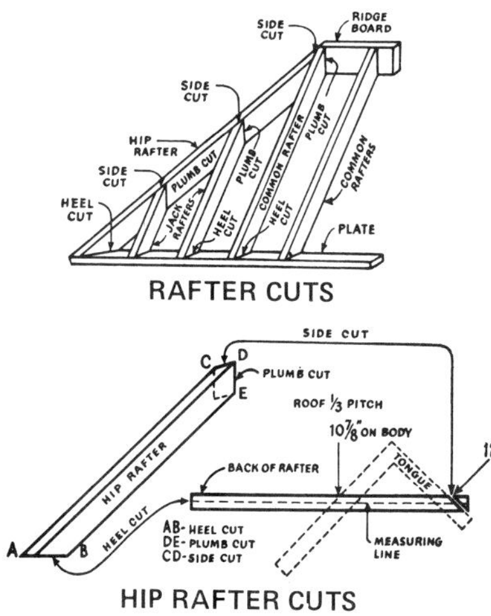

Figure 767 Rafter cuts

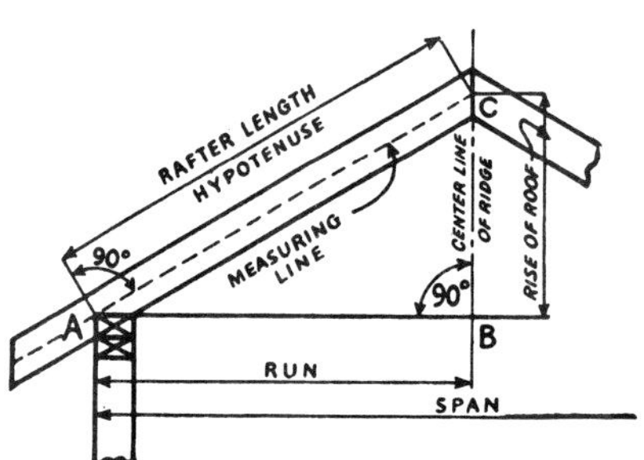

Figure 768 Rafter lengths

rafter anchor Shaped metal fastener used for holding rafters or trusses to top plates and, where appropriate, to the supporting studs. (Figure 765.) Rafter anchors are designed to resist uplift due to winds. Also called *storm clip.* See also *bridging, framing anchor, joist hanger.*

rafter angle square A small triangular square used for marking cutting angles for rafters and the like. (Figure 766.)

rafter cuts The horizontal heel cuts, vertical plumb cuts, and angled side cuts. (Figure 767.)

rafter length Length of the rafter measured from support to support. In the case of a common rafter, for example, the length is measured from where the bottom end frames against the plate (the *plumb cut* on the *bird's mouth*) to the top cut or plumb cut butted against the ridge board. (Figure 768.) See also *rafter run, rafter theoretical length.*

rafter line Line of measurement on a rafter, runs from the heel of the bird's mouth up to the plumb cut at the ridgeboard. See *pitchline.*

rafter pattern Full-scale rafter template that is used as a pattern to duplicate needed rafters on a roof system.

rafter plan A plan view showing placement and size of rafters.

rafter plate Plate on which rafters rest, commonly called a *top plate.*

rafter run Horizontal distance a rafter travels measured from support to support. A line is dropped down from the top support end to meet a line projected over from the bottom support end. (Figure 769.) The run is the distance from the lower rafter end support over to the line dropped from the upper rafter end support. Seen as a *right triangle,* the run is the distance of the base. The *rafter length* is the actual measured distance on the rafter itself. The rafter length would be the hypotenuse of the right triangle. See also *roof rise, roof run, span.*

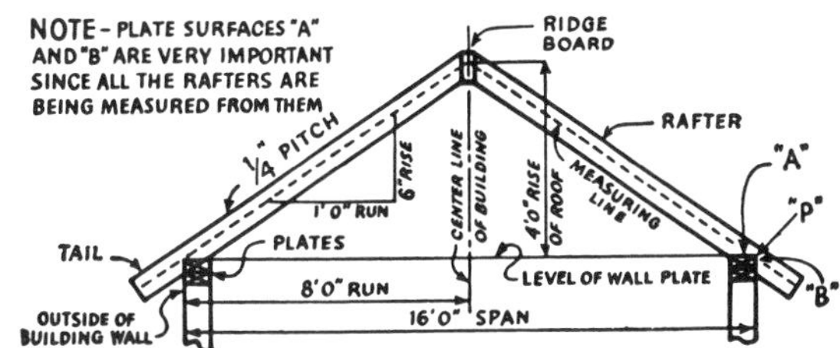

Figure 769 Rafter run

rafter seat The bird's mouth on a rafter. The *seat cut* is the horizontal part of the bird's mouth.

rafter square Right-angle framing square widely used for layout. (Figure 770.) The blade is the long arm and the tongue is the short arm on the square. The heel is the corner. Four tables are given on the square: *rafter* (gives unit length for various rafters), *essex board* (for calculating board feet), *octagon* (used for laying out eight-sided figures), and *brace* (gives common brace lengths). Also called a *framing square, steel square.* See also *rafter angle square.*

rafter table Table on a steel framing square which gives rafter length per foot of run. See *rafter square.*

rafter theoretical length Rafter length from heel cut to the center of the ridge, measured along the top of the rafter. Rafter length without adjustment for ridge reduction (one-half the thickness of the ridgeboard). Rafter is assumed to run to the center of the ridge. The unadjusted length. (Figure 771.)

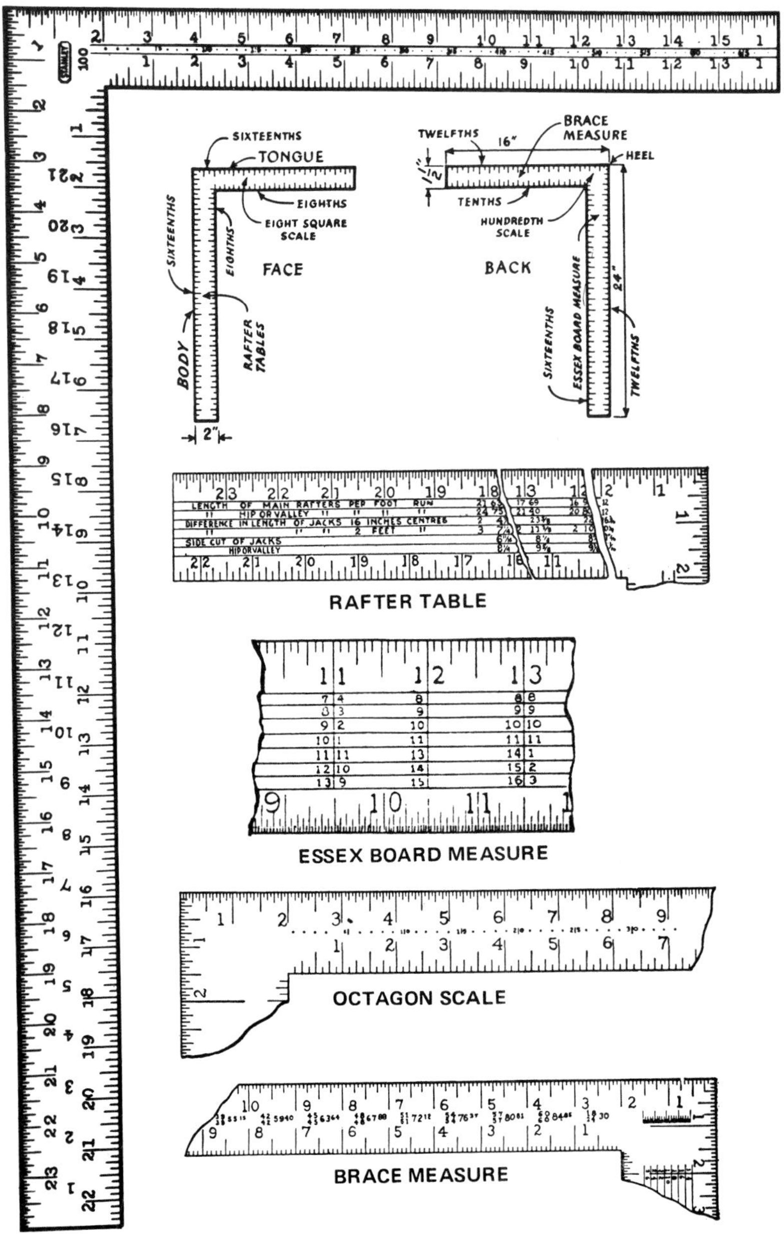

Figure 770 Rafter square

raft foundation Large slab used to support a structure. Also called a *mat.*

raggle A building unit that has a slot to receive the end of flashing. Used on a flat roof where the roof frames into a parapet wall. (Figure 772.) See *reglet.*

rail Horizontal guard at the top of a balustrade; a *hand rail.* In windows and doors, a horizontal member. In a panel door the rail at the top is called the top rail; the rail by the lock, the lock rail; the rail at the bottom, the bottom rail. The door rails fit against the vertical members called *stiles.* See *door, panel door.* In cabinetmaking, horizontal members used to frame a panel in a carcass, horizontal members used to make a panel cabinet door; top and bottom horizontal rails are joined at the two sides by vertical *stiles.* (*Figure 164*, p. 61.) In furniture construction, horizontal seat members on a chair; also cross supports between the back chair posts or legs. See also *post, splat, stretcher.*

railing Hand rail on a stairs or balustrade. (*Figure 57*, p. 23.)

rainproof Constructed so rain will not enter or interfere with operation.

rainwater pipe *Downspout* or *leader.*

raised grain Condition on wood surface where some of the wood fibers raise up from the wood. Caused by moisture swelling the porous springwood.

rake Roof projection on the gable ends of a roof. (Figure 773.) To remove mortar from a joint so the joint may be finished with an elastic caulking or other special mortar.

rake back lead In masonry work, a series of bricks or concrete blocks laid up in the center portion of a wall as a guide. Six or seven courses are laid; each higher course is raked back one-half a brick. See *mason's lead.*

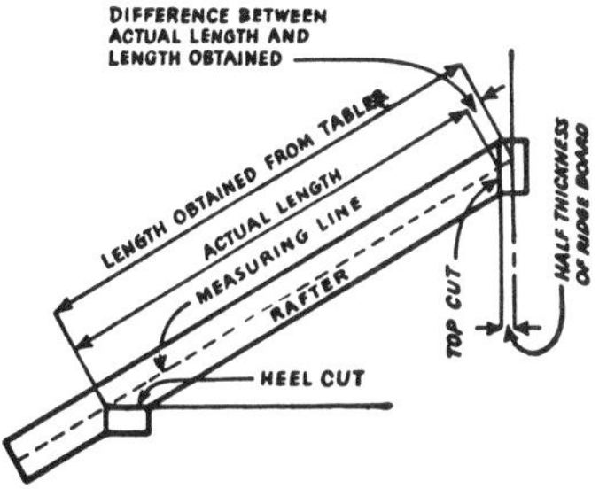

Figure 771 Rafter theoretical length and actual length

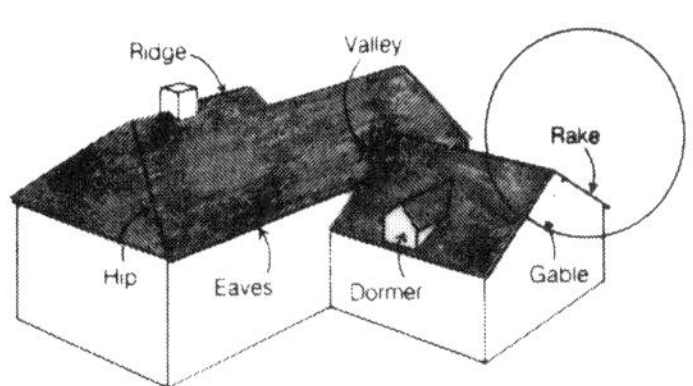

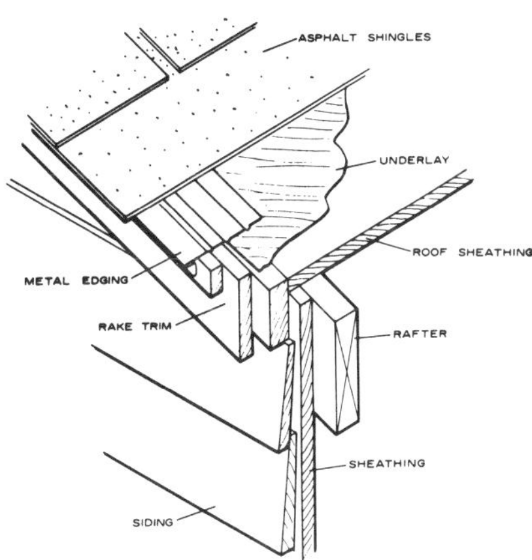

Figure 773 Rake

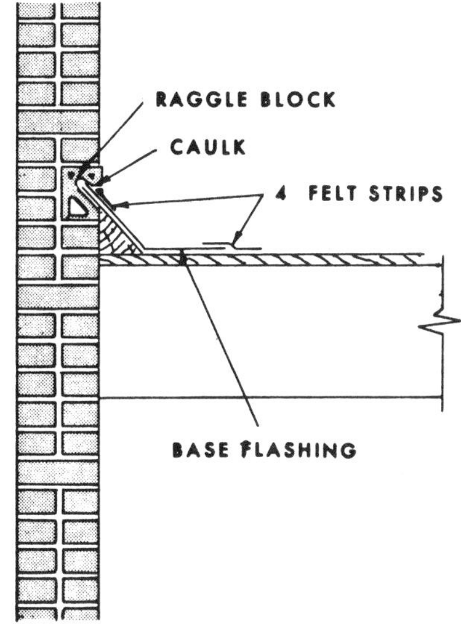

Figure 772 Raggle block

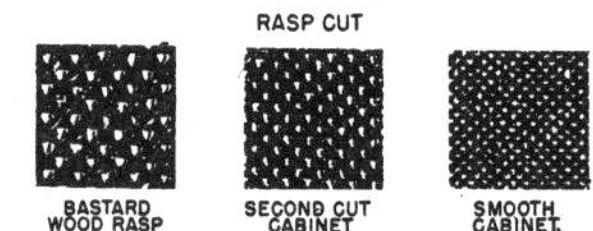

Figure 774 Rasp

raked joint Mortar joint with the mortar tooled back for a distance. See *mortar joints.* The raked joint may be finished with an elastic caulking.

rake molding Trim on the edge of a roof *rake.*

rake out To remove loose, broken mortar or to remove fresh mortar.

rakers Extensions or abutments on the ends of a bowstring truss beyond the outside walls to provide support. The ends of the trusses are extended to the ground.

raking Cleaning out weathered and decayed mortar joints before filling with fresh mortar (tuckpointing). Removing fresh mortar from a joint so it may be replaced by a special weatherproof mortar or elastic caulking.

raking bond Masonry bond pattern where the bricks are laid at an angle or diagonally to the face; a raking bond may use *headers or stretchers* at an angle. See *herringbone bond.*

raking shore Temporary timber brace; one end is fastened against a wall and one end rests on the ground. A *shore.*

rammed earth Construction that creates walls by ramming earth down into wall forms.

rammer A pneumatic device used to pack down earth preparatory to constructing a foundation.

ramp A short bend or change in direction. Also, a short sloping runway or passage, often a temporary sloping plank runway for wheelbarrows; a sloped runway for wheelchair entry.

rance Temporary shoring; timber braced at an angle against a wall.

ranch-style architecture One-story house with a rambling informal plan. A basement is normally omitted. See *architectural styles, California ranch.*

random Irregular. In masonry work, a wall with different-sized stones in no regular pattern. A random bond. See *ashlar masonry.*

random shingles Shingles of varying widths.

range hood fan Hood with exhaust fan mounted directly over the kitchen range, used to exhaust cooking fumes.

range line One of a series of north-south lines running every six miles east or west of a local *principal meridian.*

range numbers Numbering system used to locate a township. A range number is the number of rows or lines a township is located east or west of a local *principal meridian* (north-south line). The range number along with a township number is used to locate the township. See *township numbers.*

range pole In surveying, a brightly painted and striped pole used for aligning the direction of measurement.

ranger See *wale.*

rangework Stonework laid in even courses.

rasp Heavy-toothed abrading tool used for rough grinding, as for roughing out or cutting wood shape. Also used on soft metal. The cutting or abrading surface is composed of individual teeth. (Figure 774.) See also *cabinet rasp, file, half-round rasp, rasp plane, wood file.*

rasp plane A rough plane made from a slotted metal strip mounted on a frame. (Figure 775.) The metal teeth scrape or gouge off the wood.

ratchet Tool which has a reverse mechanism, as a ratchet brace, which may be turned back and forth to drive a bit in a continuous direction. A handle used to turn *sockets.* (Figure 776.) The socket is driven forward by operation of the handle back and forth. As little as a 4″ turn or sweep of the handle is sufficient to drive the socket. A reversing lever changes the ratchet direction. A square drive fitting on the end of the ratchet receives the socket head. The square drive fitting size varies: $\frac{1}{4}$″, $\frac{3}{8}$″, $\frac{1}{2}$″ and $\frac{3}{4}$″ are used. The socket head opening must be the same as the square drive fitting on the ratchet. Socket adapters are available to allow sockets of another size to fit the ratchet square drive. See *socket.*

ratchet brace A brace that has a ratchet attachment for reversing directions. The bit can be driven in the same direction by a back-and-forth motion of the handle. Used in tight corners. See *brace.*

ratchet-lever hoist Hand-operated hoist that uses a rachet action to lift or move a load. (Figure 777.) The ratchet handle is pumped up and down, lifting on the downstroke. See *hoist.*

ratchet nail Nails that have a reverse annular thread on the shank, designed to resist loosening or removal. See *nails.*

ratchet screwdriver Screwdriver with a ratchet attachment for reversing direction. See *offset ratchet screwdriver, spiral ratchet screwdriver.*

ratchet wrench See *ratchet.*

rating Electrical load capacity. Design or strength load of a material.

ratio Relation of one thing to another, as the ratio of $\frac{1}{4}$ is the relation of 4 to 1.

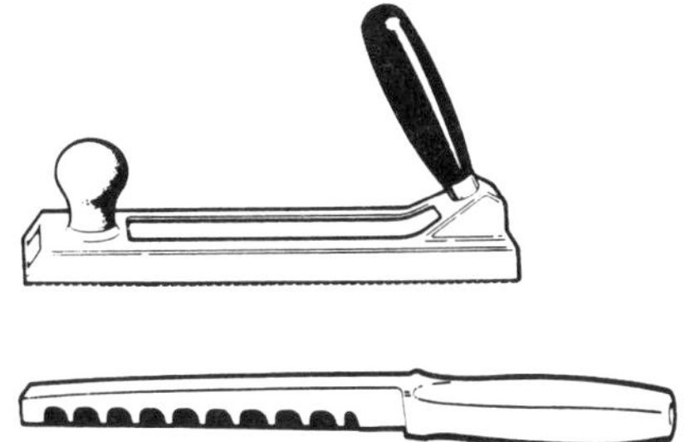

Figure 775 Rasp plane

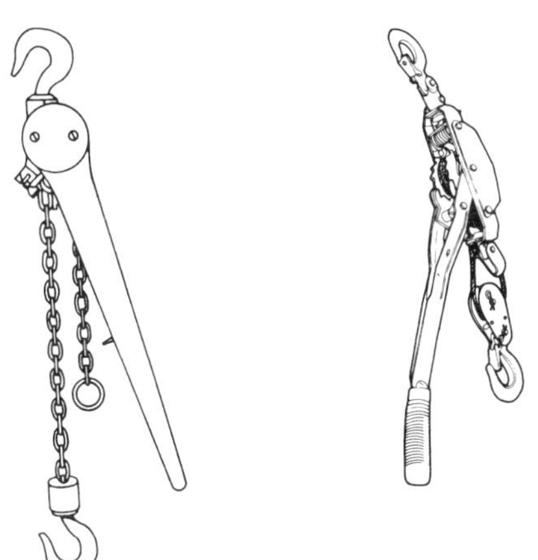

Figure 777 Ratchet-lever hoist

rattail file Small round file that tapers to a point. See also *file.*

rawhide mallet Mallet with a head made of tightly rolled rawhide.

rawlplug Short fiber cylinder lined with lead, used in masonry. The plug is inserted in a drilled hole; then the screw expands the plug to make a tight fit. (Figure 778.) See *fiberplug.*

Figure 776 Ratchet or ratchet wrench

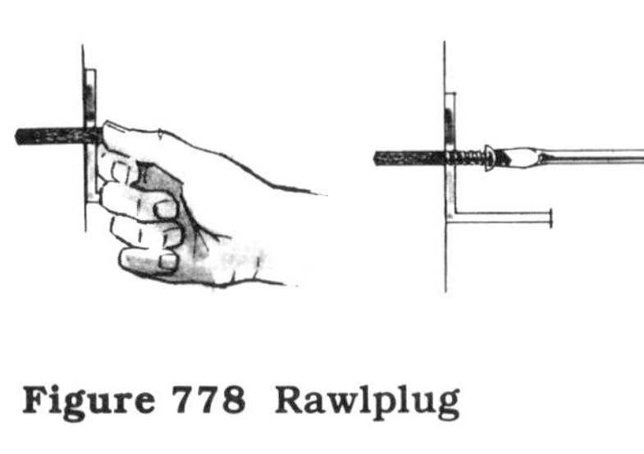

Figure 778 Rawlplug

Figure 779 Ready-mix truck

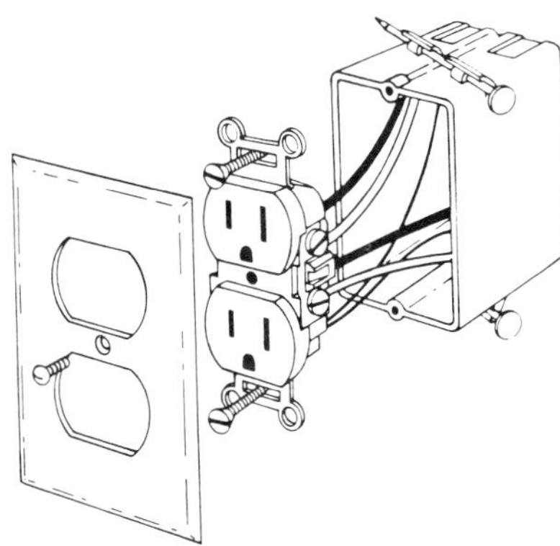

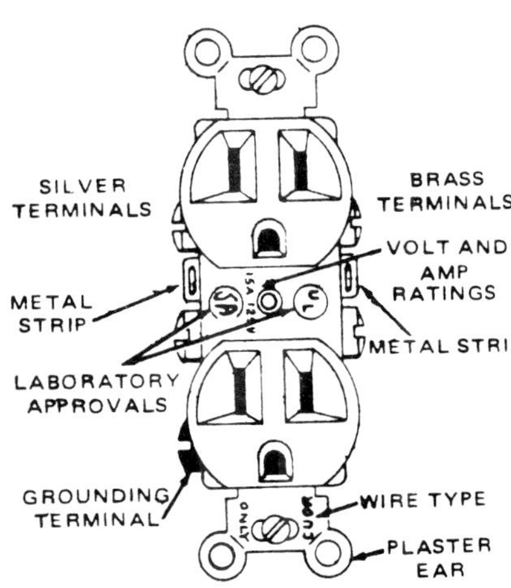

Figure 781 Receptacle: duplex

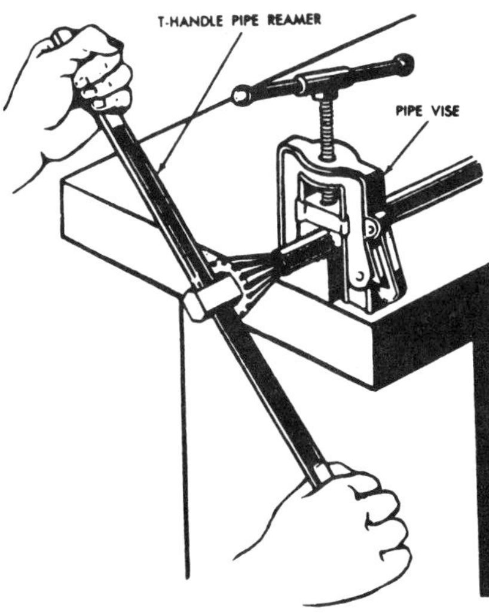

Figure 780 Reamer used to remove burrs from inside pipe

ray Wood tissue plane that extends outward from the pith; *medullary rays.* Rays appear as flakes in quartersawn wood; they are especially prominent in oak. See *wood.*

raze To demolish a building.

reaction Forces acting at a support, as the beam load on a support.

reaction wood *Tension wood* or *compression wood.*

ready mix Factory-mixed preparation, such as plasters and cements, that can be used directly out of the container. See *premix.*

ready-mix concrete Concrete mixed at the plant and carried to the job site in a truck.

ready-mix truck Truck with concrete drum that delivers ready-mix concrete to the building site. (Figure 779.) The concrete is batched or mixed at the manufacturing plant and poured in the truck. See *transit-mix concrete.*

real property Land, buildings, and crops on the land; land and everything on it that cannot readily be moved off. See *personal property.*

ream To smooth the inside of a pipe or hole. See *reamer.*

reamer Tool for removing burrs or metal pieces from the inside of a pipe or a hole. Figure 780 shows a reamer being used to remove burrs from the inside of a pipe.

rebar Reinforcing bar used for reinforcing concrete. See *bar.*

rebate Groove or recess cut into a wood member to receive another wood member or panel; a *rabbet.*

receptacle Electrical outlet that receives a plug from a light or appliance. A unit having two receptacles is called a duplex receptacle. (Figure 781.) One having three or more receptacles in a row is called a strip receptacle. In a plumbing system, a fixture that receives water or waste, as a floor drain.

receptacle outlet Electrical outlet with one or more receptacles.

recessed Placed back or behind.

recharging Adding refrigerant to an air conditioning system.

reciprocal Math term denoting the reversal of a fraction or whole number. The reciprocal of 5 is $\frac{1}{5}$, of 15 is $\frac{1}{15}$, of $\frac{1}{2}$ is 2. Used in heat-loss calculation for converting *C value* to *R value.* (The R value is the reciprocal of the C value.)

reciprocating Moving with an up-and-down or back-and-forth motion, as a reciprocating saw.

reciprocating saw Power saw that has a reciprocating (up-and-down) cutting action. (Figure 782.) Used in rough cutting on the construction site. Various types of saw blades are available for cutting wood and metal.

record drawings Blueprints and specifications which include all the changes that went into the actual structure during building. Also called *as-built drawings.*

rectangular survey system Survey system that describes land in relation to principal meridians and base lines, called *government survey system* because established by the federal government in 1785.

rectifier Direct-current welding machine that supplies DC welding current to the arc welder. See also *transformer-rectifier, welding machine.*

red knot A reddish knot in the wood, caused by a live branch or limb.

reducer Plumbing fitting with inside threads at both ends; one end is smaller. Reduces from a larger to a smaller pipe.

reeving blocks Threading a rope through the blocks to form the block and tackle.

reference line A line on a blueprint used as a reference; it is held true, and other parts of the structure are measured from it.

reference points In surveying, objects used to tie in a *hub.* Magnetic bearings are taken from the hub to various points, such as the corner of a structure or a nail in a tree, and the distance is measured.

reflected ceiling plan In more complex lighting layouts, as for commercial and industrial construction, a "plan" view is made of the lighting outlets on a ceiling. The view shows the ceiling as if it were reflected down as into a mirror.

reflective insulation Batt, blanket, or board insulation with one side covered with a foil-type sheet. The foil backing faces the living area and reflects heat.

refrigerant Fluid in an air conditioning system that absorbs heat by changing to a vapor (at the evaporator) and looses heat (at the condenser) by changing back to a fluid. A chemical called refrigerant 12 is commonly used in the system.

refrigerating circuit The path followed by the refrigerant in an air conditioning system: evaporator, compressor, condenser, and return to the evaporator. Heat is absorbed at the evaporator, causing the refrigerant to become a vapor; the compressor compresses the heated vapor back to a liquid under high pressure; the consenser allows the heat to be dissipated into the outside air.

register Grill covering for a warm-air outlet.

reglet Furrow or slot in concrete or a masonry unit designed to receive flashing or counterflashing, as where the roof meets a parapet. See *raggle.*

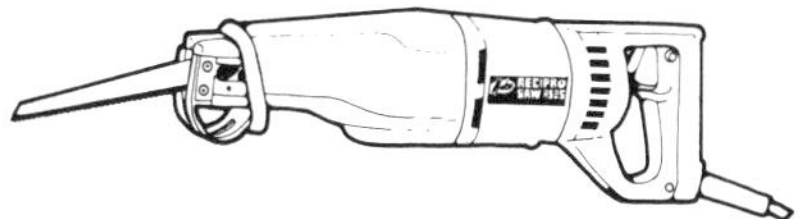

Figure 782 Reciprocating saw

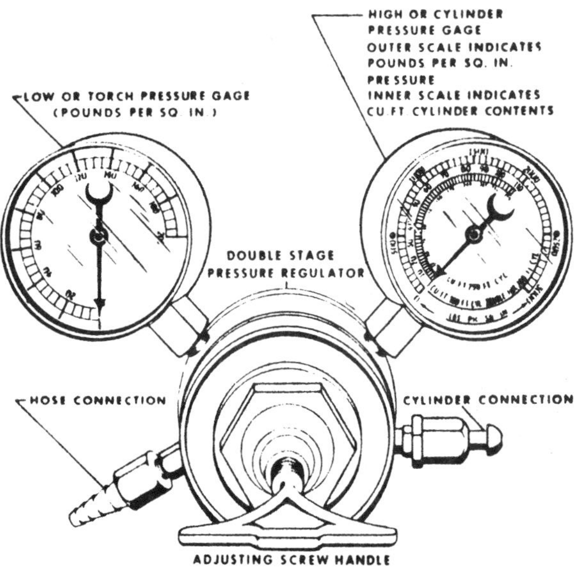

Figure 783 Regulators used for oxygen control (oxyacetylene welding)

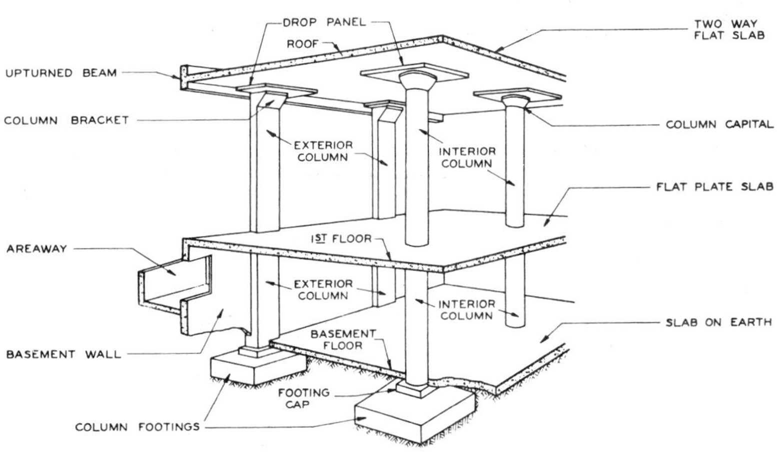

Figure 784 Reinforced concrete construction

Figure 785 Reinforced prestressed concrete construction

regulator In oxyfuel welding, control device used on gas cylinders to control the gas pressure fed into the hoses. (Figure 783.) See *gas cylinder*. See also *flowmeter*.

reinforced concrete construction Structure whose basic support is reinforced concrete. Figure 784. See *box frame construction, concrete, reinforcing bars, reinforced concrete floors, shell construction.*

reinforced concrete floors Several basic types of reinforced floors are used:

flat slab Reinforced slab with bars running in two directions and supported by columns with enlarged top supports. Two-way reinforcement forms a criss-cross pattern, and the floor is often called a *two-way flat slab.*

flat plate Designed in the same fashion as the flat slab, but column support does not have enlarged tops.

solid slab A solid reinforced slab that has beams cast into the bottom or a bearing wall to give additional support. See *one-way solid slab, two-way solid slab.*

ribbed slab The slab is hollowed out on the bottom to produce deep ribs or joists. One-way and two-way (waffle slab) rib systems are used. See *one-way floor system, two-way floor system.*

reinforced prestressed concrete construction Method of building a structure using prestressed concrete members. Individual units are precast, trucked to the job site, then hoisted into place with a crane. Every unit is numbered and goes into a specific place in the structure. Figure 785 shows a prestressed concrete building under construction. See *panel, prestressed concrete, slab.*

reinforcement schedule Table giving location, size, type, and number of reinforcement bars in a concrete slab, column, or beam.

Figure 786 Reinforcement symbols

reinforcement symbols Symbols used to show reinforcement bars used in a concrete structure. (Figure 786.)

reinforcing bars Steel bars used to give strength to concrete. Deformed bars are used. See *bar, truss bar.* The deformed bars have ridges to give a better hold in the concrete. A size number is raised up on the bar. Generally, the number can be translated into eighths. For example, a bar with a 4 on it would be $\frac{4}{8}''$ or $\frac{1}{2}''$ thick. Figure 787 shows bar sizes. Figure 788 shows layout of bars in various concrete members.

related trades All the construction trades. Sometimes used in the sense of all the trades other than the carpentry trade, such as plumbers, electricians, brickworkers, and so on.

relative humidity Percent of moisture in the air. Always expressed as a percent of the maximum water vapor that could be present at a given temperature. See *humidity.*

relief In art or ornamental work, a sculpture or figure that projects out from the background surface. Also called relievo.

relief cut Side cut made into the work piece over to the basic cut to remove material so that the saw will not bind.

Bar designation No.**	Unit weight lb/ft.	Diameter in.	Cross-sectional area, in.²	Perimeter in.
3	.376	0.375	0.11	1.178
4	.668	0.500	0.20	1.571
5	1.043	0.625	0.31	1.963
6	1.502	0.750	0.44	2.356
7	2.044	0.875	0.60	2.749
8	2.670	1.000	0.79	3.142
9	3.400	1.128	1.00	3.544
10	4.303	1.270	1.27	3.990
11	5.313	1.410	1.56	4.430
14	7.65	1.693	2.25	5.32
18	13.60	2.257	4.00	7.09

*The nominal dimensions of a deformed bar are equivalent to those of a plain round bar having the same weight per foot as the deformed bar.
**Bar numbers are based on the number of eighths of an inch included in the nominal diameter of the bars.

Figure 787 Reinforcing-bar sizes

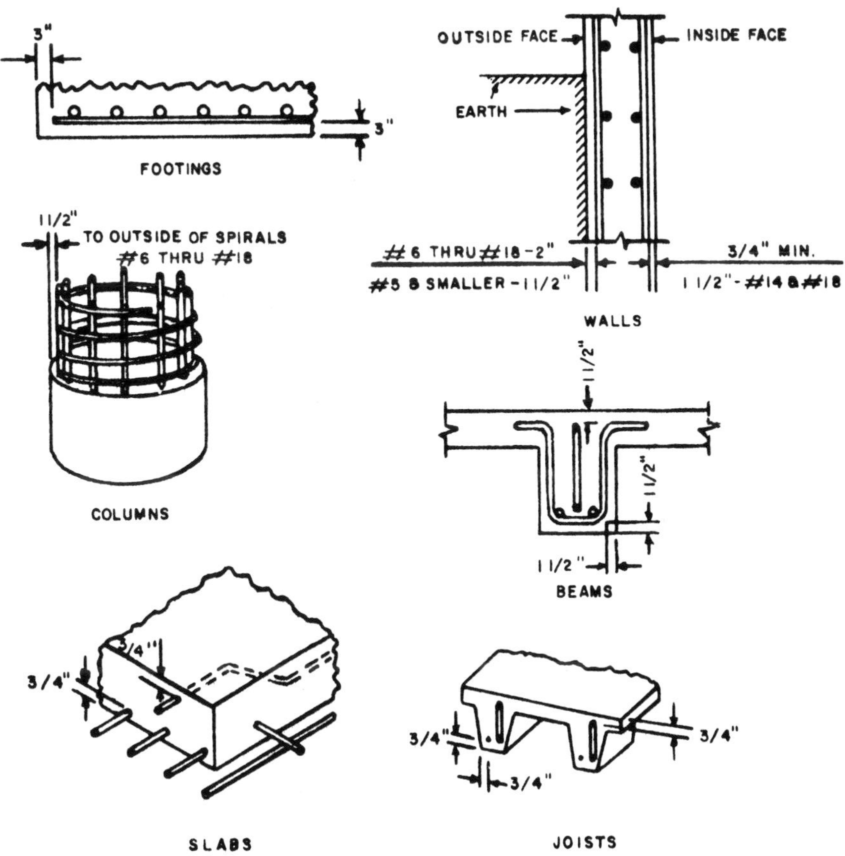

Figure 788 Reinforcement bars used in structural concrete

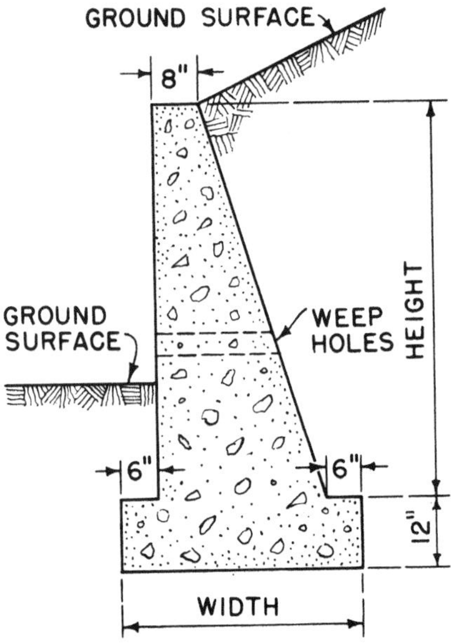

Figure 789 Retaining wall

relief valve Safety valve used with hot-water heating equipment. If the temperature or pressure gets too high, the relief valve opens to relieve the pressure.

relocatable structure Factory-built building that can be moved as needed.

remote-control wiring Low-voltage electrical wiring system that uses a relay to control devices. Doorbells and chimes are operated on this system. A transformer steps the 120-V house current down to 24 V.

rendering Perspective drawing with shading and coloring. An architect's presentation drawing to show what the completed structure will look like.

residential construction House construction; construction of single-unit dwellings.

resilient channel In drywall construction, galvanized steel channels that are attached at a right angle, 24″ O.C., to wood or metal studding or joists. Gypsum board is attached to the channel. Used to improve sound-transmission value, especially in apartment buildings and motels.

resilient-floor layer One who lays out and installs all types of floor tile and carpeting. See *floor-covering installer.*

resilient tile Asphalt, rubber, vinyl, or vinyl asbestos tile used to form a finish flooring; usually comes in 9″ × 9″ or 12″ × 12″ squares and is set into a mastic.

resin A congealed sap.

rosin Hard resin left after distilling turpentine.

rosin-core solder A solder with a rosin core; the rosin serves as a flux. Rosin-core solder is used for making electrical connections. See also *acid-core solder.*

resistance In an electrical system, the opposition to current flow, as the resistance in the conductors. Resistance to flow of electrical current is transformed into heat.

resistance welding Welding by means of the heat generated by resistance to current flow. Two pieces are pressed together and connected to electrodes. Their resistance causes them to become very hot and plastic and to blend together. The current is turned off, allowing them to cool into one welded piece. See *spot welding.*

resistance wiring Electrical wiring that heats up because of its resistance to electrical flow. Used for radiant heating when run in coils in the floor, walls, or ceiling. (*Figure 762,* p. 335.) Also used in gutters and downpipes to melt snow. Sometimes buried in concrete sidewalks or driveways to melt ice and snow.

respirator Protective device worn over the mouth and nose to protect lungs from dust and vapor in the air. Worn when sanding or paint spraying.

retaining wall Wall constructed to hold back earth, usually on a slope or hillside. (Figure 789.) Wall that receives a lateral load or pressure, such as water pressure on a *cofferdam.*

retarder Admixture added to concrete to slow down the setting, usually during hot weather.

retempering Adding more water to concrete or mortar and remixing after it has started to stiffen.

retrofit To modify an existing building, as for energy efficiency.

return To turn back on. Molding or finish that changes direction or turns back on itself, as a cornice that continues around the corner of a house. (*Figure 258*, p. 97.)

return air Inside air returned back to the heating or cooling unit. See *supply air*.

return register Cool-air return.

reveal Outside door or window side jamb between frame and outer wall surface. Also, groove in the exterior face of cast-in-place concrete or a precast panel; rustication.

reverse door Swinging door that opens outward, as opposed to a regular door that opens inward. See *hand of doors*. (A cabinet door that swings outward, however, is considered standard or regular.)

reverse polarity In arc welding, a setup with *direct-current electrode positive* (work is the negative pole, electrode is the positive pole).

revertment Facing of stone on the side of a stream or waterway, designed to protect against erosion.

revetment Concrete facing or parging on a wall to protect the surface.

revision block Area near title block on drawing to describe any change or revision made to the drawing. Number of the change, date, and initials of approving authority are given.

revolution A turning. One revolution is one 360° turn.

revolving shelf Storage cabinet shelf that revolves on an axis. Often used in the corner of a kitchen below counter-top level in a row of cabinets. A *lazy susan*.

rib Continuous projection or raised part on a concrete wall or on the underside of a concrete slab.

ribbed slab Reinforced concrete slab floor with concrete ribs (joists) on the underside. See *one-way floor system*, *two-way floor system*, *reinforced concrete floors*.

ribbon Let-in horizontal board used to help support joists in balloon framing. (*Figure 55*, p. 23.) Sometimes called a *ledger* or *girt*.

ribbon board A *ribbon* or *ledger*.

ribbon course In roofing, a double layer of shingles. Also called a *shadow course*.

ribbon figure A ribbonlike dark-and-light appearance in the wood face, found in quartersawn wood. Also, a cut on *interlocked grain*.

rice brush Brush used in plastering to press the finished surface to create a texture. *Stippling brush*.

rich mortar Mortar with a high proportion of cement.

ridge In roof construction, the top intersection of two roof slopes. The *ridgeboard* where the rafters meet at the ridge of a roof. (*Figure 763*, p. 335.)

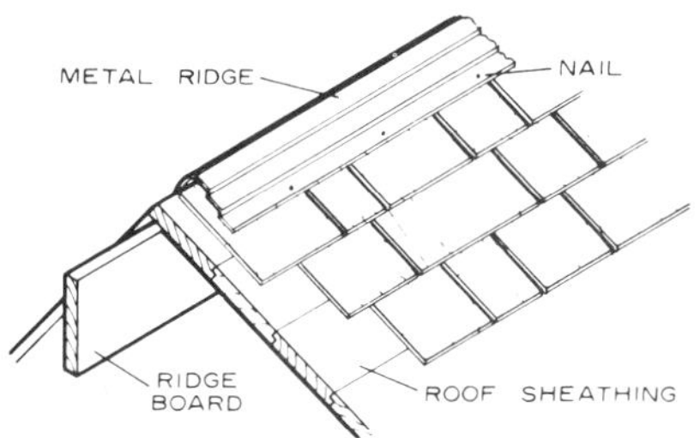

Figure 790 Ridge cap

Figure 791 Right-angle drill

ridgeboard Horizontal support member at the ridge of a roof that receives rafter ends. Also called a *ridgepole* or *rooftree.* See *rafter.* (*Figure 763,* p. 335.)

ridge cap Capping run at the roof crown where two roof slopes meet, as at a gable roof ridge. (Figure 790.) Used in metal and tile roofing. See *asphalt shingle, roof tile.*

ridge course Last course of roof tiles or shingles at the ridge.

ridge covering Capping or covering that goes over the roof ridge.

ridge cut End cut on a rafter to fit to the ridgeboard.

ridgepole Older term for the *ridgeboard.*

ridge reduction In roof construction, one-half the thickness of the ridgeboard, which is subtracted from the theoretical rafter length to get the true rafter length.

ridge roll Capping run at the roof crown where two roof slopes meet, as at a gable roof ridge. Used in asphalt shingle, metal, and tile roofing. Special curved ridge tiles are used. See *ridge cap.*

ridge strut In a rigid frame steel building with a gable roof, the steel framing member running along the apex of the roof. Corresponds to a ridgeboard in a wood frame building.

ridge ventilator Ventilator that runs along the roof ridge. See *attic ventilation.*

ridging Material for covering the roof ridge.

rift grain *Edge grain; quartersawn grain.*

rigging Rope, wire rope, or cable used at the construction site for hoisting. See *cable attachments.*

right angle Ninety-degree angle. See *angle.*

right-angle drill Drill that bores at a right angle to the body of the drill. Figure 791 shows a right-angle power drill, used in tight working conditions as in plumbing and electrical work.

righthand door Hinges are on the right when you face the door from the outside, and the door swings away from you. (If the door swings toward you, it is a reverse righthand door.) See *hand of doors.*

righthand stairs Stairs with a rail to the right as you ascend.

right of way An easement or designated strip of land for passage.

right triangle A 90° triangle with two 45° angles, used for drafting.

rigid conduit Metal piping used to run electrical conductors. See *conduit, rigid metal conduit.*

rigid frame construction Construction that uses a rigid framework as the skeleton for the structure. Reinforced concrete, reinforced prestressed concrete, and steel are used. Figure 792 shows the skeleton for a rigid frame building. Note the rigid connection of the steel columns and rafters. One rigid frame support of columns and rafters is called a *bent;* two bents together make a *bay.* (See also *Figure 750*, p. 328; *764*, p. 335.)

rigid metal conduit Metal pipe used for running electrical connectors. It is installed and then the conductors are fished through. (*Figure 232*, p. 88.)

rigid nonmetallic conduit Various vinyl, fiber, and asbestos cement materials are used for forming rigid nonmetallic conduit. The conductors are fished through the installed conduit. Used in all locations except in hazardous areas or areas of high temperatures. See *conductors.*

rigid paving Road surfacing with aggregate and cement, commonly called *concrete paving.* See also *flexible paving.*

rigid PVC conduit Unthreaded polyvinyl chloride conduit.

rim joists Joists that run on the outside of the floor platform. Also called *band joists, sill joists.* See specifically *end joists, header joists.*

rim latch Lock used on the interior side of a door. See *dead bolt.*

ring shake Lumber defect; separation of wood at the annual rings. See *windshake.*

ripper In drywall construction, narrow strips of gypsum board, used for narrow areas such as soffits.

ripping Sawing of wood in the direction of the grain. Opposed to *crosscutting.* In earthmoving, the breaking up of soil to a depth of two or three feet by running teeth through the earth. See *scarifying.*

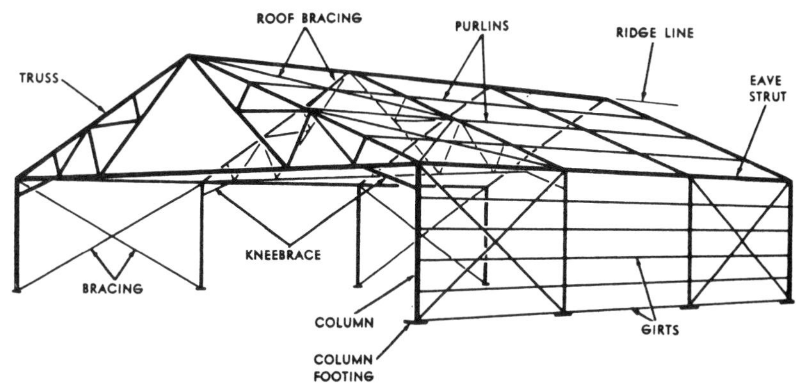

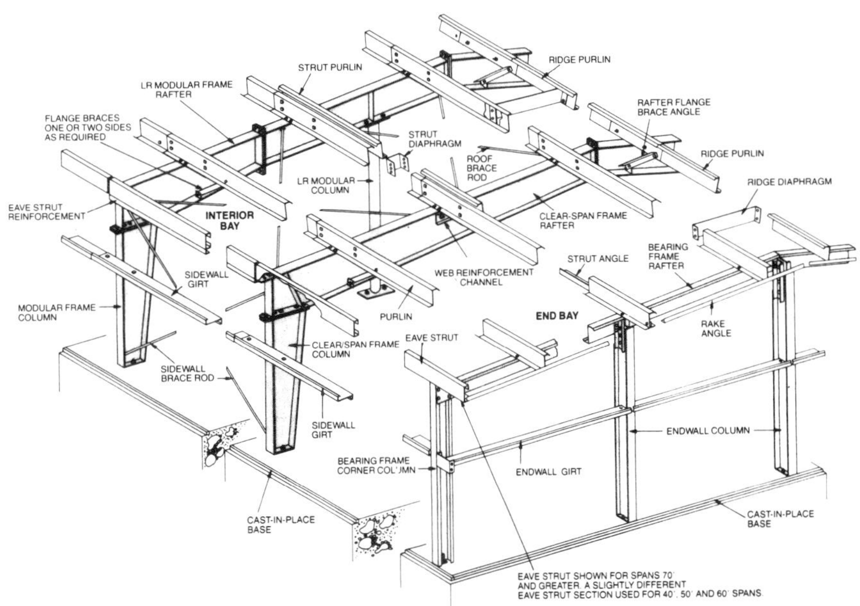

Figure 792 Rigid frame steel construction

Figure 793 Ripping bar

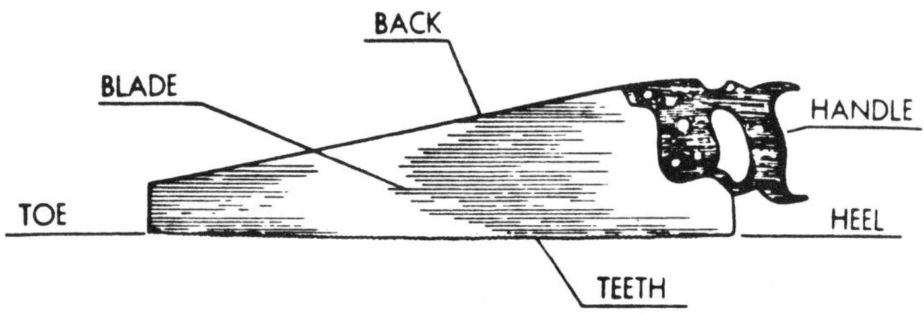

THE SIZE OF A SAW IS DETERMINED BY THE LENGTH OF THE BLADE IN INCHES
SOME POPULAR SIZES ARE 24" AND 26"
THE COARSENESS OR FINENESS OF A SAW IS DETERMINED BY THE NUMBER OF POINTS PER INCH
A COARSE SAW IS BETTER FOR FAST WORK AND FOR GREEN WOOD
A FINE SAW IS BETTER FOR SMOOTH ACCURATE CUTTING AND FOR DRY SEASONED WOOD
5-1/2 AND 6 POINTS ARE IN COMMON USE FOR RIP SAWS
7 AND 8 POINTS ARE IN COMMON USE FOR CROSS CUT SAWS
SAW TEETH ARE SET; EVERY OTHER TOOTH IS BENT TO THE RIGHT AND THOSE BETWEEN TO THE LEFT, TO MAKE THE KERF WIDER THAN THE SAW
THIS PREVENTS THE SAW FROM BINDING IN THE KERF OR SAW CUT.

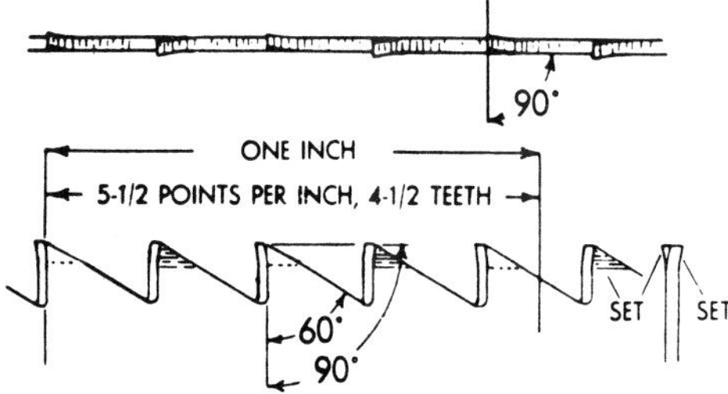

RIP SAW TEETH ARE SHAPED LIKE CHISELS. THEY CUT LIKE A GANG OF CHISELS IN A ROW.

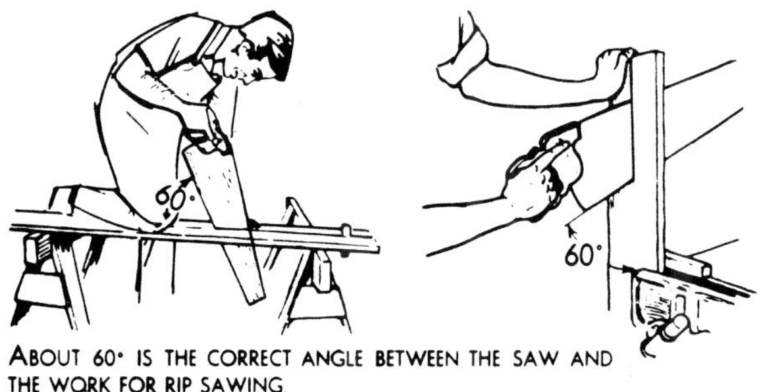

ABOUT 60° IS THE CORRECT ANGLE BETWEEN THE SAW AND THE WORK FOR RIP SAWING.

TOP VIEW OF RIP TEETH

TEETH OF RIP SAW

Figure 794 Ripsaw

ripping bar Steel bar with one end bent back and having a claw and the other end having a chisel-type edge. (Figure 793.) A *crowbar, pinch bar, prybar, wrecking bar.*

ripping chisel One-piece, all-steel chisel used for rough cutting.

ripping hammer A *straight claw hammer;* a *flooring hammer.*

ripple In a weld bead, the wave-like ridges in the bead. See *weld bead.*

riprap Haphazard arrangement of loose, irregular-sized stones or broken rock; often used along stream banks for protection against erosion. See also *revetment.*

ripsaw Hand saw with chisel-type teeth used for ripping lumber (cutting along the length of the piece parallel to the grain). A 26″ saw with 6 teeth to the inch is common. (Figure 794.) See also *crosscut saw.*

rise A vertical direction, as opposed to *run* (horizontal direction). In roofing, the height from the center of the *span* to the ridge. (Figure 795.) Also, the total stair height. (*Figure 939,* p. 414.) Height of an arch from base level to center high point (center of span). (*Figure 378, p.* 150.)

riser Vertical support board in a stair step. (*Figure 939,* p. 414.) In plumbing, vertical piping. In electricity, vertical wiring. In heating and air conditioning, vertical hot-water piping or ductwork. In masonry work, a brick or concrete block that runs vertically, as set on end.

riser pipe Vertical pipe that runs from floor to floor to carry water, steam, or gas.

riser shaft Vertical shaft that holds electrical cables.

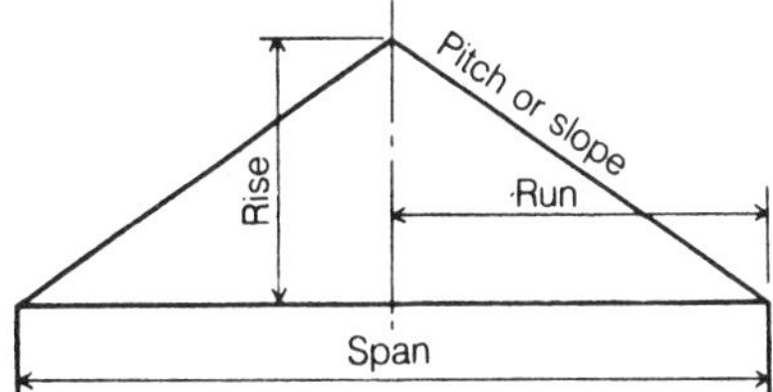

Figure 795 Rise, run, and span

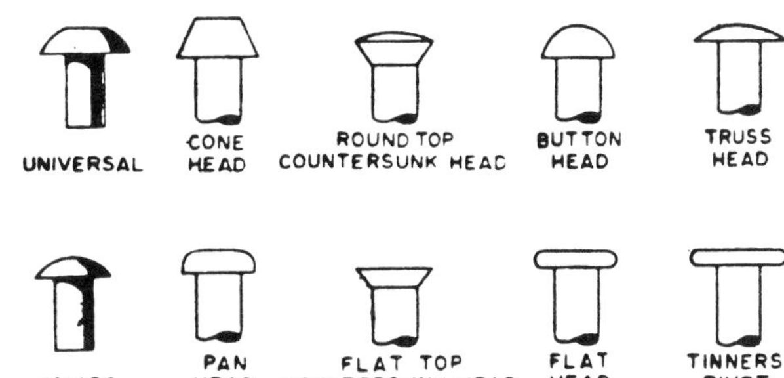

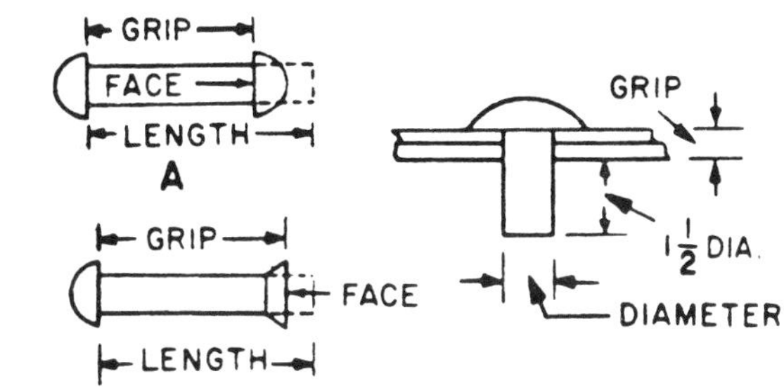

Gage of sheet metal	Rivet size (weight in pounds per 1000 rivets)
26	1
24	2
22	2 1/2
20	3
18	3 1/2
16	4

Figure 796 Rivets

rivet A kind of unthreaded bolt made of a malleable metal. (Figure 796.) The rivet is run through holes made in two or more metal pieces; then the plain end is mashed or flattened to make a second head to hold the metal pieces together. A *rivet set* may be used to form a smooth head. See also *punch-lok riveter.*

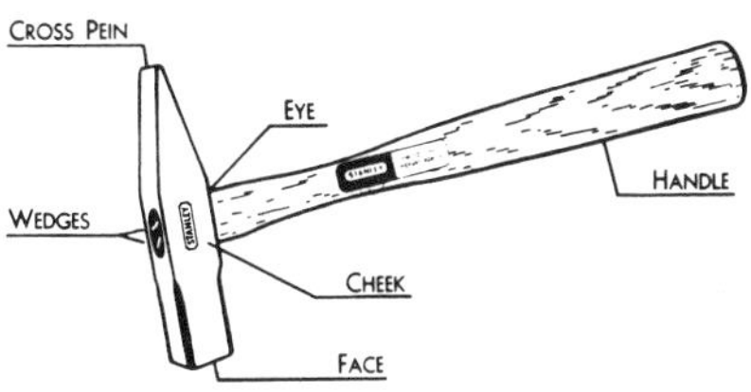

Figure 797 Riveting hammer

riveting hammer Special hammer used in sheet metal work, especially used for driving and setting rivets. (Figure 797.) See also *hammer.*

Drawing, upsetting, and heading a rivet

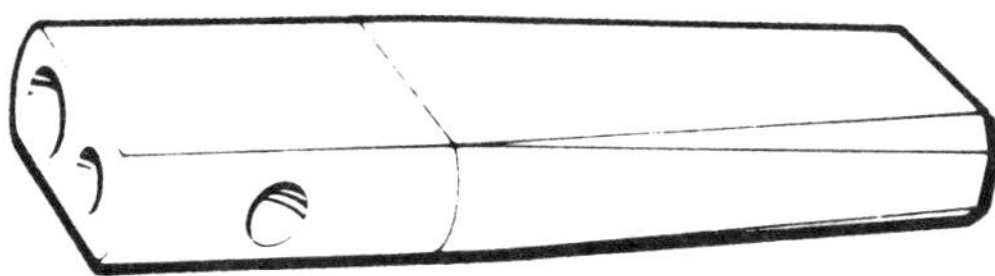

Figure 798 Rivet set

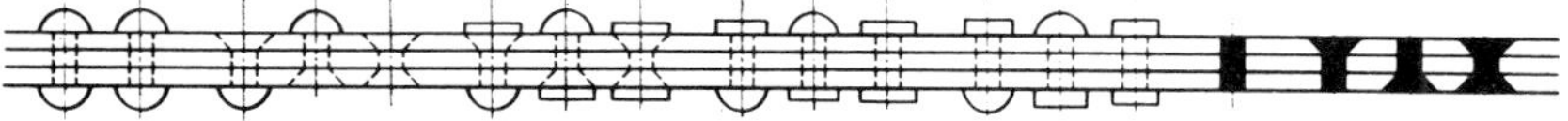

Figure 799 Rivet symbols

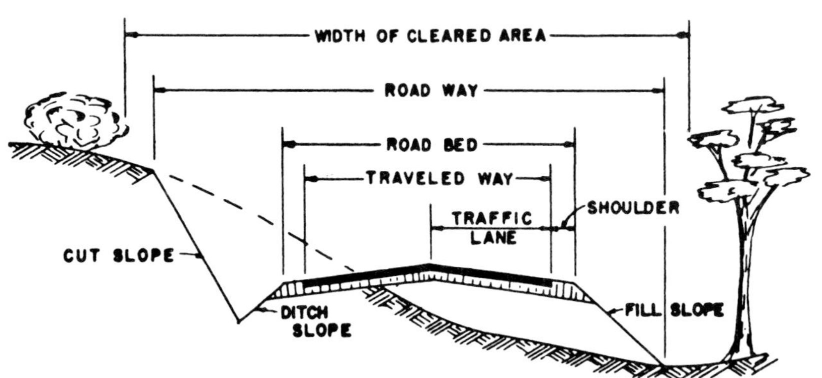

Figure 800 Road construction

rivet set Hand tool for setting or mashing down the head of a rivet, struck with a riveting hammer. (Figure 798.)

rivet symbols Symbols that show how shop rivets and field rivets are to be finished. (Figure 799.) In steel construction today most rivet fabrication is done in the shop.

roadbed In highway construction, the graded portion of the highway between the side slopes which serves as a base for the *pavement structure* and the shoulders. (Figure 800.)

roadside The area just beyond the edge of a roadway; this may include the broad area separating the divided highway.

roadside development In highway construction, the finish, landscaping, sodding, replanting, placement of ground cover, and other improvements needed to complete the highway.

roadway In highway construction, the right of way within the limits of construction. (Figure 800.)

rod A *story rod* or stick for marking set measurements when laying out framing members or parts. Also, a measure of length: $5\frac{1}{2}$ yards or $16\frac{1}{2}$ feet.

rodding In plastering, straightening (leveling) the plaster surface with a darby to make it even and of the correct thickness. In plumbing, the cleaning of drains with a flexible rod.

rod saw Wire-shaped saw blade coated with tungsten carbide, used for cutting very hard materials such as hardened steel, glass. The rod saw blade fits into a regular hacksaw. (Figure 801.)

roller In site development and highway construction, a piece of heavy equipment that compacts the earth by rolling over it. See *compacting*.

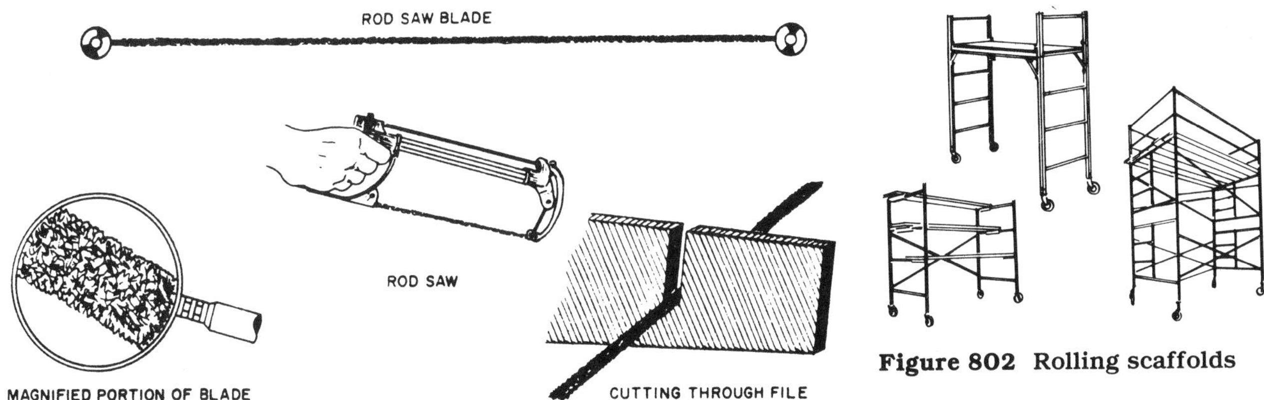

Figure 801 Rod saw

Figure 802 Rolling scaffolds

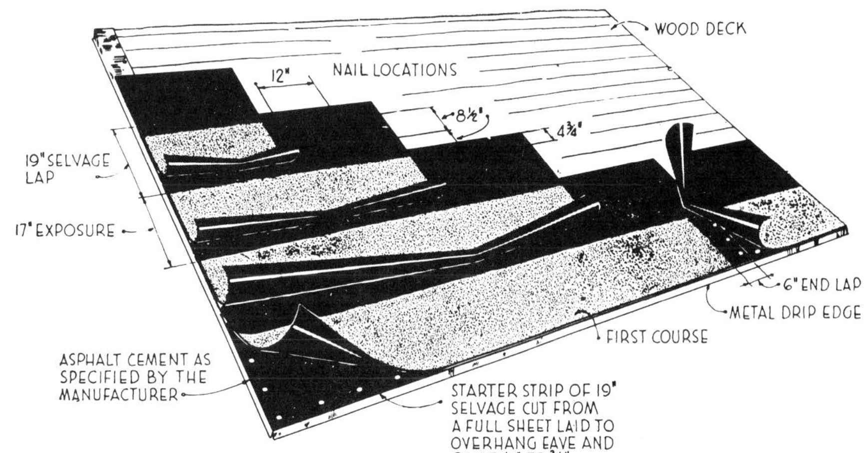

Figure 803 Roll roofing

roller catch Mechanical device for holding a cabinet door closed. A stud on the door is caught between two rollers mounted on the carcass. See *catches.*

rolling scaffold Prefabricated metal scaffold that can be rolled on wheels. Wheels are locked when the scaffold is in place. Normally used inside where a hard flat surface is available. (Figure 802.)

roll roofing Fiber and saturated asphalt roofing that comes in rolls 36″ wide, 36′ long; weights vary from 45 to 90 pounds per roll. Used in built-up roofing. The roofing is rolled out and hot-mopped down with tar. Figure 803 shows application of roll roofing.

Roman brick Kiln-baked brick that is thinner and longer than the standard building brick. The Roman brick is $1\frac{1}{2}'' \times 4'' \times 12''$.

Romex *Nonmetallic sheathed cable.*

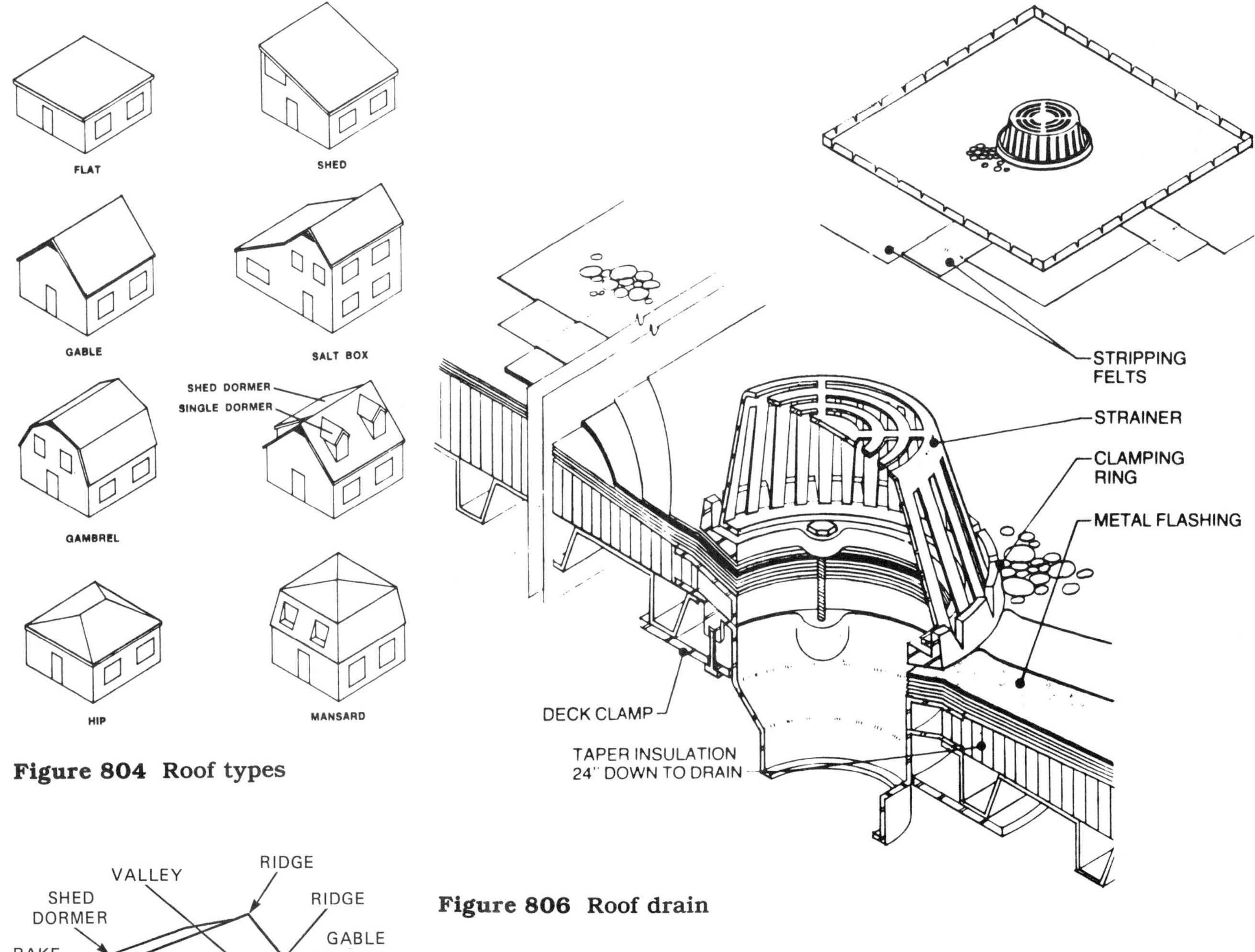

Figure 804 Roof types

Figure 806 Roof drain

Figure 805 Roof terminology

roof Covering for a structure. Figure 804 shows common roof types used in residential construction. Figure 805 shows terms used to identify roof parts.

roof assembly The total covering of a structure, including the *roof deck* and *roof system.*

roof deck Structural surface attached over rafters or trusses. See *roof assembly, roof system.*

roof dormer Small dormer built into the side of a roof. See *dormer.*

roof drain Drain on a flat roof designed to receive rain water for discharge into a downspout. (Figure 806.)

roofer One who lays asphalt shingles, wood shingles and shakes, tile, and slate to roofs of buildings. Applies composition roofing. Waterproofs and dampproofs masonry walls and other surfaces.

roofer's knife Special knife for marking and cutting asphalt shingles, roofing paper, linoleum, and the like. (Figure 807.)

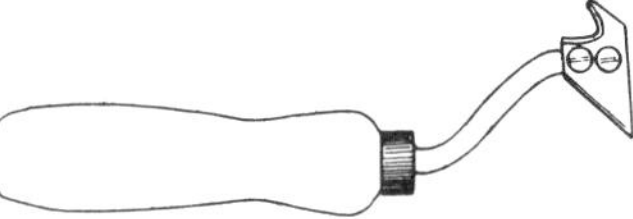

Figure 807 Roofer's knife

roof framing The erection and fastening together of all the support members in a roof. Figure 808 shows the roof framing for a simple roof layout.

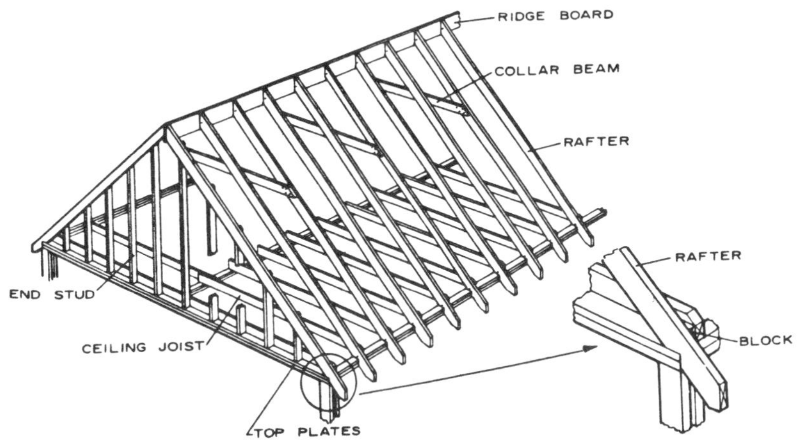

Figure 808 Roof framing

roof framing plan Plan view of the layout of all the rafters in the roof. See also *roof plan.*

roofing nail Wide-headed nail used for attaching roofing. See *nail.*

roofing stapler Air-driven stapler used to attach asphalt shingles; uses $\frac{7}{8}''$ staples, for reroofing uses $1\frac{1}{2}''$ staples. (Figure 809.) See also *sheathing stapler.*

Figure 809 Roofing stapler

roof jacket A flange that fits around the vent pipe as it exits the roof.

roof pitch Angle of the roof sides expressed as a ratio of *span* to *rise.* See *pitch.*

roof plan Top-view plan of a completely finished roof. See also *roof framing plan.*

roof rise Height from the center of the span to the ridge. See *rise.* (*Figure 795*, p. 351.) Note that the rise runs up to where the *rafter line* intersects at the ridge (one-half the ridgeboard height). See *rafter length.*

roof run Horizontal distance: one-half the span. The span runs from outside the top plate edge to outside the opposite plate edge. See *rafter length, span.*

roof sheathing Roof covering that goes on over the rafters or trusses. Plywood or hardboard panels are widely used. The roof covering, such as shingles, is nailed into the sheathing.

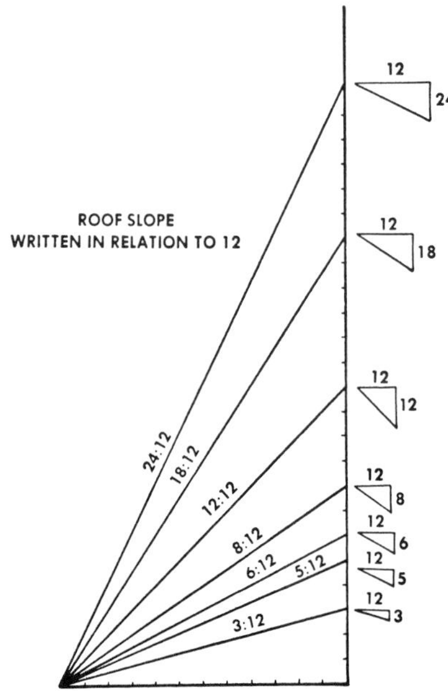

Figure 810 Roof slope

roof slope Angle of the roof sides expressed as *rise* in inches per 12 inches of *run.* (Figure 810.) See *pitch, slope.*

roof system The roof covering put down on the roof deck. See *roof assembly*. In industrial building, the roof panels. (Figure 811.)

roof tile Clay tile used as roof covering. Figure 812 shows roof tiles.

rooftree Older term for the *ridgeboard.*

roof truss Manufactured roof support that spans the roof. See *truss.*

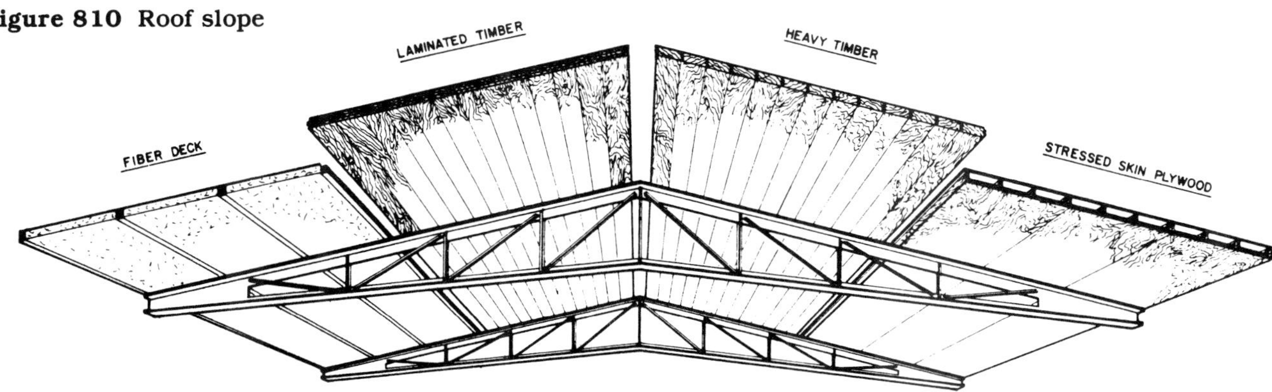

Figure 811 Roof system (industrial)

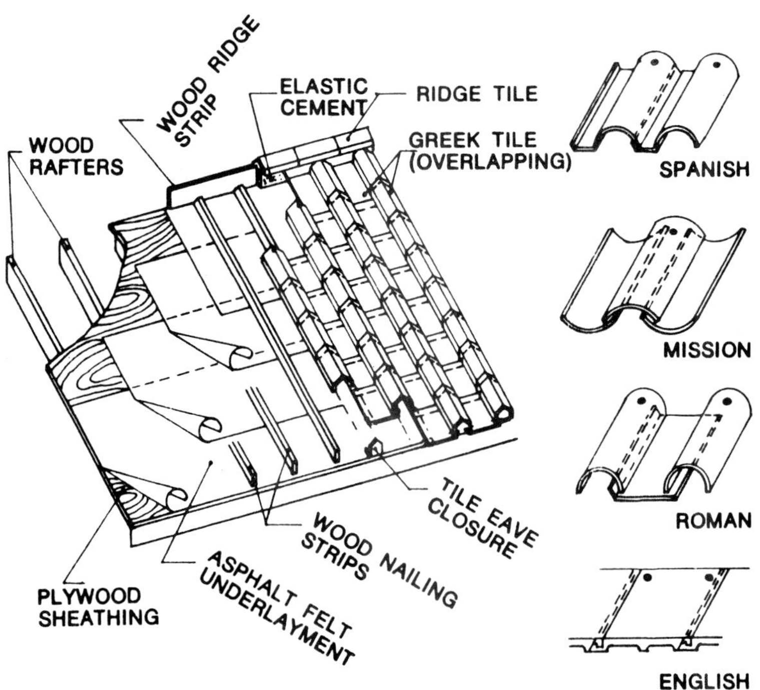

Figure 812 Roof tile

roof ventilator Ventilator in the side of the roof. See *attic ventilation.*

room finish schedule Table listing wall finish, materials, colors of walls, and information on trim for each room of the structure.

room heater Individual free-standing heating unit used to heat a single room; a *space heater.*

root of weld Point at which the back or bottom of the weld meets the base metal.

root opening In welding, the opening at the base or bottom (or at the closest part) of two pieces to be joined.

rope Twisted plant fibers that are made into a strong, flexible cord. See *rope knots* for some of the common means of securing rope. Also, a short term for steel *wire rope.*

rope knots Means of tying rope so it can be safely used to support and carry. Figure 813 shows some of the standard knots used on the job.

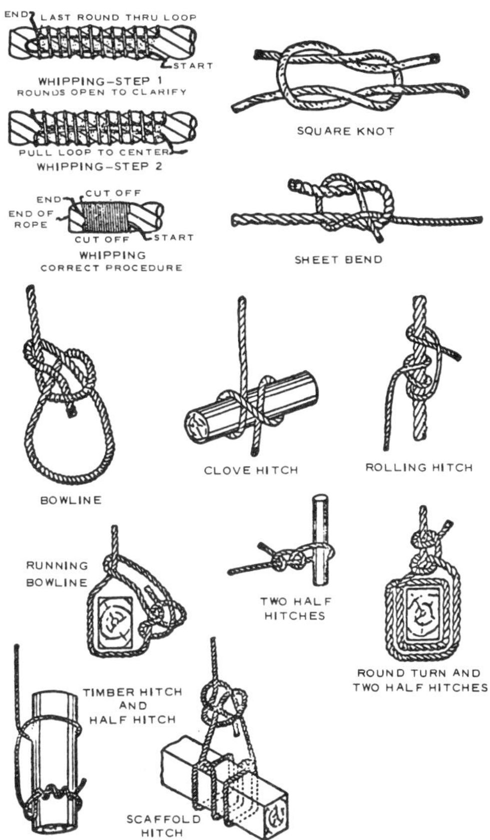

Figure 813 Rope knots

rose window Round window decorated with colored or painted glass. The design is based on the large, leaded window traditional in church design.

rosin Hard resin extruded from trees, such as pine. When dry it is used as a flux in soldering, as a drier in oil paint, in varnishes, and the like.

rosin-core solder Solder that has a rosin flux; suitable for electrical connectors.

rot Breakdown of wood by bacterial action. Breakdown of plaster caused by dampness.

rotary-cut veneer Veneer cut from a log with a knife in a continuous strip. The log or bolt is turned in a lathe.

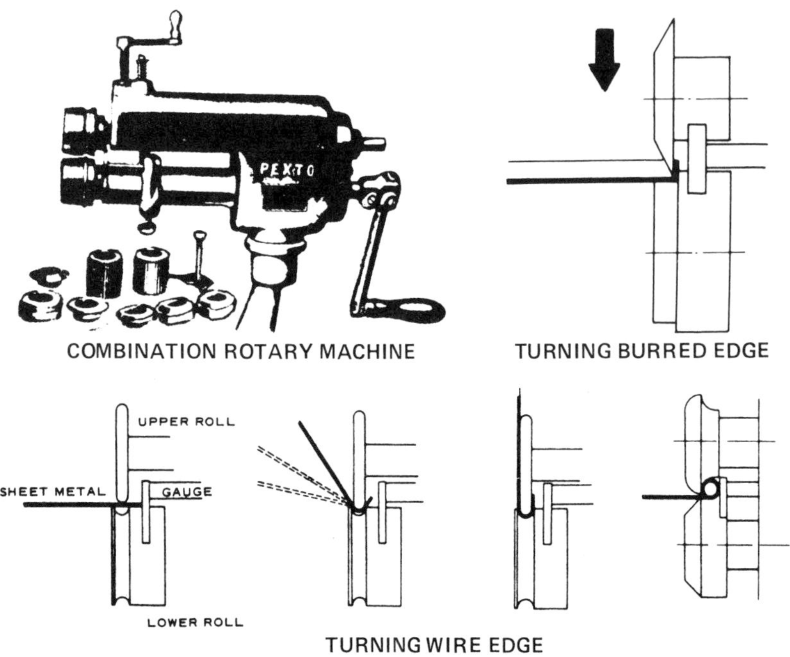

Figure 814 Rotary machine

rotary machine Sheet metal machine for making edges, beading, crimping. (Figure 814.) See also *beading, brake, cornice brake, slip roll forming machine.*

rottenstone A polishing agent (siliceous limestone) used for finishing surfaces, such as a varnished surface.

rough carpentry Erection of structural framework and the major parts of a building. See *finish carpentry.*

rough coat First plaster coat.

rough finish Completion of framing and readying of structure for finish.

rough flooring Flooring material, such as plywood or hardboard panels, that is nailed directly to the joists. The finish flooring is applied over the rough flooring.

rough hardware Concealed support metalwork used in constructing a building: nails, screws, bolts, framing anchors, metal connectors, and so on. Opposed to *finish hardware,* which is visible, such as drawer pulls.

rough-hewn Of timbers, roughly finished with an adz or axe. Used for decorative effect, as for exposed beams.

rough horse A stair *carriage* or *stringer.*

rough-in drawing A *rough-in sheet.*

rough-in sheet In plumbing, a manufacturer's sheet or drawing that gives specific information and sizing on a plumbing fitting (such as a faucet or shower head) or a plumbing fixture (such as a sink). Also called a *rough-in drawing.*

roughing-in Preliminary rough work done on a job. In plumbing, the installing of all water-supply and drain-waste-vent piping and hangers, prior to installation of fixtures.

rough lumber Undressed lumber as it comes from the mill; saw marks are clearly visible.

rough opening (R.O.) The rough-framed opening, as for a window or door.

rough plumbing Water-supply piping, waste and drainage piping, and vent piping, without fixtures. See *finish plumbing.*

rough sill The horizontal framing at the bottom of a window opening.

rough string An intermediate stringer (carriage) used for stair support. See *stairs.*

round-bend screw hook Open hook with a screw point.

round file File that is round in cross section; used for filing curved surfaces and holes. (Figure 815.) See *file, half-round file.* Also, a *drill saw.*

rout In woodworking, to cut out material with a *router.* In concrete work, to deepen a crack in preparation for patching or sealing.

routed Having had material removed with a *router.*

router A hand-operated power woodworking tool used for cutting or gouging out material from wood, to make grooves, dadoes, rabbets, edge moldings, and so on. (Figure 816.) Different bits are available for making different cuts. See *router bits, shaper.*

router bits Cutting bits used in a router for achieving different finished shapes in the wood. Figure 817 shows a variety of common router bits.

router plane A small hand plane that takes cutters of various shapes used for cleaning out cuts such as a dado or groove. (Figure 818.) See also *plane, router.*

routing Process of cutting out material with a *router.*

row house Living unit whose walls on opposite sides are party walls, shared by adjacent living units.

rowlock Masonry units set on edge (laid on the face). A rowlock stretcher is called a *bull stretcher;* a rowlock header is called a *bull header.* See *brick.*

rubber mallet Rubber-headed striking tool used for hitting finished surfaces.

rubber test plug Rubber plug used for sealing plumbing drain pipes so the system can be tested with water.

rubble Rough broken stone; *riprap.*

Figure 815 Round file

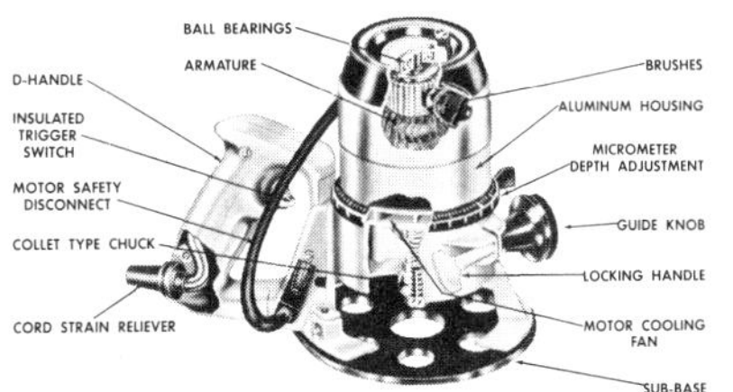

Figure 816 Router

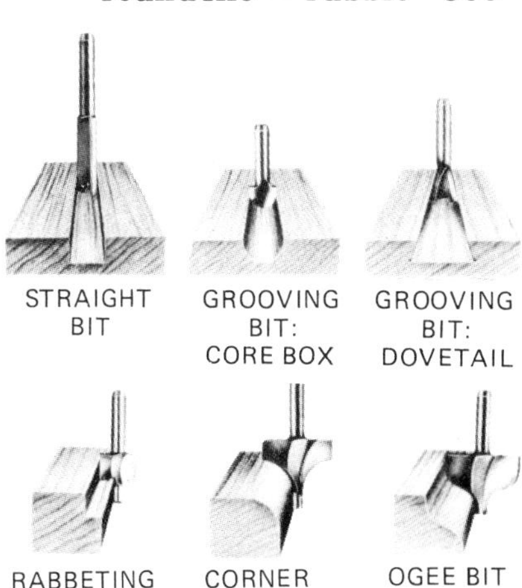

Figure 817 Router bits and typical cuts

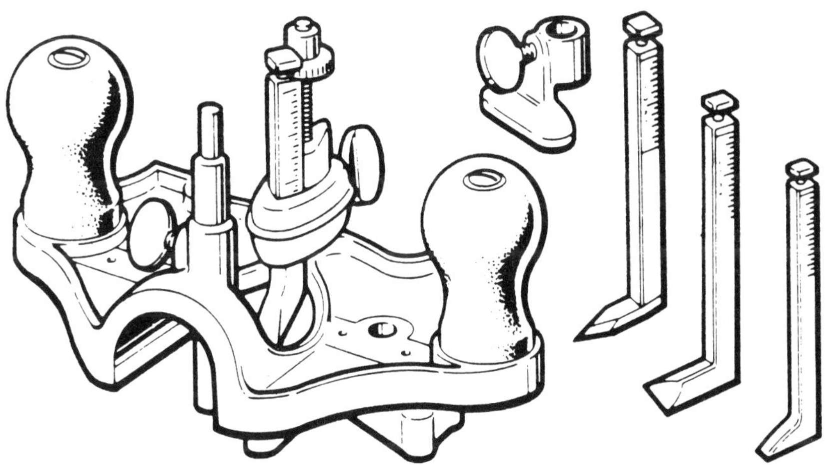

Figure 818 Router plane

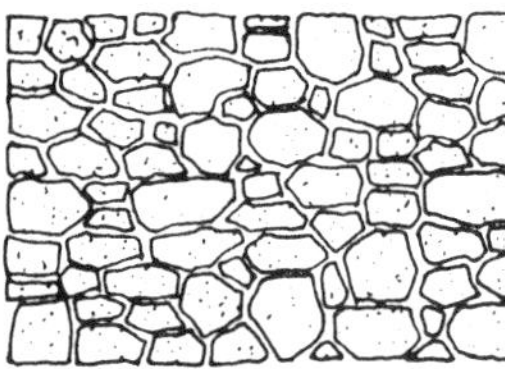

Random Rubble

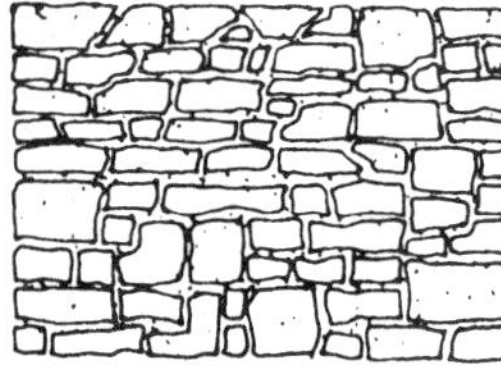

Coursed Rubble

Figure 819 Rubble masonry

rubble masonry Wall built with rough irregular stones or rubble. (Figures 819, *109*, p. 42.) Also called *rubblework*. See also *ashlar*.

rubblework Masonry work using rough, irregularly shaped stones. Also called *rubble masonry*.

rule Measuring tool marked in inches and fractions of an inch, used for layout and for measuring. See *extension rule, folding rule, push-pull rule, steel tape*.

run A horizontal direction as opposed to rise (vertical direction). In roof and floor construction and design, one-half the *span*. See *rise*. (*Figure 795*, p. 351.) In stair construction, the horizontal distance the stair goes, measured from face of top riser to face of bottom riser. See *stair*. (*Figure 937*, p. 413; *939*, p. 414.) In plumbing, horizontal piping. In electricity, horizontal wiring or sometimes the complete circuit. In heating and air conditioning, horizontal ductwork or piping; the complete distance to the outlet.

rung Horizontal bar on a ladder supported by the two side stiles; serves as a step. See *ladder*.

running bond Masonry bond pattern of all stretcher courses; joints of one course fall in the center of stretchers of other courses. See *brick bond*.

rustication Grooves in exterior concrete panel for decorative purposes.

R value Resistance to heat flow. The higher the R value, the greater the insulating value of the material. An R value of 19 to 29 may be required for an outside wall (depending on location). (The R value is the reciprocal of the *C value*. A C value of 2 would give an R value of ½ or 0.5.) See *K value, C value, U value*. Figure 820 shows recommended R values for different areas of the country. Figure 821 gives R values for different insulation materials. The R value for a typical frame wall is shown in Figure 822.

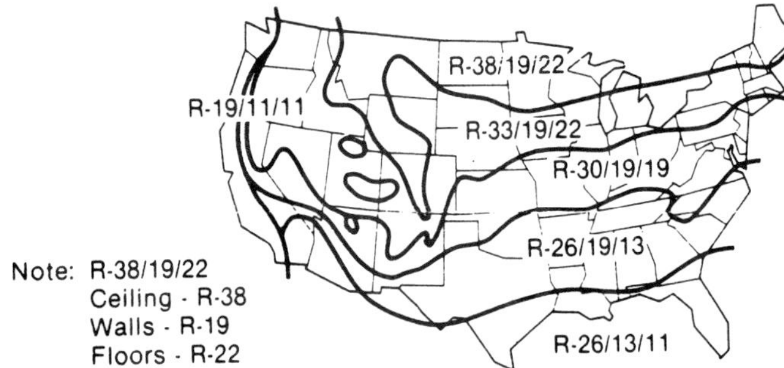

Figure 820 R values recommended for different areas of United States

Insulating Material	Approx. R/inch	R-11	R-19	R-22	R-34	R-38
			Approximate depth in inches needed for			
Loose fill						
Mineral wool and fiberglass	2.2-3.0	4	7	8	12	13½
Cellulose	3.1-3.7	3	5½	6	10	11
Vermiculite	2.2	5	9	10	15½	17
Batts or Blankets						
Fiberglass	3.1-3.6	3½	6	7	10½	12
Mineral Wool	3.2	3½	6	7	10½	12
Rigid Board						
Polystyrene beadboard	3.6	3	5½	6	9½	10½
Extruded polystyrene	5.0	2	3½	4	6½	7
Urethane & isocyanurate	7.7	1½	2½	3	4½	5
Fiberglass	4.0	3	5	5½	8½	9½
Foamed-in-place						
Urea-formaldehyde	4.8	2½	4	4½	7	8
Urethane	6.2	2	3	3½	5½	6

Figure 821 R values for typical insulations

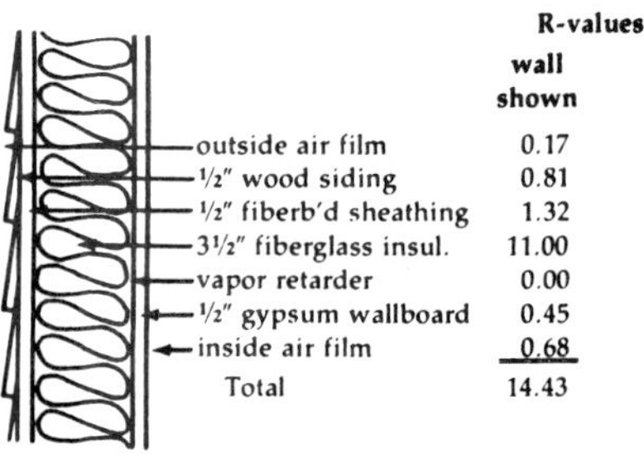

Figure 822 R value for a typical frame wall

S

saber saw Portable electric saw using a thin blade which cuts by an up-and-down reciprocating motion. The blade cuts on the up-stroke. (Figure 823.) Both wood-cutting and metal-cutting blades are used. See also *jig saw*.

sacking In concrete work, the repair of cracks or defects on a concrete surface by applying a sand-cement mixture and rubbing it in with a burlap sack or other rough material.

saddle Roof ridge covering. The small double-sloped structure on the high side of a brick chimney to divert water. (*Figure 268*, p. 100.) A *cricket*. Also, the *threshold* in a doorway.

saddle flashing Flashing used at a chimney saddle. (*Figure 268*, p. 100.)

safety color codes Various colors are by law used to symbolize different safety considerations or dangers.

Red Fire equipment; flammable-liquid container; stop button on machinery or electrical equipment.

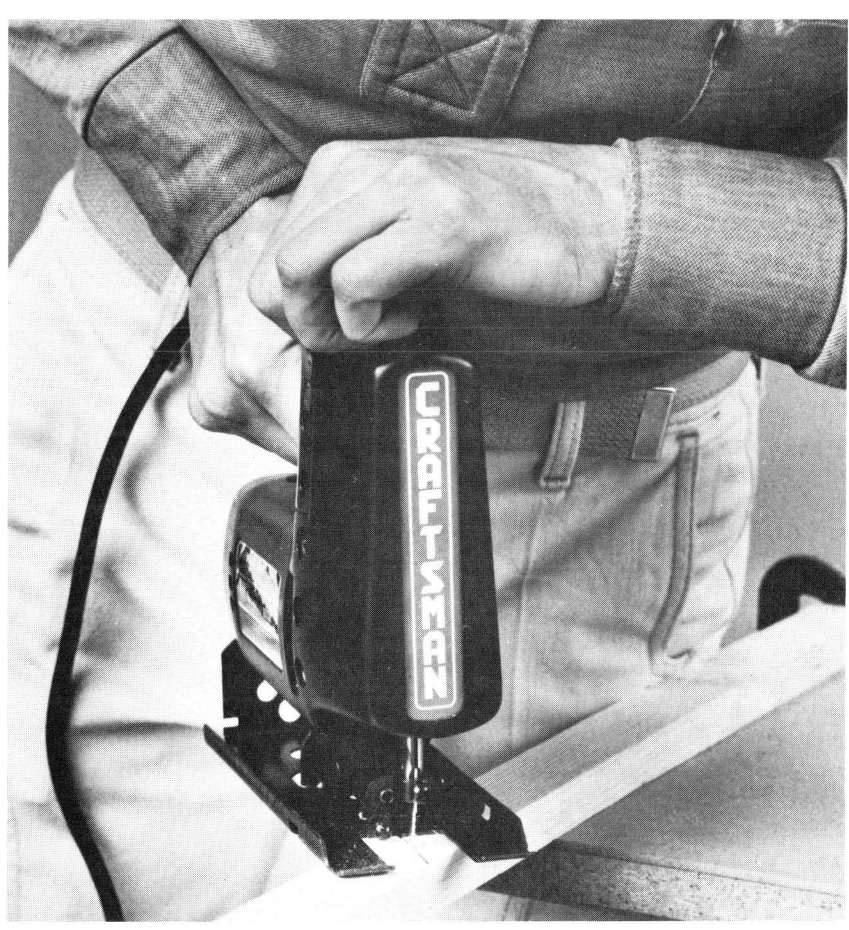

Figure 823 Saber saw

Figure 824 Safety harness

Yellow Caution; power source; waste container.
Orange Exposed cutting or moving parts; danger area on a machine; starter button.
Purple Radiation hazard.
Green Safety; first aid equipment.

safety factor The overdesign of something so its actual strength is beyond its stated strength. Something may have a stated strength of, say, 1000 pounds per square inch but actually be able to hold 3500 pounds per square inch without failure. It will actually hold 3.5 times its stated strength and has a safety factor of 3.5. In design and in practice only the stated strength is used. This results in a structure that is stronger than the actual design figures would show, so that a safety factor is introduced.

Figure 825 Salt box Colonial

safety harnesses Support slings, straps, and harnesses used to support the worker in high, exposed work areas. (Figure 824.)

safety switch A *service disconnect* for disconnecting power to a building.

sag A hanging or drooping downward; settlement.

salamander Portable heater used on the building site.

salient Projecting outward.

salt box Colonial Square or rectangular house with a gable roof that extends down to the first floor on one side; two or two-and-a-half stories. (Figure 825.) See also *architectural styles*.

sander Various power sanders are used to finish flat surfaces, as floors. See *belt sander, disc sander, orbital sander.*

sanding Finishing wood by rubbing with an abrasive, such as *sandpaper*.

sanding block Small wood piece which is used as a base for wrapping sandpaper around to create a flat sanding surface and a good gripping area. See also *sandpaper holder*.

sandpaper A paper with an abrasive, such as crushed flint or garnet particles, glued to it. See *abrasive*.

sandpaper holder Metal holder used to mount and hold sandpaper. A flat sanding surface is created. See also *sanding block*.

sandwich beam Two lumber pieces joined together with a steel plate between the two members; a *flitch beam*.

sandwich panel Plywood sheets glued on each side of a built-up honeycomb core. See *stressed-skin panel*. Also a wall unit composed of a center layer of insulation sandwiched between outside layers of concrete (a *sandwich wall*).

sandwich wall Wall panel made with insulation board set between two reinforced concrete layers.

sanitary sewer Sewer that carries only sanitary (waste) flow. See *combination sewer, storm sewer.*

sanitary unit Water closet or urinal.

sapwood Outer living wood of a tree that carries sap; called xylem. See *tree.*

sash Window frame that holds the lights (window panes). See *double-hung window, window.*

sash balance Counterbalance in a double-hung window to allow the sash to slide freely up and down and stop in any position without falling. The sash balance has a spring inside a revolving drum.

sash bar Strip of wood used for separating and holding glass panes (lights) in a window sash. See *muntin.*

sash lift Metal handle used to raise a window sash.

sash lock Lock used to fasten a window sash. Figure 826 shows a sash lock used to fasten double-hung windows. See also *lock.*

sash opening The opening the sash fits into. The distance is measured inside the window frame. See *window frame.*

sash stop Wood strip that prevents the bottom sash from falling out when completely opened. See *double-hung window.*

sash weights Weights used on older double-hung windows. Each weight hangs on a cord which is attached to the window sash. The sash weight helps to pull the window up. Weights have been replaced in all new windows by a *sash balance.*

saturated felt Felt impregnated with tar or asphalt.

saw Hand or power tool used for cutting wood or metal. A blade with a serrated or finely toothed edge is used to tear away the wood or metal. (Figure 827.) A wide variety of saws are commonly available. For hand saws see: *backsaw, coping saw, crosscut saw, dovetail saw, drywall saw, flooring saw, hacksaw, keyhole saw, ripsaw, stab saw, two-man saw.* For power saws see: *table saw, chain saw, chop saw, circular saw, concrete saw, jig saw, power miter box, radial saw, reciprocating saw, saber saw, table saw.* See also *jointer, shaper.*

saw arbor Shaft on which a circular saw is mounted.

Figure 826 Sash lock

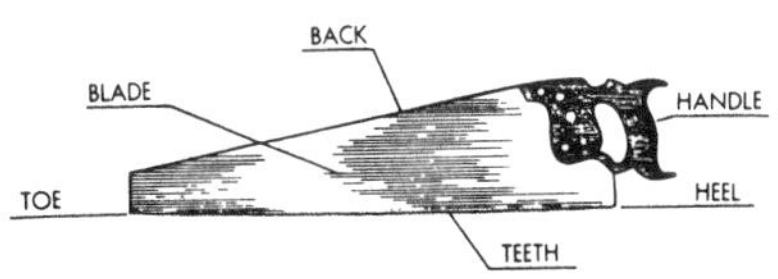

Figure 827 Saw (hand)

Blade Type	Description	Application	Blade Dia.	No. of Teeth
Chisel-tooth Combination	Flat ground, teeth set for clearance	For fast cross cuts and rip cuts on exterior plywood; construction timbers, tempered laminates, rough cut lumber and materials that tend to dull ordinary blades rapidly.	7¼" 8¼"	24 28
Regular tooth Combination	Flat ground, teeth set For clearance	For the broad range of cross and rip cuts on lumber, plywoods, laminates and other structural materials that are rough on blades.	7¼" 8¼"	40 40
Planer Miter	Hollow ground for clearance	For cross, rip or miter cuts, special design provides smooth operation. Unexcelled for interior woodwork, cabinet work, furniture, on hard or soft woods.	7¼" 8"	50 60
Flooring Cut-off	Flat ground, teeth set for clearance	For cross or miter cuts on flooring, reclaimed lumber and work where nails may be encountered. Also for aluminum, brass or copper tubing.	7¼" 8"	76 80
Plywood	Hollow ground for clearance	For smooth, splinter-free cuts in plywood, veneers and laminates. Provides a fine finish that requires little or no sanding.	7¼" 8"	160 176
Carbide Tipped	Clearance ground carbide tips	Rugged carbide tips easily cut tough materials such as asbestos, Formica, Masonite, composition siding, roofing materials. The life expectancy is 30 to 40 times that of regular blades.	7¼" 8¼"	18 24
Abrasive Cut-Off	Organic-bond abrasive, reinforced with figerglass	For cutting-off a range of ferrous and non-ferrous metals, and non-metallic materials; also for tuck-pointing masonry.	7" x 3/32" x 5/8" 7" x 1/8" x 5/8" 8" x 3/32" x 5/8" 8" x 1/8" x 5/8"	

Figure 828 Circular saw blades used in power saws

Figure 829 Sawhorse used to support work

saw blade Flat metal part of the saw with the cutting teeth. (Figure 827.) The teeth on a saw blade determine what the blade is used for. (Figure 830 compares the teeth of a hand crosscut and a ripping saw.) Figure 828 shows several circular blades used in power saws.

sawbuck A *sawhorse.*

sawed joint In concrete flatwork, a control joint made by sawing with a power *concrete saw.*

sawed veneer Veneer cut by sawing.

saw files Triangular tapering files used for sharpening saws. Sizes run from 4" to 10". See *file.*

sawhorse A support frame for holding lumber or other material while it is being worked on. Two sawhorses are used, and the material is laid across them. (Figure 829.)

saw kerf Shallow groove or cut made by a saw blade.

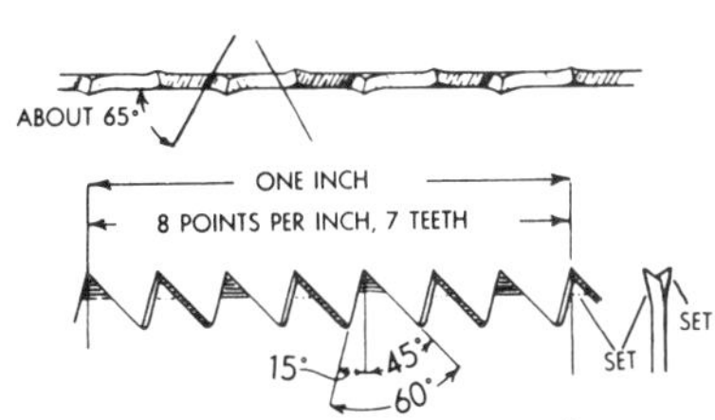

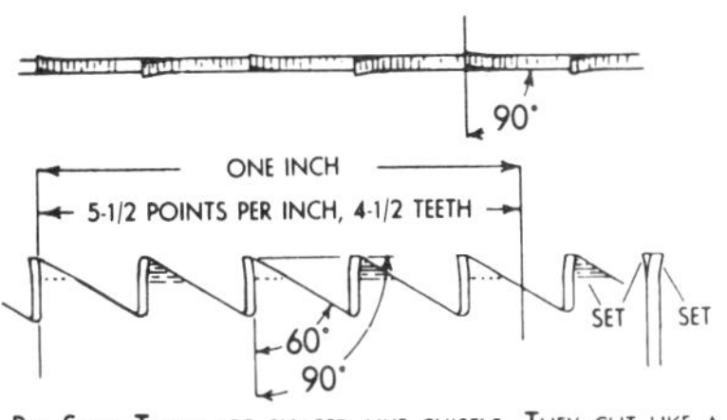

Figure 830 Saw teeth: crosscut (top) and ripsaw (bottom). The number or teeth or points per inch is specified

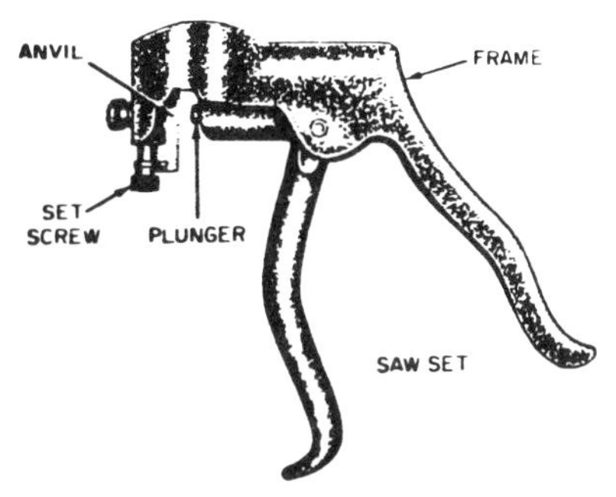

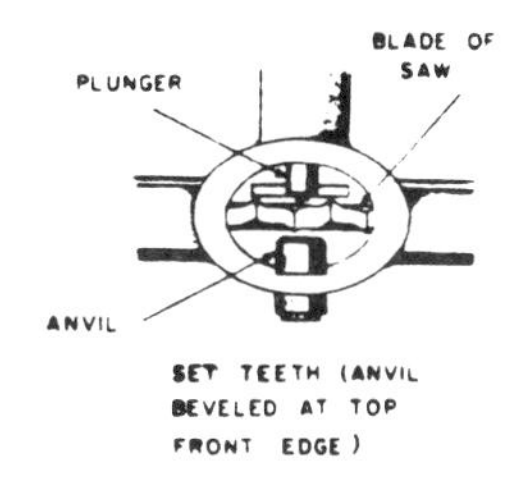

Figure 831 Saw set

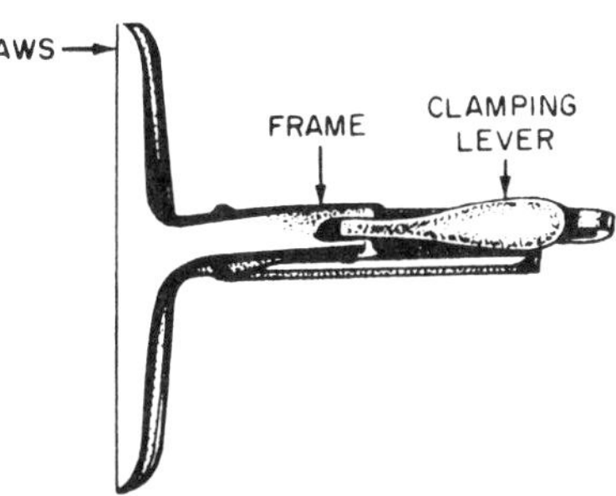

Figure 832 Saw vise

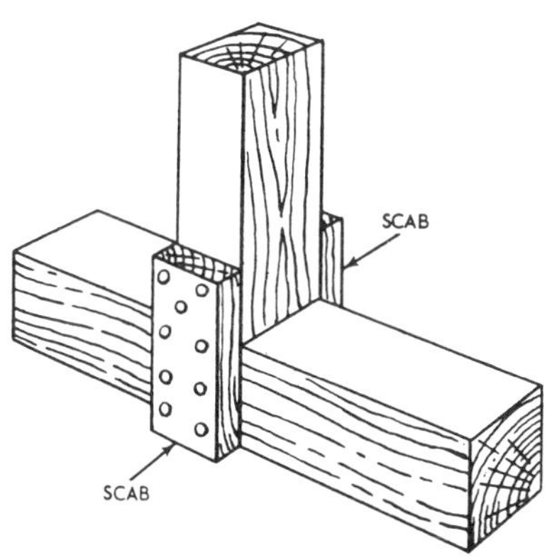

Figure 833 Scab (wood)

saw nest A set of saw blades, used with the same handle.

saw points Saws are specified by the number of points per inch. Figure 830 shows how the points are measured. A crosscut saw commonly has 7, 8, or 9 points per inch. A ripsaw commonly has 6 or 7 points per inch. Hacksaws have from 14 to 32 points per inch.

saw set The angle of the saw teeth. The instrument for setting the saw teeth. (Figure 831.) See also *set*.

saw teeth The cutting points on a saw. (Figure 830.) See *crosscut saw, ripsaw, saw blade.*

saw vise Wide-jawed plier-type vise for holding saw when setting or sharpening. (Figure 832.)

S beam A standard (S) steel beam. In standards, the S beam designation has replaced the I beam. The nominal height (distance between flanges) on S beams varies from 3″ to 24″; weight is specified as pounds per running foot.

scab Short lumber piece, used to give support or added strength. (Figure 833.)

scabbling Dressing down of rough stone.

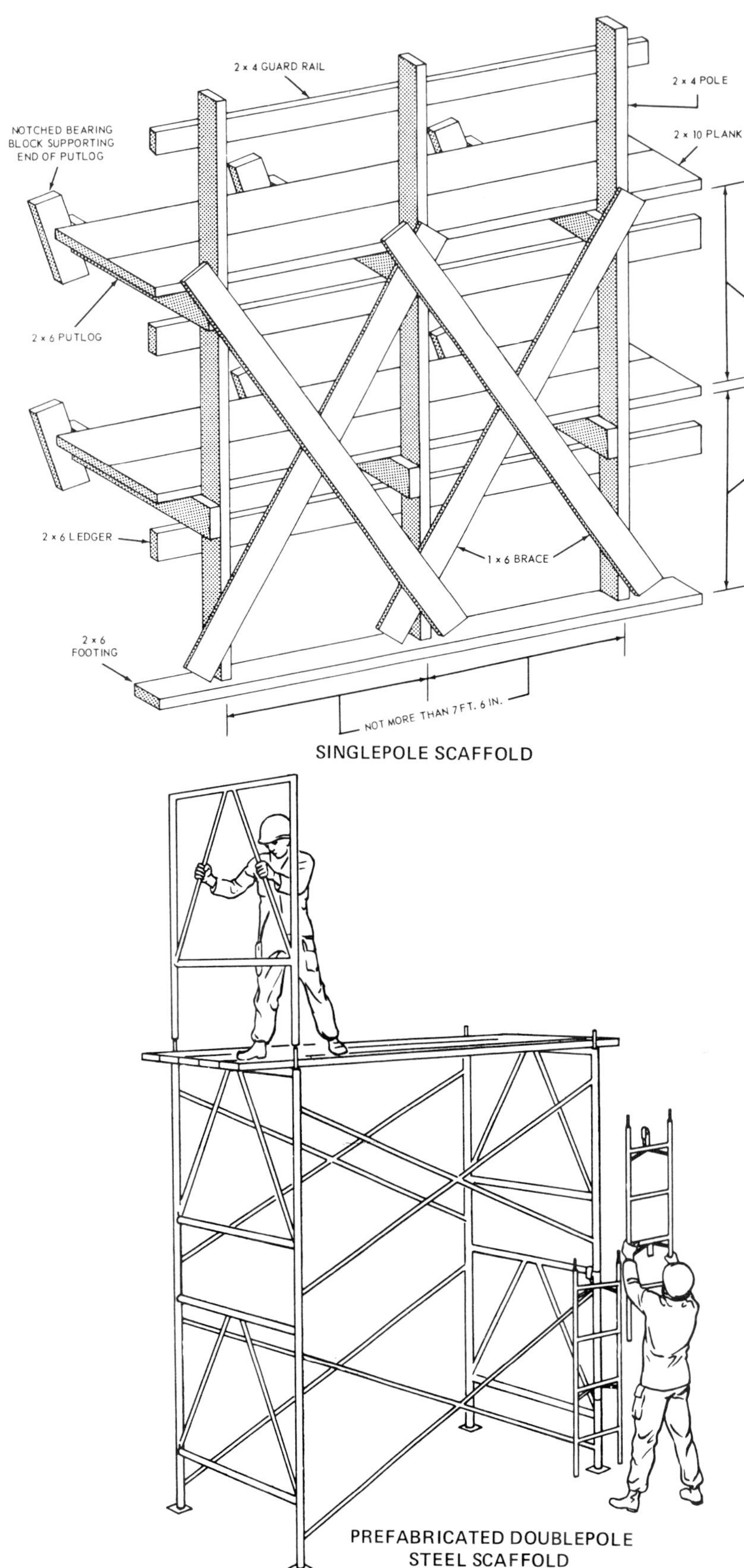

Figure 834 Scaffolds

scaffolding Temporary metal or wooden platforms constructed as work areas above ground level. (Figure 834.) Most scaffolding today is tubular steel. See *bricklayer's scaffold, hanging scaffold, rolling scaffold, swing-stage scaffold, tubular scaffold.*

scaffold nails Double-headed nails; *duplex nails*. See *nails*.

scale Measuring instrument with graduations showing measuring units, used for drawing plans. Different scales are used for drawing buildings or construction features to different sizes. See *architect's scales, draftsman's scales*.

scaled drawing Drawing made to a reduced size. Floor plans are commonly made to a scale of $\frac{1}{4}''$ = 1′–0″. Each $\frac{1}{4}''$ on the scaled floor plan represents 1′–0″ on the actual building.

scaling The breaking away of concrete or stone by the freezing and thawing cycle. In painting, the flaking or peeling away of the paint. In drafting, drawing to scale. Also, measuring a blueprint with a scale to determine building dimensions. (Not a good practice!)

scantling Lumber piece of small cross section, as a 1″ × 2″ or 2″ × 3″; a stud.

scarf joint Wood or plywood piece jointed together with a slant angle cut on each piece. (Figure 835.)

scarification In site development, a technique of abraiding or patterning an exposed, graded slope so water runoff will be retarded; the steel tracks or tires of heavy equipment, such as a bulldozer, are used to form the scarification pattern.

scarifier Wide plastering tool with long wirelike teeth used to scratch a plaster coat to form a good surface for bonding the next coat. Also called a *scratcher*. (Figure 836.)

scarifying In earthmoving, the loosening of the top soil by running blades or teeth through it. See *ripping*.

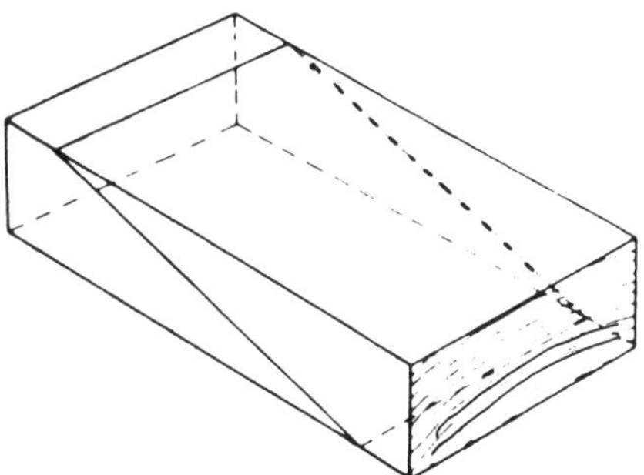

Figure 835 Scarf joint

schedule Tables used on blueprints giving number, size, type, manufacturer and location of building parts or finish. Schedules are made for : *windows, doors, room finish, beams, columns, reinforcement*, and *bending* (of reinforcement). See also *bar schedule, slab schedule*.

scissors truss A truss with a bottom chord that vees up sharply toward the center. Used to give a more open ceiling area. See *truss*.

score A cut or notch cut in something. A notch cut in hardened concrete with a masonry saw. Scratches made in plaster. See *scoring*.

scoring In plastering, horizontal scratches made in the scratch coat to give a better key for bonding on the next plaster coat. See *scarifier*. Marking wood with lines or scratches to form a better surface for a glue bond.

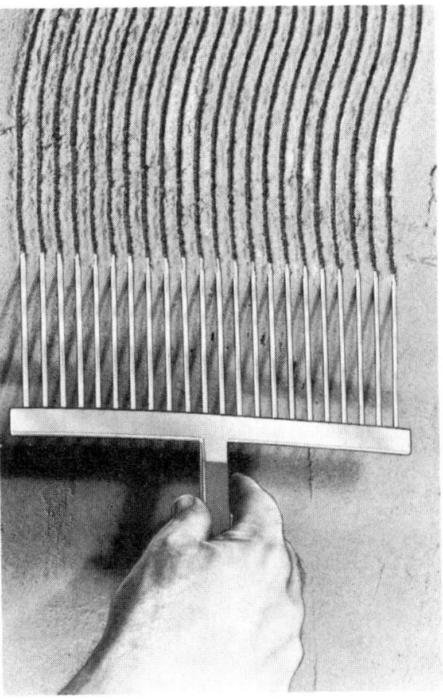

Figure 836 Scarifier

Figure 837 Scraper

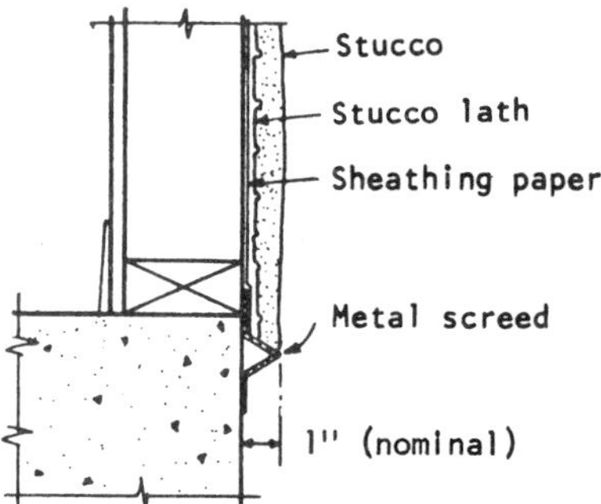

Figure 838 Screed used in stucco work

scotia A concave molding.

scrap To remove wood with a scraper; to remove a thin layer of wood.

scraper See *cabinet scraper*. Also heavy earth-moving machine used for leveling roadways and building sites. (Figure 837.) Blade sets underneath. See *grader*.

scratch awl Pointed tool used for scratching on wood or steel. The mark is used as a cutting or layout guide. See *awl*.

scratch coat First or base coat of plaster. The plaster is scratched to make a better bonding surface for the next coat.

scratcher Tool used to scratch the base coat of plaster. See *scarifier*.

SCR brick A large brick that is used in single-wythe 6" walls. The brick has a double five-core row and is $2\frac{1}{4}'' \times 5\frac{1}{2}'' \times 11\frac{1}{2}''$.

screed In concrete flatwork, a straightedge used to level (strike off) the concrete. The ends of the straightedge may ride on the tops of the forms. (The forms themselves in concrete flatwork are also sometimes referred to as screeds.) See also *power screed*. In plastering, small strips of wood or built-up strips of plaster that serve as a guide to the plaster coat depth needed and as a support for the ends of the leveling straightedge. See *grounds*. In stucco work , a metal depth guide used at the base of the stucco. (Figure 838.) In hardwood floor laying, short wood members laid in mastic on a concrete slab to support a wood floor. Wood strips 2″ × 2″ or 2″ × 3″ that are placed 16″ O.C. in a reinforced concrete slab floor. These screeds rest on top of steel stirrups; when the slab is placed, they project above its face; used for nailing wood flooring into. Also, wood boards that are laid atop the slabs for nailing a wood flooring into. See *sleepers*.

screed finish Rough finish in concrete flatwork made with a screed.

screeding Leveling of concrete flatwork; *striking off*. Leveling of plaster; *rodding*.

screw Threaded, pointed fastening device. (Figure 839.) *Wood screws* have a plain shank below the head and commonly are screwed into a bored hole. *Sheet metal screws* are threaded all the way to the head for holding metal pieces or sheets together. *Drywall screws* are driven in with a power screwdriver and cut their own hole. Some screws are *self-tapping*; they cut their own hole and have a special grooved point. Screws are available in a wide variety of shapes and sizes and are chosen to fit the specific job at hand. Screws are sized by gauge from #0 ($\frac{1}{16}$″) up to #24 ($\frac{3}{8}$″) or even larger; lengths vary. Figure 840 shows screw sizes. See also *machine screw*.

HOW TO MEASURE

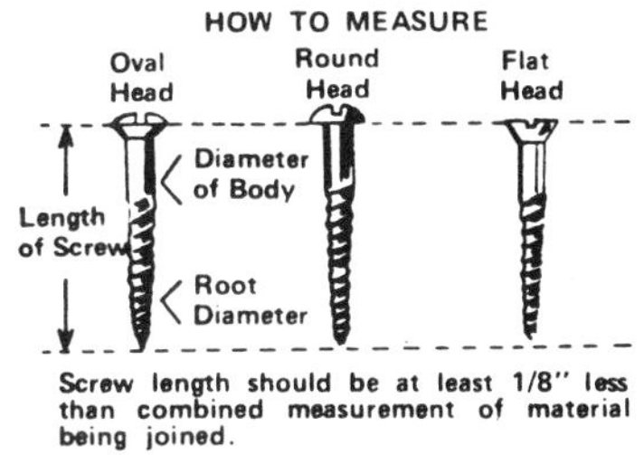

Screw length should be at least 1/8″ less than combined measurement of material being joined.

SCREW HEAD STYLES

Round Head

Oval Head

Countersink

Head structure is the only difference between round, oval, and flat head screws. They are usually available in the many lengths and gauges shown above.

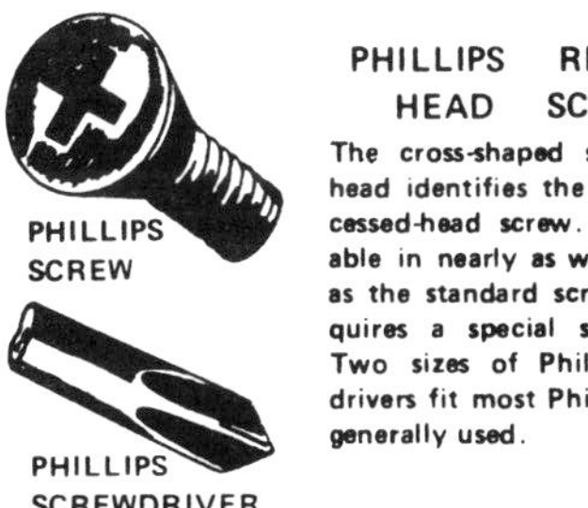

PHILLIPS RECESSED HEAD SCREWS

The cross-shaped slot in the head identifies the Phillips recessed-head screw. It is available in nearly as wide a range as the standard screw and requires a special screwdriver. Two sizes of Phillips screwdrivers fit most Phillips screws generally used.

SHEET METAL SCREWS

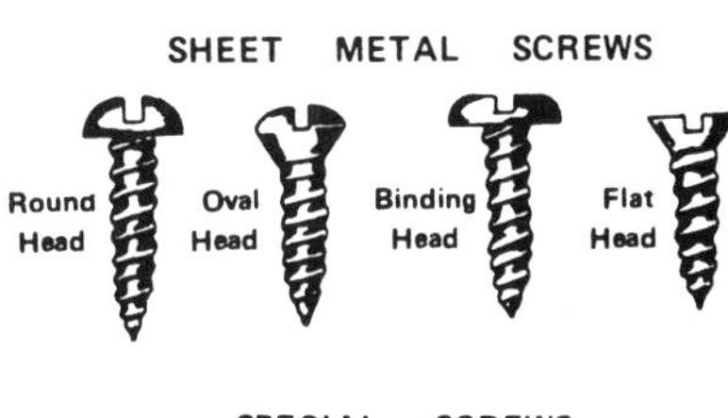

SPECIAL SCREWS

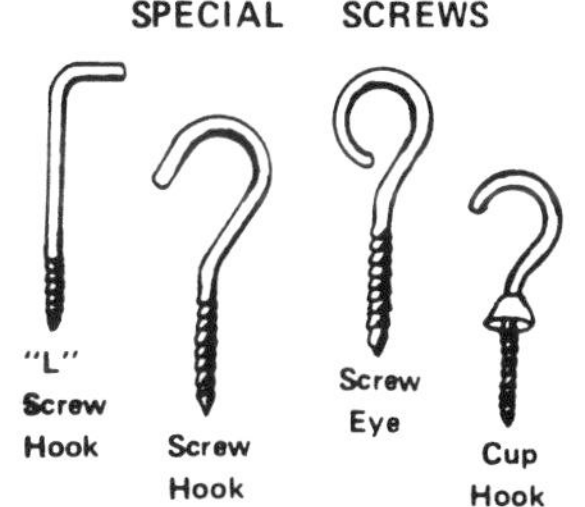

Figure 839 Screws

SCREW SIZE	HEAD SIZE	SHANK SIZE	PHILLIPS POINT SIZE
	ACTUAL SIZE		ACTUAL
0			POINT NO. 0
1			
2			POINT NO. 1
3			
4			
5			POINT NO. 2
6			
W7			
8			
W9			
10			POINT NO. 3
12			
W14			
M 1/4			
W16			
W18			POINT NO. 4
M5/16			
W20			
W24 & M 3/8			

Figure 840 Screw sizes

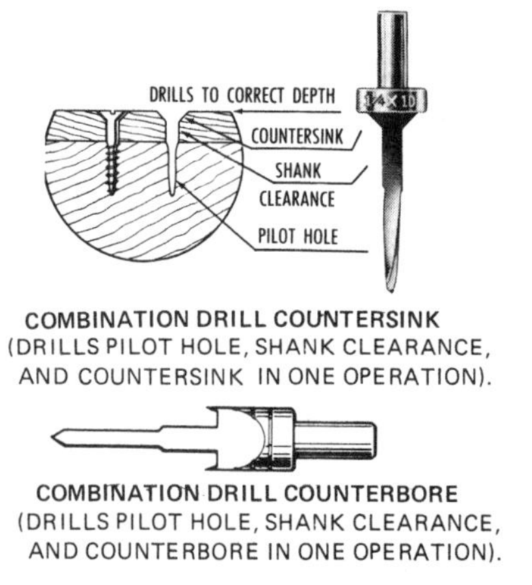

Figure 841 Screw bits

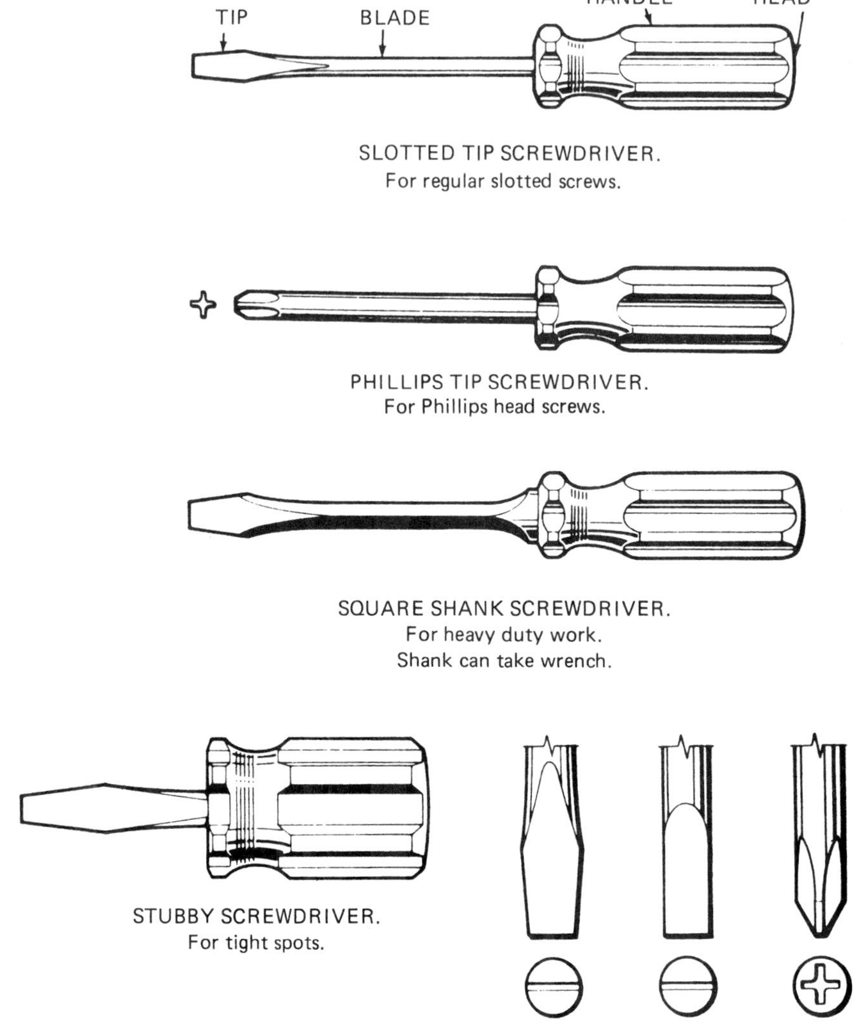

Figure 842 Screwdrivers

screw anchor Anchor that fits into a bored hole for securing a screw or bolt. Generally used in masonry. See *lead screw anchor.*

screw bit A bit, shaped to drill a hole for a wood screw, that also has a wider part for the screw shank and a countersunk part for the screw head. (Figure 841.) The bit must be chosen to fit the screw size. Also called a *drill and counterbore bit.*

screw clamp A *parallel clamp.*

screwdriver Tool designed for driving in or extracting screws. The standard screwdriver has a blade tip for driving slotted screws. (Figure 842.) Two types of standard screwdrivers are used: standard tip (flat tip with slanting sides) and cabinet tip (flat, straight tip) for driving counterbored screws. (*Figure 147*, p. 56.) The *Phillips screwdriver* has an X-shaped tip. Various sizes and lengths are available; handles are vinyl or wood. In addition to the common slotted or Phillips screwdrivers, other blade shapes are available. Square shanks are available so a locking plier can be attached for extra turning power. An *offset screwdriver* is used to drive screws in tight or hard-to-reach places, such as inside an outlet box. The screwdriver tip should always fit the head of the screw being driven. See *cabinet screwdriver, drywall screwdriver, machine screw, offset ratchet screwdriver, power screwdriver, screws, spiral ratchet screwdriver, wood screw.*

screw extractor Threaded device used for removing broken screws and bolts. A hole is drilled into the broken screw end, then the screw extractor is screwed in counterclockwise to remove the screw. (Figure 843.) Also called a *bolt extractor.*

screw eye Screw hook with a round eye on the unthreaded end. (Figure 844.) See *cuphook, L hook screw, screw hook.*

screw gauge A V-slotted gauge used for measuring the diameter or size of a screw.

screw holder Flexible grasping arms attached to a screwdriver shank for holding screw heads. Used for reaching into otherwise inaccessible areas. (Figure 845.)

screw hole plug Short wood plug that is glued into a screw hole to conceal the recessed screw. A *plug.*

screw hook Screw with a curved hook on the unthreaded end. (Figure 846.) A screw hook has an open curved end; a *screw eye* has a closed end or eye. See *cup hook, L hook screw.*

screw nail Nail with a screw-shaped shank. (Figure 847.)

screw sink See *screw bit.*

scrib To cut or scratch a line on wood or metal.

scriber Two-legged compass with two points used for scribing or marking boards against an irregular surface. (Figure 848.)

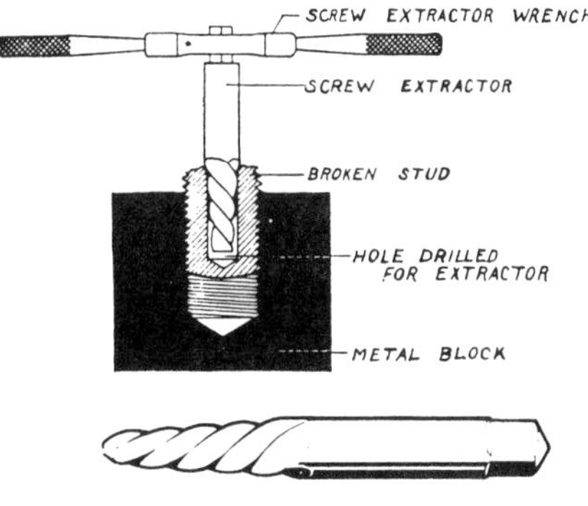

Figure 843 Screw extractor

Figure 844 Screw eye

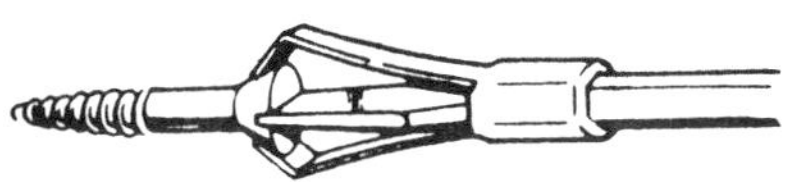

Figure 845 Screwholding screwdriver

Figure 846 Screw hook

Figure 847 Screw nail

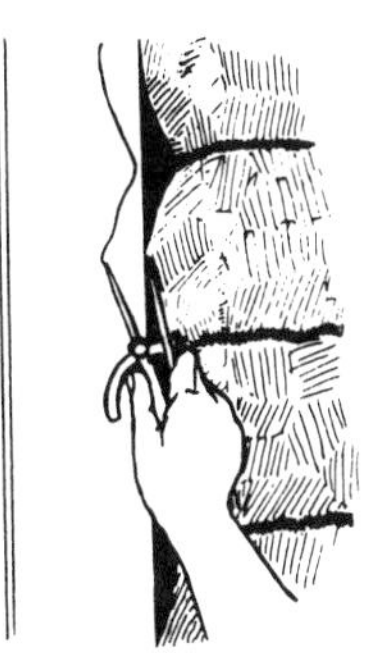

Figure 848 Scribing

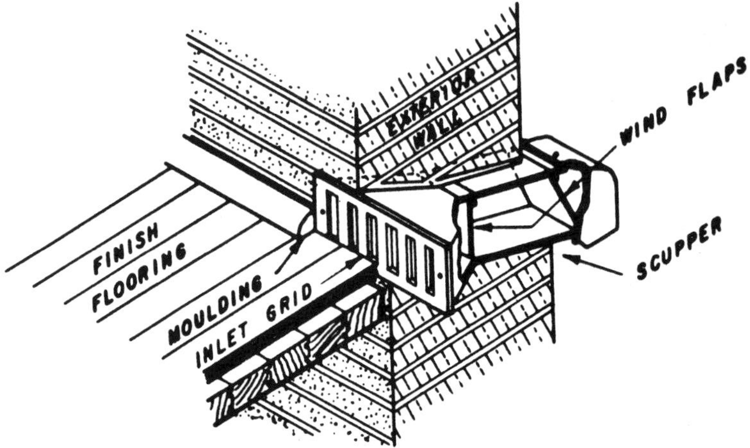

Figure 849 Scupper

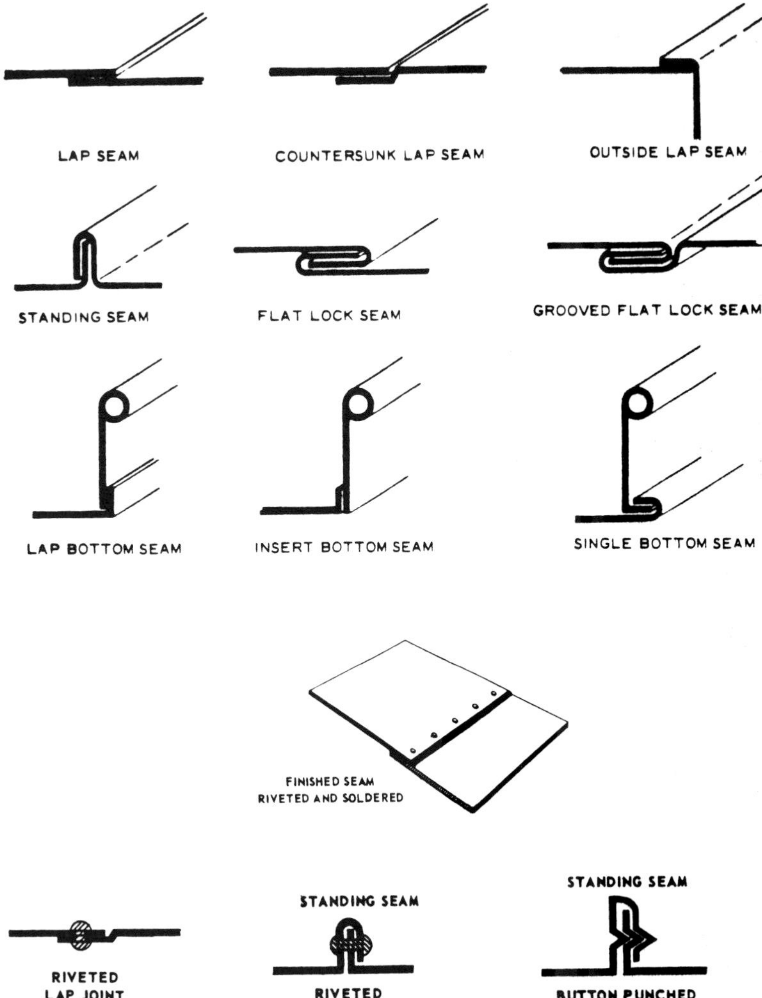

Figure 850 Seams: sheet metal

scribing Using a scriber or compass to transfer an irregular wall or surface outline to a board or panel so the edge may be sawn to fit the exact contour. (Figure 848.)

scroll Curving ornament.

scroll saw A *jig saw*.

scupper Outlet at the edge of a roof to allow for water runoff. No downspout is used. Used mainly on flat roofs. (Figure 849.)

scuttle Access opening in ceiling leading to attic area; access opening in roof.

seal cap Mortar top to a chimney.

sealed drawings Drawings with the seal of a registered architect.

sealer Various liquid sealants painted on wood or masonry surfaces to fill the pores or cracks and prevent moisture entry. Often used on wood preparatory to painting.

seal weld Weld made to obtain tightness.

seam Mechanical closure made between two parts. In sheet metal work, the joint fastening two sheets together: lapped, grooved, and standing seams are the most common. (Figure 850.) See also *drive slip, locked corner seam*. Soldering or riveting may also be used. See *edge*. Also, a sealed joint between gypsum boards.

seamer A sheet metal hand tool used for making sheet metal joints, bending and flattening seams. (Figure 851.)

seasoning The drying of lumber. Natural air seasoning or kiln drying is used.

seat cut In rafter construction, the horizontal cut in a bird's mouth. The cut rests on top of the plate. Also called a *plate cut*. See also *plumb cut*.

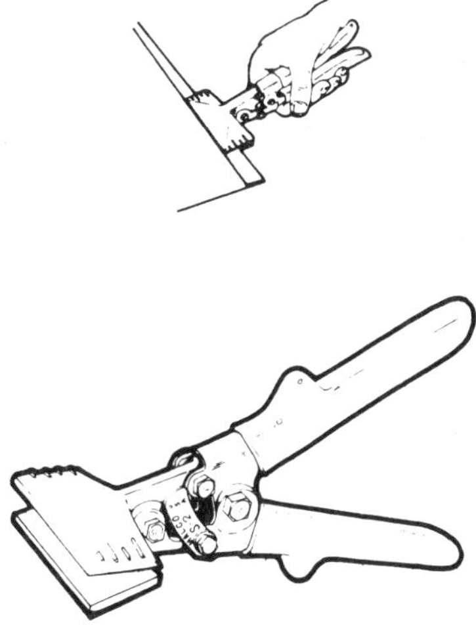

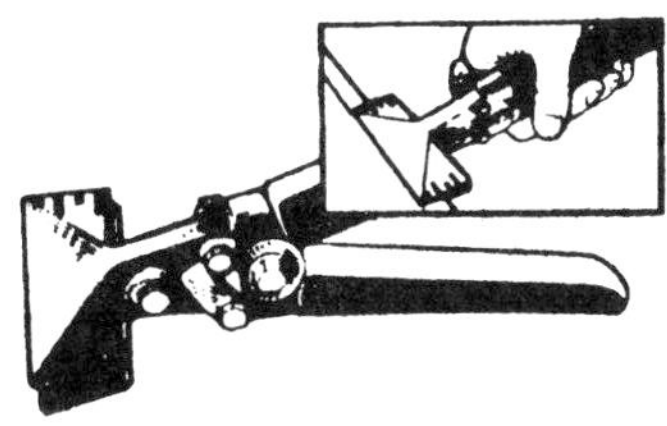

Figure 851 Hand seamer

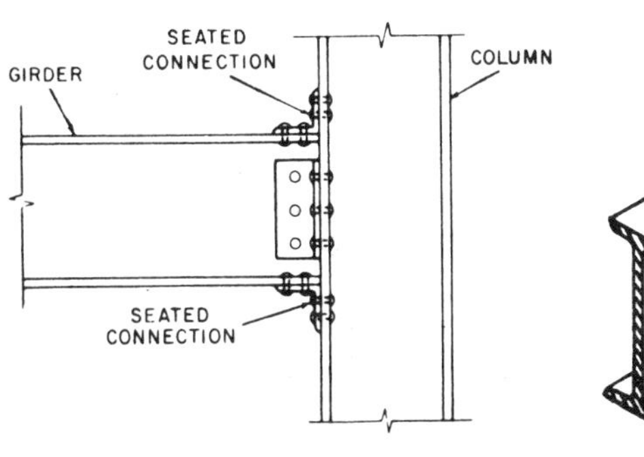

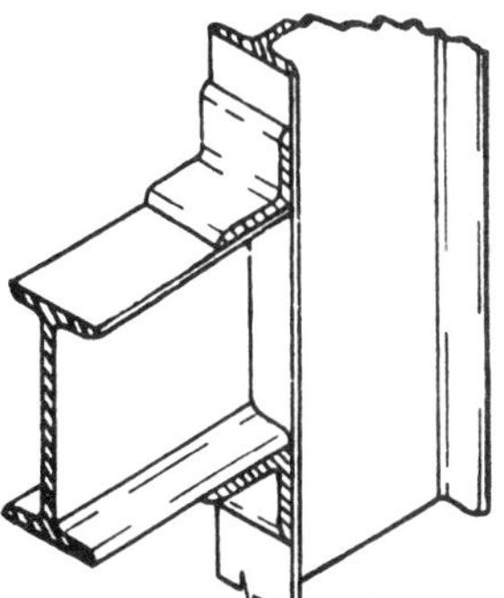

Figure 852 Seated connection

seated beam connection In steel construction, a steel beam whose end rests on a structural seat (angle) attached to a column. Seats that are attached to a column flange are called unstiffened seated connections. Seats that are attached to a column web are called stiffened seat connections. Figure 852 shows a seated beam connection; the load is transferred from the beam to the column. See also *framed beam connection*.

secondary colors Orange, green, and violet. See *primary colors*.

Second Empire architectural style See *American mansard*.

second floor plan Plan view of the second floor. See *floor plan*.

second growth Timber that grows after the *virgin growth* has been cut down.

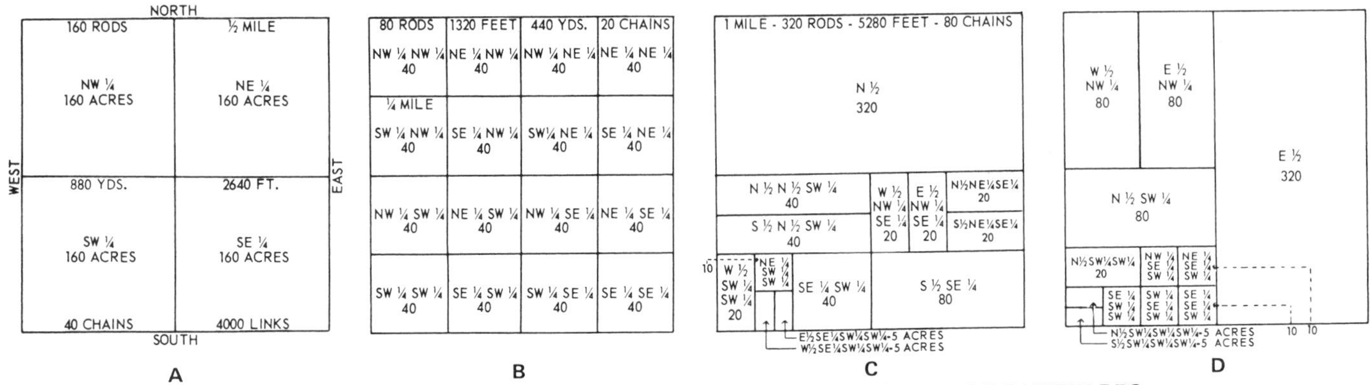

THE BEST WAY TO READ LAND DESCRIPTIONS IS FROM THE REAR OR BACKWARDS.

Descriptions of land always read FIRST from either the North or the South. In A, B, C, D, notice that they all start with N (north), S (south), such as NW, SE, etc. They are never WN (west north), ES (east south) etc.

IMPORTANT: It is comparatively simple for anyone to understand a description, that is, determine where a tract of land is located, from even a long description. The SECRET is to read or analyze the description from the rear or backwards.

EXAMPLE: Under C, the first description reads E½, SE¼, SW¼, SW¼. The last part of the description reads SW¼, which means that the tract of land we are looking for is somewhere in that quarter (as shown in A). Next back we find SW¼, which means the tract we are after is somewhere in the SW¼ SW¼ (as shown in B). Next back, we find the SE¼, which means that the tract is in the SE¼ SW¼ SW¼ (as shown in D). Next back and our last part to look up, is the E½ of the above, which is the location of the tract described by the whole description (as shown in C).

Figure 853 Sections and land descriptions

seconds Lumber that is graded as a lower quality than first.

secret nailing Nailing so the nail heads are concealed; *blind nailing*.

section In drafting, a vertical cross-cut view of part of the building. In surveying, a unit of one square mile (640 acres). Figure 853 shows how a survey section is broken into parts. A section is a mile, or 320 rods, or 5280 feet, or 80 chains on a side. When reading section land descriptions, read backward. For example, NW 1/4 SE 1/4 is located in the SE corner and the NW corner (of the SE corner).

section symbols Material symbols used to show a vertical cross-cut view. (Figure 854.) Section symbols are almost identical to *plan symbols*. See also *symbols*.

section view Vertical cross-cut view. See *section*.

sediment control See *erosion and sedimentation control plan*.

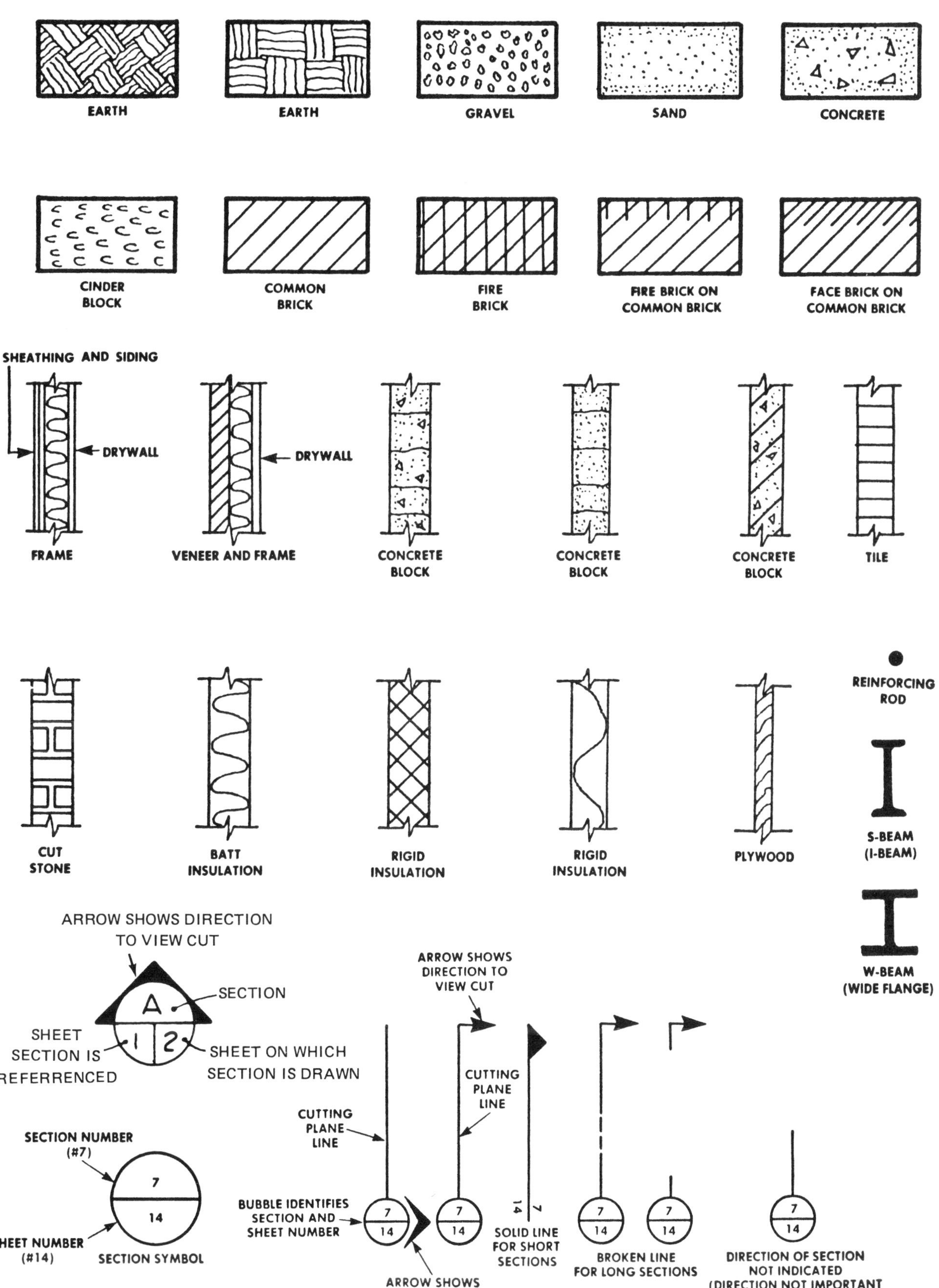

Figure 854 Section symbols

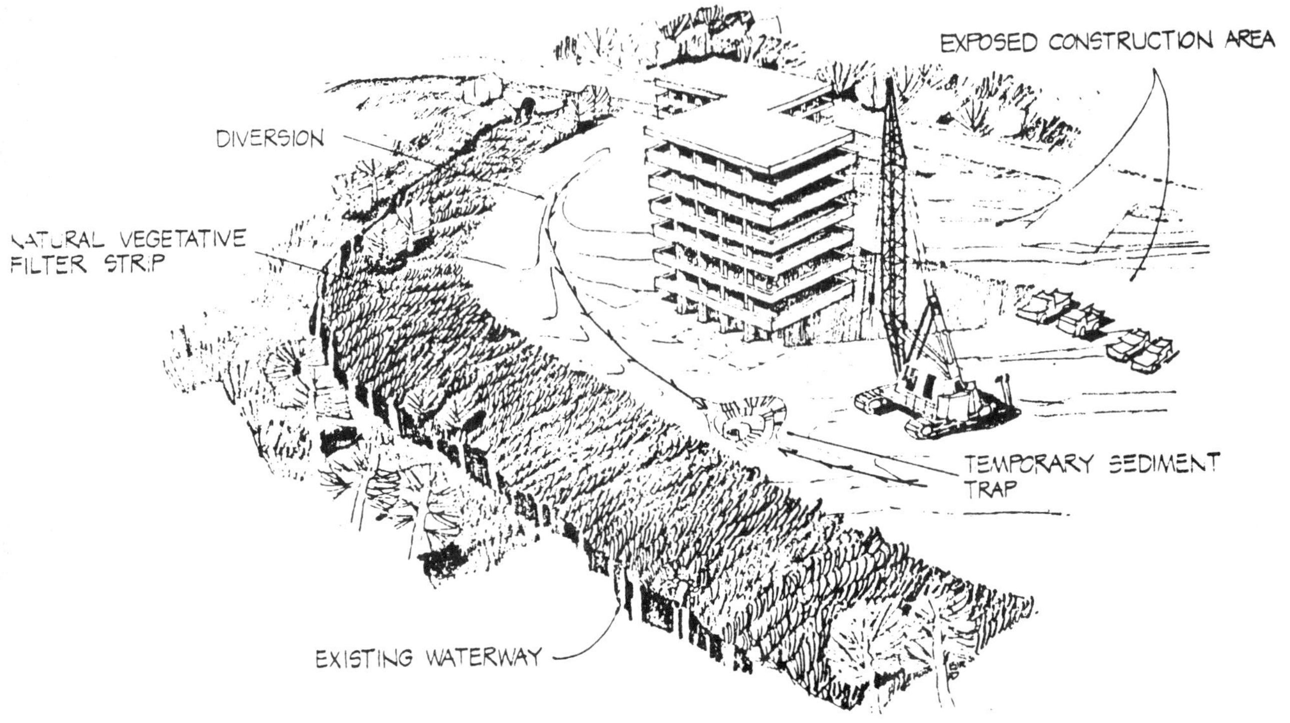

Figure 855 Sediment pool or trap holds water runoff on construction site

sediment pool Low earth trap or pool constructed to hold run-off water on the construction site so sediment may fall out and not be carried off site. Figure 855. See also *diversion, vegetative filter, erosion and sedimentation control plan.*

segment A part of the whole.

segregation In the placement of concrete, the separation or concentration of aggregates so the concrete is no longer homogeneous.

seizure In tightening a fastener, a freezing of the fastener before the full specified torque is achieved. Also called *set.*

select lumber Boards free of or with few defects.

self-centering punch Punch used to strike a hole to start a screw, as for a hinge. (Figure 855.) See *punches.*

self-drilling anchor See *snap-off anchor.*

self-feed bit A large, heavy-body bit with replaceable cutters and lead screw and shank. (Figure 856.) Sizes vary from $1\frac{3}{4}''$ to $4\frac{5}{8}''$ diameter. They are used in a power drill to drill piping holes.

self-tapping screw Screw with a special cutting point that drills its own hole into metal. (Figure 857.) *Sheet metal screws* and some *drywall screws* are self-tapping.

selvage An edge that differs from the main body of material, such as an edge of roll roofing which is bare of aggregate. Used for joining or lapping over other sheets or surfaces. Also, the unprinted edge of wallpaper.

semiautomatic welding Welding with a machine that automatically feeds the filler material but where the rate of work is manually controlled. See *manual welding, machine welding,* or *automatic welding.* See *welding procedures* for a comparison of the processes.

semicircle Half of a circle.

semicircular arch Arch shaped like a half-circle.

septic tank A holding tank for raw sewage. The sewage is broken down by bacterial action, releasing gas through vents. Solid matter settles to the bottom and must be cleaned out periodically. The liquid waste passes on to a distribution field or absorption field, where it slowly passes into the soil and is purified. (Figure 858.)

serrated Criss-cross notching, as on a hatchet face.

serration Jagged, sawtooth edge.

service conductors Supply conductors from the transformer to the house or building using electricity. Service conductors may run from the underground power street main or from overhead transformers to the building service entrance.

Figure 856 Self-feed bits

Figure 857 Self-tapping screw

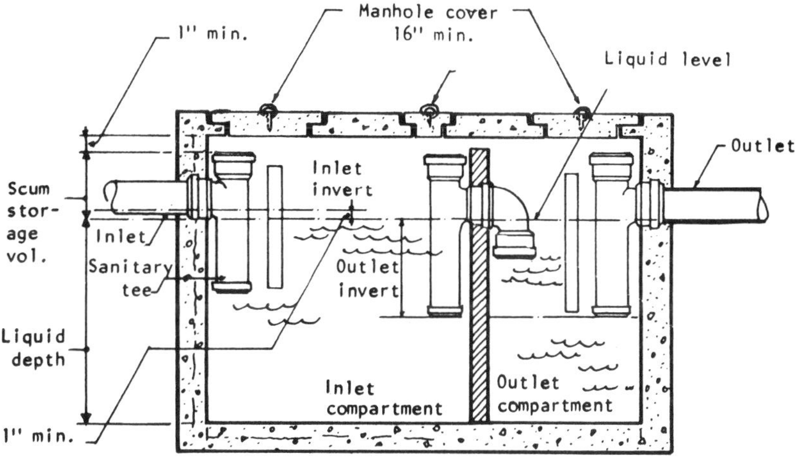

Figure 858 Septic tank (section view)

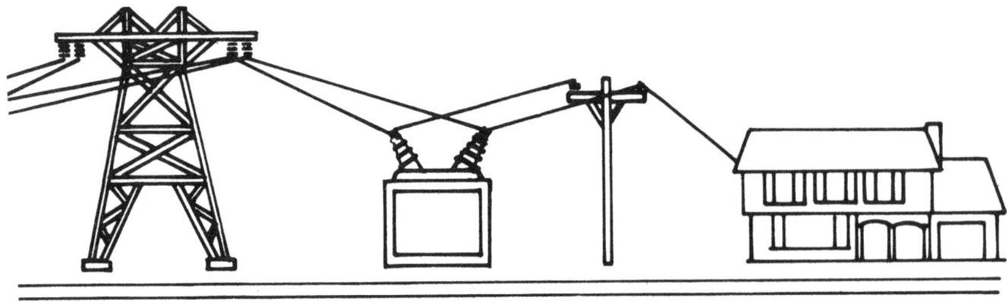

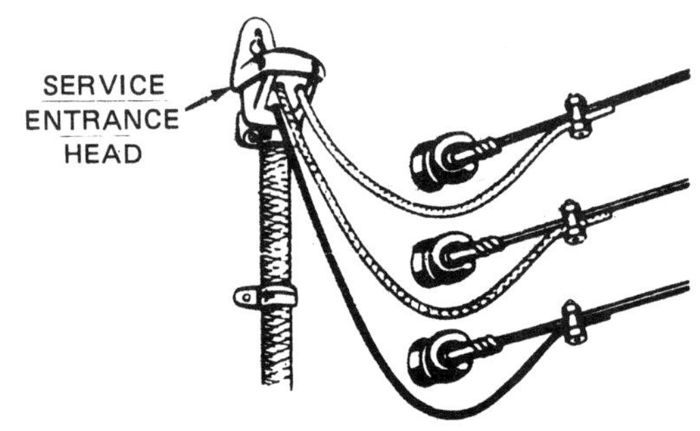

Figure 859 Service drop: overhead electrical entrance to building from power pole

service connectors Lines from the public utility power source to the *service entrance.* Two lines or conductors are run for 110-V service; three lines are run for 220 V.

service disconnect Main control switch for disconnecting power to a building. See *externally operated disconnect.*

service drop An electrical entrance where overhead power is brought into the house. (Figures *859, 123,* p. 46.) Overhead conductors or feed wires that run from the electrical power pole to the *service entrance.*

service entrance Where electrical power enters a building; power may be from a pole to a service drop or it may be underground. (Figure 859.)

service-entrance cable Factory-assembled cable. Two types are used: SE for overhead and USE for underground use.

service-entrance conductors In electrical wiring, conductors that run from the building service equipment to an overhead service drop or to a connection with the underground service lateral.

servive equipment In electrical wiring, the main control (switch and circuit breaker) that allows a disconnection between the building conductors and the outside power supply.

service lateral In electrical wiring, underground service conductors running between underground power street main and the building terminal box or meter.

service lead Wiring entrance cap at the *service drop.*

service mast In electrical wiring, an upright support for service-drop conductors.

service panel Control center that receives outside power. A main switch allows power to be turned on or off. Fuses and circuit breakers protect the system. (*Figure 656*, p. 284.) Conduit runs from the service panel to the panelboard. Sometimes the service control is located in the panelboard.

service switch Main disconnect switch for all power to a building. Normally located in the *service panel*.

set Hardening of concrete, mortar, plaster, or adhesive; the time involved. See *initial set, false set, final set, flash set.* In tightening a fastener, such as a nut on a bolt, a seizing before the full specified torque is achieved. Also called *seizure.* The angle to which saw teeth are bent. (Figure 860.)

setback A distance something is set back from some other thing, as in building placement a setback of a certain number of feet from the front lot line or street is established by local code.

setting The putting of masonry building units into place. In saw sharpening, the bending of the saw teeth to the correct angle.

setting hammer Metalworking hammer used for bending and shaping sheet metal. (Figure 861.)

settlement A sinking of the earth; a sagging or sinking down of part of a structure.

sewage Waste material that flows out of a building into the sewers or into a private septic tank and absorption field.

sewer Piping that carries waste water or sewage to a public processing plant.

sewer gas Gas that is given off by decay of organic matter in the sewers.

Figure 860 Set: the angle at which saw teeth are bent out

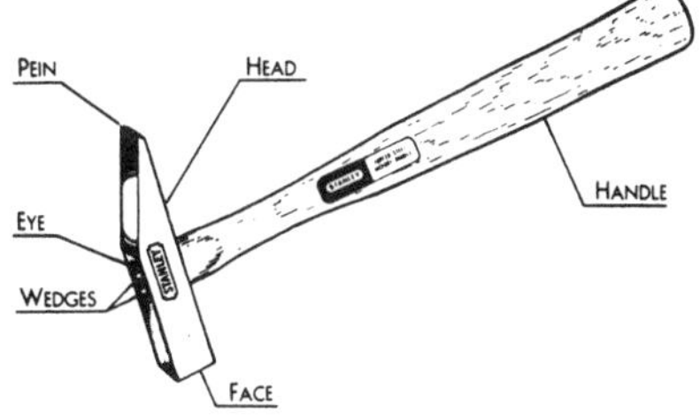

Figure 861 Setting hammer

sewer rod Flat spring-steel rod with a shaped head used for cleaning out sewers. The rod is pushed by hand into the sewer. When not in use, it is stored in a coil. See *drain auger*.

shade line Shadow made by the roof projection at the eaves. The shade line varies by the time of the year. Often the window area is directed toward the south so in the cold months, when the sun is low, direct sunlight will warm the inside of the house, but in the summer, when the sun is higher, the window is shaded by the roof projection or overhang. (*Figure 908*, p. 400.)

shadow course In roofing, a double layer of shingles. Also called a *ribbon course*.

shake Lumber defect: lengthwise separation, where the wood splits or opens around the annual rings. (Figure 862.) See *cup shake, windshake*.

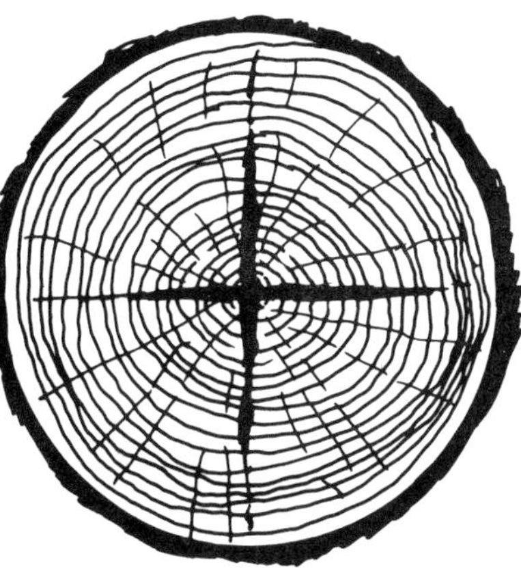

Figure 862 Shake (lumber defect)

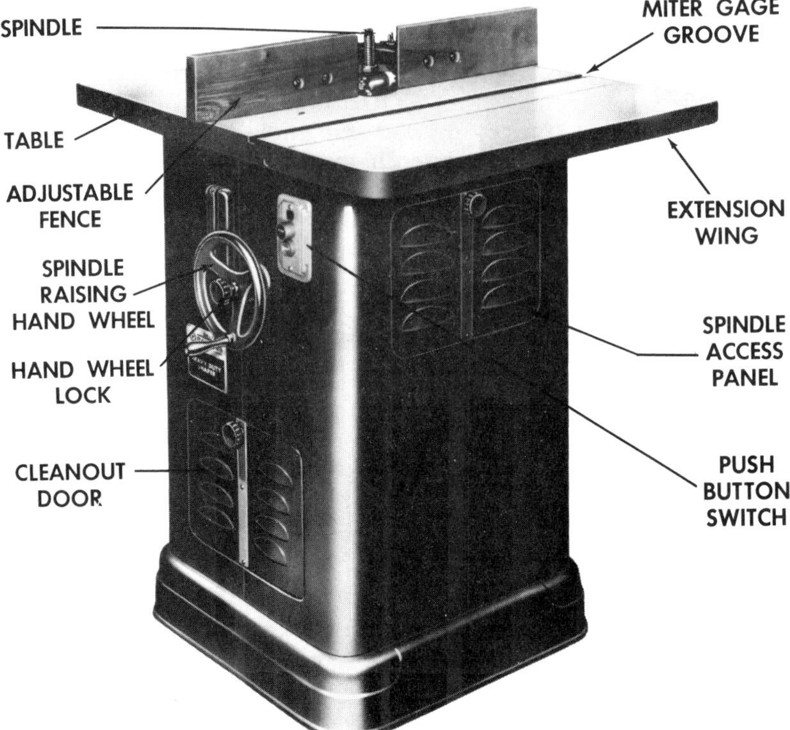

Figure 863 Shaper

shakes Wooden shingles that are split from a timber section.

shank The body of a tool below the handle or top, as the shank of a chisel. The end of a drill or bit that is inserted in the chuck.

shaper A stationary power tool used to cut joints and shape stock. Edge and end shaping of boards are commonly done on the shaper. Stock is placed on the table and pushed by the cutting head. Cutters of different shapes are used to give different shapes to the wood. (Figure 863.) See also *jointer, router*.

sharpening Putting a good cutting edge on tools. An edge is worked from coarse to fine. See *oilstone, strop, waterstone*.

shear Directly opposing forces on a member. A failure caused by opposing loads or forces. Shear that is perpendicular to a wood beam or girder is called vertical shear. Shear that is parallel to the grain is called sliding shear. A breakage caused by opposing loads or forces. See also *compression, stress, tension, torsion*. A cutting force. A hand tool for cutting sheet metal. (Figure 864.)

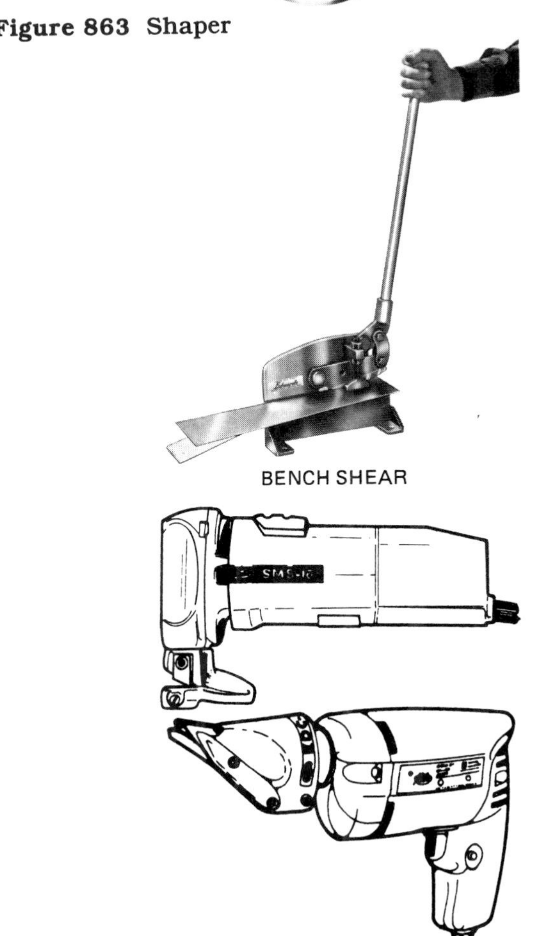

Figure 864 Shear: sheet metal

shearhead In reinforced concrete, special and extra reinforcement to strengthen a load or stress area, as a shearhead in a floor slab over a column; an assembly of shear reinforcement (bars or structional steel) around a column. Shearhead acts to give reinforcement for a column that does not have a capital or drop panel.

shear plate Metal timber connector that fits between structural members to distribute the load. (Figure 865.) See also *spike grid, timber connectors.*

shear wall Wall designed to resist lateral (shear) forces, such as wind, explosion, earthquake.

sheathing Panels that go over the outside studs; 4′ × 8′ plywood and insulation board sheets are widely used. The *siding* goes on over the sheathing.

sheathing paper Paper used over the sheathing to cut down air infiltration and moisture penetration. Also called *building paper*.

sheathing stapler Air-driven stapler used for fastening sheathing and decking. Staples from $1\frac{1}{2}''$ to 2″ are used. Also called a *decking stapler*. See also *air-driven nailer, nailer, stapler*.

sheave A grooved wheel or pulley. See *block*

she bolt In forms construction, a special bolt that threads into the ends of the tie rod. (Figure 866.) When the forms are stripped, the she bolt is removed.

shed dormer Dormer with a shed-type roof. See *dormer*.

shed roof A low-angle, one-slope roof. See *roof*.

sheeting *Sheet pilings*.

sheet metal edge See *edge*.

sheet metal locking pliers Wide-jawed pliers for bending and forming sheet metal. (Figure 867.)

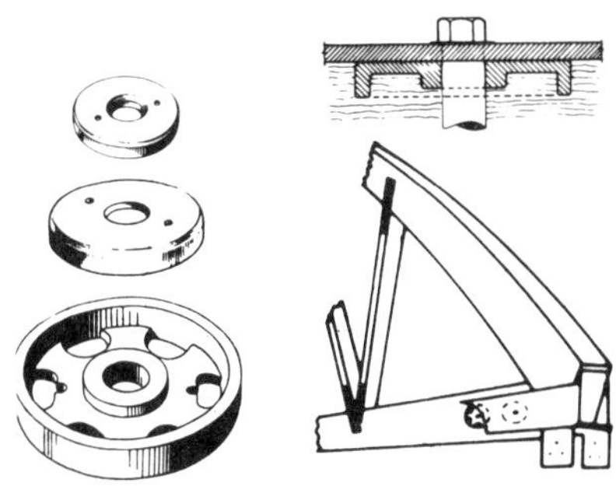

Figure 865 Sheer plates used to connect structural members

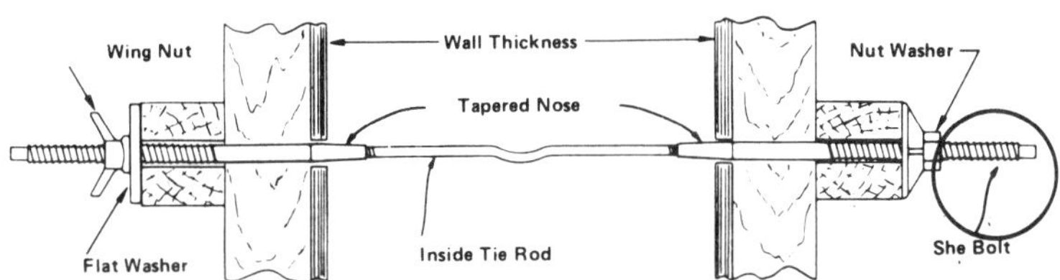

Figure 866 She bolt

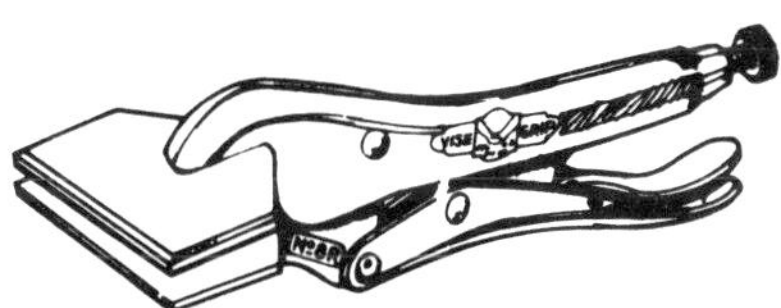

Figure 867 Sheet metal locking pliers

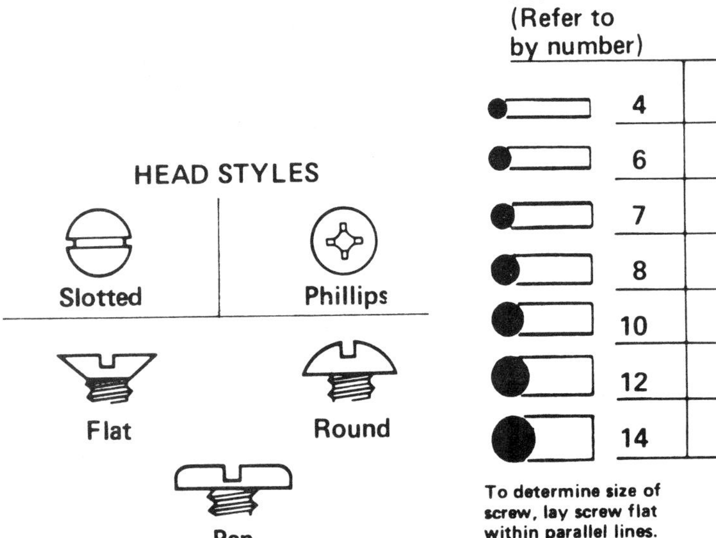

Figure 868 Sheet metal screws

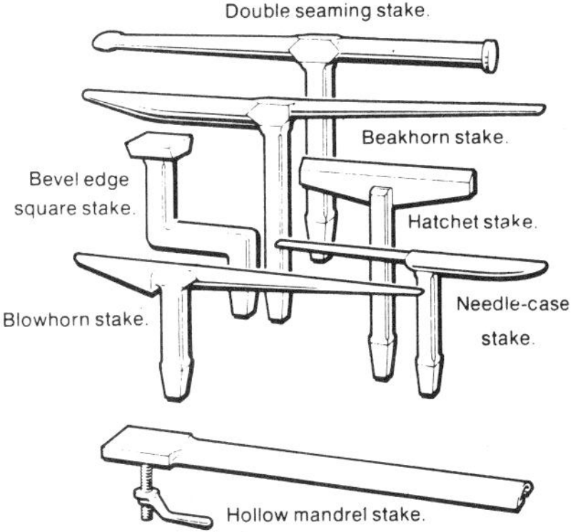

Figure 869 Sheet metal stakes

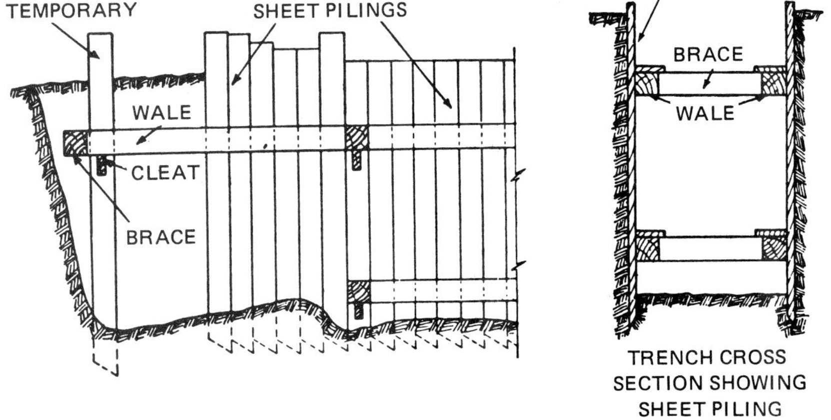

Figure 870 Sheet pilings

sheet metal punch Punch used for driving holes in sheet metal. See also *knockout punch.*

sheet metal screw Threaded fastener used for holding metal pieces or sheets together. (Figure 868.) The screw is threaded all the way to the head for maximum holding power. Round, flat, and pan heads are used. Sizes vary from gauge #4 to #14; lengths run from $\frac{1}{8}''$ up to 2″. Self-tapping points are used. See *self-tapping screws.*

sheet metal seam See *seam.*

sheet metal stakes A metal anvil that fits into a hole in the workbench; used as a support base while the workman shapes sheet metal by striking with a hammer. (Figure 869.)

sheet metal worker One who lays out, fabricates, assembles, and installs sheet metal work for air conditioning, ventilating, and heating systems. May install equipment associated with sheet metal roofing and drain pipes.

sheet pilings Vertical line of steel or wood planks used to support sides of an excavated trench. (Figure 870.) Also called *laggings, sheeting.* See *trench support.*

sheet rock *Gypsum wallboard.*

shelf life Amount of time a material can be stored and still function as intended. Often used in relation to opened packages, as a glue powder or a plaster of Paris.

shellac A surface finish made by dissolving lac (resinous secretion of the lac bug) in alcohol. The thickness of shellac is specified by cut: three- and four-pound cuts are most common. A three-pound cut, for example, has three pounds of shellac in one gallon of alcohol. Also available as a *stick shellac* in over 50 colors: comes in a solid stick or bar—colors are mixed to match the wood, then applied to cracks or holes with a hot knife.

shell construction In reinforced concrete, a thin, curved, monolithic, reinforced concrete roof or complete structure. The shell acts like a curved slab. Shell construction has also been achieved with shaped steel and hardboard or curved plywood panels. Barrel, vault, or dome shapes are formed.

shielded metal-arc welding (SMAW) Arc-welding process; arc is produced between a covered metal electrode and the work piece. Shielding comes from the electrode covering; the electrode provides filler metal. (Figure 871.) Also called *stick welding*. See *electrode, transformer, transformer-rectifier, welding machine*.

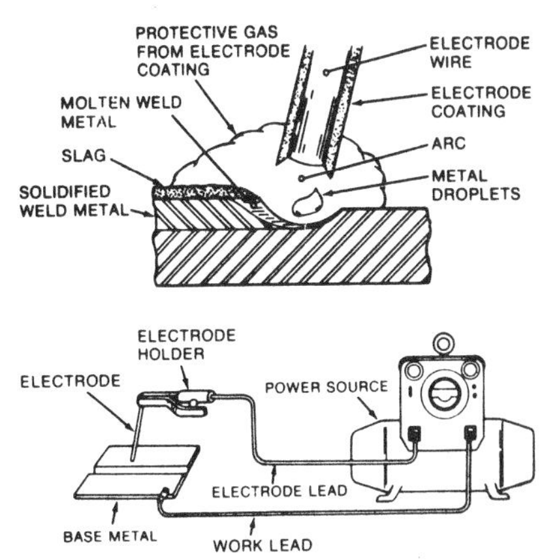

Figure 871 Shielded metal-arc welding

shielded nonmetallic sheathed cable Factory-assembled electrical cable only used in raceways. Type SNM.

shim Wood wedge used to secure, level, or fill in a part of the structure, as the shims used when installing a window frame.

shingle Individual roofing unit. Wood shingles or shakes come in 16″, 18″, or 24″ lengths; widths and thicknesses vary. (*Figure 1125*, p. 497.) Vinyl shingles, shaped to simulate wood, are also used. Various *asphalt shingles* are available. Figure 872 shows shingle types and application. See also *roof tile*.

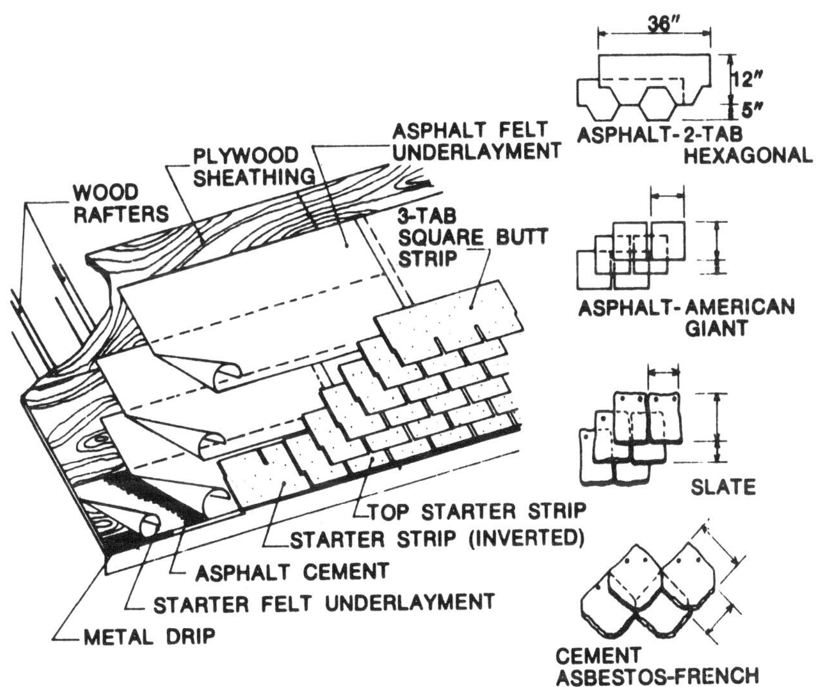

Figure 872 Shingles

shingle lap Type of joint; an asphalt shingle that keys or laps to adjacent shingles to create a tight joint.

shingle nail Wide-headed aluminum roofing nail.

shingling Application of shingles to a roof; application of felts.

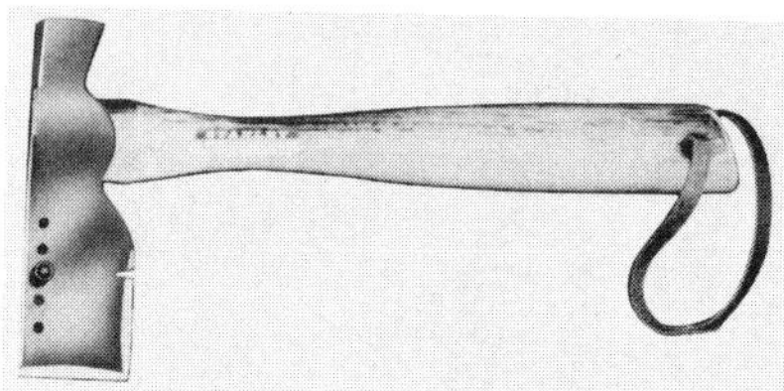

Figure 873 Shingle hatchet

Figure 874 Ship auger

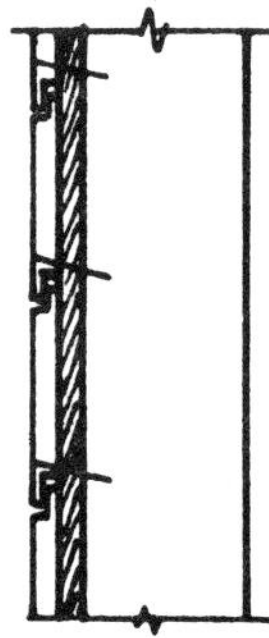

Figure 875 Shiplap joints used for siding

shingling hatchet Hatchet used for cutting, trimming, and nailing shingles. (Figure 873.) Holes are provided on the blade so the gauge can be inserted to the correct depth of exposure to weather for laying shingles. Gauge pin can be adjusted to the weather length needed. Weather is measured along the back of the blade from the striking face to the gauge pin.

ship auger Spiral-shanked single twist bit. The tip has a feed screw that draws the bit into the wood. The outside cutting lip starts the cut. (Figure 874.) The ship auger is used in a hand brace or a power drill. Generally, a long bit is used for drilling deep holes. Sizes and lengths vary. See *auger bit, bit, pole bit.*

shiplap Joint made by cutting edges back with matched rabbet cuts so two wood or plywood pieces can fit together to make an even surface. (Figure 875.) See *siding.*

shiplap lumber Tongue-and-groove lumber. See *dressed and matched, drop siding.*

shiplapped siding Siding boards that are rabbeted to fit together. (Figure 875.) See siding.

shoe Quarter-round base molding, used at the wall base. Bent joint at the end of a downspout. Metal framing connector at the base of a column. (*Figure 661*, p. 286.) Also, a term for the *sole plate.*

shoe mold Baseboard molding, a *quarter round.* A *base shoe.* See *panel molding.*

shop connection A connection made in the shop, as bolting steel beams together. Opposed to *field connection.*

shop drawings Manufacturer or contractor drawings that show in detail the fabrication and assembly of building components, or the installation details of materials or equipment.

shop lumber Lumber that is selected for further manufacturing. Also called *factory lumber.*

shop sketch Sketch or rough drawing of a part, setup, or fabrication; made in the shop to show how something is to be made or put together.

shop weld Weld made in the shop, as opposed to a *field weld.*

shore Temporary timber brace, used at an angle to brace or hold a wall. Generally any temporary member that holds or supports.

shoring Use of temporary supports or shores. The use of timbers to prevent the sliding of earth adjoining an excavation; also, the timbers and adjustable steel or wooden devices used as bracing against a wall or under decking for temporary support.

short In an electrical system, a *short circuit* caused by an accidental grounding.

short circuit Accidental grounding of an electrical circuit; a *short.*

short oil varnish Varnish with a low percentage of oil; used with interior work, such as for cabinets, where a hard, fine finish with a high gloss is needed. See *long oil varnish, varnish.*

short-term load Temporary live load, such as snow.

shotcrete Pneumatically sprayed concrete. Dry (or damp) cement and sand is blown through a nozzle where water is added to make concrete which is sprayed on a surface. There is also a wet-mix shotcrete; concrete is mixed before reaching the hose and nozzle. Also called *gunite*.

shotcreting Blowing of fresh concrete with an air gun: the architectural surface is covered with the blown concrete.

shot sledge A special heavy hammer or sledge that has lead shot in the head. This results in a recoilless sledge. When a blow is struck, there is no bounce or recoil.

shoulder The side, as the side of a highway. Defect in concrete wall when pressure of the concrete forces a misalignment between the ends of two concrete forms. A step cut in the side of a wood piece.

shoved joints In bricklaying, joints made by shoving a brick into a mortar bed so that it pushes mortar between itself and the brick already laid.

shower enclosure A complete fiberglass shower module.

shower head Fitting that sprays out water in the shower. (Figure 876.)

shrink To contract or become smaller, as the shrinkage of lumber through loss of water. Figure 877 shows lumber shrinkage for different parts of a tree.

shrink-mixed concrete Ready-mix concrete partially mixed in a central plant and then also truck mixed.

shutter Outside louvered or slatted window covering designed to close and cover the window. Today, the shutters are often decorative and remain fastened at the sides of the window.

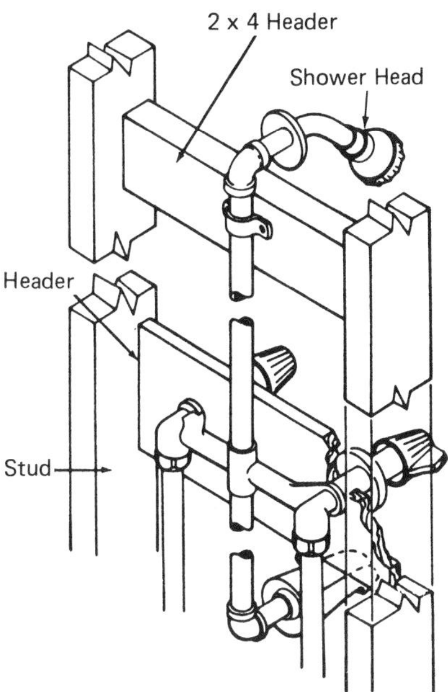

Figure 876 Shower head

side cut Cut or bevel made on the side of a rafter end to fit in at an angle, as a hip rafter framing in to the side of the ridge. A *cheek cut*. See *rafter cuts*.

side-cutting pliers Widely used heavy-jawed pliers that have cutting edges on the side. (Figure 878.) Used in electrical work for cutting wire and in reinforced steel construction for cutting rebar tie wires. Various types are available for different jobs. The handles are covered with insulative vinyl. Also called *electrician's pliers, ironworker's pliers*. See also *pliers*.

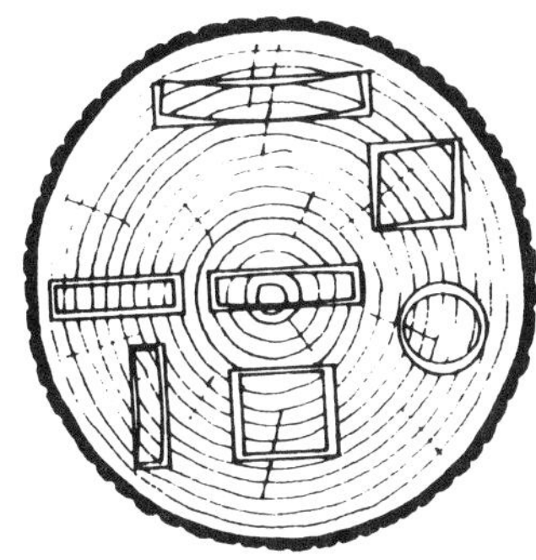

Figure 877 Shrinkage of lumber. Tangential shrinkage (with growth rings) is much greater than radial shrinkage (right angle to growth rings)

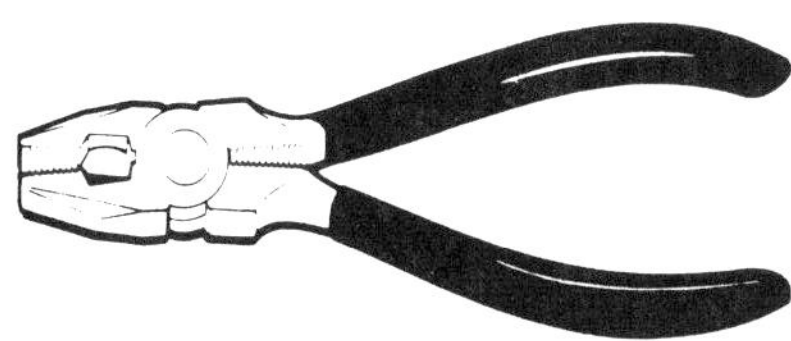

Figure 878 Side-cutting pliers

Figure 879 Sidewalk forms

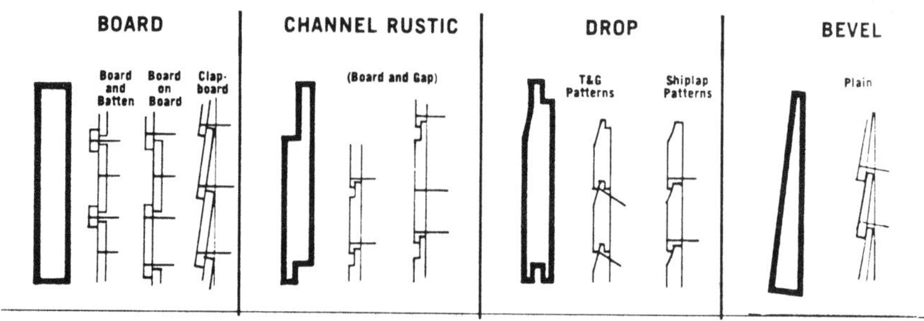

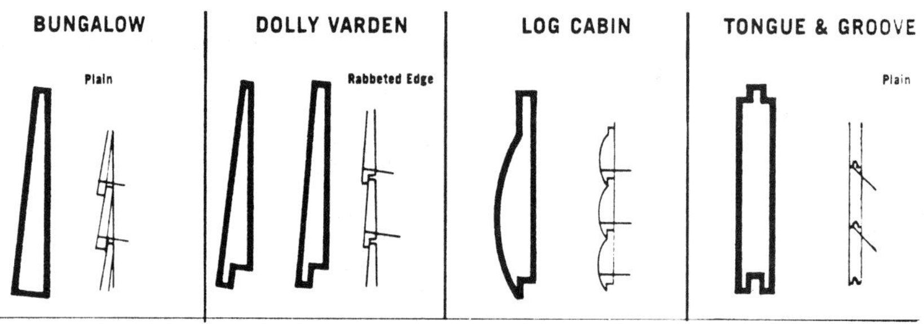

Figure 880 Siding used in residential construction

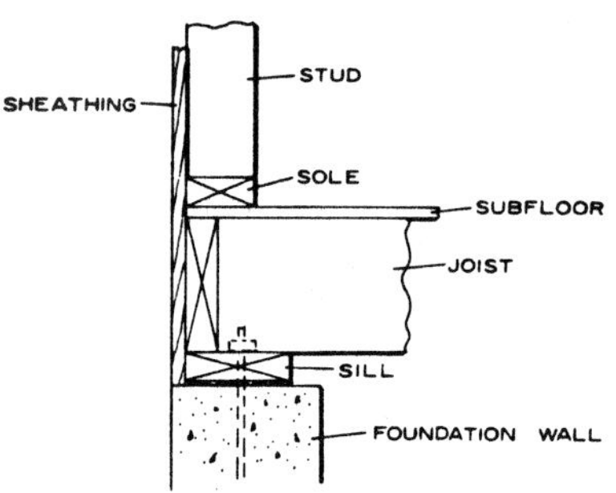

Figure 881 Sill rests on foundation wall

sidewalk forms Forms used for making sidewalks. For extensive use, metal forms are used. (Figure 879.) See *curb forms.*

siding Covering that goes on the outside of a building over the sheathing. Figure 880 shows siding types. See also *inside corner, outside corner, shakes, shingles.*

silicon carbide A very hard, synthetic abrasive mineral widely used on wet or dry abrasive papers, belts, and discs. The fine grades are widely used for wood finishing. See *abrasive.*

sill Support member (usually a 2 × 4 or 2 × 6) laid flat on the top of the foundation wall, used as the base for wall framing. Also called a *sill plate.* (Figure 881.) Also, the bottom member of a window or door frame. (Figure 882.)

sill anchor Threaded bolt imbedded in a concrete or concrete block foundation wall, used for anchoring the sill. See *anchor bolt.* See also *sill-plate anchor.*

sill caulk Caulking placed on the foundation top to form a bond with the sill.

sillcock Water faucet to which a hose may be screwed on. Also called a *bib, bibcock,* or *hose bibb.*

sill joists Joists that run on the outside of the floor platform. Also called *band joists, rim joists.* See specifically *end joists, header joists.*

sill plate The *sill.*

sill-plate anchor A metal fastener used to anchor the wood sill to the masonry or concrete foundation. (Figure 883.) They can also be used to anchor rafters to top plates.

silo trowel A rounded-end trowel used for finishing silos or swimming pools. See also *swimming-pool trowel, trowel.*

silt test In concrete work, a test made in the field to determine if the sand has too high a concentrate of fine silt or loam. A jar or bottle is half filled with sand, filled with clear water, shaken, and allowed to settle. The silt will collect at the top of the settled material. The depth is measured and compared to set standards. (Figure 884.)

silver brazing Brazing process using a filler metal that is a silver alloy.

silver soldering A soldering process using a silver-tin alloy filler metal which is melted into the joint to be bonded. A much higher temperature is used than for the regular lead-tin solder. A propane torch and a paste flux jare commonly used. A process very similar to *brazing* (which normally uses a copper-base alloy as a filler metal).

simple beam Beam openly supported at each end.

single-phase, three-wire system Standard 120-V/240-V system.

single-weld joint Weld made from one side only.

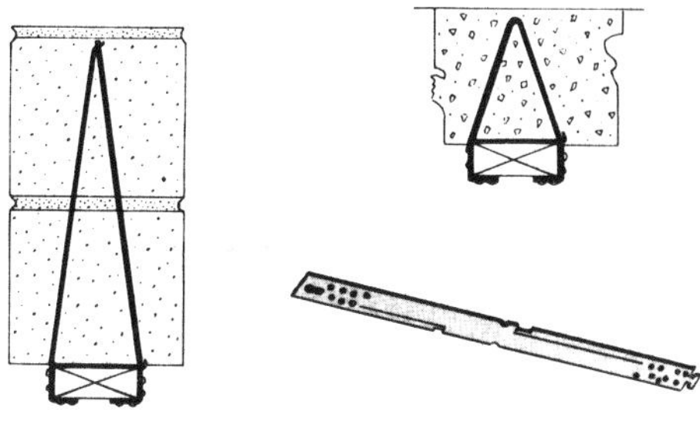

Figure 883 Sill-plate anchor

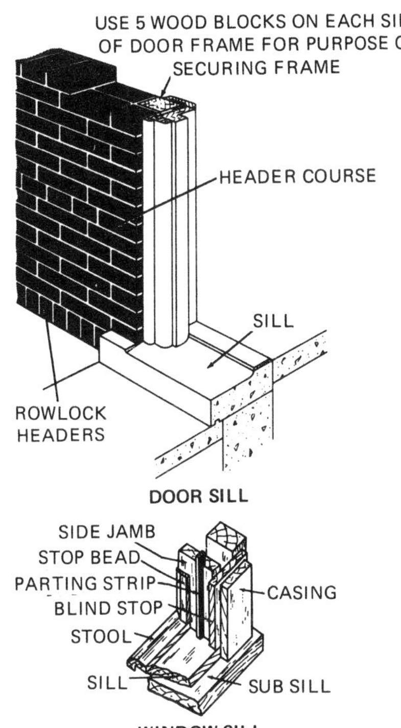

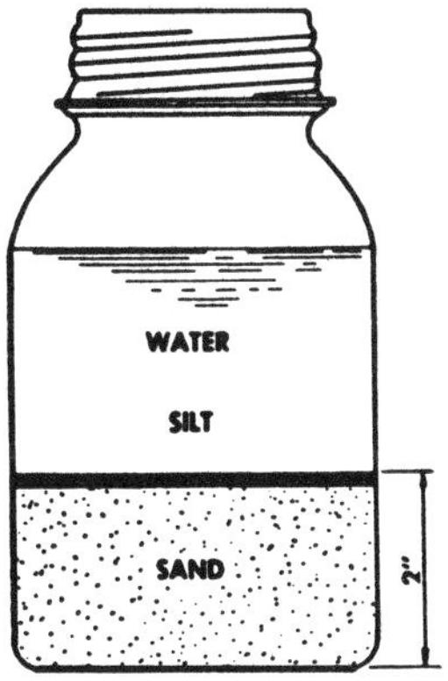

Figure 882 Sill

Figure 884 Silt test

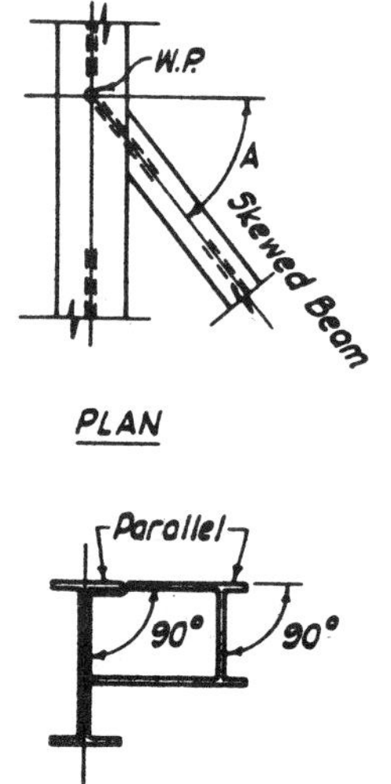

Figure 885 Skewed beam

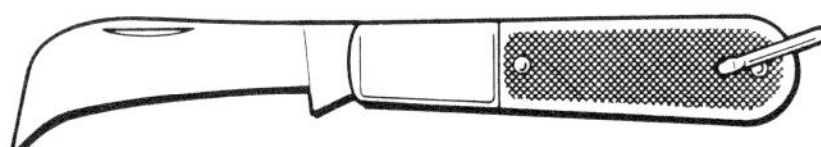

Figure 886 Skinning knife

siphoning In plumbing, loss of water in a trap by rapid flow under unequal pressures or by improper pitch on the drain pipe.

site Location where construction work is done.

site plan A *plot plan.*

six-foot rule National Electrical Code (NEC) requirement that no part of a wall base be more than six feet from an outlet.

size To fill wood pores with some material (a primer) preparatory to painting. To fill a canvas with a base before painting on it.

sizing A surface sealant. Material, such as a primer coat, that seals or fills a wood surface or other material. In fine painting, a filler is used to size the raw canvas before painting on it. Also, determining needed pipe and component sizes in a plumbing system, hot-water system, solar system. Determining ductwork size needed in a heating system.

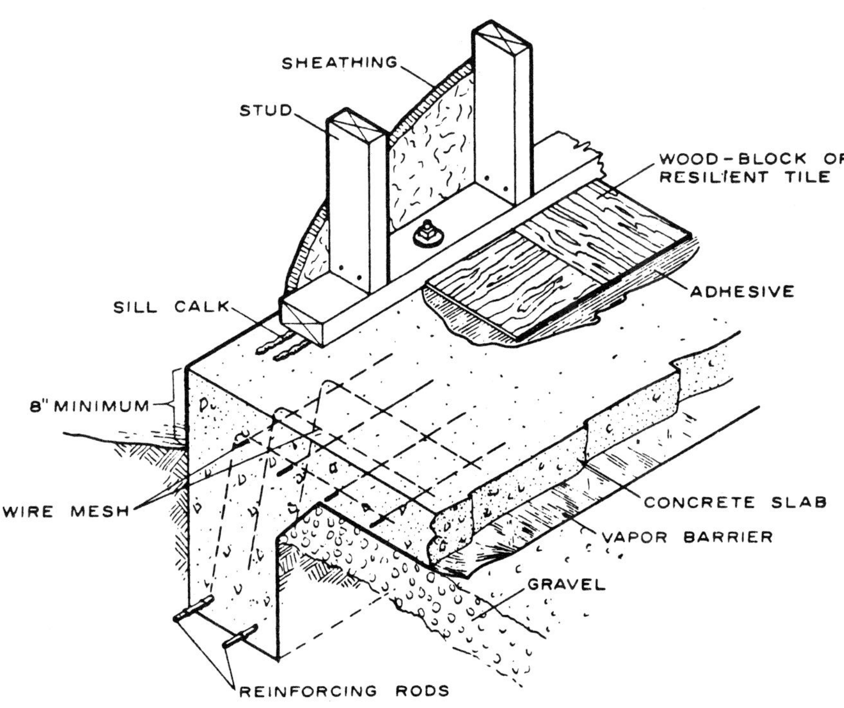

Figure 887 Slab foundation

skeleton The support framework for a building, as the steel beams for a heavy steel building.

sketch Freehand drawing of something.

skew To lean or twist.

skew chisel Woodturning tool with the cutting edge set at an angle. Used for quick cuts into the piece and for turning a shoulder. See *woodturning tools.*

skewed beam In steel construction, the framing of one steel beam into the web of another so the flanges are parallel but the web of the one is at an angle to the web of the support beam. (Figure 885.) See *canted beam, sloped beam.*

skew nailing Driving nails at an angle.

skimming In drywall construction, a light cement finish coat.

skinning knife Electrician's folding knife with a large curving blade used for skinning or cutting back conductor insulation. (Figure 886.)

skip Lumber defect; a rough area on lumber caused when an area was missed during surfacing.

skirt A rim or low border, as baseboard. Also called *skirting boad.*

skirting Baseboard.

skotch fasteners Toothed metal fasteners used to make or strengthen a corner joint. See *corrugated fastener.*

skylight Glazed roof opening used for extra lighting.

skyspace Open portion of the sky that is needed to allow a solar collector to operate effectively. Also called a *solar window.*

slab Flat concrete area. In residential construction, the concrete is normally placed on crushed aggregate and welded-wire mesh reinforcement is used. (Figure 887.) In heavy construction, a mat is made of steel reinforcing bars to strengthen the slab. (*Figure 784*, p. 344.) Also, in reinforced prestressed concrete, a horizontal hollow-core member used for flooring and roof decking.

slab band See *solid slab*.

slab bolster In heavy construction, metal supports used to hold up the reinforcing bars for the slab. See *bolsters, chairs*.

slab cut The first log cut; the cut piece has the outside bark on. From a lumber viewpoint this is a waste cut. The first board cut is made after the slab cut.

slab formwork Reusable forms used in multistory construction. Forms are stripped and hoisted to another level. (Figure 888.) See also *facade forms, forms, wall forms*.

Figure 888 Slab formwork

slab foundation Flat reinforced concrete slab resting on the ground. See *slab*.

slab schedule Table giving number, size, and location of reinforcing bars used in a slab. See *bar schedule*.

slag Cinder-type byproduct of ore smelting; it is ground up and used as an aggregate in lightweight concrete. See *slag concrete*. Also, a thin metallic crust left after arc welding; *spatter*.

slag concrete Lightweight concrete made using ground slag as an aggregate. Has good insulative and fire-resistant properties.

slag plaster Plaster that includes ground slag, used for its acoustical value.

slag strip A strip at the edge of a gravel-finished roof to catch loose gravel and prevent its falling off a *gravel strip*.

slake To hydrate quicklime by adding water to make a putty (for use as a finish coat). See *hydration, quicklime*.

slaking Addition of water to lime; causes lime to hydrate.

slate A hard metamorphic material that splits easily into thin sheets. Used for flooring and roofing.

sledge Heavy hammer used for pounding, breaking concrete, driving heavy stakes, and the like; it is operated with two hands. (Figure 889.) See also *hammer*.

Figure 889 Sledge

sled runner A type of jointer with a handle, used for finishing horizontal masonry joints. See *jointer, trowel*.

Figure 890 Sleepers

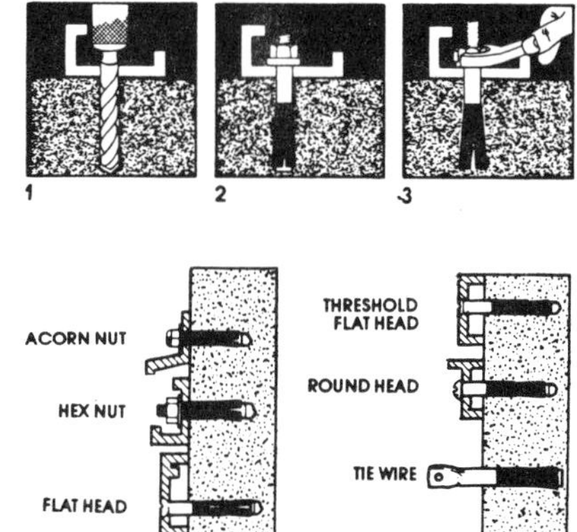

Figure 891 Sleeve anchor

sleeper Timber or wood piece laid down to support a superstructure or flooring. A board laid on a concrete slab to support the floor. (Figure 890.)

sleeve Hollow piece that fits over something. A metal tube that runs vertically through a floor slab, used to run electrical wiring or plumbing pipes. Any tube that goes through a wall or floor through which other pipes are run. Also called a *sleeve chase*. See *wall chase*.

sleeve anchor One-piece anchor used in masonry: anchor, washer and nut come as one piece. (Figure 891.) Anchor fits into bored hole, and sleeve expands to hold when tightened. See also *anchor, stud anchor, wedge anchor*.

sliced veneer Veneer that is sliced off the side of a log or bolt with a knife; comes in short sections.

slick A long-handled chisel used for paring large areas or cleaning out mortises. Slick is operated with two hands.

slicker In plastering, a long flat board used for finishing plaster. Similar to a *darby, featheredge, or plasterer's shingle*. Also, a masonry tool for tooling masonry joints; a *jointer*.

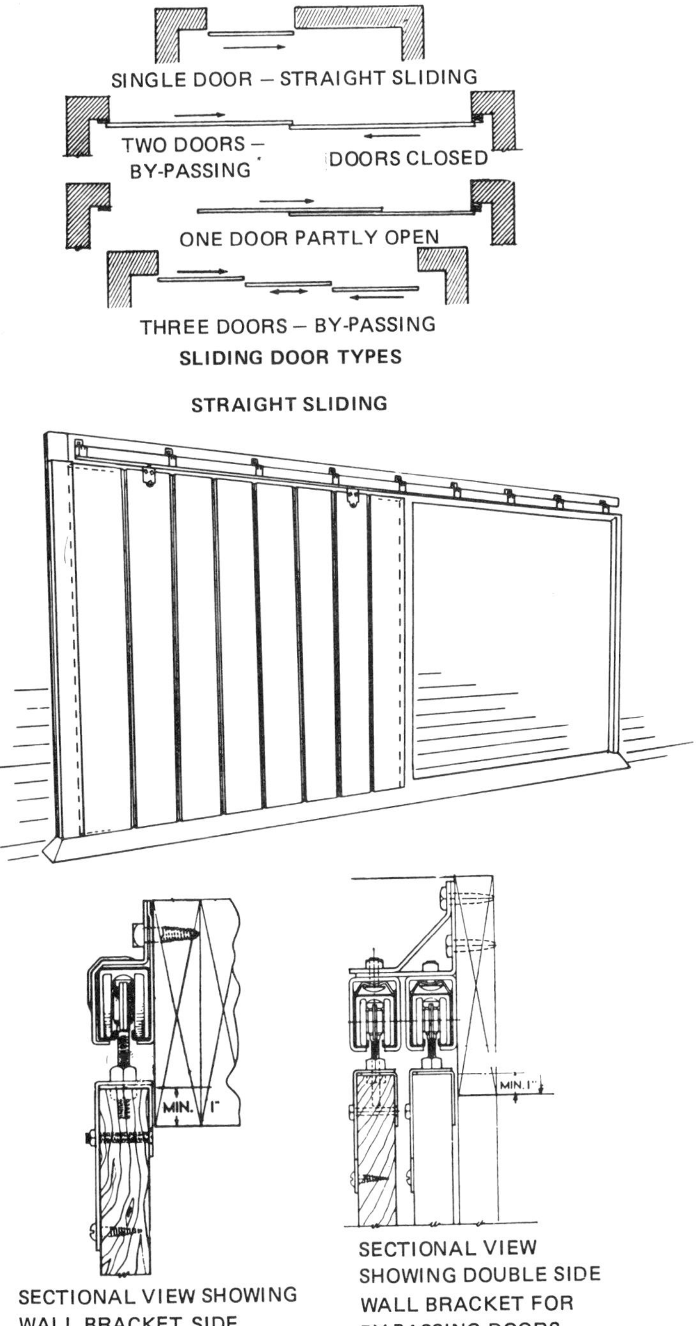

Figure 892 Sliding door

sliding doors Doors that run horizontally on tracks with door-hanger rollers. Sliding doors may be single or straight sliding, or two or more doors may bypass each other. (Figure 892.) See also *pocket door*.

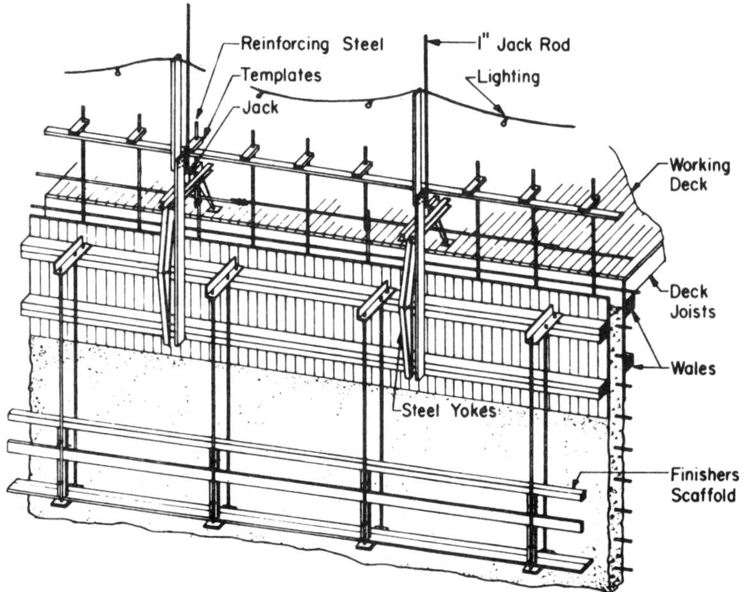

Figure 893 Slip form

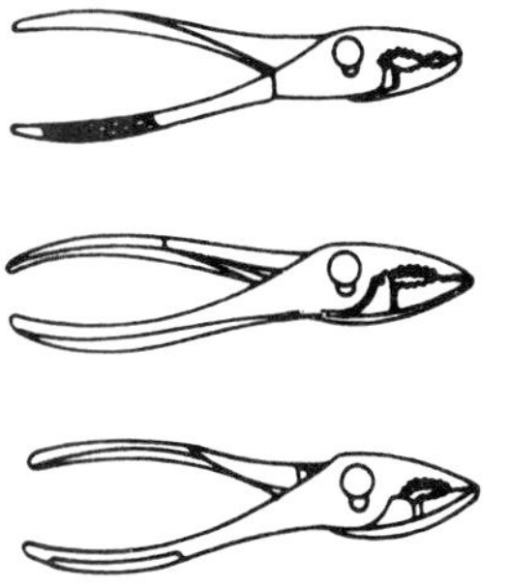

Figure 894 Slip-joint pliers

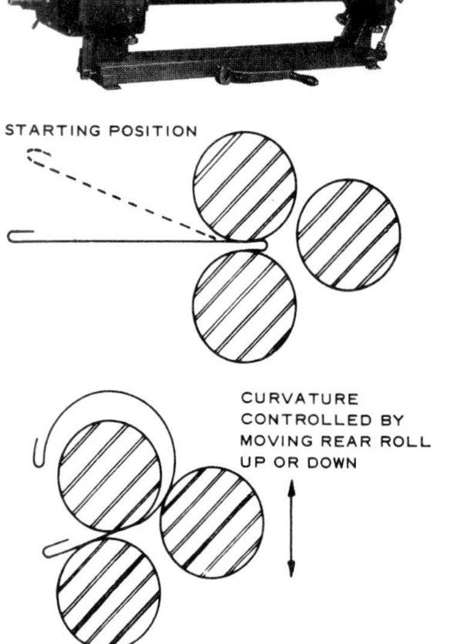

Figure 895 Slip roll forming machine

sliding windows Window sashes that run horizontally in the frame. See *window*.

sling Support loops made with a wire rope or chain, used for hoisting. See *cable attachments, tag line.*

slip form Concrete forms that are designed to be loosened and easily moved to the next higher level, commonly used in multistory reinforced concrete construction. (Figure 893.) Forms are slipped or slid from floor to floor. See *facade forms, wall forms.*

slip joint Construction joint in a wall that will let the separate wall sections move in relation to each other.

slip-joint pliers Widely used pliers with a slip joint that allows the jaws to be opened wider. (Figure 894.) Used for holding and bending. Various sizes are available. Also called *combination pliers.* See also *pliers.*

slip roll forming machine Sheet metal machine forming cylindrical and conical shapes. (Figure 895.) See also *brake.*

slip stone Rounded oilstone used to sharpen concave cutting edges, such as found on gouges.

slip tenon joint Mortise and tenon joint made in the end of a wood piece. Also called *slot mortise* or *slip mortise.* See *wood joints.*

slope Roof angle or incline; a ratio of inches of rise per running foot or 12 inches. (*Figures 697*, p. 300; *810*, p. 356.)

Level slope: less than $\frac{1}{2}$'' slope per foot.

Low slope: $\frac{1}{2}$'' to $1\frac{1}{2}$'' slope per foot.

Steep slope: Over $1\frac{1}{2}$'' slope per foot.

See *pitch, roof slope.*

sloped beam In steel construction, the framing of one steel beam into the web of another so the web faces are perpendicular but the flanges of one are at an angle up or down to the support beam flanges. (Figure 896.) See *canted beam, skewed beam.*

slot weld A weld made through a slot (long opening) in one piece through to another piece below to join the two pieces. (Figure 897.) See *plug weld.*

Sloyd knife A general-use woodworker's knife. (Figure 898.)

slump Concrete consistency; the stiffness or softness of fresh concrete as measured with a special metal tube. (Figure 899.) Checking fresh concrete is called the *slump test.*

slurry A liquid cement and water mix, used as a coating; also any very liquid mix.

slurry wall A wall built around a building foundation to hold back ground water. Trench sections are dug to a solid level, then filled with a special slurry (liquid mix of water and bentonite clay). The slurry holds the trench walls in place until reinforced concrete can be placed. Also called a *cut-off wall.*

small knot Knot between $\frac{1}{2}''$ and $\frac{3}{4}''$ in diameter.

small tools Fine ornamental plastering tools used for finishing cast plaster or for any fine plaster work. See *ornamental tools.*

smoke alarm Battery or low-voltage device that emits a sound alarm when activated by combustion fumes. See *smoke detector.*

smoke chamber Area in a fireplace just below the flue. See *fireplace.*

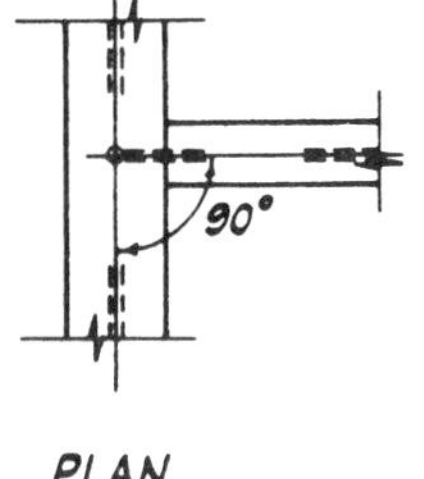

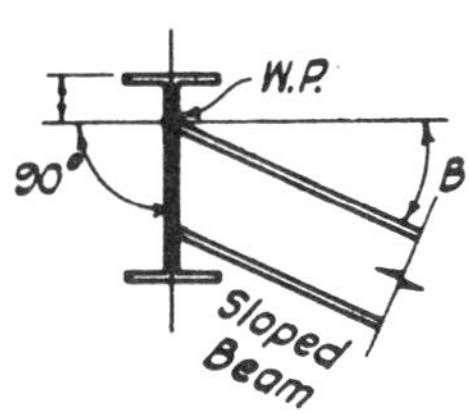

Figure 896 Sloped beam

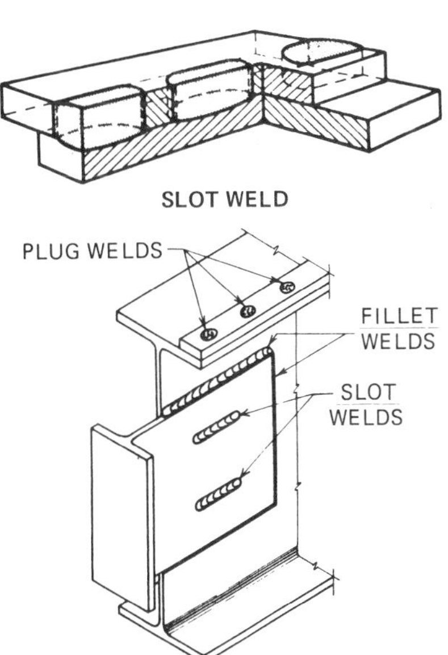

Figure 897 Slot and plug welds are made through holes in the material

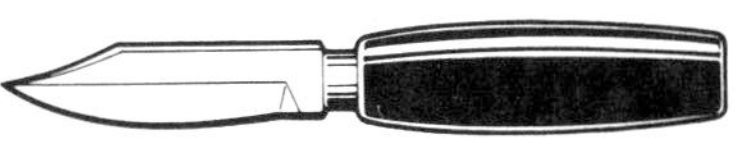

Figure 898 Sloyd knife

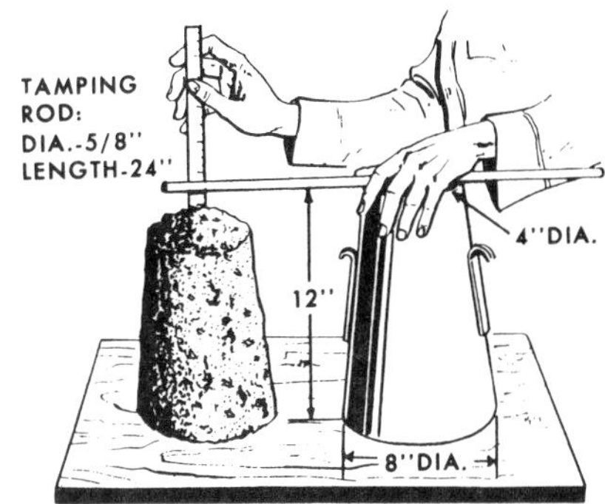

Figure 899 Slump test

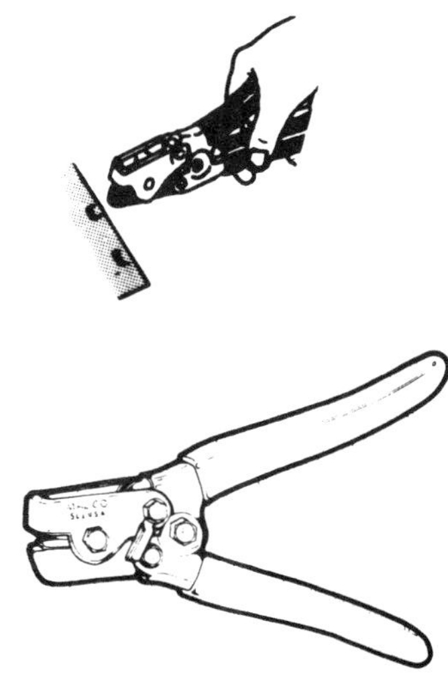

Figure 900 Snap lock punch

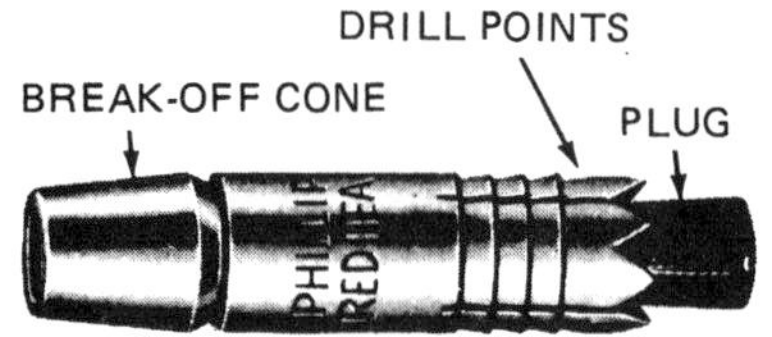

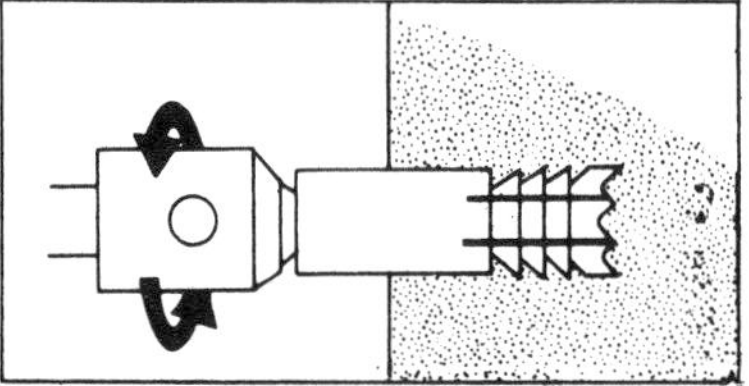

1. Using the anchor as the drill bit, drill hole until chuck holder is flush with surface of concrete. Remove anchor and clean out hole.

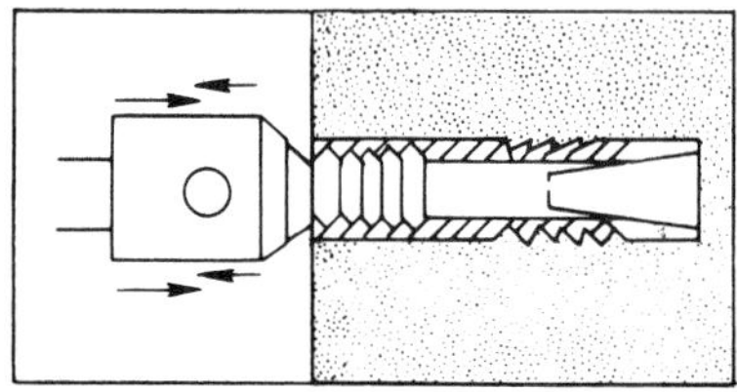

2. Insert red plug in anchor. Expand anchor by reinserting it into hole and driving it in until chuck holder is flush with the surface of the concrete. Snap off cone.

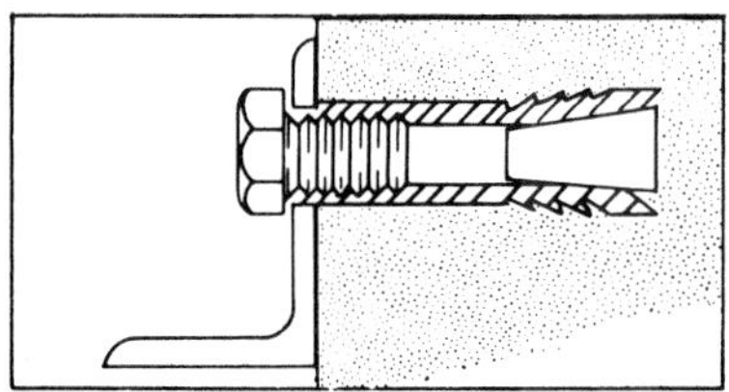

3. Bolt the object to complete the installation.

Figure 901 Snap-off anchor

smoke detector Alarm device used to detect the presence of fire in a building. Four types are used: (1) *ionization detectors*: respond to invisible combustion products; (2) *photoelectric detectors*: respond to visible smoke; (3) *infrared detectors*: respond to flame; and (4) *thermal detectors*: respond to heat.

smoke shelf Offset at the bottom of the smoke chamber in a fireplace. Designed to divert or break up air currents that might cause smoke to exit the fireplace into the living area. See *fireplace.*

smoke test In plumbing, test of the tightness of a piping system by sealing all the outlets and forcing smoke into one end of the piping.

smooth coat Plastering *finish coat*, or *putty coat.*

smoothing brush In wallpapering, a very wide, thin brush used for smoothing down the freshly laid wallpaper. See *paste brush.*

smooth plane Small hand plane, 6″ to 10″ long with $1\frac{3}{4}''$ to 2″ cutters. Used for smoothing mill and saw marks. See *plane.*

snake Plumbing device for cleaning drains; a coil spring on the end of a wire is run into the drain. See *drain cleaner*. In electrical work, a *fish tape* for running through conduit and pulling wires.

snap lock punch Sheet metal hand tool used for joining sheet metal by raising projections that lock the sheets together. (Figure 900.)

snap-off anchor Masonry anchor that fits into a drill and drills its own hole. (Figure 901.) Anchor cone end is broken off when the anchor is seated. The base serves as a fastener for screws or bolts.

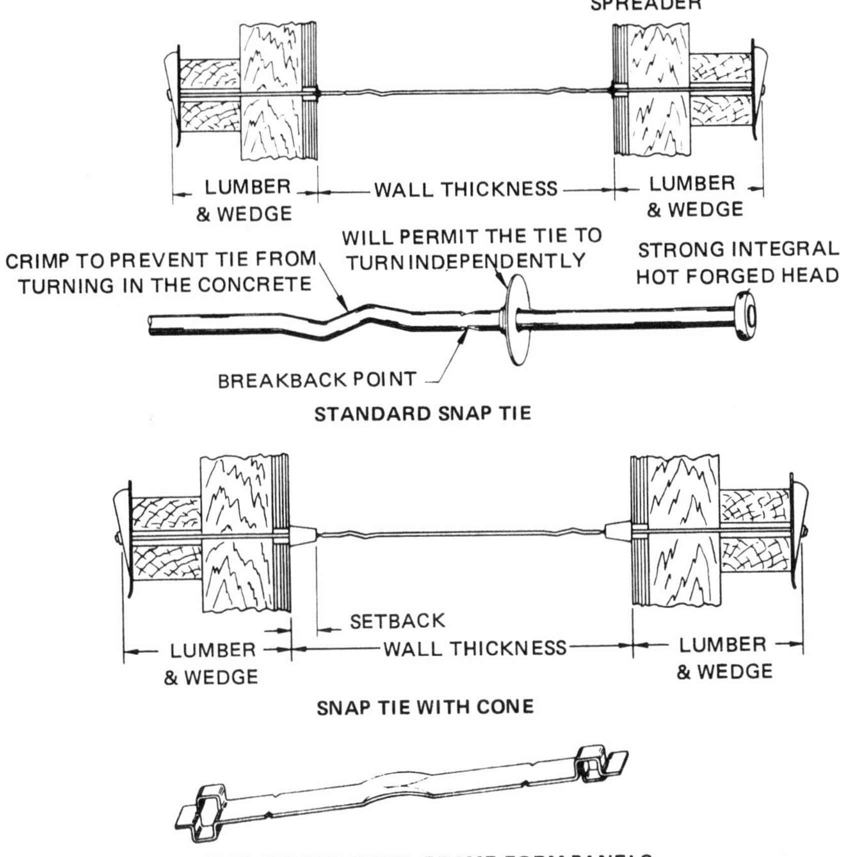

Figure 902 Snapties

snaptie Concrete form tie that snaps off when the concrete is set. (Figure 902.) Also, a simple wire loop and twist used to fasten rebars together. See ***bar ties, coil tie, tie rod.***

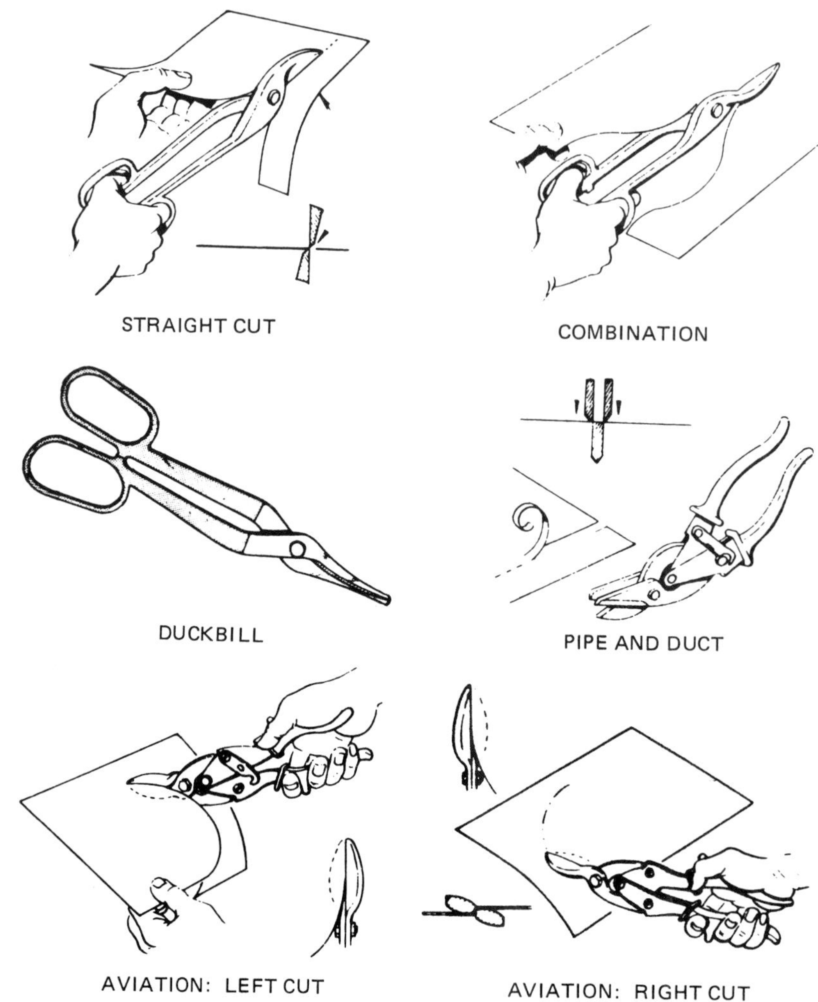

Figure 903 Snips

snips Hand shears for cutting sheet metal. (Figure 903.) Straight snips have straight cutting edges and are generally used for cutting sheet metal up to $\frac{1}{16}''$ thickness. Duckbill snips are used for cutting circles. Aviation snips have narrow blades which are operated by a compound lever action to get greater cutting power. They are used for cutting circles, squares, and irregular patterns. Aviation snips are available with curved blades (left or right curves) for cutting left or right with ease. Pipe and duct snips have double cutting blades and cut out a narrow strip of metal.

snow load Design load for the live snow load on a roof deck. Generally only a consideration on flat roofs in colder areas.

socket In plumbing, an enlarged pipe end that receives another pipe; a *bell* or *hub*. An electrical outlet that a light bulb screws into. Also, a steel jacket that seats over nuts or bolt heads. A handle or ratchet fits into one end of the socket to turn it (and the nut or bolt head). Sockets come in inch and metric sets. Regular or deep sockets are available. The inside of a socket has a number of projections or points for making contact with the nut or bolt head being turned: six-, eight-, and twelve-point sockets are available. (Figure 904.) Six-point sockets are used for hex nuts and bolts; eight-point sockets are used for square nuts only; twelve-point sockets are used for hex nuts or cap screws and very often offer a better seat than six-point sockets. Also called a *socket wrench*. See *ratchets*.

socket adapter Fitting used to allow socket wrenches with one drive size to fit a drive tool or ratchet with another size. For example, a ratchet with a $\frac{3}{8}''$ square drive can be fitted with a $\frac{3}{8}''$–$\frac{1}{4}''$ socket adapter which will allow a $\frac{1}{4}''$ square drive socket to be used.

socket chisel A heavy-duty woodworking chisel that has a socket extending up beyond the shoulder. A wood handle fits into the metal socket. See *wood chisel*.

socket wrench A *socket*.

sod A mat of growing grass imbedded in earth. Strips are cut and rolled up from where the grass is growing on the ground. Sod is then placed on a bare earth building site to form a lawn.

sodding Placement of sod for landscaping and finish of a building site. Sod is placed over bare earth to form a lawn.

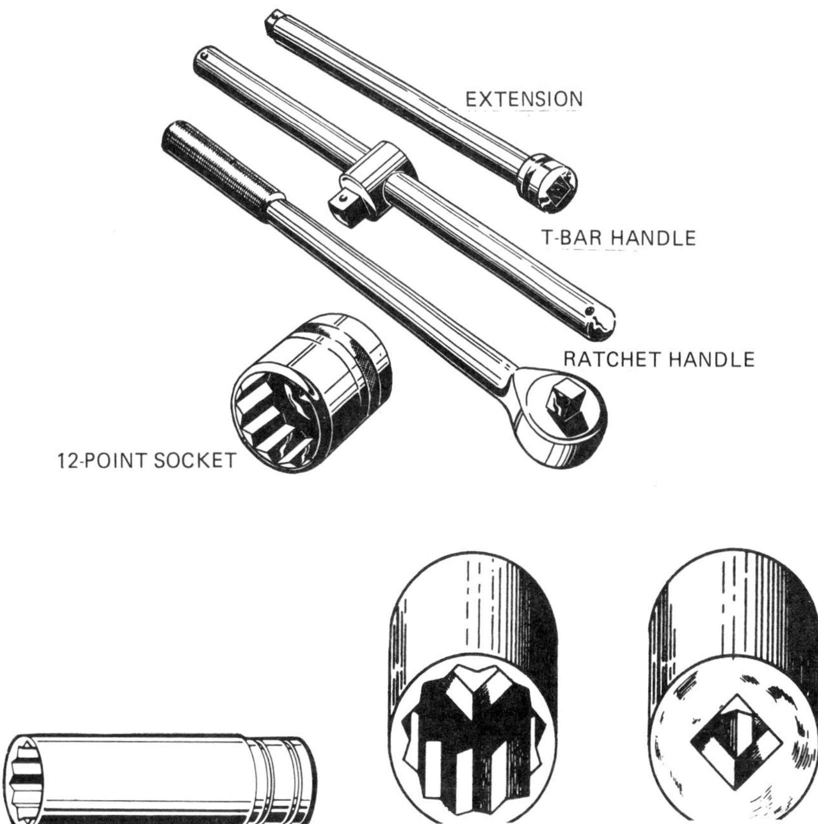

Figure 904 Sockets

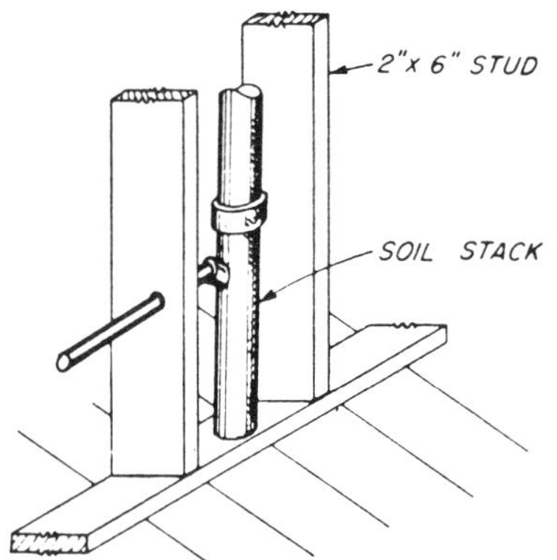

Figure 905 Soil stack

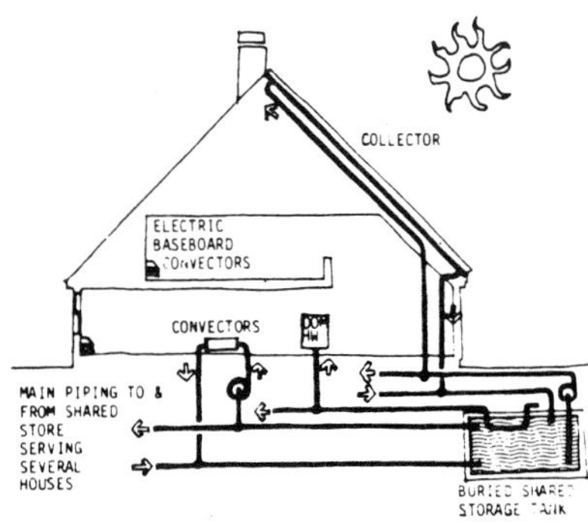

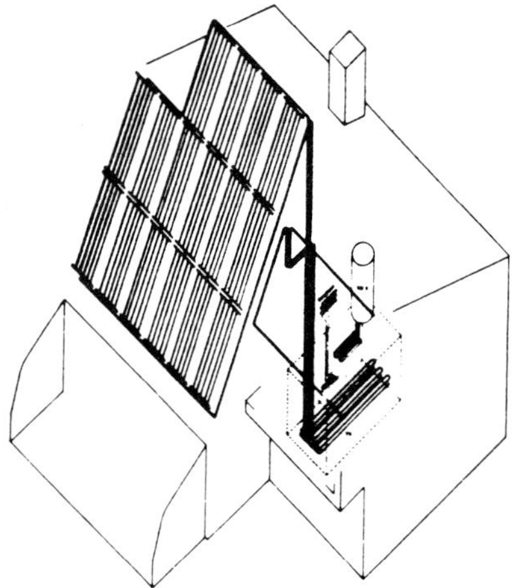

Figure 906 Solar collectors

soffit An underside area of a framing member or building part. The underside of an enclosed cornice. (*Figures 201*, p. 73; *257*, p. 97.) Also, in general, the underside of a beam or arch. In reinforced prestressed concrete, the underside of an overhang or cantilevered floor. In stair building, the underside of a flight of stairs.

soffit vent Ventilation opening, usually a screen, in a cornice soffit. (*Figure 257*, p. 73.)

soft-face hammer Hammer that has plastic faces. Used for striking wood chisels and gouges and for other sensitive work where a hard face might cause damage. See *hammer, mallet.*

soft rot Decay that occurs in the outer wood under very wet conditions; fungi attack the cellulose in the wood.

soft water Water that is mostly free of hard calcium and magnesium minerals. Soft water is easy to wash with and leaves no scum or calcium precipitate.

softwood A tree with needlelike leaves which do not fall off with the cold season; an *evergreen* or *conifer.*

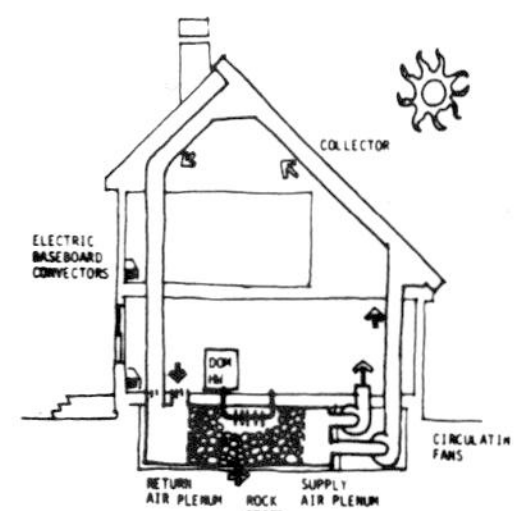

soil boring Hole drilled in the soil to determine types of soil or rock support. Cores are brought up for inspection and analysis to determine stability and load-bearing capacity.

soil pipe In plumbing, a waste pipe for carrying off waste from a water closet or similar fixture. Also, cast-iron pipe with either bell-and-spigot or no-hub ends.

soil poisoning Treatment of soil with poisons to prevent termite attack. Crawl-space areas often are poisoned.

soil stack Vertical pipe that receives waste from water closets and similar fixtures. (Figure 905.)

soil survey Analysis of subsurface soil in an area to determine composition. Soil is usually described as a percentage of sand, silt, and clay. See also *landscape plan, erosion and sedimentation control plan.*

soil test Sampling and testing of soil to determine physical properties, condition, and strength (load resistance).

soil vent That portion of a soil stack above the highest fixture discharging waste into it. See *stack vent* or *vent stack.*

solar azimuth Angle at which the sun is located, east or west, from due south. See also *altitude.*

solar collectors Panels for collecting solar heat, which is then circulated into a building. (Figure 906.) Collectors face south to catch the sun's rays. Some solar collectors circulate heat by natural processes requiring no mechanical aid. See *active solar heating, collector bank, concentrating solar collector, flat-plate solar collector, skyspace, solar window, thermosiphoning.*

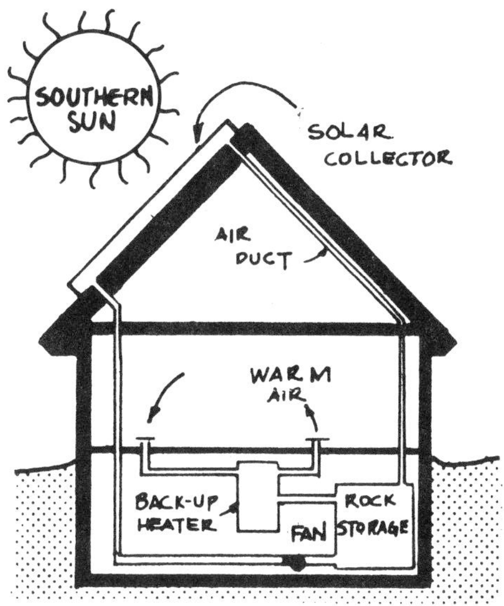

AIR-TYPE SYSTEM

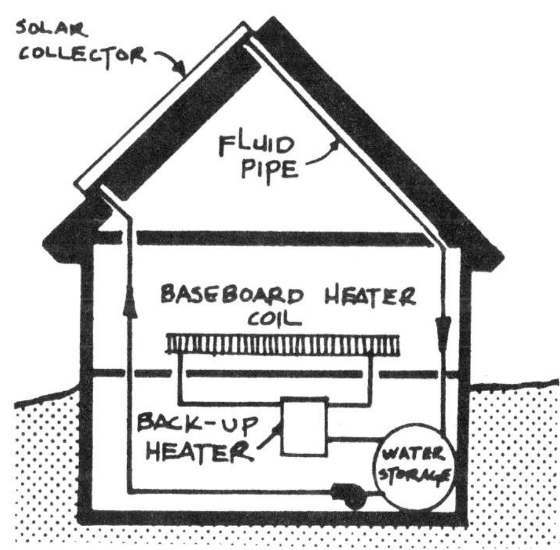

LIQUID-TYPE SYSTEM

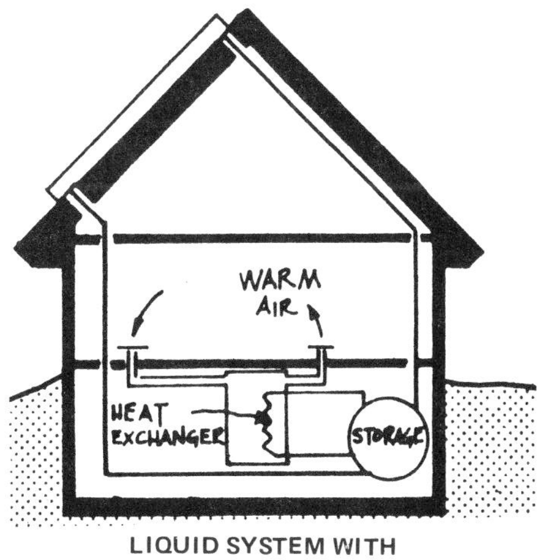

LIQUID SYSTEM WITH FORCED AIR

Figure 907 Solar heating

solar heating Heating a structure using direct (passive) or indirect (active) sunlight. (Figure 907.) Solar heat not immediately used is stored (for example in stone or water) for later use. See *active solar heating, air-type solar system, closed-loop solar system, fluid-type solar system, open-loop solar system, passive solar heating.*

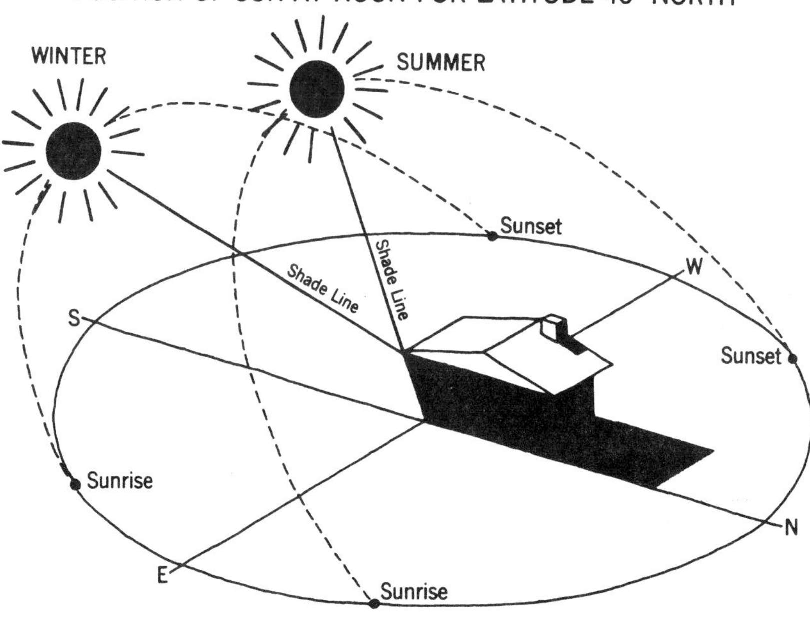

Figure 908 Solar orientation to take advantage of winter sun

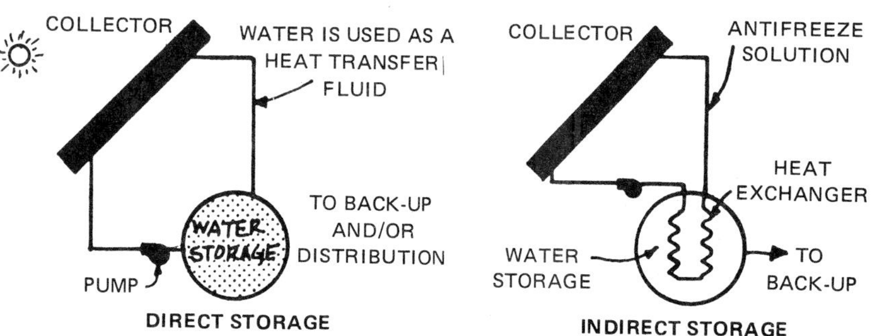

Figure 909 Solar storage

solar orientation Positioning the house to take advantage of the winter sunlight; large window area faces south to catch the low winter sun. (Figure 908.) See *orientation.*

solar panels *Solar collectors* for collecting solar heat, which is then mechanically circulated into a building.

solar radiation The sun's rays; radiation emitted from the sun. Solar radiation may be direct, diffuse (scattered by clouds or air particles), or reflected. Radiation is measured in *Langleys* (one Langley equals one calorie per cubic centimeter).

solar skyspace easement Right to enjoy access to the sun for solar heating.

solar storage Container and medium (such as water or rocks) that absorbs and holds excess heat in a solar heating system. Heat is released as needed at a later date. (Figure 909.)

solar tempering Orientation of a building to the south so the south walls receive solar heat.

solar window Large glass area positioned toward the south to take advantage of the winter sunlight. Also, open portion of the sky that is needed to allow a solar collector to operate effectively. Also called *skyspace*. (Figure 910.)

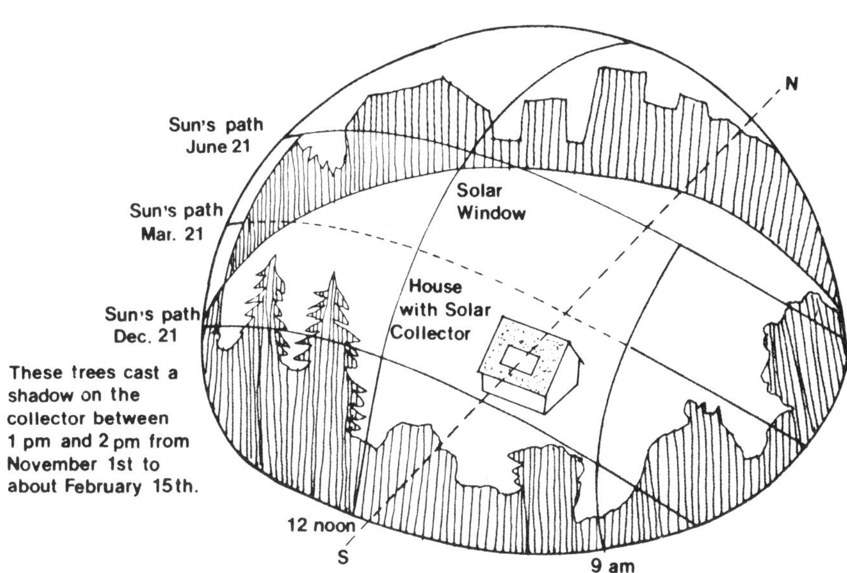

Figure 910 Solar window

solder Tin and lead alloy used, when melted, for making permanent joints between metal pieces. Available in tubular form with *rosin core* or *acid core*. For commercial use, as for plumbing joints, solder comes in bars. The bars have the tin-lead alloy percentage marked on them. A 30/70 solder has 30% tin and 70% lead. A 50/50 solder is often used in sheet metal work.

solder gun Electric heating device used to melt solder.

soldering To joint parts together with solder. (Figure 911.) See also *silver soldering*.

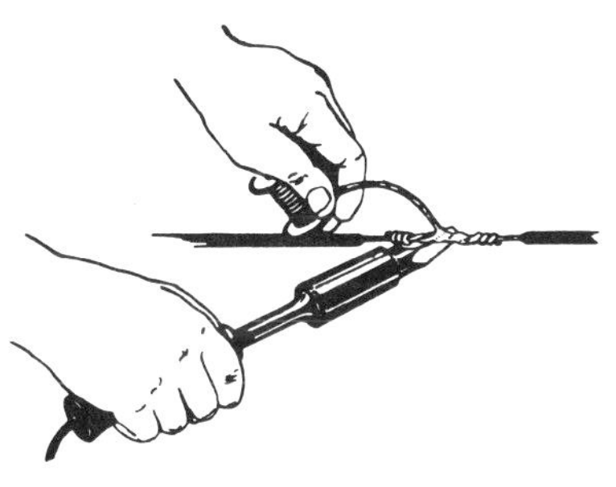

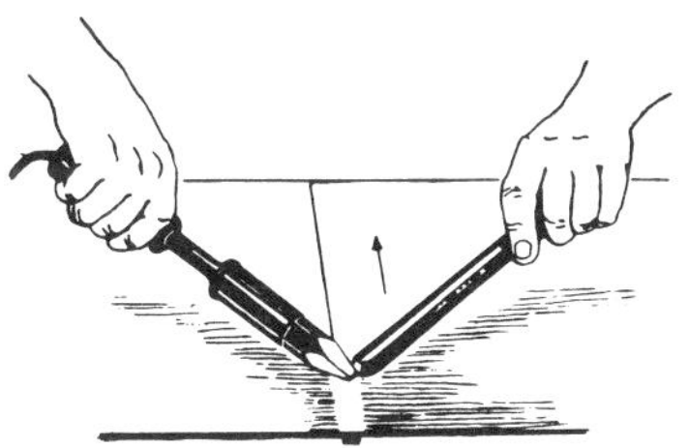

Figure 911 Soldering

soldering copper A soldering tool for melting solder; it must be heated before use. The tip or bit is copper. Still widely used by construction workers. (Figure 912.) Also called a *copper* or *soldering iron*.

Figure 912 Soldering copper

soldering iron A *soldering copper*.

soldering torch Portable torch used for heating *soldering coppers*.

solder-joint pressure fittings Copper fittings used to connect hard copper tubing for water-supply piping. See *copper tubing*.

solderless connector In electrical work, a plastic connector used for joining two electrical wire ends together. See *solderless wire connector*.

solderless wire connector A plastic nutlike connector used for connecting electrical wires together. (Figure 913.) Also called a *wire nut* or *solderless connector*.

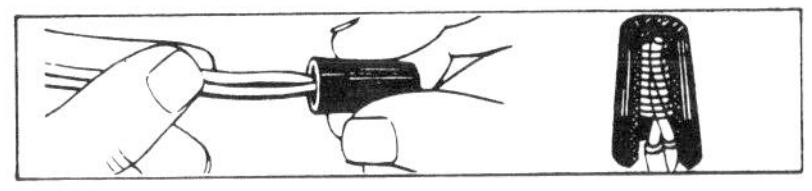

Figure 913 Solderless wire connector

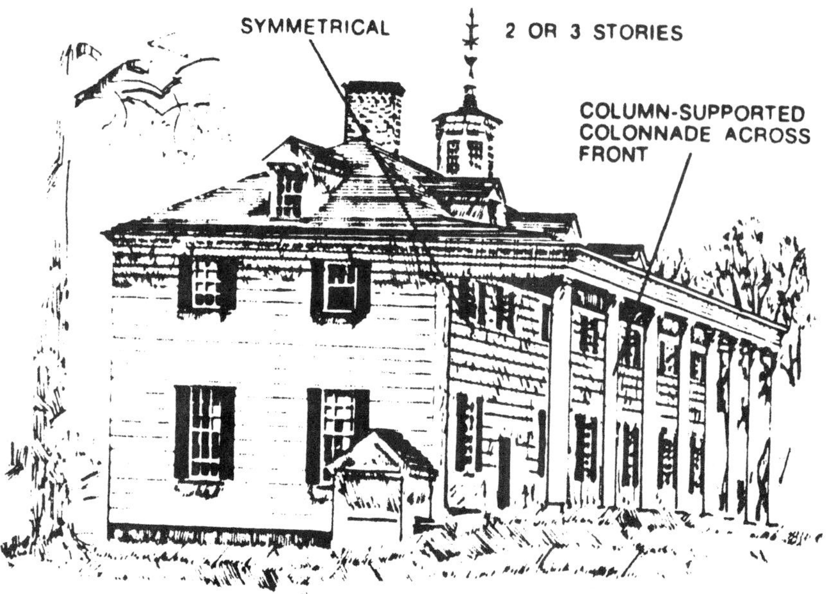

Figure 914 Southern Colonial style

soldier A brick that is set in a wall on end in a vertical position with the side facing out.

soldier course Brick course with bricks on end in a vertical position. See *course.*

sole In *platform framing*, the horizontal support member at the bottom of a wall. It rests on the floor platform. (*Figure 881*, p. 386.)

solepiece See *sole.*

soleplate See *sole.*

solid bridging Solid wood pieces that run between joists to give support. See *bridging.*

solid-core door Flush door made of a solid center, usually solid blocks glued together, with plywood glued to the faces. See *door.*

solid masonry Masonry units with voids that do not exceed 25% of the cross-sectional area (in a plane parallel to the bearing surface). See *hollow masonry.*

solid slab Reinforced concrete slab floor that has shallow beams cast in the bottom. Also called *slab band.* See *one-way solid slab, two-way solid slab, reinforced concrete floor.*

solidus In welding, the highest temperature at which a metal will still remain solid. Opposed to *liquidus.*

solvent Liquid used to dissolve another substance. In painting, a liquid that can thin or remove a paint. Common solvents are acetone, benzene, benzine, mineral spirits, turpentine.

solvent cement In plumbing, a compound used to bond rigid plastic pipe sections together.

solvent weld A type of bond made between plastic pipe and the pipe fitting using a solvent cement or primer and solvent cement. The solvent actually softens the joint, allowing the parts to fuse together, creating a permanent bond or weld.

sound In lumber, free of defect.

sound insulation Material that resists the transmission of sound.

sound knot In lumber, a knot that is firm, solid, and one with the lumber piece. Not considered a defect; it is considered to be a part of the board in strength or bearing capacity.

soundproofing Use of material that deadens the sound or prevents sound transmission.

sound-transmission class (STC) Method of assigning numerical values to walls and floors in regard to their soundproofing qualities. Loss is expressed in decibels as a reduction between two points or areas.

sound-transmission loss Reduction in sound level between two areas, measured in *decibels.*

Southern Colonial Architectural style developed in the old South, characterized by tall columns in the front and a symmetrical facade; two or three stories is common. (Figure 914.) See also *architectural styles.*

space heater Individual free-standing heating unit used to heat a single room; a *room heater*.

spackle A patching compound used for filling drywall or plaster cracks or nail holes. Comes in a powder or in a premixed vinyl paste.

spade bit Standard spade-shaped bit used in power drills. Sizes vary from $\frac{3}{8}$″ to $1\frac{1}{2}$″; bits usually come in a set. (Figure 915.) See also *bit*.

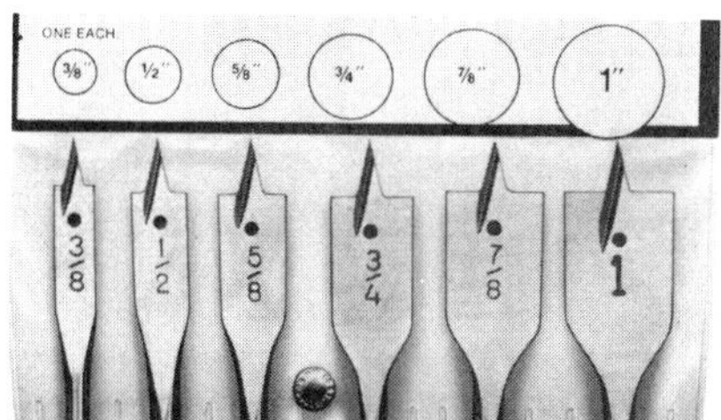

Figure 915 Spade bits

spading In concrete work, pushing a shovel or spade up and down in the freshly placed concrete to eliminate air bubbles, especially in corners and along the side of the forms.

spall In rubble masonry, a small piece of stone used to fill small spaces.

spalling In concrete work, the breaking away of chips of concrete from the hardened surface. Spalling produces pieces or chips of concrete larger than those resulting from *scaling*.

span Clear space or distance between two supports for a horizontal support member such as a beam, arch, joist, or truss. Distance is measured from clear edge of support to clear edge of support. For example, for a ceiling joist, span is measured from inside edge of top plate to inside edge of top plate. See *clear span*. See *effective span* (which is measured from center of support to center of support). See also *rafter run, run*. Roof span is measured from *outside* of plate to *outside* of plate, to correspond to the location of the *rafter line* or *pitch line*.

spandrel The triangular area at each side of a masonry or steel arch. (Figure 916.) The area may be filled in to become an integral part of the arch. In reinforced or prestressed concrete work or steel construction, the part of the wall between a window head and the window sill of the window above and between the supporting columns. See *spandrel beam*.

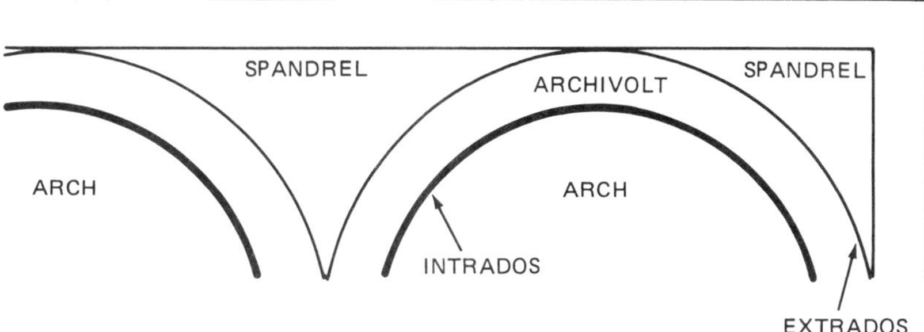

Figure 916 Spandrel

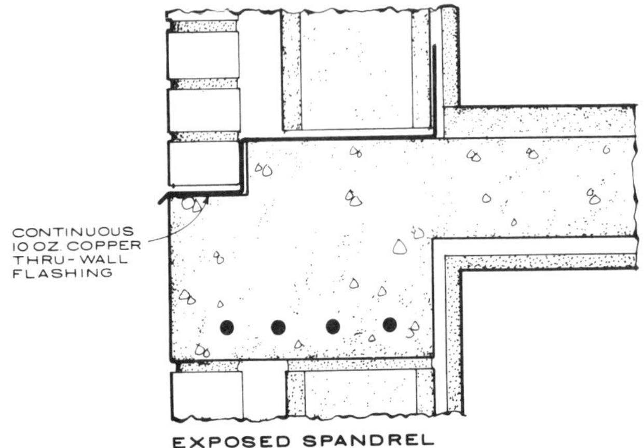

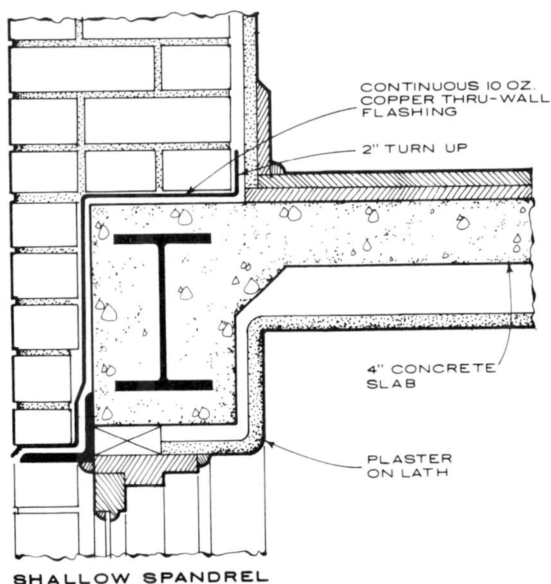

Figure 917 Spandrel beams

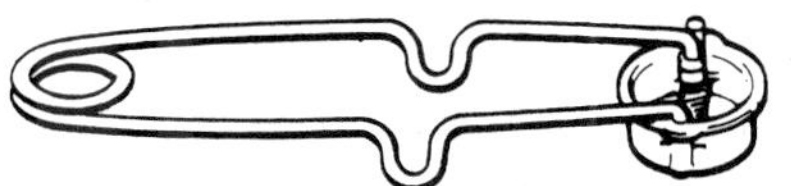

Figure 918 Spark lighter

spandrel beam Outside beam at floor level supported by columns or piers; supports the outside wall. In reinforced concrete, an edge beam; an exterior wall beam. Figure 917 shows reinforced spandrel beams cast as part of the concrete slab.

spark arrester Device set on a fireplace flue top to inhibit sparks flying out.

spark lighter A sparking device used to ignite an oxyacetylene or oxypropane torch flame. (Figure 918.)

spark test Identification of steel and iron alloys by the type of spark produced on a grinding wheel. (Figure 919.)

test \ metal	low carbon steel	medium carbon steel	high carbon steel	high sulphur steel
appearance	DARK GREY	DARK GREY	DARK GREY	DARK GREY
magnetic	STRONGLY MAGNETIC	STRONGLY MAGNETIC	STRONGLY MAGNETIC	STRONGLY MAGNETIC
chisel	CONTINUOUS CHIP SMOOTH EDGES CHIPS EASILY	CONTINUOUS CHIP SMOOTH EDGES CHIPS EASILY	HARD TO CHIP CAN BF CONTINUOUS	CONTINUOUS CHIP SMOOTH EDGES CHIPS EASILY
fracture	BRIGHT GREY	VERY LIGHT GREY	VERY LIGHT GREY	BRIGHT GREY FINE GRAIN
flame	MELTS FAST BECOMES BRIGHT RED BEFORE MELTING	MELTS FAST BECOMES BRIGHT RED BEFORE MELTING	MELTS FAST BECOMES BRIGHT RED BEFORE MELTING	MELTS FAST BECOMES BRIGHT RED BEFORE MELTING
*Spark** *For best results, use at least 5,000 surface feet per minute on grinding equipment. (Cir. x R.P.M. / 12 = S.F. per Min.)	Long Yellow Carrier Lines (Approx. .20% carbon or below)	Yellow Lines Sprigs Very Plain Now (Approx. .20% to .45% carbon)	Yellow Lines Bright Burst Very Clear Numerous Star Burst (Approx. .45% carbon and above)	Swelling Carrier Lines Cigar Shape

test \ metal	manganese steel	stainless steel	cast iron	wrought iron
appearance	DULL CAST SURFACE	BRIGHT, SILVERY SMOOTH	DULL GREY EVIDENCE OF SAND MOLD	LIGHT GREY SMOOTH
magnetic	NON MAGNETIC	DEPENDS ON EXACT ANALYSIS	STRONGLY MAGNETIC	STRONGLY MAGNETIC
chisel	EXTREMELY HARD TO CHISEL	CONTINUOUS CHIP SMOOTH BRIGHT COLOR	SMALL CHIPS ABOUT 1/8 in. NOT EASY TO CHIP, BRITTLE	CONTINUOUS CHIP SMOOTH EDGES SOFT AND EASILY CUT AND CHIPPED
fracture	COARSE GRAINED	DEPENDS ON TYPE BRIGHT	BRITTLE	BRIGHT GREY FIBROUS APPEARANCE
flame	MELTS FAST BECOMES BRIGHT RED BEFORE MELTING	MELTS FAST BECOMES BRIGHT RED BEFORE MELTING	MELTS SLOWLY BECOMES DULL RED BEFORE MELTING	MELTS FAST BECOMES BRIGHT RED BEFORE MELTING
*Spark** *For best results, use at least 5,000 surface feet per minute on grinding equipment. (Cir. x R.P.M. / 12 = S.F. per Min.)	Bright White Fan-Shaped Burst	1. Nickel-Black Shape close to wheel. 2. Moly-Short Arrow Shape Tongue (only). 3. Vanadium-Long Spearpoint Tongue (only).	Red Carrier Lines (Very little carbon exists)	Long Straw Color Lines (Practically free of bursts or sprigs)

Figure 919 Spark test

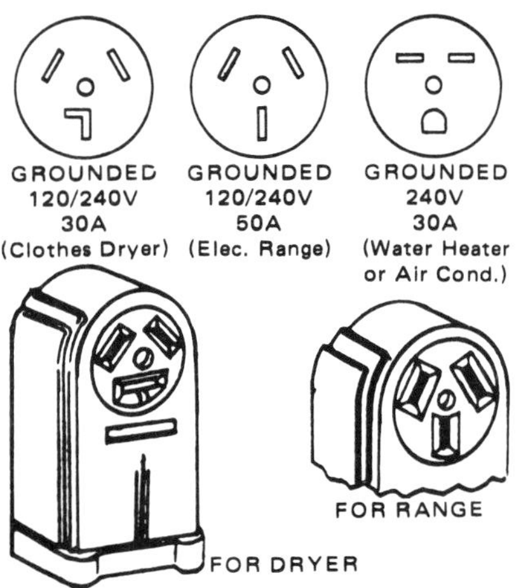

Figure 920 Special-purpose outlets

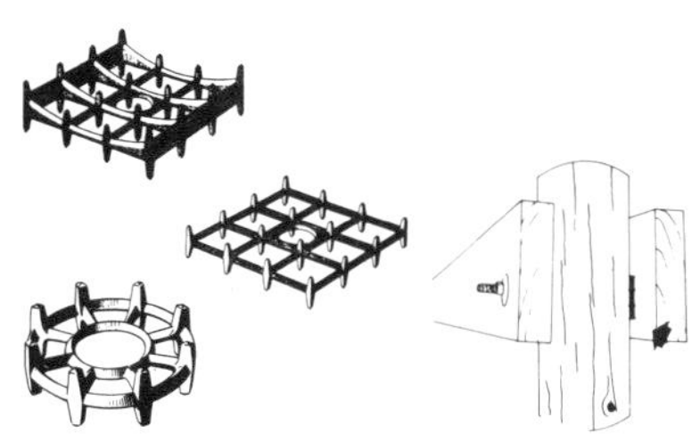

Figure 921 Spike grids

spatter In arc and gas welding, metal bits that are thrown out during the welding process.

special-purpose outlet Receptacle for special equipment such as a range or power equipment. Normally 240 V is supplied. Figure 920 shows the configuration of common special-purpose outlets.

specifications A written legal document giving information on quality and type of building materials needed. Technical standards, equipment manufacturer, and quality of workmanship are also spelled out. Specifications describe in detail how a structure is to be built. They cover three general areas: bidding requirements, general work conditions, and technical (material) specifications. Three types of technical specifications are used: *outline specifications, fill-in specifications*, and *complete specifications*. Specifications work together with the working drawings (blueprints) to give the information needed to build a structure. In case of conflict, the specifications normally take precedence over working drawings, although in bridge building the blueprints may govern. Also called *specs*. See also *Construction Specifications Institute, project manual*.

specific gravity Ratio of weight of a certain volume of material to the weight of an equal volume of water (at 39.2°F).

spigot Plain, unthreaded end of a pipe that is inserted into the enlarged end (bell or socket) of another. Sometimes used to refer to an ordinary faucet. See *bell-and-spigot, cast-iron soil pipe*.

spike A long nail used for fastening heavy timber; a nail over 60*d*.

spike grid Timber connector used in heavy construction between structural members. (Figure 921.) See also *shear plate, timber connectors*.

spike knot Knot sawed lengthwise.

spindle turning Wood turnings made on stock mounted on a wood lathe between the lathe centers. Long cylindrical pieces are made, such as a chair leg or a ballbat.

spiral A coiled reinforcement bar used in reinforced concrete columns and piers.

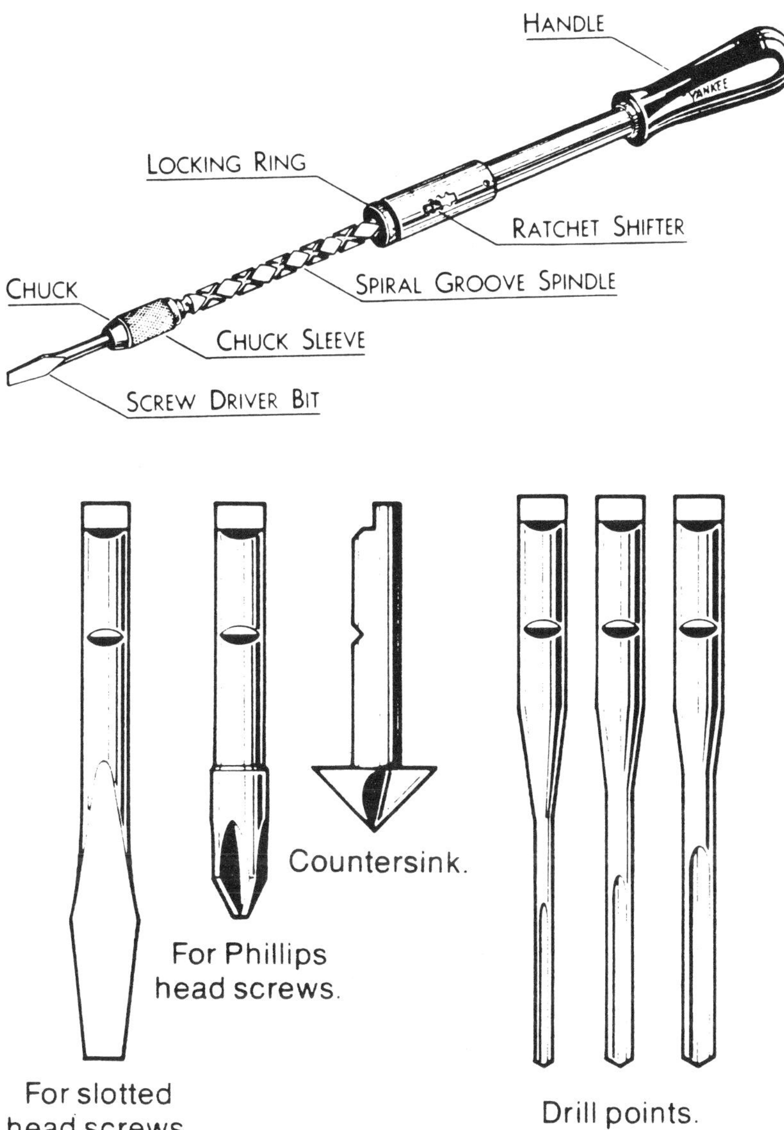

Figure 922 Spiral ratchet screwdriver and tips

spiral ratchet screwdriver A special screwdriver that drives a screw when the handle is pushed down. A spring pushes the handle back up. The rachet can be reversed so that screws can be removed. (Figure 922.) The screwdriver tips are interchangeable; both standard tips and Phillips tips are used. Special drill tips are also available.

spiral stair Stair that turns completely around as it ascends or descends.

spire Tapering tower, as on a church steeple.

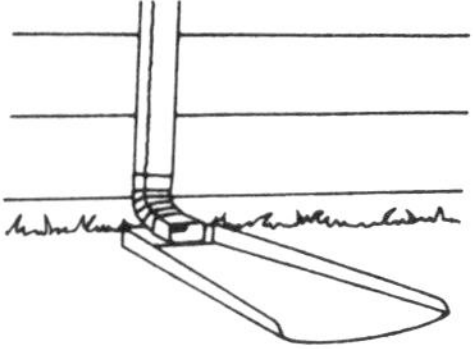

Figure 923 Splash block

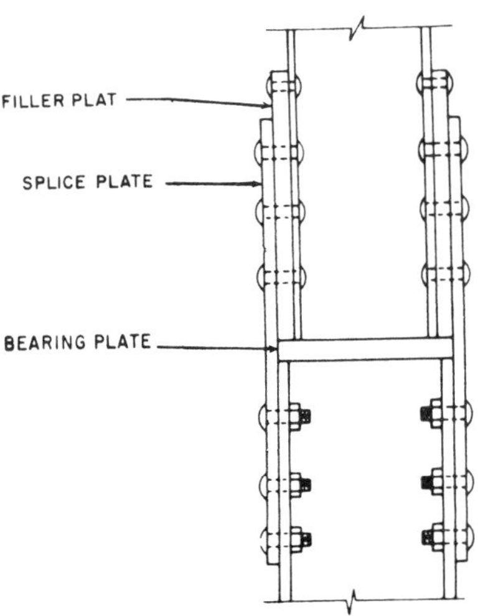

Figure 925 Splice plate (steel)

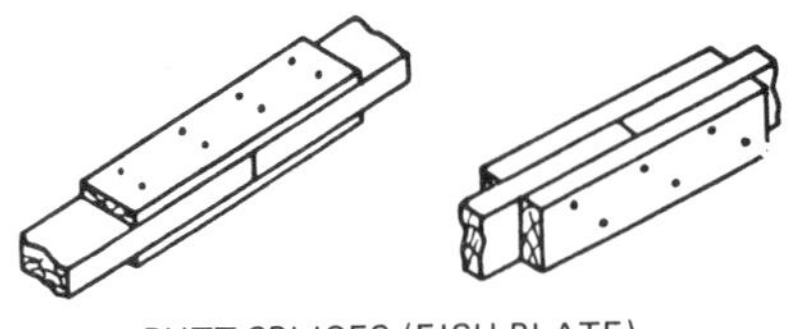

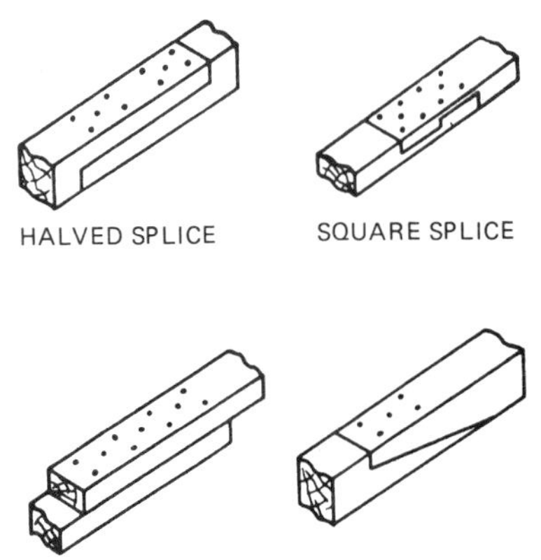

Figure 924 Splices (wood). Butt splices and halved splices are compression-resistant. Square splices and plain splices are tension-resistant. The construction splice is bend-resistant

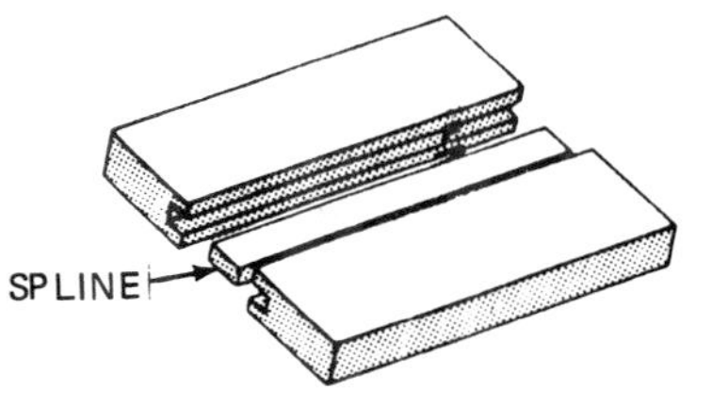

Figure 926 Spline used to join and strengthen two wood pieces

spirit level A *level*. Used for establishing true horizontal or vertical surfaces. So named for the alcohol (spirits) used in the level bubble. See *level*.

splash block Small masonry piece or concrete trough used to receive rain water runoff from the downspout. Designed to carry the water away from the edge of the building and prevent the earth from being washed away by the force of the water. (Figure 923.)

splat A thin, narrow board; a wood strip used to cover vertical wood joints on siding. Also, in furniture making, the flat, vertical, center-upright piece on the back of a chair. See also *post, rail, stretcher*.

splay A slope or bevel, as on the side of a door, to allow better closing. Slanted surface; *draft* or *batter*. An angle over 90°.

splay knot A *spike knot*.

splice A joining of two separate parts, as wood pieces, ropes, wire ropes, electrical wires, metal sheets. Figure 924 shows splices made between wood pieces. A twisted connection made between two electrical wires. See *pigtail splice, tee splice, Western Union splice*. See also *solderless wire connector*.

splice plate Flat steel plate used to splice the ends of steel members together, as the ends of multistory steel columns. (Figure 925.)

spline Narrow strip of wood running in matching grooves at the joint between two boards. Used between boards running edge-to-edge and at miter corners. Designed to strengthen the joint. (Figure 926.) See *wood joints*.

spline miter Miter that is joined with a spline. See *wood joints*.

split Lengthwise separation of wood fibers.

split bolt connector A type of U bolt used to splice heavy cable together, as in electrical wiring.

split foyer House that splits into two levels by the entryway. (Figure 927.) Also called a *split entry*.

split level Housing construction where the living areas are designed at slightly different heights but are not separated by a full floor. (Figure 928.) A house with two different levels is called a bilevel; three different levels is called a trilevel. See *split foyer*.

split ring Ring-shaped timber connector used for holding two timber pieces together, as in truss construction. Ring fits into a cut round groove. (Figure 929.) See also *timber connectors*.

split-wired receptacle Duplex receptacle with each part wired on a separate circuit.

spoil Piled-up earthen debris.

spokeshave A two-handed woodworking tool used for shaving down wood parts. (Figure 930.) The blade is drawn toward you when cutting. Historically, this tool was used for shaving down wooden spokes for wagon wheels.

spot welding Welding at localized spots. Overlapping metal parts are fused together by the heat of electrical resistance (resistance spot weld) or by an arc (arc spot weld).

spray can Pressurized canister used for applying various wood-finishing materials, such as strains, paints, and varnishes. The canister is not reusable.

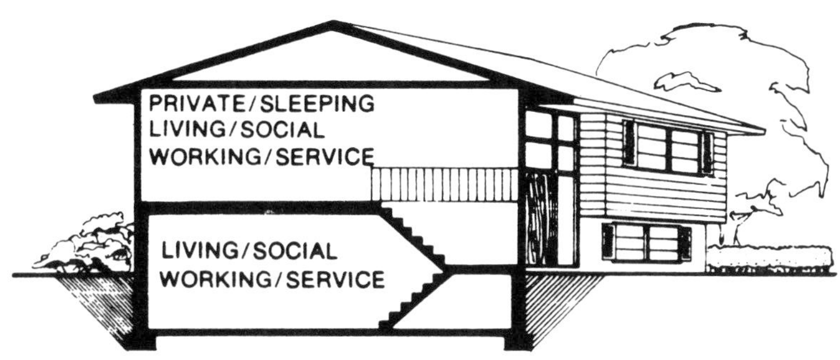

Figure 927 Split foyer: this is a bilevel ranch-style house

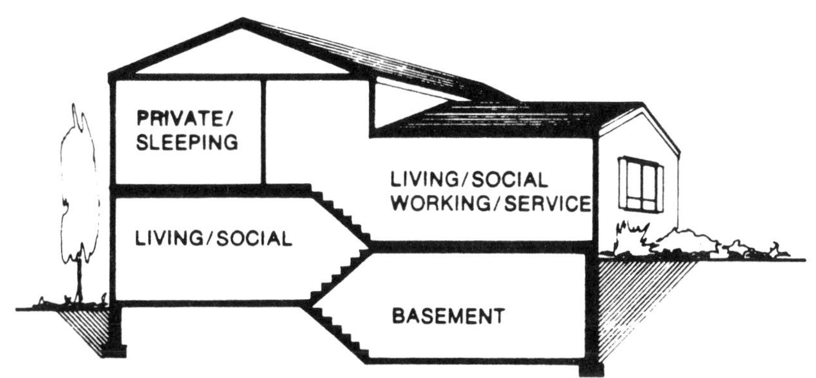

Figure 928 Split level

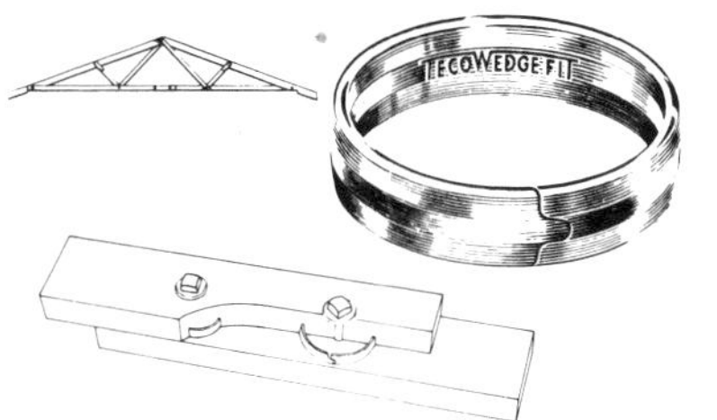

Figure 929 Split rings used to connect two wood members

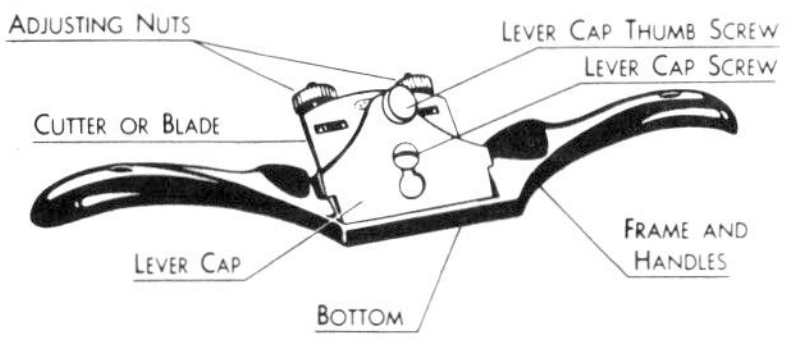

Adjusting Nuts
Lever Cap Thumb Screw
Lever Cap Screw
Cutter or Blade
Frame and Handles
Lever Cap
Bottom

Figure 930 Spokeshave

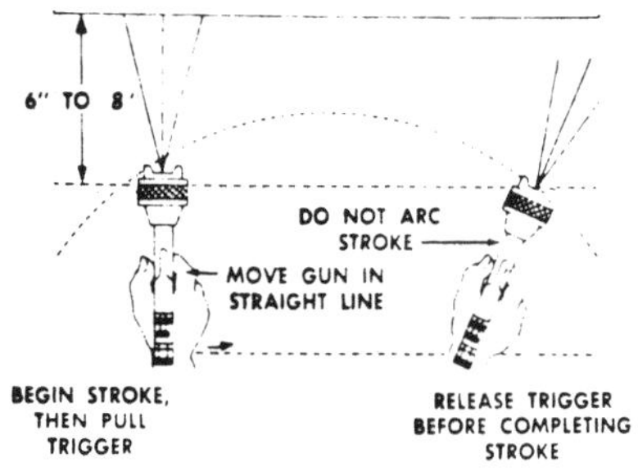

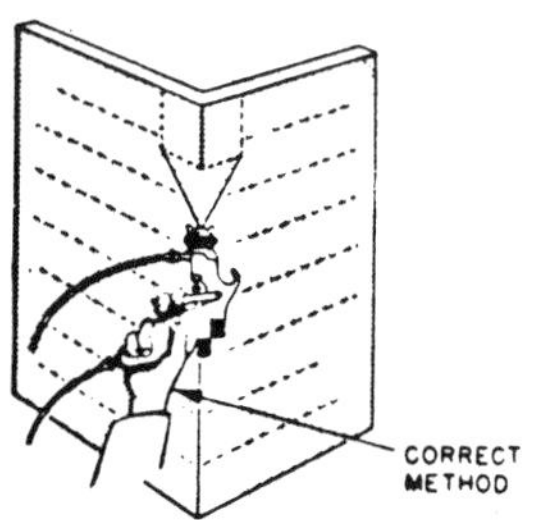

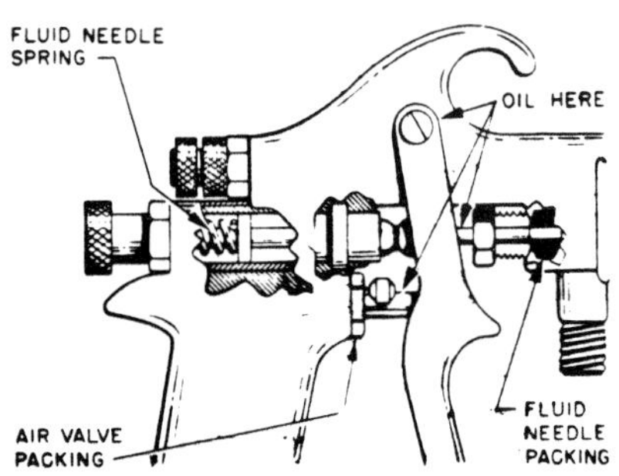

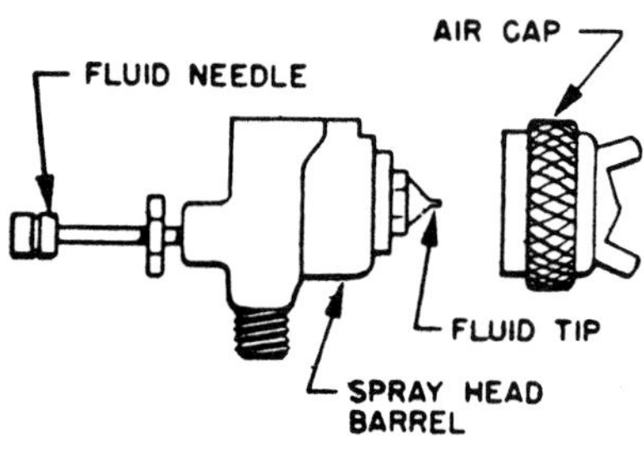

Figure 931 Spray gun

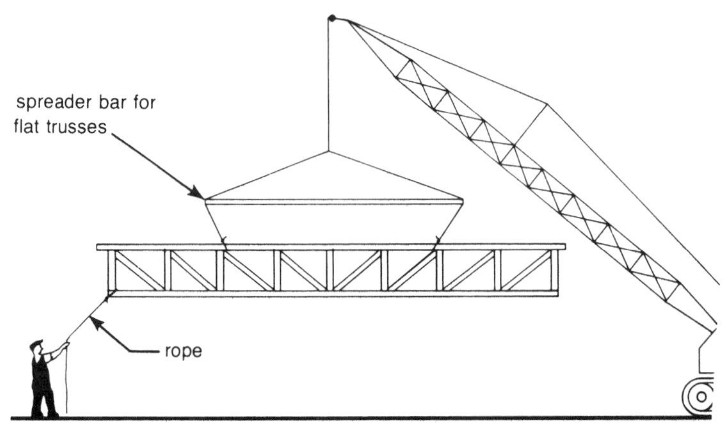

Figure 932 Spreader bar

spray gun An air-powered gun used for spraying paints, stains, lacquers, and varnishes. (Figure 931.)

spread In steel construction, the distance center-to-center of opposing gage lines of two connecting angles attached to opposite sides of the web of a steel beam. In general, distance between centers, as between the centers of two holes. See also *pitch, gage.*

spreader Rigid piece used to keep form walls separated at the proper distance until the concrete is set. See *tie wire.*

spreader bar Bar used in hoisting to create two lift points so the member being lifted will not take all the load at one point. (Figure 932.)

spread footing Footing that spreads outward at the base, has one or more unusually wide supports.

spreading Applying of glue to surfaces to be bonded together. Single spreading is applying of glue to only one surface. Double spreading is applying of glue to both surfaces.

sprig A small headless nail or dowel pin.

spring clamp In woodworking, a small clamp that is held closed with a spring, used to hold wood parts together for gluing. Sizes range from a little over 4" up to 8" long.

spring line Line running horizontally across the base of an arch over a doorway or opening.

spring-wing toggle bolt See *toggle bolt.*

spring wood In tree growth, the wood formed in the spring, usually a wide ban within the annual ring. See *tree.* Opposed to *summerwood.*

sprinkler fitter One who installs piping, equipment, and fittings for an automatic sprinkler system in commercial and heavy construction.

spud wrench See *erection wrench.*

spur center Center used on a wood lathe.

square In roofing, a bundle of shingles that will cover 100 square feet. In general, 100 square feet. A right-angle measuring tool used for checking right angles and measurements. See *steel square.*

squared Made true; at right angles.

square file File that has a square body. See *file.*

square joint Wood joint made between two wood members that are joined exactly end-to-end without any other support.

square measure An area measure in square inches, feet, or yards.

144 square inches = square foot
9 square feet = square yard

square up To make true; *squaring.*

squaring Making true; framing right angles.

stab saw Hacksaw blade set in a handle or blade holder, used for cutting in difficult-to-reach places or flush against a surface. (Figure 933.)

stack In plumbing, vertical piping for waste and vent.

stack bond Masonry bond pattern with units stacked directly on top of each other; joints line up over each other. This is a decorative pattern used in nonbearing walls. See *brick bond.*

stack cleanout Plumbing fitting that allows access to a waste or soil stack for cleaning.

stack partition Partition enclosing the soil stack (soil pipe); 2″ × 6″ studs are used to allow room for the pipe. (Figure 934.)

stack vent The portion of the soil stack above the highest horizontal drain connection from a fixture. It extends up through the roof and provides direct venting to the system. (Figure 935.)

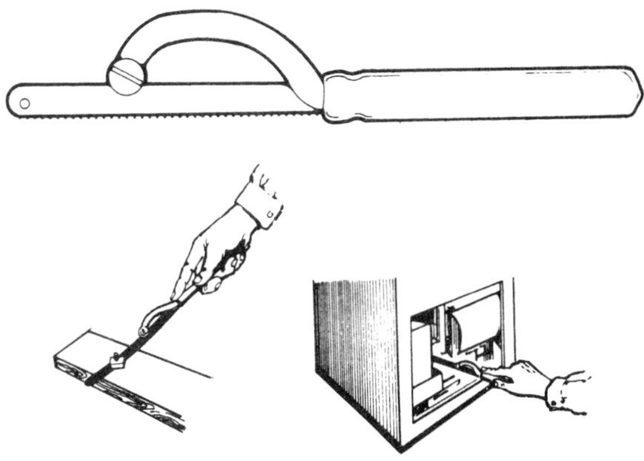

Figure 933 Stab saw

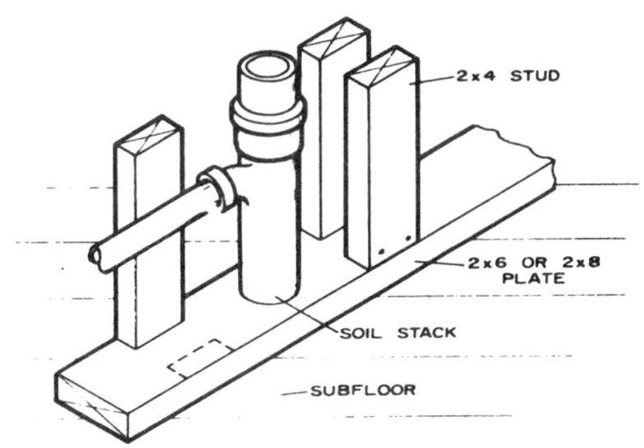

Figure 934 Stack partition

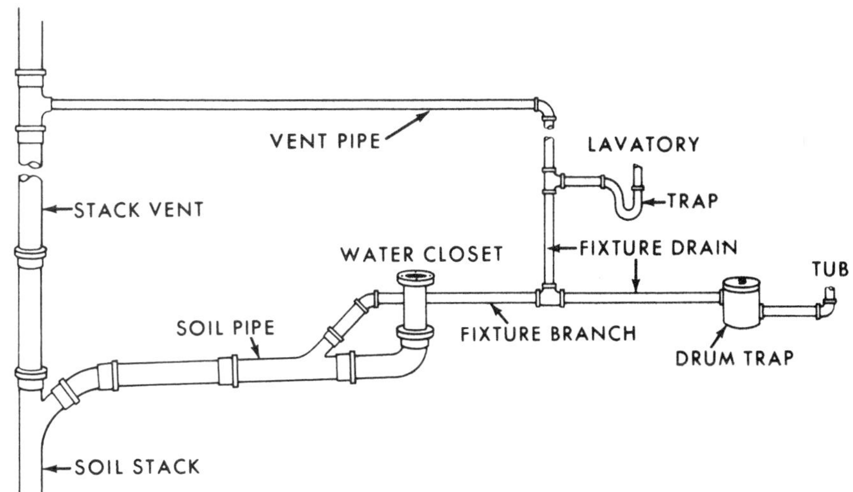

Figure 935 Stack vent

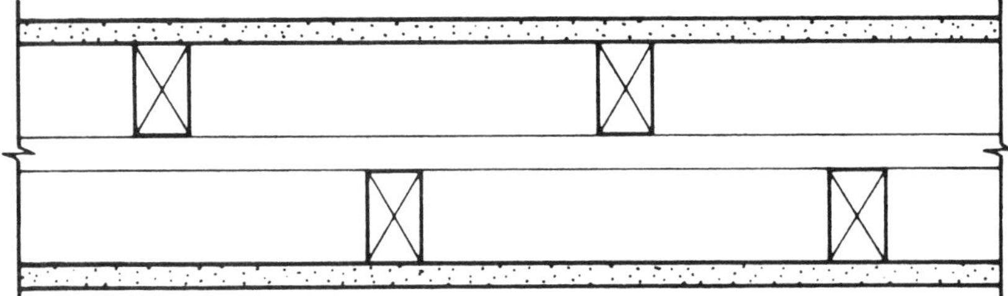

Figure 936 Staggered-stud partition

stadia rod In surveying, a rod with graduations used for estimating distances in a *stadia survey*. Different rods are available that can be used for different distances.

stadia survey Survey using the stadia method of measuring distances. Distance is estimated by using the stadia hairs on the telescope when viewing a graduated stadia rod. The number of rod graduations that fit between the stadia hairs is an indication of the distance to the rod.

staggered-stud partition A double-stud partition with studs offset from each other. (Figure 936.) Used to provide sound insulation and fire protection in party walls. See *double-stud wall.*

staging A temporary work platform; *scaffolding.*

stain a pigmented wood finish that colors the wood fibers but leaves the wood grain still visible. Also, a natural discoloration sometimes found in wood, as blue stain. The stain does not affect the strength of the wood.

stained glass Colored glass set in a lead base; widely used for church windows.

stair The run of treads and risers with supports from floor to floor. The run may be a *straight run* (without landing) or with a *landing* (platform.) Figure 937 shows basic stair construction. Also called a stairs or staircase. Also, the term stair is used to mean *stair step*. See also *circular stair, housed stair, open-string stair, platform stairs, spiral stair, straight flight stairs, straight run, winding stair.*

stair carriage A *stringer;* a support member for treads and risers. See *stair.*

staircase The complete stairs.

stair flight Run of a continuous set of stair steps in one direction, from landing to landing. See *stair.*

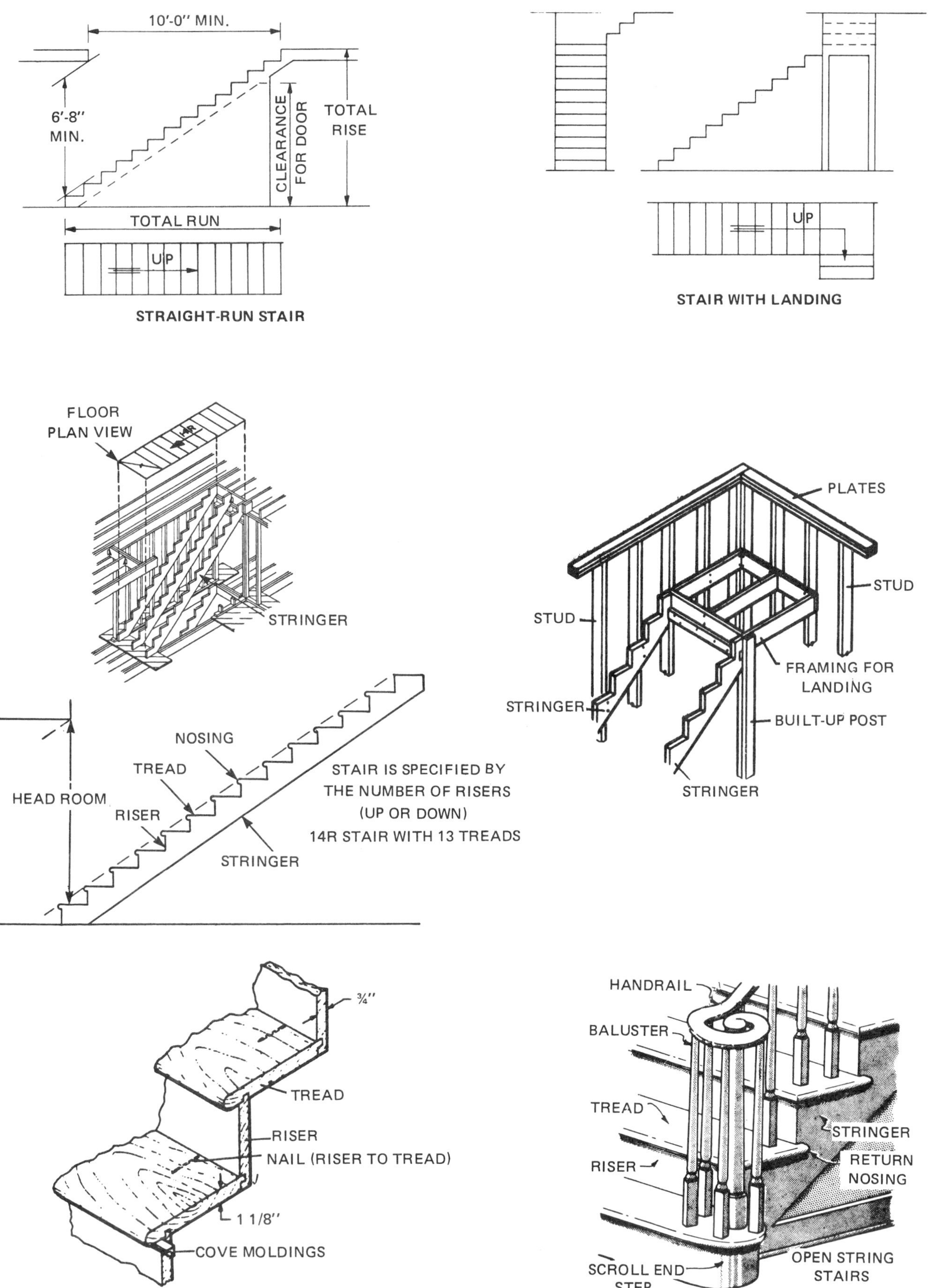

Figure 937 Stair construction

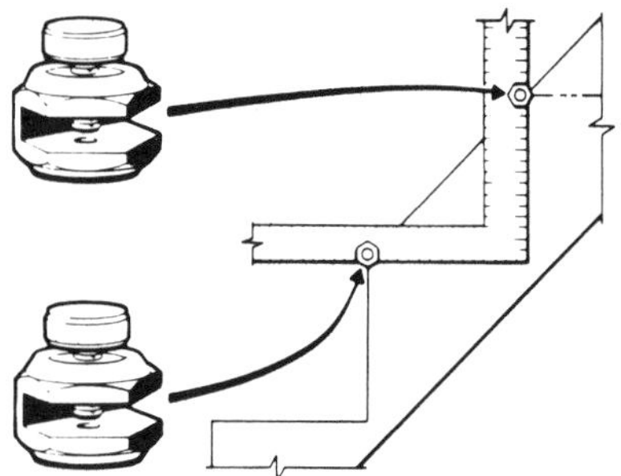

Figure 938 Stair gauges

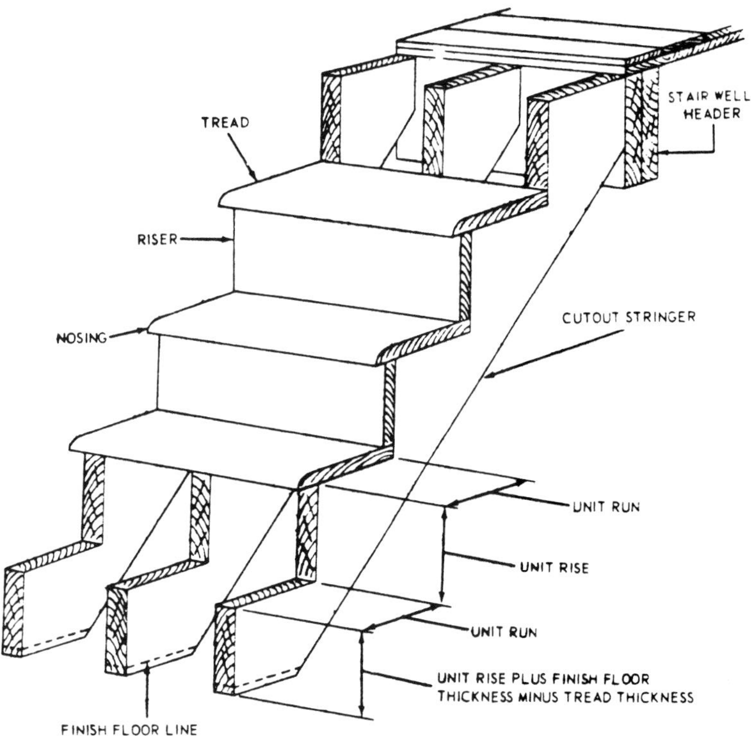

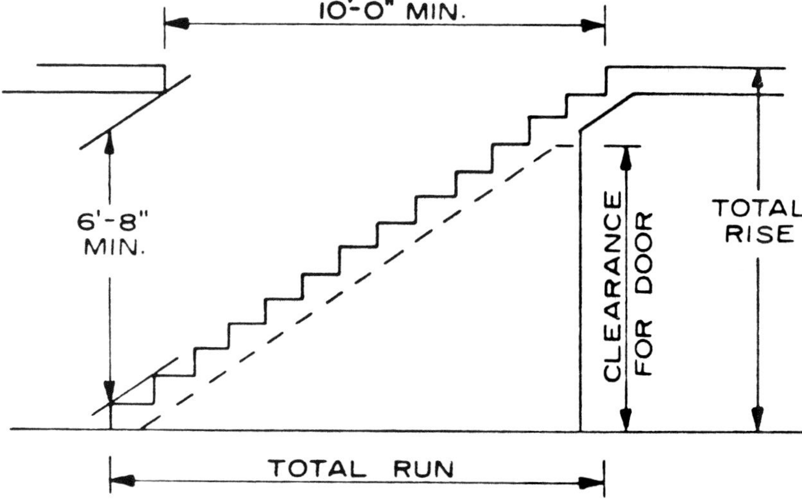

Figure 939 Stair run and stair rise

stair gauges Devices that clamp on a steel square to mark the exact rise and run for a stair layout. (Figure 938.)

stair headroom Vertical distance from top to bottom of stair flight.

stair horse A *stair stringer* or *carriage.*

stair jack A *stringer.*

stair landing Platform at beginning or end of a stair flight, often used to make a change in direction. See *stairs.*

stair rise Height from top of one tread to top of next tread. See *stair run.*

stair riser Vertical part of a stair step. See *stairs.* Abbreviated R. The number of risers are noted on the floor plan. (There is always one more riser than there are treads.) See *stair tread.*

stair run Horizontal distance of a flight of stairs, measured from face of top riser to face of bottom riser. Also, the distance of one tread (without nosing) measured from one riser to the next. (Figure 939.) Total stair run can be found by multiplying the run of one step times the number of treads. See *stair, straight run.*

stairs A *stair* or *staircase.*

stair step One step in a flight of stairs; a *tread.*

stair stringer Support member for treads and risers. Also called a *carriage.* See *stairs, stringer.*

stair tread Horizontal support member; stairstep. There is always one less tread than the number of risers, because the last step at the head of a stairs is actually the floor itself. See *stair, stair riser.*

stairwell Framed opening in which the stairs is placed.

stake Pointed lengths of wood driven into the ground to mark a boundary. Used with boards at foundation corners to mark the foundation outline. See *batterboards*. Also, in sheet metal work, an anvil that fits into a hole in the workbench. Metal is shaped over the stake by striking with a hammer. See *sheet metal stakes*.

staking out Marking lot lines and the corners of a building by driving stakes in the ground.

standards Building codes giving information on how to build or construct.

standpipe In plumbing, an open vertical pipe designed to receive water discharge from an automatic washing machine. Also, an outlet (required in tall buildings) that provides water for a fire hose.

staple Metal U-shaped fastener widely used for installing building material. (Figure 940.) Staples are driven by mechanical or air-operated staplers. Sizes vary widely, depending on the use. Smaller staples are used for tacking on building paper, vapor barriers, or other thin material. Larger staples are used to install plywood or hardboard. Also, a heavy wire staple is used to attach fencing; the heavy staples are driven in with a hammer.

stapler Mechanical, air-operated, or electric device for driving staples. (Figure 941.) See also *staple*.

stapling hammer See *stapler*.

star drill A carbide-tipped, four-sided masonry bit. (Figure 942.) A manual tool used for drilling holes; the head is struck with a hammer or light sledge.

star shake Wood defect; end grain splits from the center of the tree, and the splitting radiates outward, often opening cracks in a star shape. See *heartshake*.

starter strip In frame construction, a strip at the bottom of the outside wall, between the sheathing and the siding, to give a slant to the clapboard. (*Figure 255*, p. 96.)

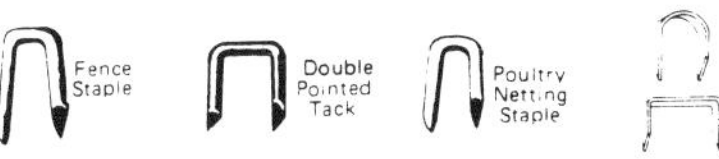

Figure 940 Staples

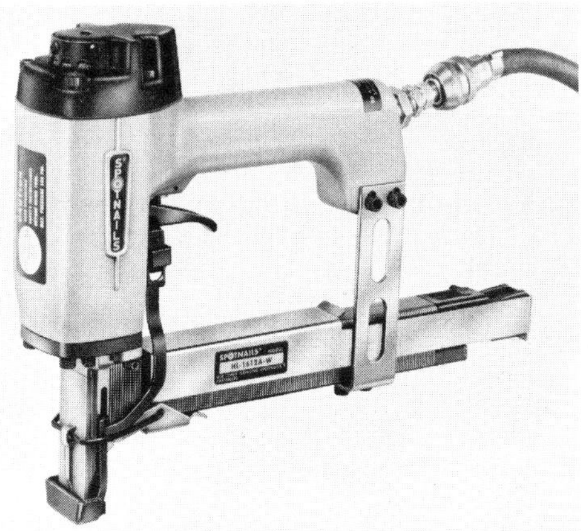

Figure 941 Pneumatic stapler

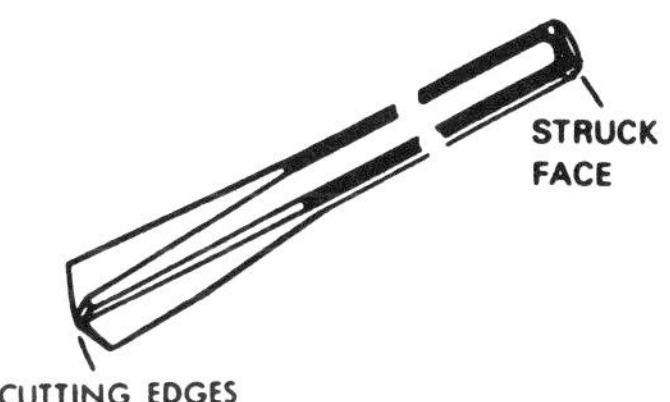

Figure 942 Star drill

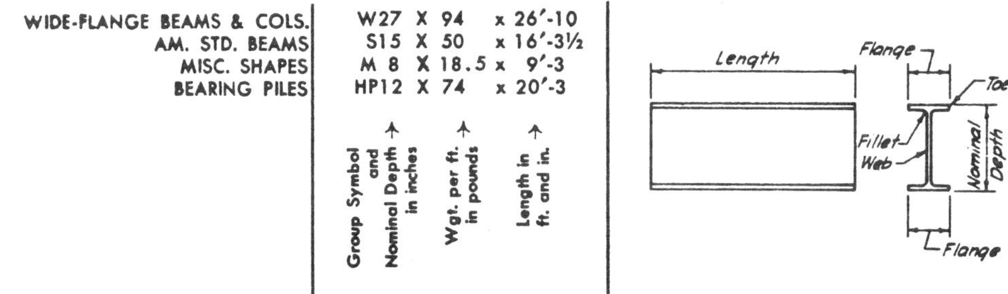

Figure 943 Steel-beam terminology and specification examples

starved joint In woodworking, a glue joint with insufficient glue. The condition can be caused by forcing the glue out by too much pressure.

station In surveying, a 100-foot distance that is marked on a line of measurement. The first station is Sta 0, the second station is Sta 1, and so on. Also, a point where the survey instrument is located at a change in direction in a traverse; a *hub*: also called *traverse station* or *traverse point*.

station point In *perspective drawing*, the viewpoint from which the drawing is made; the position of the viewer.

stay In cabinetmaking, a mechanical device, such as a folding leverlike support, used to hold drop doors in a set, open position. A brace.

steam heating Heating by the use of steam which circulates to radiators.

steel beam Standard steel beam or a wide-flange beam—refers to the type and not its use. A standard (S) or wide-flange (W) steel beam is used in all parts of construction: as a beam, a column, or as cross bracing. Figure 943 shows the parts of a typical standard steel beam, and shows how the beam is specified. See *flange, S beam, W beam, web.*

steel construction Structures framed using steel beams; the support framework is steel.

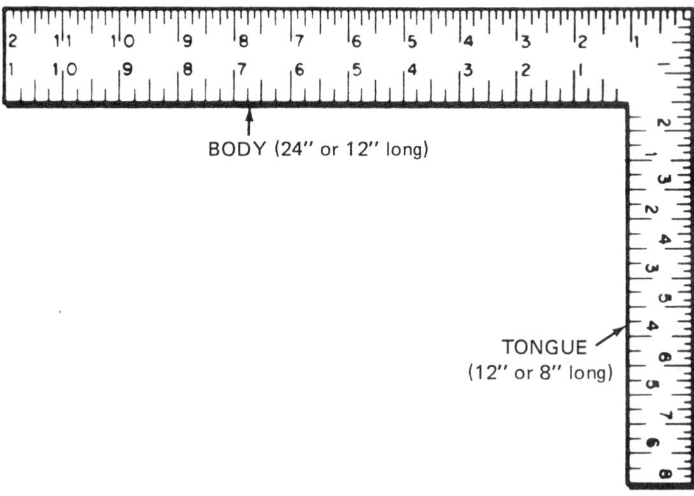

Figure 944 Steel square

steel rectangular Steel pipe that is rectangular in cross section, used for column support and steel studding in construction; sizes vary from 1½" × 1" up to 12" × 6".

steel round Cylindrical steel pipe used for column support in construction; sizes run from ⅛" up to 12". See *lally column, steel rectangular, steel square.*

steel square Layout tool used for laying out angles and for determining rafter cuts. (Figure 944.) A steel square that has tables for laying out rafters is called a *rafter square.* Also called a *framing square.* See also *carpenter's square, stair gauges.* Also, a steel pipe that is square in cross section, used for column support in construction; sizes vary from ½" square up to 4" square. See *steel rectangular, steel round.*

DESCRIPTION	PICTORIAL	MIL-STD SYMBOL	ILLUSTRATED USE	AISC SYMBOL	ILLUSTRATED USE
WIDE FLANGE SHAPE	FLANGE, WEB	WF	24 WF 76	W	W24 X 76
BEAMS				S	S15 X 14.29
AMERICAN STANDARD		I	15 I 42.9		
LIGHT BEAMS AND JOISTS		B	6B12		
STANDARD MILL		M	8M17		
JUNIOR		Jr	7 Jr 5.5		
LIGHT COLUMNS	NOMINAL DEPTH	M	8 X 8M 34.3	M	M8 X 34.3
CHANNELS				C	C9 X 13.4
AMERICAN STANDARD	DEPTH	⊔	9 ⊔ 13.4		
CAR AND SHIP		⊔	12 X 4 ⊔ 44.5		
JUNIOR		Jr ⊔	10 Jr ⊔ 8.4		
ANGLES				L	L3 X 3 X $\frac{1}{4}$
EQUAL LEG	LEG	L	L 3 X 3 X $\frac{1}{4}$		
UNEQUAL LEG		L	L 7 X 4 X $\frac{1}{2}$	L	L7 X 4 X $\frac{1}{2}$
BULB	FLANGE, WEB	BULB L	BULB L 6 X $3\frac{1}{2}$ X 17.4		
TEES				WT	WT12 X 38
STRUCTURAL	FLANGE, STEM	ST	ST5 WF 10.5		
ROLLED		T	T4 X 3 X 9.2		
BEARING PILE		BP	14 BP 73	HP	HP14 X 73
ZEE		Z	Z6 X $3\frac{1}{2}$ X 15.7		
PLATE		Pl	Pl 18 X $\frac{1}{2}$ X 2'-6"	PL	PL18 X $\frac{1}{2}$ X 2'-6"
FLAT BAR		Bar	Bar $2\frac{1}{2}$ X $\frac{1}{4}$	Bar	Bar$2\frac{1}{2}$ X $\frac{1}{4}$
TIE ROD		TR	$\frac{3}{4}$ Φ TR		
PIPE COLUMN		○	○ 6Φ	⌖	pipe 4 std

Figure 945 Steel symbols

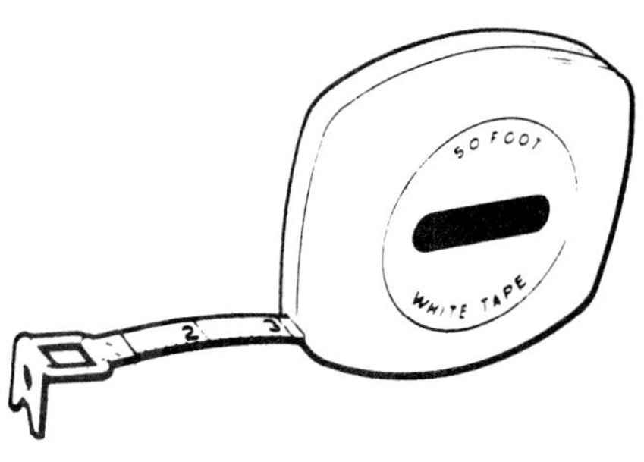

Figure 946 Steel tape

steel symbols Standard symbols used to describe structural steel shapes. Figure 945 shows both American Institute of Steel Construction and Military Standards.

steel tape Steel measuring tape; available in lengths of 50′ or 100′ Figure 946.) See also *push-pull rule*.

steel wool A ball or pad made of fine steel threads, used in woodworking for polishing; also used to clean metal surfaces.

stem In reinforced prestressed concrete, the vertical leg of a T slab or beam. See *tee*. In general, a support base.

step A stairstep

stepladder Folding ladder that opens to an "A" shape and is supported on two sets of legs. See *ladder*.

stepped flashing Flashing that follows the sloping side of a vertical surface, as the side of a chimney. (Figure 947.) Flashing units overlap each other.

stepped footings Footings that drop to different levels to follow a hillside or slope. (Figure 948.)

step trowel Right-angle trowel used for finishing the end of a concrete step. See also *corner trowel*.

stick A welding rod or electrode.

stickbuilt construction Conventional frame construction where individual pieces are cut at the site and nailed into place.

stick-electrode welding See *shielded metal-arc welding*.

sticking Setting of plaster ornament in a plaster bed. See *stick molding*.

stick shellac In woodworking, a solid stick of shellac used for filling cracks or holes in wood. Comes in over 50 colors and can be mixed to match the wood. Stick shellac is applied with a hot knife.

stick welding See *shielded metal-arc welding*.

stile Vertical side members in a door or window sash. On a panel door the stile by the hinges is called the hinge stile; the stile by the latch is called the latch stile. The door stiles fit against the horizontal door members called *rails*. See *door*. In cabinetmaking, vertical members used to frame a panel in a *carcass*; vertical members used to make a panel cabinet door — the two side stiles are joined at the top and bottom by *rails*.

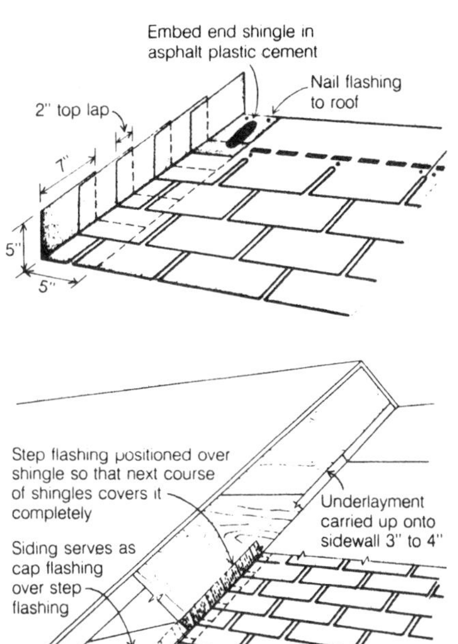

Figure 947 Stepped flashing

Stillson wrench A *pipe wrench*.

stilt house House supported on stilts or poles completely away from the ground. Used on sloping ground and to allow air circulation.

stilts Device worn by plasterers to give added height so they can reach the high parts of the wall or work on a ceiling.

stinger In concrete work, a *vibrator*.

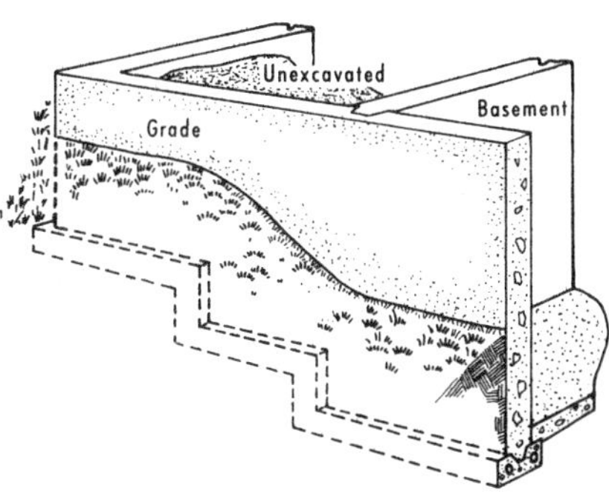

Figure 948 Stepped footings

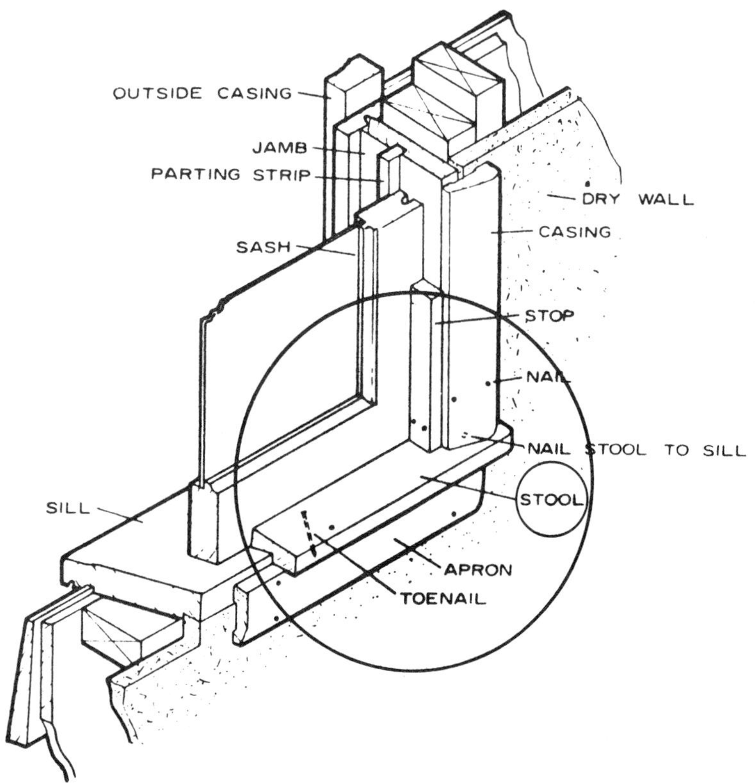

Figure 949 Stool

stippling Use of a *stippling brush* to form a texture in fresh plaster.

stippling brush Brush used to texture fresh plaster.

stirrup Metal framing anchors used to support the ends of joists, beams, girders. In reinforced concrete, U-shaped vertical reinforcing bars used near beam ends to resist diagonal stress. (*Figure 88*, p. 34.)

stock General term for lumber used in construction or woodworking.

stone carving Carving of letters, numbers, or figures into a stone face. A pneumatic gun is used to blow abrasive through a template to make the needed design.

stone cutter's chisel A toothed chisel used to cut and trim stone.

stone mason A specialist who works in stone or special masonry.

stone masonry Construction using cut or rough natural stone building units.

stool Flat interior molding over a window sill running between the two side jambs. The bottom rail of the window closes over the stool. (Figure 949.) See also *molding*. In reinforced concrete support, a *chair*.

stoop Front raised entryway.

stop Window moldings at the sides and top of the frame, designed to hold the sash in. (Figure 950.) A *door stop*.

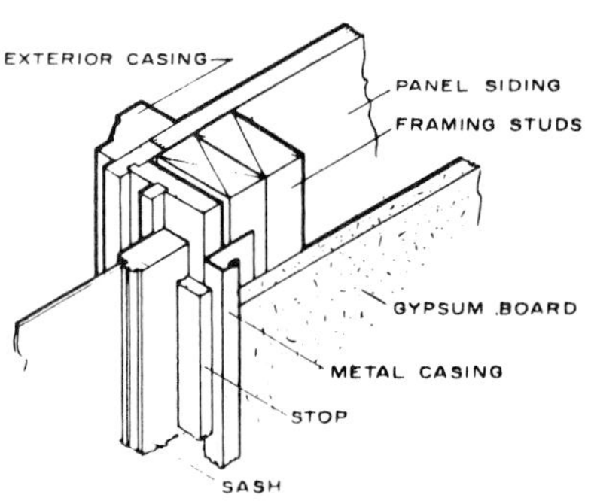

Figure 950 Stop (window)

stop box Iron boxlike shaft that holds the curb cock that controls the flow of water into a building. Also called a *buffalo box*. A small iron cover at ground level allows entry into the stop box. See *water service*.

stopcock Plumbing valve for turning water or gas on and off.

storm clip See *rafter anchor*.

storm door Extra outside door that goes in front of the regular door. Used to cut down air infiltration and heat loss.

storm sewer Sewer that carries only storm flow — that is, rain water and flood water. See *combination sewer, sanitary sewer.*

storm window Extra outside window that goes in front of the regular window. Used to cut down air infiltration and heat loss.

story A floor or level of a building; living space between two floors or levels of a building. Dwellings and other structures are referred to as one-story, one-and-a-half story, two-story, and so on. (Figure 951.)

story pole Wood piece, such as 1 × 2 or 2 × 4, that is marked off in needed dimensions, especially room heights. Used as a template in layout. Also called a *story rod.*

story rod A *story pole.*

stove bolts Widely used flat- or round-headed bolts. See *bolt.* (*Figure 106*, p. 41.)

straight claw hammer Carpenter's hammer with a straight claw. The straight claw hammer is commonly used for laying wood flooring or in electrical work. The claw is used as a sort of blade or hatchet for cutting and breaking. Also called a *rip hammer, rip claw,* or *ripping claw hammer.* (Figure 952.) See also *hammer.*

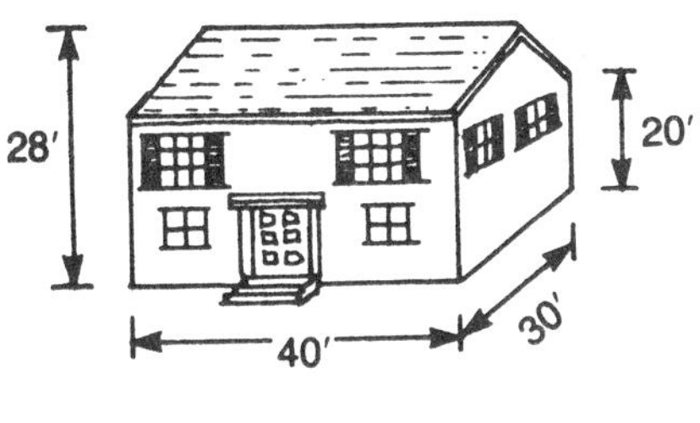

Figure 951 Story

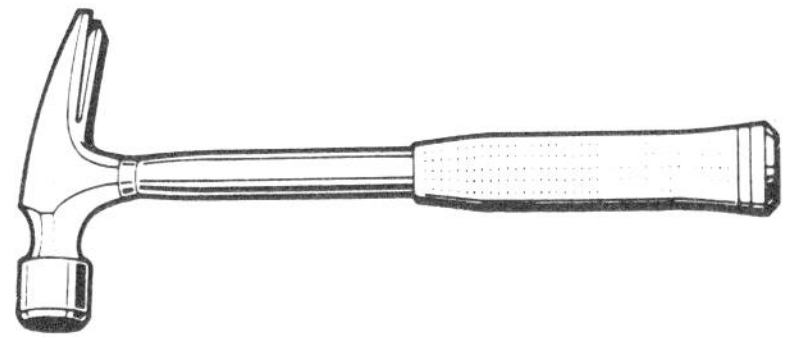

Figure 952 Straight claw or ripping hammer

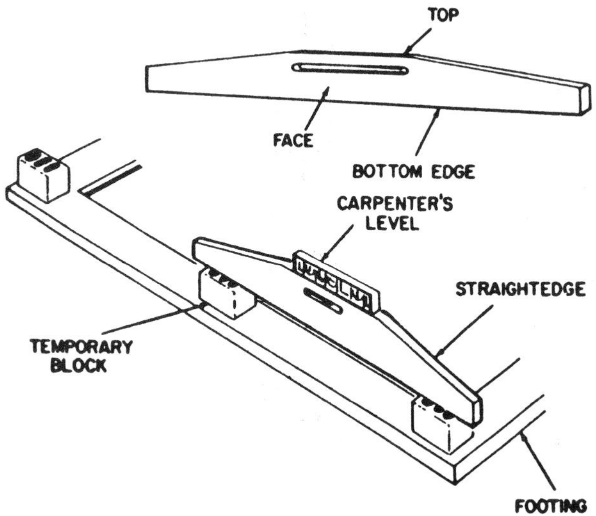

Figure 953 Straightedge (carpentry)

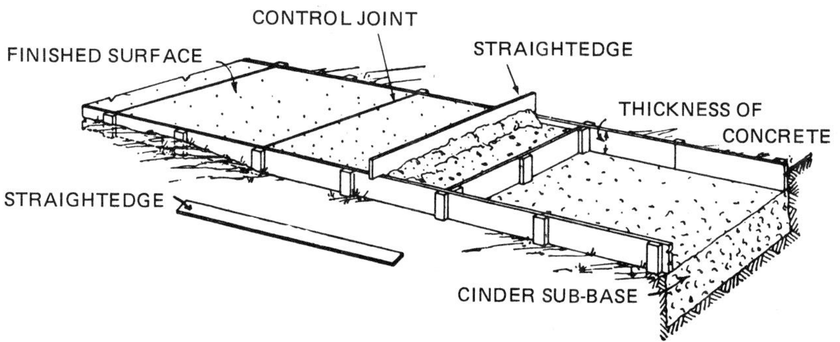

Figure 954 Straightedge used for finishing concrete surface

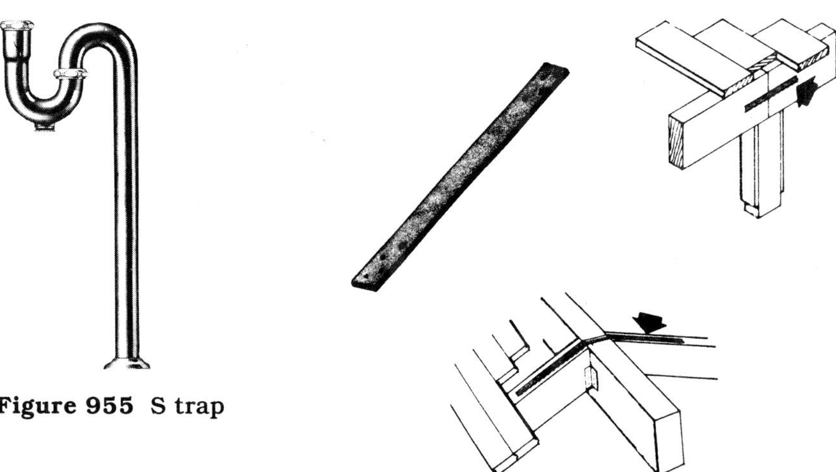

Figure 955 S trap

Figure 956 Straps used to fasten wood members

straightedge In carpentry, a long wood or steel tool used for marking or checking straight lines. (Figure 953.) In plaster and concrete work, a tool used for smoothing a plaster or concrete surface. (Figure 954.)

straight-flight stairs Straight-run stairs without a turn. See *stair, stair run.*

straight grain Wood grain that is parallel to the length of the wood piece.

straight peen hammer Metal-working hammer used for shaping and flattening metal. The peen runs in the same direction as the handle, or is straight in relation to the handle. See *ball peen hammer, cross peen hammer, hammer.*

straight polarity In arc welding, a setup with *direct current electrode negative* (the work is the positive pole, the electrode the negative pole).

straight run A straight flight of stairs, not broken with a landing. See *stair, stair run.*

straight stairway *Straight-run* or *straight-flight stairs.* See *stair, stair run.*

S trap An S-shaped water seal used on plumbing fixtures. (Figure 955.) See also *P trap.*

strap Flat metal piece used for fastening wood parts together. (Figure 956.) A *perforated strap* or band used for supporting plumbing pipes from overhead.

strap clamp See *web clamp.*

strap hinge Exposed hinge used for doors in rough framing, as for barns and sheds, and for gates. See *hinge.*

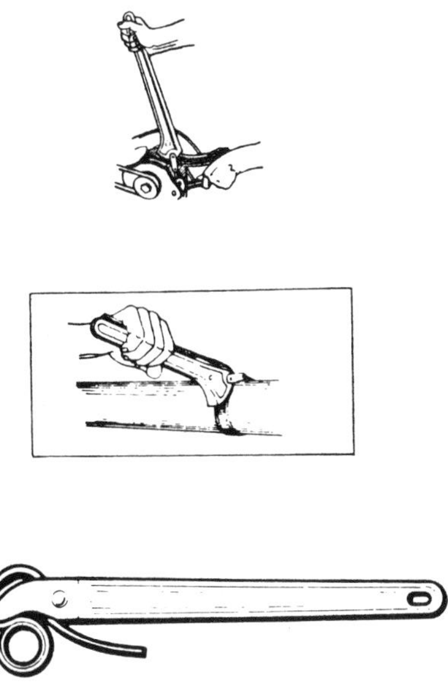

Figure 957 Strap wrench

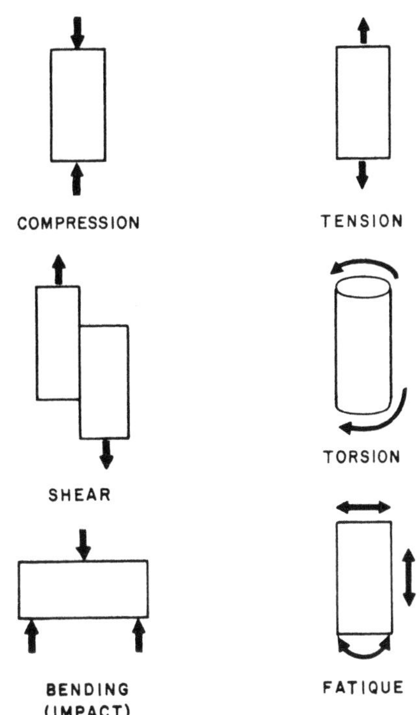

Figure 958 Stress

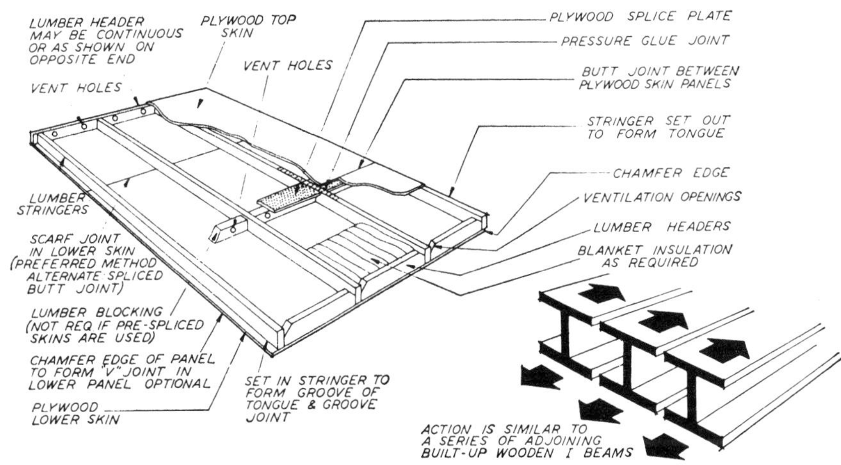

Figure 959 Stressed-skin panel

strap wrench A plumbing wrench having a strap on the end that fits over a pipe; used for turning a pipe. (Figure 957.) See also *chain wrench*.

street main Underground electrical utility line.

stress The load or force on a bearing member. Computed stress is the actual calculated load; allowable stress is the maximum stress permitted. Stress is expressed in kips (1000 pounds) per square inch. See *compression, shear, tension, torsion*. (Figure 958.)

stress grading Grading different wood types by their tested strength.

stressed-skin construction Assembly made with a plywood skin. Plywood panels are bonded to a wood framework to create an assembly that works as a unit under load. (Figure 959.)

stretcher Brick or concrete block laid lengthwise on its face; the side faces out. See *bond, brick*. In furniture construction, horizontal support members running between chair legs or posts.

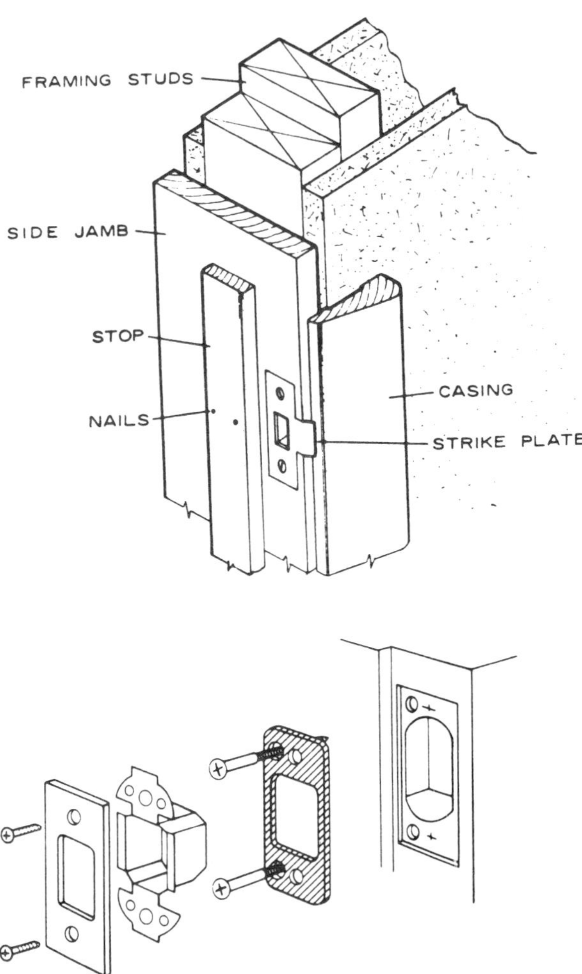

Figure 960 Strike plate

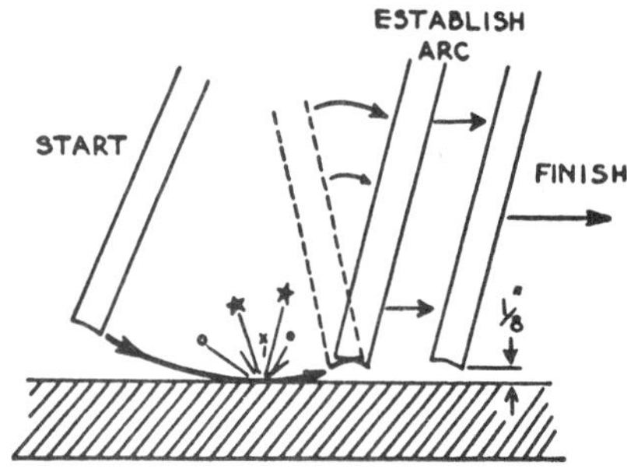

Figure 961 Striking an arc

stretchout Laying out sheet metal work with the pattern.

strike board Wide straightedge used as a *screed* to level a concrete slab.

strike-off Leveling of fresh concrete in the forms. A tool for leveling fresh concrete. *Screeding.*

strike plate Metal plate in the door jamb struck by the latch when the door closes. The latch fits through an opening in the strike plate. (Figure 960.)

striking In concrete flatwork, the leveling of the freshly placed concrete; *screeding.*

striking an arc In arc welding, establishing the electric arc between the electrode tip and the work piece by a striking motion. (Figure 961.) See *tapping an arc.*

striking iron See *jointer.*

striking off In concrete flatwork, the leveling off of the freshly placed concrete; *screeds* are used.

string A *stringer.*

string board Board fastened at the bottom of the stairs to receive and support the stair stringers. A *kick plate.* (*Figure 538*, p. 230.)

stringcourse Horizontal band on the outside of a structure; a decorative, projecting course of bricks or stone.

stringer A support for stair steps; two or three stringers are used under the stairs. Also called a *carriage* or *string*. (Figure 962.) See also *close string, face string, finish stringer, housed stringer, outer string, rough horse, rough string, string board, wall stringer.* In excavation, a horizontal support member used to hold trench sheet pilings; also called a *wale*. See *trench support*.

stringer bead In gas welding, a weld bead that is put down in a straight motion (with no weaving of the flame).

stringing mortar In bricklaying, laying out enough mortar on a bed so that several bricks can be laid.

string line Line running between stakes or boards to indicate the building outline. See *batterboards*.

strip To remove, as to remove concrete forms. To remove paint from a surface. To tear or wear off bolt threads. A small thin piece, as a short piece of tape. A thin piece of lumber.

strip flashing A type of roof flashing; see *stripping*.

stripping Taking down concrete forms after concrete is hardened. Also chemical removal of paint from a surface. In roofing, the bonding of a joint between the built-up roof and another material with several strips or plies of felt and hot asphalt; also called *strip flashing*. Removing insulation from electrical wire.

strongback Stringers nailed across ceiling joists to correct crookedness and to assure even spacing. (Figure 963.)

strop Leather strip for stropping the finely sharpened edge of a cutting tool.

stropping Pulling a cutting edge over a leather strap or piece of leather to achieve a maximum sharpness. You pull away from the edge (opposite to honing) to actually straighten the edge.

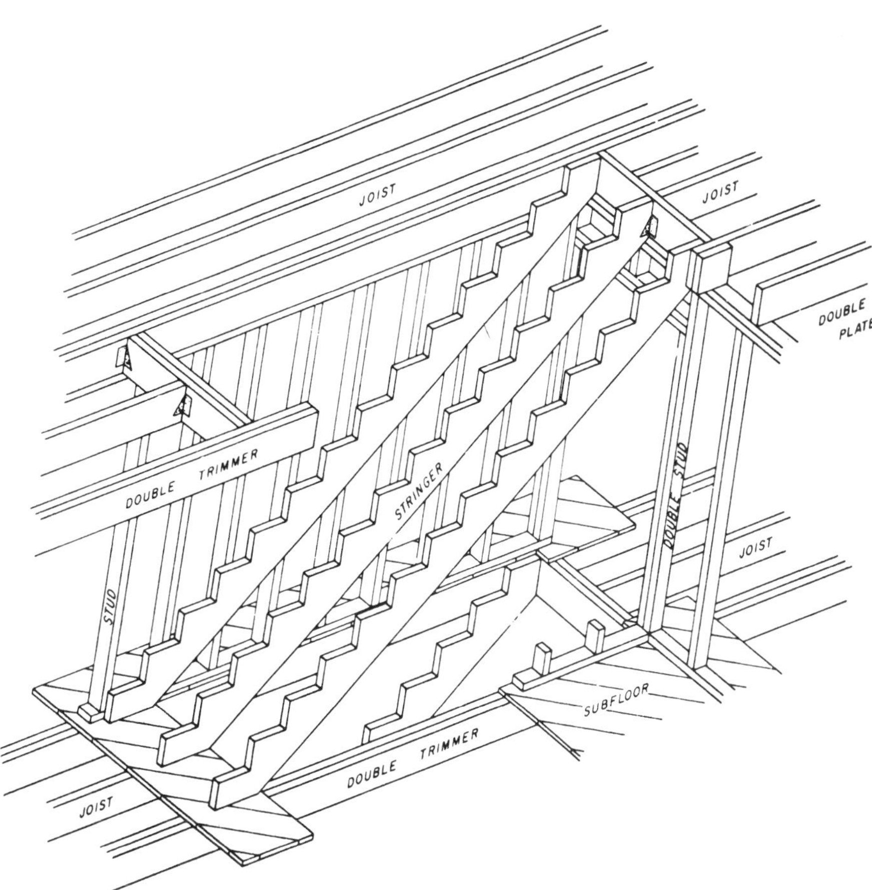

Figure 962 Stringers (or carriages) supporting the stair

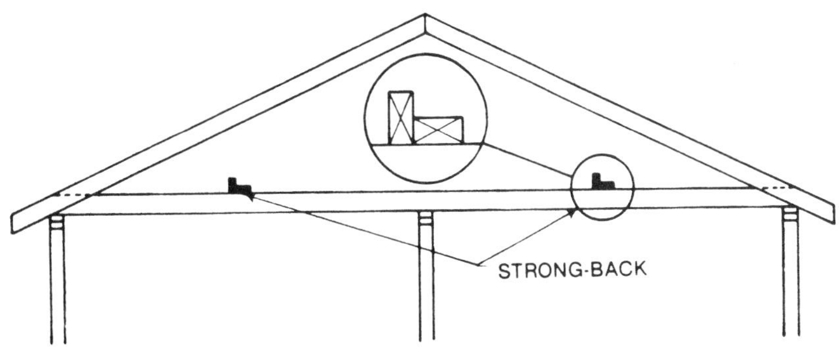

Figure 963 Strongback

Figure 964 Structural clay tile

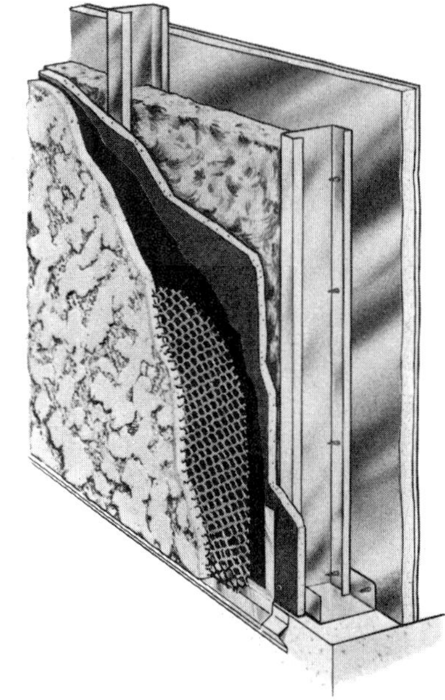

Figure 965 Stucco

struck Of masonry joints, having been tooled with a *pointing trowel*. See *tooling*. Specifically, a struck joint; see *masonry joints*. Also refers to the process of leveling concrete flatwork. See *striking off*.

struck joint Tooled mortar joint with only the bottom pushed back into the masonry joint. An inferior joint that will weather if used outdoors. See *masonry joint*.

structural bond In masonry work, a tying together of masonry units, such as brick facing on a concrete block wall, so that the entire assembly acts as a single structural unit.

structural clay tile Hollow burnt-clay building unit used for backing walls or for facing walls in various finishes. Usually 12″ long in place with mortar. (Figure 964.)

structural drawings Blueprints that show structural detail, the support of a building as the location of steel beams and columns. Identified by an S in the title box. Compare to *architectural drawings, mechanical drawings*.

structural lumber Lumber larger than 2″ × 4″.

structural timber *Structural lumber*.

strut Short support member, as a short column or solid bridging between joists.

stub mortise Mortise which is not cut completely through the wood member. See *wood joints*.

stub screwdriver Short screwdriver.

stub tenon Tenon which does not go completely through the mortise. See *wood joints*.

stucco A cement plaster covering used on the exterior of houses. (Figures 965; *838*, p. 368.) Stucco is composed of Portland cement, lime, sand, and water. See also *plaster*.

stuck molding In plastering, an ornamental cast molding which is stuck in place into fresh plaster. See *sticking*.

stud Vertical support member in a frame wall, usually 2″ × 4″ or 2″ × 6″. (Figures 966, 967.) See also *double-stud wall, metal studs*. Also, a type of bolt threaded on both ends.

stud anchor A one-piece anchor used to create an anchor stud. Anchor is driven into bored hole in masonry; the end expands to hold the stud. (Figure 968.) See also *anchor, machine-screw anchor, sleeve anchor, wedge anchor*.

studding The studs that go to make up a wall or partition.

stud driver Hand-operated device for driving threaded studs or pins into masonry or concrete. (Figure 969.) Stud or pin is held in the tube while the sliding part hammers it into the masonry. See also *powder-actuated gun*.

stud finder Magnetic device used to locate nail heads in a finished stud wall.

stud layout An elevation view showing placement of studs on one side of a house.

stud shear Device for cutting metal channels. See *channel shear*.

stump wood Wide area at the base of a tree just above the root system.

Styrofoam® A type of extruded polystyrene widely used in sheets for insulation. (Styrofoam® is a registered trademark.) See *plastic foam*.

subcontractor Someone who contracts with the prime contractor to do a part of the work on a project, such as the plumbing subcontractor.

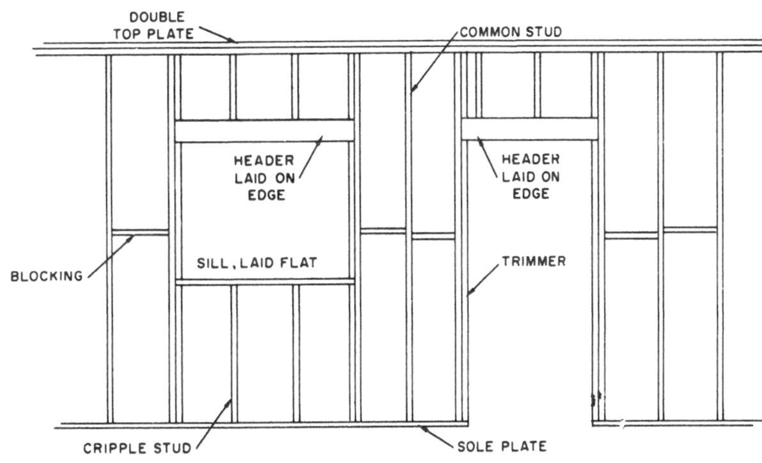

Figure 966 Stud wall

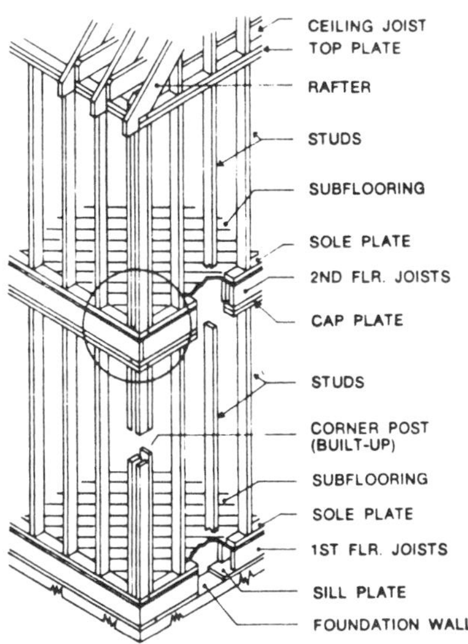

Figure 967 Studs used in platform framing

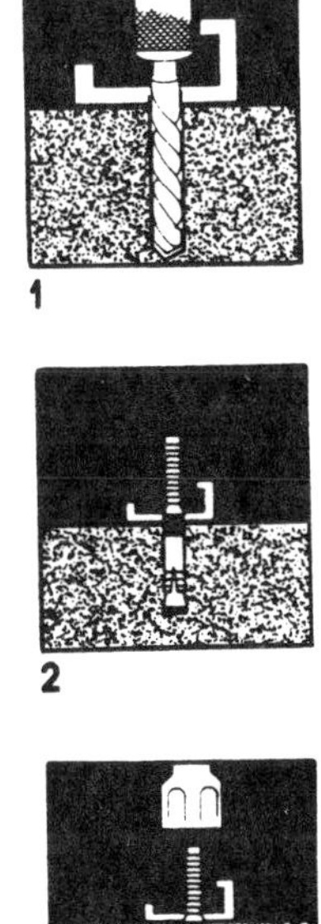

Figure 968 Stud anchor

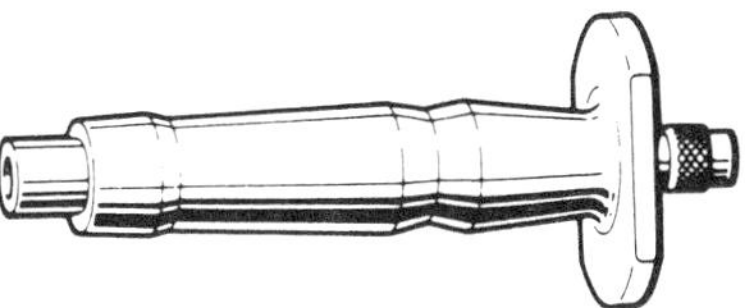

Figure 969 Stud driver

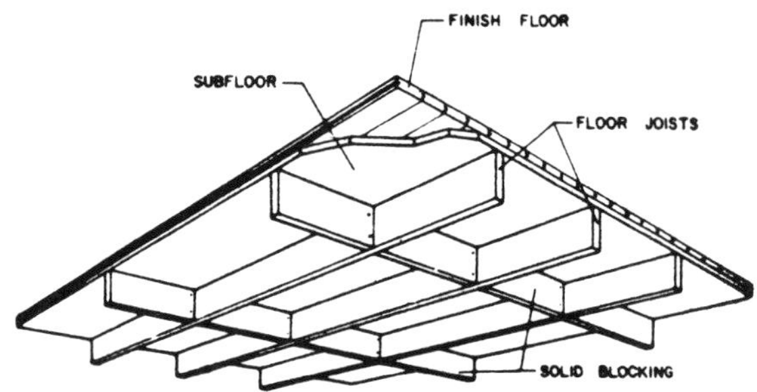

Figure 970 Subfloor

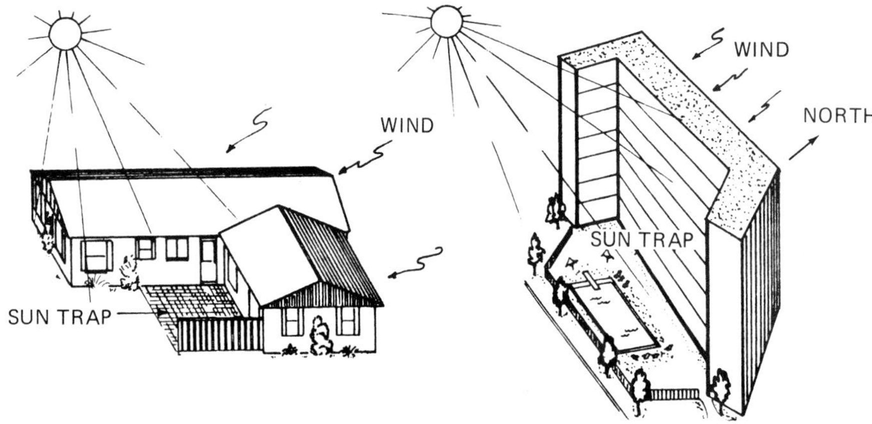

Figure 971 Sun trap

subdivision Division of land into lots and blocks for the purpose of development.

subfloor Rough flooring directly on the joists or concrete slab. The underlayment or finish floor goes over the subfloor. (Figure 970.)

subgrade Base for the concrete slab or the foundation. In highway construction, the base for the *pavement structure*.

substrate In roofing, a roof deck with or without insulation; the surface on which built-up roofing is laid.

substructure The foundation; the support for the *superstructure*.

suburban Area around a city.

summerwood Wood grown during the summer; the dark, dense part of the annual ring. See *springwood*.

sump A hole designed to collect water, as a sump in a basement to hold any seepage. The collected water is automatically pumped out with a *sump pump* when it reaches a certain level.

sump pump Pump that automatically pumps out water from a sump when it reaches a certain level. Used in basements.

sun trap Paved area next to vertical walls and with a southern exposure designed to trap sunlight for warmth. (Figure 971.)

superimposed Laid on top of.

superstructure The main structure that rests on the *substructure*.

supply air Air from the heating or cooling unit. See *return air*.

surface To finish a surface; to smooth wood, fresh concrete, or plaster. In general, the exposed face or exposed outside part.

surfacing The dressing or smoothing of the rough lumber faces.

surfaced lumber Lumber that has been smoothed by being run through a planer. Specified by the sides that have been surfaced—for example, S1S (surfaced one side) or S2S2E (surfaced two sides, two edges).

surfacing weld Weld made on a plain, unbroken surface. Several weld beads are laid down to build up the surface. Also called a *bead weld*. See *padding*.

survey To measure out a land area or building lot. Directions, angles, and distances are given. See *aerial survey, cadastral survey, construction survey, geodetic survey, plane survey, stadia survey, topographical survey.*

survey stakes Wooden stakes used to lay out construction projects and boundaries.

surveyor One who locates land boundaries and lays out a survey plot of the building site.

surveyor's arrow *Taping pin* or *chaining pin.*

surveyor's compass Magnetic compass marked in degrees that is used for taking readings. Compass is mounted on a tripod and leveled. Bearings are taken through the sighting arms. (Figure 972.) See also *needle.*

surveyor's rod See *leveling rod.*

suspended ceiling A ceiling hung on wires from structural members higher up. (Figure 973.) A *drop ceiling, false ceiling.*

swag A flaring out, as the teeth on a circular saw blade may be flared out slightly at the tip to give a wider cut and prevent binding of the blade.

swage A shaping fitting or tool used to smooth off or draw down a round bar to the required diameter, or to form metal to the shape of a groove. See also *anvil, sheet metal stakes.*

Figure 972 Surveyor's compass

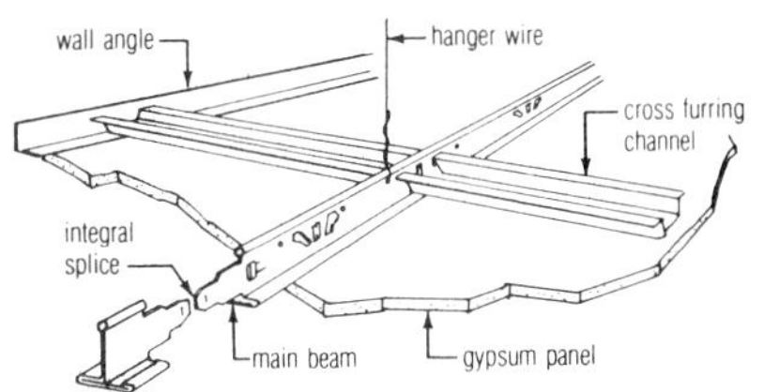

Figure 973 Suspended ceiling

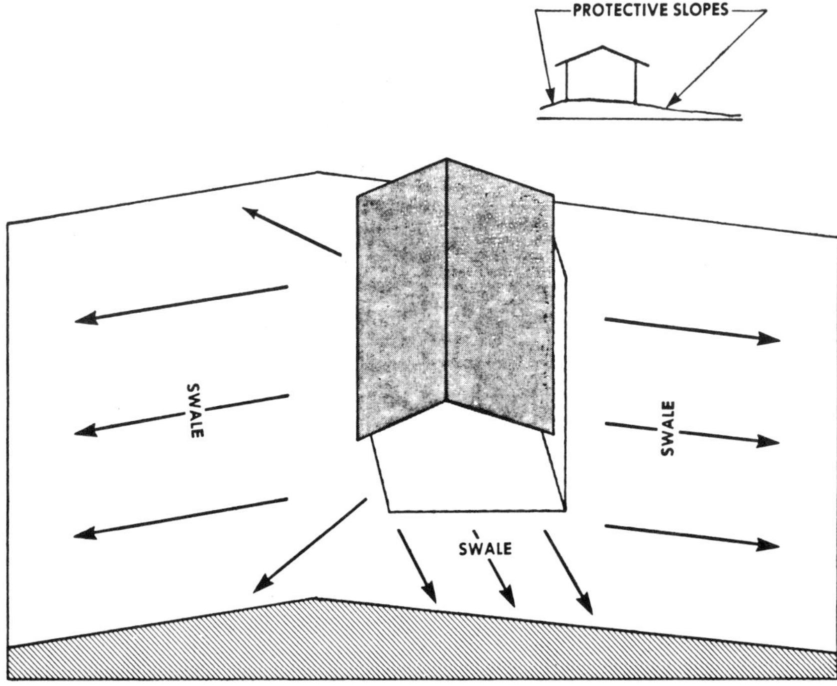

Figure 974 Swale

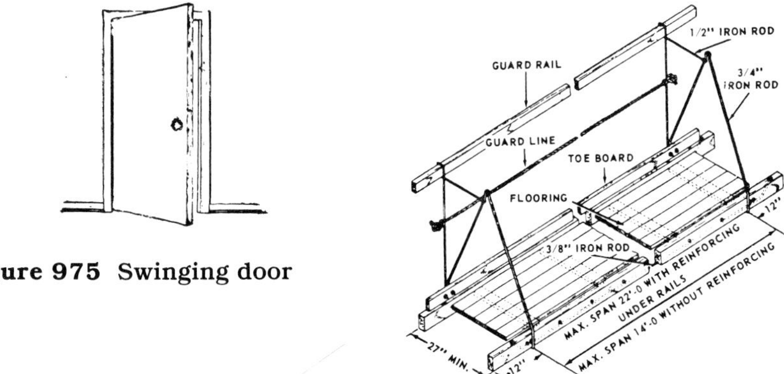

Figure 975 Swinging door

Figure 976 Swing-stage scaffold

swale Drainage flow on a building lot; low land deliberately shaped to channel off rain water. Swale is indicated on the plot plan with heavy arrows. (Figure 974.)

swarf Metal bits worn off when sharpening an edge on oilstones or waterstones; a *glaze*.

sway rod Diagonal brace used to resist the force of wind. (*Figure 642*, p. 276.)

sweating Moisture formation on a surface. Joining metal parts by *sweat soldering*.

sweat joint Copper tubing joint made by solder. The solder is not heated directly. The solder is placed on a tube end which is butted against another tube end; heat is applied to the copper tube, which in turn heats the solder, allowing a joint to be made. See *sweat soldering*.

sweatout Plaster that stays wet and does not set.

sweat soldering Joining of metal parts by heating the metal and allowing the solder to melt and flow over the joint.

sweep The curve of the cutting edge on a gouge. Curves are slight (a *flat sweep*), to semicircular (a *medium sweep*), to very sharp or U-shaped (a *quick sweep*). See *gouge*.

swimming- pool trowel Round-ended, flexible steel trowel used for finishing swimming pools. See *trowel*.

swinging door Door that swings in or out on side hinges. (Figure 975.)

swing-stage scaffold Scaffold that is lowered from above on ropes and pulleys. (Figure 976.)

swirl crotch Attractive swirl-patterned veneer cut from the crotch area of a tree. See *moonshine crotch*.

switch Electrical device for turning a circuit on and off. Figure 977 shows a common flip switch.

switch box Outlet box for electrical wall switch. See *outlet box*. (*Figure 647*, p. 278.)

swivel Coupling device that allows a part to rotate freely.

symbol A pictorial view or sign that represents something; for example, a circle is an electrical symbol for a lighting outlet. See *electrical symbols, elevation symbols, heating and ventilating symbols, plan symbols, plot plan symbols, plumbing symbols, reinforcement symbols, rivet symbols, section symbols, steel symbols, welding symbols.*

symmetrical Anything that has two matching halves; something that is artistically balanced.

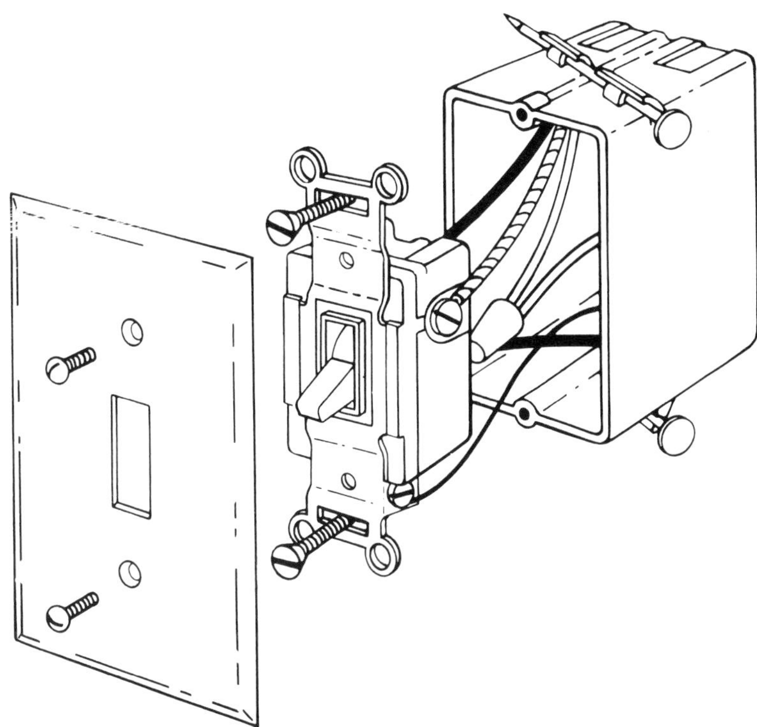

Figure 977 Switch: three-way

t

table saw Power saw used in the shop. The blade is raised or lowered to accommodate different thicknesses of stock, or can be tilted to cut at an angle. (Figure 978.) See also *saw blade*.

tack In welding or soldering, a series of small welds or solder joints made between two pieces to give a temporary joint so that the complete weld or solder joint can be made. (Figure 979.) In adhesives, the formation of an immediate bond upon contact. Also, a small short-shanked fastener. (Figure 980.)

tack cloth In woodworking, a treated cloth used for wiping wood to remove dust before finishing.

tacker See *stapler*.

tack hammer Long-necked hammer used for driving tacks and small brads. (Figure 981.) The face of the driving end is normally magnetic for holding the head of a tack.

tacking See *tack*.

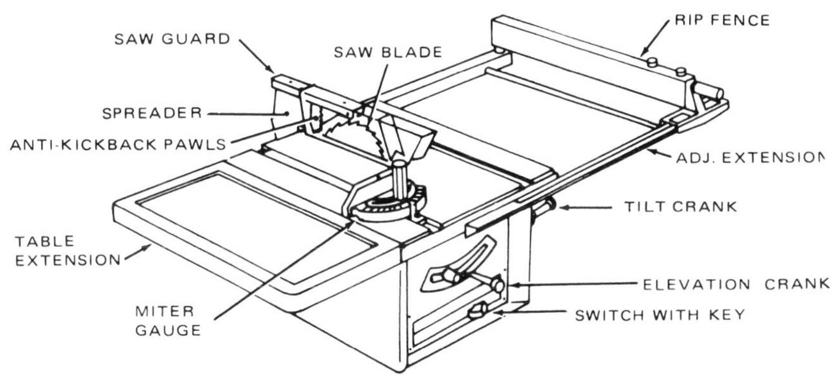

Figure 978 Table saw

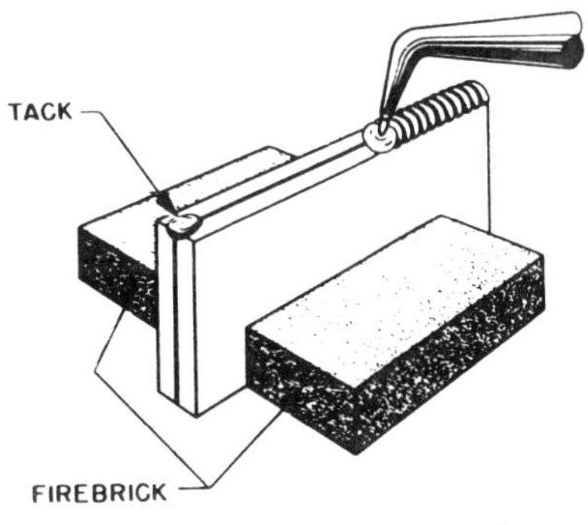

Figure 979 Tack (weld)

Figure 980 Tacks

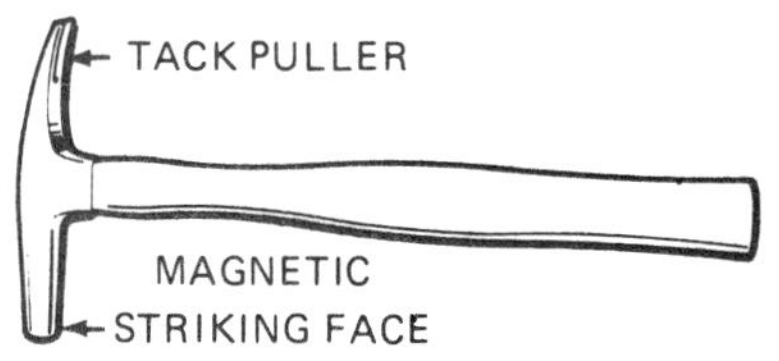

Figure 981 Tack hammer

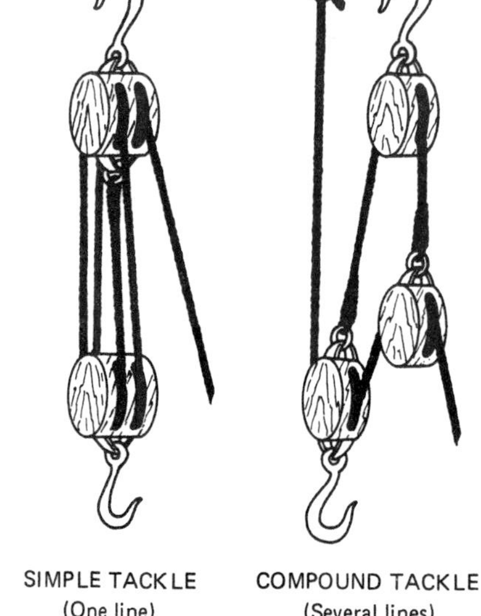

Figure 982 Tackle

tackle Assembly of blocks and lines arranged to obtain a mechanical advantage. (Figure 982.) See *block and tackle.*

tackless fitting In carpet laying, a wood or metal strip with nail points or sharp metal slivers protruding at an angle pointing away from the carpet end. The fitting is attached to the floor edge; then the carpet end is pulled over and fastened on the fitting.

tack weld Weld made on the edge of two pieces to hold them together temporarily until a complete weld is made. To join parts by a series of short welds. See *tack.*

tacky *Sticky.*

tag line A rope attached to a sling or sling-supported member to guide the member to where being moved by a crane.

tail End of a rafter over the eave. (*Figure 763*, p. 335.) In roofing, the top, thin end of a wood shingle; the part that is covered when in place on the roof. In a *dovetail joint*, the fan tail members that fit with the *pin* members.

tail bay End bay in a rigid frame steel building.

tail cut Cut of the exposed end of a rafter over the eave. (*Figure 763*, p. 335.)

tailing A stone or masonry unit that projects out of a wall.

tail joist Short joist that frames into a header and is supported by the wall at the other end. (*Figures 489*, p. 204; *531*; p. 228.) Also called a *cripple joist.*

tail length In rafter construction, the distance from the rafter tail to the heel cut measured along the top of the rafter.

tail piece Any short joist, beam, or rafter that frames into a header and is supported by the wall at the other end. See *tail joist.*

takeoff List of materials needed to build a structure. The size, lengths, and number of pieces are given.

take-off tool Sheet metal hand tool for cutting slots and for beading.

tally stick A *board rule.*

tamp To pound down the earth to create a more sound base or to compact loose earth around a foundation.

tamper Tool for tamping down the earth to create firm base for concrete. Hand tampers, a flat metal plate with a long handle, are still used for small jobs. Air-powered tampers are normally used. See *compactor.* Also, a gas-powered screed is called a tamper.

tandem Two together, one following the other.

tang The shank or shaft of a tool that goes into a handle, as the metal shaft of a screwdriver or chisel that goes into the plastic handle.

tangent Line cutting across the circumference of a circle at a right angle to the radius.

tangential surface Wood cut at a right angle to the wood rays; a lengthwise cut. Wood that is *plainsawn* or *flatsawn.*

tangent sawed See *bastard sawed.*

tap To cut an internal thread on a pipe; the tool for cutting the internal thread. (Figure 983.) In electrical wiring, an electrical connection between contacts in electrical equipment. See also *die, tap wrench.*

tap drill sizes Taps are sized by number of threads per inch and by the size of screw or bolt that will fit the hole. Figure 984 shows a gauge for tap drills and screws or bolts. A 6–32 tap would have 32 threads per inch and would take a number 6 machine screw. The 6–32 would need a hole drilled with a #28 drill and would be tapped with a #36 tap drill.

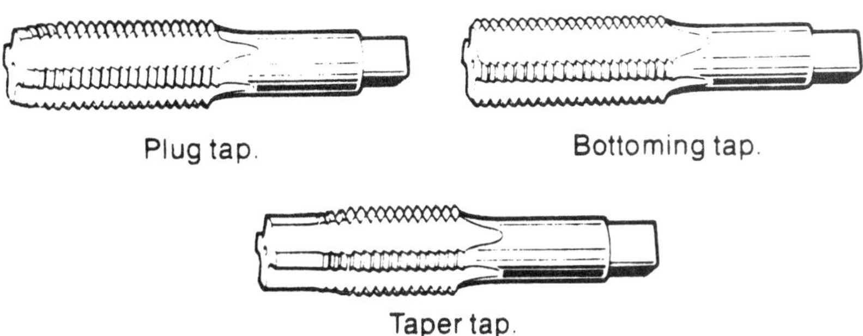

Figure 983 Taps for cutting internal threads. A tap wrench is used to drive the taps

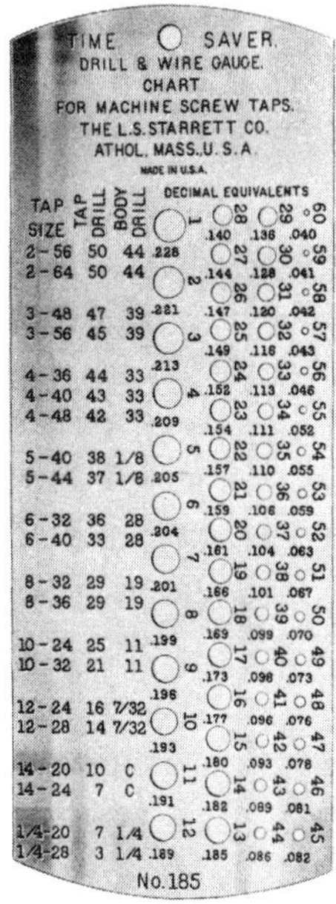

Figure 984 Tap drill sizes

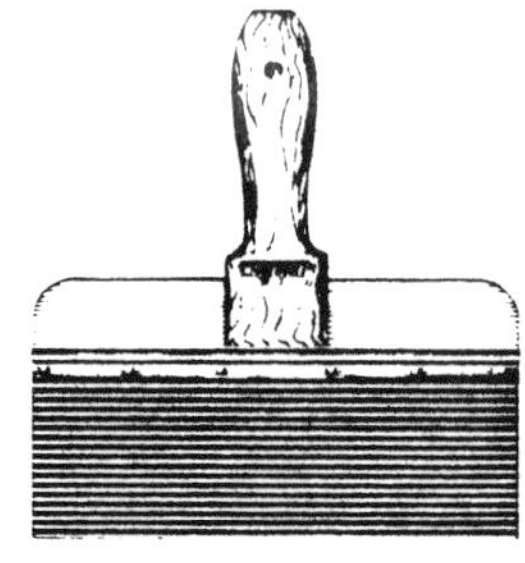

Figure 985 Taping knife

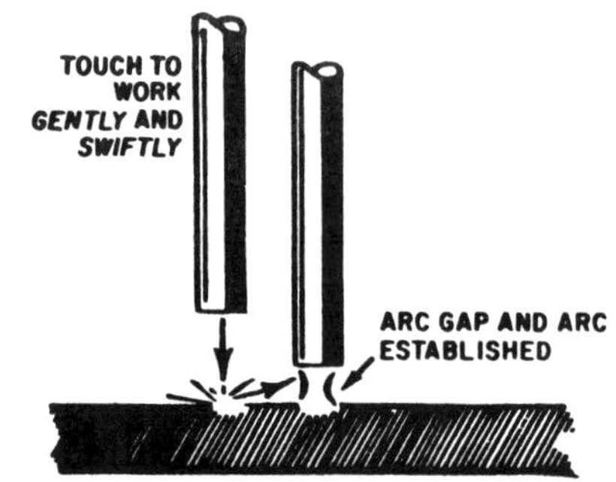

Figure 986 Tapping an arc

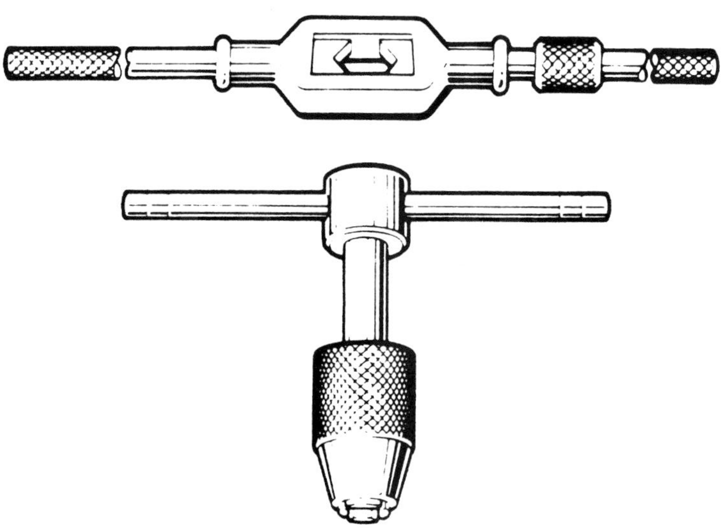

Figure 987 Tap wrench: used to hold and turn a tap

tape Measuring device. See *steel tape*. In drywalling, a flexible strip used to cover drywall sheet joints. Also, electrician's tape used to wrap splices.

tape corner tool In drywall taping, a special tool used to apply tape and joint compound to a corner in one operation. See also *banjo taper*.

tape corrections In surveying, the correction of steel tape readings for expansion and contraction caused by temperature, wind, or other factors; steel tapes give the correct length reading at 68°F.

taper Something that narrows down along the length.

tapered-edge wallboard Gypsum wallboard with tapered or thinned edges. Thin edges allow tape and joint compound to be added without causing a bulge at the joint.

taping In drywall construction, the application of tape and joint compound to the joint framed by the edges of two gypsum wallboards. (*Figure 341*, p. 132.) A thin joint compound coat is first applied to the joint, followed by the tape. A second coat and a finish coat complete the job. In roofing, the application of strip flashing. See *stripping*. In surveying, the measuring of distance with a steel tape. See also *chaining*.

taping and bedding In drywall construction, the application of drywall tape into a joint bedding cement or taping compound at the gypsum wallboard joints.

taping knife Wide flexible blade mounted on a handle used for finishing drywall joints. (Figure 985.) See *automatic taper, banjo taper, drywall knife, drywall trowel, trowel*.

taping pin In surveying, a metal pin used to mark every 100-foot length of a distance being measured. Each 100-foot mark is called a *station*. See also *chaining pin*.

tapping an arc In arc welding, establishing the electric arc between the electrode tip and the work piece by a tapping motion. (Figure 986.) See *striking an arc*.

tapping screw *Self-tapping screw*; drills its own hole.

tap splice See *tee splice*.

tap wrench Wrench used to hold and turn a *tap*. (Figure 987.)

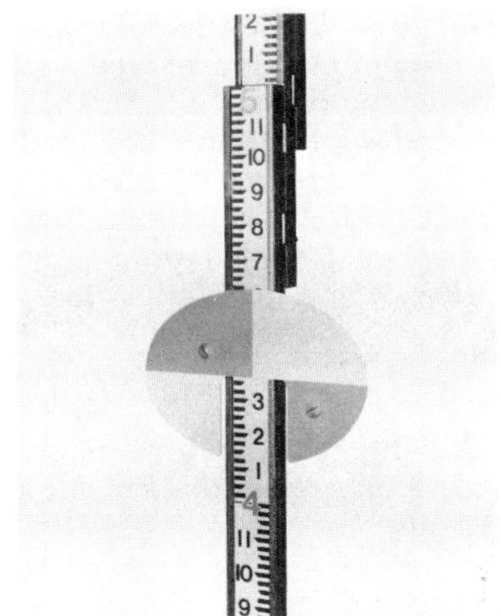

Figure 988 Target and rod

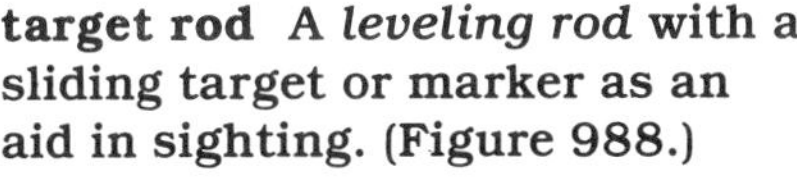

target rod A *leveling rod* with a sliding target or marker as an aid in sighting. (Figure 988.)

T beam Reinforced concrete beam shaped like a T; a steel beam shaped like a T. See *tee.*

T bevel Tool used to check and mark angles. (Figure 989.)

T brace T- shaped brace used to strengthen T intersections in light framing. (Figure 990.) See *L brace.*

technical pen Ink pen with refillable cartridge that holds ink. Widely used for drafting.

tee Reinforced prestressed beam or slab cast in the form of a T; steel beam in the shape of a T. The top horizontal part is called a flange; the vertical leg is called a stem. Figure 991 shows structural steel terminology and specifications. Also called a *T beam.* Pipe fitting shaped like the letter T. See *plumbing fittings.* In general, a T shape.

tee-joint weld Weld made between two pieces that are at right angles to each other in the form of a T. A *fillet weld.*

tee plate A T-shaped metal plate drilled and countersunk for screws; used for fastening and reinforcing in light construction. See *T brace.*

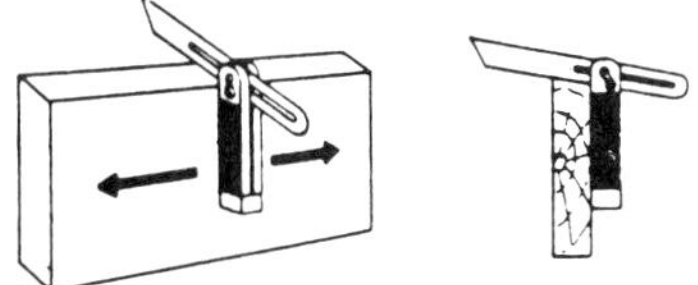

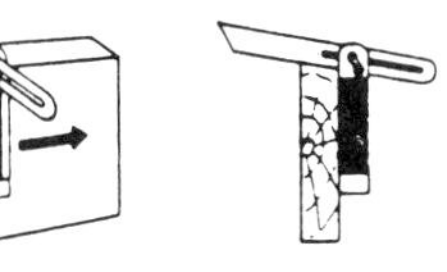

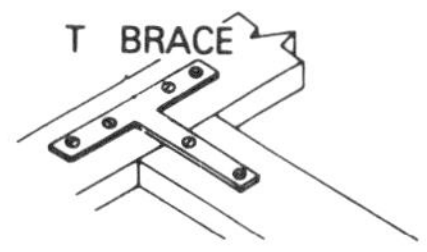

Figure 990 T brace

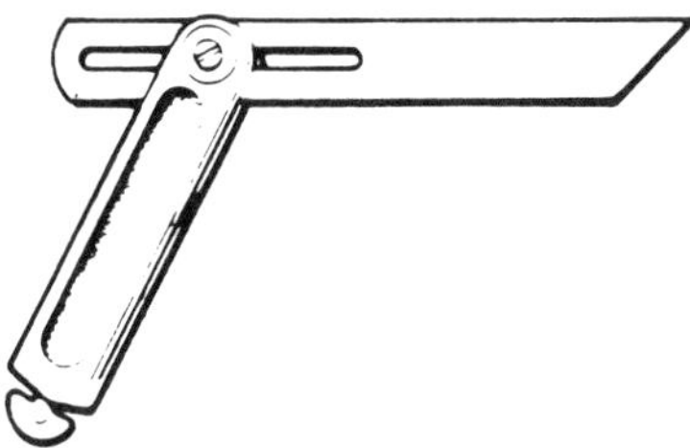

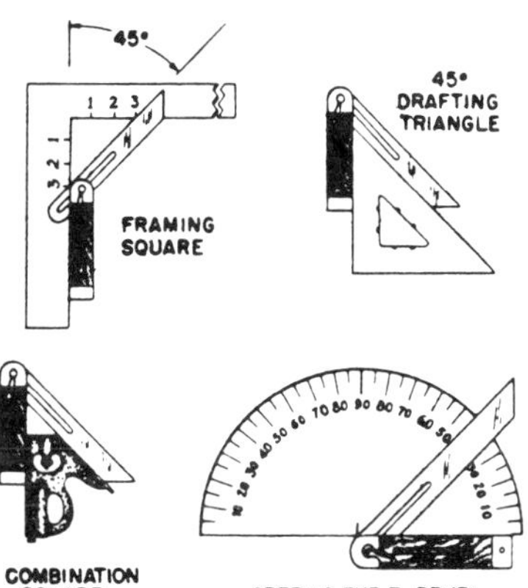

Figure 989 T bevel

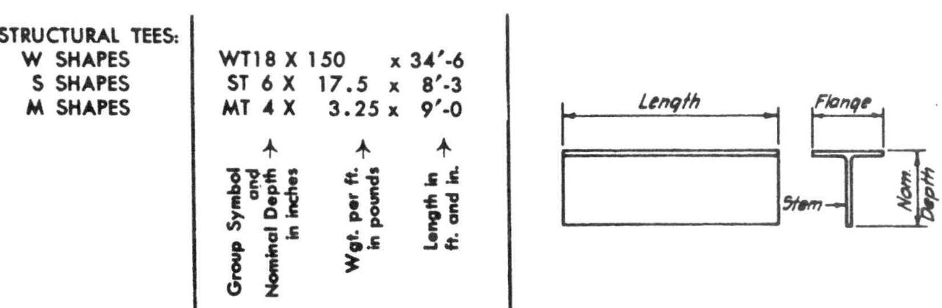

Figure 991 Structural steel tee terminology and specification examples

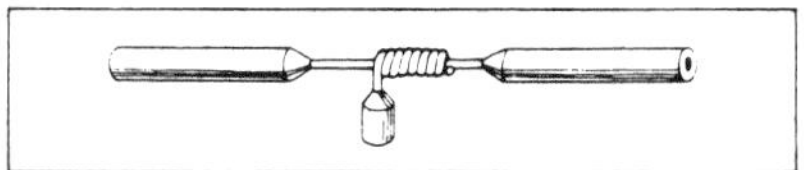

Figure 992 Tee splice

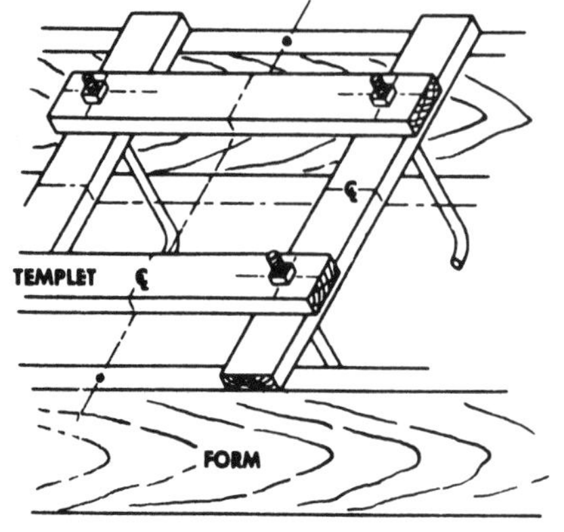

Figure 993 Template used to locate anchor bolts

tee splice Wiring splice made by wrapping one wire end around an exposed section of another wire. (Figure 992.) Solder is used. Splice is taped with an insulating tape. Also called a *tap splice.* See also *pigtail splice, Western Union splice.*

telescoping house jack Jack used as a brace to lift sagging floors. A *jackscrew.*

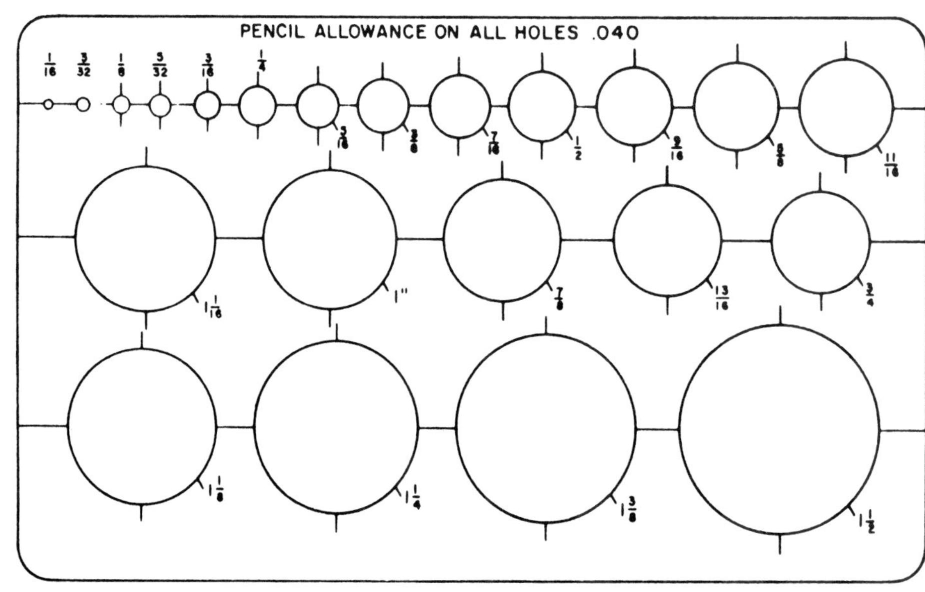

Figure 994 Template (circle)

temper To bring a metal to a specified hardness by heating and rapidly cooling (quenching).

temperature bar In reinforced concrete, special reinforcing bars designed to minimize cracking because of temperature change. They run at a right angle to the reinforcing bars and distribute the temperature stress.

tempering valve In solar heating, a *mixing valve.*

template A guide or pattern. A guide used for making some building part. A guide used to locate anchor bolts in a column footing (bolts attach to the steel column). (Figure 993.) In drafting, a guide for drawing standard parts or symbols. (Figure 994.)

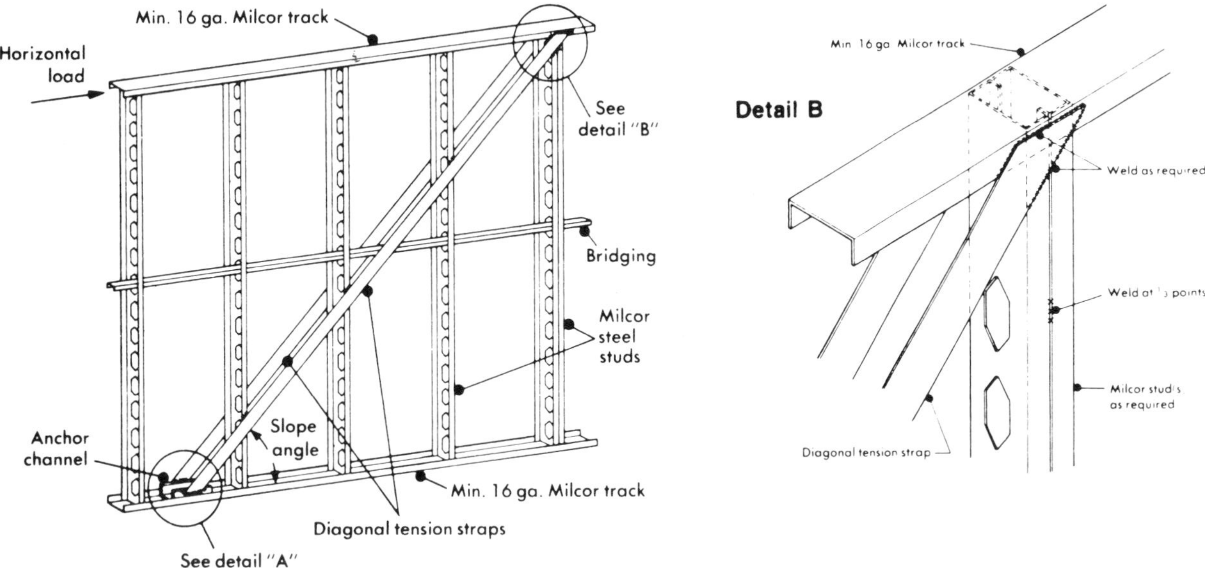

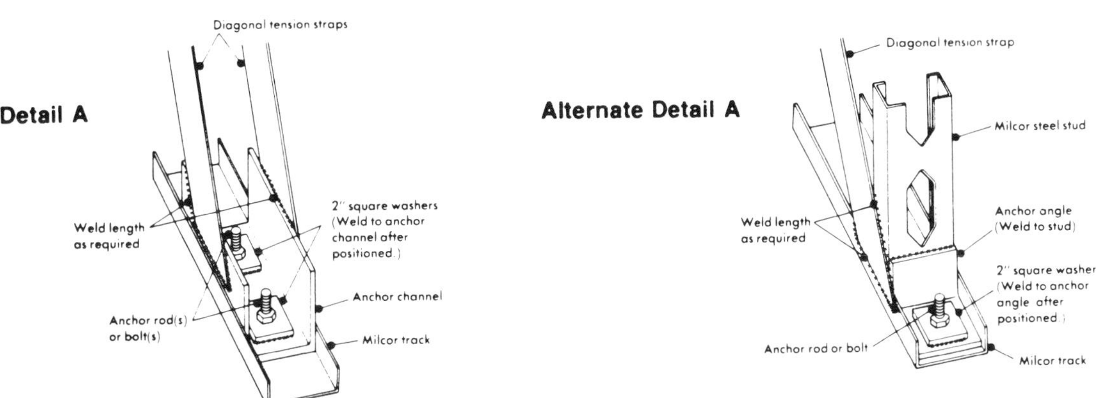

Figure 995 Tension straps

tenon Wood piece cut down to make small joint end or tongue (tenon). The tenon fits into a *mortise* to make a *mortise and tenon* joint. See *wood joints*. Also, steel wire used to create stress in prestressed concrete units.

tenon saw A *backsaw*.

tensile A pulling stress.

tensile stress A pulling or stretching force that building members are subjected to.

tension To subject to a stretching force. See *compression, shear, stress, torsion*.

tension parallel to grain (F_t) Stress calculated in pounds per square inch of tension along the grain. Values are available in published tables for various woods and are used in design calculations. See also *extreme fiber stress in bending*.

tension straps Diagonal cross bracing used in metal framing to strengthen against wind and seismic racking (lateral loading). (Figure 995.)

terminal The end or termination of something. End of vent pipe or soil pipe that projects through the roof.

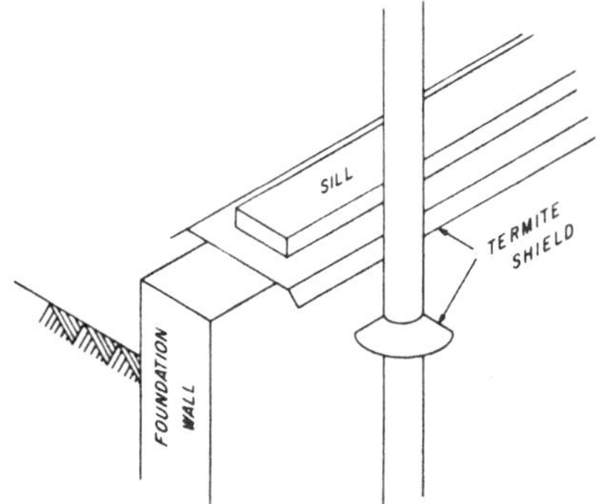

Figure 996 Termite shield

Figure 997 Theodolite

termite An insect that consumes wood.

termite poisoning Chemical treatment of earth under a building to kill termites.

termite shield Metal shield placed over foundation walls, over the pier footings, and around pipes. Shield prevents termites from moving up to reach wood members. (Figure 996.)

terne plate Steel or iron coated with a terne alloy (80% lead and 20% tin), used for roof covering, flashing, guttering.

terrace A paved or tiled area adjacent to a house, used as an outside living or recreational area. A level earth area or platform.

terra cotta Baked-clay building material. Widely used for tile. Glazing is added to the face for decorative or finish tile.

terrazzo Special cement-type flooring. Crushed stone, such as marble, is mixed with cement (often colored) to form a base; the surface is then polished.

textured finish In plastering or concrete work, a rough or decorative finish.

texture float Tool used for texturing plaster. See *float*.

T head Shore that braces into a horizontal support member at the top.

theodolite Surveying instrument for reading horizontal (level) angles and vertical (transit) angles. (Figure 997.) The instrument sets on a tripod and is leveled with four leveling screws. A plumb bob and line locates the instrument over an exact spot. Horizontal and vertical readings are simultaneously projected on a microscope located adjacent to the 30× telescope. The full name is *transit theodolite*. See also *automatic level, transit, transit laser*.

therm Heat equal to 100,000 Btu's.

thermal chimney Closure or duct designed to allow natural flow of warm air.

thermal conductor Anything that allows easy heat transfer, such as a metal.

thermal cutting Process that cuts by melting the metal. See *arc cutting, oxygen cutting*.

thermal detector *Smoke detector* that responds to heat.

thermal insulation Material that resists heat transmission. See *insulation*.

thermal mass In solar heating, capacity of a material to store heat.

thermosiphoning Upward movement by natural convection of a heated fluid, as a solar-heated fluid flows upward from a *solar collector* to a storage tank directly above.

thermostat Electrical automatic control unit for a heating/cooling system.

thimble A metal sleeve or socket, as the socket on a door lock that receives the knob spindle; the metal sleeve that fits into a chimney flue and receives a metal pipe from a heating element; a metal eye used on the end of wire rope. (*Figures 1116*, p. 491; *99*, p. 39.)

T hinge Exposed hinge used for doors in rough framing or for gates. The two parts or leaves of the hinge form a T shape. See *hinges*.

thinner Liquid used to thin paint.

thinwall conduit Metal pipe used for running electrical wiring; *electrical metal tubing* (E.M.T.). It is installed and then the conductors are fished through. (Figure 998.) The interior diameters are the same as for the *rigid metal conduit*. See *conduit*.

thread and coupling (T and C) Plumbing pipe supplied with threads on the pipe and with a threaded coupling screwed onto one end of the pipe.

threads Grooves cut in metal. External threads are cut on rods, metal studs, bolt ends, pipe. Internal threads are cut in holes and inside pipe. Most threads are specified by the *Unified Thread* standards. See also *metric threads*, *pipe threads*.

three-phase current Alternating current with three different cycles or phases displaced 120° apart from each other. Abbreviated 3ϕ. A three-phase four-wire system uses voltages 277 V/480 V and 120 V/208 V. Used with industrial equipment that employs three-phase motors.

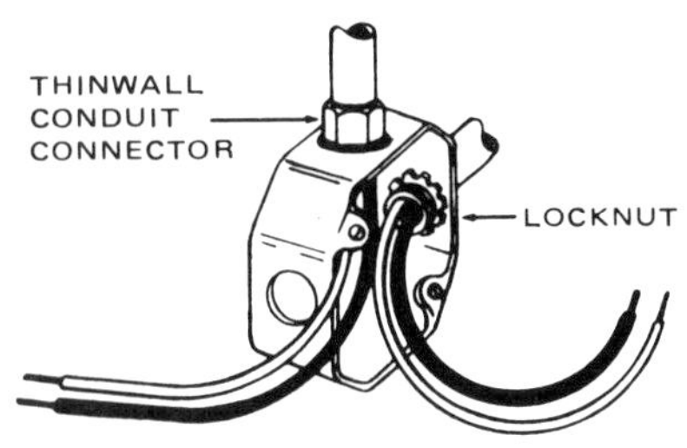

Figure 998 Thinwall conduit

three-point perspective A three-dimensional drawing that has three vanishing points. Also called *oblique perspective*. See *perspective drawing*.

three-prong plug Grounded electrical plug used with 240-v equipment or appliances.

three-way switch Electrical flip switch, of which two are used at separate places to control one light. The light can be switched on or off at either location. See *switch*.

three-wire circuit Wiring used with 240-v circuitry. Three wires run into the building from the electrical power source.

threshold Wood or metal plate that runs across the bottom of a door opening; the *sill*. Figure 999 shows a threshold with a replaceable vinyl gasket that seals against heat loss.

throat Constriction above the fireplace as it runs up to the smoke chamber. See *fireplace*.

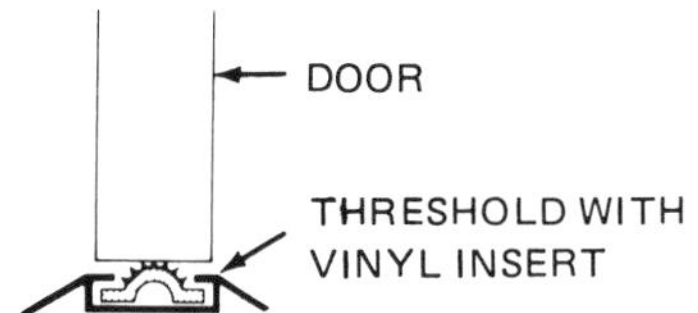

Figure 999 Threshold with replaceable vinyl gasket

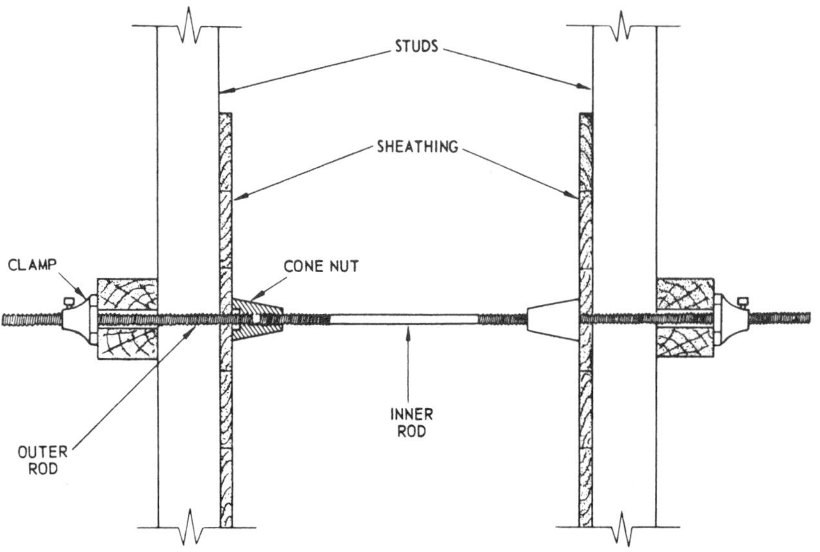

Figure 1000 Tie rod used to hold concrete forms together

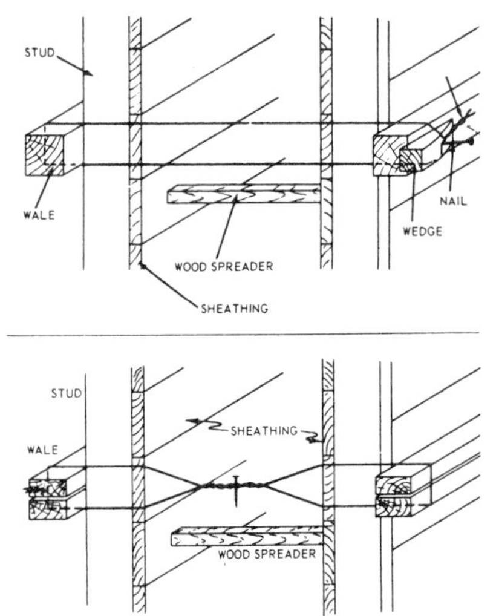

Figure 1001 Tie wire

through stone Stone that goes through a stone wall; a *bond stone.*

throw In forced-air supply, the distance the air travels after exiting the air outlet.

thrust A sideways pushing force. Thrust is both a horizontal and diagonal force, as exerted for example by a rafter on a wall top.

thumb screw Machine screw with a flat or enlarged head that can be held between the thumb and a finger. Generally used as a clamping screw on tools or as an adjusting screw.

tie Something that holds two separate parts to each other, as a metal tie that holds brick veneer to the frame wall. See *anchor, form ties, metal ties, tie rod.* In concrete reinforcement, a horizontal wire support loop made around reinforcing bars for vertical members, such as a column. See *bar tie.*

tie bars In concrete reinforcement, bars that run at right angles to the main support bars. They are wired together with the main support bars to form a mat.

tie beam A *collar beam.*

tie in In roofing, to join shingles together from separate roofing areas.

tier A floor level.

tie rod In concrete forms construction, a steel rod or tie that runs between forms to hold them together after concrete is placed. (Figure 1000.) Tie rods run through the form to the outside of whalers. Their ends are broken off after concrete is set. See also *coil tie, forms, form tie, snaptie, tie wire, wall spacer.*

tie wire In concrete forms, wiring that holds the two sides of the form together so they will not spread when filled with concrete. (Figure 1001.) Any wire that is used to fasten parts together, as wire used to hold reinforcing bars together. See also *form tie, tie, tie rod.*

tig *Gas tungsten-arc welding (GTAW).*

tiger grain In maple wood, a striped type of grain figure.

tig welding *Gas tungsten-arc welding (GTAW).*

tile A hard ceramic or plastic unit used for flooring or covering walls. See *ceramic tile, plastic tile.* Also, a curved roof unit made of fired clay. See also *drain tile, roof tile.*

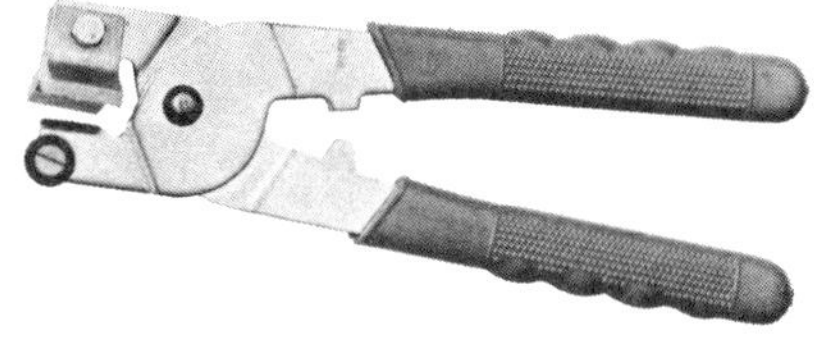

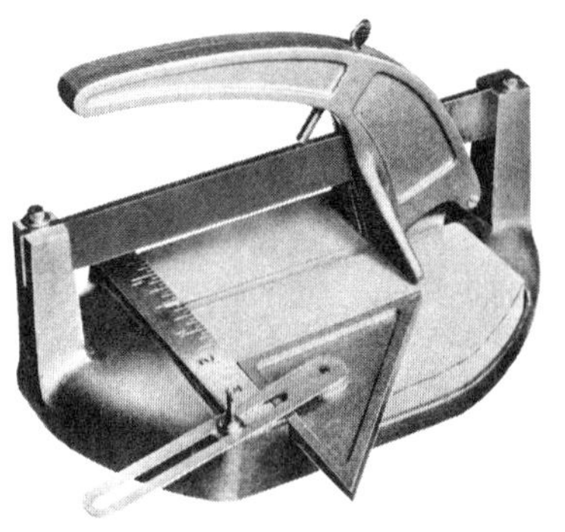

Figure 1002 Tile cutters for ceramic tile

Figure 1003 Tilt-up construction

Figure 1004 Timber carrier

tile cutter Special tool for breaking or cutting ceramic tile (Figure 1002.)

tilt-up construction Precast concrete units, normally walls, that are cast beside the structure, then tilted up into vertical position. (Figure 1003.)

tilt-up door *Roll-up doors* used for garages.

timber Heavy lumber pieces 5″ × 5″ or larger. Also, a growing stand of trees.

timber carrier Two-handled tongs used to carry timbers or logs. (Figure 1004.) See *cant hook, peavy.*

timber connectors Metal anchors, ties, straps, and fasteners used in assemblying timber pieces together—they are designed to transmit stress. Figure 1005 shows typical heavy timber connectors. See also *framing anchor, joist anchor, joist hanger, plywood clip, post anchor, post cap, rafter anchor, shear plate, spike grid, split rings.*

DESCRIPTION	SYMBOL	ILLUSTRATED USE	PICTORIAL
SPLIT RING	SR	$2\frac{1}{2}$SR	
TOOTHED RING	TR	2TR	
CLAW PLATE, MALE	CPM	$2\frac{5}{8}$CPM	
CLAW PLATE, FEMALE	CPF	$3\frac{1}{8}$CPF	
SHEAR PLATE	SP	4SP	
BULLDOG, ROUND	BR	$3\frac{3}{4}$BR	
BULLDOG, SQUARE	BS	5BS	

DESCRIPTION	SYMBOL	ILLUSTRATED USE	PICTORIAL
CIRCULAR SPIKE	CS	$3\frac{1}{8}$CS	
CLAMPING PLATE, PLAIN	CPP	5x5CPP	
CLAMPING PLATE, FLANGED	CPFL	5x8CPFL	
SPIKE GRID, FLAT	SGF	$4\frac{1}{8}$x$4\frac{1}{8}$SGF	
SPIKE GRID, SINGLE CURVE	SGSC	$4\frac{1}{8}$x$4\frac{1}{8}$SGSC	
SPIKE GRID, DOUBLE CURVE	SGDC	$4\frac{1}{8}$x$4\frac{1}{8}$SGDC	
WOOD SPLICE PLATES			

Figure 1005 Timber connectors

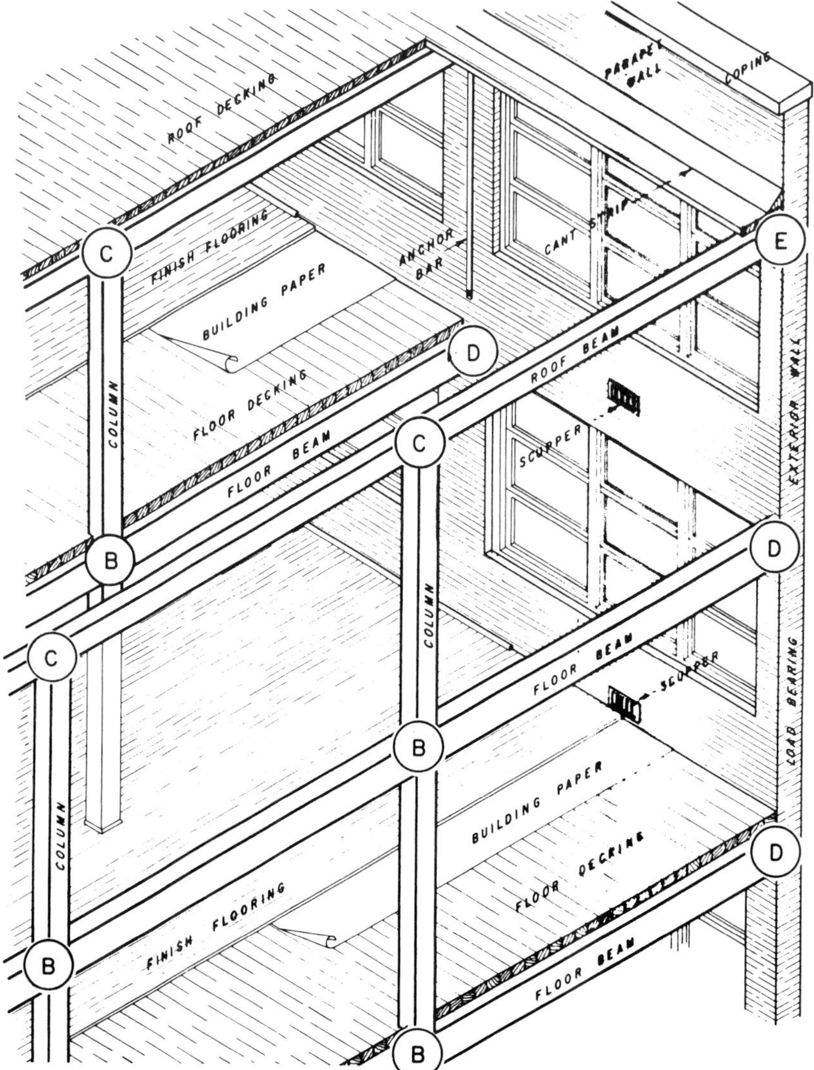

Figure 1006 Timber construction with exterior brick walls

timber construction Framing with timbers. Sometimes the walls may be brick or concrete block but the internal frame is timber. Figures 1006 and 1007 show typical timber construction and framing details.

tin snips Hand shears for cutting sheet metal. See *snips*.

title block On a blueprint, the blocked-off area where the drawing is identified, usually the lower righthand corner or right side.

title sheet The cover sheet in a set of blueprints. It identifies the structure and sometimes gives a table of contents for the sheets.

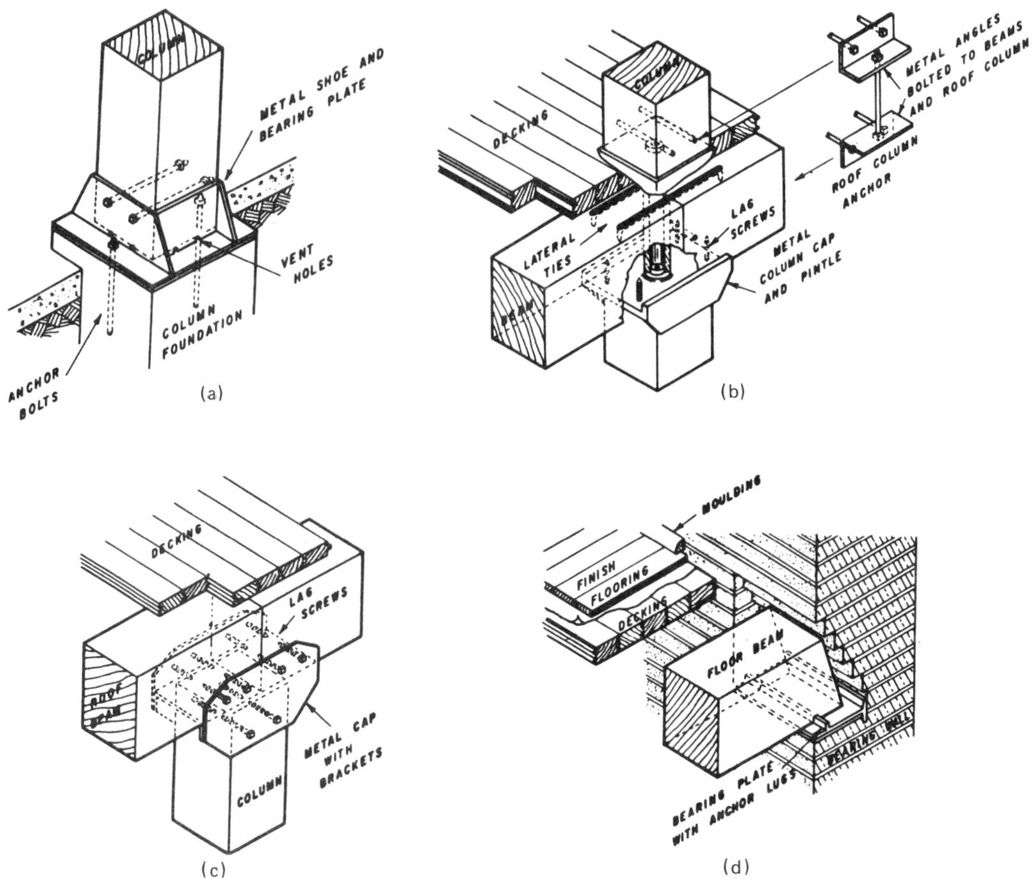

Figure 1007 Timber construction framing details (refer back to Figure 1006)

T joint A plumbing tee. See *plumbing fitting, pipe fittings.* (*Figure 691*, p. 297.) Two lumber pieces joined so as to form a "T" shape.

T nail Nail with a T-shaped head, used in nailing guns. T nails come in strips that are inserted in the electric or pneumatic nailer. (Figure 1008.)

toe Underside of rafter where it meets the top plate (but does not bear on the plate) at the bird's mouth. See *bird's mouth.* (*Figure 763*, p. 335.)

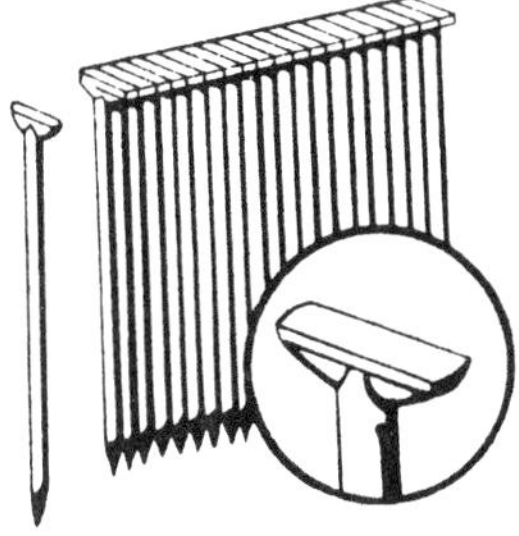

Figure 1008 T nails

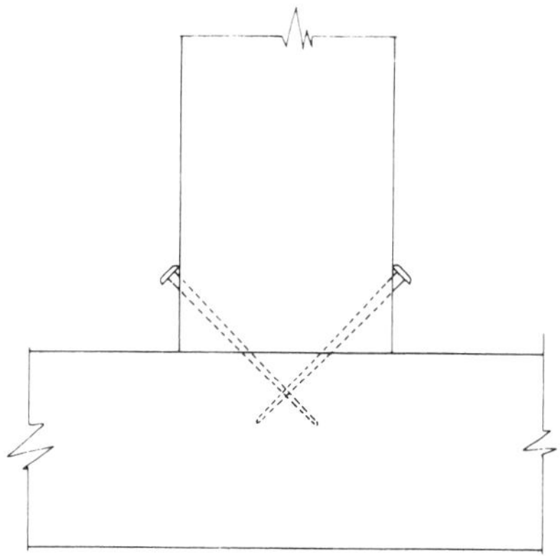

Figure 1009 Toenailing

Figure 1010 Toggle bolt

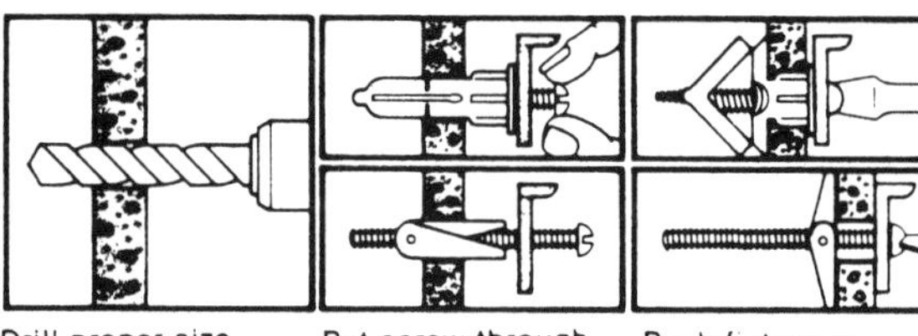

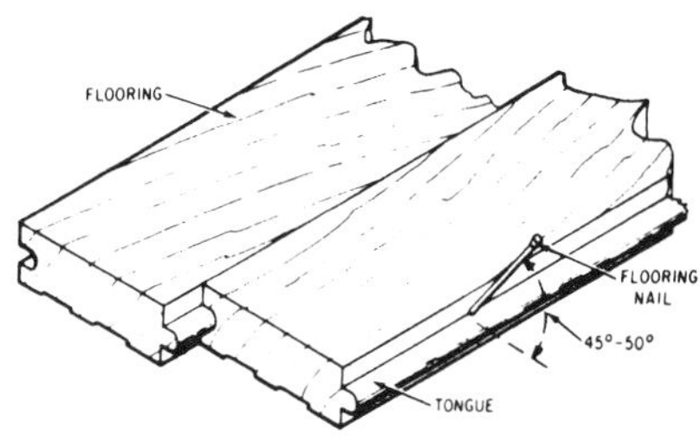

Figure 1011 Tongue and groove

toe board Board at the base on the outside edge of scaffolding; designed to stop tools from falling off. See *scaffold.*

toeing Driving nails in at an angle; *toenailing.*

toenailing Driving nails in at an angle; *toeing.* (Figure 1009.)

toggle bolt Holding device with spring-out wings, used in panel or drywall construction. Device is inserted through bored hole, then the wings spring open on the other side of the panel or wallboard. Bolt is then tightened into the holder. (Figure 1010.) See *hollow-wall anchor, plastic toggle.*

tongs In blacksmithing, a long-handled tool used for holding hot metal while working with the metal.

tongue and groove Joint made by a tongued edge of a board or panel fitting into a groove in another board or panel. (Figure 1011.) See *dressed and matched, siding, wood joints.*

tongue-and-groove pliers See *adjustable-joint pliers.*

ton of refrigeration Removal of heat at the rate of 12,000 Btu's per hour.

tool To round a concrete edge while still plastic; to joint. A device or instrument for doing mechanical work.

tool brush Small brush used in plastering to work water into ornamental or detail work.

tooled joint Masonry joint that has been jointed. The mortar is pushed back into the joint to protect the joint against weather. See *jointing, masonry joint.*

tooling Finishing or jointing a masonry joint; the mortar is pushed back into the joint. Also called *jointing.* See *masonry joints.*

tooth chisel *Stonecutter's chisel.*

toothed rings Timber-fastening device used to strengthen timber joints. See *timber connectors.*

toothing Roughening a wood surface to provide a better key for gluing. Also, roughening a plaster coat to provide a better key for the next coat. See *scarifier, scratcher.*

top chord Top support members of a truss. See *truss.*

top cut Vertical (plumb) cut at the upper end of a rafter.

top dressing In site development, the spreading of a thin soil coat over an area. See also *filling.*

topographical map Contour map showing location of all natural and manmade features on an area of land.

topographical survey Land mapping; survey of land to determine contours and location of natural land areas. See also *aerial survey, construction survey, geodetic survey, plane survey.*

topping Finish concrete or asphalt coat put down on a floor, as a 2-inch concrete topping on a prestressed concrete roof deck.

top plates Two 2 × 4's running horizontally over a stud wall or partition to tie the studs together. Top plates below rafters are sometimes referred to as *rafter plates.* Top plates below joists are sometimes referred to as *joist plates.* (*Figure 967*, p. 427.)

top rail Top horizontal member of a window sash or panel door. See *door, window sash.*

torch A *welding torch.*

torch brazing Welding process where metal is melted and allowed to flow into the joint by capillary action. See *brazing.*

torpedo level See pocket level.

torque A twisting force, as the force of a nut being tightened on a bolt.

torque wrench Wrench having a measuring device on it that indicates the number of pounds of pressure exerted. Used in tightening bolts to specification. (Figure 1012.)

torsion A twisting force. See *compression, shear, tension.* See also *stress.*

torus A semicircular molding.

to weather Distance a shingle is exposed to the elements.

tower crane Tall mobile or stationary-erected crane used at the building site to hoist building parts. Crane may work from the ground or from the floor of the building under construction. (Figure 1013.) See *climbing crane.*

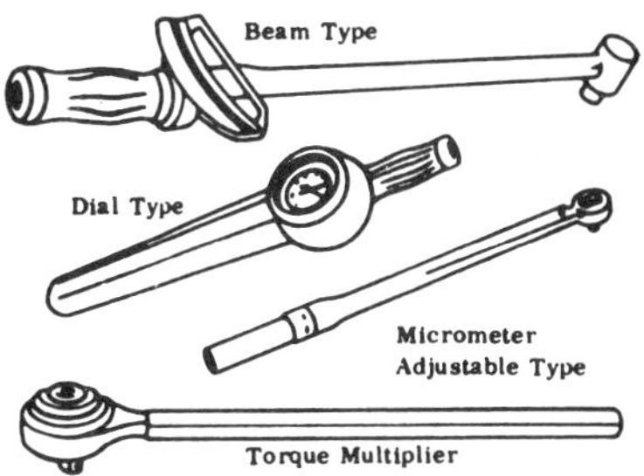

Figure 1012 Torque wrench. Various sockets are used

Figure 1013 Tower crane

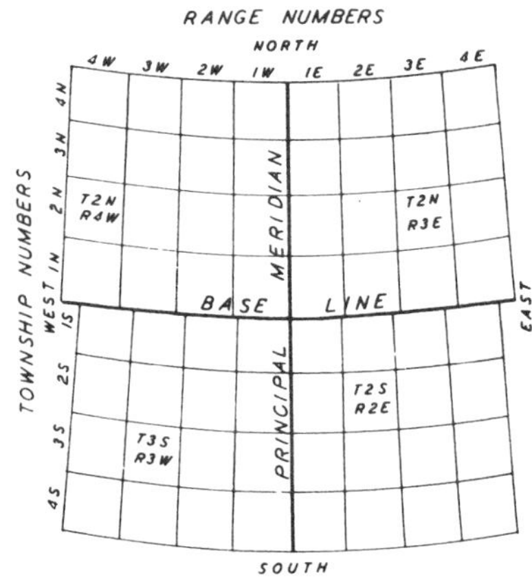

Figure 1014 Township numbers are east and west of the principal meridian. Range numbers are north and south of the base line

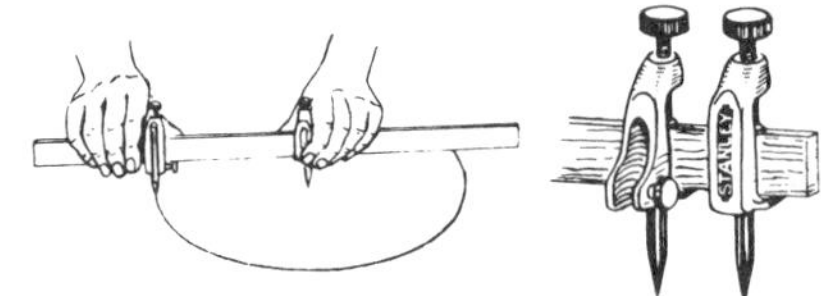

Figure 1015 Trammel

townhouse Multi-unit construction where dwelling units share common walls. See also *Western row house*.

township Land laid out in a square six miles on each side, an area of 36 square miles; contains 36 *sections*, each 640 acres.

township line East-west lines running every six miles north or south of a local *base line*.

township number Numbering system used to locate a township. A township number is the number of rows or lines a township is located north or south of a local *base line* (east-west line). The township number along with a range number is used to locate the township. See *range number*. Figure 1014 shows how townships are located north or south from a base line and east or west from a principal meridian (range numbers). For example, township T2N/R4W is located two up going north from the township base line, and four over going west from the range principal meridian.

T plate A flat T-shaped metal piece used for strengthening T joints. Holes are bored and countersunk in the metal so wood screws may be used. See *T brace*. See also *corner brace, flat corner iron, L brace, mending plate*.

track Metal base used to secure metal studding.

tract Parcel of land.

tract development A planned subdivision. All housing is planned, developed, and built following an overall plan.

tractor See *bulldozer*.

trade An organized craft, such as electrical workers or plumbers; a construction speciality.

trade associations Groups of manufacturers or contractors who join together to promote safety and high standards. Different construction areas have different trade associations—for example, the Portland Cement Association or the American Institute for Steel Construction.

trammel Device for scribing large circles. The trammel points slide on a bar. (Figure 1015.)

trammel points See *trammel*.

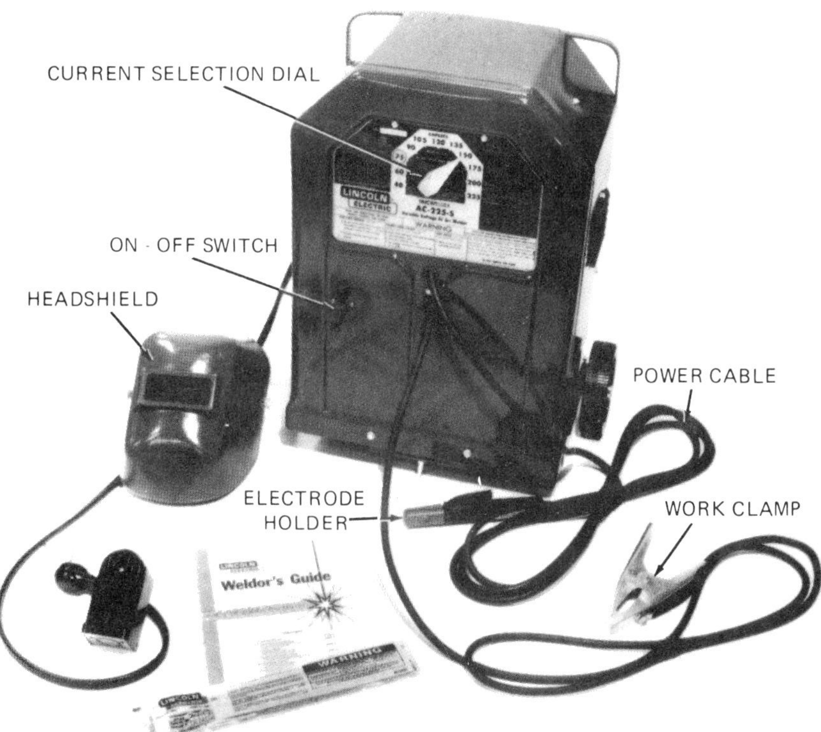

Figure 1016 Transformer: arc welding

transformer Electrical device used to decrease or increase voltage. A transformer is used to decrease voltage from power lines so it can be used in the home. A transformer that decreases voltage is called a step-down transformer. (One that increases voltage is called a step-up transformer.) In arc welding, a power supply that steps the available AC voltage down for welding. (Figure 1016.) The voltage is stepped down to a low voltage and a high amperage, as required. See also *transformer-rectifier*.

transformer-rectifier An AC-DC welding machine that supplies both AC and DC power. The transformer steps down the supplied 220 volts and increases the amperage. The rectifier changes AC to DC. (Figure 1017.)

Figure 1017 Transformer-rectifier (230 amp ac/150 amp dc)

Figure 1018 Transit

Figure 1019 Transit laser

Figure 1020 Transit-mix truck

transit A transit theodolite. Surveying instrument for reading vertical (transit) angles and horizontal (level) angles. (Figure 1018.) The instrument sets on a tripod and is leveled by four leveling screws. A plumb bob and line locates the instrument over an exact point. A vertical 360° circle and vernier is located on the side. A horizontal 360° circle and vernier is located at the base around the compass. The telescope is commonly 20× and can be flipped 180° end-to-end (transited). See also *alidade, automatic level, builder's level, plunging, theodolite, transit laser, vernier.*

transiting In surveying, turning a telescope end-for-end around its horizontal axis. Also called *plunging.*

transition piece Sheet metal fixture that makes a change from one size of ductwork to another size.

transit laser Surveying transit that uses a laser beam for taking readings and for establishing a constant reference line. (Figure 1019.) See also *builder's level, transit.*

transit-mix concrete Concrete mixed in transit to a construction site. The ingredients are measured and placed in a truck for mixing. The mixer revolves to mix the water, cement, and aggregates together. Figure 1020 shows a transit-mix truck. See also *ready-mix truck.*

transit theodolite See *theodolite.*

transom Hinged window over a door. A lintel.

transverse Crosswise; at a right angle to a long piece or member.

transverse section Cross-cut view of a structure made from side to side. See *longitudinal section* (from end to end).

trap In plumbing, a bend in a drain pipe that holds water to prevent sewer gases from entering the building. A *P trap* is most commonly used. See also *S trap*.

trap door A small door opening into attic or cellar areas; an *access door*.

travelers Electrical connections between terminals on separate control switches.

traverse A lateral, side-to-side, movement or direction. In surveying, a line or series of connecting lines surveying across a plot or lot area. The line lengths are measured, as are the angles where the lines meet. See *closed traverse, open traverse*.

traverse point See *station*.

traverse station See *station*.

tread Horizontal flat portion of a stair step; the surface you step on. (*Figure 939*, p. 414.) See *stair*.

tree Hardwood or softwood trees are used to produce lumber. Figure 1021 shows the parts of a tree. Each year annual rings are deposited: light for the springwood and dark for the summerwood. See *wood*.

treenail Well-seasoned wooden peg; when in place, it gains moisture and expands in the hole.

tremie Tube used for carrying concrete under water; concrete flows through the tube directly into the form. (Figure 1022.)

trench Narrow excavation made in the earth.

trench jack Horizontal screw or hydraulic jack used as cross bracing to hold trench shoring. (Figure 1023.)

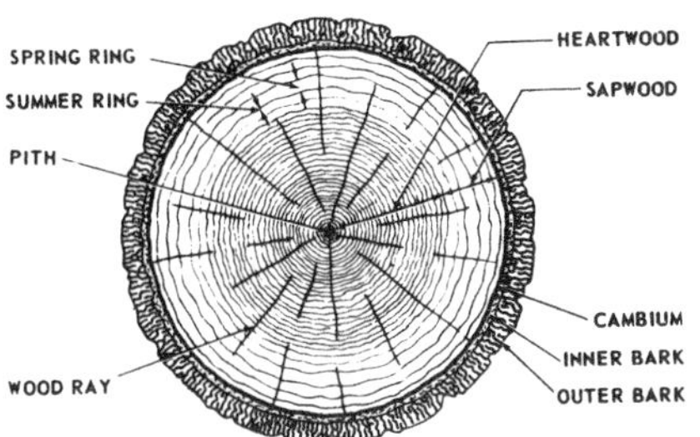

Figure 1021 Tree terminology: shown in cross section

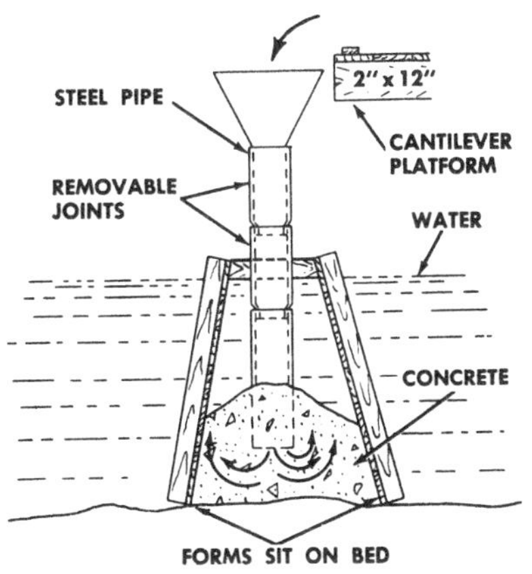

Figure 1022 Tremie

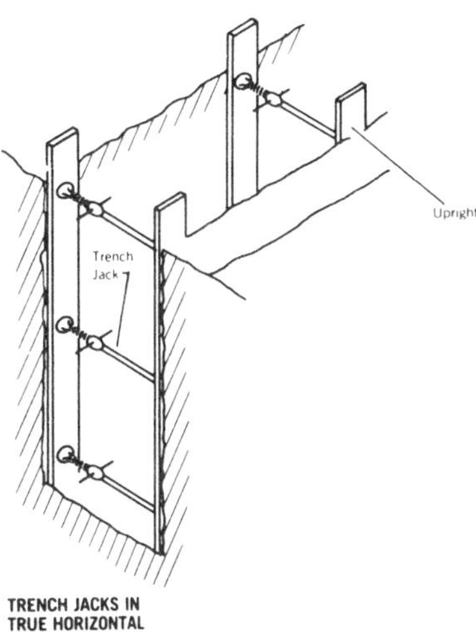

Figure 1023 Trench jacks

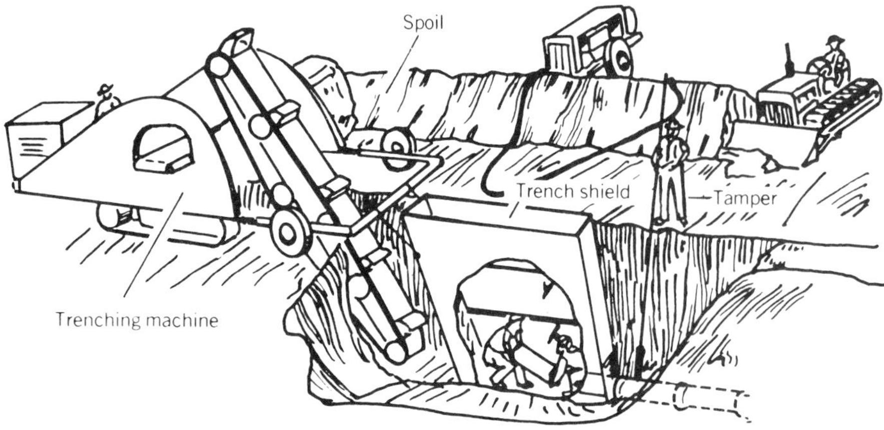

Figure 1024 Trench shield

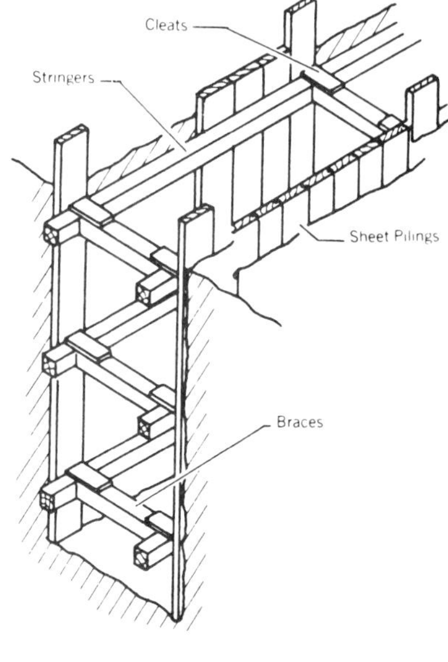

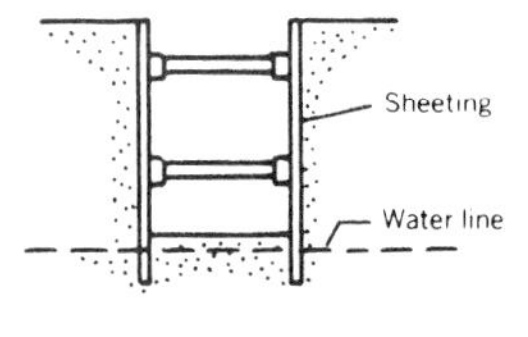

Figure 1025 Trench support

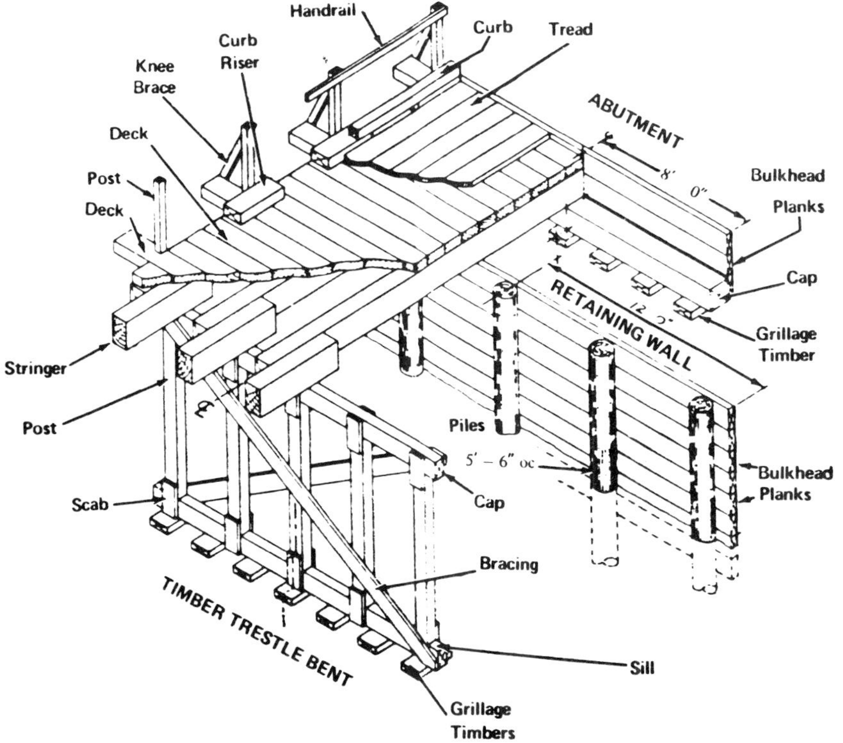

Figure 1026 Trestle bridge

trench shield Prefabricated movable steel box or shield used as protection in trench excavation. (Figure 1024.)

trench support Braced walls (sheeting or sheet pilings) used to hold trench walls in place. (Figure 1025.) See also *trench shield*.

trestle Braced framework of steel, heavy timber, or piles, used to carry a load such as a bridge or roadway. Figure 1026 shows a trestle bridge.

triangle Closed, three-sided plane figure. (Figure 1027.) A right triangle has one corner with a 90° angle. An equilateral triangle has three sides of equal length. An isosceles triangle has two equal sides. A scalene triangle has all sides and all angles unequal. A 45° or 30°–60° triangle used in drafting. (Figure 1028.)

triangular file File that is triangular in section; used for filing metal corners. Also called a three-square file. (Figure 1029.)

triangulation In surveying, checking stations along a traverse by taking sightings that form triangles. Since the sum of the interior angles of any triangle is 180°, errors can be detected and adjustments made. By using a *base line* (line of known length) and two angles of the triangle, the length of the other two sides can be computed.

trim Interior finish; includes wood casings, moldings, metal hardware, and so on. In general, any wood casings, interior or exterior. (*Figure 612*, p. 263.) See also *base trim*.

trimmer Studs, joists, beams, or rafters that frame into the ends of a double header at an opening. (*Figures 489*, p. 204; *531*, p. 228; *962*, p. 425; *966*, p. 427.)

triple glazing Window sash with three panes of glass with air spaces sealed in. Used for the insulative value. (Figure 1030.) See *double glazing*.

tripod Three-legged stand, as for a theodolite or level. (Figure 1031.)

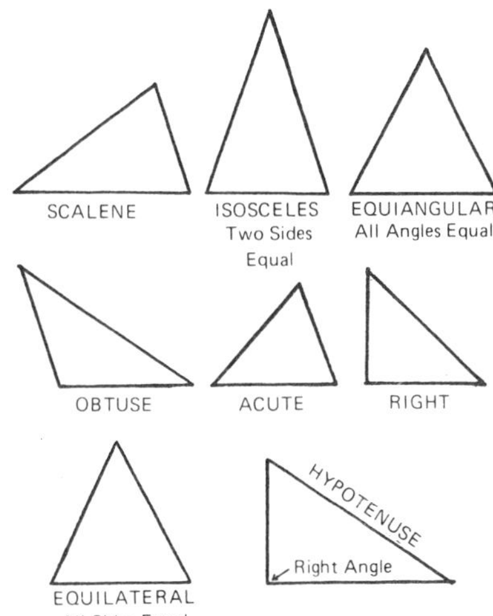

Figure 1027 Triangles

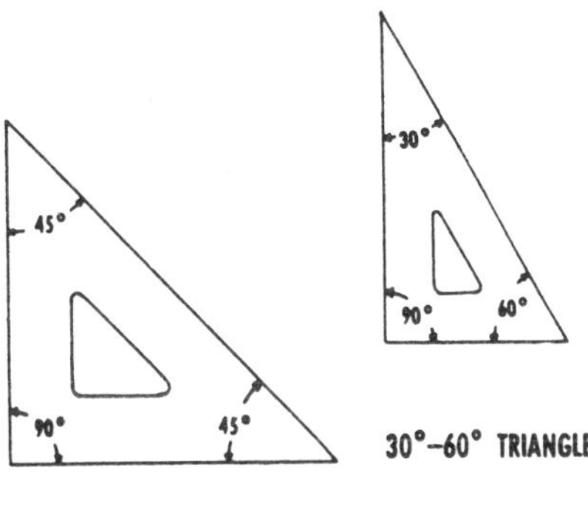

Figure 1028 Triangles used by drafters

Figure 1029 Triangular file

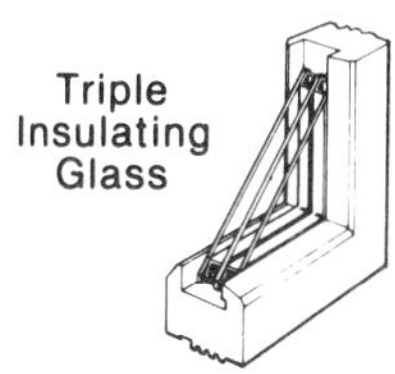

Figure 1030 Triple glazing

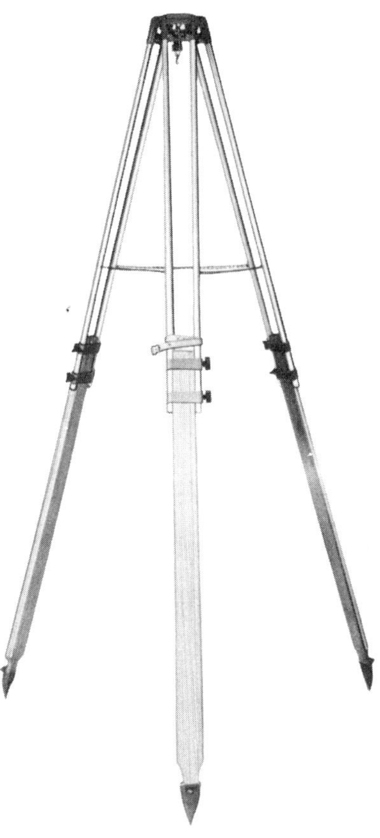

Figure 1031 Tripod for surveying instrument

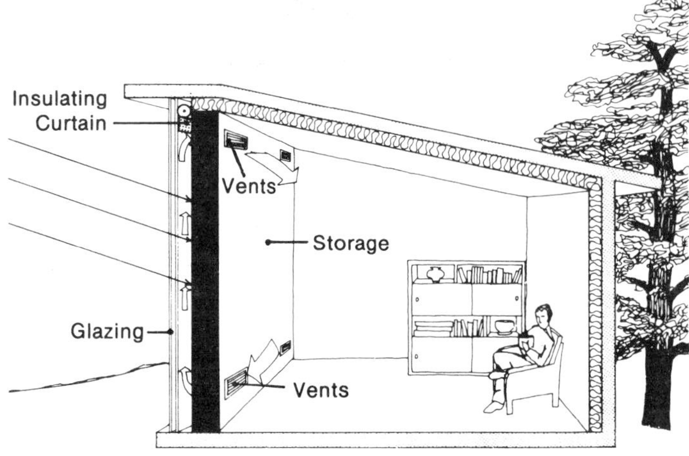

Figure 1032 Trombe wall

Trombe wall In passive solar heating, a masonry wall designed to collect direct sunlight. (Figure 1032.) Originally developed by Felix Trombe. See *solar heating*.

trowel A versatile, generally flat-sided tool for applying, working, and finishing cement, mortar, plaster, and drywall compound. Figure 1033 shows some of the basic trowels used. Variations from these tools are available. *Cement mason's trowels* are used in finishing flatwork. Specialty trowels are available, as the flexible swimming-pool trowel or the pipe trowel, which can work behind plumbing pipes. Edgers finish the edge of the flatwork; groovers make a groove for a control joint. Angled trowels are used to finish concrete steps and gutters. Floats are used for finish. See also *bull float, concrete placer, concrete rake, darby, float, kumalong, power trowel, screed. Brick mason trowels* are used for placing mortar and finishing joints. Joints are finished with joint fillers on jointers. See *jointers, pointing, tuckpointing. Plasterer's trowels* are used for placing and finishing plaster. See also *angle plane, darby, featheredge, ornamental tools, plasterer's shingle, slicker*. Drywall trowels are used to apply and finish drywall compound over taped joints. *Drywall knives* are used to spread the joint compound; the flexible trowel is slightly curved to feather the joint compound. See also *drywall construction*. See also *bucket trowel, buttering trowel, duckbill trowel, gauging trowel, mastic trowel*.

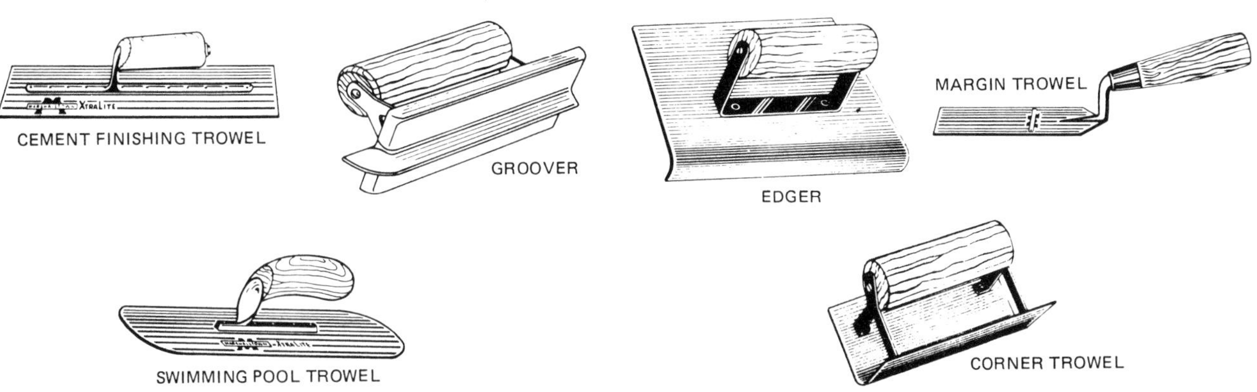

CEMENT MASON'S TROWELS

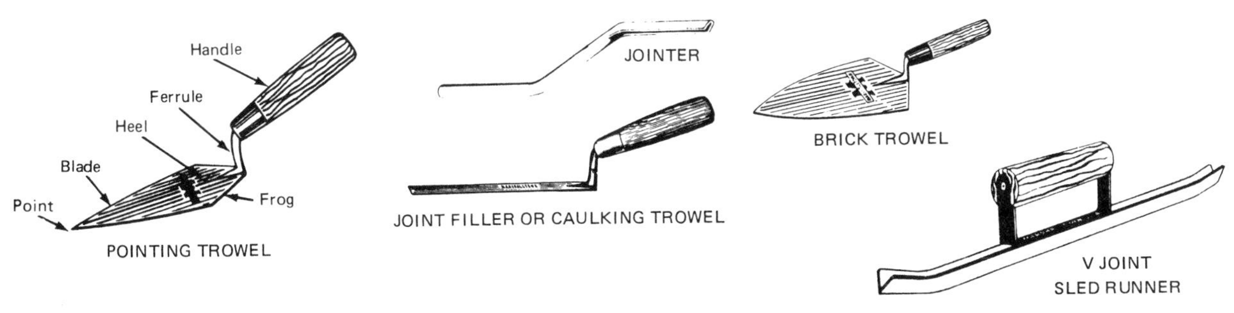

BRICK MASON'S TROWELS

PLASTERING TROWELS

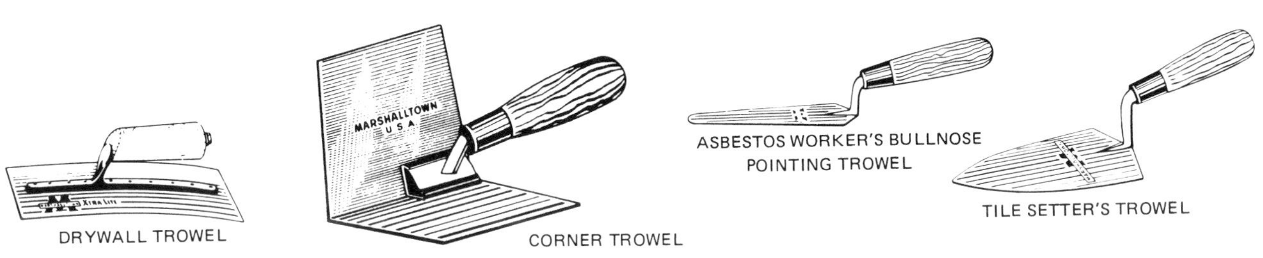

DRYWALL TROWELS

Figure 1033 Common trowels

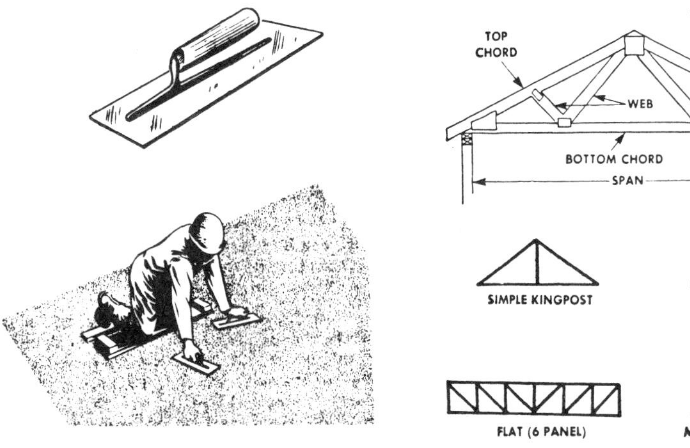

Figure 1034 Troweling

Figure 1036 Trusses (wood)

Figure 1035 Truck crane

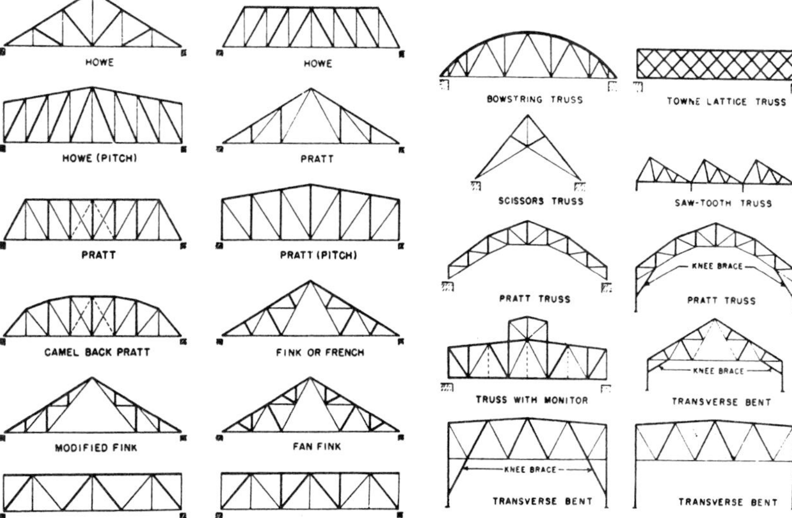

Figure 1037 Trusses (steel)

troweling In concrete flatwork, finishing of the concrete. Troweling is done after concrete has been screeded, surface water has disappeared, and the concrete has started to stiffen. A wood or steel trowel is used. (Figure 1034.) Either hand troweling or power troweling is done. See also *floating*.

truck crane Crane mounted on the bed of a truck. Used at the building site to hoist building parts into place. (Figure 1035.)

truck mixer Truck-mounted concrete mixer.

true bearing Angle between a survey line and true north.

true meridian True north direction, as opposed to magnetic north or *magnetic meridian*. The true meridian is normally shown on a map by a line with a full arrowhead. The angle between true meridian and magnetic meridian is indicated.

truss Manufactured roof support member. Internal support is given by cross braces called *webs*. Figure 1036 shows common wood roof trusses. Trusses can be designed to fit any roof situation. Flat floor and roof trusses are also widely used. Figure 1037 shows trusses used in steel construction. See *floor truss*.

truss bar A reinforcing bar that is bent up at an angle so the top of the bar runs near the top of the concrete member.

truss clip Metal connector used in fastening the parts of a truss together. (Figure 1038.) Clips are also used for joining and strengthening other framing members. See *truss plate*.

truss joist See *open-web joist*.

truss plate Perforated metal plate nailed across truss joints to hold the joint. Also called a *truss clip*.

try square Layout tool used for constructing and testing right angles. (Figure 1039.) See also *miter try square*.

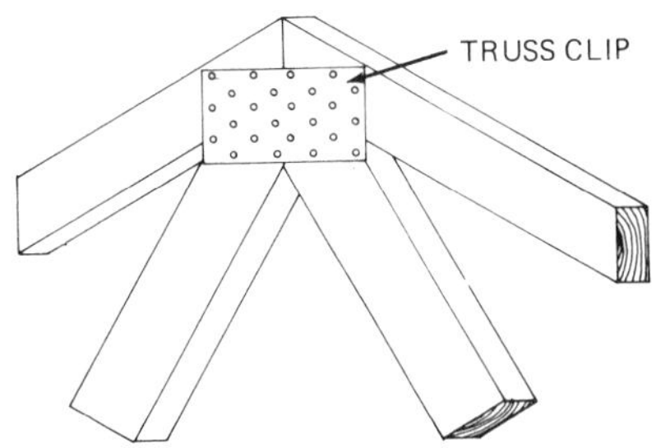

Figure 1038 Truss plate

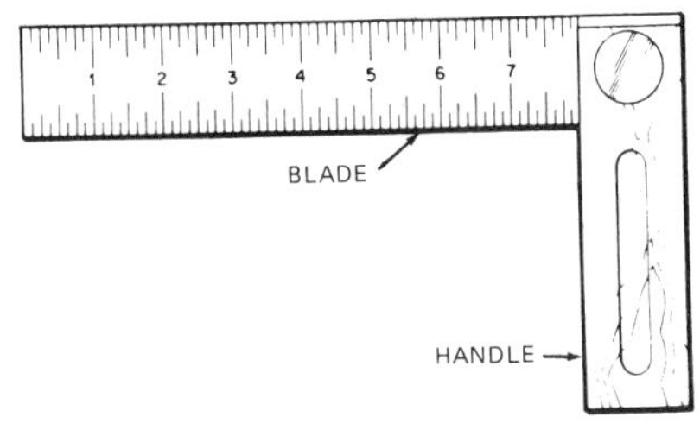

Figure 1039 Try square

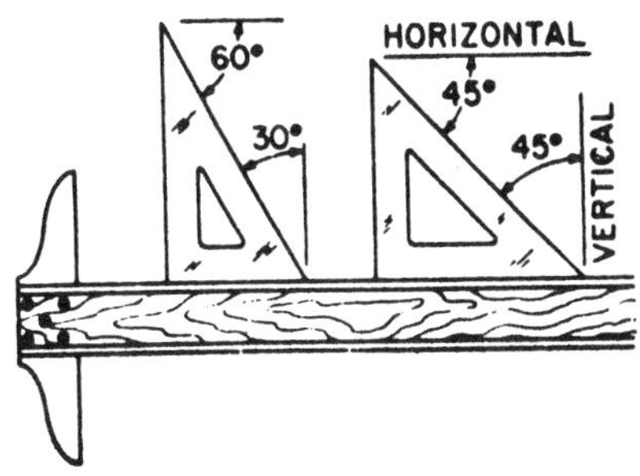

Figure 1040 T square and triangles

Figure 1041 Tube-cutting tool

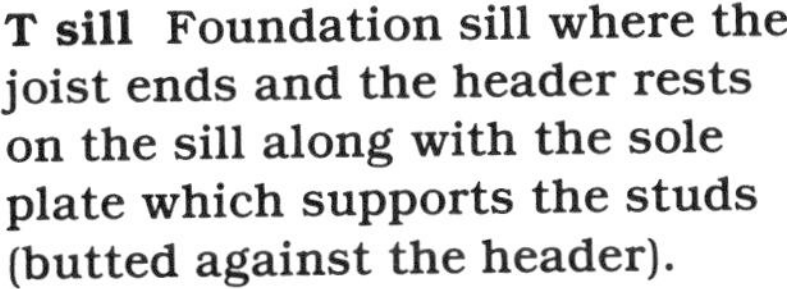

T sill Foundation sill where the joist ends and the header rests on the sill along with the sole plate which supports the studs (butted against the header).

T square Drafting tool used for drawing straight lines. The "T" part (head) slides along the edge of a drafting board. Triangles are used along the tongue or blade to draw angles and lines. Figure 1040 shows a T square with a 30°-60° and a 45° triangle.

tube-cutting tool Tool for cutting *tubing*. (Figure 1041.)

tube steel Structural hollow steel members: square, rectangular, and circular tubes are available. Symbol: TS. See *steel rectangular, steel round, steel square.*

tube system In heavy construction, the use of closely spaced columns around the perimeter of the building. Columns are connected by *spandrel beams.*

tubing A kind of thinwall pipe. Both copper and plastic tubing are used in the water and gas supply system. See also *tube-cutting tool.*

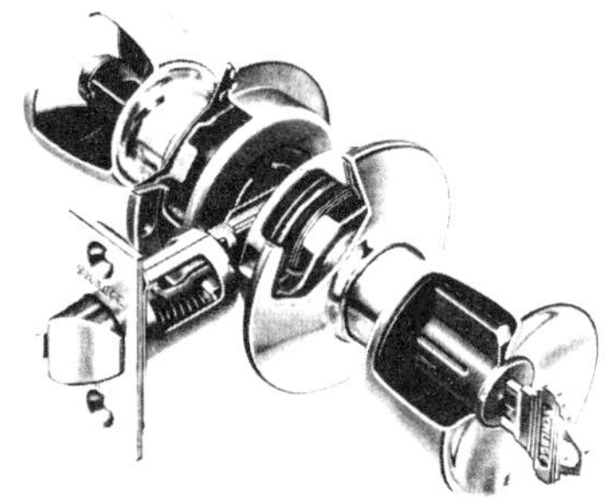

Figure 1042 Tubular lock

Figure 1043 Tudor style

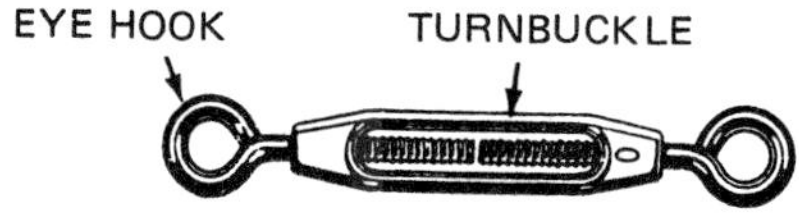

Figure 1044 Turnbuckle

tubular lock A common lock in a tubular case used in interior doors, as for bathrooms or bedrooms. (Figure 1042.) See *cylindrical lock.*

tubular scaffold Prefabricated scaffold made of tube steel; the scaffold is assembled at the site as needed and dismantled when the job is completed. See *scaffolding.*

tuckpointing Finishing of masonry joints. Repair of broken or weathered mortar joints with fresh mortar. Old broken mortar is removed from the joint and fresh mortar is troweled in and then jointed. See *pointing.*

tuckpoint trowel See *joint filler.*

Tudor architecture Stone and brick house, usually two-and-a-half stories. Developed in the late fifteenth century. (Figure 1043.) See also *architectural styles.*

tung oil A drying oil developed from the tung nut, used in varnishes to make wood finishes.

tungsten carbide A very hard synthetic abrasive mineral used on belt and disc sanders. See *abrasives.*

turfing See *sodding.*

turnbuckle Fastener or coupling for joining the ends of threaded rods. The rods can be tightened together by turning the turnbuckle. With eyebolts the turnbuckle is used to tighten wires and lines. (Figures 1044; *479*, p. 197.) See also *clevis.*

turning Shaping of wood piece with woodworking tools while piece is turning in a wood lathe. (Figure 1045.)

turning gouge Woodworking tool for cutting down work that is turning in a lathe. See *woodturning tools.*

turning point In surveying, a point from which two readings are taken on the leveling rods: a *backsight* on the rod at a known point, and a *foresight* forward on a rod at a new, unknown point. Abbreviated T.P. (Figure 1046.)

turn-of-the-nut method Method of tightening high-strength bolts and nuts by turning a certain rotational distance beyond a tight or snug fit.

Tuscan order A classical Roman order. See *orders.*

twentieth- century American architecture Architecture characteristic of the early part of the twentieth century. See *adobe, international.* See also *architectural styles, post World War II American architecture.*

twist Lumber defect; a board or lumber piece warps out of flat so that the whole board is twisted. (Figure 1047.)

twist drill See *drill.*

two-man saw Two-handled saw used by two persons for felling trees, cutting heavy timber. (Figure 1048.)

two-pipe system In hot-water or steam heating, a system that uses one pipe to carry heat to the outlets and another pipe to return cooled water or steam back to the furnace. See also *one-pipe heating system.*

two-point perspective A three-dimensional drawing that has two vanishing points. Also called *angular perspective.* See *perspective drawing.*

two-way flat slab See *flat slab concrete floor.*

Figure 1045 Turning in a wood lath

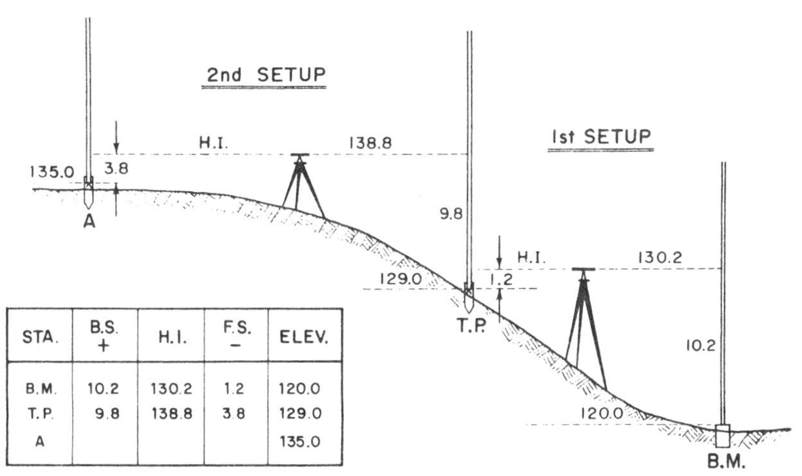

STA.	B.S. +	H.I.	F.S. −	ELEV.
B.M.	10.2	130.2	1.2	120.0
T.P.	9.8	138.8	3.8	129.0
A				135.0

Figure 1046 Turning point and level notes

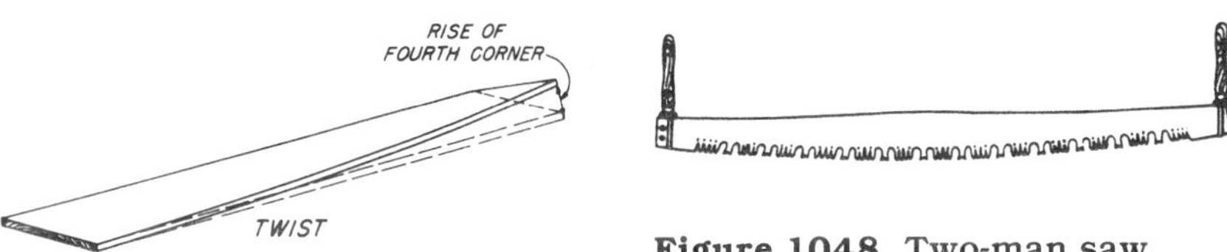

Figure 1047 Twist (lumber defect)

Figure 1048 Two-man saw

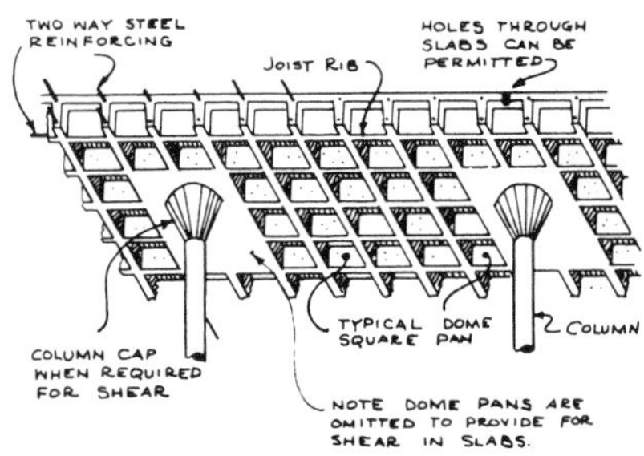

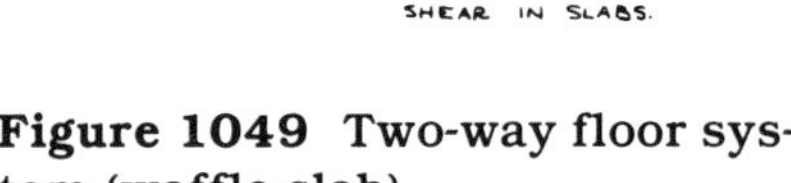

Figure 1049 Two-way floor system (waffle slab)

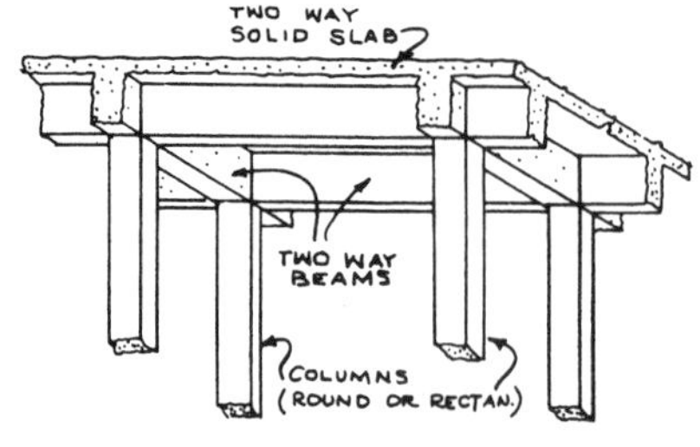

Figure 1050 Two-way solid slab

two-way floor system In reinforced concrete, a floor slab that has support beams running in two directions at right angles to each other. Also called a two-way joist system or a *waffle slab*. (Figure 1049.) See *one-way floor system, reinforced concrete floor*.

two-way joist system See *two-way floor system*.

two-way solid slab Solid reinforced concrete slab floor with beams underneath running two ways. (Figure 1050.) See *one-way solid slab*.

u

U bolt Fastener shaped like a "U" with threads on each end. (*Figure 106*, p. 41; *1116*, p. 491.)

ultimate stress Point of extreme stress on a support member just before it fails.

undercoat Base or primer paint.

undercut To cut back underneath. In arc welding, a cut back under the weld bead.

under-floor plenum Warm-air diffusion system that uses crawl space to carry warm air to perimeter outlets. See *crawl-space plenum*.

underground construction Structures built partially or completely underground. *Berms* (earthen slopes) may be built up around the sides. *Earth-sheltered construction* is partly or wholly underground and is often combined with *solar heating* (through open glass areas).

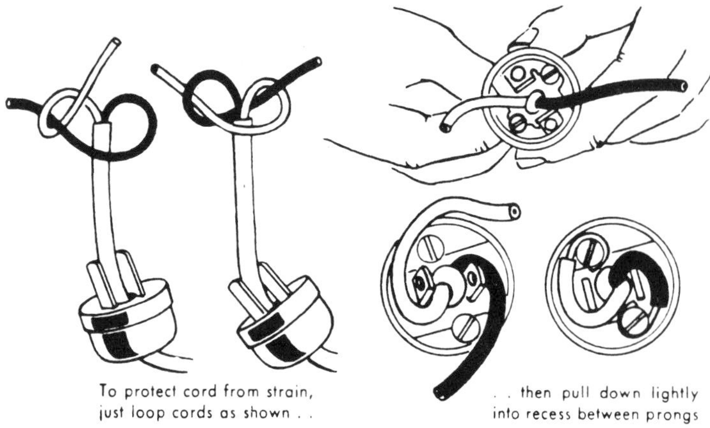

Figure 1051 Underwriter's knot

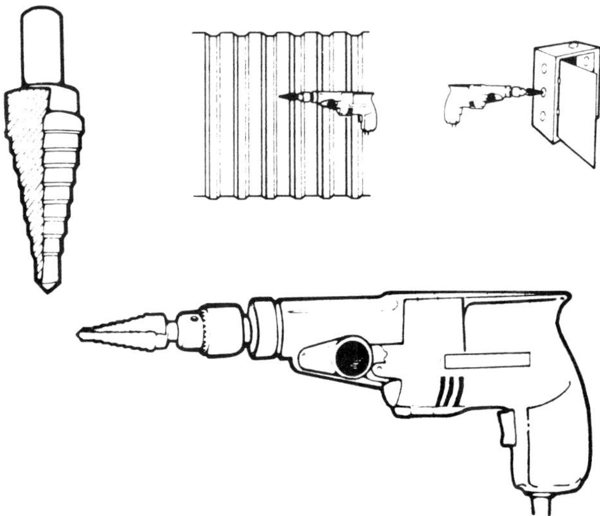

Figure 1052 Unibit

Diameter			Threads Per Inch		
No.	Inch	Decimal equivalent	UNC	UNF	UNEF
0	----	.0600	---	80	---
1	----	.0730	64	72	---
2	----	.0860	56	64	---
3	----	.0990	48	56	---
4	----	.1120	40	48	---
5	----	.1250	40	44	---
6	----	.1380	32	40	---
8	----	.1640	32	36	---
10	----	.1900	24	32	40
12	----	.2160	24	28	---
---	1/4	.2500	20	28	36
---	5/16	.3125	18	24	32
---	3/8	.3750	16	24	32
---	7/16	.4375	14	20	28
---	1/2	.5000	13	20	28
---	9/16	.5625	12	18	24
---	5/8	.6250	11	18	24
---	3/4	.7500	10	16	20
---	7/8	.8750	9	14	20
---	1	1.0000	8	14	20

Figure 1053 Unified threads

underlayment A support surface put down to smooth or level a rough floor so the finish floor may be installed. Also, a base laid on the roof deck so the roof finish may be installed.

Underwriter's knot Knot used to secure the two ends of an electrical cord when wiring into an electrical plug. (Figure 1051.)

Underwriters Laboratories (UL) Organization that tests materials, products, equipment, constructions, methods, and systems to establish safety with respect to hazards affecting life and property. All products that pass UL standards are given a UL seal of approval.

unibit Bit that can cut a hole of varying size, depending on the depth to which the bit is run. (Figure 1052.)

Unified Coarse Thread (UNC) Unified Thread series that specifies general-purpose threads, as for screws, bolts, nuts, holes. Figure 1053 shows the UNC series. A unified screw thread is specified by nominal size, number of threads per inch, thread series, and class of screw thread. For example: ".250–20UNC–2A" would be read as .250 inch ($\frac{1}{4}$"), 20 threads per inch, in the Unified Coarse series with an external (A) general-purpose (2) thread. The designation "A" means external. The designation "B" means internal. Class 1 has a wide allowance, used for rapid assembly fasteners. Class 2 is a general purpose thread used on bolts, nuts, etc. Class 3 is a close tolerance thread.

Unified Extra Fine Thread (UNEF) Unified Thread series that is used to specify threads used where precision or fine adjustment is needed, as for adjusting screws. (Figure 1053.)

Unified Fine Thread (UNF) Unified Thread series that specifies special machine fasteners, as those used in automotive and air craft production. (Figure 1053.)

Unified Threads A widely used standard for metal threaded fastening devices and drilled holes. Three series are used: *Unified Coarse (UNC)*, widely used for screws, bolts, nuts, holes; *Unified Fine (UNF)*, used on automotive and aircraft fasteners; and *Unified Extra Fine (UNEF)*, used where fine precision is needed. (Figure 1053.)

uniform load Load distributed evenly, opposed to *concentrated load*.

union: Plumbing fixture for joining two pipes end-to-end. See *pipe fitting*. (*Figure 691*, p. 297.)

unit heater Suspended heater used to heat commercial and industrial spaces.

unit lock Type of heavy-duty lock used in commercial construction. (Figure 1054.)

universal joint A right-angle fitting used with a *ratchet* or *flex handle* to drive sockets from a 90° angle. (Figure 1055.)

upflow furnace Furnace where air flows upward and is discharged at the top. See *downflow furnace*.

upright Vertical, as wall framing members.

upset To flatten a rivet or rod end by hammering.

urban Within a city.

urban renewal Rebuilding of older areas in cities.

urinal Plumbing fixture for receiving and draining away urine; a water flush is used for flushing and cleaning the urinal.

utility knife Cutting knife with a replaceable blade, used for scoring drywall, cutting insulation board, and the like. (Figure 1056.)

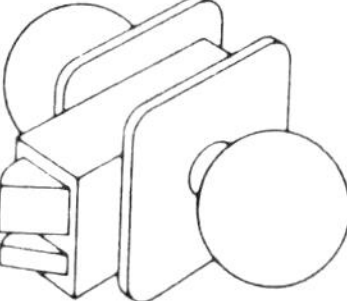
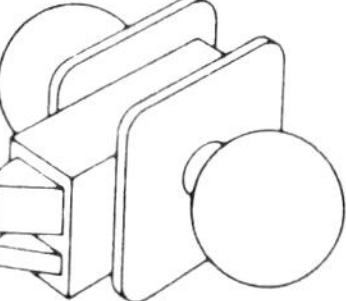

Figure 1054 Unit lock

U value Rate of heat flow, as through a wall. The lower the U value, the greater the insulating value. Derived by adding together all the *C values*, or by taking the reciprocal of all the *R values* for the area. U value indicates the number of Btu's lost per square foot per hour (assuming 1°F difference between the sides of the wall or part).

Figure 1055 Universal joint

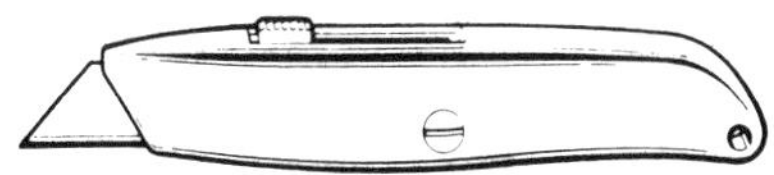

Figure 1056 Utility knife

VAC Volts, alternating current. Term is commonly used on power equipment, such as a panelboard.

valence A short panel or curtain over and across a window area. Designed to give esthetic balance and control light entrance. Often a fluorescent lamp is concealed behind a panel valence.

valence lighting Lighting that is shielded by a panel or skirt, as a light over a window with a short plywood panel or skirt in front of it.

valley Surfaces that join together to make a "V". The "V" made by two sloping roofs joining together. See *roof*.

valley flashing Metal piece that is laid in a roof valley to prevent water and moisture from leaking in. Roof shingling or finish runs over the flashing. (*Figure 402*, p. 163.)

valley rafter Rafter that runs in a roof valley. See *rafter*.

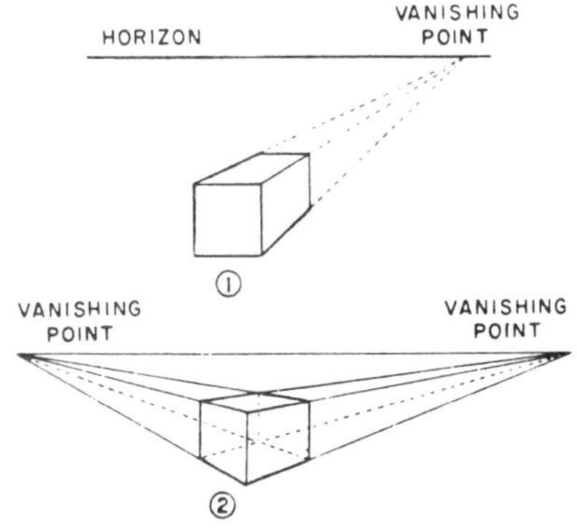

Figure 1057 Vanishing points

valve Device for regulating the flow of water or gas in a pipe.

vanishing point In *perspective drawing*, a point (in the distance) where object lines converge together. (Figure 1057.)

vanity A lavatory (wash basin) with an enclosed cabinet below. A *counter-top lavatory*.

vapor barrier Impermeable material, normally a plastic sheet, fastened on the inside surface of a wall before the interior wallboard or drywall is installed. The material serves as a barrier to prevent moisture from the living area entering the wall. (Figure 1058.) Cutouts are made for openings.

Figure 1058 Vapor barrier being installed

varnish Clear, waterproof preparation of oil, resin, solvent, and drier used as a wood finish. Boiled linseed or tung oil is used as a base in natural varnishes. Modified varnishes are available that have a polyurethane base. The drier causes the oil to oxidize quickly and harden. *Tung oil* varnishes have a high percentage of oil and are used for exterior work where a tough finish is needed. Spar varnish and marine varnish are examples. *Short oil varnishes* have a low percentage of oil and are used for interior fine finishes, such as for cabinets, where a hard finish with a high gloss is needed.

vault Arched masonry structure that forms a ceiling or roof. Commonly found in older church architecture.

vegetative filter In site development, ground cover, such as grass or other planting, used around the construction site to retard water runoff and prevent erosion. (*Figure 855*, p. 376.) See also *landscaping plan*.

vehicle Medium that holds the paint pigment.

velocity Rate of motion in a specific direction.

veneer In woodworking, a thin, fine sheet or ply of wood that is bonded to an inferior wood. Also, in cabinetmaking, a plastic or vinyl sheet that is bonded to a wood or pressed wood surface, used to make counter tops. In brick masonry, a low nonbearing wall attached to the house frame. See *rotary-cut veneer, sawed veneer, sliced veneer.* Also see *brick veneer*.

veneering Bonding a veneer to a base surface.

veneer wall Masonry wall that is tied to, but may not be a load bearing structural part of, a building framework, See *brick veneer*. (*Figure 129*, p. 49.) Also see *faced wall*.

vent Opening for air flow.

vent flashing Flashing collar and base that fits over exposed vent pipe and is attached to the roof. Roof cover is fastened over the edges of the flashing. (Figure 1059.)

ventilating brick Metal, louvered device sometimes used in a brick wall to allow air to circulate into a crawl space.

ventilation Circulation of fresh air into and through the living area. Circulation of air into and through the attic or crawl space. See *attic ventilation, crawl-space ventilation.* Also, air access into the plumbing system through a *vent stack* or *vent pipe.*

vent increaser Enlarged pipe section at top of vent. (Figure 1059.) Allows more efficient dispersion of vent fumes.

venting Removal of gases from a system. Also, removal of gases and equalization of pressure in a plumbing system.

vent pipe Flue or pipe that allows fumes to exit or air to come in, as for a gas hot-water heater or plumbing waste system. (Figure 1060.) Also, a *vent stack*. See *gas venting*.

vent stack A vertical vent pipe that vents the drainage system. The vent stack is separate from the *stack vent* (which is an extension of the soil pipe or soil stack). (*Figure 326*, p. 126.)

vent system In plumbing, the piping that opens through the roof to allow air flow into the drain and waste system. The total system is called the *drain-waste-vent system*. (*Figure 326*, p. 126.)

verge An edge, as the roof edge projecting over a gable roof. The whole area where the roof meets the gables is called the *rake*.

verge board Board running along the gable ends of a roof. Also called *barge board*.

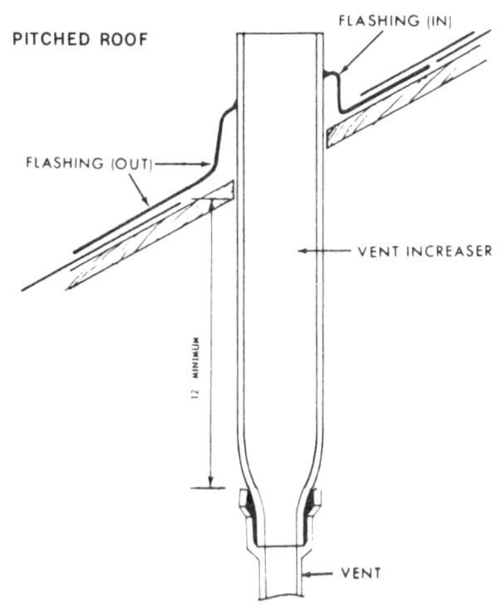

Figure 1059 Vent increaser

vermiculite Loose insulation made from expanded bits of mica. Used as a fill between attic joists.

vernier Auxiliary scale used with the base scale for measuring lengths and angles. A very high degree of accuracy is attained. In surveying, a vernier scale is used to take a final reading on angles, or on a leveling rod. Figure 1061 shows a vernier scale used with horizontal circle graduated from 0° to 360° (reading two ways). The figure reads 17° + 25′ = 17°25′ from left to right and 340°30′ + 05′ = 342° 35′ from right to left.

vertical Straight up and down. Opposed to *horizontal* (level to the horizon). Vertical is at a right angle to the horizontal.

vertical cut Cut that is straight up and down when the framing member is in place; a *plumb cut*.

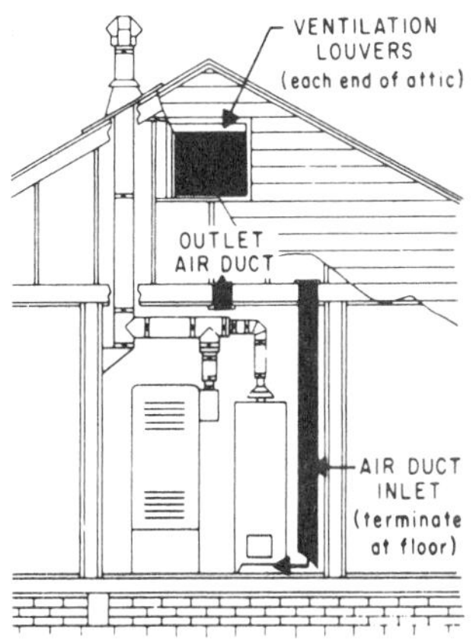

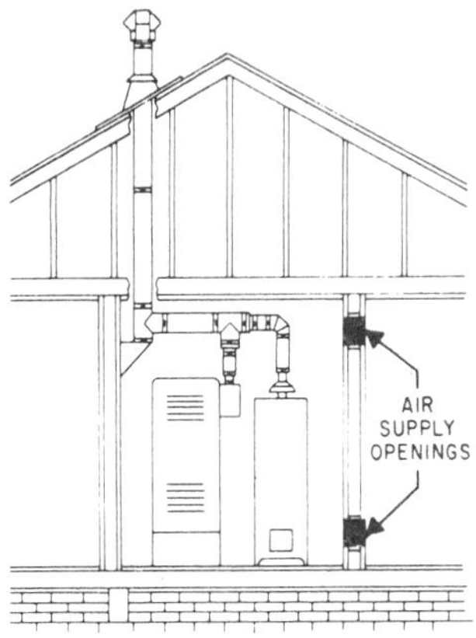

Figure 1060 Vent pipe

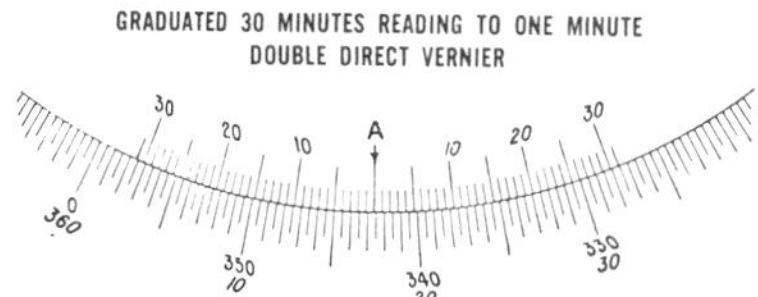

Figure 1061 Vernier

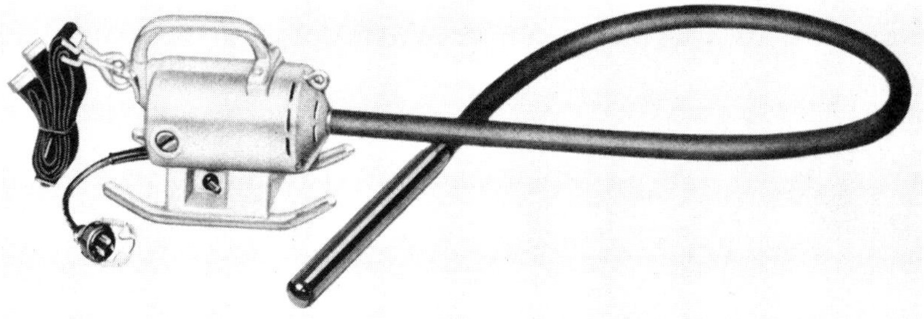

Figure 1062 Concrete vibrator

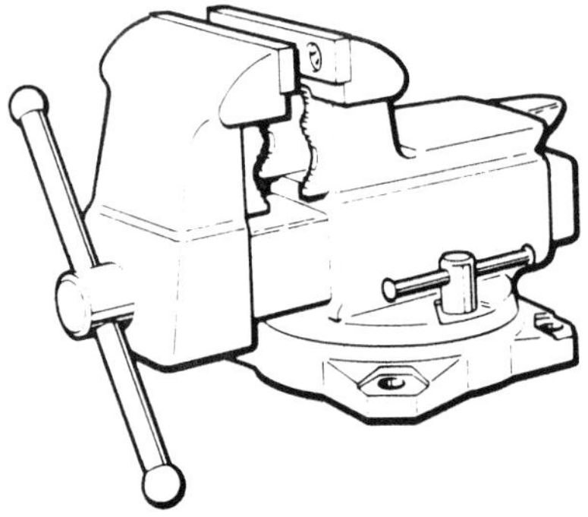

Figure 1063 Vise

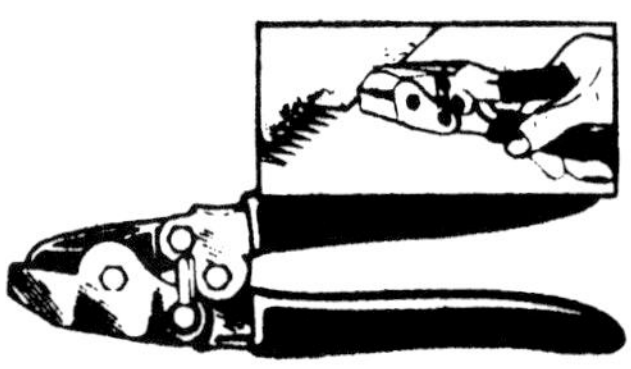

Figure 1064 V notcher

vertical grain *Edge grain; quartersawn grain.*

vertical-position welding Weld that runs up and down. See *welding position.*

vessel In wood, a tubelike cell that runs lengthwise in the tree trunk or branch. When seen cut endwise, it is called a *pore.* See *wood.*

vibration An agitation of fresh concrete to consolidate it in the forms and remove air voids. See *vibrator.*

vibrator A mechanical vibrating tool for agitating fresh concrete to consolidate the concrete in the forms and remove air voids. Sometimes called a *stinger.* (Figure 1062.)

Vicegrip[R] Proprietary term sometimes used to refer to *locking pliers.*

view On floor plans, the identification of which side of the building is shown: front view, rear view, North view, and so on. See also *orthographic views, perspective drawing.*

vinyl-coated nail Nail with a complete vinyl coating to resist temperature changes and prevent rusting.

vinyl-headed nail Nail with colored vinyl head, used for attaching prefinished, colored interior panels.

vinyl-tile cutter Tile cutter for vinyl and asphalt floor tile.

virgin growth First or original timber growth; mature trees that have grown naturally without disturbance or culture by man.

vise Tool for holding or clamping; jaws tighten to grasp the work. (Figure 1063.) See also *clamp, woodworking vise.*

visualization Seeing the whole from parts. When you can see what a three-dimensional object or structure looks like by viewing two-dimensional drawings, you are visualizing.

vitreous Glasslike.

V joint Masonry joint that has been tooled to form a "V". See *jointing, masonry joints.*

V notcher Sheet metal tool for cutting V notches. (Figure 1064.)

volt (V) Electromotive force that causes current to flow in a circuit. This force can be thought of as similar to water pressure in a plumbing system. In an electrical system a power supply provides the force or voltage. Two voltages are normally used in the home: 110 volts for convenience outlets and 220 volts for equipment and heavy appliances. Industry uses 440 volts.

$$\text{volts} = \text{amperes} \times \text{ohms}$$

$$\text{volts} = \frac{\text{watts}}{\text{amperes}}$$

voltage drop A drop or lessening of electromotive force due to resistance, usually because of the distance from the power supply. In welding, if long cables are used, the voltage may drop below the amount needed for welding.

voltage tester Electrical tester used for checking circuits to locate a dead wire or fuse.

voltmeter Electrical device for measuring voltage (electrical pressure). See *circuit tester*.

volume A space; the cubic content of something. Generally used in relation to liquid contents in a container. See *cubic measure*.

voussoir Wedge-shaped stone or masonry unit used to create a arch. See *keystone*. (Figure 1065.)

V weld A weld joining two angled edges. See also *bevel weld*.

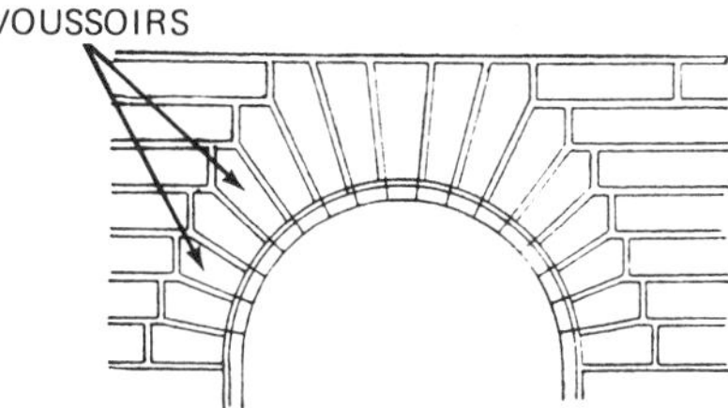

Figure 1065 Voussoirs

waferboard A manufactured structural panel made with pressed wood chips and resin. It has a strength similar to that of plywood and is used for roof and wall sheathing.

waffle slab Two-way reinforced concrete slab; joist ribs run in two directions. See *two-way floor system*.

wainscot Interior wall covering at the base of a wall. (*Figure 612*, p. 263.)

wainscoting Interior wall base area, specifically the lower three or four feet when covered with paneling or other material different from the rest of the wall. The *wainscot*.

wale Horizontal support member running along the outside of concrete forms. Support ties run through the forms and the wales and are fastened on the outside of the wales. (Figure 1066.) See *forms*. Also called *rangers, walers, whalers*. In excavation, a horizontal support member used to hold trench sheet piling. (*Figure 870*, p. 382.) Also called a *stringer*. See *trench support*.

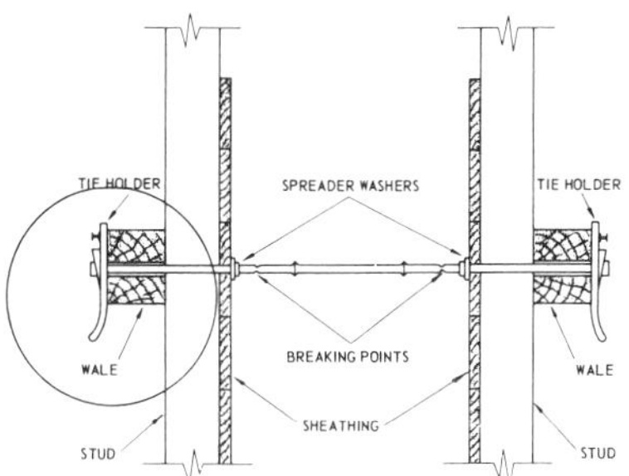

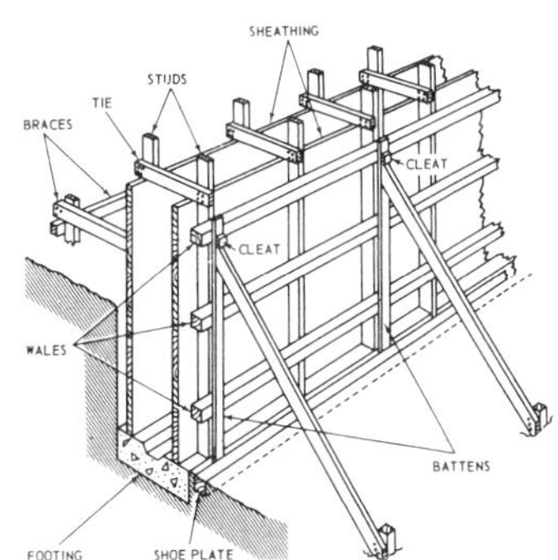

Figure 1066 Wale

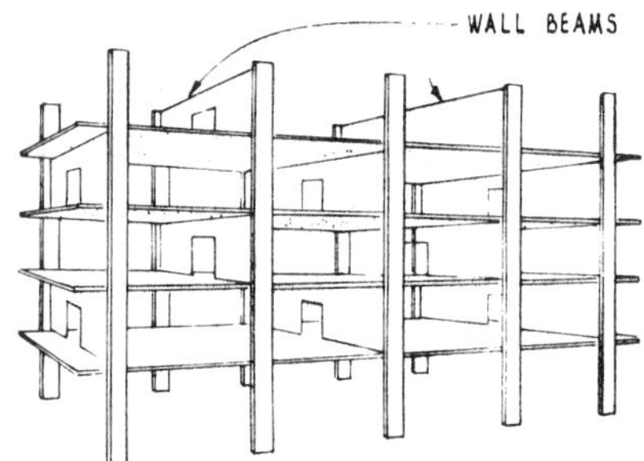

Figure 1067 Wall beams

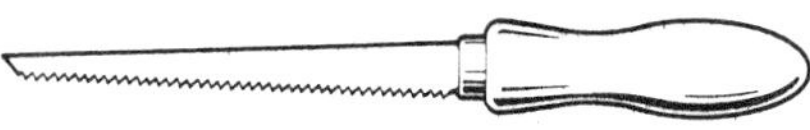

Figure 1068 Wallboard saw

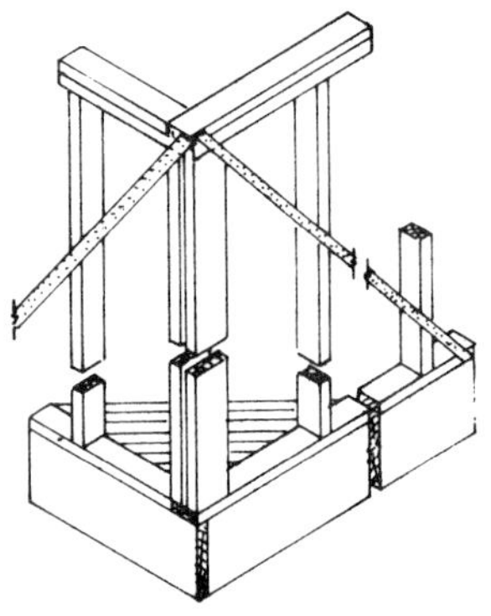

Figure 1069 Wall braces

Figure 1070 Wall forms

waler See *wale.*

wall A load-bearing, enclosing side or interior part of a building. Non-load-bearing interior walls are called partitions. A structure that holds back or retains earth, a *retaining wall.* See *bearing wall, brick veneer, cavity wall, curtain wall, faced wall, firewall, frame wall, foundation wall, load-bearing wall, masonry wall, nonbearing wall, party wall, tilt wall, veneer wall, wall system.*

wall anchor See *joist anchor.*

wall beam In reinforced concrete construction, a full-story, reinforced bearing wall that spans the full width of the building from column support to column support. (Figure 1067.)

wallboard saw A short, thin-bladed saw used for cutting gypsum board, insulating board, masonite, hardboard, and the like. (Figure 1068.)

wallboard square See *drywall T square.*

wall braces Metal braces used between studs at the corner to stiffen the wall. (Figure 1069.) See also *cut-in brace, let-in brace, tension straps.*

wall chase Opening through a masonry wall. See *chase.*

wall coping Covering for the top of a masonry wall. See *capping, coping.*

wall fan Exhaust fan mounted in the wall to vent air directly to the outside; used in kitchen areas and bathrooms. See *ceiling fan, range hood fan.*

wall formwork Formwork used to make concrete walls. In heavy construction, reusable forms are hoisted into place. (Figure 1070.) See *facade forms, forms.*

wall heater Heater recessed in the wall.

wall line Outside edge of a wall; face line; foundation edge. A line laid out with *batter boards* and string.

wall panel Precast concrete unit, often used to form a *curtain wall*. A wood or fabricated wall unit. See *precast concrete, preengineered building, wall system.*

wallpaper hanging Applying wallpaper to walls. Several types of wallpaper today come prepasted; others must have paste applied to the back. Some "wallpapers" are not actually paper. Vinyl, various fabrics, and even foil wall coverings are available. See *paste brush, smoothing brush.*

wall plate General term sometimes used to mean *top plate.*

wall socket Electrical outlet; *receptacle.*

wall spacers Wood or metal separators used to hold the form walls apart until the concrete is poured. (*Figures 426*, p. 172; *826*, p. 363; *1001*, p. 442; *1066*, p. 471.) The wood spacers are removed (by a wire pull) as the form is filled with concrete. The metal rod spacers or ties are left in the concrete wall; the ends are broken off when the forms are removed. See *form ties, tie rod.*

wall spreader In reinforced concrete, a Z- or U-shaped separator used in a wall to separate two curtains or mats or reinforcement bars.

wall stringer Stair stringer that is located next to a wall.

wall symbols Symbols used on floor plans to identify the type of wall in a building. See *plan symbols.*

wall system In industrial construction, the complete wall covering or wall panel system. (Figure 1071.)

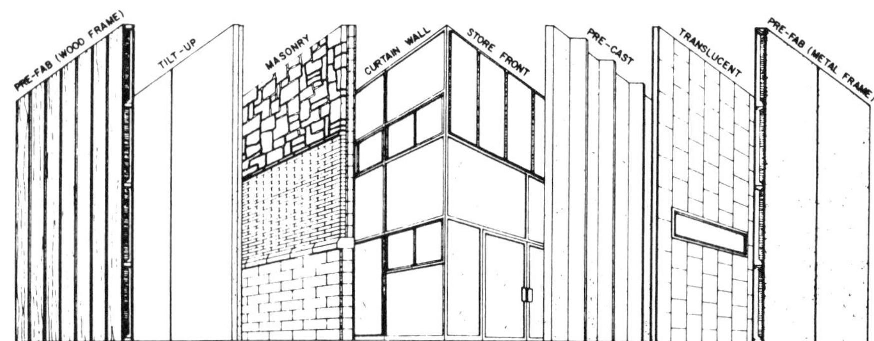

Figure 1071 Wall systems

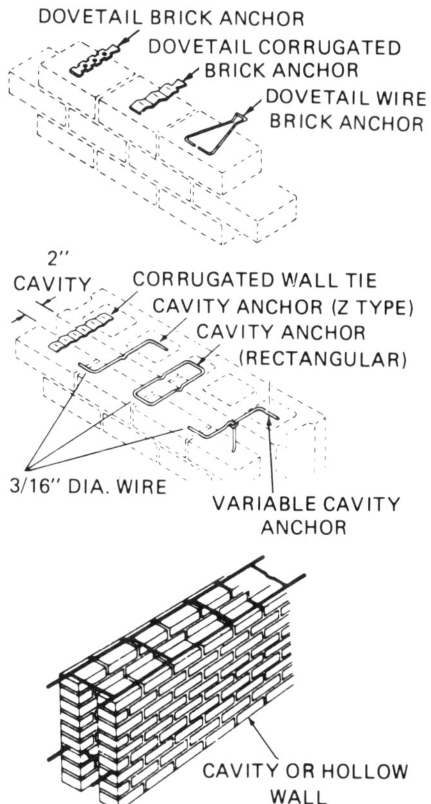

Figure 1072 Wall ties

wall tie Metal pieces used to tie hollow wall wythes together or to tie veneer to the frame. (Figure 1072.) See *brick veneer, cavity wall, hollow wall.* Also, reinforcement used between courses of brick or concrete block.

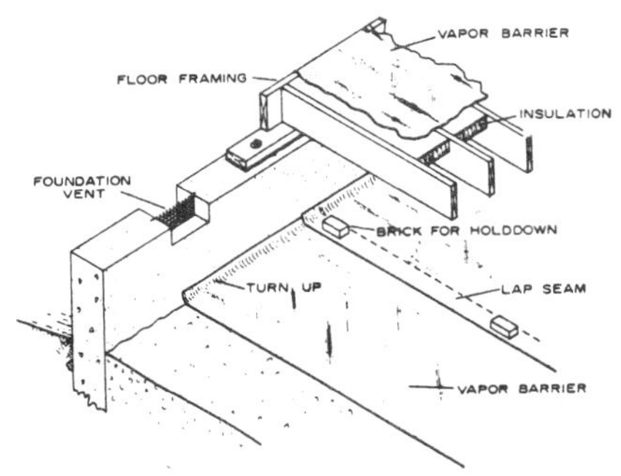

Figure 1073 Wall or foundation vent

wall vent A screened opening or metal ventilator used in concrete foundations or brick or concrete block walls to allow air to circulate in the wall or into a crawl-space area. (Figure 1073.) See *crawl-space ventilation, ventilator brick*.

wane Lumber defect: wood missing, usually on the edge, because of bark coverage or peel-off.

warm-air heating See *forced warm-air heating*.

warp Lumber defect: a distortion caused by the wood twisting or bending out of shape.

wash Downward slope on exposed masonry surface to allow water runoff; slope on exposed wood framing member, as a window sill, to allow water runoff.

SIZES: Washers are size-labeled according to the sizes of screws and bolts. Therefore, for a snug-fitting washer, select the same size washer as the screw or bolt you're using.

TYPES

FLAT: Use whenever you want to distribute load – prevents material from buckling or screw/bolt from pulling through.

SPLIT-LOCK: Ends of washer serve as "friction points" to prevent assembly from loosening. Made of spring metal so constant pressure. Use to prevent loosening if tension varies in assembly, space is not critical and/or you may occasionally disassemble.

TOOTH-LOCK: Teeth serve as points of friction to prevent assembly from loosening. Made of thin spring metal. Use if tension doesn't vary significantly, space is critical and/or you don't often disassemble.

Inside Diameters (Refer to by number)	Standard Outside Diameters			
	Flat	Split-Lock	Tooth-Lock	Finishing
6	.375	.250	.305	.375
8	.438	.293	.365	.438
10	.500	.334	.395	.500
3/16	.562			
12	.562	.377	.460	.562
1/4	.625	.489	.494	.625
5/16	.688	.586	.588	.688
3/8	.812	.683	.670	
7/16	.922	.779	.740	
1/2	1.062	.873	.880	

Figure 1074 Washers

washcoat A sealer coat put over an existing finish to stop bleeding. Thinned shellac is used as a washcoat.

washer Round collar used with a bolt and nut to give an enlarged bearing surface under the nut. (Figure 1074.) Washers are choosen to fit the bolt size. See *bolt, nut*, Washers are also sometimes used with machine screws.

waste Unusable material cut away from the main part, as the end of a rafter piece. Also, material and waste water carried away by the plumbing waste system.

waste pipe Plumbing pipe that carries waste into sewer system. See *drain-waste-vent system, plumbing system*.

waste stack Vertical pipe that receives waste from fixtures such as sinks, lavatories, showers or bathtubs, but not from a water closet or similar units. See *soil pipe, soil stack.*

waste trap See *trap.*

waste vent See *vent stack* or *stack vent.*

water closet Toilet fixture; abbreviated W.C. Figure 1075 shows the major parts of the tank.

water cooling tower Tower for cooling water by allowing it to lose heat to the surrounding air.

water hammer Banging noise in a water supply system caused by improper design and layout; occurs when tap is turned off.

water heater Plumbing appliance that heats water for personal use.

water level Rubber tube or hose with glass tubes at each end. When the level is filled with water, the glass tubes, when held vertically at the same height, will have exactly the same water level. Used at the construction site for establishing the same height at widely separated spots.

water main Muncipal water supply pipe.

water meter Device for measuring use of water in gallons or cubic feet.

waterproof cement Portland cement with a waterproofing additive.

waterproofing Protection against water penetration or seepage.

water-reducing admixture Admixture added to concrete that reduces the amount of water needed to mix the concrete.

water-repellent admixture Admixture added to concrete to produce a concrete that is resistant to water or moisture penetration.

water seal Water in the "U" of a plumbing trap. See *trap.*

water service Entry of fresh water to a structure. Water is taken from the water main and is measured by the water meter in the structure. Water to the building is turned on and off by a curb cock at the bottom of a stop box. The water company controls the water by a corporation cock located by the water main.

watershed See *water table.*

water softener Plumbing appliance that is attached to the water supply system to remove hard minerals from the *hard water.* The unit replaces calcium and magnesium in the water with sodium, which results in a *soft water.*

waterstone Manufactured Oriental sharpening stone that uses water as a base (rather than oil). Three stones are used: coarse (800 grit), medium (1200 grit) and fine (6000 grit). See *abrasives, oilstone.*

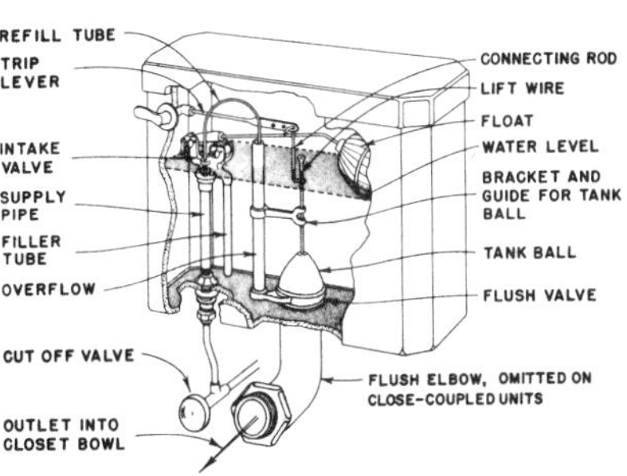

Figure 1075 Water-closet tank

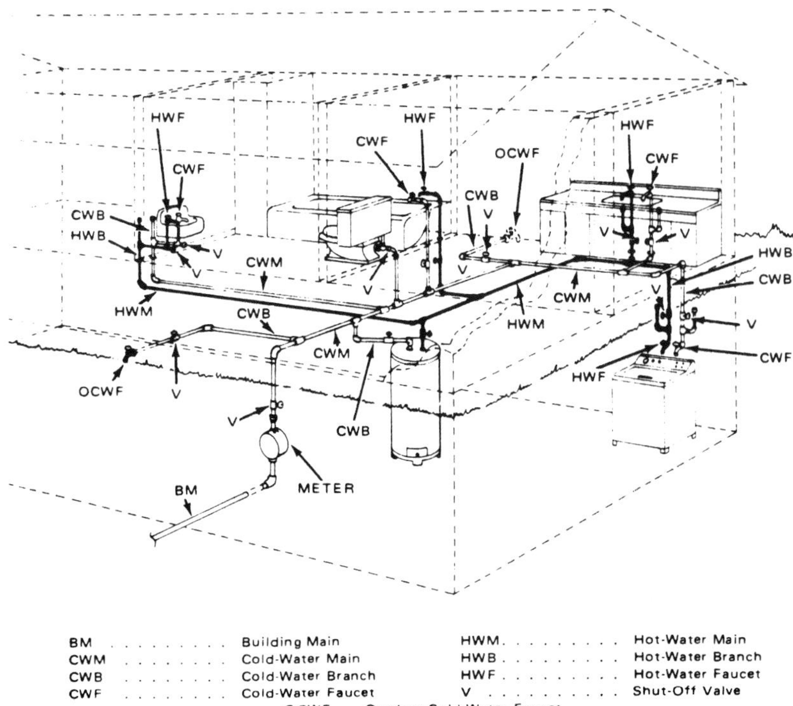

Figure 1076 Water supply system

water supply system Piping that brings in the fresh water. (Figure 1076.)

water table Projection just above the foundation top and at the bottom of the wall. Allows water to flow away from the side of the house. (*Figure 337*, p. 130.) Sometimes called a *watershed* or *drip cap*. Also, level of subsurface ground water.

watertight Constructed so moisture will not enter or interfere with operation.

watt (W) Measurement of electrical power. Volts and amperes together give the total electrical power or wattage. When one volt causes one ampere of current to flow, one watt of power is used. Electrical appliances and light bulbs are rated in watts (for example, a 60-watt bulb). Equipment or light bulbs normally state what wattage they use. The wattage is a function of the amps (rate of flow) times the 120-volt current supplied. A 100-watt light bulb draws approximately 0.83 amp (0.83 × 120V = 100 watts).

amps × volts = watts
1 amp at pressure of 1 volt = 1 watt
1 watt used for 1 hour = 1 watt-hour
1000 watt-hours = 1 kilowatt-hour (kWh)

wax sticks In woodworking, soft colored sticks that are used for filling nail holes and other small holes in fine hardwoods.

W beam A steel beam that has wide flanges. The designation W has replaced the old designation WF (wide flange) in current standards. The nominal height (distance between flanges) in W beams varies from 6″ to 36″; weight is specified as pounds per running foot. See *S beam*, *steel beam*.

weatherproof Constructed so that exposure to weather will not interfere with operation.

weatherproof outlet Receptacle used outdoors; the device is covered when not in use. (Figure 1077.)

weatherstripping Seals used around windows, doors, or other openings to reduce air passage. (Figure 1078.)

weather-struck joint Masonry joint that is *struck* or *pointed*; the mortar is pushed back into the joint. See *tooling*. (*Figure 588*, p. 253.)

weave bead In welding, a weld bead that is laid down by a back-and-forth or zigzag motion of the rod or electrode.

weaving In welding, enlarging a weld bead by moving the electrode in a sideways pattern. (Figure 1079.) Figure 1080 shows typical weaving techniques.

web In truss construction, the cross support members between the bottom chord and top chord. See *truss*. On a steel beam, the steel between the top and bottom flanges. See *open-web joist, steel beam*

webbing Cross bridging in metal studs, flat trusses, or manufactured joists. See *metal studs, trusses*.

web clamp Long belt of nylon webbing used for holding wood parts together for gluing. (Figure 1081.) Belt goes around the wood pieces and is pulled tight with a buckle. A tightening nut is used to make a tight fit. See *clamp*.

web connection Framing into the web of a steel beam; holes are drilled for the connection.

web frame In cabinetmaking, the four-sided internal horizontal frames used to strengthen a carcass and provide support for drawers.

web leg In steel construction, the leg of an angle attached to the web of a beam. The other leg (outside leg) in attached to another beam in the field. See *outside leg*.

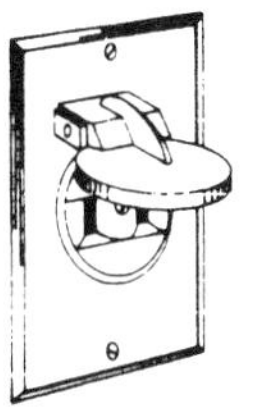

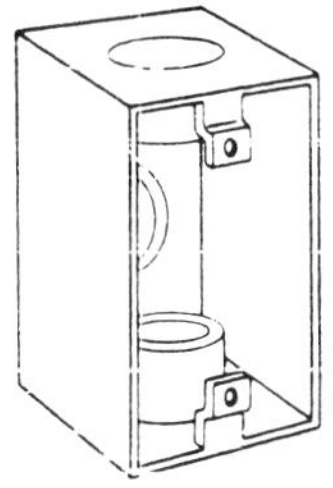

Figure 1077 Weatherproof outlet and box

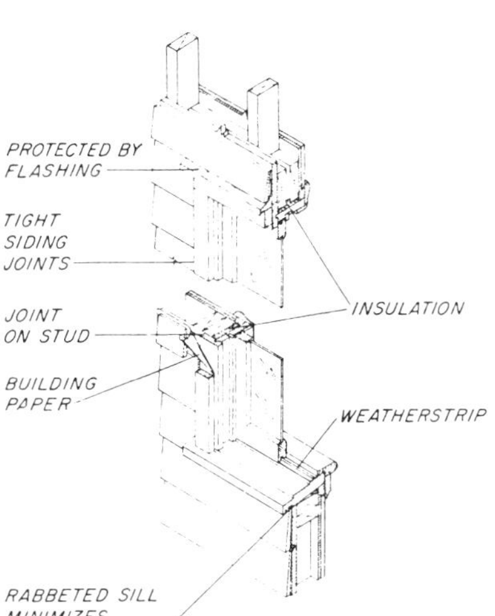

Figure 1078 Weatherstripping and insulation

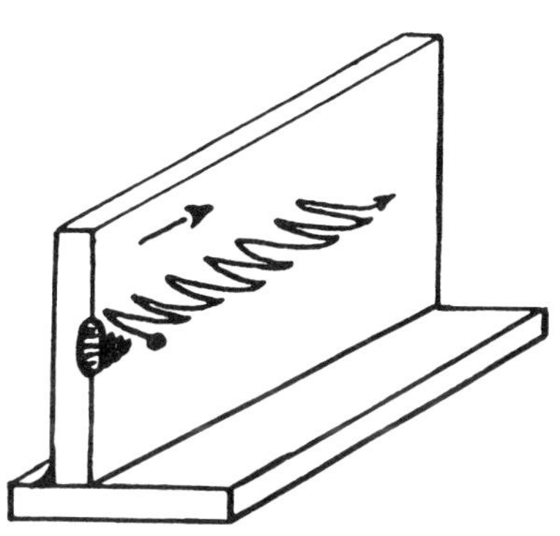

Figure 1079 Weaving

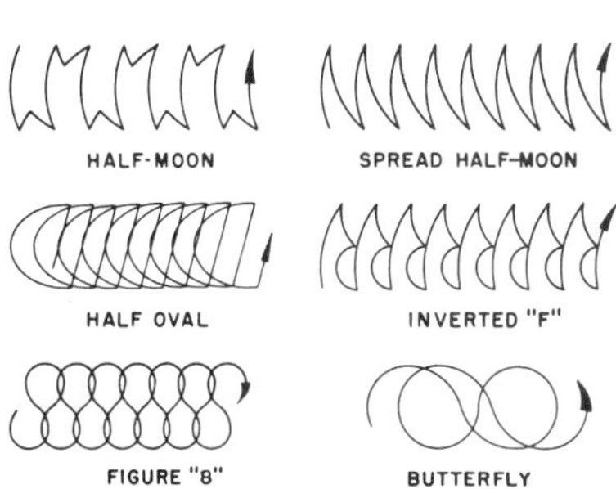

Figure 1080 Weaving techniques

Figure 1081 Web clamp

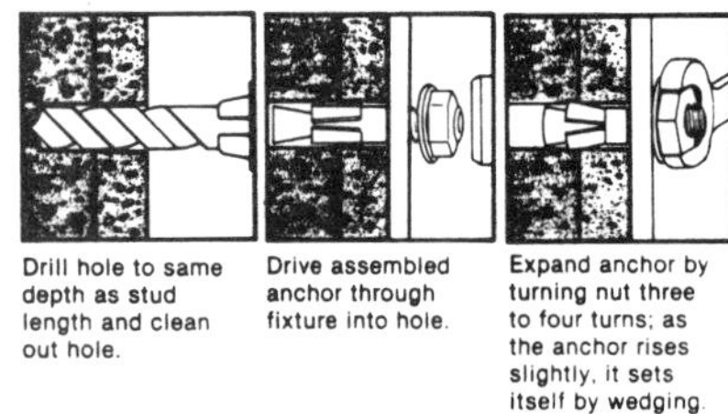

Figure 1082 Wedge anchor

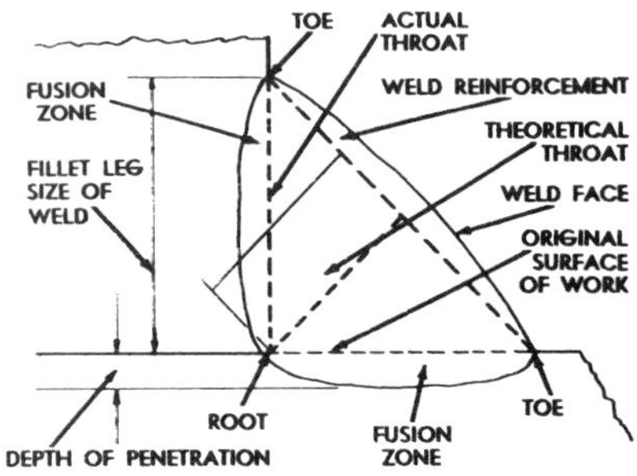

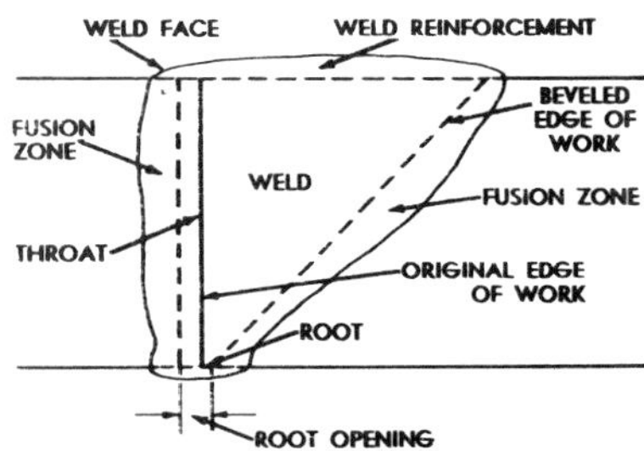

Figure 1083 Weld-bead nomenclature

STYLE DESIGNATION		Steel Area Sq. In. Per Ft.		Weight Approx. Lbs. Per 100 Sq. Ft.
New Designation (By W-number)	Old Designation (By Steel Wire Gage)	Longit.	Transv.	
ROLLS				
6x6–W1.4xW1.4	6x6–10x10	.03	.03	21
6x6–W2xW2	6x6–8x8*	.04	.04	29
6x6–W2.9xW2.9	6x6–6x6	.06	.06	42
6x6–W4xW4	6x6–4x4	.08	.08	58
4x4–W1.4xW1.4	4x4–10x10	.04	.04	31
4x4–W2xW2	4x4–8x8*	.06	.06	43
4x4–W2.9xW2.9	4x4–6x6	.09	.09	62
4x4–W4xW4	4x4–4x4	.12	.12	86
SHEETS				
6x6–W2.9xW2.9	6x6–6x6	.06	.06	42
6x6–W4xW4	6x6–4x4	.08	.08	58
6x6–W5.5xW5.5	6x6–2x2**	.11	.11	80
4x4–W4xW4	4x4–4x4	.12	.12	86

*Exact W-number size for 8-gage is W2.1.
**Exact W-number size for 2-gage is W5.4.

Figure 1084 Welded-wire fabric: new and old designations

wedge Any metal or wooden piece shaped like a sharp V, used to expand or split wood. Metal piece driven into the end of a hammer handle to expand the wood in the hammer eye.

wedge anchor A one-piece anchor used for fastening into masonry: bolt, nut, washer, and anchor come as a single unit. Anchor fits into drilled hole; tip sleeve is expanded by tightening nut. (Figure 1082.) See also *anchor, machine-screw anchor, sleeve anchor, stud anchor.*

weephole Opening left in a brick veneer or solid masonry wall to allow for moisture drainage and evaporation. (*Figures 129*, p. 49; *789*, p. 346.)

weld bead The deposit put down by one weld pass. See *bead weld, stringer bead, weave bead.* Figure 1083 shows the nomenclature of a weld bead.

welded-steel fabirc Wire mesh welded together in a rectangular pattern; *welded-wire fabric.*

welded-wire fabric (WWF) Steel wires or bars welded together to form a mat, used for concrete slab reinforcement. (*Figure 887*, p. 388.) Spacing of the bars is specified in inches — for example, a 10 × 10 fabric is spaced in a criss-cross pattern with the wires or bars spaced 10" apart in both directions. The size of the wire or bars is specified by gauge or, in the new standards, by cross-sectional area in hundredths of a square inch. Figure 1084 compares the old gauge sizes to the new cross-sectional W sizes. Welded-wire fabric (also called *mesh) is available in rolls or sheets.*

welded-wire mesh (WWM) See *welded-wire fabric.*

weld groove Groove where the two pieces to be welded are joined. See *groove weld.*

welding Process of fusing metal parts together with heat; a filler rod may be used to supply metal to make the joint. Welding gives a permanent bond between metal parts. Construction welding commonly uses gas (*oxyacetylene* or *oxyfuel*) or *electric-arc welding*. See *arc welding, oxyacetylene welding*. See also *brazing, pipe welding, welding processes.*

welding clamp Locking plier-like clamp used to hold metal pieces together for welding. (Figure 1085.)

welding flame In gas welding, the flame made by the combustion of oxygen and acetylene (or other gas such as propane or natural gas). Three different basic flames are made by adjusting the proportion of oxygen and acetylene: neutral, carburizing, and oxidizing. (Figure 1086.) *Neutral flame*: Generally used for welding. Roughly, equal amounts of oxygen and acetylene are combined to produce a flame of around 5900°F. A white inner cone is produced. *Reducing* or *carburizing flame*: This flame has a high acetylene mixture and adds carbon to the weld. Note in the illustration the acetylene feather. This flame leaves a brittle surface. Also called a carbonizing flame. *Oxidizing flame*: This flame has an excess of oxygen and produces a very hot flame, around 6300°F. It is not commonly used for welding because it forms oxides and a brittle surface. With the addition of more *oxygen* it is used for cutting metals.

welding helmet A protective device worn over the head when welding. (Figure 1087.) The front welding plate lifts up. Different lenses are used depending on the type of welding. (Figure 1088.)

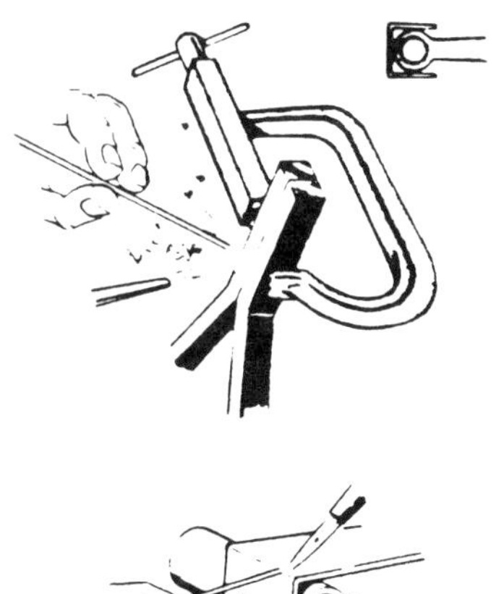

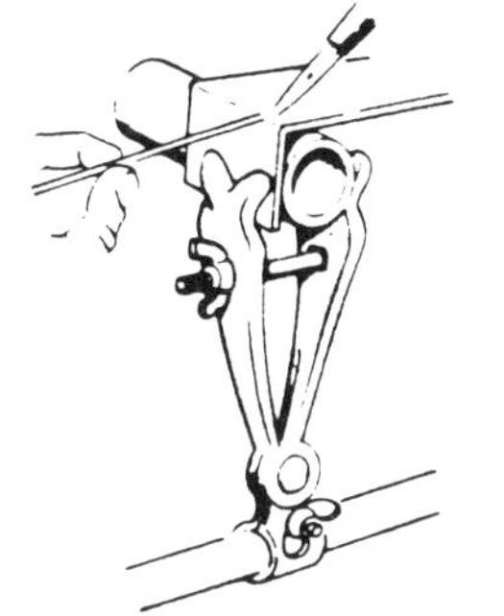

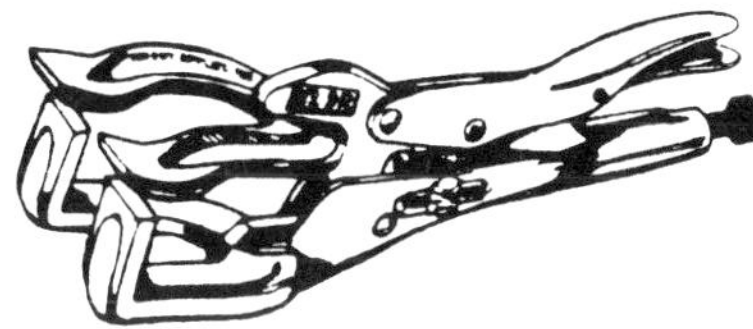

Figure 1085 Welding clamps

Figure 1087 Welding helmet

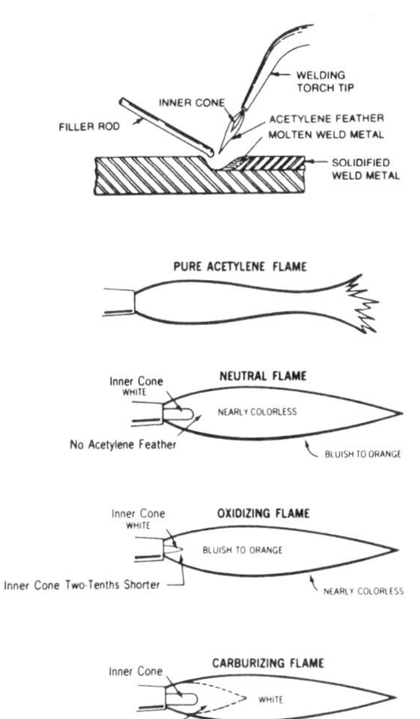

A neutral flame is used for almost all gas welding. The oxy-acetylene flame consumes all oxygen in the air around the weld area. This leaves an uncontaminated weld area resulting in maximum weld strength. An oxidizing flame is rarely used, and a carburizing flame is occasionally helpful when flame hardening, or brazing.

Figure 1086 Welding flame

lens shade selector

OPERATION	SHADE NUMBER
SOLDERING	2
TORCH BRAZING	3 or 4
OXYGEN CUTTING up to 1 inch 1 to 6 inch 6 inch and over	 3 or 4 4 or 5 5 or 6
GAS WELDING up to 1/8 inch 1/8 to 1/2 inch 1/2 inch and over	 4 or 5 5 or 6 6 or 8
SHIELDED METAL-ARC WELDING 1/16, 3/32, 1/8, 5/32 inch electrodes	 10
GAS TUNGSTEN-ARC WELDING (Nonferrous) GAS METAL-ARC WELDING (Nonferrous) 1/16, 3/32, 1/8, 5/32 inch electrodes	 11
GAS TUNGSTEN-ARC WELDING (Ferrous) GAS METAL-ARC WELDING (Ferrous) 1/16, 3/32, 1/8, 5/32 inch electrodes	 12
SHIELDED METAL-ARC WELDING 3/16, 7/32, 1/4 inch electrodes 5/16, 3/8 inch electrodes	 12 14
ATOMIC HYDROGEN WELDING	10 to 14
CARBON-ARC WELDING	14

Figure 1088 Welding helmet: lens-shade selector

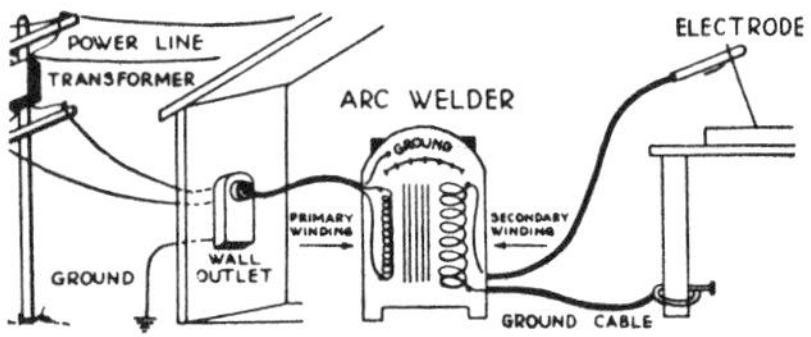

Figure 1089 Welding machine: AC transformer

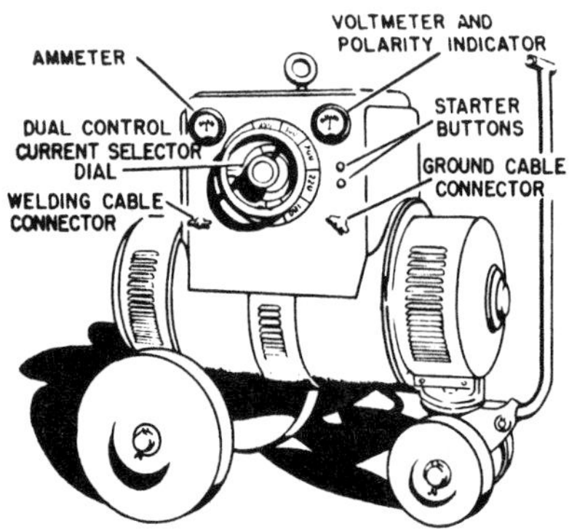

Figure 1090 Welding machine: DC generator (dual control)

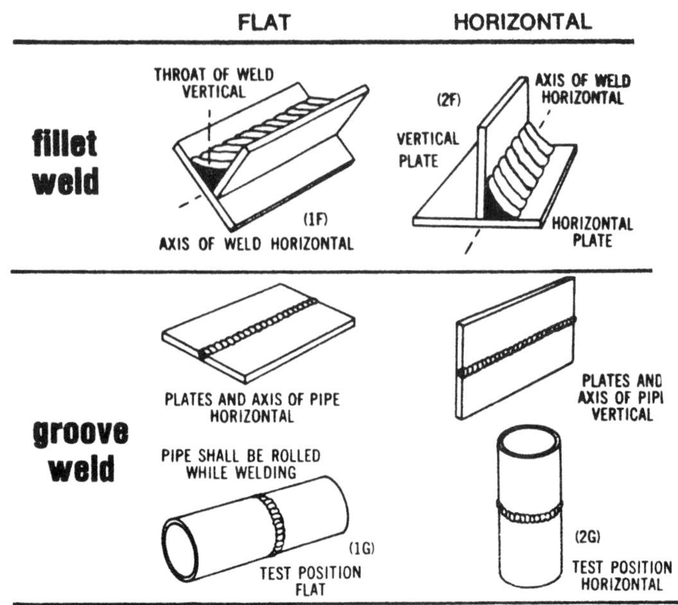

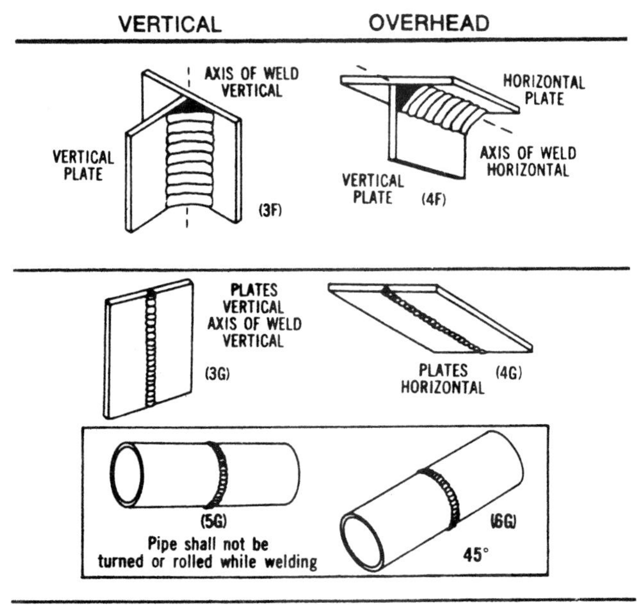

Figure 1091 Welding positions

welding machine In arc welding, the AC or DC power supply or welder. Alternating-current welders are *transformers* that change the available 220V or 240V electrical power to low-voltage, high-amperage welding current. (Figure 1089.) Direct-current power is supplied by a *rectifier* or *transformer-rectifier* which transforms AC to DC. The transformer-rectifier supplies both AC and DC. An electric- or gasoline-operated generator can also provide AC or DC power. (Figure 1090.) Power is rated in amps. For example: 150–200 amps (light welding), 250–300 amps (standard welding), and 400–600 amps (heavy-duty).

welding, plastic See *plastic welding*.

welding position Four basic welding positions are used: horizontal, flat, overhead, and vertical. (Figure 1091.)

Method of Application → / ↓ Welding Process	MA Manual	SA Semiautomatic	ME Machine	AU Automatic
Shielded metal arc*	Most popular	Not used	Not used	Special
Gas tungsten arc*	Most popular	Possible—rare	Used	Used
Plasma arc*	Most popular	Not used	Used	Used
Submerged arc*	Not possible	Little used	Most popular	Popular
Gas metal arc*	Not possible	Most popular	Used	Popular
Flux cored arc*	Not possible	Most popular	Used	Popular
Electroslag	Not possible	Possible—rare	Most popular	Used
Torch brazing*	Most popular	Used	Used	Used
Oxyfuel gas*	Most popular	Not used	Little used	Little used
Thermal cutting*	Most popular	— —	Popular	Popular

* *Manipulative skills required for these processes when applied manually or semiautomatically.*

Figure 1092 Welding procedures

welding procedures Various procedures or methods of application are used for making welds: *manual* (completely operator controlled), *semiautomatic* (some operator manual control), *machine* (operator supervision), and *automatic* (no operator). Figure 1092 compares these different methods of welding. Figure 1092 illustrates how different welding processes relate to the four methods of application.

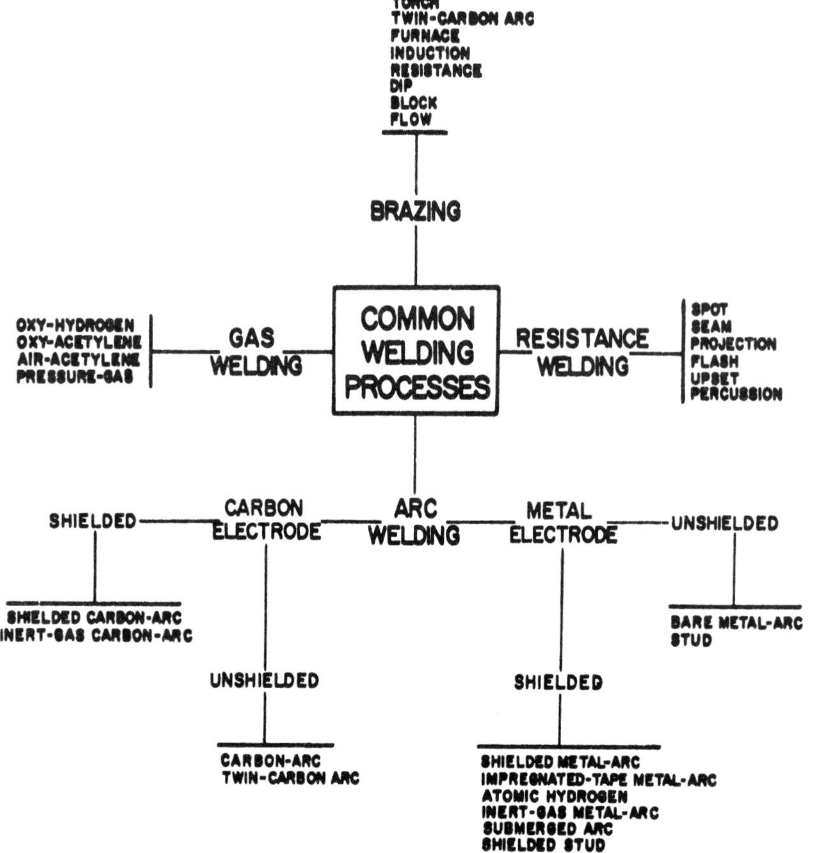

Figure 1093 Welding processes

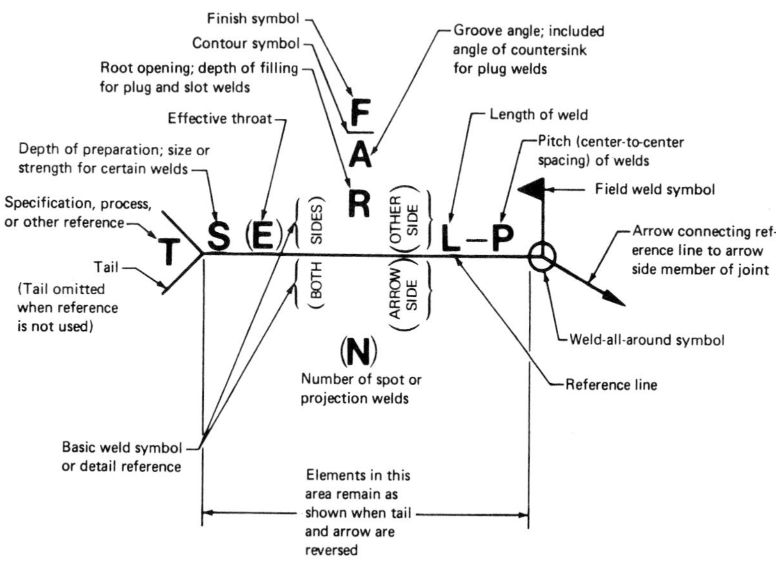

Figure 1095 Basic welding symbol

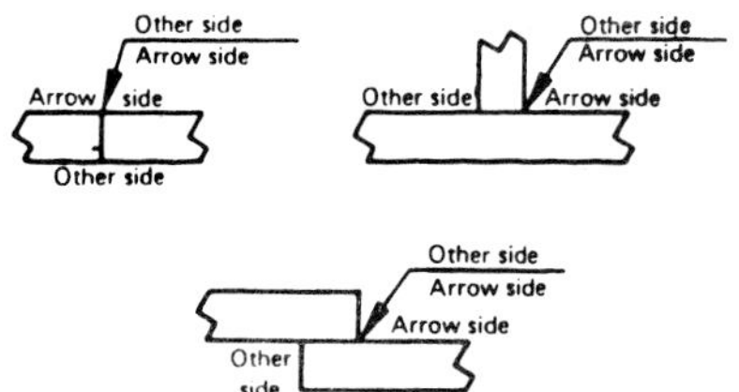

Figure 1096 Welding-arrow location significance

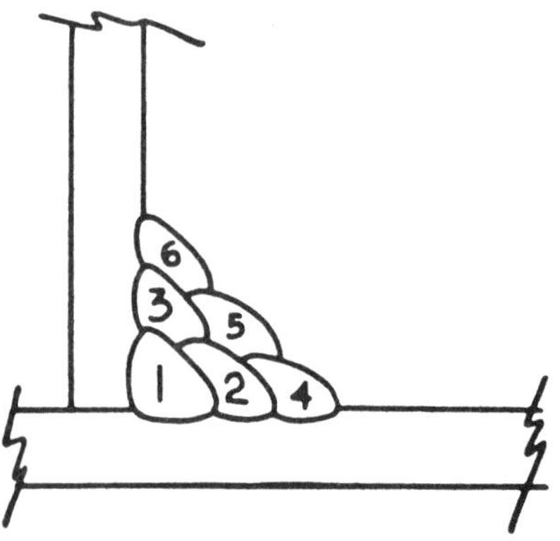

Figure 1094 Welding sequence for a fillet weld

welding processes Various processes are used in welding. In construction, *gas welding* and *arc welding* are used. *Brazing* is also used for making joints. Figure 1093 shows some of the more common welding processes.

welding rod In gas welding, a stick of filler metal that is used to fill in and bond the joint being welded or brazed. The rod is melted by the heat of welding.

welding sequence Order of welds at a joint, as the placement and sequence of the weld beads. Figure 1094 shows a typical weld sequence for a fillet weld.

welding symbols Standard signs used to give information on type and size of weld. The symbols are developed and standardized by the American Welding Society. Figure 1095 shows the basic welding symbol used on welds. A weld symbol on *top* of the arrow reference line indicates the weld is made on the *other side* from where the arrow is pointing. A weld symbol on the *bottom* of the arrow reference line indicates the weld is made on the *same side* as the arrow. Symbols on both sides of the reference line indicate welds on both sides of where the arrow is pointing. Figure 1096 shows how this concept is applied. The AWS standard weld symbols are shown in Figure 1097. A distinction is made between the *welding symbol* (Figure 1095) and the *weld symbols* (Figure 1097), which indicate the type of weld.

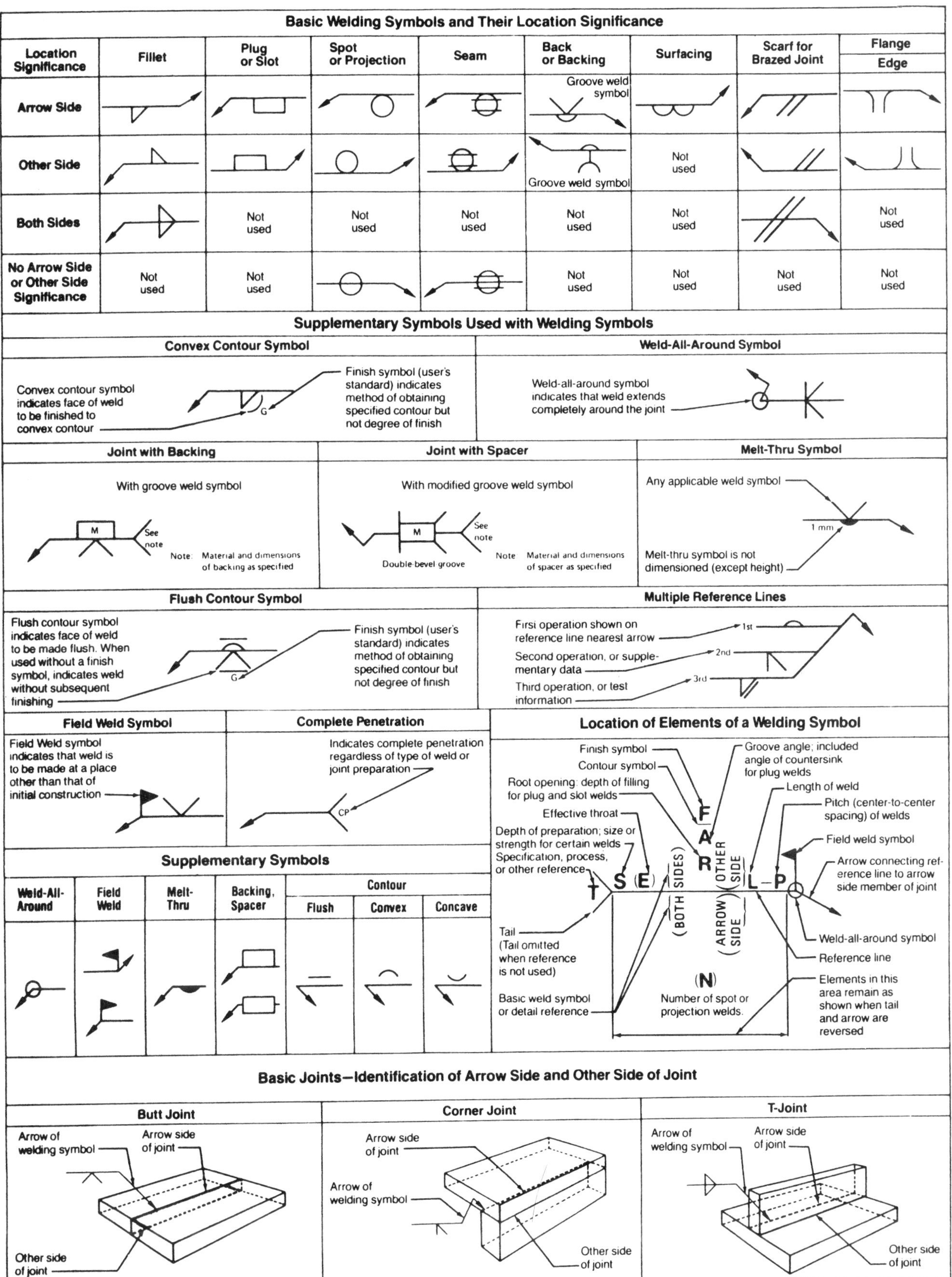

Figure 1097 AWS basic welding symbols

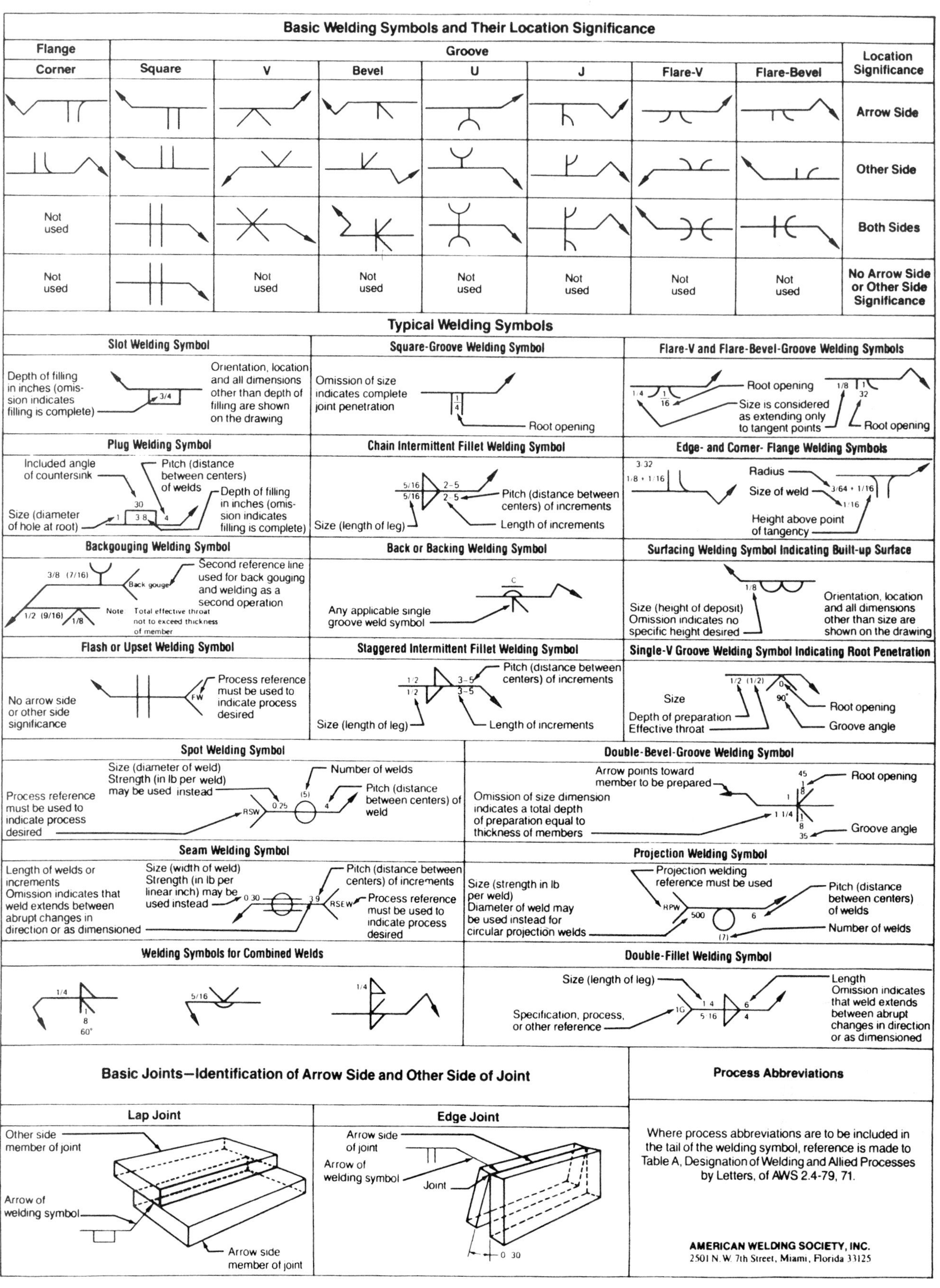

Basic Welding Symbols and Their Location Significance
Flange
Corner
Groove
Square
V
Bevel
U
J
Flare-V
Flare-Bevel
Location Significance
Arrow Side
Other Side
Not used
Both Sides
No Arrow Side or Other Side Significance
Typical Welding Symbols
Slot Welding Symbol
Depth of filling in inches (omission indicates filling is complete)
3/4
Orientation, location and all dimensions other than depth of filling are shown on the drawing
Square-Groove Welding Symbol
Omission of size indicates complete joint penetration
Root opening
Flare-V and Flare-Bevel-Groove Welding Symbols
Root opening
Size is considered as extending only to tangent points
Root opening
Plug Welding Symbol
Included angle of countersink
Pitch (distance between centers) of welds
Depth of filling in inches (omission indicates filling is complete)
Size (diameter of hole at root)
Chain Intermittent Fillet Welding Symbol
Pitch (distance between centers) of increments
Size (length of leg)
Length of increments
Edge- and Corner- Flange Welding Symbols
Radius
Size of weld
Height above point of tangency
Backgouging Welding Symbol
Second reference line used for back gouging and welding as a second operation
Back gouge
Note Total effective throat not to exceed thickness of member
Back or Backing Welding Symbol
Any applicable single groove weld symbol
Surfacing Welding Symbol Indicating Built-up Surface
Size (height of deposit) Omission indicates no specific height desired
Orientation, location and all dimensions other than size are shown on the drawing
Flash or Upset Welding Symbol
No arrow side or other side significance
FW
Process reference must be used to indicate process desired
Staggered Intermittent Fillet Welding Symbol
Pitch (distance between centers) of increments
Size (length of leg)
Length of increments
Single-V Groove Welding Symbol Indicating Root Penetration
Size
Depth of preparation
Effective throat
Root opening
Groove angle
Spot Welding Symbol
Size (diameter of weld) Strength (in lb per weld) may be used instead
Number of welds
Pitch (distance between centers) of weld
Process reference must be used to indicate process desired
RSW
Double-Bevel-Groove Welding Symbol
Arrow points toward member to be prepared
Root opening
Omission of size dimension indicates a total depth of preparation equal to thickness of members
Groove angle
Seam Welding Symbol
Length of welds or increments Omission indicates that weld extends between abrupt changes in direction or as dimensioned
Size (width of weld) Strength (in lb per linear inch) may be used instead
Pitch (distance between centers) of increments
Process reference must be used to indicate process desired
RSEW
Projection Welding Symbol
Projection welding reference must be used
Size (strength in lb per weld) Diameter of weld may be used instead for circular projection welds
RPW
Pitch (distance between centers) of welds
Number of welds
Welding Symbols for Combined Welds
Double-Fillet Welding Symbol
Size (length of leg)
Specification, process, or other reference
Length Omission indicates that weld extends between abrupt changes in direction or as dimensioned
Basic Joints—Identification of Arrow Side and Other Side of Joint
Lap Joint
Other side member of joint
Arrow of welding symbol
Arrow side member of joint
Edge Joint
Arrow side of joint
Arrow of welding symbol
Joint
Process Abbreviations
Where process abbreviations are to be included in the tail of the welding symbol, reference is made to Table A, Designation of Welding and Allied Processes by Letters, of AWS 2.4-79, 71.
AMERICAN WELDING SOCIETY, INC.
2501 N.W. 7th Street, Miami, Florida 33125

welding tip In oxyacetylene (oxyfuel) welding, a tip that is fitted into the end of the torch. Different types are used for different jaws. (Figure 1098.)

welding torch In oxyacetylene (oxyfuel) welding, the tube that mixes the gases. (Figure 1098.) In arc welding, the gun that holds the electrode. See *gas metal-arc welding gun, oxyacetylene welding, oxyfuel welding, oxygen cutting torch.*

weld joints Welds made between pieces joined together at a butt, corner, lap, edge, or tee. (Figure 1099.) See also *groove weld, weld types.*

weld length Length of weld bead specified in inches. Length is specified on the welding symbol reference line. The weld length and length between welds (in intermittent welding) is sometimes specified.

weldment The completed weld joint; the complete project with all parts welded together.

weld pass One continuous weld operation along the joint to be welded; a bead or weld deposit is put down.

weld root See *root of weld.*

weld size Size of weld bead specified in fraction of an inch, as $\frac{1}{4}$, $\frac{3}{16}$, $\frac{3}{8}$, and so on. Size is specified on the welding symbol reference line.

weld types Welds are made in the following ways: plug, slot, arc seam, bead, groove, edge, and fillet. (Figure 1100.)

weld voltage The *arc voltage.*

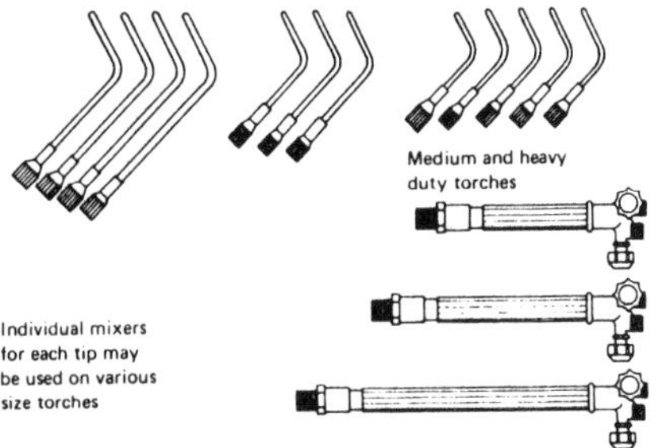

Figure 1098 Welding tips and torches

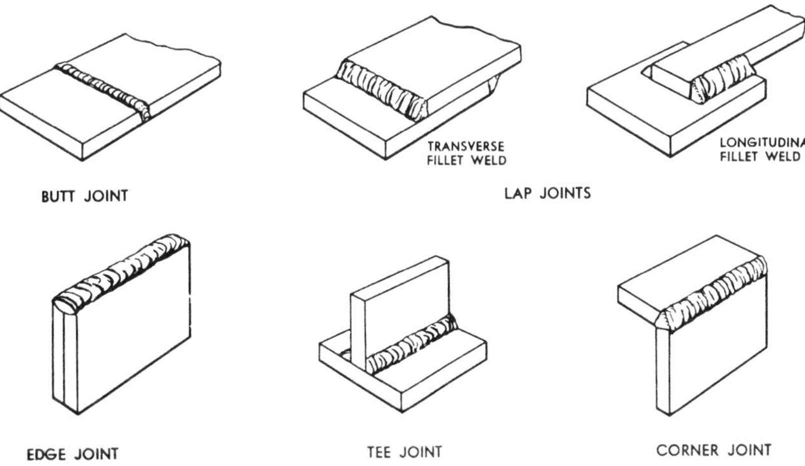

Figure 1099 Weld joints

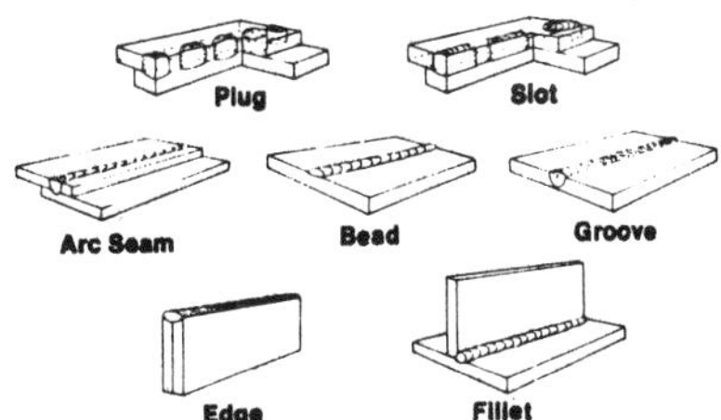

Figure 1100 Weld types

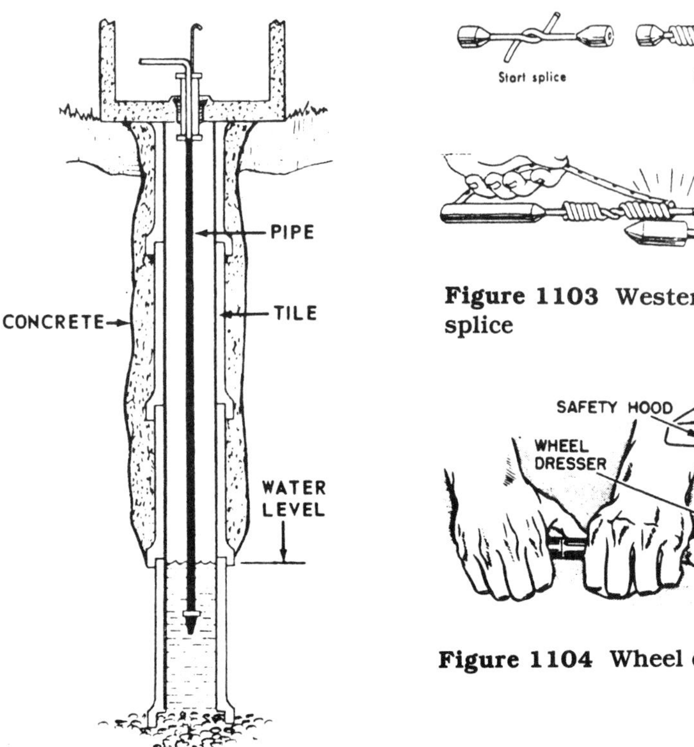

Figure 1101 Well for fresh water

Figure 1103 Western Union splice

Figure 1104 Wheel dresser

Figure 1102 Western row house or townhouse

well Open shaft running from floor to floor, as for a stairwell, elevator shaft, air vent; a well-hole. Also, a cylindrical hole bored or dug into the earth to reach water. (Figure 1101.)

Western framing *Platform framing.*

Western row house Series of houses with common side walls, usually extending for an entire block. (Figure 1102.) Different styles are used. See also *architectural styles.*

Western Union splice Double twisted splice used to join electrical wires. Solder is used. (Figure 1103.) Splice is taped with an insulating tape. See also *pigtail splice, tee splice.*

wet-pipe sprinkler system Automatic sprinkler system in which there is water under pressure. Heat from a fire will melt a locking device on each sprinkler head, releasing water. See *dry-pipe sprinkler system.*

wetwood Heartwood, which has a high moisture content. Also the inner sapwood.

whaler See *wale.*

wheelbarrow Steel container mounted over a single forward wheel and controlled by two hand holds at the rear. Used at the construction site to move small loads. An open wheelbarrow is used to carry brick and tile.

wheel dresser Device for cleaning glazed surface of a circular grinding wheel. (Figure 1104.) The wheel dresser is applied while the grinding wheel is revolving.

whetstone Stone for sharpening a cutting edge.

whipping Securing the end of a rope by wrapping and trying with a cord. Also, in arc welding, a manipulative technique used to control the size of the weld puddle. The electrode is moved away slightly and then back to control the heat and the rate at which the filler metal is added. (Figure 1105.) (Do not confuse with *weaving*, a side-to-side motion.)

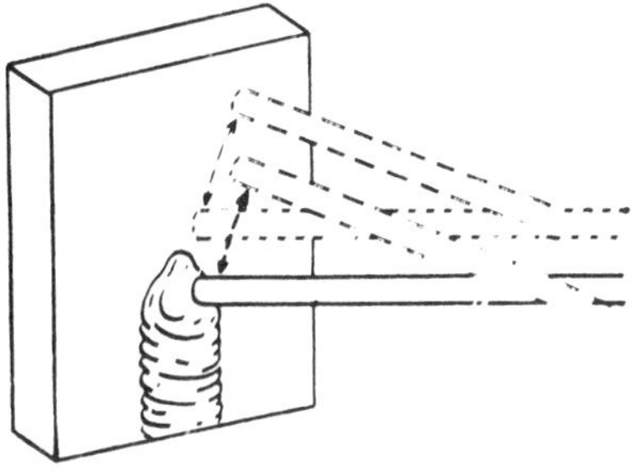

Figure 1105 Whipping

white coat In plastering, a finish coat. Specifically, a finish coat with plaster of Paris.

white rot A whitish fungus decay in wood produced when both cellulose and lignin are broken down.

wide-flange (WF) beam Steel beam with wide flanges; the designation WF has been replaced by W in current standards. See *W beam*.

winch Mechanical hoisting device. A steel rope rolled around a drum can the played out or reeled in to lift or pull. See *windlass*.

wind (Rhymes with "kind.") A twist or bend as in a lumber piece.

wind brace Diagonal bracing in a building designed to resist the lateral wind load. Brace used to reinforce a column to resist wind pressures or wind loading.

windbreak Something that breaks up the wind and shelters a structure. Fencing, hedges, and other plantings are normally used as windbreaks.

wind chill Cooling caused by the action of wind blowing over something and carrying away heat.

wind columns Extra columns often used in rigid frame steel construction to strengthen the structure against wind uplift or lateral force. Two columns are used at opposite sides of the structure.

winders Stairsteps that turn a curve. (Figure 1106.) The treads are cut narrower at one end to swing around the axis. Local building codes restrict and control the use of winders.

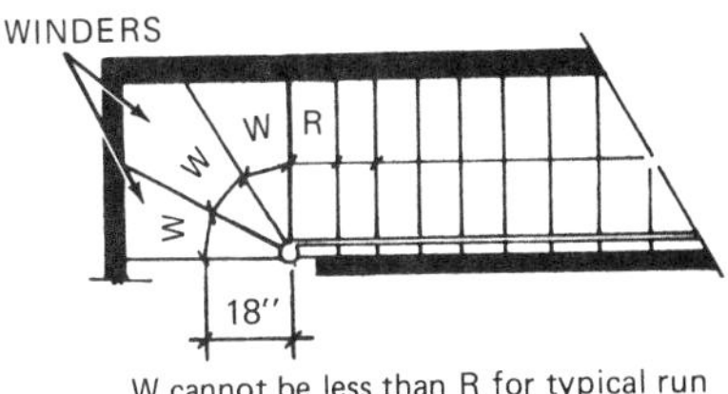

Figure 1106 Winders

winding stair A curving stairs that uses *winders*. See *circular stairs*.

windlass A mechanical hoist. A *winch*.

windload The total effect of wind on a structure. Wind is assumed to affect a building horizontally, creating a positive pressure on the windward side and a negative pressure on the side away from the wind. The windload is the total of these two pressures. The windload is calculated into the design and support for the roof. It is specified in pounds per square foot on the roof. See *wind pressure*.

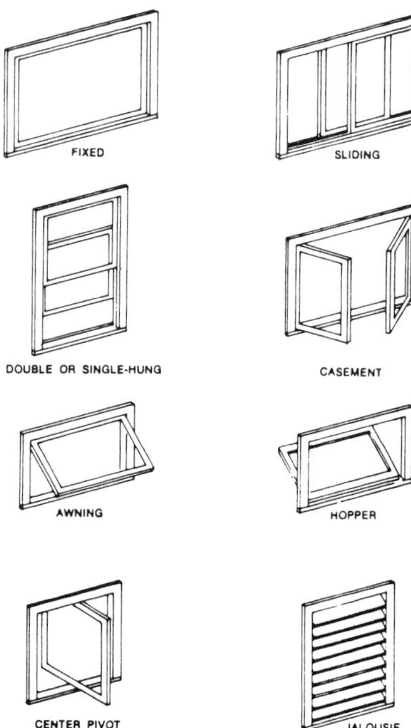

Figure 1107 Windows

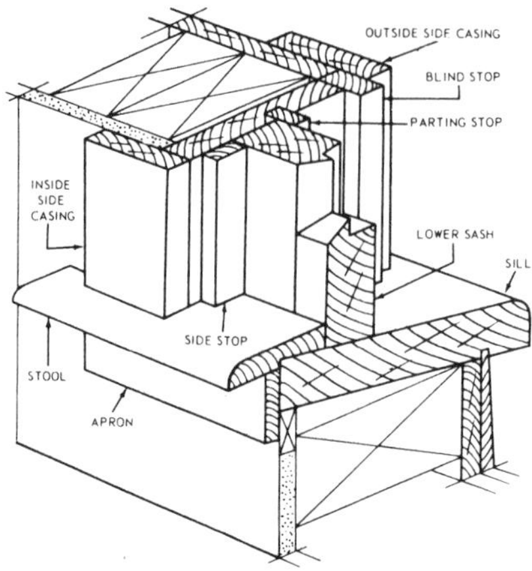

Figure 1108 Window apron and stool

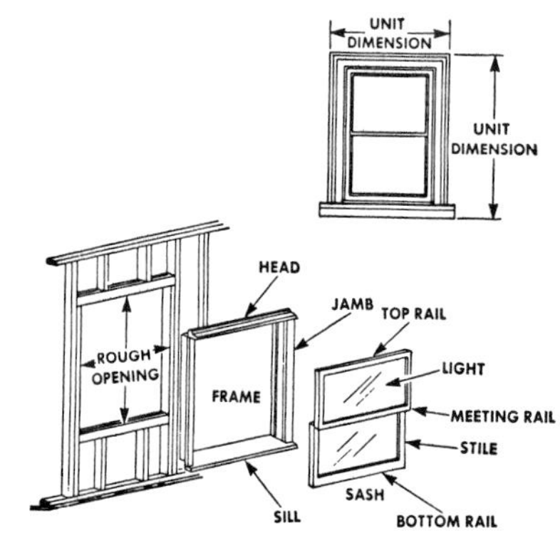

Figure 1110 Window frame and sashes

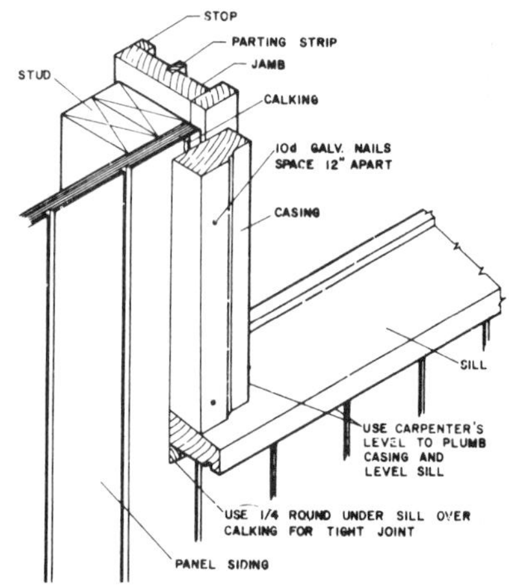

Figure 1109 Window casing

window A glazed opening in a wall of a building. Figure 1107 shows the common window types.

window apron Inside molding or trim below the window stool. (Figure 1108.)

window bar Member that supports separate lights in a window sash. A *muntin.*

window casing Trim around the window frame. (Figure 1109.)

window catch Device for latching a window sash.

window details *Window sections.*

window frame Wood frame that holds the window sash. (Figure 1110.)

window head Top area of a window. See *window sections.*

window jamb Side area of a window. See *window sections.*

window opening Rough opening in the wall that receives the window frame. (Figure 1110.)

window sash Frame that holds the window lights. The sash fits into the window frame. (Figure 1110.)

window schedule Table giving number, size, rough opening, type, manufacturer, light size, hardware, finish, and so on for every window in a structure. Each type of window is identified by a symbol (letter or number).

window sections Vertical cross-cut views through a window to expose the parts. The three major areas of a window are shown: head, jamb, and sill. *Note*: The jamb section is actually a horizontal section view (of the side of the window) that by tradition is revolved to line up with the head and sill sections. (Figure 1111.) Sometimes a window section is referred to as a *window detail*.

window sill Bottom area of a window. (Figures 1109 and 1111.)

window stool Interior projection at the bottom of the window frame. (Figure 1111.)

window stop Wooden piece at the sides and top of a double-hung window to prevent the sash from falling out. (*Figures 950*, p. 420; 1108, 1109.)

window symbols Symbols used on floor plans and elevations to identify the window type. (Figure 1112.)

window trim Interior window finish.

window wall Outside building wall made up of nonstructural window units set into a structural framework. See *curtain wall*.

window well See *areaway*.

wind pressure Wind is assumed to affect a building horizontally, creating a positive pressure on the windward side and a negative pressure on the side away from the wind. A *windload* is the total of these two pressures.

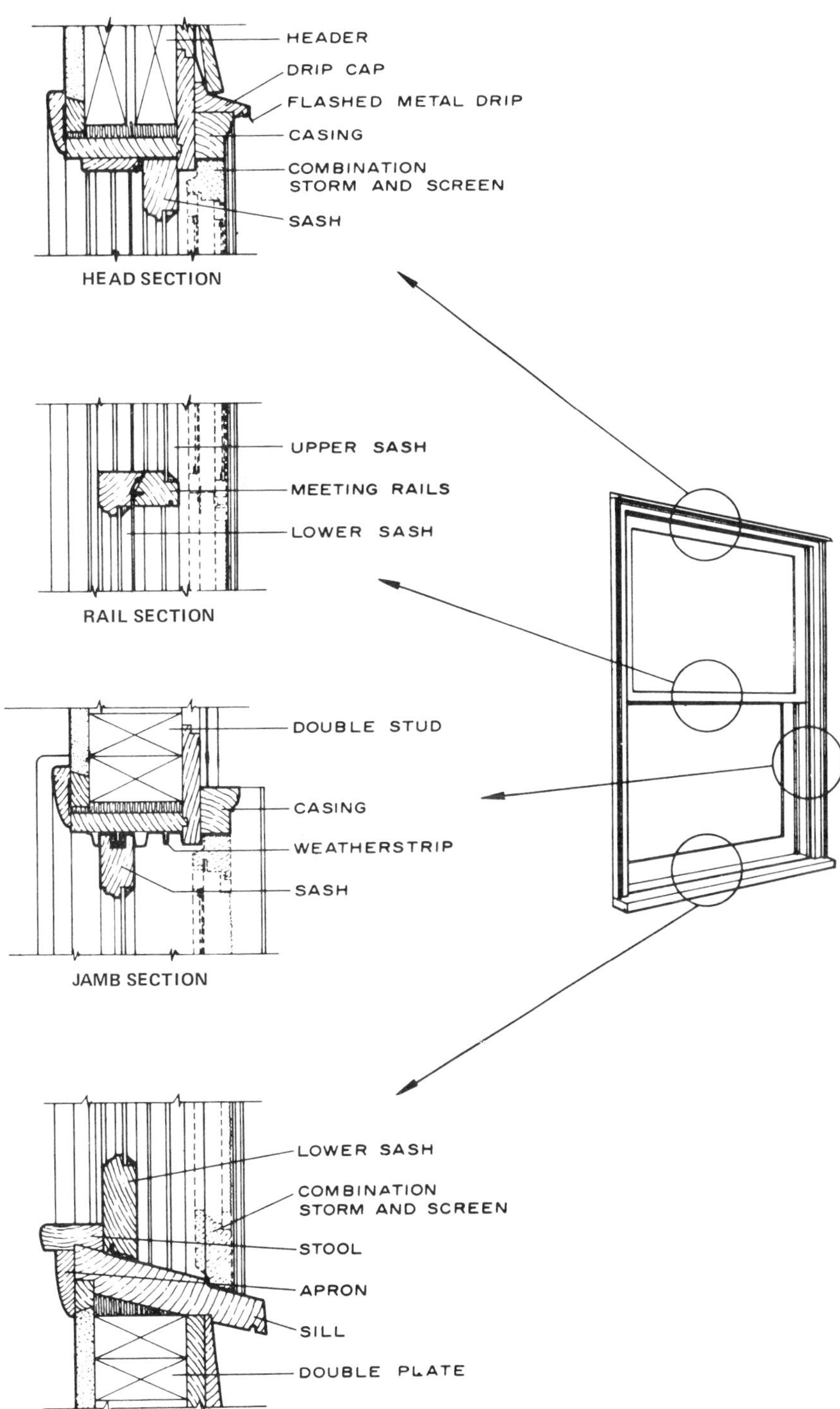

Figure 1111 Window sections

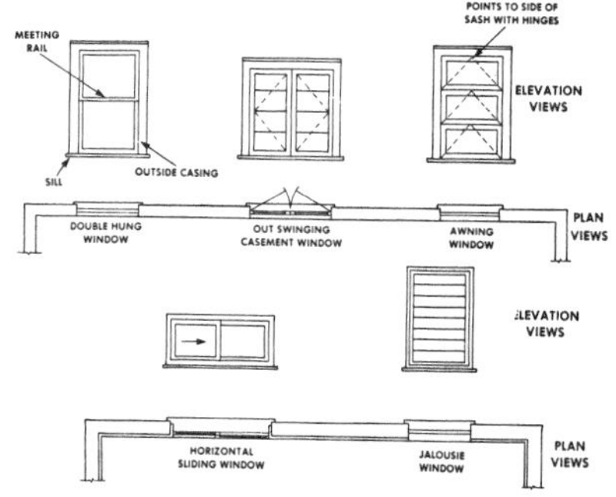

Figure 1112 Window symbols

Figure 1113 Wire brush

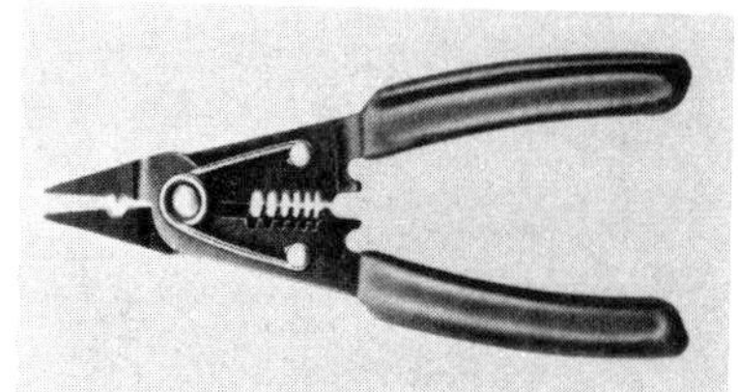

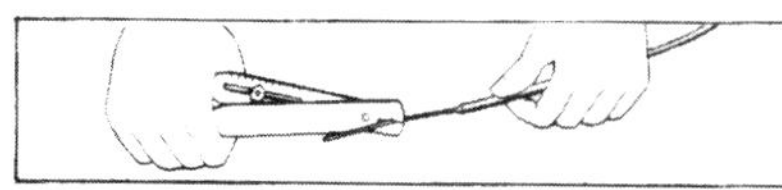

Figure 1114 Wire cutter and stripper

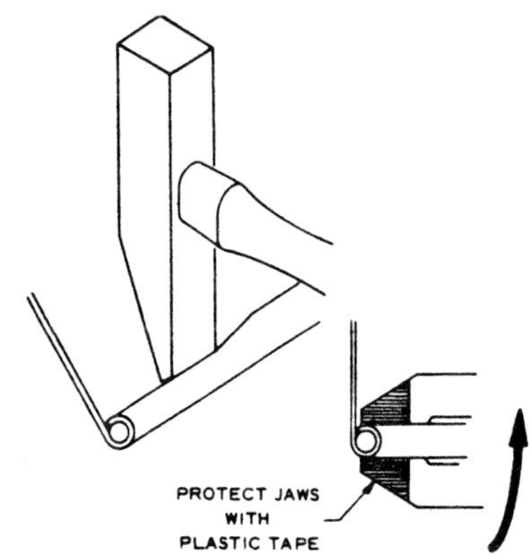

Figure 1115 Wired edge

windshake Lumber defect where the annual rings of a tree open and separate from each other, thought to be caused by the wind shaking the tree. (See *cupshake, shake.*

winter degree day Unit used in estimating fuel consumption and determining heating loads. Unit is based on days when the mean temperature is less than 65 °F. There are as many winter degree days as there are Fahrenheit degrees difference in temperature between the mean temperature for the day and 65 °F. See *degree day.*

winterizing Preparing a home for cold weather by putting in storm windows and doors, sealing cracks with caulking, and adding weatherstripping.

wiped joint Solder joint made by dripping or wiping solder into a joint. A cloth is used to shape (wipe) the solder joint.

wire brads Small nails running from $\frac{1}{2}$'' to $\frac{3}{4}$''.

wire brush Wooden-handled brush with wire bristles. (Figure 1113.) Used for removing rust, dirt, paint, and so on.

wire connectors Plastic connector used to twist two electrical wires together. Also called a *solderless connector, wire nut.*

wire cutter Tool for cutting and stripping solid and stranded electrical wire. (Figure 1114.)

wired edge In sheet metal work, an edge that is rolled or wrapped over a wire. (Figure 1115.) See *edge, rotary machine.*

wire edge A burr created on a cutting edge by oversharpening; the edge becomes so thinned that it turns over. Also, in sheet metal work, an edge that is rolled over a wire; a *wired edge.*

wire gauge Flat circular gauge used to measure or check wire diameter or sheet metal thickness. See *American standard wire gauge.*

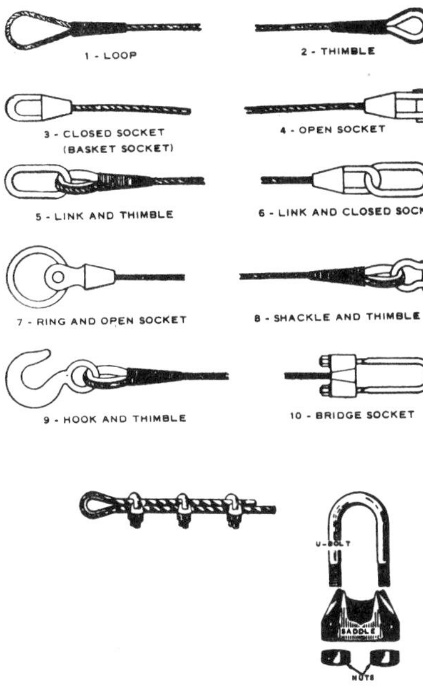

Figure 1116 Wire rope fittings

wire lath See *metal lath.*

wire loop Steel wire loop cast into precast concrete panels, used for lifting by the crane. Loop is cut off after panel is in place.

wire mesh See *welded-wire mesh.*

wire nail Standard nail; opposed to *cut nail.* See *nail.*

wire nut Plastic nut used for connecting electrical wires together. See *solderless connector.*

wire rope Cable made from steel wires twisted together. Figure 1116 shows wire rope fittings.

wire sizes Standard gauge sizes are developed for all wiring. (Figure 1117.) The larger the number, the smaller the size. See also *conductor.* See *American standard wire gauge.*

wire stripper A special electrician's tool for cutting conductor insulation and stripping it off. See *wire cutter and stripper.* See also *crimping tool, multipurpose tool.*

wire ties See *wall ties.*

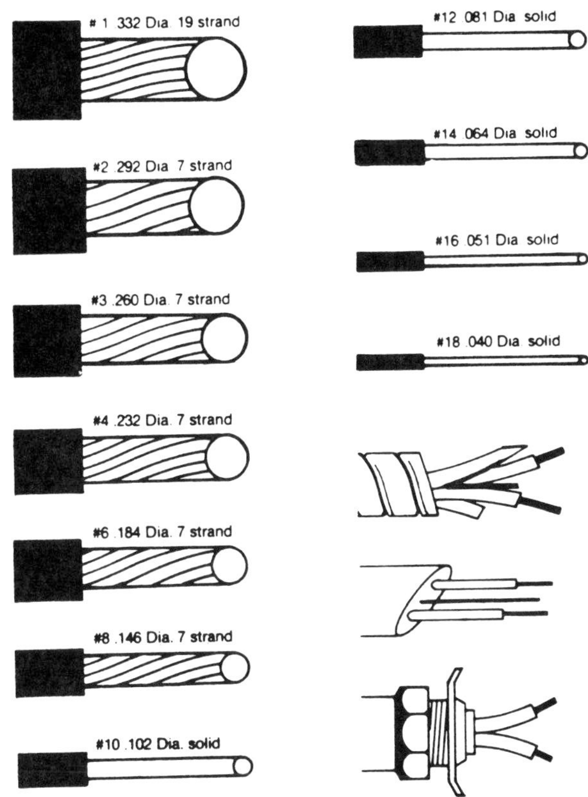

Circuit Breaker or Fuse Ampere Rating	Minimum Copper Wire Size	Minimum Aluminum Wire Size	Circuit Breaker or Fuse Ampere Rating	Minimum Copper Wire Size	Minimum Aluminum Wire Size
FEEDERS AND BRANCH CIRCUITS					
15	14	—	60	4	3
20	12	—	70	4	2
25	10	—	100	1	0
30	10	8	125	1	00
40	8	6	150	0	000
50	6	4	200	000	250MCM
THREE WIRE SERVICE CONDUCTORS OR OTHER CONDUCTORS WHICH CARRY ALL THE CURRENT FOR THE BUILDING					
100	4	2	150	1	00
125	2	0	200	00	0000

Figure 1117 Wire sizes

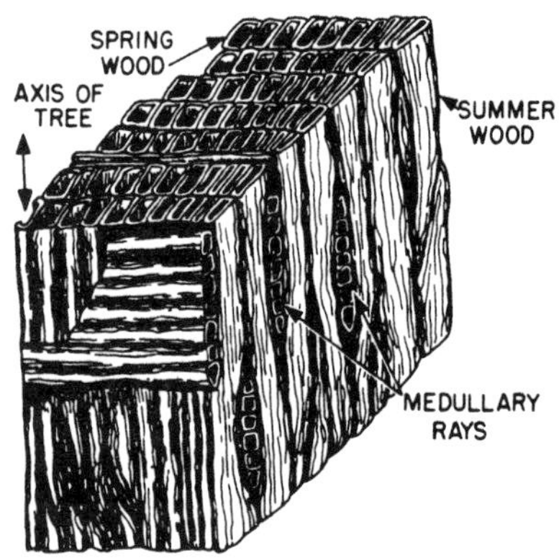

Figure 1118 Wood structure (highly magnified)

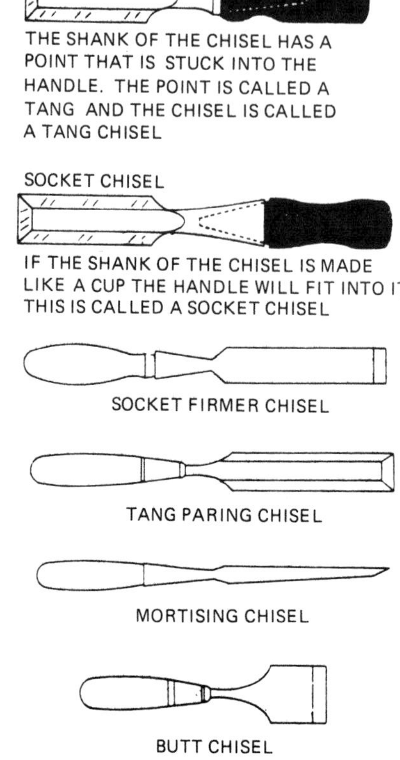

Figure 1119 Wood chisels

wireway Sheet metal trough with hinged or removable covers, used for housing electrical cable and wires.

wiring diagram Schematic showing the wiring layout and connections.

withe A *wythe*.

Wolmanized Of wood members, having been treated or processed to resist decay.

wood A fibrous material produced by trees; natural timber. Both *softwood* and *hardwood* is used in building construction. Figure 1118 shows a small section of wood highly magnified to show the individual wood cells. See *tree*.

wood chisel Wood chisels are used for removing sections of wood either by hand pressure or by tapping with a mallet or soft hammer. Rough cutting is done with the bevel side down; fine or finish cutting is done with the bevel side up. There are two basic types of woodworking chisels: socket and tang. A socket chisel has a socket at the end of the metal chisel that receives the wooden handle. Socket chisels are designed to be struck with a mallet. A tang chisel has a thin pointed end or tang on the end of the metal chisel. The tang fits into a wooden or plastic handle. (Figure 1119.) Long-handled chisels are used for cutting on turning wood pieces in a lathe. For carpentry work the firming or firmer chisel is used. The framing chisel is a heavy wood chisel very similar to the firming chisel and is used in rough carpentry work. The paring chisel is thin and has a side bevel for paring work. A mortising chisel is used for cutting mortises for mortise and tenon joints. The butt chisel is used for chiseling gains for butt hinges on doors. Wood chisels are held with their flat side or back against the work. For rough cuts the bevel is held against the work. Various sizes are available. See also *chisel, gouge, woodturning tools.*

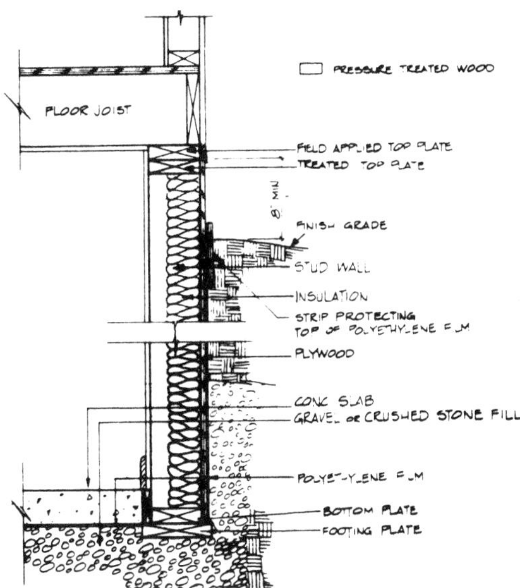

Figure 1120 Wood foundation (note wood footing)

wood file File used for smoothing out tool marks or file marks left by a wood rasp. See *file*.

wood filler Sealer used to fill cracks and pores in wood, colored to match different woods; *wood putty* or *plastic wood*.

wood foundation Support base for a house made using wood framing members. (Figure 1120.) Special treated wood is used. Note that the footing consists of a flat wood member laid on gravel.

wood frame construction See *frame construction*.

wood fungi Microscopic plants that feed on wood, causing decay and stains.

wood hand screw See *parallel clamps*.

wood joint A fastening together of two separate pieces of wood. Figure 1121 shows common wood joints. Note that the wood pieces are shaped so that one fits another. Gluing or nailing may be used to produce a firm joint. See also *finger joint, fishplate, scarf joint*.

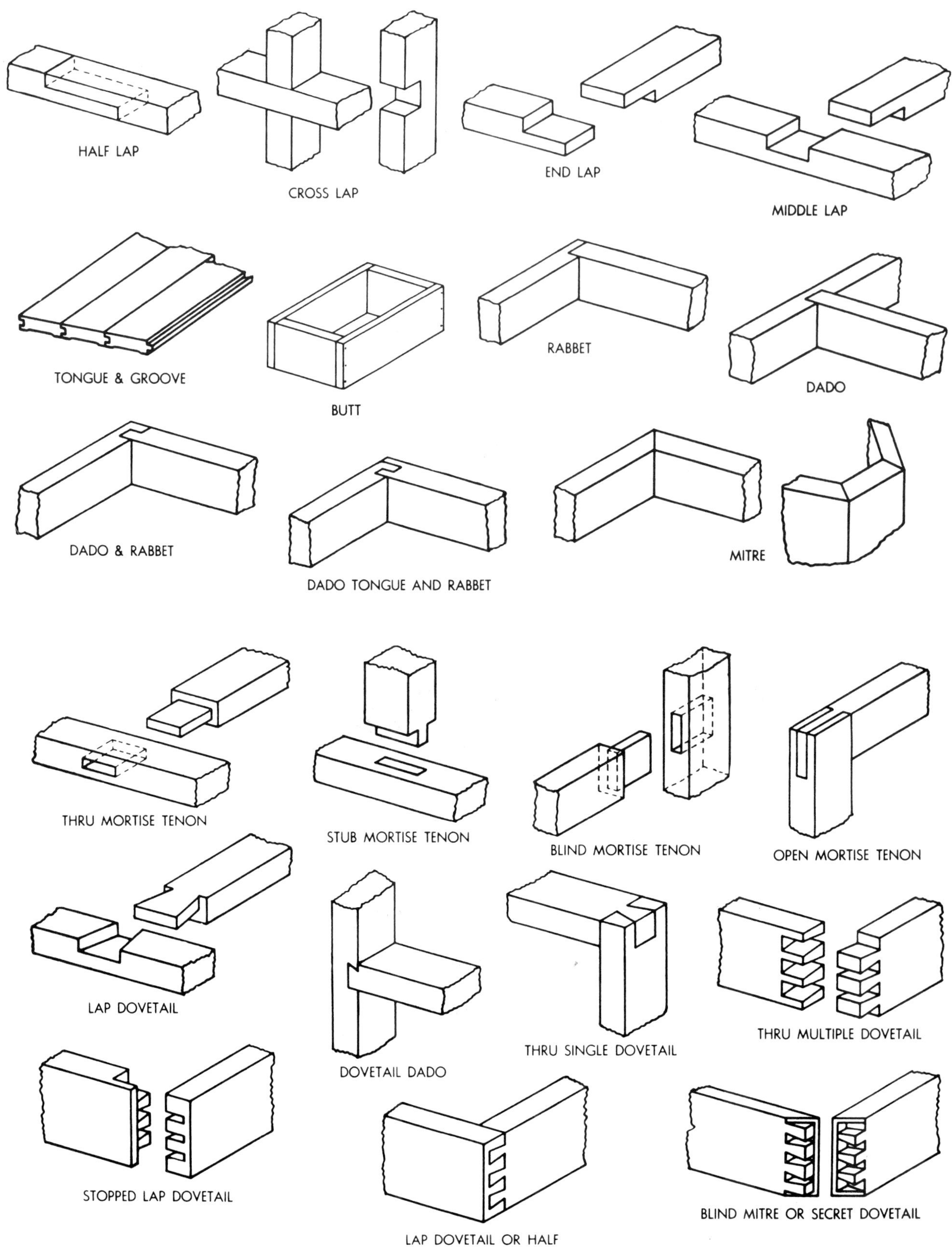

Figure 1121 Wood joints

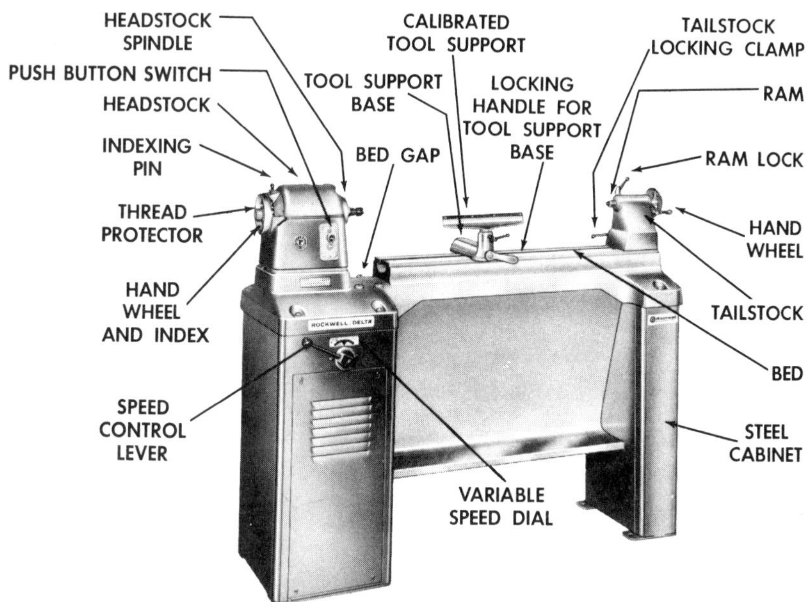

Figure 1122 Wood lathe

wood lathe Power machine used for turning wood pieces. (Figure 1122.) Wood stock is mounted between the fixed head stock and the movable tail stock. Woodturning tools (a type of long-handled chisel) are used to cut the wood to shape as it turns. The tool support or tool rest is used as a rest while the cutting tool is held against the revolving wood. (Figure 1123.) See also *faceplate turning, turning.*

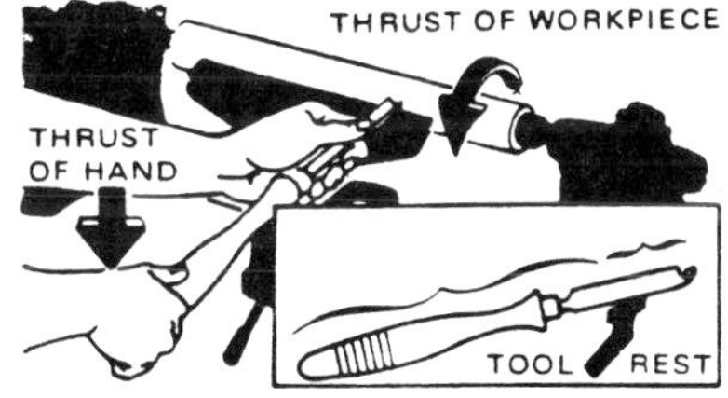

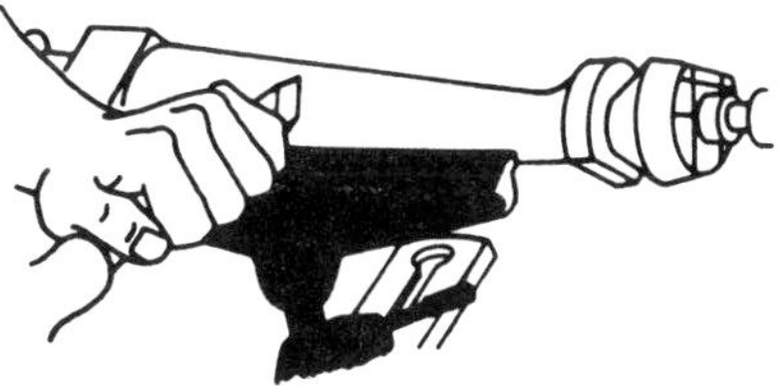

Figure 1123 Wood lathe: turning

wood mallet Wooden tool for striking a wood chisel or for driving wood cabinet parts together. See *mallet.*

wood preservative Chemical used to treat wood to prevent decay. Preservatives are applied by *dipping* or by *pressure treatment.*

wood putty Powder mixture of wood powder and an adhesive. Mixed with water and used for filling cracks or holes in wood.

wood rasp See *rasp.*

wood rays *Medullary rays.*

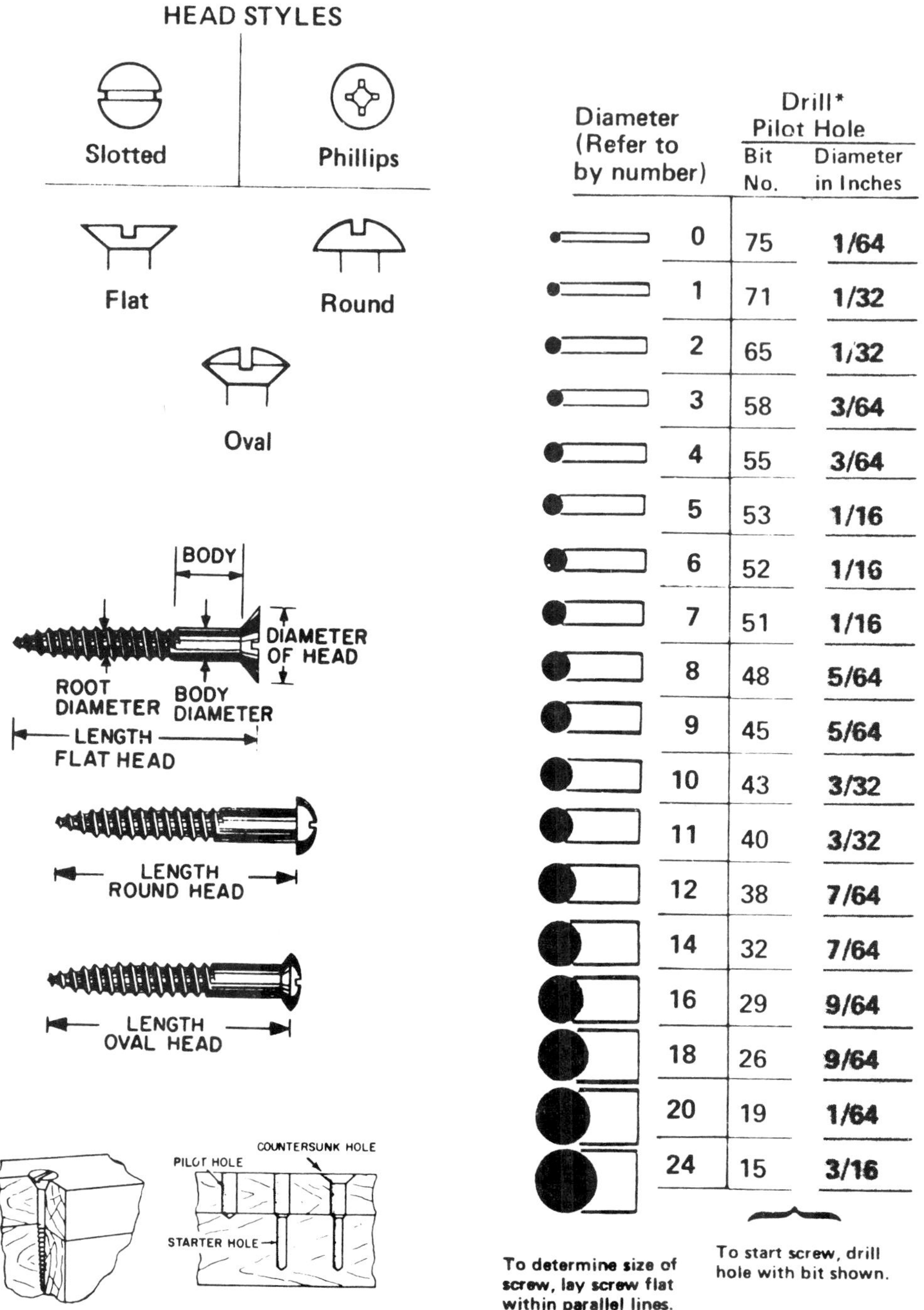

Diameter (Refer to by number)	Drill* Pilot Hole Bit No.	Diameter in Inches
0	75	1/64
1	71	1/32
2	65	1/32
3	58	3/64
4	55	3/64
5	53	1/16
6	52	1/16
7	51	1/16
8	48	5/64
9	45	5/64
10	43	3/32
11	40	3/32
12	38	7/64
14	32	7/64
16	29	9/64
18	26	9/64
20	19	1/64
24	15	3/16

Figure 1124 Wood screws

wood screw Threaded, pointed fasteners used for holding wood members together. (Figure 1124.) A hole is normally drilled to accept the screw. The straight, unthreaded shank under the head allows one wood member to be driven down to a base member without damage to the outside member. Oval, flat, or round heads are used. Most wood screws have a slotted opening in the head, although Phillips heads are also widely used. Screws are sized by diameter gauge number from #0 up to #24 ($\frac{1}{16}$″ up to $\frac{3}{8}$″). *Figure 840*, p. 269, shows the wood screw sizes. Lengths vary from $\frac{1}{8}$″ to 2″. See also *drywall screws, machine screws, sheet metal screws.*

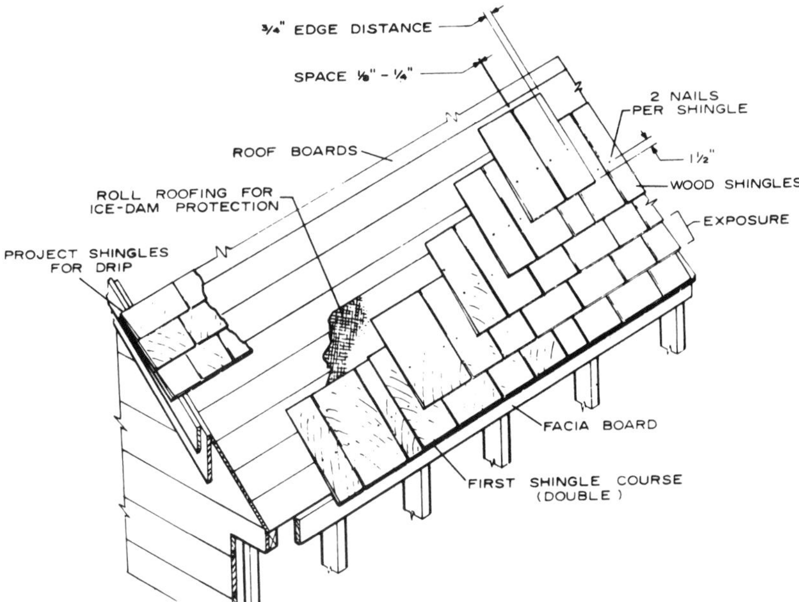

Figure 1125 Wood shingle application

wood shingle Individual roofing unit. Wood shingles or shakes come in 16″, 18″, or 24″ lengths; widths and thicknesses vary. The wood shingle is sawn. Shakes are split off. Figure 1125 shows how wood shingles are applied to the roof. Shingles are sold in *squares* (bundles that will cover 100 square feet).

woodturning Shaping wood pieces on a wood lathe. (Figure 1123.) See *turning, woodturning tools.*

woodturning tools These are edged, long-handled cutting or gouging tools used on turning wood pieces in a lathe. (Figure 1126.) *Gouges* and *roundnose chisels* are used for cutting curves or long curves. A *parting tool* is used for cutting straight into the piece as deep as needed. A *spear-point chisel* is used for cleaning out corners and for cutting shallow vees and beads. *Skews* are used to cut deep vees and beads. Variously shaped chisels are used for *faceplate work* (one end of the wood is glued to a turning faceplate, the other is open and free).

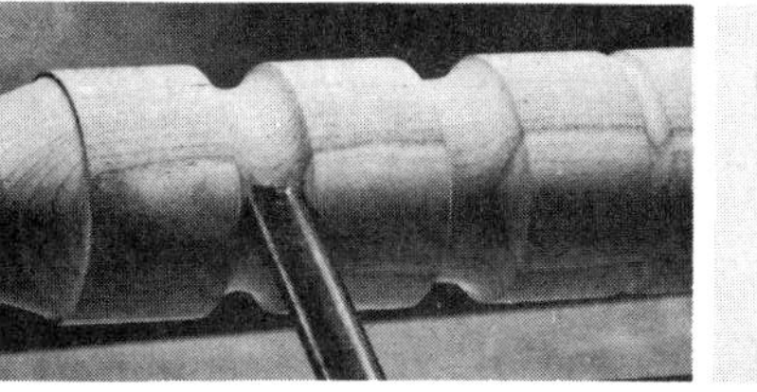
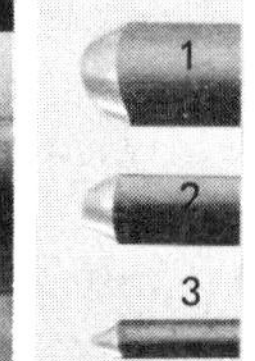
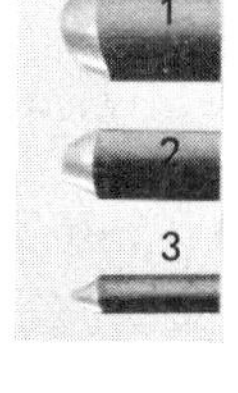

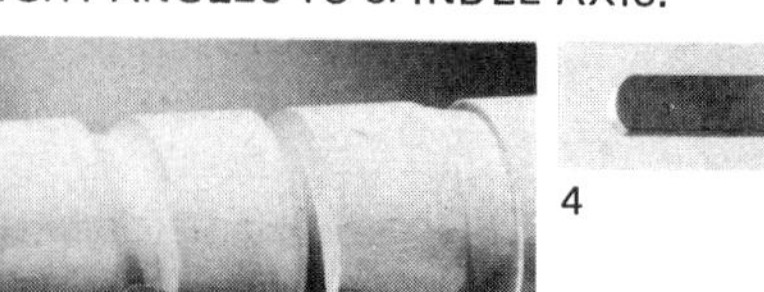
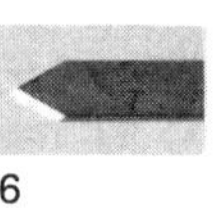
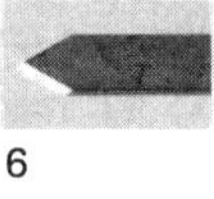
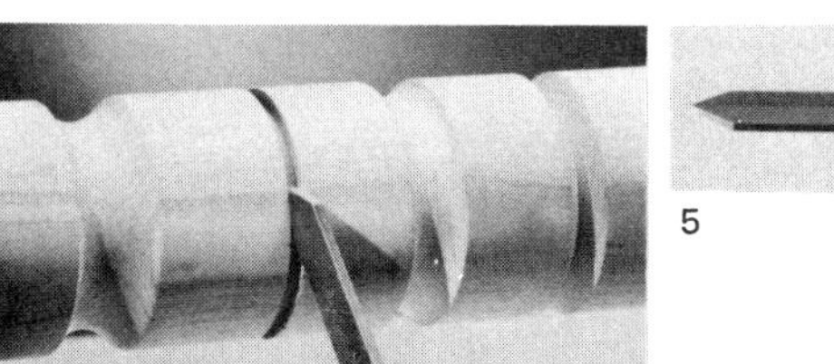
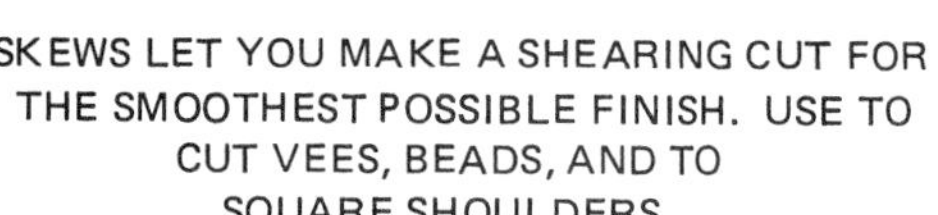
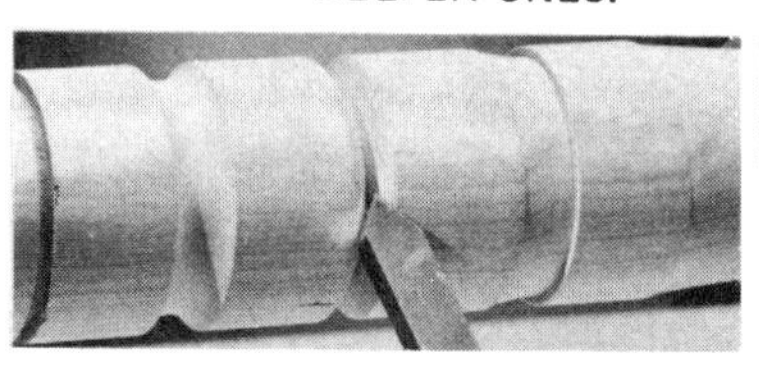

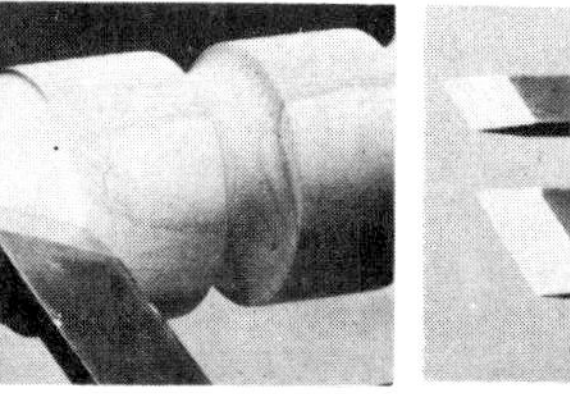
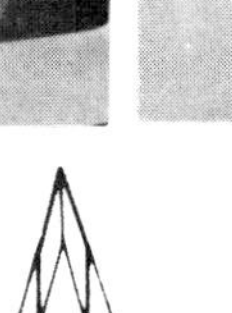
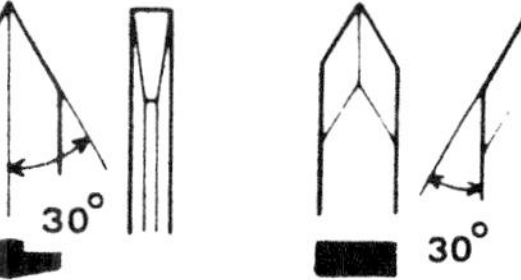

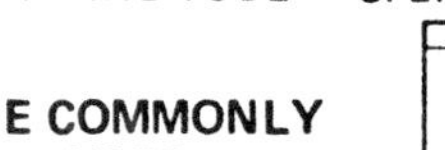
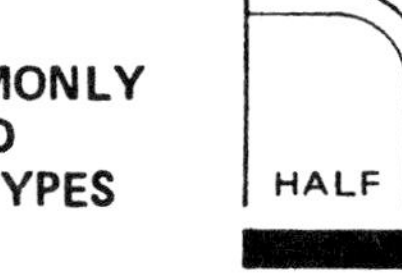

Figure 1126 Woodworking tools

woodworker's bench Wooden bench laid out for woodworking. Vices and clamping bars are built in. (Figure 1127.)

woodworking clamp See *clamp*.

woodworking vise Large-jawed vise used for holding lumber pieces. The vise is mounted on the side of the woodworker's bench so the top edges of the jaws are flush with the bench. (Figure 1128.) See *woodworker's bench*.

work angle In welding, the angle the torch, electrode, or welding rod makes with the work.

worked lumber Dressed lumber that has been further machined, as tongued and grooved, shiplapped, or patterned lumber.

working line In steel construction, a key line from which other dimensions or distances are taken; a *reference line*.

working drawings The architect's final drawings used to build a structure. Working drawings are reproduced in the form of *blueprints* for use on the job.

working load See *dead load, live load, snow load, wind load*.

work lead In arc welding, a conductor between the power supply and the work.

work triangle In kitchen layout and design, the triangle made between the three work areas or centers: (1) refrigerator (storage) area, (2) sink (preparation) area, and (3) range (cooking) area. (Figure 1129.)

woven-wire fabric *Welded-wire fabric*.

wrecking bar A *crowbar* or *pinch bar*. Also called a *ripping bar*.

Figure 1127 Woodworker's bench

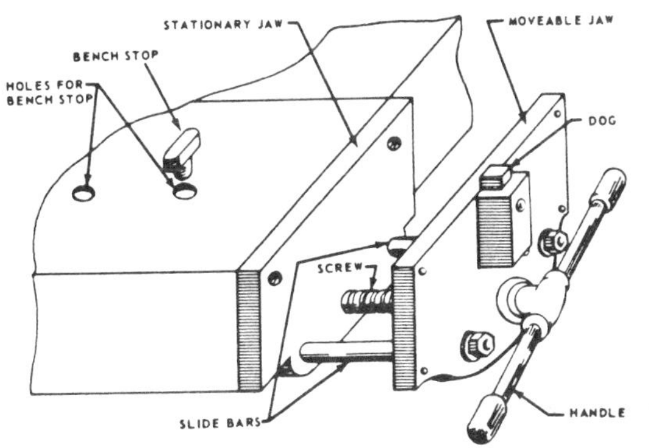

Figure 1128 Woodworking vise

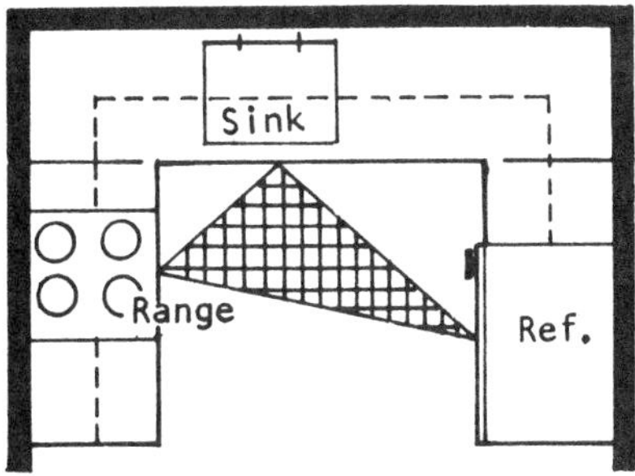

Figure 1129 Work triangle

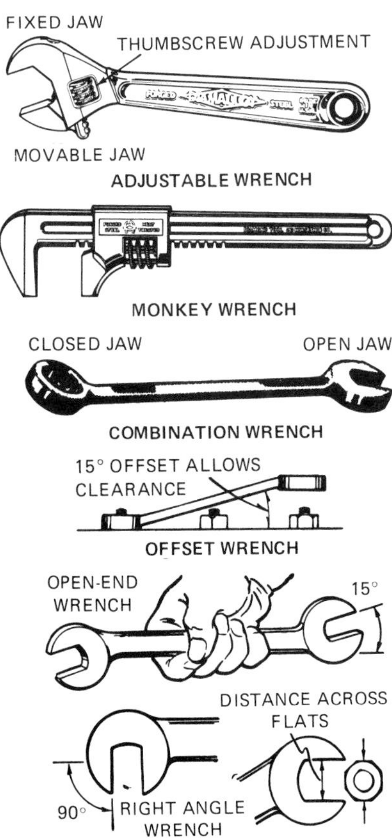

Figure 1130 Wrenches

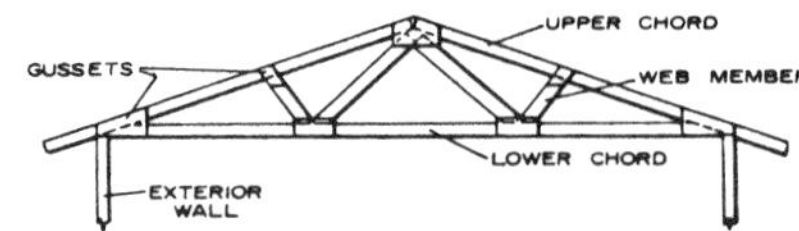

Figure 1131 W truss

wrench A turning device used for tightening nuts on bolts, piping, or other heavy fasteners. Wrenches have fixed open or closed jaws or adjustable jaws. (Figure 1130.) See also *adjustable wrench, Allen wrench, box wrench, chain tongs, chain wrench, combination wrench, crescent wrench, erection wrench, impact wrench, locking pliers, monkey wrench, offset wrench, open-end wrench, pipe wrench, ratchet, socket, strap wrench, torque wrench.*

wrought iron Malleable iron used for railings, balustrades, and the like. Abbreviated W.I.

W truss Common roof truss used in residential construction. (Figure 1131.) Also called a Fink truss. See *truss.*

wye Pipe fitting shaped like the letter Y. See *pipe fittings.*

wythe Single vertical course of masonry units, such as brick or concrete block. A vertical layer of concrete, as in a *sandwich panel.*

xyz

X brace Double diagonal cross bracing.

X joint Two lumber pieces joined so as to form an X shape.

yardage Short for cubic yard.

yard lumber Standard lumber designed for ordinary frame construction.

year ring *Annual ring* in wood. See *tree*.

Y fitting Plumbing fitting shaped like the letter Y; a *wye*.

Z bar A wire bent in the form of a Z used to tie together the two wythes of a *hollow wall*. Also a special channel, shaped like a Z in cross section, used for hanging ceiling panels.

zeebar Extruded, shaped metal used to support ceiling panels.

zero-clearance fireplace Metal woodburning fireplace unit that is insulated so that no clearance ("zero clearance") is needed between it and other wood members in the house.

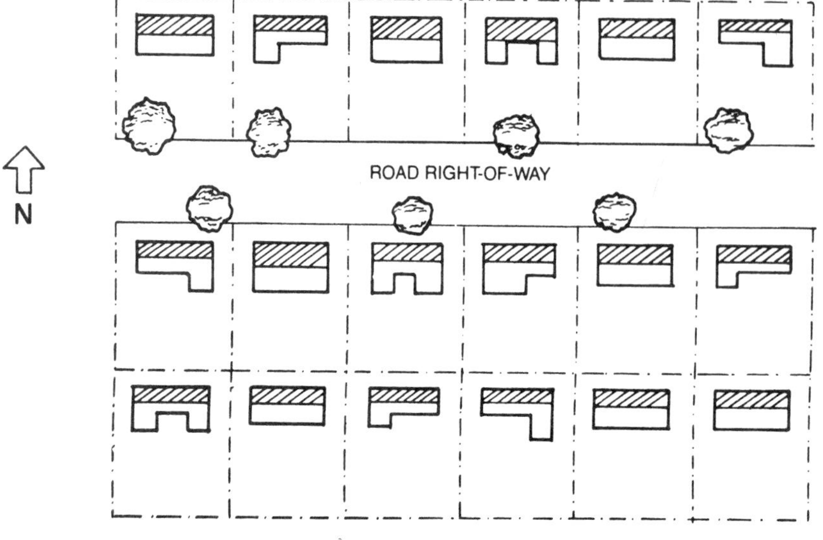

Figure 1132 Zero-lot-line siting: northward

zero-lot-line siting Placement of structure so it abuts on property line. Figure 1132 shows zero-lot-line siting to the north to allow maximum exposure and yard area to the south and to maximize solar access.

zigzag rule A *folding rule.*

zone control Heating system with separate areas of heat control; a separate thermostat is used in each zone.

zone heating A heating system divided into independent sections, each with its own controls—used in warm-air and hydronic systems.

zoning Official regulation restricting or regulating use or development of land in a specific area.

zonolite concrete A special insulating concrete.

legal, real estate and management terms

abeyance Condition in which ownership is undetermined.

abstract A summary of the history of the legal title to a piece of property; summary of a document; itemized list of building materials.

abstract of judgment Summary of a court decision.

abstract of title Short history of the title to a tract of land; includes description of land, all deeds, mortgages, wills, judgments, liens, foreclosures, and releases.

accretion Increase of land area through natural causes, such as the build-up of sediments.

acquittal Verdict of not guilty.

acre Land area containing 43,560 square feet (160 square rods); if square, each side is 208.71 feet long.

action Lawsuit.

act of God A natural act of violence, such as a storm, flood, earthquake, or other cataclysmic act of nature.

addendum Something added. Change made on the drawings or specifications.

ad hoc Latin: for this special purpose.

ad infinitum Latin: forever.

ad valorem Assessment of taxes against a property.

adverse possession Possession by open, continuous physical occupation of property: the actual owner must be notified of the assertion of ownership by the person occupying the property. See *squatter's rights*.

advertisement for bids Published notice asking for bids on a construction project. Fulfills legal requirement that bidding be open.

A/E Architect/engineer.

affidavit Signed and sworn statement that is witnessed and notarized.

affirm Promise to tell the truth; assertion of truth.

affirmative action plan Plan that contractor has to meet equal employment requirements.

agent Someone who acts for another.

agreement Understanding between two or more parties.

air rights Legal right to air space above a property.

alienation Transfer of property to another.

alienation clause Clause in a mortgage or deed stating that the total debt becomes due if the property is sold or transferred.

allotment Subdivision; division of land into smaller parts.

amenities Attractive or desirable features on or near a property.

American Institute of Architects Professional society of architects.

American National Standards Institute (ANSI) Group that publishes voluntary safety and building standards.

American Society for Testing and Materials (ASTM) Group that establishes standards for materials and products, often referenced in specifications.

amortization Scheduled liquidation (satisfaction) of a long-term debt by set amounts over a period of time. For a mortgage, monthly payments are made over a set number of years; payments include prorated interest, principal, taxes, insurance. As each payment toward principal is made, the mortgage amount is reduced or amortized by that amount.

ancillary Subordinate to.

annual percentage rate Interest calculated on a yearly basis.

answer Written statement of defendant's position.

appeal Legal procedure of bringing a decision of a lower court to a higher court.

appraisal The setting of a value on something. An evaluation of the property to determine its value. An appraisal is concerned chiefly with market value — what the property would sell for in the marketplace.

appraisal inventory Complete list and evaluation including all items relating to a property.

appreciation Increase in value of a property.

apprentice Someone who works for a number of years (usually three or four years) under a formal training system to learn a trade.

apprenticeship Agreement under which an apprentice works for a number of years to learn a trade.

approved Of construction, conforming to code or legal requirements. Construction or material that has been inspected and found to be in conformity to standards.

appurtenance Belonging to; a part of real property as a house or an easement.

arbitration Settlement of a dispute through the process of submitting claims to a third party for decision.

arrow diagram method (ADM) Work-event sequence used in network analysis: sequence is shown with arrows linking *nodes*. See *network analysis system, precedence diagram method.*

asking price The price at which the seller is offering something.

assemblage Bringing parts together to form a new whole, as parcels of land combined.

assent Agreement.

assessment Imposition of a tax or charge against a property; valuation for tax purposes.

asset Item of property.

assignment Written transfer of an interest or ownership. Transfer of rights from one person or institution to another.

Associated General Contractors of America An association of contractors engaged in construction.

assumability See *mortgage assumption.*

attachment Legal seizure of property; taking of property by judicial order to hold for application against debt.

auction Public sale to the highest bidder.

balance Amount outstanding; amount still owed.

balloon payment Mortgage loan that provides for unscheduled payments.

bankruptcy Insolvency in which the property of a debtor is taken over by a receiver or trustee.

batture Land formed by water action, as the land at the edge of a lake or river, especially when the water level is low. See also *reliction.*

bench mark An object of known elevation used as a reference to establish elevations of buildings; it is a permanent reference mark established by surveyors.

beneficiary The recipient of something; person receiving income from a *trust.*

bid An offer to buy or do something. A written offer to perform work for a price.

bid bond Dollar guarantee by the bidder (contractor called the "principal") for a contract, backed by a *surety*, that if his bid is accepted he will enter into a contract and faithfully perform and complete the work and pay for all labor and materials. See *performance bond.*

bidding documents Documents that relate to the bidding; notice to bidders (invitation or advertisement to bid), bid instructions, bid form, contract documents, and any addenda prior to bid acceptance.

bid solicitation Invitation to bid on a project. Qualified bidders may be searched out and contacted.

bid time Specified time when all bids must be in.

binder A receipt for money paid to secure the right to purchase real estate upon agreed terms. An offer to purchase.

blockbusting Practice of preying on the fears of local residents by stressing local rapid change in ethnic composition.

bona fide Latin: in good faith; without deceit.

bond A guarantee to pay a set amount or cover certain costs if specified conditions are not met, as a contractor's *surety bond* to guarantee that work is performed on schedule to specifications. See *performance bond.*

book depreciation Amount reserved on an annual basis to provide for replacement.

breach of contract Violation of the terms of a contract; failure to meet the requirements.

brief Written arguments presented to a court.

broker Licensed person or corporation who deals (selling and buying) in real estate; one who buys and sells for a commission.

building code Laws or regulations that control the construction of buildings.

building permits Authorization given by local governing body to construct a building.

building restrictions Prohibitions by a local governing body against construction of certain types of structures in certain locations of the lot.

bylaws Rules adopted by a corporation to govern the conduct of its business.

call loan A loan that must be paid on demand.

capital expenditure Investment in *real property*.

cash flow Net spendable income from an investment, determined after deducting all expenses.

caveat Latin: let him beware. A notice or warning.

certificate of title Document that signifies ownership of a house and property. It usually contains a legal description of the house and its land.

change order Written change issues by the owner (or his authorized agent) for modification of a contract after the execution of the contract. Change orders affect work, contract sum, and contract time.

charge account Buying on credit and paying later.

chattel Article of personal property.

civil action A court case that is *not* criminal.

claim A demand or assertion of a right; asking for something due.

Clean Air Act Federal legislation passed in 1970 requiring individual states to develop pollution control standards.

client Person or organization who engages the services of a builder or contractor, developer, or architect; one for whom the professional builder or designer works.

closed construction A building that is structurally closed or finished so it cannot be inspected at site without damage or disassembly.

closed shop Requirements that prospective employee be a member of a union before he can be hired by an employer. See *open shop*.

closing costs Sometimes called *settlement costs*. Costs in addition to the price of the house, including mortgage service charges, title search and insurance, and transfer-of-ownership charges.

closing day The date on which the title for property passes from the seller to the buyer and/or the date on which the borrower signs the mortgage.

code An organized set of laws or regulations governing something—for example, building code, plumbing code, electrical code, fire code.

codicil Supplement to a *will*.

collateral Real property pledged as security for a loan; security.

collective bargaining Negotiation between employers and unions.

commercial property Property used for business; property used to produce a income.

common law Body of law based on custom and past court decisions.

common property Tract of land that is the property of the public.

completion bond Guarantee by owner to lender or mortgager that project will be completed and free of lien. *Not a performance bond.*

component Large building part, such as a room or series of rooms, completely manufactured off-site in a factory; building assembly.

compromise An agreement or settlement based on mutual agreement.

condemnation Legal declaration stating that a structure is unfit for use or occupancy; taking of private property for public use. See *eminent domain*.

condominium Individual ownership of a living unit in a larger multiunit structure; common areas and facilities are shared.

consequential damage Indirect damage to property; damage incidental to an act or occurrence.

construction documents Working drawings and the specifications.

construction management Integrated management system, from planning and design to construction, designed to give maximum quality at greatest savings in time and cost. Organized around a *construction manager*, who coordinates the total project from design through construction, usually utilizing computer-assisted evaluation and scheduling. See also *critical-path method, network analysis system*.

construction management control system (CMCS) Developed by the General Services Administration to control material delivery and work on the site. See *network analysis system*.

construction manager (CM) One who manages, coordinates, and schedules the total project, including the work of specialty constructors. See *contract manager, project manager*.

construction team Under the construction management concept: the construction manager (CM), the owner, and the architect/engineer (A/E).

contract Oral or written agreement between two or more parties.

contract bond Security furnished by the contractor and his surety as a guarantee that work will be done in accordance with the contract.

contract documents Information prepared by the architect covering construction of a building. Material covered includes owner-contractor agreement, conditions of contract, drawings, specifications, and all modifications and addenda issued after the contract is executed. Contract documents are the basis from which bids are made, the contract executed, and the construction completed. See also *bidding documents.*

contract manager Person in charge of a construction project.

contractor A *general contractor.* A company or an individual who agrees to build a structure or project for a set amount of money; he coordinates all work and hires, supervises, and pays *subcontractors.*

contract purchase Real estate sale wherein seller holds title until purchase is complete; buyer is in possession while completing purchase.

contract time Number of days allowed for completion of a contract, including authorized extensions.

contributory negligence Want of care; omission that contributes to an injury.

conveyance Written document that passes ownership of property.

cooperative housing An apartment building or a group of dwellings owned by residents and operated for their benefit by their elected board of directors. The resident occupies but does not own his unit; rather, he owns a share of stock in the total enterprise.

corporation Legal entity with limited liability to persons who form the organization. *Public corporations:* government bodies or government-owned bodies, such as cities. *Not-for-profit corporations:* corporations not organized for monetary gain, such as religious, educational, charitable, and fraternal institutions. *Corporations for profit:* business corporations with capital stock organized to make a profit.

cosigner A person who signs for another's loan and assumes equal liability.

cost plus Agreement to pay the construction cost plus a set percent over the costs.

covenant Written agreement between parties; a promise.

credit Charging of goods or services with promise to pay later.

creditor An individual or a business that lends money or sells goods on credit.

credit rating Evaluation of someone's ability to pay debts.

critical-path method (CPM) Scheduling of jobs by shortest time; work sequences are worked out graphically before starting the project. Normally computer-assisted analysis and evaluation is used. See *network analysis system.*

CRT Cathode ray tube; viewing screen, similar to a TV, used to project computer-generated information and graphics.

damages Assessed monetary amount for an injury.

datum Horizontal plane of zero elevation, used for establishing elevation.

debtor Someone who owes money.

dead end street Public street with one end closed.

deadline Agreed completion time.

deed Written legal document that transfers title and ownership of real property from one person to anther.

de facto Latin: in fact, actually.

default Failure to perform something required by law or agreement; failure to make agreed-on payment.

defeasance Condition in a written document, such as a deed, that when performed renders the document void.

demand loan Loan due on demand.

density zoning Ordinances restricting number of living units in a area, generally used in relation to subdivisions.

depreciation Decline in value of an asset by use, aging, obsolescence, adverse changes in a neighborhood.

descent Passing of an estate through inheritance.

design-built Design and construction of a project under one management.

developer Someone who prepares land for construction; one who builds.

development The planned construction of a number of living units; a whole community laid out and built from a master plan.

devise Property bequeathed through a will.

discharge To free from a legal obligation.

domicile Place of permanent habitation.

down payment Amount of money paid at time of purchase toward the total purchase price.

due-on-sale Mortgage clause requiring balance of home loan to be paid at lender's option if property is sold. Forces new buyer of previously financed property to obtain new financing at current interest rates.

due process The regular progress of any legal procedure through the courts.

dwelling A building containing one or more *dwelling units* that are used for living purposes.

dwelling unit Single living unit providing facilities for living, eating, sleeping, cooking, and sanitation: a *living unit.*

earnest money Deposit money given to the seller by the potential buyer to show serious interest. The earnest money is applied to the purchase price if the deal goes through. It may be forfeited if the potential buyer decides not to complete the purchase.

easement rights A right of way granted to a person or company authorizing access to or over the owner's land. Electric companies often have easement rights across property.

eminent domain Right of a government or municipal body

to acquire property for public use through a condemnation process.

encroachment Building onto or using someone else's property.

encumbrance A restriction on a property, such as an easement.

environmental impact Analysis of the effect construction will have on the surrounding area.

Environmental Protection Agency (E.P.A.) Federal agency charged with the protection of the environment; monitors state pollution control plans.

Equal Employment Opportunity Federal executive order requiring that employment and material purchase be made on an equal basis without discrimination.

equity ownership A buyer's initial and increasing ownership rights in a house as he pays off the mortgage. When the mortgage is fully paid off, the buyer has 100% equity in the house.

equivalent Equal to or approved equal.

escheat Reversion of land to the state, commonly for failure to pay taxes.

escrow funds Money, or papers representing financial transactions, which is given to a third party to hold until all conditions in a contract are fulfilled.

estate Property; rights in a property.

estimator's criteria sheet Outline giving information and financial conditions on a project, used in preparing job estimate.

eviction Dispossession of a tenant from use of property.

evidence Facts or information presented in court.

exclusive listing Listing of real estate with one broker only, opposed to an *open listing*.

executor Someone appointed to carry out the terms of a will.

extra work Work not provided for in the contract but essential to the completion of the job.

fair housing laws Guarantees that no person will be discriminated against in regard to housing on the basis of race, color, religion, sex, or national origin.

Federal Housing Authority (FHA) Operated under the Development of Housing and Urban Development (HUD) to insure housing purchases; guarantees loans made for purchase, but the real estate must be approved by FHA and loans must meet interest requirements.

fee simple absolute Holding an estate without restrictions.

fee simple limited Estate in fee controlled by a set condition.

fee tail Conditional inheritance of estate — for example, restriction of inheritance to the male heirs.

fire resistance Property of construction materials that prevents or retards passage of excessive heat, hot gases, or flames.

fire-resistance ratings Time in hours or parts of an hour that a material, construction, or assembly will withstand fire exposure before failure.

fire-resistive material See *incombustible building material.*

fire wall Wall with qualities of fire resistance; resists the spread of fire.

first mortgage The basic contract for property. Obligation to the first mortgage must be satisfied before any payments can be made on a *second mortgage.*

fixture Something attached or built into real property so it becomes part of the property, such as a toilet or light fixture.

float Amount of time between earliest scheduled start date and latest scheduled start date. *Or* amount of time between earliest scheduled completion date and latest scheduled completion date. Also called *slack.*

foreclosure Legal process that cancels an ownership or mortgage.

foreman Experienced tradesman in charge of work on the job; a *superintendent.*

forfeit To lose a right to something.

freehold Unencumbered tenure or ownership.

frontage The part of a lot or structure running along a public street, highway, waterway, or other thoroughfare.

front of lot The part of a lot that borders on a street or highway; in case of uncertainty, one side is designated as the front.

general contractor Overall contractor who is responsible for the construction of the total project; works with subcontractors, such as the plumbers, HVAC, and electrical subcontractors, to coordinate and complete the building.

general overhead Costs other than labor and materials, such as taxes, administrative costs, finance charges, and rental.

goodwill Intangible value added to a business by successful continuation of reputable performance. A dollar value for goodwill can be assigned to the business only upon sale and transfer of title.

government survey system Rectangular survey system.

graduated-payment mortgage Holds down homeowner payments in early years to below market interest rates and raises payments in later years.

grantee One who takes a grant.

grantor One who grants.

grid A 40-acre parcel; one-sixteenth of a section.

gross area Area of a building based on outside dimensions.

gross lease Fixed rental; renter does not pay fixed expenses and maintenance. See *ground lease, net lease, percentage lease, variable lease.*

ground lease Long-term lease of the land only; tenant may build or own a building on the land.

growing equity mortgage Ties monthy payments to general wage or price index; additional payments are made as income rises.

guaranty A promise or agreement to answer for someone else's performance.

hazard insurance Insurance to protect against damages caused to property by fire, windstorm, and other common hazards.

hidden costs Costs or expenses that are not easily seen or apparent, such as a special insurance cost or requirement.

high-rise Tall, many-storied structure that provides living area combined with in-building shopping and convenience areas.

impact insulation class (IIC) Number rating which gives an estimate of the impact sound-insulating performance of a floor-ceiling assembly.

improvement Change made on a structure or site that increases its value; this may be the addition of a building to a site or the addition of a room to an existing building.

incombustible building material Material that will not burn in a standard $2\frac{1}{2}$-hour test in a furnace.

idemnify To secure against loss; to reimburse or compensate. To undertake an obligation to pay for any loss or damage suffered by another person.

indemnity Security or protection against loss; immunity. Compensation for loss.

individual retirement account (IRA) Money that may be set aside at interest for retirement; no taxes are paid until money (or part of it) is removed from the account, usually at retirement when the income tax is presumably lower.

industrialized building Construction that uses a high degree of *prefabricated construction* and/or building *components*.

industrial park A development created especially for commercial and industrial use: structures are built for industry, roads are designed for heavy use, rail spurs are brought in.

industrial property Land zoned for industrial use, as for factories or for warehousing.

inspection Examination of a structure to see that code restrictions or building plans and specification requirements have been followed; examination of premises to see that it conforms to health standards, fire safety, electrical code, and so on.

insolvency Inability to pay debts.

installment A regular payment on a debt.

instrument Written document; any legal writing such as a deed, will, or lease.

intangible property Value established by doing business, such as a good reputation, as opposed to real property.

interest Additional money paid for money borrowed.

intestate Not leaving a valid will. When a person dies intestate, ownership to any property will pass on to heirs as determined by the state.

in toto Latin: completely.

invitation for bids Advertisement for proposals for work or materials.

ipso facto Latin: the fact itself.

joint tenancy Estate held by two or more persons at the same time.

joint venture Undertaking of a project by two or more companies.

journeyman A craftsman who has completed an apprenticeship in a trade.

judgment by default Judgment in favor of plaintiff's evidence because defendant did not appear.

jurisdictional dispute Disagreement between two separate unions or crafts as to who should do a particular job.

labor and material payment bond Dollar guarantee (at least one-half of contract price) put forward by the principal (contractor) and *surety* guaranteeing that the contract will be completed in accordance with its conditions, or if not completed then sufficient funds will be available to pay for the cost of completion (with a different contractor).

labor union An association of workers banded together for the purpose of bargaining for their rights.

laches Failure to make a claim within a reasonable length of time; court doctrine preventing assertion of a right because of undue delay or failure to assert a right.

landlord The owner of property who rents or leases to a *tenant.*

land-use planning Long-range planning that prescribes, usually through zoning, how land is to be developed and used.

lease Rental or conveyance of real estate for a set period of time.

leaseback Facility built for the specific use of another; the user does not purchase but leases from the builder.

legal instrument A formal written document, such as a lease or deed.

lessee Someone (tenant) who takes a lease (from the *lessor*).

lessor The person or institution who conveys real estate by lease (to a *lessee*).

letter of intent Letter outlining intention to do something — before the actual contract is completed or signed.

leverage Use of borrowed money to finance something, such as the purchase of a home.

liability Something owed; costs outstanding.

license Legal permit to do something.

license bond Permit to perform

something: contractor is responsible for obtaining any needed license bonds from the state or municipality.

lien Legal obligation or claim made against property, as a mechanic's lien for payment for services rendered; a legal judgment made against the property, such as for taxes; a type of *encumbrance.*

lien bond Bond taken by contractor to guarantee owner aginst loss due to liens against the property.

light and air easement Easement guaranteeing continued use of light and air around a structure; prevents other developers from constructing a building that will interfere with existing light and air use, such as a tall building that will block sun and cut off air flow.

listed Equipment that is included on a list published by an approved testing agency. Real estate that is offered through a broker. See *exclusive listing, open listing.*

listing See *listed.*

litigation Suit at law.

littoral Right of landowner to water rights of bodies of water adjacent to his property; ownership up to high-water mark of land bordering water bodies. See *riparian rights.*

living unit Complete residential unit that includes facilities for one family, with provisions for living, sleeping, eating, cooking, and sanitation. A *dwelling unit.*

lockout Refusal of an employer to allow employees to work in a work place.

lot Plot of ground, as in a subdivision.

lot and block description Description of property by reference to lot and block number, usually within a subdivision.

lot line A legal boundary of a building site, determined by survey.

lump-sum contract Agreement to make payments in predetermined sums at set times during construction.

management contract System wherein the contractor's main responsibility is not to build but to organize and manage the subcontractors.

mandamus Latin: we command. A legal endowment ordering something to be done.

mandatory Shall be done; used in codes.

manufactured building Closed finished structure or building (not a mobile home) assembled in a manufacturing facility.

maturity Date a negotiable security or note comes due.

mechanic A skilled craftsman or worker.

mechanic's lien Legal claim against a building or other structure for service performed to cover labor and materials; claim is made by those who did the work.

mediation Use of a third party to settle a dispute.

metes and bounds Description of property by reference to a known point and by a description using directions and distances from this point; description follows all sides of the land parcel and returns to the beginning.

mobile home Residential unit that may be moved on its own support system.

model home Completed and furnished home used for showing prospective buyers what a completed unit looks like; used by developer to sell, rent, or lease homes (often before they are built).

mortgage assumption Taking over existing mortgage at existing interest rate.

mortgage discount "points" Discounts (points) are a onetime charge assessed by a lending institution to increase the yield from the mortgage loan to a competitive position with the yield from other types of investments. A point is 1% of the mortgage loan.

mortgage loan A special kind of long-term loan for buying a house. There are three main kinds of mortgage financing for single-family homes in the United States: the conventional mortgage; the VA (Veterans Administration), sometimes called the GI mortgage; and the FHA (Federal Housing Administration) insured loan.

mortgagee The bank or lender who loans the money to the mortgagor.

mortgagor The owner who is obligated to repay a mortgage loan on a property he has purchased.

net assets Material worth remaining after all liabilities are subtracted from the whole.

net earnings Profit remaining after all costs are deducted from the total earnings.

net lease Commercial property: tenant pays fixed rent plus other fixed expenses and the maintenance. See *gross lease, ground lease, percentage lease, variable lease.*

network analysis system Graphic process chart and mathematical analysis of all sequential activities and work on a project, including design and drawing preparation, material and equipment procurement and delivery, and all on-site work. Flow charts and diagrams show sequence and time for tasks and events; arrows or numbers show sequence of activities. See *critical-path method, construction management control system, precedence diagram method, project evaluation and review technique.*

man day (MD) Full project work day; used in estimating and scheduling.

networking Diagraming of construction work activities. See *network analysis system.*

node Event or activity space; used in programming, as in network analysis. Nodes are represented by circles or boxes.

notary public Someone licensed to officially witness and seal legal documents.

note Written promise to repay borrowed money at a stated interest and time; instrument of credit attesting to a debt.

notice to bidders Offical notice to bidders.

novation Substitution or replacement of an existing agreement or contract with a new contract.

oath A swearing to the truth.

obligee The owner of a building that has been contracted by a principal (contractor): a term used in a bond to denote to whom an obligation will be due if the principal fails to perform as contracted. See *bid bond, labor and material payment bond, performance bond.*

off-site improvements Improvements made near a property that add to its value, such as paved streets, sewer system.

open construction Building that can be inspected at site without damage or disassembly.

open listing Listing of real estate with several brokers, opposed to an *exclusive listing.*

open shop Requirement that, once hired, an employee must join a union within a set period. See *closed shop.*

option A right to buy or sell that is kept open for a set period of time.

ordinance An official rule or law.

orientation Placement of structure on lot in relation to sun, winds, and privacy.

owner Person or institution who has legal right to property.

partnership Two or more people organized together in a business association. *Limited partnership:* each partner liable for his interest only. *General partnership:* all partners liable for the entire interest.

pay item Specifically described work for which a price is provided in the contract.

percentage lease Retail or commercial property rent based on gross or net income. See *gross lease, ground lease, net lease, variable lease.*

perform To do; to discharge a duty.

performance bond Dollar guarantee by principal (contractor) and *surety* that a contract will be faithfully performed and all labor and materials costs will be paid.

permit Official document allowing work to be done.

personal property Personal, movable possessions, such as household goods. See *real property.*

picketing Right of organized labor to present grievances by physical presence.

plat A map or plan showing location and boundaries of individual properties.

points In mortgage charges, a point is 1% of the total amount of the loan.

power of attorney Delegation of authority by one person to another person so that the other person may act for him.

precedence diagram method (PDM) System for scheduling and control of work progress on a project: Work-event sequence is shown by numbered or lettered sequence by each *node.* See *arrow diagram method, network analysis system.*

prefabricated construction Structure that uses building parts manufactured off-site, usually in a factory.

prepaid expenses The initial deposit at time of closing for taxes and hazard insurance, and the subsequent monthly deposits made to the lender for that purpose.

prescriptive easement Easement awarded by court because of long, unchallenged use.

prevailing wages Standard wage for a job, including benefits, in any area, often required on federal jobs.

principal Amount of a loan as distinguished from interest or other costs; a main party to a transaction or agreement; the main part.

prima facie Latin: at first sight; on the face.

probate Establishing the validity of a will in a court of law; adjudication of any question concerning estates, as for a minor, for someone incompetent, or on behalf of a deceased person.

probate court Court with jurisdiction over estates — as in the case of a deceased person, minor, or someone incompetent.

process Judicial summons; legal proceedings.

progress chart Chart or table recording the progress of the various building operations: it shows the start date, amount completed, and finish date. The chart commonly is based on and checked by *critical-path* analysis.

project evaluation and review technique (PERT) Method developed by the U.S. Navy for *critical-path* scheduling of a job, normally done today using computer-assisted evaluation, analysis, and checking. The shortest work schedule in relation to material delivery is determined.

project manager Coordinator and supervisor of all work on a project.

promissory note Written promise to pay a set amount at a certain time or on demand.

proof Establishment of something as a fact (for evidence).

property line A *lot line.*

proposal Offer of a bidder to perform work and furnish labor and material at the quoted prices.

proposal guaranty Security furnished with a bid to guarantee that the bidder will accept a contract if his bid is accepted.

proration In a real estate transaction, the division of financial responsibility between buyer and seller.

prospectus Brochure outlining the details of an investment.

proxy Authorization to vote for someone else at a meeting.

public housing Low-rent housing developed with government aid.

public space Legal open space accessible to a public way or street and devoted to public use.

purchase agreement Document spelling out conditions of sale.

purchase order Company authorization to buy material and supplies. Abbreviated: P.O.

quantity takeoff Detailed listing of materials needed to do a job.

quitclaim A deed or document granting ownership of property without any guarantees.

quorum Necessary number of people present to conduct business.

real estate Land and property.

real property Land, buildings, and crops on the land; land and everything on it that cannot readily be moved off.

realtor Real estate broker; one who sells real estate.

rebate Discount or reduction in price or cost.

reclamation Claiming of unusable land, as by draining, for development.

recording Filing real estate documents with the appropriate authority, such as the registrar of deeds, so as to become a matter of record.

rectangular survey system Survey system that describes land in relation to principal meridians and base lines, called *government survey system* because established by the federal government in 1785.

release Waiving of a claim.

reliction Creation of usable land by permanent recession of water, as the permanent lowering of a lake. See also *batture.*

replacement cost Expense of replacing something calculated at current prices.

required Mandatory.

rescind To terminate, as a contract or agreement.

rescission Cancellation of a contract.

revocation Rejection, annulment, or cancellation of agreement or promise.

rider A part added to a document.

right of way An easement or designated strip of land for passage.

right-to-work Laws that specify that no person can be denied work because of membership or nonmembership in a labor organization.

riparian rights Land owner's rights to water and land under water on water area bordering his property; right to access and use of water on bordering property. Rights extend to the midpoint of body of water. See *littoral rights.*

road Public thoroughfare, often used in reference to rural roads.

rod Unit of linear measurement used in surveying, equals $16\frac{1}{2}$ feet.

schedule Time sequence for performance of set tasks or objectives to do a job.

second mortgage Mortgage made on property that has an existing *first mortgage.* If there is a foreclosure, the second mortgage cannot be satisfied until the first mortgage obligations are satisfied.

section Land area one mile square (one mile on each side of a square); contains 640 acres.

security Property pledged to the creditor in case of a default on a loan; collateral.

sentence Judgment of the court in a criminal case.

setback Distance house must be set back from street; distance is established by zoning.

shall Indicates that which is required.

site Location of a structure.

slack Spare time available to perform a task.

small claims court A limited-jurisdiction court for small claims, usually around $300 to $800; cases are usually heard without lawyers.

solvency Ability to pay debt.

special assessment A tax for a specific purpose such as providing paved streets or new sewers. People whose properties abut the improved streets or tie into the new sewer system must pay the tax.

specifications Legal document that details how structure is to be built. Specifications describe work to be done and spell out specific materials to be used. Bidding requirements and general work conditions, including insurance, codes, permits, bonds, and tax requirements, are also covered. See also *contract documents.*

squatter's rights Extralegal rights to property based on use, occupancy, or development of the property outside of any legal right to it; recognition based on presence, often in relation to public lands. See *adverse possession.*

standby pay Pay for being on the job ready for work, even though the work may not materialize, as an electrician on standby to provide temporary power in case it is needed. Sometimes a part of the union contract.

statute A law; legislative act.

statute of limitation Law limiting the time during which an action, such as a lawsuit or criminal action, may be brought to court.

statutory bond Bond guaranteeing that compliance will be made to statute.

stipulation Condition or article in an agreement.

stop payment Demand that the bank not pay on a check.

stop work order Ban on further work on a project.

story Part of a building between two floors or between a floor and the roof.

street Public thoroughfare.

strike Refusal to work by the employees, usually an organized response in pursuit of better working conditions and/or increased wages and benefits.

structure Building or construction on a site such as a bridge or dam.

subcontract Order or agreement with the general contractor for work to be done that is a part of the general work.

subcontract bond Bond taken by subcontractor guaranteeing (to the general contractor) that the subcontract will be faithfully executed.

subcontractor Individual or company who contracts with the general contractor to do work.

subdivision Division of land into lots and blocks for the purpose of development or sale.

sublease Leasing of already leased property by the tenant.

subrogation Substitution of one creditor for another.

subsidy Grant to lower cost of a project. Made by a governmental body, often made by the federal government to aid in the development of low-rent housing.

suburban Around a city.

summons Legal notice to appear or to answer a complaint.

superintendent Someone in charge of the construction on a site.

surety Someone who stands behind a guarantee; an insurer.

surety bond A bond taken by a contractor with a surety company (a sort of insurance company) that work will be performed as contracted.

system Interrelationship or assembly of parts to form a whole. Often used to include the planning and work to produce the finished whole.

takeoff List of materials needed for a project; amounts and sizes are specified. A takeoff is made from analysis of drawings and specifications.

tenant One who rents or leases property from a *landlord.*

testate Leaving a valid will. See *intestate.*

time-shared occupancy Several buyers purchase a vacation home and share the use.

title The evidence of a person's legal right to possession of property, normally in the form of a deed; documents that prove ownership.

title company A company that specializes in insuring title to property.

title insurance Special insurance which usually protects lenders against loss of their interest in property due to unforeseen occurrences that might be traced to legal flaws in previous ownerships; an owner can protect his interest by purchasing separate coverage.

title search A check of the title records, generally at the local courthouse, to make sure of ownership and that there are no liens, overdue special assessments, or other claims or outstanding restrictive covenants filed in the record.

tort Civil wrong committed against someone.

townhouse Multiunit construction where dwelling units share party walls.

township Land laid out or in a six-mile square (six miles on each side — an area of 36 square miles); contains 36 *sections,* each 640 acres.

tract Parcel of land.

trade associations Groups of manufacturers or contractors who join together to promote safety and high standards. Different construction areas have different trade associations — for example, the Portland Cement Association or the American Institute for Steel Construction.

trust Property held by someone (*trustee*) for the benefit of another (*beneficiary*).

trustee Administrator of a trust; one who holds property on behalf of another person (a *beneficiary*).

turn key Completion of a structure to the point where it is ready for the client to move in (turn the key in the lock and move in).

Underwriters Laboratories (UL) Organization that tests materials, products, equipment, construction, methods, and systems to establish safety with respect to hazards affecting life and property. All products that pass UL standards are given a UL seal of approval.

urban Within a city.

urban renewal Rebuilding of older areas in cities.

VA loans Loans for housing guaranteed by the Veterans Administration for eligible veterans.

value engineering Total evaluation of a system or project aimed at achieving maximum cost effectiveness within the limitations of performance, quality, and safety objectives.

variable lease Lease at a fixed rental that may vary with the cost-of-living index. See *gross lease, ground lease, net lease, percentage lease.*

variance Permission by zoning authorities to make an exception to the zoning law; written order allowing a code change or change of a contract.

vendee The buyer.

vendor The seller.

venue Geographic area where a court has authority.

verify To prove to be true.

void Having no legal force; null.

waiver Abandonment of a legal right.

warranty Affirmation of something by the seller.

warranty deed Deed affirming title to real property; gives the grantee the right to sue the grantor if anyone later makes a claim against the property.

way Street, driveway, thoroughfare, or easement for passage of people or vehicles.

will Legal document providing for transfer of property after death.

work Labor and all materials and supervision needed to complete a contract.

working drawings Architects' drawings showing graphically, to scale, the parts and details of the structure: plans, elevations, sections, and details are used. Parts are dimensioned, and notes give specific construction information.

workmen's compensation Protection granted to workers and dependents against injury and death in the course of employment. Insurance is paid by the employer.

writ Court order requiring someone to do something.

zoning Official regulation restricting or regulating use or development of land in a specific area.

credits

1 Sears, Roebuck and Co.
2 AEG Power Tool Corp.
3 Merit Abrasive Products, Inc.
4 Sears, Roebuck and Co.
5 HUD: *Manual of Acceptable Practices*
6 Realtors National Marketing Institute, National Association of Realtors
7 USDA
8 Diamond Tool Co.
9 *Top:* Diamond Tool Co. *Bottom left:* U.S. Navy. *Bottom right:* Klein Tools, Inc., Hand Tools Institute
0 Realtors National Marketing Institute, National Association of Realtors
1 U.S. Navy
2 USDA
3 (The Williamson Co.)
4 Duo-Fast Fastener Corp.
5 Keuffel & Esser Co.
6 U.S. Navy
7 Comteck
8 Realtors National Marketing Institute, National Association of Realtors
9 U.S. Navy
0 Graves Humphreys, Inc.
1 U.S. Army
3 The Stanley Works
4 American Institute of Steel Construction
5 U.S. Dept. Labor: OSHA
6 Goldblatt Tool Co.
7 U.S. Navy
28 Northwestern Steel and Wire Co.
29 Sears, Roebuck and Co.
30 *Top:* The Stanley Works. *Bottom:* Milwaukee Tool & Equipment Co.
31 Concrete Reinforcing Steel Institute
32 American Institute of Timber Construction
33 American Institute of Timber Construction
34 Comteck
35 U.S. Navy
36 Advance Drainage Systems, Inc.
37 Sears, Roebuck and Co.
38 Midwest Plan Service
39 Goldblatt Tool Co.
40 USDA
41 *Top left:* Asphalt Roofing Manufacturers Assoc. *Top right:* National Retail Hardware Assoc. *Bottom:* USDA
42 *Top:* Illinois Dept. of Energy and Natural Resources. *Bottom:* USDA
43 *Top:* The Irwin Co. *Bottom:* The Stanley Works
44 Keuffel & Esser Co.
45 United States Gypsum
46 Klein Tools, Inc., Hand Tools Institute
47 U.S. Navy
48 U.S. Navy
49 U.S. Army
50 U.S. Navy
51 Ford Tractor Operations
52 U.S. Navy
53 American Welding Society: *Welding Handbook*, Vol. 2
54 U.S. Army
55 Realtors National Marketing Institute, National Association of Realtors
56 The Stanley Works
57 USDA
58 Rockwell International Corp., Power Tool Division
59 Marshalltown Trowel Co.
60 Concrete Reinforcing Steel Institute
61 Klein Tools, Inc. Hand Tools Institute
62 Roper Whitney, Inc.
63 U.S. Navy
64 Sears, Roebuck and Co.
65 Concrete Reinforcing Steel Institute
66 U.S.E. Diamond, Inc., Expansion Bolt Division, York, Pa.
67 Concrete Reinforcing Steel Institute
68 Concrete Reinforcing Steel Institute
69 U.S. Navy
70 U.S. Navy
71 USDA
72 Comteck
73 U.S. Navy
74 HUD
75 Gypsum Assoc.

76 Sears, Roebuck and Co.
77 Illinois Dept. of Energy and Natural Resources
78 U.S. Navy
79 USDA
80 Butler Manufacturing Co.
81 National Forest Products Assoc.
82 U.S. Navy
83 U.S. Navy
84 U.S. Navy
85 U.S. Army
86 Canadian Wood Council
87 American Institute of Timber Construction
88 U.S. Navy
89 USDA
90 U.S. Army
91 Sears, Roebuck and Co.
92 Rockwell International Corp., Power Tool Division
93 U.S.G.S., U.S. Army
94 Butler Manufacturing Co.
95 U.S. Navy
96 National Solar Heating and Cooling Information Center
97 The Irwin Co.
98 U.S. Navy
99 U.S. Navy
100 U.S. Navy
101 *Top:* Porter Cable Corp. *Bottom:* U.S. Navy
102 U.S. Navy
103 Bruning, Itasca, IL
104 U.S. Navy
105 U.S. Army
106 *Top:* U.S. Army. *Bottom:* Graves-Humphreys, Inc.
107 U.S. Navy
108 HUD: *Minimum Property Standards*
109 USDA
110 U.S. Navy
111 Asphalt Roofing Manufacturing Assoc.
112 USDA: Forest Products Laboratory
113 U.S. Navy
114 U.S. Navy
115 Roper Whitney, Inc.
116 American Plywood Assoc.
117 U.S. Navy
118 Klein Tools, Inc., Hand Tools Institute
119 The Stanley Works
120 U.S. Navy
121 U.S. Navy
122 Tennsmith, Inc.
123 Comteck
124 Hobart Brothers Co.
125 U.S. Army
126 U.S. Army
127 U.S. Army
128 *Top:* Klein Tools, Inc., Hand Tools Institute. *Bottom:* U.S. Army
129 National Forest Products Assoc.
130 *Top:* Teco. *Center and bottom:* National Forest Products Assoc.
131 Marshalltown Trowel Co.
132 *Top:* Keuffel & Esser Co. *Bottom:* U.S. Navy
133 USDA
134 U.S. Army
135 Fiat Allis North American, Inc.
136 U.S. Army
137 Goldblatt Tool Co.
138 The Stanley Works
139 *Top:* U.S. Navy. *Bottom:* The Stanley Works
141 The Stanley Works
142 U.S. Navy
143 U.S. Navy
144 U.S. Navy
145 The Stanley Works
146 The Stanley Works
147 Klein Tools, Inc. Hand Tools Institute
148 Sears, Roebuck and Co.
149 U.S. Army
150 U.S. Navy
151 United States Gypsum
152 Realtors National Marketing Institute, National Association of Realtors
153 U.S. Navy
154 U.S. Navy
155 American Institute of Steel Construction
156 U.S. Navy
157 USDA
158 Realtors National Marketing Institute, National Association of Realtors
159 Realtors National Marketing Institute, National Association of Realtors
160 Asphalt Roofing Manufacturers Assoc.
161 HUD: *Manual of Acceptable Practices*
162 U.S. Navy
163 Hobart Brothers Co.
164 Realtors National Marketing Institute, National Association of Realtors
165 The Stanley Works
166 U.S. Army
167 Sears, Roebuck and Co.
168 Ajax Industries, Hardware Division
169 *Top:* HUD. *Bottom:* Illinois Dept. of Energy and Natural Resources
171 Klein Tools, Inc., Hand Tools Institute
172 U.S. Dept. Commerce
173 Ajax Industries, Hardware Division
174 U.S. Navy
175 Homelite Division of Textron, Inc.
176 *Top:* Diamond Tool Co. *Bottom:* U.S. Navy
177 U.S. Army
178 Wood Molding and Millwork Producers Assoc.
179 U.S. Army
180 American Institute of Steel Construction
181 United States Gypsum Co.
182 Gold Bond Building Products
183 HUD: *Manual of Acceptable Practices*
184 *Top:* Lenco, Inc. *Bottom:* U.S. Navy
185 Black & Decker Industrial/Construction Division
186 U.S. Navy
187 *Top:* Marshalltown Trowel Co. *Bottom:* United States Gypsum Co.
188 General Electric Co.
189 General Electric Co.
190 *Top:* Black & Decker, Inc. *Bottom:* U.S. Navy
191 Porter-Cable Corp.
192 U.S. Army
193 U.S. Army
194 The Stanley Works
195 U.S. Army
196 USDA
197 Illinois Dept. of Energy and Natural Resources
199 American Society of Civil Engineers
200 Teco
201 USDA
202 Asphalt Roofing Manufacturers Assoc.
203 Dayton Superior Corp.
204 U.S. Navy
205 National Forest Products Assoc.
206 HUD: American Planning Assoc.
207 American Institute of Timber Construction
208 United States Gypsum Co.
209 *Top:* Gates & Sons, Inc. *Bottom:* National Forest Products Assoc.

210 U.S. Army
211 U.S. Navy
212 U.S. Navy
213 U.S. Navy
214 Sears, Roebuck and Co.
215 Klein Tools, Inc., Hand Tools Institute
216 American Rental Association
217 Keuffel & Esser Co.
218 U.S. Navy
219 Western Wood Products Assoc.
220 Western Wood Products Assoc.
221 Western Wood Products Assoc.
222 U.S. Navy
223 Midwest Plan Service: *Home and Yard Improvement Handbook*
224 USDA
225 Midwest Plan Service: *Home and Yard Improvement Handbook*
226 Goldblatt Tool Co.
227 Goldblatt Tool Co.
228 Morgan Manufacturing Co.
229 Goldblatt Tool Co.
230 Marshalltown Trowel Co.
231 *Top:* Southwire Co. *Bottom:* Sears, Roebuck and Co.
232 Realtors National Marketing Institute, National Association of Realtors
233 Montgomery Ward
234 Sears, Roebuck and Co.
235 Sears, Roebuck and Co.
236 U.S. Army
237 Realtors National Marketing Institute, National Association of Realtors
238 General Electric Co.
239 Vermont American, Hardware Tool Division
240 HUD: American Planning Assoc.
241 U.S. Army
242 *Top:* Midwest Plan Service: *Home and Yard Improvement Handbook* *Bottom:* U.S. Army
243 HUD: *Manual of Acceptable Practices*
244 United States Gypsum Co.
245 USDA
246 USDA
247 National Forest Products Assoc.
248 U.S. Navy
249 Sears, Roebuck and Co.
250 Sears, Roebuck and Co.
251 *Top:* United States Gypsum Co. *Bottom:* U.S. Navy
252 U.S. Navy
253 U.S.E. Diamond, Inc., Expansion Bolt Division, York, Pa.
254 Adjustable Clamp Co.
255 Masonite Corp., Central Hardboard Division
256 Sears, Roebuck and Co.
257 USDA
258 U.S. Navy
259 U.S. Navy
260 USDA
261 U.S. Navy
262 U.S. Navy
263 *Left:* U.S. Army. *Right:* American Hoist & Derrick Co.
264 U.S. Army
265 American Institute of Steel Construction
266 U.S. Navy
267 USDA
268 USDA
269 Malco Products, Inc.
270 *Left:* Vaco Products Co. *Right:* General Electric Co.
271 National Forest Products Assoc.
272 USDA: Forest Products Laboratory
273 U.S. Navy
274 *Left:* The Stanley Works. *Right:* U.S. Navy
275 The Stanley Works
276 U.S. Navy
277 HUD: *Manual of Acceptable Practices*
278 USDA: Forest Products Laboratory
280 Jaquith Industries, Inc.
281 U.S. Army
282 U.S. Army
283 *Top:* Schlage Lock Co. *Bottom:* U.S. Navy
284 Rockwell International Corp., Power Tool Division
285 U.S. Army
286 U.S. Navy
287 HUD
288 Harlock Products Corp.
289 U.S. Navy
290 Asphalt Roofing Manufacturers Assoc.
291 U.S. Army
292 HUD: Office of Policy Development and Research
293 Klein Tools, Inc., Hand Tools Institute
294 U.S. Navy
295 U.S. Army
296 U.S. Army
297 United States Gypsum Co.
298 American Welding Society: AWS A3.0–80
299 American Welding Society: AWS A3.0–80
300 Sears, Roebuck and Co.
301 Rockwell International Corp., Power Tool Division
302 HUD: *Manual of Acceptable Practices*
303 *Top:* The Stanley Works *Bottom:* U.S. Navy
304 J. Thomas Gussel, Wisconsin Dells, Wis.
305 The Ceco Corp.
306 Realtors National Marketing Institute, National Association of Realtors
307 *Top:* Comteck. *Bottom:* U.S. Navy
308 U.S. Navy
309 USDA
310 USDA
311 Ajax Industries, Hardware Division
312 U.S. Navy
313 USDA
314 U.S. Navy
315 USDA
316 Pella/Rolscreen Co.
317 U.S. Army
318 Western Wood Products Assoc.
319 U.S. Navy
320 Franklin Chemical Industries
321 Franklin Chemical Industries
322 The Stanley Works
323 USDA
324 Bruning, Itasca, IL
325 American Rental Assoc.
326 Genova, Inc.
327 Sears, Roebuck and Co.
328 U.S. Navy
330 U.S. Navy
331 U.S. Navy
332 Omark Industries
333 U.S. Navy
334 Rockwell International Corp., Power Tool Division
335 The Stanley Works
336 U.S. Navy
337 USDA
338 U.S. Navy
339 U.S. Navy
340 Gypsum Assoc.
341 USDA
342 Vaughan & Bushnell Mfg. Co.
343 United States Gypsum Co.

344 Marshalltown Trowel Co.

345 United States Gypsum Co.

346 Duo-Fast Fastener Corp.

347 Marshalltown Trowel Co.

348 Marshalltown Trowel Co.

349 Sears, Roebuck and Co.

350 U.S. Navy

351 USDA

352 Asphalt Roofing Manufacturers Assoc.

353 U.S. Navy

354 Marshalltown Trowel Co.

355 James F. Lincoln Arc Welding Foundation

356 Adjustable Clamp Co.

357 Sears, Roebuck and Co.

358 Comteck

359 General Electric Co.

360 Sears, Roebuck and Co.

361 The Stanley Works

362 Klein Tools, Inc., Hand Tools Institute

363 *Top:* James F. Lincoln Arc Welding Foundation *Bottom:* American Welding Society: *Welding Handbook*, Vol. 2

364 Hobart Brothers Co.

365 *Top:* James F. Lincoln Arc Welding Foundation. *Bottom:* U.S. Army

366 James F. Lincoln Arc Welding Foundation

367 Comteck

368 Realtors National Marketing Institute, National Association of Realtors

369 Klein Tools, Inc., Hand Tools Institute

370 U.S. Navy
Tuscan order

372 *Top:* Butler Manufacturing Co. *Bottom:* U.S. Navy

373 Fiat Allis North American, Inc.

374 *Top:* The Irwin Co. *Bottom:* U.S. Navy

375 U.S. Army

376 USDA

377 *Top:* The Stanley Works *Bottom:* U.S. Navy

379 Outinord Universal Co.

381 Sears, Roebuck and Co.

382 Klein Tools, Inc.

383 USDA

384 U.S. Navy

385 Goldblatt Tool Co.

386 Realtors National Marketing Institute, National Association of Realtors

387 U.S. Navy

388 *Top:* U.S. Navy *Bottom:* Sears, Roebuck and Co.

389 U.S. Army

390 USDA

391 United States Gypsum Co.

392 U.S. Army

393 American Insurance Assoc.

394 Walter Kidde, Division of Kidde, Inc.

395 USDA

396 USDA

397 USDA

398 *Top:* U.S. Navy. *Bottom:* Square D Co.

399 The Wm. Powell Co.

400 The Ceco Corp.

401 U.S. Navy

402 *Top:* USDA. *Bottom:* National Forest Products Assoc.

402 *Top left:* USDA. *Bottom:* National Forest Products Assoc.

403 U.S. Navy

404 U.S. Navy

405 American Society of Civil Engineers

406 American Society of Civil Engineers

407 U.S. Army

408 *Top:* Marshalltown Trowel Co. *Bottom:* U.S. Navy

409 U.S. Army

410 American Plywood Assoc.

411 Klein Tools, Inc.

412 Goldblatt Tool Co.

413 Comteck

414 Steelplex Corp.

415 Comteck

416 U.S. Army

417 American Welding Society: *Welding Handbook*, Vol. 2

418 American Welding Society: *Welding Handbook*, Vol. 2

419 American Plywood Assoc.

420 The Stanley Works

421 *Top:* The Stanley Works *Bottom:* U.S. Navy

422 USDA

423 U.S. Navy

424 U.S. Navy

425 Dayton Superior Corp.

426 *Top:* USDA *Bottom:* Outinord Universal Co.

427 USDA

428 U.S. Navy

429 U.S. Navy

430 Georgia Pacific

431 Teco

432 U.S. Navy

433 Comteck

434 Realtors National Marketing Institute, National Association of Realtors

435 Realtors National Marketing Institute, National Association of Realtors

436 U.S. Navy

437 USDA

438 United States Gypsum Co.

439 Realtors National Marketing Institute, National Association of Realtors

440 General Electric Co.

441 General Electric Co.

443 U.S. Navy

444 Sears, Roebuck and Co.

445 Sears, Roebuck and Co.

446 American Welding Society: *Welding Handbook*, Vol. 2

447 American Welding Society: *Welding Handbook*, Vol. 2

448 American Welding Society: *Welding Handbook*, Vol. 2

449 American Welding Society: *Welding Handbook*, Vol. 2

450 American Welding Society. *Welding Handbook*, Vol. 2

451 HUD: *Manual of Acceptable Practices*

452 Sears, Roebuck and Co.

453 Marshalltown Trowel Co.

454 Homelite Division of Textron, Inc.

455 Realtors National Marketing Institute, National Association of Realtors

456 U.S. Navy

457 Red Devil, Inc.

458 Red Devil, Inc.

459 *Top:* Graves-Humphreys, Inc. *Bottom:* USDA

460 Sears, Roebuck and Co.

461 *Top:* Sellstrom Mfg. Co. *Bottom:* Klein Tools, Inc.

462 Greenlee Tool Co.

463 USDA

464 Galion Dresser: Galion Manufacturing Division

465 Galion Dresser: Galion Manufacturing Division

466 U.S. Army

467 U.S. Army

468 USDA

469 Rockwell International Corp., Power Tool Division

470 Concrete Reinforcing Steel Institute

471 Marshalltown Trowel Co.
472 U.S. Navy
473 U.S. Army
474 Square D Co.
475 Sears, Roebuck and Co.
476 Sears, Roebuck and Co.
477 Sears, Roebuck and Co.
478 USDA
479 U.S. Navy
480 U.S. Navy
481 U.S. Navy
482 U.S. Navy. *Top left:* The Stanley Works
483 The Stanley Works
484 *Top and third from top:* Schlage Lock Co. *Bottom and third from bottom:* U.S. Navy
485 U.S. Navy
486 Sellstrom Mfg. Co.
487 U.S. Navy
488 U.S. Navy
489 USDA
490 USDA
492 Conteck
493 Commonwealth Edison
495 U.S. Navy
496 American Institute of Steel Construction
497 Midwest Plan Service: *Home and Yard Improvement Handbook*
498 U.S.E. Diamond, Inc., Expansion Bolt Division, York, Pa.
499 Malco Products, Inc.
500 Malco Products, Inc.
501 Black & Decker, Industrial/Construction Division
502 U.S.E. Diamond, Inc., Expansion Bolt Division, York, Pa.
503 U.S. Navy
504 USDA
505 USDA
506 USDA
507 Truss Joist Corp.
508 AEG Power Tool Corp.
509 Comteck
510 Sears, Roebuck and Co.
511 USDA
512 Owens-Corning Fiberglas Corp.
513 Realtors National Marketing Institute, National Association of Realtors
514 Gilpin, Inc.
515 Klein Tools, Inc., Hand Tools Institute
516 HUD: *Manual of Acceptable Practices*
517 U.S. Army
518 U.S. Navy
519 U.S. Navy
520 USDA
521 U.S. Navy
522 Rockwell International Corp., Power Tool Division
523 USDA
524 Rockwell International Corp., Power Tool Division
525 The Stanley Works
526 U.S. Navy
527 U.S. Navy
528 Goldblatt Tool Co.
529 USDA
530 Inryco, Inc.
531 U.S. Navy
532 *Top:* Teco. *Bottom:* National Forest Products Assoc.
533 Neptune/Elkay Disposer Corp.
534 Sears, Roebuck and Co.
535 U.S. Army
536 U.S. Navy
538 U.S. Navy
539 USDA
540 Amercian Institute of Steel Construction
541 USDA
542 Sears, Roebuck and Co.
543 U.S. Navy
545 American Rental Assoc.
546 National Forest Products Assoc.
547 U.S. Navy
548 U.S. Navy
549 U.S. Navy
550 U.S.E. Diamond, Inc., Expansion Bolt Division, York, Pa.
551 U.S. Army
552 Schlage Lock Co.
553 HUD: Office of Policy Development and Research
554 United States Gypsum Co.
555 Goldblatt Tool Co.
556 USDA
557 USDA
558 The Rawlplug Co.
559 Sears, Roebuck and Co.
560 U.S. Navy
561 U.S. Army
562 U.S. Navy
563 David White Instrument
565 Marshalltown Trowel Co.
566 U.S. Navy
567 Fiat Allis North American, Inc.
568 J.I. Case
569 Schlage Lock Co.
570 HUD: *Manual of Acceptable Practices*
571 U.S. Navy
572 U.S. Navy
573 Kwikset Division, Emhart Hardware Group
574 The Irwin Co.
575 Klein Tools, Inc., Hand Tools Institute
576 National Forest Products Assoc.
577 USDA
578 U.S. Dept. of Commerce, National Bureau of Standards: PS 20–70
579 U.S. Dept. of Commerce, National Bureau of Standards: PS 20–70
580 D.R.I. Industries, Inc.
581 U.S.E. Diamond, Inc., Expansion Bolt Division, York, Pa.
582 U.S. Army
583 *Top:* Klein Tools, Inc., Hand Tools Institute *Bottom:* U.S. Navy
584 U.S. Navy
585 U.S. Navy
586 Klein Tools, Inc.
587 U.S. Navy
588 U.S. Army
589 Morgan Manufacturing Co.
590 Marshalltown Trowel Co.
591 U.S. Navy
593 United States Gypsum Co.
594 Inryco, Inc.
597 U.S. Navy
599 U.S. Navy
600 Realtors National Marketing Institute, National Association of Realtors
601 U.S. Army
602 U.S. Navy
603 U.S. Navy
604 U.S. Navy
605 The Stanley Works
606 Morgan Manufacturing Co.
607 National Forest Products Assoc.
608 U.S. Navy
609 Wood Molding and Millwork Producers Assoc.
610 Gossen Division of U.S. Gypsum Co.
611 USDA
612 Diamond Tool Co.
613 U.S. Army
614 Klein Tools, Inc.
615 Graves-Humphreys, Inc.
616 Duo-Fast Fastener Corp.
617 D.R.I. Industries, Inc.
618 Duo-Fast Fastener Corp.
619 D.R.I. Industries, Inc.
620 U.S. Navy

621 *Top:* U.S. Navy. *Bottom:* The Stanley Works
622 D.R.I. Industries, Inc.
623 Duo-Fast Fastener Corp.
624 Realtors National Marketing Institute, National Association of Realtors
625 Malco Products, Inc.
626 Fernco, Inc.
627 Sears, Roebuck and Co.
628 Malco Products, Inc.
629 D.R.I. Industries, Inc.
630 U.S. Navy
631 U.S.E. Diamond, Inc., Expansion Bolt Division, York, Pa.
632 U.S. Navy
633 U.S. Navy
634 U.S. Navy
635 Comteck
636 HUD: *Manual of Acceptable Practices*
637 American Society of Civil Engineers
638 American Society of Civil Engineers
639 USDA
640 *Top:* U.S. Navy. *Bottom:* Klein Tools, Inc., Hand Tools Institute
641 Asphalt Roofing Manufacturers Assoc.
642 Trus Joist Corp.
643 Porter Cable Corp.
644 Realtors National Marketing Institute, National Association of Realtors
645 Goldblatt Tool Co.
646 U.S. Navy
647 Sears, Roebuck and Co.
648 American Institute of Timber Construction
649 U.S. Navy
650 American Welding Society: *Welding Handbook*, Vol. 2
651 American Welding Society: *Welding Handbook*, Vol. 2
652 Hobart Brothers Co.
653 American Welding Society: *Welding Handbook*, Vol. 2
654 BernzOMatic Corp.
655 U.S. Navy
656 General Electric Co.
657 Teco
658 Comteck
659 U.S. Navy
660 United States Gypsum Co.
661 Wood Molding and Millwork Producers Assoc.
662 U.S. Navy
663 *Top:* Adjustable Clamp Co. *Bottom:* U.S. Army
664 National Forest Products Assoc.
665 Illinois Dept. of Energy and Natural Resources
666 U.S. Navy
667 U.S. Navy
668 USDA
669 U.S. Navy
670 USDA
671 Midwest Plan Service: *Home and Yard Improvment Handbook*
672 HUD: *Manual of Acceptable Practices*
673 USDA
674 Comteck
675 *Top and bottom:* U.S. Navy *Middle:* U.S. Army
676 U.S. Navy
677 Klein Tools, Inc., Hand Tools Institute
678 U.S. Navy
679 U.S. Army
680 HUD: *Minimun Property Standards*
681 Comteck
682 Sears, Roebuck and Co.
683 Comteck
684 U.S. Army
685 U.S. Navy
686 Graves-Humphreys, Inc.
687 U.S. Navy
688 Adjustable Clamp Co.
689 U.S. Navy
690 U.S. Navy
691 Sears, Roebuck and Co.
692 Sears, Roebuck and Co.
693 Klein Tools, Inc., Hand Tools Institute
694 Klein Tools, Inc., Hand Tools Institute
695 U.S. Navy
696 U.S. Navy
697 U.S. Navy
698 Portland Cement Assoc.
699 *Top:* U.S. Forest Products Laboratory *Bottom:* U.S. Navy
700 *Top:* The Stanley Works. *Bottom:* U.S. Navy
701 Rockwell International Corp., Power Tool Division
702 Comteck
703 Hobart Brothers Co.
704 Sears, Roebuck and Co.
705 Goldblatt Tool Co.
706 U.S.E. Diamond, Inc., Expansion Bolt Division, York, Pa.
707 Illinois Dept. of Energy and Natural Resources
708 Sears, Roebuck and Co.
709 Genova, Inc.
710 American Plywood Assoc.
711 Wheeling Corrugating Co.
712 U.S. Navy
713 Federal Housing Administration: HUD
714 Comteck
715 U.S. Army
716 Sears, Roebuck and Co.
717 U.S. Navy
718 Genova, Inc.
719 U.S. Navy
720 U.S. Dept. of Commerce: PS 51–71
721 American Plywood Assoc.
722 The Stanley Works
723 U.S. Navy
724 USDA
725 Northeast Regional Agricultural Engineering Service
726 U.S. Navy
727 Teco
728 Realtors National Marketing Institute, National Association of Realtors
729 Teco
730 *Top:* Amca International *Bottom:* U.S. Navy
731 U.S. Navy
732 Amca International
733 Sears, Roebuck and Co.
734 Sears, Roebuck and Co.
735 Goldblatt Tool Co.
736 Skil Corp., Subsidiary of Emerson Electric Co.
737 *Top:* Superior Featherweight Tool Co. *Bottom:* Marshalltown Trowel Co.
738 U.S. Navy
739 HUD: *Building Value into Housing, 1980 Awards*
740 Prestressed Concrete Institute
741 U.S. Navy
742 U.S. Navy
743 U.S. Navy
744 Teledyne Post
745 U.S. Army
746 The Stanley Works
747 *Top:* Genova, Inc. *Bottom:* USDA
748 U.S. Navy
749 Malco Products, Inc.
750 U.S. Navy
751 *Top:* The Stanley Works. *Bottom:* U.S. Navy
752 U.S. Navy
753 Sears, Roebuck and Co.
754 U.S. Navy
755 Rockford Map Publishers, Inc., Brown Country Farm Bureau
756 U.S. Forest Products Laboratory

758 U.S. Army
759 The Stanley Works
760 Realtors National Marketing Institute, National Association of Realtors
761 Rockwell International Corp., Power Tool Division
762 USDA
763 Comteck
764 Stran Builders
765 Teco
766 Swanson Tool Co.
767 The Stanley Works
768 The Stanley Works
769 The Stanley Works
770 The Stanley Works
771 The Stanley Works
772 U.S. Army
773 *Top:* Asphalt Roofing Manufacturers Assoc. *Bottom:* USDA
774 U.S. Navy
775 U.S. Navy
776 Diamond Tool Co.
777 *Left:* U.S. Navy. *Right:* Malco Products, Inc.
778 The Rawlplug Co.
779 U.S. Army
780 U.S. Navy
781 *Top:* Square D Co. *Bottom:* Sears, Roebuck and Co.
782 Malco Products, Inc.
783 U.S. Navy
784 Concrete Reinforcing Steel Institute
785 The Flexicore Co., Inc.
786 U.S. Army
787 Concrete Reinforcing Steel Institute
788 U.S. Navy
789 USDA
790 USDA
791 Black and Decker Industrial/Construction Division
792 *Top:* U.S. Army. *Bottom:* Stran Builders
793 The Stanley Works
794 *Top:* The Stanley Works. *Bottom:* U.S. Navy
795 Asphalt Roofing Manufacturers Assoc.
796 U.S. Navy
797 The Stanley Works
798 U.S. Navy
799 U.S. Army
800 U.S. Army
801 U.S. Navy
802 American Rental Assoc.
803 USDA
804 Realtors National Marketing Institute, National Association of Realtors
805 Comteck
806 National Roofing Contractors Assoc.
807 Malco Products Inc.
808 USDA
809 Duo-Fast Fastener Corp.
810 Comteck
811 Steelplex Corp.
812 Realtors National Marketing Institute, National Association of Realtors
813 U.S. Navy
814 *Top left:* Roper Whitney, Inc. *Top right and bottom:* U.S. Navy
815 U.S. Navy
816 Porter-Cable Corp.
817 Black and Decker, Inc.
818 U.S. Navy
819 Midwest Plan Service
820 USDA
821 Illinois Dept. of Energy and Natural Resources
822 Illinois Dept. of Energy and Natural Resources
823 Sears, Roebuck and Co.
824 Klein Tools, Inc.
825 Realtors National Marketing Institute, National Association of Realtors
826 Ajax Industries, Hardware Division
827 The Stanley Works
828 AEG Power Tool Corp.
829 The Stanley Works
830 The Stanley Works
831 U.S. Navy
832 U.S. Navy
833 U.S. Navy
834 U.S. Navy
835 USDA
836 Goldblatt Tool Co.
837 Terex: Division of General Motors
838 HUD: *Manual of Acceptable Practices*
839 Graves-Humphreys, Inc.
840 The Stanley Works
841 *Top:* The Stanley Works. *Bottom:* U.S. Navy
842 U.S. Navy
843 U.S. Navy
845 Klein Tools, Inc.
847 Northwestern Steel and Wire Co.
848 The Stanley Works
849 National Forest Products Assoc.
850 U.S. Navy
851 Malco Products, Inc.
852 U.S. Navy
853 Rockford Map Publishers, Inc., Brown County Farm Bureau
854 Comteck
855 Michigan Department of Natural Resources
856 Black and Decker Industrial/Construction Division
857 Buildex Division of Illinois Tool Works Inc. Teks®
858 HUD: *Manual of Acceptable Practices*
859 Square D Co.
860 U.S. Navy
861 The Stanley Works
863 Rockwell International Corp., Power Tool Division
864 *Top:* Edwards Mfg. Co. *Bottom:* Malco Products, Inc.
865 Teco
866 Dayton Superior Corp.
867 Peterson Manufacturing Co., Inc.
868 D.R.I. Industries, Inc.
869 U.S. Navy
870 U.S. Navy
871 Hobart Brothers Co.
872 Realtors National Marketing Institute, National Association of Realtors
873 The Stanley Works
874 The Irwin Co.
875 HUD
876 Sears, Roebuck and Co.
877 U.S. Forest Products Laboratory
878 U.S. Navy
879 Jaquith Industries, Inc.
880 Western Wood Products Assoc.
881 USDA
882 *Top:* U.S. Army *Bottom:* U.S. Navy
883 Teco
884 U.S. Navy
885 American Institute of Steel Construction
886 U.S. Navy
887 USDA
888 Outinord Universal Co.
889 U.S. Navy
890 USDA
891 ITT Phillips Drill Division
892 The Stanley Works
893 Concrete Reinforcing Steel Institute
894 Klein Tools, Inc., Hand Tools Institute
895 *Top:* Roper Whitney Co. *Bottom:* U.S. Navy
896 American Institute of Steel Construction

897 *Top:* American Welding Society: AWS A3.0–80 *Bottom:* American Institute of Steel Construction
898 U.S. Navy
899 U.S. Army
900 Malco Products, Inc.
901 ITT Phillips Drill Division
902 Dayton Superior Corp.
903 Klein Tools, Inc.
904 U.S. Navy
905 USDA
906 HUD: *Solar Dwellings Design Concepts*
907 Illinois Dept. Of Energy and Natural Resources
908 USDA
909 Illinois Dept. of Energy and Natural Resources
910 HUD: Franklin Research Center
911 The Stanley Works
912 U.S. Navy
913 Sears, Roebuck and Co.
914 Realtors National Marketing Institute, National Association of Realtors
915 The Irwin Co.
917 Copper Development Assoc., Inc.
918 U.S. Navy
919 Hobart Brothers Co.
920 Sears, Roebuck and Co.
921 Teco
922 *Top:* The Stanley Works. *Bottom:* U.S. Navy
923 USDA
924 U.S. Army
925 U.S. Navy
926 Franklin Chemical Industries
927 Realtors National Marketing Institute, National Association of Realtors
928 Realtors National Marketing Institute, National Association of Realtors
929 Teco
930 The Stanley Works
931 U.S. Navy
932 Truss Plate Institute, Inc.
933 Malco Products, Inc.
934 USDA
935 USDA
936 Gypsum Assoc.
937 *Top:* USDA *Center left:* Comteck *Center right:* USDA *Bottom:* USDA
938 U.S. Navy
939 *Top:* U.S. Navy. *Bottom:* U.S. Army
940 Graves-Humphreys, Inc.
941 Spotnails: Division of Swingline, Inc.
942 Klein Tools, Inc., Hand Tools Institute
943 American Institute of Steel Construction
944 U.S. Navy
945 U.S. Navy
946 U.S. Navy
947 Asphalt Roofing Manufacturer's Assoc.
948 USDA
949 USDA
950 USDA
951 HUD: American Planning Assoc.
952 U.S. Navy
953 U.S. Army
954 USDA
955 Dearborn Brass
956 Teco
957 *Top:* Klein Tools, Inc. *Bottom:* U.S. Navy
958 U.S. Navy
959 American Plywood Assoc.
960 *Top:* USDA *Bottom:* Schlage Lock Co.
961 James F. Lincoln Arc Welding Foundation
962 National Forest Products Assoc.
963 Gypsum Assoc.
964 U.S. Navy
965 United States Gypsum Co.
966 U.S. Navy
967 Realtors National Marketing Institute, National Association of Realtors
968 ITT Phillips Drill Division
969 U.S. Navy
970 USDA
971 HUD: *Manual of Acceptable Practices*
972 Warren-Knight Instrument Co.
973 United States Gypsum Co.
974 Comteck
975 Comteck
976 U.S. Navy
977 Square D Co.
978 Sears, Roebuck and Co.
979 U.S. Navy
980 Graves-Humphreys, Inc.
981 U.S. Navy
982 U.S. Navy
983 U.S. Navy
984 L.S. Starrett Co.
985 Marshalltown Trowel Co.
986 U.S. Navy
987 U.S. Navy
988 David White Instruments
989 U.S. Navy
990 USDA
991 American Institute of Steel Construction
992 Sears, Roebuck and Co.
993 U.S. Army
994 U.S. Navy
995 Inryco, Inc.
996 National Forest Products Assoc.
997 Keuffel & Esser Co.
998 Sears, Roebuck and Co.
999 USDA
1000 U.S. Navy
1001 U.S. Navy
1002 Red Devil, Inc.
1003 Dayton Superior Corp.
1004 U.S. Navy
1005 U.S. Army
1006 National Forest Products Assoc.
1007 National Forest Products Assoc.
1008 Duo-Fast Fastener Corp.
1009 U.S. Navy
1010 U.S.E. Diamond, Inc., Expansion Bolt Division, York, Pa.
1011 U.S. Navy
1012 Klein Tools, Inc., Hand Tools Institute
1013 Morrow Crane Co.
1014 Rockford Map Publishers, Inc., Brown County Farm Bureau
1015 The Stanley Works
1016 The Lincoln Electric Co.
1017 Miller Electric Mfg. Co.
1018 Keuffel & Esser Co.
1019 Keufel & Esser Co.
1020 Barber-Greene
1021 U.S. Navy
1022 U.S. Army
1023 U.S. Dept. Labor: OSHA
1024 U.S. Dept. Labor: OSHA
1025 U.S. Dept. Labor: OSHA
1026 U.S. Army
1028 U.S. Navy
1029 U.S. Navy
1030 Pella/Rolscreen Co.
1031 David White Instruments
1032 U.S. Dept. of Energy
1033 Marshalltown Trowel Co.
1034 U.S. Army
1035 FMC Corp. Construction Equipment Group
1036 Comteck
1037 U.S. Army
1038 Teco
1039 U.S. Navy
1040 U.S. Navy
1041 U.S. Navy
1042 Schlage Lock Co.
1043 Realtors National Marketing Institute, National Association of Realtors

1044 U.S.E. Diamond, Inc., Expansion Bolt Division, York, Pa.
1045 Rockwell International
1046 U.S. Army
1047 USDA: Forest Products Laboratory
1048 U.S. Navy
1049 American Society of Civil Engineers
1050 American Society of Civil Engineers
1051 Sears, Roebuck and Co.
1052 Malco Products, Inc.
1053 U.S. Navy
1054 Schlage Lock Co.
1055 U.S. Navy
1056 U.S. Navy
1057 *Top:* U.S. Army *Bottom:* U.S. Navy
1058 Owens-Corning Fiberglas Corp.
1059 USDA
1060 Sears, Roebuck and Co.
1061 Keuffel & Esser Co.
1062 Goldblatt Tool Co.
1063 U.S. Navy
1064 Malco Products, Inc.
1066 U.S. Navy
1067 American Society of Civil Engineers
1068 U.S. Navy
1069 Teco
1070 Outinord Universal Co.
1071 Steelplex Corp.
1072 *Top:* Dayton Superior Corp. *Bottom:* Dur-O-Wall
1073 USDA
1074 D.R.I. Industries, Inc.
1075 USDA
1076 Sears, Roebuck and Co.
1077 Sears, Roebuck and Co.
1078 USDA
1079 James F. Lincoln Arc Welding Foundation
1080 U.S. Navy
1081 Klein Tools, Inc., Hand Tools Institute
1082 U.S.E. Diamond, Inc., Expansion Bolt Division, York, Pa.
1083 U.S. Army
1084 Concrete Reinforcing Steel Institute
1085 *Top:* Klein Tools, Inc. Bottom: Peterson Manufacturing Co., Inc.
1086 *Top:* Hobart Brothers Co. *Bottom:* Victor Equipment Co.
1087 Sellstrom Manufacturing Co.
1088 Hobart Brothers Co.
1089 James F. Lincoln Welding Foundation
1090 U.S. Navy
1091 Hobart Brothers Co.
1092 Hobart Brothers Co.
1093 U.S. Army
1094 James F. Lincoln Welding Foundation
1095 American Welding Society: AWS A2.4–1979
1096 American Welding Society: AWS A2.4–1979
1097 American Welding Society: AWS A2.4–1979
1098 American Welding Society: *Welding Handbook*, Vol. 2
1099 U.S. Army
1100 Hobart Brothers Co.
1101 Sears, Roebuck and Co.
1102 Realtors National Marketing Institute, National Association of Realtors
1103 Sears, Roebuck and Co.
1104 U.S. Navy
1105 James F. Lincoln Arc Welding Foundation
1106 HUD: *Manual of Acceptable Practices*
1107 Realtors National Marketing Institute, National Association of Realtors
1108 U.S. Navy
1109 USDA
1110 Comteck
1111 USDA
1112 Comteck
1113 U.S. Navy
1114 *Top:* Easco Hand Tools, Inc. *Bottom:* Sears, Roebuck and Co.
1115 U.S. Navy
1116 U.S. Navy
1117 Square D Co.
1118 U.S. Navy
1119 U.S. Navy
1120 National Forest Products Assoc.
1121 The Stanley Works
1122 Rockwell International Corp., Power Tool Division
1123 Sears, Roebuck and Co.
1124 *Top and right:* D.R.I. Industries, Inc. *Bottom and left:* U.S. Navy
1125 USDA
1126 Sears, Roebuck and Co.
1127 Sears, Roebuck and Co.
1128 U.S. Navy
1129 HUD: *Manual of Acceptable Practices*
1130 *Top:* Diamond Tool Co. *Bottom:* U.S. Navy
1131 USDA
1132 HUD: American Planning Assoc.

Other Practical References

▲ National Construction Estimator

Current building costs for residential, commercial, and industrial construction. Estimated prices for every common building material. Manhours, recommended crew, and labor cost for installation. Includes *Estimate Writer*, an electronic version of the book on computer disk, with a stand-alone estimating program — free on 5¼" high density (1.2Mb) disk. The National Construction Estimator and *Estimate Writer* on 1.2Mb disk cost **$26.50**. (Add $10 if you want *Estimate Writer* on 5¼" double density 360K disks or 3½" 720K disks.) **576 pages, 8½ x 11, $26.50. Revised annually**

▲ Construction Estimating Reference Data

Collected in this single volume are the building estimator's 300 most useful estimating reference tables. Labor requirements for nearly every type of construction are included: site work, concrete work, masonry, steel, carpentry, thermal & moisture protection, doors and windows, finishes, mechanical and electrical. Each section explains in detail the work being estimated and gives the appropriate crew size and equipment needed. **368 pages, 8½ x 11, $26.00**

▲ Building Cost Manual

Square foot costs for residential, commercial, industrial, and farm buildings. Quickly work up a reliable budget estimate based on actual materials and design features, area, shape, wall height, number of floors, and support requirements. Includes all the important variables that can make any building unique from a cost standpoint. **240 pages, 8½ x 11, $16.50. Revised annually**

▲ Berger Building Cost File

Labor and material costs needed to estimate major projects: shopping centers, hospitals, educational facilities, office complexes, industrial and institutional buildings, and housing projects. All cost estimates show both manhours required and typical crew needed. Figure the price and schedule the work quickly and easily. **304 pages, 8½ x 11, $30.00. Revised annually**

▲ Carpentry for Residential Construction

How to do professional quality carpentry work in homes and apartments. Illustrated instructions show you everything from setting batterboards to framing floors and walls, installing floor, wall and roof sheathing, and applying roofing. Covers finish carpentry, also: how to install each type of cornice, frieze, lookout, ledger, fascia and soffit; how to hang windows and doors; how to install siding, drywall and trim. Each job description includes the tools and materials needed, the estimated manhours required, and a step-by-step guide to each part of the task. **400 pages, 5½ x 8½, $19.75**

▲ Estimating Home Building Costs

Estimate every phase of residential construction from site costs to the profit margin you should include in your bid. Shows how to keep track of manhours and make accurate labor cost estimates for footings, foundations, framing and sheathing finishes, electrical, plumbing and more. Explains the work being estimated and provides sample cost estimate worksheets with complete instructions for each job phase. **320 pages, 5½ x 8½, $17.00**

▲ Estimating Plumbing Costs

Offers a basic procedure for estimating materials, labor, and direct and indirect costs for residential and commercial plumbing jobs. Explains how to interpret and understand plot plans, design drainage, waste, and vent systems, meet code requirements, and make an accurate take-off for materials and labor. Includes sample cost sheets, manhour production tables, complete illustrations, and all the practical information you need to accurately estimate plumbing costs. **224 pages, 8½ x 11, $17.25**

▲ Estimating Tables for Home Building

Produce accurate estimates in minutes for nearly any home or multi-family dwelling. This handy manual has the tables you need to find the quantity of materials and labor for most residential construction. Includes overhead and profit, how to develop unit costs for labor and materials, and how to be sure you've considered every cost in the job. **336 pages, 8½ x 11, $21.50**

▲ Painter's Handbook

Loaded with "how-to" information you'll use every day to get professional results on any job: the best way to prepare a surface for painting or repainting; selecting and using the right materials and tools (including airless spray); tips for repainting kitchens, bathrooms, cabinets, eaves and porches; how to match and blend colors; why coatings fail and what to do about it. Thirty profitable specialties that could be your gravy train in the painting business. Every professional painter needs this practical handbook. **320 pages, 8½ x 11, $21.25**

▲ Electrical Blueprint Reading, Revised

Shows how to read and interpret electrical drawings, wiring diagrams and specifications for construction of electrical systems in buildings. Shows how a typical lighting plan and power layout would appear on the plans and explains what the contractor would do to execute this plan. Describes how to use a panelboard or heating schedule and includes typical electrical specifications. **208 pages, 8½ x 11, $18.00**

▲ Paint Contractor's Manual

How to start and run a profitable paint contracting company: getting set up and organized to handle volume work, avoiding the mistakes most painters make, getting top production from your crews and the most value from your advertising dollar. Shows how to estimate all prep and painting. Loaded with manhour estimates, sample forms, contracts, charts, tables and examples you can use. **224 pages, 8½ x 11, $19.25**

▲ Carpentry in Commercial Construction

Covers forming, framing, exteriors, interior finish and cabinet installation in commercial buildings: designing and building concrete forms, selecting lumber dimensions, grades and species for the design load, installing materials selected for their fire rating or sound transmission characteristics, and how to plan and organize the job to improve production. Loaded with illustrations, tables, charts and diagrams. **272 pages, 5½ x 8½, $19.00**